AF528571

Normal and Therapeutic Nutrition

THE AUTHORS

Professor Neelam Khetarpaul, presently working as ICAR Emeritus Professor (Foods and Nutrition), also former Dean (COHS), Associate Director (Farm Advisory Service) and (Home Science) and HOD (Foods and Nutrition) at CCS Haryana Agricultural University, Hisar is the recipient of many awards *viz.*, Distinguished Teacher Award, Young Investigator Award, Ms. Manju Utreja Gold Medal and cash award for doing the best research work in the University and Best Research Paper Awards on many occasions by many scientific organizations. She is the recipient of various Visiting Fellowship Abroad funded by different national and international agencies and visited USA, Australia, UK, Hungary, the Netherlands, Ghana, Benin, Ecuador *etc.* for academic pursuits. She was the Country Coordinator of Indo-Tailoring Biotechnologies Project Programme (2007-2012).

Dr. Sudha Katyal, Professor and Principal of Government Home Science College, Chandigarh, is State Awardee for her outstanding performance, dedication and contribution in successful implementation of Poshan Abhiyaan in Chandigarh. She has been felicitated with a number of awards for Appreciation/Outstanding Contribution/ Excellence by various Govt. Departments of Chandigarh Administration *viz.* Deptt. of Social Welfare of Women and Child Development, Deptt. of Election, Environment, Chandigarh Commission for Protection of Child Rights and State Legal Services Authority besides a number of educational societies of Chandigarh and Punjab. She is the member of various State level Committees of U. T. Chandigarh and Punjab in the field of Higher Education, Early Childhood Education, Mid-Day Meal, Right to Education Act, to name a few. She has authored a number of books, chaired various national and international seminars. Under her dynamic leadership, Govt. Home Science College Chandigarh topped amongst Chandigarh Colleges twice in a row and bagged 72nd and 86th position at National Level in NIRF Rankings released by MHRD in 2019 and 2020, respectively.

Dr. Vipul Khetarpaul graduated with distinction in the MBBS degree from KMC, Manipal University, and scored the highest marks in Manipal University. He is the recipient of various prestigious gold medals, prizes and awards. He did Masters in Surgery and Fellowship in Vascular Surgery from Washington University, St. Louis, USA. Presently, he is working as Assistant Professor, Department of Surgery, Washington University, School of Medicine in St. Louis, MO, USA. He has keen interest in human nutrition and is the co-author of two more books on Human Nutrition and presented various research papers on role of nutrition in degenerative diseases at various platforms.

Normal and Therapeutic Nutrition

Authors

Neelam Khetarpaul

Sudha Katyal

Vipul Khetarpaul

2022

Daya Publishing House®

A Division of

Astral International Pvt. Ltd.

New Delhi – 110 002

ISBN 978-93-5461-238-1 (PB)

Published by : **Daya Publishing House®**
A Division of
Astral International Pvt. Ltd.
– ISO 9001:2015 Certified Company –
47636/23, Ansari Road, Darya Ganj
New Delhi-110 002
Ph. 011-43549197, 23278134
E-mail: info@astralint.com
Website: www.astralint.com

Digitally Printed at : **Replika Press Pvt. Ltd.**

Preface

Food is the necessity of life. Everybody eats food for existence and many people enjoy eating food. There is a famous saying 'you are what you eat'. Right from the beginning, scientists were curious about the food they consumed, its passage in the body and its effect on health. This curiosity led to the science of nutrition and dietetics.

As we all know that good nutrition is an important part of leading a healthy lifestyle. Combined with physical activity, well balanced diet can help us to reach and maintain a healthy weight, promote overall health and reduce the risk of chronic and degenerative diseases related with obesity like heart disease, diabetes and cancer the prevalence of which are increasing tremendously all over the world including India due to changing life style *viz.*, sedentary physical activity and faulty food habits. Smart nutrition and choosing right kind of food choices in the daily diet can help the body cope more successfully with an ongoing illness too. Hence, understanding good nutrition and paying attention to what one should eat normally or in diseased condition can help to maintain or improve the health.

The risk factors for adult chronic diseases, like hypertension, CHD and type 2 diabetes, are increasingly seen in younger ages too, often a result of unhealthy eating habits and increased weight gain. Dietary habits established in childhood often carry into adulthood, so it is important to teach children right from beginning at home and then at school and college levels; how to plan and eat healthy diet right from the young age that will help them stay healthy throughout their life.

Presently, people have become very much nutrition and health conscious. Print and electronic media have also started pouring out nutrition messages to the public. Super markets are flooded with foods of varied types, natural, processed, and ready to eat. Under such situation, there is a need to provide scientific knowledge to the public about nutrition through nutritionists and dietitians. Moreover, keeping in view the relationship between intake of right kind of nutrition and maintenance of healthy life through life span, teaching of Human Nutrition and Dietetics course has also become a part of the course curricula of secondary schools and university level undergraduate and postgraduate courses. Now a days, it is becoming a well-sought course at school and college levels. Looking into the great demand of this course, many colleges have started undergraduate and postgraduate degree/diploma in Human Nutrition and Dietetics. The career after pursuing a degree in nutrition and dietetics is quite prolific too because people are becoming more aware of the health and fitness day by day. There is a great scope for Nutritionists and Dietitians now a days as they can get jobs in teaching, research, health clubs, hospitals or work as private consultants *etc.* Scope of doing practice in nutrition and dietetics is gaining momentum in India which was not so many years ago.

Looking at the present changing scenario in selection of academic courses by the students and public awareness about interrelationship between diet, health and disease, it has become very important to write text and reference books providing latest information about type of nutrition to be provided across the life span and type of diets to be provided in various diseased conditions including the newer diseases being faced recently for promoting and maintaining good health to them for reading and learning.

This book on Normal and Therapeutic Nutrition has been organized into two units. Unit one has thirteen chapters including basic principles of meal planning, planning meals for individual and family, food groups and their use in meal planning, use of food guide pyramids in formulation of balanced diets for various age groups, recommended dietary allowances, food exchanges, nutrition across life span (infancy till old age). The Unit two has17 chapters *viz.*, principles of diet therapy, therapeutic adaptations of normal diet, nutrition screening and nutrition assessment of hospitalized patients, enteral feeding, parenteral feeding, diets in infections, fever, COVID-19, HIV/AIDS, diabetes, cardiovascular diseases, gastrointestinal disorders, diseases of liver, pancreas, kidney, obesity and underweight, cancer, food hypersensitivity, inherited metabolic disorders, COVID-19 and sports nutrition. The authors have collected latest proved and updated detailed authentic information from journals, books, Web Med *etc*. Efforts have been made to include the recent topics and diseases too in this book.

We strongly feel that this book shall be a very informative and updated information source for nutritionists, dietitians and health professionals in teaching graduate and postgraduate students. Private diet consultants can also refer to it for getting updated knowledge about any topic related to normal and therapeutic nutrition while counselling their clients. It shall also be of great use for any reader who is interested to gain information about the role of human nutrition, clinical nutrition, therapeutic nutrition *etc*. in health and disease. We hope to get great response from our readers as in our previous publications. We are grateful to our students, colleagues, friends and medical doctors who directly or indirectly helped us in writing this book.

Neelam Khetarpaul

Sudha Katyal

Vipul Khetarpaul

Contents

Unit 2

Unit 1

1

Basic Principles of Meal Planning

Meal planning involves deciding what to eat and how much to eat each day at each meal. Generally, each day we eat three meals and perhaps an additional snack. The various members of the family partake of these meals. To ensure that these meet the physiological and nutritional needs of each family member, we need to plan the meals.

Meal planning means planning of well balanced nutritious meals which are colourful, attractive, appetizing, palatable, satisfying, enjoyable within the budget of the individual or family concerned. A meal plan proves to be a time, energy and money saver. It also avoids unwanted, unappetizing and unsatisfying meals. The meal planned should be flexible enough to take into consideration changes due to price fluctuations, seasonal foods, family tastes and desires, holidays, fasts, special family functions and guests.

To build and maintain healthy bodies, the knowledge of basic nutrition must be combined with imagination when planning meals. Appetite appeal is as important as nutritive value of meal planning because the best food is nutritious only when it is eaten. Hence, meal planning is both a science and an art. It is a science because planning meals in a nutritious way requires the meal manager to have scientific knowledge of various nutrients, their functions in the body, and food sources; while art is involved in presenting the nutritious meal in an attractive, colourful, appetizing, palatable and satisfying manner.

Planning for meals is planning which begins before shopping and preparation. It often does not end until the meal is over, for a person may revise the plan for dessert while planning the main course. In a more restricted sense, meal planning involves nothing but menu planning, that is, selecting the foods for a given meal. In the broader sense, meal planning includes menu making in accordance with goals plus plans for shopping, preparing and serving meals. With experience, it is possible to do all this without apparent plans or shopping lists. But in the beginning, a home maker has to make many rational decisions especially with limited resources of time, money, energy, knowledge and skill.

Goals of Meal Planning

There are four basic goals of meal planning for an individual or family:

1. Achieving nutritional adequacy
2. Matching meals to the budget for food
3. Likes and dislikes of family members
4. Matching meals to available time, energy and other resources available

1. Achieving Nutritional Adequacy

Good nutrition means more than subsistence. It means a diet that will contribute to health, vigour and happiness. Hence, it is the first prerequisite of a good meal plan. The meal planned should meet the nutritional needs of the individual and the family as a whole. Following statements regarding foods and nutrition may be helpful in meal planning:

- ✰ Nutrient needs can be met through food.
- ✰ No one food contains all the nutrients. Even milk which is considered to be a wholesome food, is deficient in iron and vitamin C.

- All foods except some that are highly refined or milled, contain several nutrients like carbohydrate, fat, protein, one or more vitamins and one or more mineral elements.
- Some foods are especially rich in one or more nutrients, for example, milk in protein, calcium, riboflavin and vitamin A; citrus fruits in vitamin C; green leafy vegetables in β-carotene and iron; sugars in carbohydrate; pulses in protein; and so on.
- Some foods are miserly in their nutrient content; for example, sugars and some fats provide only energy. Cereals and some roots and tubers contain only small quantities of protein, minerals and vitamins.
- To provide nutrient needs for calcium, iron, vitamin A and C, it is necessary to introduce specific foods into meals.
- Meals that provide enough protein, calcium and vitamin A will provide sufficient riboflavin and niacin.
- If foods containing protein, calcium, vitamin A and C are included in the meal, the remaining nutrient needs will be provided. However, when calories are restricted, this does not always hold true.
- Good protein quality can be satisfactorily supplied by mixtures of legumes or cereals combined with small quantities of milk or milk products in the vegetarian meal. Non-vegetarians can take meat or eggs to obtain good quality and quantity of protein.
- Include a variety of foods from different food groups every day in each meal in sufficient quantities to meet the needs for different nutrients.

For convenience in meal planning, different foods have been grouped depending upon their function and major nutrient contributions which (Figure 1.1) include:

- Energy giving foods
- Body building foods
- Protective foods

Include some of these foods from each food group in the meal so as to make it a balanced diet.

Hence, while planning the meal, foods from all the above mentioned food groups in right proportion should be selected and included to have a wholesome well balanced nutritious meal.

Individuals in a family differ in their nutrient needs according to their age, sex, size, physical activity, physiological status and pathological conditions. For example, if a family, consists of an infant, a school going child, lactating mother, hardworking adult man and grandparents, the nutritional needs of all these family members are different. The housewife planning the meals for such a family must ensure that nutritional requirements of all the family members are adequately met through judicious selection of foods from different food groups as mentioned above. This does not necessarily mean that different meals are to be planned for each family member. The same family meals, can be served in different amounts as per dietary allowances recommended by ICMR (2011) for various age groups with different levels of physical activity and physiological conditions. The pregnant woman, lactating mother, growing child or an adolescent girl can be given extra amount of milk, legumes or eggs to meet extra protein needs.

With slight modifications, in texture, consistency and ingredients, the same meal can cater to the individual nutritional needs of all family members. For example, a salad consisting of radish and carrot can be served in a grated form to the old family members while others can take in the stick form.

2. Matching Meals to the Budget for Food

A large part of the family budget is generally spent on food. The amount of money required for food for any given family depends on the size of the family, the number of children and their ages, the activities of various family members, special needs for pregnancy and lactation, and possible therapeutic requirements, if any. It costs more to feed some persons than others because their food needs are greater. For example, it costs more to feed men than women of the same age unless the latter are pregnant or lactating in which situation its costs more to feed her than him; it costs more

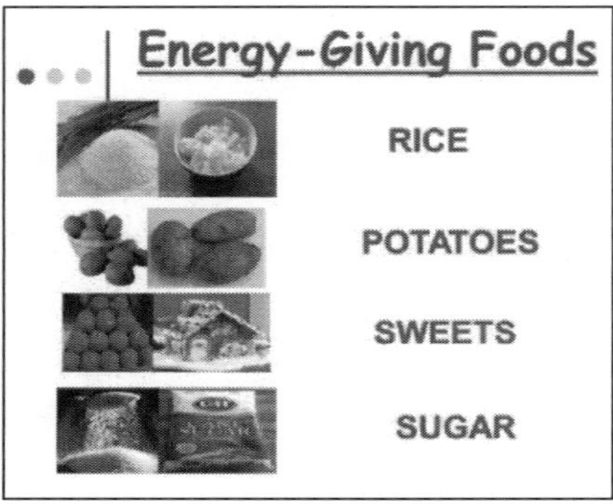

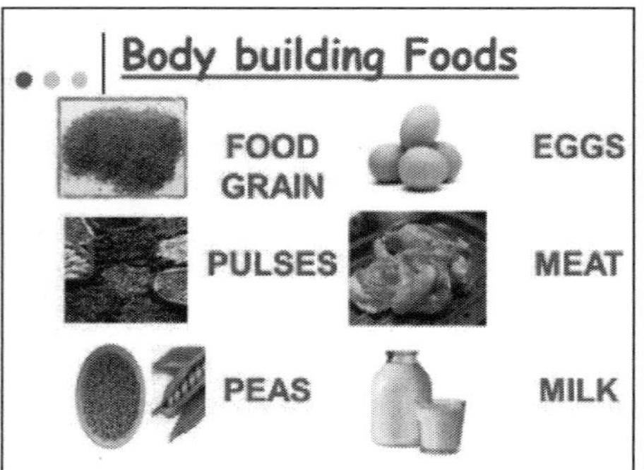

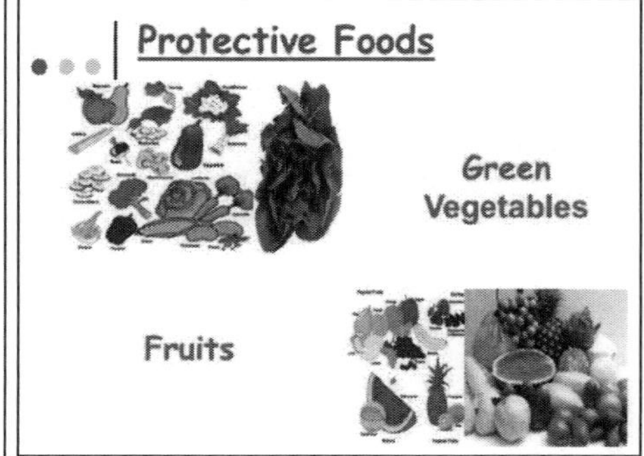

Figure 1.1 Basic Food Groups.

to feed boys than girls after they reach age nine, and it costs more to feed a six-year old than a three-year old. Family food needs are the composite of individual food needs. Significant determinants of the budget for food are the quantity and kind of foods, the family must buy.

The amount of money spent for food does not indicate how well the food has supplied nutrient requirements. The price of a food is ruled by the law of demand and supply. Food may seem more expensive than it really is because it is purchased more often than many other items or services.

Family income is a powerful determinant of how much can be spent for food. If one is spending one-fourth of the paycheck for food, one can easily obtain nutritious and palatable meals at a more reasonable price. If the income is average and one spends less than one-sixth of it, for food, the meals may lack healthful qualities or variety. It is a well established fact that although the absolute amount of money spent on food increases with increase in family's total income, the proportion of expenditure on food decreases. On the other hand, the proportionate expenditure on food increases with a decrease in the total income. In such circumstances, one spends a higher proportion of the food budget to procure staple foods rather than purchasing protective foods such as milk, fruits and vegetables.

It is important to plan and buy food items wisely within the given food budget so that maximum benefit is derived from the money spent. It cannot be definitely specified that how much amount or money should be spent on food. In a typical family, roughly the following division of expenditure on food may work as a guideline:

- ☆ 25 per cent of food money on meat, fish, eggs, cheese, poultry, legumes and pulses
- ☆ 20 per cent on milk
- ☆ 20 per cent on fruits and vegetables
- ☆ 20 per cent on cereals, breads and biscuits
- ☆ 15 per cent on fats, sweets and miscellaneous items such as spices, coffee, tea *etc.*

To economize and get more for the food cost in terms of nutrient value, satisfaction and convenience, one should consider the following:

- ☆ Plan what one to buy, and buy only that and no more. Consult Table 1.1 for an alternative, if a certain food item is not available, in order to see what else one can buy without upsetting the menu, budget and food value.
- ☆ Carefully select the brand of dry and tinned foodstuffs that suit the budget.

Table 1.1: Food Groups: Choose Foods from Every Group Each Day for a Balanced Diet

Five Food Groups	*Food stuffs*	*Main Nutrient Contribution*
1.	Milk, curds, paneer, cheese, skim milk powder Pulses, grams, *dals*, dried beans and peas Meat, fish, poultry, eggs, liver	Protein Calcium Riboflavin
2.	Fruits: Oranges, tomato, mango, papaya, amla, lemon, lime, guava, grapefruit *etc.*	β-carotene, Vitamin C
	Green leafy vegetables: *Sag*, cabbage, carrot, radish tops, beet root leaves, amaranths, spinach, mustard leaves, mint, coriander leaves	Calcium Iron
3.	Other vegetables: Brinjal, gourds, fresh beans, pumpkin, lady finger, cauliflower, carrots	Vitamins and minerals
4.	Cereals: Rice, wheat, corn, ragi, bajra, jowar Roots and tubers: Yams, potato, colocasia, green banana *etc.*	Carbohydrates B-vitamins Protein (in cereals)
5.	Fats and oils: Vegetable oils, butter, ghee	Energy, Essential fatty acids, vitamin A (in animal fat only)
6.	Sugar, Jaggery *etc.*	Carbohydrates in sugar only

Table 1.2: Daily, Weekly or Monthly Shopping List and Price Record of Grocery Store Purchases

	Groceries (Food items)			*Dairy Products*			*Vegetables*			*Fruits*			*Meat, Fish or Poultry*		
Date	*Item*	*Size or Weight*	*Cost*	*Item*	*Weight*	*Cost*	*Item*	*Weight*	*Cost*	*Item*	*Size or Weight*	*Cost*	*Item*	*Weight*	*Cost*

Total for month = Rs. ____________________

Number of guests and approximate amount they added to food costs = ____________________

Total cost of food, including meals away from home = ____________________

- ☆ Keep a price record of grocery store purchases and other fresh food items of last month (Table 1.2) to compare with current market prices so as to make best buy for the money spent.
- ☆ One should buy foods from fair price shops, super markets and smaller stores to compare quality, selection, service and price *etc.*
- ☆ Buy unperishable foods like cereals, pulses, spices, tea leaves, coffee powder, sugar and fats and oils *etc.* once a month if possible, but no more than once a week.
- ☆ One should be selective while purchasing fresh foods like fruits, vegetables, eggs, poultry, bread and butter. Use seasonal foods as they are both economical and nutritious. Buying foods that are in good supply generally yields good returns for the money spent.
- ☆ Store and cook the food properly as it can make a big difference in cost and as in nutrients retained and satisfaction received.
- ☆ If storage space is less, purchase smaller amounts, even when a larger number or size costs less per serving. Make proper use of left-overs and the commonly discarded foods *e.g.* leaves of radish and carrots.
- ☆ Store the right kind of food at required optimum temperature to avoid food wastage.
- ☆ Use all fresh foods within 24 hours of purchase unless one has the refrigeration facility and or similar storage facility.
- ☆ If facilities are available, one should maintain kitchen garden and/or poultry farms.

3. Likes and Dislikes of Family Members

Each family is different. Two families with similar nutrient requirements may not enjoy the same meals.

There are many psychological factors which influence our eating habits which get well established right from the childhood. Family should develop sound food habits among the children as childhood food habits have long lasting effect. The association of a food with a pleasant or unpleasant experience may cause a person to like or dislike that food all his life. Likes and dislikes for food may also be based on national traditions and customs, regional supply of foods, religious practices or even on publicity for or against a food. Another influence can be attributed to fads and fallacies.

This goal of achieving meals that the family wants and likes is not easy as it has to be achieved within the limits of provided resources *i.e.* budget and time. In fact, the same meal may not be accepted, eaten and enjoyed with the same degree of enthusiasm by all members of a family. While meal planning, likes and dislikes of all the family members should be kept in mind. There is no set guidelines for planning meals which assure the acceptance and enjoyment of meals by all individuals. However, some guidelines are based on knowledge of food preferences and sensory perceptions; others on recognition that man is not entirely rational or logical about his food and meals. This objective can be achieved by keeping the following factors into consideration:

☆ Make Food Appealing

Sense perceptions affect food preferences in proportion to the sensitivity of the individual. The senses of sight, smell, taste and touch are important in determining food preferences. One can make the simplest meal satisfying if one knows how does food affect the senses and how to create a pleasing effect.

The meal can be made appealing and satisfying if one keeps in mind the following factors in mind while doing meal planning:

i) Variety

Variety means better acceptability. It can be introduced through variation in colour, texture, flavour and method of preparation. Even favourite foods become less interesting when they are prepared day after day without variation. It is important to consider variety within each of the basic four food groups each day.

ii) Appearance

Food should look good to eat as the first appeal to the appetite is through the eyes. The initial reaction to food is based on its appearance, it is essential to consider the colours and shapes of food when planning meals. Avoid a menu of all white or neutral coloured foods. In a menu, colours and shapes should vary and blend in a harmonious way.

Some colours stimulate the autonomic nervous system of which the digestive system is a part. These colours are yellow, orange, orange-red, bright green and brown. Purple, red-purple, yellow green, olive and grey are not pleasing. Colour should be added in a meal through inclusion of foods having colours from different food groups.

Colours naturally associated with food have more appeal than artificial colours. Yellow tomatoes and watermelon do not look as tempting as the more familiar red and pink varieties.

Food should not be coloured artificially. For example, beet root or red radish which is added in the salad to introduce red colour – but many a times these are an artificial red replaced by white radish coloured red with colour which is in fact disliked.

Use of unusual colours can arouse suspicion *e.g.* green bread, red mashed potatoes *etc.* Artificial colour should be used wisely and sparingly *e.g.* few drops of green colour to mint or coriander chutney make it more like fresh one. The colour used should blend well with the flavour of recipe.

Select the food combinations in a meal which are colourful. If the menu for lunch includes wheat *chapati*, boiled rice, curd, cauliflower and potato vegetable, salad (onion, radish) and *kheer* (rice pudding), it would be monotonous in its colour scheme and nor appealing to eye. It can be made more colourful, appealing and attractive by making slight modifications *e.g.*, wheat *chapati*, tomato rice, mint *raita*, mixed vegetable (cauliflower, potato, peas, carrot, capsicum), salad (onion, lettuce leaves, cucumber, radish, beet root) and chocolate pudding. Use of garnishes such as chopped coriander, grated *paneer*, cream, onion and tomato rings, fried onions *etc.* can also add colour to the dish.

Some thought can also be given to the colour of dishes used for serving the prepared dish. For instance, orange juice served in blue glass would not be as appealing to the eyes as when served in a clear glass. Now-a-days modern ceramic and plastic dinnerware available in the market are in bright in colour and bold in design. Their wise selection and use can further add and enhance colour appeal to the food served in them.

iii) Shapes

Shapes also influence our feelings about food. A variety of shapes is more tempting than a similarity. Irregular shapes are often more attractive than smooth, definite ones. Different shapes in a dish make it more attractive. Round, flat, shredded, sliced or diced, cubes, spheres *etc.* are the different shapes that may be used. The arrangement in which food is served can be appealing or disturbing. Orderliness is more appealing to the appetite than a jumbled mingling of foods on a plate.

iv) Texture

Mechanical stimulation of sense organs present in the tongue, gums and hard and soft palate results in the judgement of mouth feel, that is, the texture of food or beverage. Texture variation is equally important in meal planning. To assure appealing texture variations in meals, the foods should be checked for contrasts of soft and firm, pliable and crisp, smooth and rough, thin and thick, moist and dry. Some texture qualities such as grittiness, fibrousness, toughness, mushiness, lumpiness and dryness lower the food acceptability. Textures like smoothness, crispness, softness, lightness, moistness, flakiness, fluffiness *etc.* are generally liked. Undercooked and overcooked foods lack the desired textures. Depending on the food, staling results in dryness, hardness, toughness and sogging which is not acceptable to the human palate.

In menu planning, recipes selected should have acceptable and varied textures. For instance, children and elderly persons will not like to eat well if all the foods in a meal require lot of chewing, while adults do not like all soft foods. Meal planned should have variations in textures to satisfy the requirements of different age groups in a family.

v) Taste

Taste of food means the sensation which the food imparts to taste buds in the tongue. Food must taste good otherwise it will not be eaten. No matter how appealing the food looks or how appetizing it smells, it will not be eaten unless it tastes good. The four basic tastes are sweet, sour, salty and bitter. In planning meals, one can take advantage of the effects of these four tastes to make the foods more satisfying. Most of foods are blends of two or more tastes. If in a menu, tastes are combined and contrasting tastes are used, it shall give more enjoyment to eating. Salt accents the sweet taste of food, as when used on orange, sweet lime, grapefruit, watermelon, apple *etc.* Similarly, sweet reduces the salty taste. Salt and sweet lessen sourness, so a little salt mixed with the sugar in cooking tomato sauce or ketchup.

Different tastes have different effects on our appetite. The effect of these tastes on the appetite should determine the type of food served at different courses of a meal. A tart, tangy, spicy food should be used for an appetizer course. Sweets such as fruit custard, bread pudding, rice pudding *etc.* should be served as dessert, not an appetizer.

vi) Flavour

Flavour means more than taste as it includes all *i.e.* odour, taste and feel or texture of the food. When the sense of smell is lost, all food tastes alike.

A meal should contain variety but a balance of flavours. A blend of seasonings and flavourings enhances the savour of a meal. Serve flavours that go well together in a meal. Food flavours range from the very intense to the very delicate. Foods that have intense flavours include garlic, onions, ginger, many herbs and spices. Foods that are delicate and bland in flavour include bread, wheat, rice, potatoes, white sauce, skim milk powder, peas, cucumber, papaya, banana *etc.*

The real art of menu planning is in the blending of flavours within a meal. Spicy foods and marked flavours should not dominate the meal, but should provide the accent. Do not serve more than one or two strong tasting foods in one course of a meal. Green onions, chilly, chutneys, garlic and pickles have distinct, strong flavours. For balance, they should be used with mild-tasting flavours. The overuse of herbs and spices in one dish or several dishes in one meal destroys the pleasing flavour experience. Flavour variety precludes using the same food in two or more forms at one meal; for example, tomato soup and tomato in salad or boiled rice and rice pudding *etc.* would not good choices. Combine foods in accordance with preferred flavour combination.

vii) Temperature

A contrast of hot and cold adds variety and attraction to a meal, have something hot with a cold meal and something cold with a hot meal. Even on hot days, one hot food enhances the pleasure derived from cold foods. Often, the service of a hot bread (*chapati, puri, nan* or *parantha*) may be a key factor in increasing the enjoyment of a meal. For most foods there is a right temperature at which flavour qualities are at their peak. Serve hot food hot and cold food cold for optimum perception of food flavours and the tastes. For example, the best temperature for ice cream is 16°F; serving it colder subdues the flavour and a temperature of 20°F or more may make the ice cream taste soured. The flavours of banana and pineapple reach full strength with warmth. In this regard, there are individual preferences too. Only those foods can be chosen in a meal which can be served at the most suitable temperature and this is possible if meal planning has been done wisely.

viii) Season

Season of the year also requires some consideration during meal planning. During cold winter day, hot soup, hot puddings (*halwas*), spiced hot curries may seem especially desirable. Summer meals also require that the same nutrients be provided, but one is likely to prefer less of heavy and rich foods.

In northern India, besides rice and wheat, other cereals *viz.* maize and bajra are used for making *'roti'* in winter season but not in summer as people consider maize and bajra as hot foods. Mustard *sag* with plentiful of butter or ghee is consumed in winter but is not liked in summer months. Fried and spiced foods are prepared in winter rather than in summer season. These considerations must be taken into account while planning meals for a family.

ix) Methods of Preparation

This can also add variety to the meal. The foods to be served may be cooked using different methods such as boiling, roasting, deep fat frying, shallow fat frying, baking, stewing, broiling, steaming *etc.* For example, eggs may be used in omelets of various kinds, soft cooked, hard boiled, poached or half fried. Similarly, tomato can be eaten fresh in salads or juice. It can be used in preparation of soup by boiling. Tomatoes with different kinds of stuffings can be baked. Hence, various kinds of dishes can be prepared from one type of food using different methods of cooking.

x) Satiety Value

A meal should be satisfying. It should allay the sense of hunger until almost time for the next meal. Fats, dietary fibre and proteins have a higher satiety value than carbohydrates. Including all foods rich in proteins or carbohydrates or fats make the meal monotonous. A combination of carbohydrate and protein foods interspersed with fat add variety to the meal.

Variety in the meal is the sum total of many kinds and classes of foods served in pleasing colour combinations with a judicious mixture of soft and crisp foods, of bland and sharp flavours, and hot and cold dishes. If one plans varied meals, the diet is more likely to be nutritionally adequate than it would be if just a few foods are used day after day.

xi) Sociological Preferences

These include religious and socio-culture traditions and customs which influence our choice of food. For instance, in Hindu religion, beef is prohibited while the same is true for pork in Muslim and Jew society. In Jain religion, generally onion and garlic are not included in the meal. These customs and beliefs also influence the food that is eaten on particularly some religious occasions and during fast days. It may be necessary to alter the plan on such days and use an appropriate substitute. For example, during some fasts, cereals are not to be consumed and they can be substituted by *sawank, kuttu ka atta,* sago *etc.* These factors should be kept in mind while planning meals.

xii) Food Fads and Fallacies

Fallacious beliefs concerning food are common. Most of these trickle down unquestioned, from generation to generation. Some of them, however, have their origin in coincidental illness following the eating of a particular food or food combination. Some of the common examples of dietary misconceptions prevalent among Indian society are:

a) Milk consumed along with fish in any form may cause Vitiligo. Entirely erroneous as such a combination is consumed daily by many.

b) Pregnant woman should not eat papaya as it can cause miscarriage. In fact, it is far from truth. Papaya is a very good source of β- carotene and should be taken in the diet.

c) If you drink buttermilk at night with the meal, it can cause knee joint pains all your life. But at the same time, it is alright to have yoghurt. This has no scientific basis.

d) Dry fruits have 'hot' properties and are best eaten only in winters. Whereas the truth is that dry fruits are very healthy and a handful of them should be taken daily throughout the year. There is nothing like 'hot' or 'cold' foods.

e) Skim milk has little nutritive value. Skim milk is whole milk with just butterfat removed. It remains a superior food with all its protein, calcium and vitamins.

f) Foods cooked in aluminum utensils can cause cancer. A small amount of aluminum certainly gets dissolved in food while cooking, but little if any is absorbed from the gastrointestinal tract.

g) Fruit juice is an important part of a slimmer's diet. Juice of any fruit has all its calories. The only thing that is lost is fibre and some vitamins, which could rather be helpful.

h) Vegetable oils are light and not fattening when compared to *desi* ghee. Vegetable oils are comparable to animal fats in their caloric value and are as fattening.

One should be cautious of fad diets or for that matter any diet that encourages limiting or leaving out essential food groups.

xiii) Individual Preferences

It is especially important to consider the individual likes and dislikes of various family members. If a particular food is disliked by everyone, it may be possible to substitute its nutritional equal. If a food is disliked by only one member of the family, it is advisable to serve it during that person's absence or that food can be tried out in some other acceptable dish. For example, if a child in a family does not like spinach *ka sag*; he may be served spinach soup or nutritious parantha containing wheat flour, spinach puree, left over cooked dal or *besan*. If a person does not like to drink milk, he may be given milk in other form such as banana milk shake or mango shake, cold coffee, *kulfi*, *kheer* (rice, sago or *paneer*), *paneer bhuji*, curd or any *raita etc.* There are many ways to prepare several dishes from a single food.

Old people cannot eat crunchy foods that involve lot of chewing. Instead of serving carrot in the form of carrot stick in the salad, grated carrot in salad or carrot juice can be planned for them. Young children generally do not tolerate highly spiced food preparations. For them before adding spices, *dal* or vegetable or curries can be taken out from the family pot.

xiv) Family Size and Composition

This has a great role to play in meal planning. It is known that if the family income remains constant but the family size increases, the money spent for food per person decreases. Such is the case in a family where there are several closely spaced children. As the money available remains unchanged even with increase in family size, a larger part of it shall be spent for purchasing staple foods (wheat, rice *etc.*) and expenditure on purchase of protective foods (fruits, vegetables, milk *etc.*) may decrease which affects the meals adversely. The case is different if money available also increases with increase in family size. In that case, meal planning is not that tedious as in the former case where meal plan has to be a judicious and careful one so that a nutritious and satisfying meal is provided to all family members within a limited budget.

Family composition affects the kind and amount of food needed and the pattern of meals served. For example, a family consisting of young children (below 5 years of age) requires more milk. Adolescents in the family require more food to support growth and their activities than adults. The older persons in the family require less food but with different consistency due to lack of teeth and poor chewing capacity. For school going children in the family, packed lunch is required during school time. Therefore, family size and composition both have to be considered while meal planning.

4. Matching Meals to Available Time, Energy and Other Resources Available

In our modern civilization, with each family member, including the homemaker, busy in many and varied activities, meals that can be prepared and served in comparatively short period of time, are often desirable. It is very important to make wise use of available resources *i.e.* time and energy while planning meal for the family. By doing so, a homemaker can avail leisure time for herself and can plan some leisure time activities for her family members too. Like money, time and energy also require budgeting for their best and efficient use. In this regard, some important tips are:

- ☆ Menus should be planned for a week at a time or at least 3 or 4 days in advance. By doing so, trips to the market may be limited to once or twice a week. This in itself, is time and energy saving as it ensures having the necessary food supplies to work with and eliminates numerous trips to the grocery store.
- ☆ Shopping should be done when market is less crowded.
- ☆ The kitchen should be arranged in efficient working units, keeping utensils and supplies near the centre where they are to be used.
- ☆ Time and energy saving kitchen equipments and appliances should be used, if possible and affordable within the family budget. They all affect the working schedule in the home since their efficient use will cut the time spent in food preparation as well as other operations in the home.
- ☆ Convenience foods, though expensive, can be used when time saving is more important than the slightly higher cost incurred on food.
- ☆ Plan a work schedule of each food.

2

Planning Meals for Individuals and Family

Meals a day should be balanced and properly spaced. Do not skip meals, plan them. If any of the meals in a day is neglected, omitted or skipped, there is too much nutritional load put on the remaining meals and snacks in a day's intake. For example, in case of skipping breakfast, the total intake of nutrients for the day shall be inadequate or it shall be compensated by overeating of some other meal which shall consequently result in concentration of nutrients in a meal later in the day instead of being divided equally over 24 hours. There are also chances that skipping of a meal may result in more of snacking of 'empty' calorie fast foods and drinks which may be deleterious for optimum health and nutrition in the long run. Hence, one must have the knowledge of the following:

- ☆ How many meals must be prepared each day?
- ☆ At what time, the meals should be served?
- ☆ What pattern should be followed in selecting types of food in different meals?
- ☆ How to schedule meal preparation work to fit in with other activities?

How Many Meals a Day

In Indian households, the conventional plan of three or four meals a day is generally followed. It includes:

- ☆ Breakfast
- ☆ Lunch
- ☆ Dinner

In between these three basic meals, especially between lunch and dinner, evening tea with light snack is also an established custom in majority of our households. There is no apparent physiological reason to serve the day's food at three meals. Some people prefer four or even six smaller meals. Many people today have between-meal foods. There is some evidence of an advantage to eating smaller meals at shorter intervals, if the total amount of food satisfies the body needs. The mid-morning tea or coffee during office hours or at work place, a snack during the recess period in the school *etc.* are part of the overall meal plan. A typical meal pattern usually followed by an Indian urban family is shown below:

Typical Meal Pattern for an Urban Family

Breakfast	Milk/tea/coffee Bread butter/bread jam/bread omelet or Simple *parantha*/Stuffed *parantha* with butter Curd Tea or Milk/tea/coffee *Vada*/*idli* with chutney or *samber* Seasonal fruit
Lunch	*Chapati*/Rice Dry seasonal vegetable or legume preparation Curd/*Raita* Salad Seasonal fruit
Evening tea	Tea or coffee Biscuit/*Chiwda*/Sprouted *dal chat*

Dinner	*Chapati*/Rice *Dal* or vegetable curry Seasonal vegetable preparation Salad
At night	Milk

In general, planning meals in advance helps to balance the required nutrients. It ensures balanced, attractive and palatable meals. It is also more economical with regard to time and money.

Begin with a Good Breakfast

It is important to have breakfast and to avoid the temptation to snack between meals. As breakfast usually follows a 12-15 hours period of starvation (the word 'breakfast' means to break a fast), it is important to provide energy to the body early in the morning. Start the day with a good breakfast. Ideally, the whole family should eat breakfast together. However, group breakfast is impractical as different family members may have to leave for work place, school or college at different hours.

Eating breakfast has been found to be essential for maximum efficiency – both physical and mental, and avoiding breakfast results in decreased output and decreased work concentration and mental alertness. Studies have shown that skipping breakfast results in:

- ✰ Attitude towards school work and scholastic achievement is poorer.
- ✰ Efficiency in physiologic performance decreases in late morning hours.
- ✰ Greater hunger during later part of the day.

A good carefully planned breakfast should supply one-fourth or the days nutrient needs. On the contrary, the general observation is that breakfast is either skipped or taken in a hurry and most family members carry packed lunch to work or study place. Adolescent girls or those who are on weight reduction diets are apt to develop a 'no-breakfast' fad. Children may prefer fifteen minutes more sleep than to take breakfast. If the house lady is non-working, she has time to plan, prepare and serve the nutritious breakfast and packed lunches to all family members. But in case of working women and no additional household help, breakfast is generally a neglected meal. It may consist only of a cup of milk/tea/coffee with biscuits or bread toast with butter or jam. An ideal breakfast pattern as shown under should be nutritious as well as easy and quick to prepare. If it furnishes 1/4 to 1/3 of the nutrients especially energy and protein needed in the day, it is adequately nourishing.

Breakfast is the time to serve foods that awaken the appetite and appeal to the sight, smell and taste. Colourful foods and table settings, good flavour and daily change in the menu makes the breakfast more enjoyable. There is no doubt that variety is desirable but a radical change in foods or introduction of a strong flavour may not be accepted by the family. At this first meal, most people want familiar foods. The breakfast pattern should be such that adults will need no-mid morning foods. Children, adolescents, pregnant and lactating women may need extra energy food between breakfast and the lunch.

Many combinations of foods are possible for breakfast. It should include protein rich foods such as egg or milk, cereal or breadstuff or both and a beverage. Breakfast cereals may be hot or cold; breads or *paranthas* may vary from simple ones to stuffed ones. If citrus fruit, a good source of vitamin C is included at breakfast, not only the day's allowance for this nutrient is assured but it also awakens the appetite. A breakfast may be light or heavy depending upon the individual's activity and preferences. Sunday and holiday breakfast should be a special occasion.

Different Breakfast Menus

I	*II*
Nutritious *paranthas* (Whole wheat flour + *Besan*/ cooked *dal* if leftover + GLV + onion + salt) Curd Orange/Apple juice	Stuffed *parantha* (Cauliflower or radish or boiled potato or *paneer*) Butter Banana/Apple/Orange Tea/Coffee/Milk
III	***IV***
Bread toast Egg omelet Milk Any seasonal fruit	Instant breakfast cereal Milk (hot or cold) Orange juice
V	***VI***
Porridge (Broken wheat or rolled oats) Milk Boiled egg Any seasonal fruit	*Suji Upma* with seasonal vegetables Banana/Apple/Orange Tea/Coffee/milk
VII	***VIII***
Pohe (Flaked rice with seasonal vegetables) Any seasonal fruit Tea/Coffee/Milk	*Idli/Vada Samber* or chutney Any seasonal fruit Tea/Coffee/Milk
IX	***X***
Bread sandwich (Cucumber slice, tomato slice, cheese slice, butter, green chutney) Any seasonal fruit Milk/Tea/Coffee	Bread toast with jam Boiled egg Any seasonal fruit Tea/Coffee/Milk

Lunch

In the metros, most of the workers carry packed lunches from home or depend upon lunches available in the canteens at work places. In big cities, tiffin service is also available. In small cities, distances are generally not much between home and work places, in such cases people prefer to visit home for taking lunch in break period.

Even at home, the noon meal may be coarsely assembled and hurriedly eaten especially when both the parents are working. Perhaps even more neglected than the worker and school children are the homemaker and preschool children. If the husband and children carry packed lunches, the homemaker is alone during lunch time and generally has a tendency to not to plan and prepare the elaborate lunch but use the leftovers.

Occasionally food choice for the worker and school children is so limited that there is little one can do to improve the lunch pattern other than carrying a lunch. A heavy noon meal is not desirable for adults doing sedentary work. Whatsoever may be the case, the noon meal is often slighted and neglected. The 'snack' lunch is apt to be monotonous and dry.

Nutritionally, lunch should provide 1/4 to 1/3 of the essential nutrients, including protein, energy, minerals and vitamins. Lunch can be inexpensive, easy to prepare, tasty and nutritionally adequate with a little foresight and planning. Examples of a few good lunch patterns are given below. These basic patterns offer a wide variety from which to choose the one that suits age and activity requirements.

Basic Pattern for Packed Lunch

Some basic patterns are given as under:

Pattern I	*Pattern II*
☆ Boiled rice/*pulav*/*chapati*/*parantha* ☆ Dry seasonal vegetable *e.g.* cauliflower and potato or ☆ Dry bean preparation *e.g.* dry *channa* ☆ Pickle/green chutney ☆ Seasonal fruit like guava, *ber*, banana, apple or orange	☆ *Pav* ☆ *Bhaji* ☆ Seasonal fruit
Pattern III	***Pattern IV***
☆ Sandwiches containing a wedge of lettuce, carrot sticks, or radish, cucumber rings, cheese slice ☆ Fruit	☆ Vegetable patties ☆ Fruit
Pattern V	
☆ Masala *dosa* ☆ Coconut chutney ☆ Seasonal fruit	

Nutritionally balanced packed lunches	
Pattern VI	***Pattern VII***
☆ *Paushtik roti* (wheat + *besan* + GLV + oil) ☆ Green chutney ☆ Seasonal fruit	☆ Rice-*dal cheela* ☆ Potato vegetable ☆ Groundnut *chikki*
Pattern VIII	***Pattern IX***
☆ Sprouted moong sandwiches ☆ Tomato/Cucumber ☆ Fruit	☆ Broken wheat pulao (*Dalia*, sprouted *dal* + vegetable) ☆ *Channa – murmura ladoo* ☆ Fruit

Some Important Tips for Packed Lunch

- ☆ Curry vegetables should not be included in tiffin as they can be messy.
- ☆ Instead of a dry *chapati*, *chapati* smeared with a little butter/ghee in summer or vegetable oil in winter should be preferred as fat or oil used keeps the *chapati* soft which is easy and palatable to consume.
- ☆ *Roti* stuffed with seasoned vegetable mixture can also be an ideal choice. To make *roti* nutritious as well as manageable in the tiffin, it is better to prepare *roti* from the dough containing the mixture of wheat flour, besan or leftover cooked dhal or boiled and mashed potatoes and pureed spinach or *methi* and milk. Such a *roti* packed in the tiffin along with pickle or curd with fruit is a complete nutritious meal which is easy and quick to prepare and pack.
- ☆ Addition of pickle (mango, mixed vegetable, teent *etc.*) makes the packed meal more interesting and palatable.
- ☆ In summer season, one should avoid taking curd in the tiffin.
- ☆ Children prefer noodles in the tiffin. To make noodles more nutritious, seasonal vegetables can be mixed in it which otherwise children do not prefer.
- ☆ For children and adolescents, packed sandwiches are favourites for the noon meal. Care should be taken that sandwich filling is neither so moist that it soaks the bread nor so dry that it is difficult to eat.
- ☆ In the packed lunch, only those fruits which carry well, are crunchy, tart and easy to eat *e.g.* apple, banana, orange, guava *etc.* should be included. Fruits like mango, papaya, sapota *etc.* are not desirable.
- ☆ Vegetable patties, brown bread slices stuffed with cooked dry vegetables or *paneer* preparation and toasted in non-stick sandwich maker are often liked by children and are easy to carry.

Basic Pattern for at Home Lunch

At home, some broader choices than the packed one can be offered at noon meal even though the individual is in a hurry to return to work place. Examples of some basic patterns for at home lunch are mentioned below:

Menu for Lunch at Home

Pattern I

- Boiled rice/vegetable *pulav* or *chapati* or both
- Curry preparation (*Rajmah* curry or *Kabuli channa* or black *channa* or vegetable *kofta* preparation *etc.*)
- Dry seasonal vegetable preparation (Brinjal *bhurta*/cauliflower and potato/*methi-aloo*/peas-mushroom/*palak*-paneer/Ghia vegetable/Dry *bhindi etc.*
- Curd/*Raita* (mint/*boondi*/*ghia*/*aloo etc.*)
- Salad (Onion, tomato cucumber, carrot, radish, beetroot *etc.*)
- Green chutney/Pickle
- Any seasonal fruit/Dessert

Pattern II

- *Makki ki roti*
- *Sarson ka sag*
- Butter
- Buttermilk
- Salad
- Seasonal fruit

Pattern III

- Boiled rice/*pulav*/*Chapati*
- *Besan* curry
- Seasonal vegetable
- Salad
- Seasonal fruit

Different food items to be served in the noon meal can be served at appropriate temperature at home which makes the meal more interesting and palatable. A non-vegetarian family may substitute *dal* with egg, fish or meat preparation. In routine, soups and desserts are not normal during the noon meal in an urban household where both the partners are working.

Dinner

The chief meal of the day may be served at noon or at night, depending upon family custom, working pattern of family members, availability of time *etc.* Generally, it has been observed that dinner time is family time in Indian households. It is the meal at which all the family members of the family join together. Therefore, more emphasis is to be laid on dinner to supply the necessary nutrients to the family members according to their requirements.

Dinner is often a good time to balance the calories for the day. In terms of minerals and vitamins, it must make up any lack that may have occurred in the other meals. The custom of making the dinner meal much larger than the other two, however, is not a good practice. When one meal provides a large share of the day's carbohydrates and calories, the normal metabolic mechanisms are overloaded. As a result, alternate metabolic pathways may be used. These alternate pathways favour the synthesis of adipose tissue and cholesterol, both of which are not desirable.

Soups, salad and dessert if not included in the noon meal, should be served at dinner. Appetizers or soups are a good way to get the family together and ready for the meal and to fill the time when children become restless, waiting for dinner. Dinner is the time for vegetables. Depending upon the family customs, some prefer to take *dal* preparation at dinner only *e.g.* Punjabis while others prefer *dal* only at noon meal *e.g. marwaris.* Any way, *dal* preparation if not served at lunch must be included in the dinner plan. Non-vegetarians can replace '*dal*' preparation with meat, chicken or fish curry. The dessert for dinner can be a substantial one, such as *kheer*, pudding, fruit custard, fruit cream, *halwa etc.* An example of typical menu for dinner generally preferred in Northern Indian households is given as under:

Dinner Menu

- *Dal* preparation (Moong/*channa*/*arhar etc.*)
- Seasonal vegetable preparation (Cabbage and peas, peas and potato, *aloo-bangen*, mixed vegetables, *sag*, *methi-aloo*, *ghia kofta*, stuffed *karela*, stuffed *bhindi etc.*).
- *Chapati*
- Salad
- Pickle/*Papad*
- Dessert (*Kheer*/pudding/custard/*kulfi*/ice cream/*halwa*/Fruit *etc.*

Whatever the dinner menu for the day, a few well-chosen, good foods make a more satisfactory meal than a large collection, indifferently prepared.

Snacks

There can be no rigid rule concerning between meal eating. Active children often benefit by having a mid-morning or mid-afternoon snack provided that it is of such a nature that the appetite at mealtime is not suppressed. Workers in factories, industries, hospitals, teaching profession, students and others experience a 'lift' with a snack. Pregnant and lactating women should also take a snack in between the meals. It is the quality of snack which is more important. Fried *samosas*, *bondas*, french fries, *patties*, carbonated drinks and concentrated sweets may contain little other than carbohydrate and may destroy mealtime appetite; but fruits, fruit juices, fruit *chat*, buttermilk, light snack like sandwich *chiwda, Bhel puri etc.* depending upon the individuals' activity and the interval between meals, carry many valuable nutrients. Between-meal snacks, properly chosen, may actually aid some persons to maintain weight by reducing the tendency to overeating at mealtime.

Teens often obtain one-fourth to one-third of all their energy and major nutrients from snacks. Teenagers can obtain many nutrients from snacking. Even fast food restaurants offer some good food choices. By choosing wisely, one can eat at fast food restaurants and still consume a very good diet. Snacks and fast food restaurants in and of themselves are not the problem: poor food choices are.

Selection of Snacks

No snack can be a perfect one. However, some snacks are nutritious than the others. Most of the snack foods are rich in energy but contain fewer amounts of protein, vitamins and minerals. The nutritive value of some of the snacks is given in Table 2.1. By simple planning well in advance, nutritious snacks can be prepared and given to children when they return from school as they require something to nibble in order to satisfy their hunger. Fruit yoghurt, fruit milk shake, *patties* containing seasonal vegetables, cheese sandwich, vegetable *rava upma, dahi wada*, fruit *chat*, sprouted *dal* chat, *Atta* noodles with vegetables, mushroom pizza, egg sandwich, stuffed *masala dosa*, groundnut *chikki*, vegetable burger, *channa-gur murmura etc.* are examples of some of the nutritious snacks which can be served in between the main meals. Many people in the urban areas prefer fast food since it is relatively low cost food available at a food joint which provides quick service and is convenient. These should be eaten only once in a while and care should be taken that a snack or fast food should not replace a good wholesome meal.

Table 2.1: Nutritive Value of Snacks

Preparation	*One Serving Size*	*Calories (Kcals)*
Savoury snacks		
Pakora	8 Nos.	280
Samosa	1 No.	200
Pizza (cheese and tomato)	1 slice	200
Potato bonda	2 Nos.	200
Besan ka pura	1 No.	220
Batata wada	2Nos.	190
Dahi vada	2 Nos.	180
Vada	2 Nos.	140
Cheese balls	2 Nos.	250
Masala dosa	1 No.	200
Masala vada	2 Nos.	150
Sukhadi	2 pieces (2'x2")	159
Rava upma	¾ cup	150
Uttapam	1 medium	205
Medu wada	2 medium	200
Cheese sandwich	2 pieces	230
Egg sandwich	2 pieces	230
Vegetable sandwich	2 pieces	150
Chiwda (rice flakes and peanuts)	¾ cup	210
Chiwda (puffed rice and peanuts)	¾ cup	154
Chat (*dahi-pakori*)	5 pieces	220
Potato chips	10 pieces	100
Pea-*kachori*	2 Nos.	380
Sago *wada*	2 Nos.	210
Chutneys		
Coconut/groundnut/*til*	2 tbsp	120
Tomato	1 tbsp	10
Tamarind (with jaggery)	1 tbsp	60
Sweets and desserts		
Besan burfi	2 pieces	400
Ice cream	½ cup	200
Groundnut *chikki*	2 pieces	290
Fruit cake	1 piece	270
Jam	1 tsp	20
Halwa	½ cup	320
Sandesh	2 Nos.	140
Beverages		
Cold drink	1 bottle (200 ml)	150
Fresh lime juice	1 glass (200 ml)	60
Sharbat	1 glass (200 ml)	200
Buttermilk (lassi) with 2 tsp sugar)	1 glass (200 ml)	110
Tea (50 ml toned milk + 2tsp sugar)	1 cup	75
Coffee (2 tsp sugar + 100 ml toned milk)	1 cup	110

Source: Dietary Guidelines for Indians: A Manual. NIN, ICMR, 2011.

Typical Meal Plan for North Indian Urban Family

Family		
Husband	38 years	
Wife	33 years	
Son	10 years	
Daughter	4 years	
Recommended dietary intake		
	Energy (Kcal)	***Protein (g)***
Husband	2425	60
Wife	1875	50
Son	2190	54
Daughter	1690	30

Table 2.2: Balanced Diet for the Family

Sl.No.	Food Group	g/ Portion	Husband		Wife		Son		Daughter	
			V	NV	V	NV	V	NV	V	NV
1.	Cereals and millets	30	14	14	10	10	11	11	7	7
2.	Pulses	30	2	1	2	1	2	1	1.5	0.5
3.	Milk	100 ml	3	3	3	3	5	5	5	5
4.	Roots and tubers	100	2	2	1	1	1	1	1	1
5.	Green leafy vegetables	100	1	1	1	1	1	1	0.5	0.5
6.	Other vegetables	100	1	1	1	1	1	1	0.5	0.5
7.	Fruits	100	1	1	1	1	1	1	1	1
8.	Sugar	5	5	5	4	4	7	7	6	6
9.	Fats and oils (visible)	5	4	4	4	4	5	5	5	5
10.	Egg/Meat/Chicken/Fish	30	-	1	-	1	-	1*	-	1*

V = Vegetarian; NV = Non-vegetarian.

* One portion of pulse may be exchanged with one portion (50 g) of egg/meat/chicken/fish in case of son and daughter.

Family Menu Plan (Vegetarian)

A sample menu plan for a family is given as under:

Sl.No.	Name of Meal	Husband	Wife	Son	Daughter
1.	**Morning tea**	1cup	1 cup	-	-
2.	**Breakfast**				
	Bread toast with butter and cheese slice	2 slices	2 slices	2 slices	1 slice
	Apple	1	1	1	½
	Milk	1 cup	1 cup	1 cup	1 cup
3.	**Lunch (packed)**				
	Paushtik chapati	3	2	-	-
	Stuffed *parantha*	-	-	One	One
	Dry mixed vegetable	1 cup	1 cup	½ cup	½ cup
	Mango pickle	1 slice	1 slice	1 slice	1 slice
	Groundnut *chikki*	1 piece	½ piece	½ piece	¼ piece
4.	**Tea**	1 cup	1 cup	-	-
	Biscuits	2	2	2	2
	Banana milk shake	-	-	1 glass	½ glass
5.	**Dinner**				
	Tomato soup	1 cup	1 cup	1 cup	½ cup
	Channa dal	1 katori	1 katori	½ katori	¼ katori
	Palak-aloo	1 cup	1 cup	½ cup	¼ cup
	Salad				
	Chapati	3	2	One	One
6.	**Bed time milk**	1 cup	1 cup	1 cup	1 cup

3

Food Groups and their Use in Meal Planning

Different foods contain different nutrients and other healthful substances. Foods contain most of the nutrients but in different amounts. No single food can supply all the nutrients in the amounts one needs. For example, oranges provide vitamin C and folate but no vitamin B_{12}, cheese provides calcium and vitamin B_{12} but no vitamin C.

While planning meals for an individual or a family one must ensure that the meal is nutritious and a balanced one. In a balanced diet, all the essential nutrients should be present in required amounts and proper proportions to preserve and promote good health. Now, the question arises as to how one might attain a balanced diet. If all the time during ordinary meal planning, one has to consult the tables of Recommended Dietary Allowances for Indians as suggested by ICMR (2011) and Indian Food Composition Tables (Longvah *et al.*, 2017), it would be quite an inconvenient, ardous and time consuming task for the meal planner. Fortunately, nutritionists have devised an elementary method that simplifies planning a balanced diet. This is based on Basic Food Groups. A thorough knowledge of these food groups makes planning nutritious meals easy and fast.

What are Food Groups?

On the basis of nutrients present, foods have been classified into eleven, three, four or five basic groups.

Eleven food groups: Foods have been broadly categorized into 11 groups depending upon their nutritive value. These are:

- ✰ Cereals and millets
- ✰ Pulses
- ✰ Nuts and oilseeds
- ✰ Vegetables
- ✰ Fruits
- ✰ Milk and milk products
- ✰ Eggs
- ✰ Meat, fish and other animal foods
- ✰ Fats and oils
- ✰ Sugar and other carbohydrate foods
- ✰ Spices and condiments

The 11-Food Group plan as suggested by US Department of Agriculture, 1964 is given in Table 3.1.

Table 3.1: 11-Food Group Plan

Sl.No.	*Food Group*	*Main Nutrients*
1.	a) Milk and cheese	Calcium, phosphorus, vitamins and protein
	b) Ice-cream	Fat and carbohydrates
2.	Meat, poultry and fish	Proteins, phosphorus, iron, B-vitamins
3.	Eggs	Proteins, fats, vitamins, iron and phosphorus
4.	Dry beans, peas and nuts	Proteins and B-vitamins
5,	Flour, cereals and baked products (whole grains enriched restored)	Thiamine, niacin, riboflavin, iron, carbohydrates and fibre
6.	Citrus fruits and tomatoes	Ascorbic acid and potassium
7.	Dark green and deep yellow vegetables	β-carotene, ascorbic acid and iron
8.	Potatoes	Carbohydrates and ascorbic acid

Sl.No.	Food Group	Main Nutrients
9.	Other vegetables and fruits	Ascorbic acid and fibre
10.	Fats and oils	Essential fatty acids and vitamin E
11.	Sugar, syrup and preserves	Carbohydrates

However, on the basis of functions, foods may also be classified into three groups (NIN, 2011) as given in Table 3.2

Table 3.2: Food Groups on the Basis of their Functions in the Human Body

Food Group	Major Nutrients	Other Nutrients
Energy rich foods	**Carbohydrates and fats**	
	Whole grain cereals, millets	Protein, fibre, minerals, calcium, iron and B-complex vitamins
	Vegetable oils, ghee, butter	Fat soluble vitamins, essential fatty acids
	Nuts and oilseeds	Proteins, vitamins/ minerals
	Sugar	Nil
Body building foods	**Proteins**	
	Pulses, nuts and oilseeds	B-complex vitamins, invisible fat, fibre
	Milk and milk products	Calcium, vitamin A, riboflavin, vitamin B_{12}
	Meat, fish, poultry	B-complex vitamins, iron, iodine, fat
Protective foods	**Vitamins and minerals**	
	Green leafy vegetables	Antioxidants, fibre and other carotenoids
	Other vegetables/fruits	Fibre, sugar and antioxidants
	Eggs, milk and milk products and flesh foods	Protein and fat

Basic Four Good Groups

As per NIN (1998), foods are conveniently grouped as:

1. Cereals, millets and pulses
2. Vegetables and fruits
3. Milk and milk products, egg, meat and fish
4. Oils and fats, nuts and oilseeds and sugar

Basic Five Food Groups

On the basis of nutrients present, the following five basic food groups have been suggested by Thimmayamma and Pasricha (1996).

Table 3.3 Basic Five Food Groups

Sl.No.	Food groups	Nutrients
1.	Cereals and millets	Energy, iron, thiamine, niacin and folic acid
2.	Pulses, legumes, nuts and oilseeds	Protein, energy and calcium
3.	Milk, egg and flesh foods	Protein, calcium, vitamin A
4.	Vegetables and fruits	Calcium, iron, β-carotene, vitamin B-complex, vitamin C
5.	Fats and sugar	Energy, essential fatty acids (fats only)

These foods have been discussed in detail as given below:

1. Cereals and Millets

Foods made from grains (wheat, rice, oats *etc.*) help to form the foundation of a nutritious diet as they provide vitamins, minerals, carbohydrates (starch and dietary fibre) and other substances that are important for good health.

Rice and wheat are the major cereals consumed in India. Wheat is the staple food in North India while rice in South India. Other millets used are sorghum (*jowar*), pearl millet (*bajra*), ragi and corn. They contribute 70 to 80 per cent of the daily energy intake of most of the Indian population. They are the cheapest and widely available source of energy especially among the poor income families. As the income increases, the amount of calories derived from cereals and millets decreases, but still in the highest income groups, they contribute 50 to 60 per cent of the daily energy intake. Besides energy, they also provide protein, fibre, calcium, iron and B-complex vitamins (Table 3.2).

Cereals provide more than 50 per cent of daily protein intake as they contain 6 to 12 g protein per 100 g of edible portion. However, they lack in an essential amino acid lysine which is present in sufficient amount in pulses. That is why nutritionists recommend mutual supplementation of cereals and pulses which is very common in Indian dietary pattern *viz., khichdi, idli, dosa etc.* is recommended to improve the protein quality of vegetarian diets. Among cereals, the protein quality of rice is superior to other cereals and millets.

Grain products are low in fat, unless fat is added in processing, in preparation or at the table. The total fat content in cereals may vary from 2 to 3 per 100 g. Considering the amount of fat present in cereals in Indian diets, it is estimated that fat present in cereals in our diets can meet more than 50 per cent of our essential fatty acid requirement. Cereals along with legumes can nearly meet the requirement of EFAs in adults.

All unmilled cereals are also sources of minerals like calcium, phosphorus and iron. Among cereals, polished

Facts About Nutritive Value of Cereals

- ☆ All cereals are excellent sources of energy (125-397 kcal/100 g) which is present in the form of carbohydrates.
- ☆ Of the soluble carbohydrates, starch is the most important carbohydrate in all cereals.
- ☆ They are also a significant source of protein (4.7-29.2 per cent) in diet although this protein has relatively low biological value as compared to animal protein because cereals lack in certain essential amino acids like lysine, leucine *etc.*
- ☆ Cereal like refined wheat flour has an insoluble protein called gluten which has a property of elasticity and is good for breadmaking.
- ☆ All unmilled cereals contain good amount of fibre and minerals including calcium, phosphorus and iron.
- ☆ B-group vitamins except niacin are present in all cereals.
- ☆ Cereals do not contain vitamin A, C and D.
- ☆ The amount of minerals and vitamins are greatly reduced by the milling process, so we should use whole cereals as far as possible

rice however, is poorer in these minerals. Although these minerals are not present in large amounts but they contribute significantly due to more amount of cereals consumed in our daily diets. Millets on the whole are good sources of minerals and fibre but higher amount of antinutrients like phytic acid and polyphenols present in them limit the bioavailability of minerals.

Whole cereal grains contain good amount of B-complex vitamins. Since most of these vitamins are present in the outer bran, processing *viz.* milling, polishing and refining drastically (high extraction wheat flour (*maida*) pearl millet and polished rice *etc.*) reduces their contents of B-vitamins. The extent of loss depends upon the degree of refining. On the contrary, parboiling results in seeping the vitamins present in outer layer into the grain. Hence, milled and polished parboiled rice contains more of B-complex vitamins than the milled and polished ordinary rice.

Cereals do not contain vitamin A and C. But yellow maize and some varieties of pearl millet and sorghum do contain small amounts of β-carotene.

Whole grains differ from refined grains in the amount of fibre and nutrients they provide and different grain foods differ in nutrient content, so choose a variety of whole and enriched grains in a balanced diet. Eating plenty of whole grains or their combination such as whole wheat *chapati* or bread, oatmeal, barley, parboiled rice *etc.* may help to protect against many chronic diseases. The consumption of which as mentioned under, is healthy as its consumption lowers blood cholesterol and sugar levels.

Composition of Nutritious *Atta*

Whole wheat	7 ½ kg
Barley	1 kg
Soybean flour*	½ kg
Chickpea	1 kg
Total	10 kg

* Soybean should be soaked overnight in water, dehulled manually, dried and ground into flour.

An adult man and woman, doing sedentary activity should daily take 12.5 and 9 portion sizes (@ 30 g per portion size) of cereals and millets, respectively. In case of elderly persons, reduce 3 portions of cereals and millets and add an extra serving of fruits.

2. Pulses

Pulses or legume seeds are second to cereals as a source of human food. They are foods of choice worldwide and provide variety to the human diet. The term 'Legume' is derived from a latin word 'legumen' meaning seeds harvested in pods. Alternative terms for the edible seeds include 'grain legumes' and 'pulse', from the latin 'puls' meaning pottage. The importance of food legumes, especially in the diets of developing nations like India is well-established.

India is the largest pulse growing country in the world, both in terms of area and production. Despite being the largest producer, India is also the largest importer and consumer (23-24 million tonnes) of pulses in the world. In India, during 2017-18, pulses were cultivated over > 29 million ha (Mha) of area and recorded the highest ever production of 25.23 million tonnes (Mt) at a productivity level of 841 kg/ha. For the first time, the country witnessed near self-sufficiency in pulses. But still per capita availability/day of pulses is 47.2 g against the daily requirement of 60 g by an adult woman.

The major pulses grown and consumed in Indian meals are pigeonpea or red gram (*arhar*), bengal gram (*channa*), blackgram (*urd*), green gram (*moong*) and lentils (*masur*), cowpeas, kidney beans (*rajmah*), peas *etc.*

The food legumes contribute a good source of several important nutrients including protein, calorie, vitamins and minerals. Pulses have high protein content which ranges from 20-40 per cent and this is important in human diet from the point of nutrition. Since most of the Indian population is vegetarian and moreover, animal foods are very expensive, so pulses are the good source for meeting the protein requirement of poor

Nutrient Composition of Pulses

- Pulses are good sources of protein (7.2-43.2 per cent).
- They have good amount of carbohydrates (15.9-60.9 per cent).
- Pulses like soybean are rich in fat content (19.5 per cent).
- Mineral profile of pulses is also very good as they are the source of calcium, magnesium, zinc, iron, potassium and phosphorus.
- Pulses contain small amount of carotene.
- They are also fairly rich in thiamine and niacin but are poor in riboflavin.
- Dry legumes are almost devoid of ascorbic acid.

people. Hence, they are rightly said as 'poor man's meat' in India.

Unlike cereals, pulses are rich in lysine, leucine and isoleucine but deficient in another essential amino acid – methionine which is present in sufficient amount in cereals. Pulses are also deficient in tryptophan amino acid which is the precursor of niacin whose deficiency leads to pellagra. To enhance the protein quality of a diet, a judicious combination of cereal grains and pulses in 7:1 or 8:1 (or 4 parts cereal protein + 1 part pulse protein) is recommended in the balanced diet. Different pulses have 1-5 per cent fat content and along with cereals contribute to meet essential fatty acid requirements of an adult. Pulses are fairly rich sources of B-vitamins especially thiamine and niacin (Table 3.2). They contain small amount of carotene but like cereals, they do not contain vitamin A and C. But germinated pulses do contain vitamin C. Mineral profile of pulse is also good as they are the source of calcium, magnesium, zinc, iron, potassium and phosphorus.

The total carbohydrate of pulses ranges from 24 to 68 per cent, with total soluble sugars representing only a small percentage. Oligosaccharides such as raffinose, stachyose, verbicose *etc.* have been implicated in flatulence problem associated with intake of pulse. These sugars account for about 31 to 76 per cent of total sugars. Pulses contain relatively high level of soluble and insoluble fibre and significant level of pectin.

Besides nutrients, certain antinutrients are also present in pulses. These are:

- Enzyme inhibitors
- Phytic acid
- Saponin
- Phenolic compound
- Flatulence causing substances (raffinose, stachyose and verbicose)
- Oxalate content

These antinutritional factors have an adverse effect on the nutritive value of pulses, hence should be removed through the processing methods as they upgrade the nutritional quality of pulses. Various processing and cooking methods *viz.* soaking, dehulling, ordinary and pressure cooking, germination and fermentation *etc.* have beneficial effects on the nutritive value of pulses which are as under:

- Save fuel, time and energy used in cooking
- Reduce the contents of antinutrients
- Improve the digestibility of starch and protein
- Enhance the bioavailability of dietary essential minerals
- Improve the shelf life of the product
- Improve the flavour, palatability and acceptability of food
- Germination increases the ascorbic acid content.

A vegetarian adult man and woman (sedentary activity) should have 2.5and 2 servings of pulses (30 g per serving) in the daily diet, respectively. A non-vegetarian adult man or woman can replace one serving of pulses with one serving (30 g per serving) of egg/meat/chicken/fish.

Vegetables and Fruits

Vegetables are edible forms of plant tissues. They are served whether raw or cooked as part of the main course of a meal. Besides providing minerals and vitamins, vegetables also add colour, variety, texture and flavour to our food. The texture and appearance of a meal can be varied by the way the vegetable is served. They have a special and important role to play in the vegetarian diet of Indians. Vegetables are generally classified into three categories:

- Green leafy vegetables
- Roots and tubers
- Other vegetables

Green Leafy Vegetables

The commonly consumed green leafy vegetables are spinach, mustard leaves, coriander leaves, mint, amaranth, fenugreek *etc.* They are important sources of calcium, iron, β-carotene, riboflavin, vitamin C, folic acid and fibre. They are inexpensive sources of various vitamins and minerals in the cereal based diets of poor

Facts about Nutrient Composition of Vegetables

- ☆ Vegetables especially GLVs are high in moisture content (58.3-96.5 per cent).
- ☆ Most of the vegetables like cucumber, radish, ash gourd, tomato, cauliflower and cabbage contain very less calories *i.e.* 10-30 Kcal/100 g.
- ☆ Roots and tubers contain more energy *i.e.* 97-157 Kcal/100 g.
- ☆ Vegetables contain less amount of protein (0.2-8.4 g/100 g). Leguminous vegetables like peas and beans provide protein but that is incomplete protein due to lack of certain essential amino acids.
- ☆ Vegetables are very low in fat content (0.1-1.7 per cent), hence low in calories.
- ☆ Vegetables are good sources of vitamins and minerals.
- ☆ Orange and dark green leafy vegetables are rich in β-carotene – precursor of vitamin A.
- ☆ GLVs are rich in calcium, iron, vitamin C and riboflavin.
- ☆ In vegetables, carbohydrate is present in the form of cellulose, starch and sugars.
- ☆ Non-volatile acids *viz.* citric acid, malic acid, oxalic acid and succinic acid present in vegetables contribute flavour.
- ☆ Dietary fibre of vegetables is required to avoid constipation, appendicitis, diverticular diseases *etc.*

vegetarian population. The rural as well as urban community should be encouraged to grow green leafy vegetables in backyard as kitchen gardens, pots and school gardens so as to have their availability throughout the year.

Roots and Tubers

The commonly consumed roots and tuber vegetables are potato, sweet potato, carrots, colocasia (*arbi*), tapioca and yam (*zimikand*). They are generally rich in carbohydrate and can form important source of energy in the diet. Carrots and yellow varieties of yam are rich sources of β-carotene. Potato contains significant amount of ascorbic acid. Tapioca and yam are rich in calcium.

Other Vegetables

Vegetables which are not covered under GLVs and roots and tubers are covered in this category. They can be:

- ☆ Stem vegetables: Lotus
- ☆ Fruit vegetables: Tomato, cucumber, pumpkin, brinjal, ladies finger, various gourds
- ☆ Pods and seeds: Pea, bean, fresh bean, French bean, guar bean
- ☆ Flower vegetables: Cauliflower, drumstick

These vegetables not only add variety to the meal but also provide certain minerals, vitamin C and dietary fibre.

Fruits

A fruit is a matured ovary of a flower, including its seeds and adjacent parts. In India, agroclimatic conditions are favourable for growth of wide range of tropical and temperate fruits including mango, apples, pears, papaya, pineapple, orange, sweet lime, plums, banana, *jamun*, *lichi*, grapes *etc.* Fruits are generally sweet

Some Facts about Nutrient Composition of Fruits

- ☆ Fruits contain high amount of water (15.3-95.8 per cent). Dates contain the minimum (15.3 per cent) whereas melon contains the maximum (95.8 per cent) amount of moisture.
- ☆ Fruits contain very less amount of protein (0.1-7.1 per cent) and fat (0.1-22 per cent) except avocado or butter fruit which contains 25 per cent fat.
- ☆ Carbohydrate content ranges from 0.8-75.8 per cent in various fruits.
- ☆ Fruits contain free organic acids like malic acid, citric acid, tartaric acid and oxalic acid.
- ☆ Amla is the richest source of vitamin C (600 mg/100 g). Apples, grapes and watermelon contain only 1 mg vitamin C per 100 g. Citrus fruits have 30-63 mg vitamin/100 g whereas guava (212 mg/100 g) has 3-5 times more vitamin C than that in citrus fruits. About 100 g of orange juice or 20 g guava can meet the RDA of vitamin C of an adult man.
- ☆ Yellow fruits like papaya, mango, apricot and orange are the good sources of β-carotene.
- ☆ Generally, fruits contain very low amounts of B-vitamins.
- ☆ Fruits like dates, figs, amla, lime, *etc.* are good sources of calcium (50-130 mg/100 g).
- ☆ Fresh fruits contain good amount of sulphur, phosphorus and iron.
- ☆ Fruits are good sources of roughage which avoids constipation and gall bladder stones.

having a wide range of flavours, colour and texture. The characteristic colour of the fruit is derived from the pigments present in it *e.g.* blue, purple, red, and pink colour in plums, cherries, grapes and berries are due to anthocyanin pigment; carotene gives yellow colour to fruits; banana, peaches and citrus fruits contain the pigments lycopene and xanthophylls.

Fruits contain high amount of water (15.3-95.8 per cent). They are good sources of vitamin C. Yellow and orange red coloured fruits *viz.* papaya, mango are rich sources of β-carotene. Banana is a good source of carbohydrates and hence, energy. Banana and papaya contain pectins which provide bulk to the diet and aid in bowel movement. Dried fruits *e.g.* date and raisins provide iron.

Why Eat Plenty of Fruits and Vegetables

To promote health, an adult man or woman should eat a variety of fruits and vegetables – at least one serving of fruits (100 g per serving) and two servings each of roots and tubers and other vegetables and one serving of green leafy vegetables @ 100g per serving each day.

- Eat a variety of fruits and vegetables to protect against chronic diseases and for promoting good health.
- Different fruits and vegetables are rich in different nutrients.
- Low in fat and calories and are filling.
- Some are excellent sources of carotenoids while others may be rich in vitamin C, folate or potassium.
- Fruits and vegetables especially dry beans and peas also contain fibre and other substances that are associated with good health. Fibre promotes healthy bowel function.
- Choose whole or cut up fruits and vegetables rather than juices most often. Juices contain little or no fibre.
- Eat variety of fruits and vegetables to protect against chronic diseases and for promoting good health.

Which Fruits and Vegetables Provide the most Nutrients?

The list given as under shows different fruits and vegetables contain various vitamins and minerals.

Sources of Vitamin A (Carotenoids)

- Orange vegetables like carrots, sweet potatoes, pumpkin.
- Dark green leafy vegetables such as spinach, mint, fenugreek, amaranth, radish leaves, carrot leaves.
- Yellow and orange fruits like mango, apricots, papaya.
- Tomatoes

Sources of Vitamin C

- Amla, guava, citric fruits and juices, straw berries
- Tomato, cabbage
- Leafy greens

Sources of Folate

- Cooked dry beans and peas, peanuts
- Oranges, orange juice
- Dark green leafy vegetables like spinach, mustard greens, lettuce
- Green peas

Sources of Potassium

- Sweet potato, cooked greens, sweet lime
- Bananas, dried fruits such as apricots and prunes, orange juice, coconut water
- Cooked dry beans and lentils

3. Milk, Egg and Flesh Foods

a) Milk and Milk Products

All types of milk and milk products are included in this group. Some of the commonly milk and milk products in India are liquid milk, skimmed milk, milk powder, curd, buttermilk, *paneer* or *channa*, ice cream, *khoya etc.* Butter and ghee contents are excluded because of high fat content and proportionately low protein and calcium content. This group provides proteins, fats, carbohydrates, calcium, phosphorus, vitamin A, riboflavin and niacin.

Milk has always been considered an ideal and complete food for infants and children and a good supplementary food for all other age groups. Milk contains good quality protein and it varies from 1.1 to 4.7 per cent in milk obtained from different species. Lactose is the main carbohydrate (1.5-6.0 per cent) of milk. In some individuals, due to deficiency of enzyme lactase, milk sugar is not digested, producing gas and diarrhoea which is called lactose intolerance. Such persons should take fermented milk product *e.g.* curd and buttermilk in which lactose is converted into lactic acid. Milk contains good amount of calcium, phosphorus, zinc, magnesium, potassium, sodium, chloride and sulphur. Milk is a good source of vitamin A. Riboflavin is present in a higher concentration than the other B-vitamins, but this vitamin is lost rapidly on exposure to light. Milk is deficient in iron and vitamin C. Vitamin D is present in small amount in milk.

The number of servings varies according to age and physiological condition. An adult man or woman should take 3 servings (100 g per serving) of milk and its products while two additional servings are required each day during pregnancy and lactation in the meal. Children require more servings of milk and its products than the adults to meet the additional requirements for good quality protein, calcium and riboflavin due to rapid growth.

b) Egg

In Indian meals, hen's egg is more common than that of duck. Eggs are rich in all the nutrients except vitamin C. An egg has two distinct portions *i.e.* egg white and yolk. Egg white contains mainly protein known as egg albumen. Egg protein is superior hence it is also known as reference protein to any other dietary protein source. The yellow part *i.e.* yolk contains fat, vitamins and minerals.

Raw egg should not be consumed as it contains two antinutritional factors namely avidin which binds biotin and makes it unavailable; and ovomucoid which can inhibit trypsin activity. Duck's egg white has more content of ovomucoid than hen's eggs. Heat processing of any kind whether boiling or cooking destroys these antinutritional factors.

c) Flesh Foods

Flesh foods like meat, poultry and fish are sources of good quality protein and B-vitamins. Vitamin B_{12} not present in plant foods can be obtained from flesh foods. Liver of flesh foods is a good source of vitamin A. Fish contains Omega-3-fatty acids which are known to protect against cardiovascular diseases. One serving portion (30 g) of fish/chicken/meat/egg can be a substitute for pulses in cereal based diets. Non-vegetarians can take one serving (30 g portion size) of flesh foods each day in the diet to have good quality protein, some fat, iron, copper, sodium, potassium, chloride, phosphorus, magnesium, zinc, vitamin A, B-complex and D.

Meat alternates such as pulses, beans and nuts can be good sources of protein and are usually less expensive than meats, fish and poultry.

4. Fats and Oils and Nuts and Oilseeds

Dietary fats are derived from both plant and animal sources and are classified as 'visible' or 'invisible' types. The visible fats generally consumed in Indian households are butter, ghee, hydrogenated oils (*Vanaspati*), and various vegetable oils such as groundnut oil, soybean oil, corn oil, cottonseed oil, safflower oil, mustard oil, coconut oil *etc.* Fats that are integral part of different foods are referred to as 'invisible' fat. Besides protein, nuts and oilseeds and most of the animal foods provide high amounts of invisible fat too. The small amounts of invisible fat present in cereals, pulses, nuts and oilseeds add up to substantial fat level in our daily diet and can nearly meet the essential fatty acid requirement of an adult.

Fats and oils in our diets are the nutrients that probably concern us most. They contain more than twice the Kcalories per gram (9 instead of 4) when compared with proteins and carbohydrates. Diets that provide excess of calories, fats and cholesterol elevate blood lipids. Excessive fat in the diet increases the risk of obesity, heart disease, stroke and cancer.

Depending upon the physical activity and physiological status, the total fat (invisible + visible) in the daily diet should provide between 15-30 per cent total calories. An adult with sedentary habits should consume about 4 servings (5 g per portion size) or should consume about 20 g visible fat while individuals performing hard physical work require 30 g higher amounts of visible fat.

Depending upon the fatty aid composition, various fats and oils consumed in our daily diet can be classified as under:

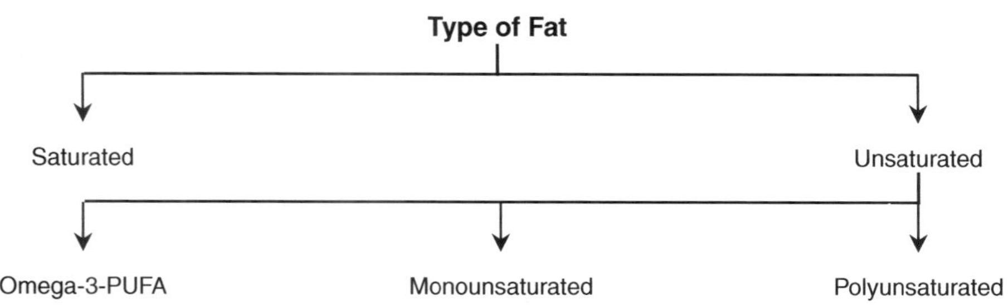

Saturated	*Monounsaturated*	*Polyunsaturated (Omega-6-PUFA)*	*Omega-3-PUFA*
☆ Coconut oil ☆ *Vanaspati* ☆ Ghee ☆ Butter ☆ Animal foods (Meat, milk and milk products)	☆ Olive oil ☆ Almond oil ☆ Red palm oil ☆ Groundnut oil ☆ Rice bran ☆ Sesame oil ☆ Canola	☆ Safflower oil ☆ Sunflower oil ☆ Cottonseed oil ☆ Corn oil ☆ Soybean oil	☆ Fish and fish oil Salmon, oil tuna, and Mackerel oil, Halibut oil and oils from cold water fish

Food high in saturated fats tend to raise the levels of total LDL cholesterol and are therefore, atherogenic. These foods include high fat dairy products like cheese, whole milk, cream, butter, ice cream *etc.*, fatty fresh and red meat, the skin and fat of poultry, lard and coconut oil. The intake of such foods should be kept low. Foods that are high in cholesterol also tend to raise blood cholesterol. These foods include liver and other organ meats, egg yolks and dairy fats.

On the other hand, unsaturated oils reduce total and LDL cholesterol, and, hence, are anti-atherogenic. Unsaturated oils include both monounsaturated fats and polyunsaturated fats. Olive, canola and mustard, sunflower, sesame (*til*) and peanut oils are some of the oils high in MUFA content. Vegetable oils such as soybean oil, corn oil and cottonseed oil and many kinds of nuts are good sources of polyunsaturated fats.

Though both PUFA and MUFA reduce LDL cholesterol, PUFA intake exceeding 10 per cent of total energy intake may also be detrimental, as it may reduce HDL-cholesterol. Therefore, oils with very high PUFA content alone are not so desirable. Hence, goal is to strike a balance between PUFA and MUFA intake, and minimize saturated fat intake. One should use moderate amounts of foods high in unsaturated fats, taking care to avoid excess calories. Some fish such as salmon, tuna and mackerel that contain omega-3-fatty acids are more beneficial for prevention of inflammation, accumulation of fatty material in blood vessels (atherosclerosis) and clotting of blood (thrombosis).

Rich Sources of Cholesterol in Diet

Food Sources	***Cholesterol (mg per 100 g)***
Brain	2000
Egg, whole	500
Kidney	375
Liver	300
Butter	250
Heart	150
Cheese, cheddar	100
Lard and other animal fat	95
Milk, dried, whole	85
Beef	70
Pork	70
Cheese spread	65
Mutton	65
Chicken	60
Ice cream	45
Whole milk	34

Cholesterol Rich Foods

- ✰ Red meat
- ✰ Liver, kidney, brain and other organ meats
- ✰ Lard
- ✰ Egg yolk
- ✰ Whole milk and milk products *e.g.* cream, ice cream, processed cheese, milk based sweets *etc.*

Foods high in trans fatty acids tend to raise blood cholesterol and promote abnormal blood clotting too. These foods include those high in partially hydrogenated vegetable oils, such as many hard margarines, *vanaspati* ghee and shortenings. Even low fat butter and butter substitutes have trans fatty acids. Foods with a high amount of these ingredients include some commercially fried foods and nearly all commercial bakery goods. In small amounts, butter is healthier than *vanaspati* or margarines.

Table 3.4: Upper Limit on Fat for the Calories One Consumes

Total Calories per Day	***Saturated Fat (g)***	***Maximum* Total Fat (g) (No more than 30 per cent of energy)***	
1600	18 or less	53	
2000	20	65	67
2200	24	73	
2500	25	80	83
2800	31	93	100

* About 1/3 of this amount is provided by invisible fat in cereals, pulses, dairy products, meat and poultry

One should prefer a blend of oils rather than taking a single oil. A blend of sunflower/corn/soybean/safflower with groundnut in equal proportions is a good choice. This should be supplemented by occasional mustard oil and ghee. Other suitable alternatives are canola or sesame (*til*) oil alone or a blend of rice bran and sunflower oils. Olive oil is also a good choice. The percentage of monounsaturated, polyunsaturated and saturated fat of different oils has been given in Table 3.5.

Dry fruits and nuts are rich sources of mono-unsaturated fats and vitamin E which are beneficial for the heart. They are devoid of cholesterol also. However, they are rich in energy too. Therefore, their intake should be restricted to not more than 15-20 g a day. Obese people should avoid dry nuts and oilseeds.

Table 3.5 Percentage of Saturated and Unsaturated Fat in different Cooking Oils

Oil	***Saturated Fat (SFA)***	***Mono-unsaturated Fat (MUFA)***	***Poly-unsaturated Fat (PUFA)***
Almond	9	73	18
Butter	65	30	4
Canola	7	62	31
Coconut	92	6	2
Corn	13	25	62
Cottonseed	27	19	54
Groundnut	19	51	30

Oil	Saturated Fat (SFA)	Mono-unsaturated Fat (MUFA)	Poly-unsaturated Fat (PUFA)
Mustard	8	70	22
Palm	51	39	10
Safflower	9	13	78
Sesame	15	42	43
Soybean	15	24	61
Sunflower	11	20	69
Olive	10	82	8
Vegetable ghee	28	45	27

Desirable Ratio of Saturated, Monounsaturated and Polyunsaturated Fats in the Diet

The cholesterol raising fat *i.e.* saturated fat should be reduced in the diet as much as is practical to effect a substantial lowering of serum cholesterol concentrations. This is possible by reducing intake of animal fats *i.e.* fat from meat and milk sources and by gradually reducing intakes of trans fatty acids through consumption of commercial fried and baked foods.

Two-Step Dietary Therapy as Recommended by American Heart Association to Reduce Blood Cholesterol Level

Nutrients	Step I	Step II
Total fat	< 30 per cent of total calories	< 30 per cent of total calories
Saturated fatty acids	< 10 per cent of total calories	< 7 per cent of total calories
Polyunsaturated fatty acids	Upto 10 per cent of total calories	Upto 10 per cent of total calories
Monounsaturated fatty acids	10 to 15 per cent of total calories	10 to 15 per cent of total calories
Cholesterol	< 300 mg/day	< 200 mg/day

Saturated fats should provide 7-8 per cent of the total energy intake. There are concerns about increase in polyunsaturated fat in the diet too. There is no doubt that polyunsaturated fat lowers the blood total cholesterol and LDL-cholesterol (bad cholesterol) but at the same time excess of everything is bad for health. Studies done on laboratory animals have indicated that high intakes of polyunsaturated fat can promote chemical carcinogenesis, and suppress the immune system. Limited epidemiologic data further suggests that high polyunsaturated fat consumption can increase the risk factors for human cancer, arthritis and can accelerate aging. But the same is not true for monounsaturated fats. Hence, because of the potential harmful effects of high intake of polyunsaturated fats, their consumption should not exceed 7 per cent of total energy.

Thus, the ratio of saturated to polyunsaturated should be about one to one, with total intakes of each being in the range of 7 per cent of total energy. The overall ratio of MUFA, PUFA and saturated fat in the diet should be 1.5:1:1.

The desirable intake of monounsaturated fat should be 15-16 per cent of total energy. A diet of this type is recommended currently by the American Heart Association for the general public. However, for individuals, depending on personal preferences, intakes of MUFA can vary from 10 to 20 per cent of total energy. This range will allow for a ratio of MUFA to PUFA (or to saturated fat) to vary from 1:1 to 3:1. For most people, this range appears to be healthy.

To achieve the above desirable ratio of different fats, tips given as under should be followed while selecting the cooking media.

- ☆ An average of no more than 30 per cent of energy from fat should be taken.

Method for Calculating Energy Intake from Fat

- ☆ Calculate 30 per cent of daily energy intake of an individual. For example, his or her daily energy intake should be 1800 Kcal.

  Daily energy intake from fat =

  $$\frac{1800 \times 30}{100} = 540 \text{ Kcals}$$

- ☆ Divide total requirement by 9 as to get total fat in grams per day because we get 9 Kcals from 1 g of fat or oil.

  $$\text{Total fat intake} = \frac{540}{9} = 60 \text{ g}$$

- ☆ One third of 60 g *i.e.* 20 g is provided by invisible fat present in various food items *e.g.* cereals, pulses, dairy products, meat and poultry.
- ☆ Rest of the daily fat intake *i.e.* 60-20 = 40 g should come from visible sources of fats or oils.
- ☆ In this case, visible fat *i.e.* 40 g per day should come from a variety of cooking oil sources to get ideal proportion of saturated, MUFA and PUFA intake.

This should also be taken from a variety of vegetable oil and fats instead of a single source. This is because of the fact that most of vegetable oils do not have an ideal mix of saturated fat, MUFA, PUFA and omega 3-PUFA.

- ☆ In Indian conditions, one should mix PUFA rich oil *i.e.* any of the vegetable oils such as safflower, sunflower, soybean, or corn oil with MUFA rich groundnut oil in 1:1.5 proportions.
- ☆ Occasionally, mustard oil can be used in preparation of dry vegetables or for frying purpose to get omega-3 PUFA. Frequent consumption and use of this oil as the sole

cooking medium is not desirable as it contains erucic acid which is detrimental for health.

- ✰ Canola (rapeseed) oil extracted from a variety containing less erucic acid alone can be used as the sole cooking medium as it contains saturated, MUFA and PUFA in an ideal and desired proportion.
- ✰ Sesame (*til*) oil contains almost equal amounts of MUFA and PUFA, hence can be used alone.
- ✰ Rice bran oil (mainly MUFA) blended with any PUFA rich vegetable oil such as sunflower oil is also a good choice.
- ✰ Small amounts of butter and ghee should also be used as they do not have detrimental effect on health.
- ✰ Excess intake of any fat or oils whether vegetable oils or animal fats in the diet is bad for health as the energy provided by one gram of ghee or vegetable oil is same. Hence, it is not wise to consider only animal fats as 'fatty'.
- ✰ For frying purpose, oils with high smoking point (> 200°C) should be chosen. Most of the vegetable cooking oils have adequately high smoking point. Butter and coconut oil should not be preferred for frying purpose as they have low smoking point. Ghee has an intermediate smoking point.

5. Sugar

Sugars are carbohydrates and a source of energy. Their sources include cane sugar, jaggery, glucose, honey and syrup. The commonly used sugar as sweetening agents in beverages and various food preparations to enhance palatability include sucrose or cane sugar. Sugars occur both naturally and added as an ingredient in many foods. They are present in natural foods like fruits, some vegetable, milk and honey. Refined or table sugar (sucrose) provides 'empty' calories. Jaggery may contain some iron. Excess consumption of foods containing added sugars like soft drinks, candied chocolates *etc.* can promote tooth decay, may lead to obesity and elevated blood lipids.

Major Sources of Added Sugar

- ✰ Sweets
- ✰ Soft drinks
- ✰ Cakes, cookies, pastries
- ✰ Dairy desserts such as ice creams
- ✰ Candy

Names of Added Sugar that Appear on Food Labels

✰ Corn syrup	✰ Molasses
✰ Dextrose	✰ Raw sugar
✰ Glucose	✰ Sucrose
✰ High-fructose corn syrup	✰ Table sugar
✰ Honey	✰ Corn sweetener
✰ Lactose	✰ Fructose
✰ Invert sugar	✰ Fruit Juice concentration

An adult man doing sedentary work requires four serving portions of added sugar (@ 5g per portion size).

4

Use of Food Guide Pyramids in Formulation of Balanced Diets

What is Balanced Diet?

The nutrients such as proteins, carbohydrates, fats, vitamins and minerals are the nourishing substances contained in the food. A balanced diet is one which provides all these essential nutrients in required amounts and proper proportions, so as to meet all the body's needs. Such a diet contains the various groups of foodstuffs such as energy yielding, body building and protective foods in the correct proportions and the individual consuming this diet is assured of obtaining the minimum requirement of all the nutrients. It helps to keep the body in a state of good health.

In order to plan a balanced diet a variety of foodstuffs should be included. Greater the variety of foods, better are the chances of consuming an adequately balanced diet. We can have variety by choosing some foods from each of the five groups (Table 3.3) every day.

However, it is not sufficient to eat a variety of foodstuffs, unless the selected foods contain right amount and correct proportions of various nutrients. A correct combination of various amounts of foods is also very important to get the desired results for proper growth and development, body building and repair and maintenance of various body functions. The quantities of foods needed to meet the nutrient requirements vary with age, gender, physical activity and physiological status. For instance, a manual worker needs more of carbohydrate and protein rich foods than a person who does sedentary work. A pregnant or a lactating woman needs more calcium, protein, iron and vitamin rich foods. This means that a balanced diet for different groups of persons will contain different amounts of various nutrients.

While formulating a well-balanced diet, the ratio of the energy contribution from carbohydrates, proteins and fats is usually calculated. These ratios are not well-defined as much depends on the types of foods available in a region, economic status of the population, eating habits of the communities *etc.* Depending upon the prevailing conditions in India, it has been proposed by NIN, Hyderabad that a balanced diet should provide around 50-60 per cent of total calories from carbohydrates, preferably complex carbohydrates, about 15 per cent from proteins and 20-30 per cent from visible and invisible fat.

In addition, a balanced diet should provide non-nutrients such as dietary fibre, antioxidants and phytochemicals which have positive health effects. Antioxidants *viz.*, vitamin E, selenium, β-carotene, ascorbic acid and riboflavin rich foods protect the human body from free radical damage. Spices like turmeric, ginger, garlic, cloves and cumin contain good amounts of antioxidants. Phytochemicals such as polyphenols, flavones *etc.* also protect against oxidative damage.

Importance of Balanced Diet during Various Stages of Life

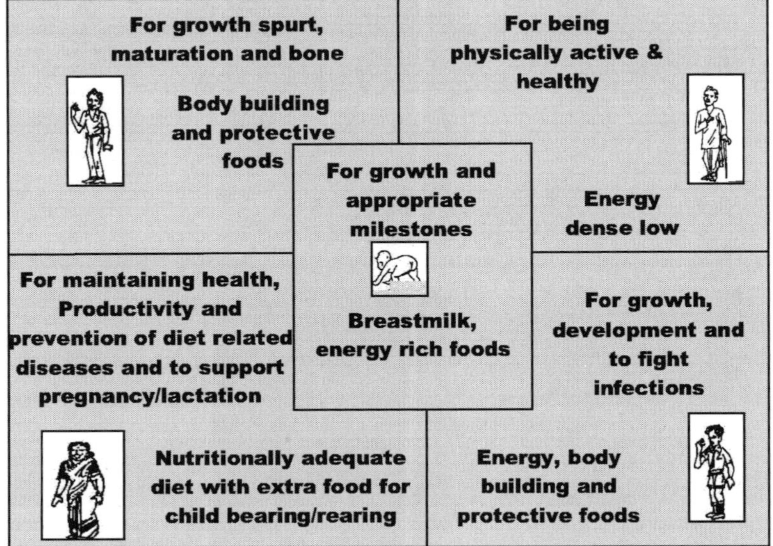

Figure 4.1a: Importance of Balanced Diet for Various Age Groups.

Balanced Diet at Various Costs

High cost: Such diets shall include liberal amounts of expensive foods such as animal foods – milk, egg, meat, fish, poultry *etc.* and fruits and moderate amounts of cereals, pulses, nuts and fats.

Moderate cost: These diets shall include liberal amounts of cereals, pulses, nuts and green leafy vegetables but moderate amounts of animal foods and fats.

Low cost: Such diets contain more amounts of cereals, pulses, nuts and green leafy vegetables but only small amount of animal foods and fats.

Food Guide Pyramid in Formulation of Balanced Diets

Different foods contain different nutrients and other healthful substances. No single food can supply all the required nutrients in desired amounts we need. To get all the nutrients and other substances required for optimum health, one should build a healthy base by using the Food Guide Pyramid (Figure 4.1b) as a starting point.

What is the Food Guide Pyramid

The pyramid is an outline of what to eat each day. It is not a rigid prescription, but a general guide that helps one to choose a healthful balanced diet needed for promoting and maintaining healthy life style.

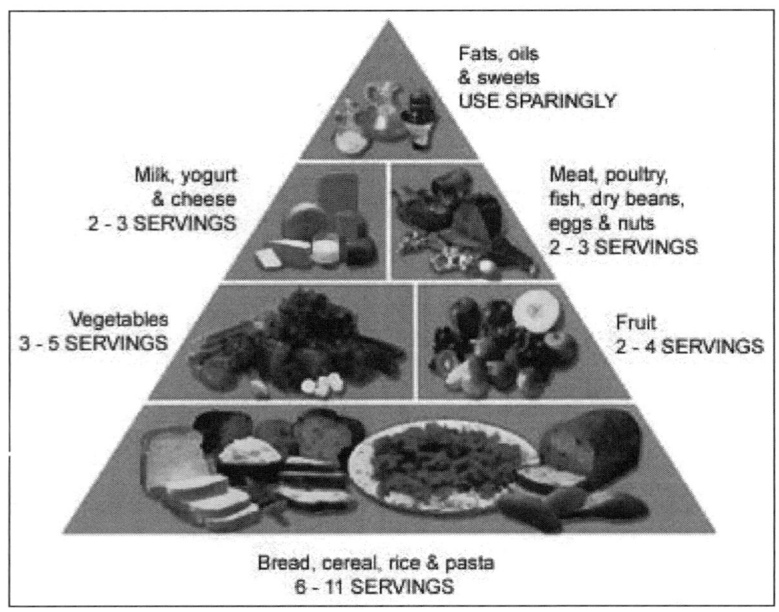

Figure 4.1b: Different Components of Food Guide Pyramid.

The Pyramid calls for eating a variety of foods to get the nutrients one needs and at the same time the right amount of calories to maintain or improve body weight.

Different Components of Food Pyramid

The Food Guide Pyramid emphasizes foods from the five major food groups shown in the three lower sections of the Pyramid. Choose the recommended number of daily servings from each of the five major food groups. Each of these food groups provides, some, but not all, of the nutrients required by us. Foods in one group cannot replace those in another. No one food is more important than another as for good health, we need them all.

The small tip of the Pyramid shows fats, oils, sugar and sweets. These foods include ghee, vegetable oils, butter, margarine, cream, salad dressings, sugars, jaggery, honey, candies, soft drinks and *sherbats*, sweets and desserts. Such foods provide mainly calories and little else nutritionally, hence these foods should be used sparingly.

The next level below small tip of the Pyramid contains two groups of foods that come from animals: milk and milk products; meat, poultry, egg and fish for non-vegetarians and meat alternatives/substitutes *viz.* pulses and nuts for vegetarians. Such foods should be taken in moderate amounts. These foods provide protein, calcium, iron and zinc. Growing children, teenagers, pregnant and lactating women and adults over 50 years have especially high need for calcium, but most people need to eat plenty of good sources of calcium for healthy bones throughout life. When selecting dairy products to get enough calcium, one should choose those that are low in fat or fat free to avoid getting too much saturated fat.

Next level of Food Guide Pyramid includes foods that come from plants – vegetables and fruits. These foods should be taken in liberal amounts as most people need to eat more of these foods for the vitamins, minerals and fiber they supply. Fruits and vegetables should be the key part of our daily diet. Eating plenty of fruits and vegetables of different kinds may help to protect us against many chronic diseases. It also promotes healthy bowel function. Hence, to promote good health, one should eat a variety of fruits and vegetables. The list given below shows which fruits and vegetables are the best sources of which nutrients (Table 4.1).

Why Eat Plenty of Vegetables and Fruits

- Different fruits and vegetables are rich in different nutrients.
- Some fruits and vegetables are excellent sources of carotenoids including those which form vitamin A, while others may be rich in vitamin C, folate or potassium.
- Dark green leafy vegetables and deep coloured fruits are especially rich in many nutrients.
- Some are high in dietary fibre which is associated with good health.
- Choose whole or cut-up fruits and vegetables rather than juices most often as juices contain little or no fibre.

Table 4.1: Fruits and Vegetables as Rich Sources of Vitamins and Minerals

Sources of Vitamin A (carotenoids)
☆ Deep red, yellow and orange coloured fruits and vegetables *viz.*, carrots, tomatoes, pumpkin, mango, apricots, papaya *etc.*
☆ Dark green leafy vegetables such as spinach, mustard leaves, coriander leaves, mint, fenugreek leaves *etc.*
Sources of Vitamin C
☆ Citrus fruits and juices, strawberries
☆ Tomatoes, cabbage, potatoes
☆ Amla
☆ Guava
☆ Leafy green vegetables
Sources of Folate
☆ Fresh green peas and beans
☆ Oranges
☆ Dark green leafy vegetables
Sources of Potassium
☆ Cooked greens, sweet potato
☆ Banana, dried fruits such as apricots and prunes
☆ Citrus fruits (sweet lime)

At the base of the Food Guide Pyramid is grain group – comprising of various cereals and millets, breakfast cereals, noodles and pasta *etc.* Foods from this group should be taken in adequate amount. These foods provide complex carbohydrates (starches) which are an important source of energy, especially in low fat diets. They also provide vitamins, minerals and fibre.

What Counts as a Serving Portion?

The serving portions of various food groups and their nutritive value are given in Table 4.2. If you eat a large portion, count it more than one serving portion. For example, if a *chapati* is prepared from 45g wheat flour, it shall count as one and a half serving cereal. If you eat a smaller portion, count it as a part of one serving portion.

Table 4.2: Serving Portion Sizes of Foods (Raw) and their Nutritive Value (NIN, 2011)

Food Groups	*g/ portion*	*Energy (Kcal)*	*Protein (g)*	*Carbo-hydrate (g)*	*Fat (g)*
Cereals and millets	30	100	3.0	20	0.8
Pulses	30	100	6.0	15	0.7
Eggs	50	85	7.0	-	7.0
Meat/chicken/fish	50	100	9.0	-	7.0
Milk (ml)* and milk products	100	70	3.0	5	3.0
Roots and tubers	100	80	1.3	18	-
Green leafy vegetables	100	46	3.6	-	0.4
Other vegetables	100	28	1.7	-	0.2
Fruits	100	40	-	10	-
Sugar	5	20	-	5	-
Fats and oils (visible)	5	45	-	-	5.0

* Toned milk.

Table 4.3: Quantity of One Serving of Various Food Preparations

Preparation	*Quantity for One Serving*
Cereals and millets	
Bread	1 slice
Rice	1 cup
Phulka/Chapati	1 No.
Parantha	1 No.
Puri	1 No.
Poha	1 cup
Upma	1 cup
Idli	2 Nos.
Khichdi	1 cup
Wheat porridge	1 cup
Ready-to-eat cereal	1 cup
Pulse	
Plain *dhal*	¾ cup
Sambar	1 cup
Non-vegetarian foods	
Boiled egg	1 No.
Omelet	1 No.
Mutton curry	¾ cup
Chicken curry	¾ cup
Fish cutlets	2 Nos.
Vegetable	
Dry	1 cup
Curry	1 cup
Fruit	
Apple/Banana/Orange	1 medium
Fruit juice	1 glass

How to make the Food Pyramid work for you

The Pyramid shows a range of serving portions for each major food group for Indians (Figure 4.2). The number of serving portions required a day depends on how many calories per day are needed, which in turn depends on age, sex, size and activity level of an individual.

The number of serving portions required in balanced diets for various age groups are depicted in the following chapters.

Aren't 4-23 Servings of Cereals a Lot?

It may sound like a lot, but it is really not. For example, a slice of bread is one serving, so a sandwich for breakfast would equal two servings. One *chapati* prepared from 30 g wheat flour equals one serving. If at lunch, three *chapatis* are eaten, they are equivalent to three servings. One cup of cooked rice is another serving. Two small biscuits add yet another serving.

Any Need for Measuring Servings?

There is no need to measure servings. Use servings only as a general guide. For mixed foods, one can estimate the food group servings of the main ingredients. For example, a *samosa* served in evening tea would count in grain group (crust), the vegetable group (potato and peas filling) and fat group (used for frying). But be sure to eat at least the lowest number of servings from the five major food groups in the Food Guide Pyramid. To reduce weight, one should choose the lowest fat choices from the food group. To gain weight, increase the amounts of foods you eat from all the food groups especially sugars and fats and oils.

Advice for Today

- ✰ Eat a variety of foods from Food Guide Pyramid to get the energy, protein, vitamins, minerals and fibre you need for good health.
- ✰ Use plant foods as the foundation of your meals. There are many ways to create a healthy eating pattern but they all start with the three food groups at the base of the Pyramid.
- ✰ Choose a diet with plenty and variety of grain products, vegetables and fruits which provide

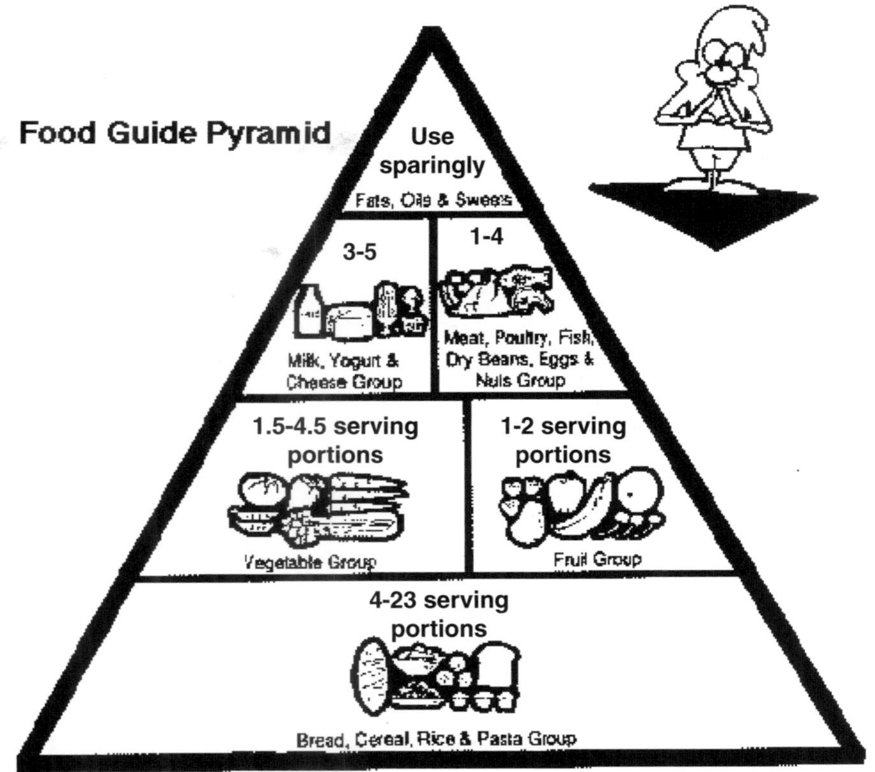

Figure 4.2: Food Guide Pyramid Showing Number of Serving Portions for Indians of Various Age Groups.

needed vitamins, minerals, fibre and complex carbohydrates, and can help you lower your fat intake.

- ☆ If you usually avoid all foods from one or two of the food groups, be sure to get enough nutrients from other food group. For example, if you choose not to eat milk products because of intolerance to lactose or for other reasons, choose other foods that are good sources of calcium *e.g.* dark green leafy vegetables and be sure to get enough vitamin D.
- ☆ Choose a diet low in fat, saturated fat and cholesterol to reduce risk of heart attack and certain types of cancer and to help maintain a healthy weight.
- ☆ Choose a diet moderate in sugars. A diet with lots of sugars has too many calories and too few nutrients for most people and can contribute to tooth decay.
- ☆ Choose a diet moderate in salt and sodium to help reduce the risk of high blood pressure.
- ☆ If you drink alcoholic beverages, do so in moderation. Alcoholic beverages supply calories, but little or no nutrients. It is otherwise also the cause of many health problems and accidents and can lead to addiction.
- ☆ Keep an eye on servings.
- ☆ Balance the food you eat with physical activity to maintain or improve your weight to reduce your chances of having degenerative diseases.

5

Recommended Dietary Allowances

We need a wide range of nutrients to lead a healthy and active life. These nutrients are derived from the diet we consume daily. For maintaining good health and physical efficiency, the components of the diet must be chosen judiciously to provide all the nutrients one needs in adequate amount and in proper proportions. Hence, for designing balanced diets for various age groups, we must know the daily requirements of different nutrients which vary with the age and physiological status. Adults need nutrients for maintaining constant body weight and ensuring proper body functions. On the other hand, infants and young children require nutrients not only for maintenance but also for growth. As compared to adults, infants and children require relatively more nutrients (2-3 times) per kg body weight. Similarly, under special physiological conditions like pregnancy and lactation, women need additional amount of some nutrients to meet the extra demand for foetal growth and maternal tissue expansion in pregnancy and milk secretion during lactation. Allowances for different nutrients for various age groups and physiological conditions have been recommended by various National Committees and some International Committees such as National Expert Group, ICMR, India; Food and Nutrition Board, National Research Council, USA; Panel on recommended allowances of nutrients, Department of Health and Social Services, U.K. and FAO/WHO Expert Group.

Recommended Dietary Allowances for Indians

History

RDA refers to the average daily dietary nutrient intake level sufficient to meet the nutrient requirement of nearly all (97-98 per cent) healthy individuals in a particular life stage and gender group. The first attempt to recommended dietary allowances of energy, protein, iron, calcium, vitamin A, thiamine, ascorbic acid and vitamin D for Indian was made by the Nutrition Advisory Committee of the Indian Research Fund Association (now ICMR) in 1944, following the recommendations of the League of Nations in 1937. Along side, a typical balanced diet based on habitual Indian diet was also formulated to provide all the nutrients to meet the Recommended Dietary Allowances (RDA) of a reference man. It means that the then existing diets of Indians particularly of the poor were inadequate in several respects and could be improved by inclusion of certain protective foods.

In the wake of recommendations of energy and protein requirements by FAO in 1950 and 1957, an attempt was made by the Indian Council of Medical Research (ICMR) in 1958 to define protein and calorie requirements of Indians based on data available at that time. In 1968, the requirements of all nutrients except energy for Indians were reviewed and recommendations on dietary requirements were made by an Expert Committee constituted by the ICMR. In arriving at those recommendations, International data provided by the Joint FAO/WHO Experts Groups and those generated in India were made use of. In 1978, the RDAs were again reviewed by another Expert Group of the ICMR and the RDAs for several nutrients were revised. In both the recommendations made in 1968 and in 1978, a wide range of balanced diets for several nutrients were formulated which could ensure a daily intake of nutrients at the recommended levels.

Energy allowances for Indians recommended in 1958 were not revised till 1989 whereas RDAs for proteins and other nutrients were revised in 1968 and 1978. In view of these developments and of newer data generated in India on energy and protein requirements of Indians, it was felt that the 1978 recommendations required a review. Accordingly an Expert Committee was constituted in 1988 by the ICMR to review the 1978 recommendations and update the RDA wherever necessary. This Expert Group constituted by the ICMR along with several resource persons met in 1988 and discussed RDAs for Indians. Reference body weights adopted in 1978 for an Indian reference man and woman and for children, were also re-examined. The Group in its final recommendations had modified the reference body weights for Indian adults and revised the RDAs in respect of energy, fat, vitamin D and vitamin A. No substantial changes were made in the RDAs for protein, B-complex, vitamins, iron, and calcium. Among nutrients considered for the first time by this expert group were certain trace elements, electrolytes, magnesium and phosphorus, two other fat soluble vitamin *i.e.* vitamins E and vitamin K and dietary fibre. Provisional recommendations on some trace elements were made on basis of limited available data.

The FAO/WHO/UNU Expert Consultation considered the revision of human nutrient requirements again after 2000. One Committee revised the micronutrients requirements in 2001 and energy in 2004 and protein in 2007. In its revision, the International Expert Group considered several other micronutrient requirements of humans. Based on newer developments on human nutrient requirements reported internationally and some Indian research reported for the last two decades, ICMR constituted an Expert Group to revise and upgrade the earlier RDA of nutrients for Indians. In this regard, some of expert group members and resource persons from NIN drafted report in 2009 on revised RDA of all nutrients including some of the newer nutrients, dietary components like selenium, B_6 and dietary fibre and antioxidants which was released to the public by NIN/ICMR in 2011. The revised recommendations of the Group with respect to a number of nutrients are given in Tables 5.1 and 5.2.

General Principles of Deriving RDA

There are certain general guiding principles in arriving at nutritional requirements of an individual and the RDA for a population which are distinctly different. The former depends upon the age, body weight and the physiological and metabolic status of the individual. The latter must take into consideration individual variation within the group, the quality of the diet, the effect of cooking and processing and the bioavailability of its nutrients. However, the general principles are:

i) **Dietary intakes:** The approach is used in arriving at the energy requirements of children. Energy intakes of normally growing healthy children are utilized for this purpose.

ii) **Growth:** The requirement of any particular nutrient or the breast milk intake, for satisfactory growth has been utilized for defining requirements in an early infancy.

iii) **Nutrient balance:** The minimum intake of nutrient for equilibrium (intake = output) in adults, and nutrient retention consistent with satisfactory growth in children have been used widely for arriving at the protein requirements.

iv) **Obligatory loss of nutrients:** The minimal loss of any nutrient or of its metabolic product through normal routes of elimination, *viz.*, urine, faeces and sweat, is determined on a diet devoid of or very low in the nutrient to be consumed daily through the diet to replace the obligatory loss. In infants and children, growth requirements are added to the above maintenance requirements.

v) **Factorial approach:** The requirements for different functions are assessed separately and added up to arrive at the total daily requirement. This is the basis of arriving at energy requirements.

vi) **Nutrient turnover:** Data from turnover of nutrients in healthy persons, using isotopically labelled nutrients have been employed in arriving at requirements of certain nutrients. Requirements of vitamin A, C, iron and vitamin B_{12} have been determined on this basis. Earlier, radioisotopic labelled compounds were utilized and currently compounds labelled with a stable isotope are being increasingly used to determine the turnover or nutrients in the body. Stable isotopes are particularly useful in infants, children and in women during pregnancy and lactation where radioisotopes are contradicted.

vii) **Depletion and repletion studies:** This approach has been used to determine the requirements of water-soluble vitamins. Requirements of vitamin C, thiamine, riboflavin, niacin and pyridoxine have been established employing this approach. The subjects are first fed a diet very low in nutrient under study till the biochemical parameter reach a low level after which response to feeding grade doses of the nutrient is determined. The level at which response increases rapidly is an indication of requirement.

Important Considerations in Arriving at RDA

Individual Variability

The RDA takes into consideration the variation that exists between individuals and a level of intake is chosen which includes an allowance for this variation.

Table 5.1: Recommended Dietary Allowances for Indians (Macronutrients and minerals)

Group	*Particulars*	*Body wt. kg*	*Net Energy Kcal/d*	*Protein g/d*	*Visible Fat g/d*	*Calcium mg/d*	*Iron mg/d*
Man	Sedentary work		2320		25		
	Moderate work	60	2730	0	30	600	17
	Heavy work		3490		40		
Woman	Sedentary work		1900		20		
	Moderate work	55	2230	55	25	600	21
	Heavy work		2850		30		
	Pregnant woman	55	+350	+23	30	1200	35
	Lactation						
	0-6 months	55	+600	+19	30	1200	21
	6-12 months		+520	+13	30		
Infants	0-6 months	5.4	92 Kcal/kg/d	1.69 µg/kg/d	-	500	-
	6-12 months	8.4	80 Kcal/kg/d	1.69 µg/kg/d	19		46µg/kg/d
Children	1-3 years	12.9	1060	16.7	27		09
	4-6 years	18.0	1350	20.1	25	600	13
	7-9 years	25.1	1690	29.5	30		16
Boys	10-12 years	34.3	2190	39.9	35	800	21
Girls	10-12 years	35.0	2010	40.4	35	800	27
Boys	13-15 years	47.6	2750	54.3	45	800	32
Girls	13-15 years	46.6	2330	51.9	40	800	27
Boys	16-18 years	55.4	3020	61.5	50	800	28
Girls	16-18 years	52.1	2440	55.5	35	800	26

Source: ICMR (2011).

Table 5.2: Recommended Dietary Allowances for Indians (Vitamins)

Group	*Particulars*	*Vit. A µg/d*		*Thiamine mg/d*	*Riboflavin mg/d*	*Niacin Equivalent mg/d*	*Pyridoxin mg/d*	*Ascorbic Acid mg/d*	*Dietary Folate µg/d*	*Vit. B_{12} µg/d*	*Mg mg/d*	*Zn mg/d*
		Retinol	*β-carotene*									
Man	Sedentary work			1.2	1.4	16						
	Moderate work	600	4800	1.4	1.6	18	2.0	40	200	1	340	12
	Heavy work			1.7	2.1	21						
Woman	Sedentary work			1.0	1.1	12						
	Moderate work	600	4800	1.1	1.3	14	2.0	40	200	1	310	10
	Heavy work			1.4	1.7	16						
	Pregnant woman	800	6400	+0.2	+0.3	+2	2.5	60	500	1.2	310	12
	Lactation											
	0-6 months	950	7600	+0.3	+0.4	+4	2.5	80	300	1.5	310	12
	6-12 months	950	7600	+0.2	+0.3	+3	2.5	80	300	1.5	310	12
Infants	0-6 months	-	-	0.2	0.3	710 µg/kg	0.1	25	25		30	-
	6-12 months	350	2800	0.3	0.4	650 µg/kg	0.4	25	25	0.2	45	-
Children	1-3 years	400	3200	0.5	0.6	8	0.9		80		50	5
	4-6 years	400	3200	0.7	0.8	11	0.9	40	100	0.2-1.0	70	7
	7-9 years	600	4800	0.8	1.0	13	1.6		120		100	8
Boys	10-12 years			1.1	1.3	15	1.6	40	140	0.2-1.0	120	9
Girls	10-12 years			1.0	1.2	13	1.6				160	9
Boys	13-15 years	600	4800	1.4	1.6	16	2.0	40	150	0.2-1.0	165	11
Girls	13-15 years			1.2	1.4	14	2.0				210	11
Boys	16-18 years			1.5	1.8	17	2.0	40	200	0.2-1.0	195	12
Girls	16-18 years			1.0	1.2	14	2.0				235	12

Source: ICMR (2011).

It thus aims at meeting the requirement of nearly all individuals in a population. In practice, a level of intake corresponding to the mean plus 2 standard deviation, which covers the requirements of 97.5 per cent of population is chosen to define RDA. This has been termed as safe level of intake and the chance of findings this level inadequate is not more than 2.5 per cent. This safe level approach is not used in case of energy since excess energy intake is as undesirable as inadequate intake. Only the average requirement of energy is defined as RDA. In case of other nutrients the RDA is 25 per cent (+2SD) higher than the mean requirement, 12.5 per cent being considered as the extent of individual variability in the requirements of all those nutrients.

Normal, Healthy Population

RDA is applicable only to a healthy normal population living under normal conditions.

Knowledge of Average Requirement of Nutrients

RDA is derived from knowledge of the average requirement of nutrients for each physiological group. Based on the nutrient requirements arrived at using one or more of the approaches as mentioned above, the RDA is computed after taking into consideration the individual variability and nutrient bioavailability from the diet. In designing the RDA for a specific nutrient, it is assumed that the requirements of all the other nutrients are adequately met.

Bioavailability of Nutrients

Bioavailability depends upon the nutritive value of particular nutrient and on the quality and components of the whole diet. For certain nutrients, bioavailability from diet is a critical factor *viz.*, iron, protein, Ca, Zn, vitamin B_{12}, β-carotene. For protein, the essential amino acid composition *i.e.* quality of the dietary protein is also important.

Pro-Vitamins

In case of vitamin A and niacin, pro-vitamins are present in foods which can be converted into the respective vitamins in the body. For example, β-carotene can give rise to vitamin A and tryptophan can yield niacin. The extent of absorption of β-carotene depends upon its source in the diet, and the amount of tryptophan available for conversion depends on the quality of protein in the diet. These factors need to be taken into account in defining the RDA for vitamin A and niacin.

Age, Sex and Body Weight

All these factors determine the nutrient requirements of an individual. In recommending dietary intakes of Indians in 1978, the body weights of well-to-do Indian children and adolescents were used in computing their nutritional requirements. But since 1944 for adults, reference body weight of 55 kg for men and 45 kg for women were used. In 1988, the Committee recommended that reference weights for man and woman should be 60 kg and 50 kg, respectively but now these have been revised to 60 kg and 55 kg, respectively by the Expert Committee (ICMR, 2011).

Nutrient requirements vary with the age, sex and physical activity. Hence, the RDAs are different for man and woman performing different level of physical activities *i.e.* sedentary, moderate or heavy.

Other Factors

Other factors like climate, body size and composition, physiological state *etc.* also influence the determination of RDA.

6

Food Exchanges

In meal planning, nutritive value of the planned diet is generally calculated using the Food Composition Tables as by National Institute of Nutrition, Indian Council of Medial Research, New Delhi. This is very laborious and time consuming. Moreover, normally such detailed calculations are also not needed in day-to-day practice. The use of food exchange list makes the job or meal planning quite easy and quicker and one gets fairly an accurate idea of nutritive value of diets. Such exchanges are used to calculate various proximate nutrients *i.e.* energy, carbohydrates, protein and fat contents of the meals. The other advantages of using food exchanges are:

- ✰ If followed diligently, the exchange lists aids in planning a wide variety of menus within the daily food allowances.
- ✰ One can choose foods of his or her liking and has not to stick to a rigid diet. Hence its use makes possible to eat well, the right kind of food and enjoy eating.
- ✰ Its use avoids monotony in the diet by providing variety of foods.
- ✰ Depending upon the regional and seasonal availability of foods, patients can select the foods of choice which are not only palatable but also inexpensive too as they are available in abundance in season.

Food exchange lists have been developed in several countries to help the patients to choose foods suitable for them. These exchange lists were first published in 1950 by a Joint Committee of the American Dietetic Association, the American Diabetes Association and the United States Public Health Service and were revised in 1976.

In the food exchange list, foods having similar chemical composition are grouped together. In that group, specified amounts of all the listed foods have approximately similar nutrient composition especially with regard to macro-nutrients. The nutrient composition of the various specified foods listed in the exchange group may vary to some extent from the average value for that food exchange. But it does not make a much difference when a variety of foods from that exchange list are chosen daily in the meal.

In the American Food Exchange Lists, foods have been broadly grouped into six groups as:

- ✰ Bread list
- ✰ Milk list
- ✰ Meat list
- ✰ Vegetable list

- ✰ Fruit list
- ✰ Fat list

Within a given list any food can be exchanged with another thereby broadly providing same amount of specific nutrients *i.e.* calories, carbohydrates, protein and fat contents in specified portions. For instances, one cereal exchange equals 18 g carbohydrate, 2.5 g protein and 85 calories. One large slice of bread is equal to one cereal exchange, as also two Marie biscuits or 3 tablespoons rice flakes. Hence, foods in any one group can be substituted for foods within the same group though the portion/size may not be the same.

In Great Britain, there are three groups of food exchange lists:

- ☆ Carbohydrate exchange list
- ☆ Protein exchange list
- ☆ Fat exchange list

The dietary pattern of Indians is quite different from that of Western population. Here, most of the people are vegetarian and basically consume cereal-based diets. The consumption of animal foods *i.e.* milk, meat, fish and poultry is low in Indian dietaries. Pulses, nuts and oilseeds are important sources of protein and fat. The carbohydrate content of different vegetables also varies to a great extent. Keeping in view the above facts, a modified version of the American Food Exchange list has been developed for Indians. This exchange list is based on three food groups according to functions of food.

The exchange list is applicable to all types of food groups. The eight food exchange lists as mentioned in Figure 6.1 are used. The meal planning is simplified by grouping foods into exchange lists.

Table 6.1: Approximate Measures of some Foodstuffs

Sl.No.	Food item	Weight
1.	A small size *chapati*	20 g
2.	A medium sized *chapati* (12" in diameter)	25 g
3.	A medium slice of sponge cake	25 g
4.	Bread slice	
	a) Medium size	25 g
	b) Large size	35 g
5.	A medium sized *parantha*	35g
6.	A cup of milk	250 ml
7.	A level teaspoon of sugar	5 g
8.	A level teaspoon of fat or vegetable oil	5 g
9.	One egg	40 g

Use of Weighing Scale in Exchange List

The use of weighing scale is necessary. However, it is easier to express quantities in terms of household measures than to use an accurate weighing scale. Commonly used home measurements are:

- ☆ One level teaspoon 5 g
- ☆ One level tablespoon 15 g
- ☆ One cup 150 ml
  (Diameter of top 2 ",
  diameter of bottom 2" and
  depth 4")

It is advisable to weigh and measure proportions till one is able to judge the serving size visually.

Different Food Exchange Lists

The different food exchange lists are described as under:

I. Carbohydrate Exchanges Containing Protein

Cereal/Starch exchange: Each exchange supplies

- ☆ 15 g carbohydrate
- ☆ 2-3.5 g proteins
- ☆ Negligible fat and
- ☆ 70 Kcals

Exchanges based on rice or ragi supply 2 g protein while those based on wheat and pearl millet provide about 3.5 g protein. Sorghum or corn based exchanges give 2.8 g protein. When roots and tubers are selected from this exchange, they provide less than 1 g of protein. The basis of this exchange is one big slice of bread (30 g) or small size *chapati* prepared from 20 g wheat flour as these provide 15 g carbohydrates. Hence, one cereal exchange is equivalent to specific portion of various cereals and their products which provide 15 g carbohydrates. This exchange includes:

- ☆ Cereals and their products *viz.* wheat, rice, corn, bajra, sorghum, ragi, semolina, maida, corn flour, bread, biscuits, noodles, pasta, macroni, spaghetti, rice flakes, cake, cookies, vermicelli, puffed rice, breakfast cereals *etc.*

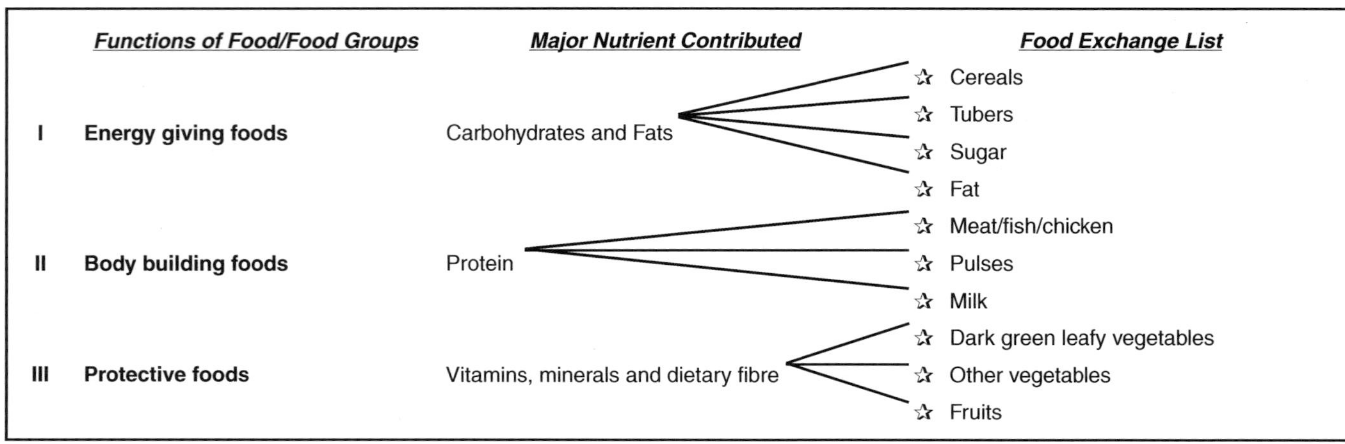

Figure 6.1: Food Exchange Lists Based on Food Groups.

Cereals supply about 70-80 per cent of the carbohydrates and calories in the diets of low income groups. As the income increases, generally the amount of carbohydrates decreases in the diet.

- ☆ Starches (arrowroot, sago)
- ☆ Roots and tubers *e.g.* potato, sweet potato, colocasia, yam, tapioca. Tapioca along with rice is consumed as staple food.

Table 6.2: Cereal/Starch Exchange List (On an average, one cereal exchange provides 15 g carbohydrate, 2 g protein, negligible fat, 70 kcal)

Foodstuff	*Amount (g)*	*Carbo-hydrate (g)*	*Protein (g)*	*Fat (g)*	*Energy (kcal)*
Bajra	22	15	2.6	-	79
Barley	22	15	2.5	-	74
Bread, white	29	15	2.3	-	71
Bread, brown	31	15	2.7	-	76
Biscuit, salty	28	15	1.8	-	156
Biscuit, sweet	21	15	1.3	-	156
Jowar	21	15	2.2	-	73
Maize, dry	23	15	2.6	-	79
Maize, tender	61	15	2.9	-	76
Oatmeal	24	15	3.3	-	90
Ragi	21	15	1.5	-	69
Rice, raw milled	19	15	1.3	-	66
Rice flakes	19	15	1.3	-	66
Rice, puffed	20	15	1.5	-	65
Semolina	20	15	2.1	-	70
Bulgar wheat	19	15	1.6	-	68
Wheat flour, whole	22	15	2.7	-	75
Wheat flour, refined	20	15	2.2	-	70
Macaroni Vermicelli Noodles	19	15	1.7	-	67
Lotus stem*	75	15	1.2	-	70
Colocasia, fresh	71	15	2.1	-	69
Potato	66	15	1.0	-	64
Potato, sweet	53	15	0.6	-	64
Yam	58	15	0.8	-	64

* 100 g of fresh lotus stem provides 10 g fibre and 24.2 mg iron.

Other key nutrients provided by cereal exchange are iron, thiamine, niacin and dietary fibre which are mainly obtained from whole cereals and their products

Source: Khanna *et al.* (1997)

II. Carbohydrate Exchanges not Containing Proteins

Sugar Exchange

This exchange includes sugar, jaggery, syrups, honey, jams, jellies, marmalades, canned fruits, squashes, *etc.* and each exchange provides:

- ☆ Carbohydrate 5 g
- ☆ Calories 20 Kcal

One teaspoon (5 g) of sugar is the basis of this exchange. In case of other products such as jams, jellies *etc.*, 7 to 8 g (1 ½ tsp) is taken as one sugar exchange, as this much amount provides 5 g carbohydrates.

III. Protein Exchanges

These exchanges include milk exchange, meat exchange and legume exchange.

Milk Exchange

This exchange supplies:

- ☆ Protein 8 g
- ☆ Carbohydrate 12 g
- ☆ Fat 10 g
- ☆ Calories 170 Kcal

Table 6.3: Milk Exchange List (On an average, one milk exchange provides 8 g protein, 12 g carbohydrate, 10 g fat, 170 kcal)

Foodstuff	*Amount (g)*	*Protein (g)*	*Carbo-hydrate (g)*	*Fat (g)*	*Energy (kcal)*
Milk, cow	250	8	11.5	103	168
Milk, buffalo	185	8	9.3	16.3	216
Skim milk, liquid	320	8	14.7	Neg*	93
Whole milk powder	31	8	11.8	8.0	154
Skim milk powder	21	8	10.7	Neg*	75
Butter milk	1000	8	5.0	11.0	155
Curds	258	8	7.7	10.3	155
Channa, cow	44	8	0.5	9.2	117
Channa, buffalo	60	8	4.7	13.8	175
Cheese	33	8	2.1	8.2	115
Khoa, buffalo	55	8	11.3	17.2	231
Khoa, cow	44	8	10.0	10.4	165
Khoa, skim milk	36 (3 +)	8	8.7	.5	70

*Neg; Negligible.

Other key nutrients provided by milk exchange are retinol, calcium and riboflavin

Source: Khanna *et al.* (1997).

The basis for this exchange list is one cup (250 ml) of cow's milk (toned milk) with about 4 per cent fat. The amount of protein *i.e.* 8 g provided by 1 cup (250 ml) of cow's milk is taken as the constant for calculating this exchange. For one milk exchange, the amount of other milk *viz.* skim milk powder, whole milk powder, *paneer*, *khoa*, *channa*, cheese, curd *etc.* Butter and cream are milk products but they are not included in this exchange as they provide negligible protein.

When milk exchanges are used as sources of protein, the amount of carbohydrate and fat should also calculated and noted. For example, buffalo milk exchange provides more fat and more calories while skim milk does not contain any fat and so provides less calories too. To make isocaloric diet, 2 fat exchanges can be added if skim milk has been chosen.

Pulse Exchange

This gives out the types and amount of raw legumes and pulses for substituting food of equal value in exchange. The basis for this exchange is 30 g raw pulse providing 7 g of protein. Hence, in this exchange 7 g protein content is taken as the constant *i.e.* any amount of pulse/legume providing 7 g of protein is considered as one pulse exchange. The quantity of carbohydrate should also be noted. One exchange of pulse/legume contains the following:

- ✰ Protein 7 g
- ✰ Fat Negligible
- ✰ Carbohydrate 17 g
- ✰ Energy 100 Kcal

Exception is soybean as 16 g of it provides 7 g protein, more of fat and carbohydrates as compared to other pulses.

This exchange list includes all types of whole, split and dehusked pulses namely chickpea, urd bean, mung bean, kidney beans, lentils, dry peas, soy bean, cowpea, moth bean *etc.* (Table 6.4). Sprouted pulses are also included in this list. Sprouts are easier to digest. They provide bulk to the diet.

Table 6.4: Pulse Exchange List
(On an average, one pulse exchange provides 7 g protein, negligible fat, 17 g carbohydrate, 100 kcal)

Foodstuff	*Amount (g)*	*Protein (g)*	*Fat (g)*	*Carbo-hydrate (g)*	*Energy (kcal)*
Bengal gram (roasted)	31	7	1.6	18	114
Bengal gram (whole)	41	7	2.2	25	148
Bengal gram dal	34	7	1.8	20	125
Black gram dal	29	7	0.4	17	101
Cowpea (lobia)	29	7	0.3	16	94
Green gram (whole)	29	7	0.4	16	97
Green gram dal	28	7	0.3	17	101
Lentil	29	7	0.2	16	96
Peas, dry	35	7	04	20	110
Rajmah	30	7	05	18	105
Red gram dal	31	7	0.5	18	104
Soybean	16	7	32	34	70
Moth beans	30	7	0.3	17	99

Note: When 100 g of whole green grain is sprouted, the sprouts provide 0.06 mg thiamine, 0.66 mg riboflavin, 1.5 mg niacin and 82 mg ascorbic acid

Other key nutrients provided by the pulses exchange are iron, vitamins of B-group and dietary fibre which mainly comes from whole pulses

Source: Khanna *et al.* (1997)

Meat, Fish and Egg Exchange

The basis for this exchange is 40 g of edible portion of mutton muscle providing 7 g protein. This protein content is taken as the constant in this exchange, hence, each meat exchange provides 7 g protein. One meat/fish/egg exchange contains:

- ✰ Protein 7 g
- ✰ Fat 5 g
- ✰ Carbohydrate Negligible
- ✰ Energy 70 Kcal

When used as a source of protein, the quantity of fat provided by each meat exchange should be noted. Beef (tender, ribs, steak), lamb (excluding breast), poultry (excluding goose), lean mutton, fish and sea foods (lobster, prawn, crab *etc.*) and cheese (<2 per cent butter fat) contain less than 4 per cent fat while pork, mutton, liver (goat), pigeon, seafood (shrimps, small crab), egg (hen), egg yolk and cheese made from whole milk have 4-8 per cent fat. Breasts of lamb, veal and beef are high fat meat, their fat content varies between 8 and 15 per cent. Like meat, *paneer* and cheese are good sources of animal protein with low amount of carbohydrate as lactose is lost in whey water during *paneer*/cheese preparation. By listing these milk products in the meat exchange, one can plan vegetarian diets.

Table 6.5: Meat Exchange List
(On an average, one meat exchange provides 7 g protein, 5 g fat, negligible carbohydrate, 70 kcal)

Foodstuff	*Amount (g)*	*Protein (g)*	*Carbo-hydrate (g)*	*Fat (g)*	*Energy (kcal)*
Egg	53	7	-	7.0	81
Fowl	27	7	-	0.2	29
Goat meat	32	7	-	1.2	38
Goat liver	35	7	-	1.1	38
Sheep liver	35	7	-	2.6	53

Foodstuff	Amount (g)	Protein (g)	Carbo-hydrate (g)	Fat (g)	Energy (kcal)
Mutton muscle	38	7	-	5.1	73
Pork	37	7	-	1.7	43
Crab, small	63	7	-	6.1	106
Hilsa	32	7	-	6.2	87
Katla	35	7	-	0.8	89
Pomfret, white	41	7	-	0.5	33
Prawn	37	7	-	0.4	32
Sardines	32	7	-	0.6	34
Rohu	42	7	-	0.6	41
Singhara	31	7	-	1.0	53
Cheese	29	7	1.8	7.3	101
Channa, cow	38	7	0.46	7.9	101
Channa, buffalo	52	7	4.11	1.2	152

Other key nutrients provided by meat exchange are retinol, niacin, thiamine and iron

Source: Khanna *et al.* (1997)

IV. Fat Exchange

This list gives the amount of fat/oil containing foods for substituting equal value in exchange. One teaspoon or 5 g of fat or oil is taken as the basis for this exchange and each fat exchange contains:

- ☆ Fat 5 g
- ☆ Energy 45 Kcal

Butter, ghee, vegetable oils, margarine, cream (light), cream (heavy), hard salad dressings and nuts and oilseeds are included in this exchange list. Besides fat, nuts and oilseeds also contribute good amount of protein, minerals like iron and calcium and B-group vitamins. Hydrogenated fats contain fortified fat soluble vitamins *viz.* A, D, E and K whereas butter, cream and ghee contain good amount of retinol (vitamin A).

Table 6.6: Fat and Nuts Exchange List (On an average one fat exchange provides 5 g fat, 45 kcals)

Foodstuff	Weight (g)	Fat (g)	Protein (g)	Carbo-hydrate (g)	Energy (kcal)
Butter	6	5	-	-	45
Ghee	5	5	-	-	45
Hydrogenated oil	5	5	-	-	45
Cooking oil	5	5	-	-	45
Cream	25	5	-	-	50
Almonds	8	5	1.7	0.8	53
Cashewnuts	11	5	2.3	2.5	66
Coconut, dry	8	5	0.3	1.5	53
Coconut, fresh	12	5	0.5	1.6	53
Gingelly seeds	12	5	2.2	3.0	58
Groundnuts, roasted	13	5	3.4	3.5	74
Walnuts	8	5	1.3	0.9	55

Other key nutrients provided by fat exchange are fat soluble vitamins. Butter and cream are a source of retinol, whereas most of the hydrogenated oils and vegetable oils (solid in packaged form) are fortified with vitamins A and D. Nuts and oilseeds also contribute significant amount of minerals like iron and calcium, and B-group vitamins in the diet

Source: Khanna *et al.* (1997).

V. Vegetable Exchange

On this basis of carbohydrate content, vegetable exchange is divided into 2 groups:

- ☆ Vegetable A exchange
- ☆ Vegetable B exchange

Vegetable A Exchange

This exchange list contains all the vegetables with 3 per cent or less carbohydrate; 100 g or ½ cup of vegetable makes one exchange which contains the following:

- ☆ Carbohydrate 3 g or less
- ☆ Protein Negligible
- ☆ Fat Negligible
- ☆ Energy 12 Kcal

Table 6.7: Vegetable A Exchange List (1 exchange of vegetable A is 100 gms and provides upto 3 g carbohydrate, negligible protein and fat)*

Foodstuff	Per cent Carbohydrate
Bathua leaves	2.9
Bottle gourd	2.5
Cucumber	2.5
Ghosala (ghia tori)	2.9
Lettuce	2.5
Mustard leaves	3.2
Parwal	2.2
Radish leaves	2.4
Radish, white	3.4
Ridge gourd	3.4
Snake gourd	3.3
Spinach	2.9
Tinda	3.4

* 2 exchanges of vegetable A=1 exchange of vegetable B

Other key nutrients provided by vegetable A exchange are β-carotene, ascorbic acid, riboflavin, vitamin K, iron, calcium and dietary fibre

Source: Khanna *et al.* (1997).

The vegetables included in this exchange list are raddish, spinach, *bathua* leaves, cucumber, ash gourd, lettuce, mustard leaves, ridge gourd *etc.* (Table 6.7). These vegetables also provide significant amount of vitamins, minerals and dietary fibre.

Vegetable B Exchange List

This exchange includes all the vegetables which are not included in vegetable A exchange or cereal/starch exchange.

One exchange of vegetable B exchange list contributes:

- ✰ Carbohydrate 7 g
- ✰ Protein 2 g
- ✰ Fat Negligible
- ✰ Energy 40 Kcal

Table 6.8: Vegetable B Exchange List
(On an average, one vegetable B exchange provides 7 g carbohydrate, 2 g protein, negligible fat and 40 kcal)

Foodstuff	Weight (g)	Carbo-hydrate (g)	Protein (g)	Energy (kcal)
Amaranth	115	7	4.6	52
Beetroot	80	7	1.4	34
Bittergourd	167	7	2.7	42
Brinjal	175	7	2.7	42
Broad beans	97	7	4.4	47
Cabbage	152	7	2.7	41
Capsicum	162	7	2.1	39
Carrot	66	7	0.6	32
Cauliflower	175	7	4.5	53
Colocasia leaves	103	7	4.3	58
Coriander leaves	111	7	3.7	40
Drumstick	192	7	4.8	48
Fenugreek leaves	117	7	3.1	57
French beans	156	7	2.6	41
Jack fruit	75	7	1.9	38
Knol khol	184	7	22.1	38
Lady's finger	109	7	2.0	38
Mint leaves	121	7	5.0	58
Onion (small)	56	7	1.0	33
Plantain, green	50	7	0.7	32
Peas	44	7	3.2	41
Pumpkin	152	7	2.1	38
Tomato, ripe	194	7	1.7	30
Turnip	113	7	0.6	33
Mushroom	163	7	2.3	47

Other key nutrients provided by vegetable B exchange are riboflavin, β-carotene, ascorbic acid, folic acid, vitamin K, iron, calcium and dietary fibre

Source: Khanna *et al.* (1997)

A carbohydrate content of 7 g is taken as the constant in this exchange. Vitamins, minerals and dietary fibre are also exchanged by vegetables included in this exchange. One exchange of vegetable B is equal to two exchanges of vegetable A exchange. Table 6.8 depicts various vegetables and their amounts to be taken as one exchange.

VI. Fruit Exchange

This list gives the amount of raw fruit for substituting equal value in exchange. One exchange of fruit contains:

- ✰ Carbohydrate 10 g
- ✰ Protein Negligible
- ✰ Fat Negligible
- ✰ Energy 40 Kcal

Table 6.9: Fruit Exchange List
(On an average one fruit exchange provides, 10 g carbohydrate, negligible protein and fat, 40 kcal)

Foodstuff	Amount (g)	Carbo-hydrate (g)	Energy (kcal)
Amla	74	10	42
Apricot, fresh	86	10	46
Apple	75	10	44
Banana	37	10	42
Cherries	72	10	46
Dates, fresh	30	10	43
Dates, dry	13	10	41
Grape fruit	143	10	46
Grapes	61	10	41
Guava	80	10	45
Lemon (sweet)	92	10	51
Lichi	73	10	44
Lime, sweet (Malta)	128	10	46
Lime, sweet (Musambi)	107	10	46
Loquat	104	10	44
Mango	59	10	44
Orange	92	10	44
Papaya	139	10	44
Pear	84	10	44
Peaches	95	10	48
Pineapple	93	10	43
Plums	90	10	47
Pomegranate	69	10	44
Rasberry	80	10	48
Raisins	13	10	41
Sapota	47	10	46
Water melon	303	10	49
Musk melon	286	10	49

Other key nutrients provided by fruit exchange are dietary fibre, ascorbic acid mainly from citrus fruits, guava and amla, β-carotene mainly from deep yellow and orange coloured fruits like mango, papaya, orange and iron mainly from dates and water melon

Source: Khanna *et al.* (1997).

The carbohydrate content of 10 g is taken as constant and the amount of fruit providing this much amount of carbohydrate is taken as one fruit exchange (Table 6.9). Fruits also provide vitamins, minerals and dietary fibre.

Use of Food Exchange List in Meal Planning

To illustrate how one can plan meals in a simple, easy and quicker way by using food exchange list, the following example is being elicited:

Aim

Plan a full day diet for an adult vegetarian woman, 25 year, sedentary worker and belonging to moderate income group. Calculate energy, protein, carbohydrate and fat in her diet.

Step I

Write down the specifications as given under:

Specifications

1. Age 25 years
2. Sex Female
3. Physical activity Sedentary
4. Socio-economic status Middle income group
5. Food habit Vegetarian

Step II

Using ICMR tables, write down the Recommended Dietary Allowances (RDA) for the nutrients to be calculated for the above mentioned specifications.

Nutrients	*RDAs*
Energy (Kcal)	1875
Protein (g)	50
Carbohydrate (g)	280
Fat (g)	60
(Visible and Invisible)	

Step III

Write down the amount of different food exchanges required in a balanced diet for an adult woman (sedentary) so as to provide the desired amount of above mentioned nutrients. Her food preferences should also be taken into consideration. In this case, it is as given as under:

Food Exchange	*No. of Exch-anges*	*Energy (Kcal)*	*Protein (g)*	*Carbo-hydrate (g)*	*Fat (g)*
Milk	2	340	16	24	20
Meat	-	-	-	-	-
Pulse	2	200	14	34	-
Cereal/Starch	14	980	28	210	-
Vegetable A	2	30	2	6	-
Vegetable B	1	40	2	7	-
Fruit	1	40	-	10	-
Fat	4	180	-	-	20
Sugar	3 ¼	65	-	16.25	-
Total		**1875**	**62**	**307.25**	**40**

Keeping in view the following steps, the above mentioned food exchange plan is prepared:

- The number of protein exchanges (milk, meat and pulse exchanges) are planned first on keeping in mind the food habits and socio-economic status of the individual. Protein rich foods should be eaten moderately.
- Next the number of cereal/starch exchanges are estimated as these are the staple foods of our diet and mainly meet the energy and protein requirements. Cereals/starch should be eaten adequately.
- Depending upon the socio-economic status, the remaining energy needs are met by the judicious selection of vegetables, fruits, fats and sugar exchanges. Fats and sugar exchanges should be used sparingly as their excess use adversely affects the health and lowers the intake of other exchanges. Vegetables and fruits should be eaten liberally.

In case of low income group, number of expensive food exchanges *viz.* milk, meat, fruits, fat, vegetable A and B *etc.* are decreased while those of cereal/starches and pulse are increased.

For high income group, number of milk, meat, fruits, fat and sugar exchanges are increased while those of cereal/starch, pulse and vegetable B are decreased.

Step IV

In this step, the above planned food exchanges are to be distributed among different meals. Normally in Indian households, daily meal pattern mainly consists of three main meals *i.e.* breakfast, lunch and dinner. Evening tea is also a general custom. While distributing different food exchanges among the meals planned, one should be careful that each of the three main meals fulfils one third of the daily nutritional requirements especially of energy and protein.

Exchange	*Total Number*	*Break-fast*	*Lunch*	*Evening Tea*	*Dinner*
Milk	2	0.6	0.4	0.2	0.8
Meat	-	-	-	-	-
Pulse	2	-	1	-	1
Cereal/Starch	14	4	5	1	4
Vegetable A	2	0.2	0.4	-	1.4

Exchange	Total Number	Break-fast	Lunch	Evening Tea	Dinner
Vegetable B	1	-	0.5	-	0.5
Fruit	1	1	-	-	-
Fat	4	1.0	1.5	-	1.5
Sugar	3 ¼	1	-	1	1 ¼

Step V

Keeping in view the principles of meal planning, plan the menu. Depending upon the menu, select the food from different food exchanges.

Mean	Menu	Food Exchange	No. of Exchanges	Food Item	Amount (g)
Breakfast	Stuffed radish *parantha* (2)	Milk	0.6	Cow's milk	150
		Cereal/Starch	4	Wheat flour	90
		Vegetable A	0.2	Radish	20
		Fruit	1	Apple	75
	Butter	Fat	1	Oil	5
	Apple	Sugar	1	Sugar	5
	1 cup tea				
Lunch	Boiled rice	Milk	0.4	Curd	100 ml
	Wheat flour	Pulse	1	Kidney bean	30
	Rajmah curry	Cereal/Starch	5	Rice	40
		Vegetable A	0.4	Wheat flour	40
	Cabbage and peas vegetable	Vegetable B	0.5	Potato	60
		Fat	1.5	Cucumber	10
	Curd			Lettuce	10
	Salad (cucumber, lettuce)			Cabbage	10
				Peas	5
				Onion	15
				Oil	7.5
Evening tea	Tea	Milk	0.2	Milk	50
	Biscuit (salty)	Cereal/Starch	1	Biscuit	30
		Sugar	1	Sugar	5
Dinner	Mixed veg. (Tomato, spinach, bottlegourd)	Milk	0.8	Milk	200
		Pulse	1	Moong dal	30
		Cereal/Starch	4	Wheat flour	65
		Vegetable A	1.4	Potato	65
				Spinach	105
				Bottle gourd	25
	Chapati	Vegetable B	0.5	Capsicum	40
	Moong dal	Fat	1.5	Onion	10
	Stuffed capsicum	Sugar	1.25	Tomato	20
				Oil	7.5
	Rice pudding			Sugar	6.25

From a single exchange plan as given in Step III depending upon the socio-economic status and seasonal availability, number of menus can be planned while maintaining almost similar nutrient composition.

7

Nutrition during Adulthood

Adulthood represents the steady state of life, when a person would have completed his/her growth in terms of body size. Adulthood can be divided into three categories:

- ✰ Early adulthood
- ✰ Middle adulthood
- ✰ Late or older adulthood

Early adulthood is considered as the time period between 18 and 30 years of age and middle adulthood from 30 to 65 years of age. In this chapter, nutrition during early and middle adulthood shall be discussed.

During early adulthood, physical growth and development, particularly in terms of height and body stature, stops to a certain extent. Correspondingly, the amount of specific nutrients required by an adult decreases to maintenance level. Moreover, as the age increases, the basal metabolic rate also decreases. On the other hand, psycho-social development continues which varies among individuals. The adult may follow any one of a number of lifestyles, each of which has a unique influence on food habits and nutrient needs. Hence, the nutritional needs are for maintenance of body functions as breakdown and repair of body tissues continues though there is no apparent growth during adulthood. Adulthood also represents the productive stage of life. To keep up the vitality and positive attitude in life, good nutrition is highly essential for optimum.

An adult man should take a diet which provides all the essential nutrients in right amount and proportion in order to keep the body in a state of good health. For planning such a nutritious well balanced diet for maintaining good health and physical efficiency of an adult, it is essential to know his or her recommended dietary allowances. From time to time, Indian Council of Medical Research has suggested recommended dietary allowances (RDAs) for different nutrients for various age groups. In case of adults, ICMR has recommended dietary allowances for reference man and woman which are defined as below:

Reference Indian Adult Man

Reference man is between 18-29 years of age and weighs 60 kg with a height of 1.73m with a BMI of 20.3. He is free from disease and physically fit for active work. On each working day, he is employed for 8 hours in occupation that usually involves moderate activity. While not at work, he spends 8 hours in bed, 4-6 hours sitting and moving about, and 2 hours in walking and in active recreation or household duties.

Reference Indian Adult Woman

Reference woman is between 18-29 years of age, non-pregnant, non-lactating, free from disease and healthy, and weighs 55 kg with a height of 1.61 m and a BMI of 21.2. She may be engaged in 8 hours of occupation which generally includes moderate activity. Apart from 8 hours in bed, she spends 4-6 hours sitting or moving around in light activity, and 2 hours in walking or active recreation or household chores.

Nutritional Requirements and Recommended Dietary Allowances for Reference Adult Man and Woman

The nutritional requirements for reference adult man and woman as suggested by ICMR (2011) are given in Table 7.1 and discussed as under:

Table 7.1: Recommended Daily Dietary Allowances for an Indian Adult Man and Woman (ICMR, 2011)

Nutrient	Man			Woman		
	SW*	MW*	HW*	SW*	MW*	HW*
Energy (kcals)	2320	2730	3490	1900	2230	2850
Protein (g)	60	60	60	55	55	55
Visible fat (g)	25	30	40	20	25	30
Calcium (mg)	600	600	600	600	600	600
Iron (mg)	17	17	17	21	21	21
Retinol (μg)	600	600	600	600	600	600
Thiamine (mg)	1.2	1.4	1.7	1.0	1.1	1.4
Riboflavin (mg)	1.4	1.6	2.1	1.1	1.3	1.7
Niacin equivalent (mg)	16	18	21	12	14	16
Pyridoxin (mg)	2.0	2.0	2.0	2.0	2.0	2.0
Ascorbic acid (mg)	40	40	40	40	40	40
Dietary folate (μg)	200	200	200	200	200	200
Vitamin B_{12} (μg)	1.0	1.0	1.0	1.0	1.0	1.0
Magnesium (mg)	340	340	340	310	310	310
Zinc (mg)	12	12	12	10	10	10

* SW: Sedentary worker; MW: Moderate worker; HW: Heavy worker.

Energy

Energy is required for maintaining body temperature, metabolic activity, supporting growth and for doing physical activity. The energy requirement of an individual is defined as the level of energy intake from food that will balance energy expenditure when the individual has a body size and composition and level of physical activity, consistent with long term good health, and that will allow for maintenance of economically necessary and socially desirable activity.

The energy requirement of an individual depends on various factors such as:

- Age and sex
- Body size and composition (Body weight)
- Physical activity

Climate is not normally important in determining energy requirements. Any effect, if present, is small and minimized by changes in clothing, housing and the use of external heat.

Requirements for energy are completely different from the requirements for other nutrients because there is no satisfactory means of adaptation to an energy intake in excess of their individual energy requirements. If an individual eats more or less energy than his requirement, his body weight will slowly change; any flexibility in his metabolism is surprisingly small and can be neglected for the purpose of specifying requirements. The relative fixed nature of each individual's energy requirement and therefore, the need to eat a specified amount of food is unique to energy. Currently, it is recommended that energy requirement must be assessed in terms of energy expenditure rather than in terms of energy intake as it may vary from day-to-day, on some days, it may be above or below the energy expenditure. The assessment of energy expenditure is therefore, a more logical approach, where one can specify the energy requirements in terms of energy output for productive work and leisure activity of adults and tissue deposition in infants, children and during pregnancy and milk secretion during lactation. An individual tends to maintain energy balance and constant body weight over a long period of time. Hence, energy requirement is best determined by measurement of energy expenditure.

Energy expenditure from physiological point of view is made of three major components (traditional approach).

- Basal metabolic rate
- Regulatory thermogenesis
- Physical activity

According to simplified new approach only two components *i.e.* physical activity and basal metabolic rate (BMR) are taken into consideration for estimating energy expenditure. Regulated thermogenesis includes the metabolic response to food ingested *i.e.* the use of energy in digesting, absorbing, storing and disposing of ingested nutrients as well as the response of the body to stimuli such as cold, stimulants and drugs. For all practical purposes, energy expenditure involved in regulatory thermogenesis is well taken care of while calculating energy cost of various physical activities.

Basal Metabolic Rate

Even when the body is in a complete state of physical and mental rest and is in postabsorptive stage, a number of processes such as heart beating, breathing, regulation of body temperature, glandular activities *etc.* go on in our body to ensure the continuance of life without any conscious effort. These processes are known as basal metabolic processes and the energy spent to keep these going is known as basal energy need. It is measured as energy used per unit of body weight per unit of time (Kcals/kg/hr) and is called the Basal Metabolic Rate (BMR). It is determined when a person is resting quietly awake in a comfortable environment and is in post absorptive (12-15 hours after eating) state. Nearly half the daily energy expenditure is accounted for by the BMR.

BMR is affected by a number of factors such as body size and composition, age, growth, state of climate and certain ailments *etc.* The requirement for energy decreases with advancing age. This is because the basal metabolism of an adult is reduced accompanied by decrease in his physical activity. Thus after the age of 23 years, energy requirement is decreased by 7.5 per cent, for every 10 years. If an adult maintains the normal body weight, it means his or her energy requirement is being met.

The modified equations for predicting BMR of Indians as recommended by ICMR (2011) are given in Table 7.2.

Using these equations, BMR for Indian reference man and reference woman will be 1515 Kcal and 1241 Kcals for 24 hours, respectively. The basal energy accounts for about 60 per cent of the total energy requirement for most people. The energy expenditure during sleep is almost equal to BMR and BMR factor for sleep also known as Resting Energy Expenditure (REE) is taken as 1.0. Actual measured BMR of Indians by using these equations (Table 7.2) is 5 per cent lower than that predicted by the FAO/WHO/UNU proposed for International use. Climate and environmental factors rather than genetic or ethnic factors appear to be responsible for this small difference of about 5 per cent seen between better nourished Indians and North American/European subjects.

Table 7.2: Equations as Proposed by ICMR Expert Group for Predicting BMR (Kcal/24 h) of Indians

Sex	*Age (yr)*	*Prediction Equation*	*Correlation Coefficient*	*S.D.*
Male	18-30	14.5 x B.W. (kg) + 645	0.65	151
	30-60	10.9 x B.W. (kg) + 833	0.60	164
	> 60	12.6 x B.W. (kg) + 463	0.79	146
Female	18-30	14.0 x B.W. (kg) + 471	0.72	121
	30-60	8.3 x B.W. (kg) + 788	0.70	108
	> 60	10 x B.W. (kg) + 565	0.74	108

Source: ICMR (2011)

Physical Activity

Based on the occupation, ICMR (1990) has categorized the physical activities as:

- Sedentary
- Moderate
- Heavy

Classification of Activities Based on Occupation (ICMR, 1990)

A. Sedentary

Male

Teacher, Executive, Peon, Postman, Barber, Tailor, Priest, Shoemaker, Landlord *etc.*

Female

Teacher, Housewife, Nurse, Tailor, Executive *etc.*

B. Moderate

Male

Potter, Industrial worker, Coolie, Driver, Weaver, Rickshaw puller, Welder, Carpenter, Goldsmith, Agricultural labourer, Mason *etc.*

Female

Maidservant, Agricultural labourer, Weaver, Basket maker *etc.*

C. Heavy

Male

Stonecutter, Mineworker, Blacksmith

Female

Stonecutter

Assessment of energy requirement is based on specifying the needs for BMR and then considering all physical activity costs as effectively occurring in the fed state *i.e.* inclusive of the energy costs of thermogenesis.

Energy Cost of Various Physical Activities in Terms of PAR and PAL

Traditionally, the energy cost of an activity has been adjusted to take account of the average weight of the group. This was necessary because the total amount of energy used each minute is greater in large than in small people. However, to simplify the calculations by expressing the energy cost of different occupations as physical activity ratios (PAR) are expressed in terms of the estimated BMR rather than weight of the group. The physical activity ratio expresses the energy cost of an individual activity per minute as a ratio of cost of BMR per minute. The PAR is unit less and the decided advantage of expressing energy expenditure in terms of PAR values is quite obvious. PAR values of some activities are given in Table 7.3.

$$\text{Physical Activity Ratio (PAR)} = \frac{\text{Energy cost of an activity per minute}}{\text{Energy cost of basal metabolism per minute}}$$

Studies have shown that the cost of women and men undertaking a specific task is very similar for the two sexes once it is expressed as a PAR. This approach has also the merit that people with very different body weights, and therefore very different rates of total energy expenditure, have the same activity ratio. Thus, the energy cost of walking on the level at 1-2 km/h for an 80 kg man is 2 Kcal/min compared with a cost of 2.5 Kcal/min for a 50 kg man. When expressed in activity ratios, however, the cost in both cases is similar and amounts to more than twice the BMR. Thus, by calculating the BMR first, one can assign an energy cost ratio of, for example, walking, sewing or hoeing in the garden to groups of either sex and of different weights.

Table 7.3: Comparison of Energy Cost of some Common Daily Activities in Terms of PAR Values

Activities	*Energy Cost of Daily Activities in PAR Values*	
	Indian Data	*International Data*
Sitting quietly	1.20	1.25
Standing quietly	1.40	1.33
Sitting at desk	1.30	1.36
Standing + doing lab work	2.0	1.95
Harvesting	3.6	3.5
Hand saw	7.4	7.5
Typing (sitting)	1.58	1.69
Walking at 3 MPH	3.71	3.77

PAR values for activities performed in a day can be aggregated over that period to yield the Physical Activity Level (PAL), which is the ratio of the energy expenditure for 24 hours and the BMR over 24 hours. For example, a person spending 8 hours in sleep (with a PAR of 1), 8 hours in domestic and leisure activity (with an average PAR of 2) and 8 hours at work (with an average PAR of 3), would have a total PAR-hour value= (8x1) + (8x2) + (8x3)= 48 PAR-hours.

The PAL (for the day) = Total PAR-hours/Total time = 48/24=2.0

TEE (Total energy expenditure) = Predicted BMR x PAL

Expressing daily energy requirements in terms of PAL values would automatically take care of the problem of lower BMR of certain population groups such as Indians. PAL values for Indians engaged in different activities are given in Table 7.4. PAL value > 2.40 is difficult to maintain over a prolonged period.

Table 7.4: PAL Values Proposed by ICMR Expert Group (2009), ICMR (1989) and Compared to the Figures Proposed by FAO/WHO/UNU Consultation, 2004

Level of Activity	*ICMR (1989)*	*ICMR (2009)*	*FAO/WHO/ UNU*
Sedentary work	1.6	1.53	1.40-1.69
Moderate work	1.9	1.8	1.70-1.99
Heavy work	2.5	2.3	2.0-2.40

Source: ICMR (2011).

Computation of Energy Requirement of Indian Adults in Terms of PAL Units

Body weights of normal healthy Indian adult individual man and woman are taken as 60 kg and 55 kg, respectively. Factorial computation of energy expenditure by adult Indian population is given in Table 7.5.

Adjustments for Body Weight and Age

Energy expenditure at three levels of activity at different age groups and at different body weights are given in Tables 7.6 and 7.7 (NIN/ICMR, 2011). Using the factorial method, BMR factors for computing the daily energy requirements of reference Indian men and women engaged in light, moderate and heavy activities have been derived.

It is very clear from the tables that energy requirement is roughly related to body weight. Energy requirement decreases with age beyond 30 years. Males are taller and heavier as compared to females hence, their energy requirement is also more.

Table 7.5: Factorial Computation of Energy Expenditure of Adult Indian Population

Main Daily Activities	*Duration (h)*	*Major Lifestyles, Energy Expenditure (PAR values)*		
		Sedentary	*Moderate Active*	*Heavy or Vigorous Active*
Sleep	8	1.0	1.0	1.0
Occupational activity	8	1.5	2.3	3.8
Non-occupational activity	8	2.1	2.1	2.1
Mean		**1.53**	**1.80**	**2.30**
Non-occupational activity details				
Personal care	1	2.3	2.3	2.3
Eating	1	1.5	1.5	1.5
Commuting to work by bus or by vehicle or by walk	1	2.0	2.0	2.0
General household or other activities	2	2.5	2.5	2.5
Walking at various speeds without load	1	3.2	3.2	3.2
Light leisure activity	2	1.4	1.4	1.4
Mean non-occupational activity	8	2.1	2.1	2.1
Men: body wt. (60 kg), BMR (1515 kcal)	-	2318	2727	3485
Women: body wt. (55 kg), BMR (1241 kcal)	-	1899	2234	2854

Table 7.6: Energy Requirements of Adult Males and Females Aged 18-30 Years with different Body Weights (Kcal/24 h)

Body Weight (kg)	Male				Female			
	BMR	Sedentary	Moderate	Heavy	BMR	Sedentary	Moderate	Heavy
40	–	–	–	–	1030	1577	1856	2371
45	1298	1986	2336	2985	1101	1685	1982	2532
50	1370	2096	2466	3151	1171	1792	2108	2693
55	1443	2208	2597	3319	1241	1899	2234	2854
60	1515	2318	2727	3485	1311	2006	2360	3015
65	1588	2430	2858	3652	1381	2113	2486	3176
70	1660	2540	2988	3818	1451	2220	2612	3337
75	1753	2716	3191	4078	–	–	–	–

Table 7.7: Energy Requirement at different Ages of Adult Males and Females with different Body Weights (Kcals/24 h)

Sex	Body Weight (kg)	Age 30 -60 years Activity				Age > 60 years Activity		
		BMR	Sedentary	Moderate	Heavy	BMR	Sedentary	Moderate
Male	45	1324	2026	2383	3045	1039	1590	1870
	50	1358	2078	2444	3123	1103	1688	1985
	55	1433	2196	2579	3296	1167	1786	2101
	60	1487	2275	2671	3420	1231	1883	2216
	65	1542	2359	2736	3547	1295	1981	2331
	70	1596	2442	2873	3671	1359	2079	2446
	75	1651	2526	2971	3797	1423	2177	2565
Female	40	1120	1714	2016	2576	965	1477	1737
	45	1160	1778	2092	2673	1015	1553	1627
	50	1203	1841	2165	2767	1065	1630	1917
	55	1245	1905	2241	2864	1115	1706	2007
	60	1286	1966	2315	2958	1163	1782	2097
	65	1328	2032	2390	3054	1215	1860	2187
	70	1369	2095	2464	3149	1265	1936	2277

Protein

In adults, protein is required not for growth but for wear and tear of body tissues. The rate of protein synthesis decreases every year as age advances. No new tissues are formed except that there is maintenance of worn out tissues during adulthood.

Daily protein requirement is expressed on a g/kg body weight basis. The safe level of requirement is given by the Mean ± 2 SD of the mean requirement.

Further, the actual amount of protein to be consumed daily to meet the requirement depends upon the quality of dietary protein. The higher the quality, lower the requirement and vice-versa. Egg protein is used as a reference protein against which the quality of other proteins is compared. Vegetable proteins are generally of poorer quality than animal proteins. The relative biological quality of the proteins of Indian mixed vegetarian (cereal-pulse based) diet is 65 while that of egg protein is 100. Therefore, for calculating the protein requirement, NPU of Indian diets is considered to be 65 per cent.

Protein requirement can be determined by several methods as given under:

- Nitrogen balance
- Obligatory nitrogen loss
- Factorial method

Nitrogen Balance

Nitrogen balance data obtained under appropriate conditions forms the basis for arriving at minimum maintenance nitrogen requirement.

Data on nitrogen balance studies in Indian adults indicate that the minimum dietary protein intake to maintain nitrogen equilibrium should be on an average 0.58 g per kg/day (0.5-0.66 g/kg/day), which after allowing for the integumental loss of N 911 mg N/ kg/day), works out to be 0.65 g protein per kg/day. Assuming a cv of 12.5 per cent, the corresponding safe level of intake (Mean + 1.96 x SD) will be computed as 0.81 g per kg body weight/day. The corresponding requirement based on a mixed Indian diet will be 0.98

g protein/kg/day and this figure is very close to 1g protein/kg/day. This figure is also very close to the RDA derived by the factorial method.

Factorial Method

Based upon this approach, the average daily protein requirement of an Indian adult in terms of egg/milk protein at the physiological level is estimated to be 0.5 g/kg. This requirement figure is based on:

- An obligatory loss of urinary (37 mg/kg/d) and faecal (20 mg/kg/d) nitrogen.
- Cutaneous N loss (11 mg/kg/d)
- Total N loss = 68 mg N/kg/d = 0.4 g protein/kg/d
- The factorial value as obtained above is increased by 50 per cent to obtain the requirement at the physiological level; this works out to be 0.64 g protein/kg/day.
- Assuming a coefficient of variation of 12.5 per cent, the safe level of intake

  (Mean + 1.96 SD) for an adult will be 0.8 g of protein per kg/d.
- Corrected for amino acid score and digestibility (PDCAAS of 82.5 per cent) = 0.97 g/kg/d

Therefore, this method also yields a figure of about 1.0 g protein/kg/day.

The protein requirement is not affected by physical activity and it is the same (1 g/kg/day) for both males and females. ICMR (2011) has recommended daily protein requirement of 60 g for reference adult man and 55 g for reference adult woman (Table 7.8).

Table 7.8: Protein Requirements of Adults

Group	*Body Weight (kg)*	*Requirement (g)*	
		g/kg/d	*per day (g)*
Man	60	1.0	60
Woman	55	1.0	55

Fat

Fat as a concentrated source of energy increases the energy density of the diet. It is also required for the absorption of fat soluble vitamins. Minimum amount of fat has to be present in the diet to meet the minimum essential requirement of the two unsaturated essential fatty acids *i.e.* linoleic and linolenic acid. At the same time, excess of fat intake is not desirable for adults as it can lead to obesity and related degenerative diseases. Hence, according to ICMR (2011), the minimum level, the desirable level and the safe upper limit of fat intake should be considered. Fat requirement of an adult has to be considered from the point of view of total fat requirement and essential fatty acid (EFA) requirement.

Total fat and essential fatty acids intake is through the following two sources:

- **Visible**: It is also known as added fat. This is in the form of ghee, butter, *vanaspati*, vegetable oils *etc.* generally used as cooking medium. Many cooking oils such as mustard, corn, safflower, groundnut, sunflower *etc.* are rich in linoleic acid. Mustard/rapeseed, soybean and linseed oils are good sources of linolenic acid.
- **Invisible**: A significant amount of fat present in cereals, pulses, milk, egg, meat, nuts and oilseeds is known as invisible fat. The small amounts of invisible fat present in different foods add up to a substantial level in the daily diet.

The total fat comprising of both visible and invisible fat in the diet should provide between 15-30 per cent of total calories.

Requirement of Essential Fatty Acids (EFAS)

The desirable amounts of linoleic acid to be consumed by adults is placed at 3 en per cent *i.e.* 3 per cent of total dietary energy. The requirement of linolenic acid is only 0.03 en per cent. Besides total EFA content, ratio of linoleic (n-6) to linolenic (n-3) acids is also important; desirable ratio is 5-10 and not higher than it.

The invisible fat content of Indian diets based on cereals and legumes provides nearly 10 en per cent and meets about 2/3 of EFA requirement *i.e.* 2 en per cent. The remaining 1 per cent of EFA requirement is to be met through visible fat. Any of the common vegetable oils used for cooking can readily provide 1 en per cent of linoleic acid.

Fat Requirement

On the basis of above considerations, the minimum visible fat requirement of Indian adult has been placed at 5 en per cent (12 g/day). In addition, fat is also required to satisfy other functions *viz.* to impart palatability and energy density to diets. Hence, recommended desirable visible fat intake of 10 en per cent (20 g/day) has been suggested to provide energy density and palatability to the diet for both reference man and woman (ICMR, 2011). The recommendations for dietary fats in Indians have been revised taking into account FAO and WHO recommendations for i) total fat, individual fatty acids and health promoting non-glyceride components ii) sources of dietary fats in Indians and iii) availability of fats.

An adult doing sedentary physical activity should desirably consume about 20 g visible fat while individuals doing moderate and hard physical work require 25 and 30 g higher amounts of visible fat, respectively. The visible fat should preferably be in the form of edible oil containing 20 per cent linoleic acid.

Table 7.9 Recommendations for dietary fat intake in Indians (ICMR, 2011)

Gender	Physical Activity	Minimum Level of Total Fat (per cent E)	Fat from Foods other than Visible Fat[a] per cent E	Visible Fat[b]	
				Per cent E	g/p/d
Adult man	Sedentary	20	10	10	25
	Moderate				30
	Heavy				40
Adult woman	Sedentary	20	10	10	20
	Moderate				25
	Heavy				30

a: If higher than 10 per cent E, visible fat requirement proportionately reduces; b: Cooking oils, butter, ghee and margarine

For sedentary adult, the upper safe limit of fat should not exceed 30 per cent of the total energy intake (10 en per cent from visible fat and 20 en per cent visible fat) to avoid obesity and related degenerative diseases. Hence, an adult should not consume less than 50 g visible fat per day.

The total fat intake (30 per cent of the total energy intake) should comprise of the saturated, polyunsaturated and monounsaturated fat in the ratio of 1:1:1.5.

- ☆ Monounsaturated fatty acids (upto 15-16 per cent of total energy)
- ☆ Polyunsaturated fatty acids (less than 10 per cent of total energy intake)
- ☆ Saturated (less than 10 per cent of energy intake)

Sources of the above fatty acids are given in detail in chapter 3.

An adult should consume less than 300 mg cholesterol per day.

Minerals and Trace Elements

Minerals are inorganic elements found in body fluids. Macro-minerals like calcium, phosphorus, and magnesium and the electrolytes (sodium, potassium and chloride) are of great importance in human nutrition. Besides the macrominerals, several trace elements are also required which are essential for many enzyme reactions and body functions.

I. Microminerals

i) Calcium and Phosphorus

ICMR (2011) has considered the requirements for calcium and phosphorus together because of their closely linked functions and requirements. Calcium is required for the formation and maintenance of skeleton and teeth. An adult man of 60 kg has nearly 1 kg of calcium which is mostly present in the bone. Non-skeletal calcium is required for normal contraction of muscle to make limbs move, contraction of heart for its normal function, nervous activity and blood clotting. The calcium levels in cells and plasma are well maintained. In case of calcium deficiency, skeletal calcium plays a role in maintaining plasma calcium level.

Calcium requirement of adult Indians has been determined on the basis of some long term calcium balance and turnover studies. Calcium is required by an adult man for replacing calcium lost from the body through urine, stool, bile and sweat and this loss has been estimated to be 700 mg per day. But the body can reduce this loss by reduced excretion on a low calcium intake through the process of adaptation. The average rate of calcium absorption of Indian mixed diets is about 20-30 per cent; it is enhanced by vitamin D. The dietary levels of both animal protein and sodium are known to limit calcium loss in urine. It has been estimated that reduction in animal protein from 60 g to 20 g or sodium from 150 to 50 m mol/day could decrease calcium requirement by about 200 m g/d and a combination of both could be additive accounting up to 400 mg/d.

Long term calcium balance studies among population groups consuming moderate levels of calcium indicate that for maintaining positive Ca balance, 500-600 mg/d calcium intake is required. In view of this, ICMR (2011) has recommended a daily intake of 600 mg calcium per day for both adult man and women. To achieve this level of intake, a minimum of 200 ml of milk/d would be essential on a cereal- legume diet.

The utilization of phosphorus is closely related to that of calcium since most of the calcium is deposited as calcium phosphate in the bone and teeth. Phosphorus is also a component of nucleic acids. It has also an important role to play in cellular metabolism of nutrients *viz.*, carbohydrate, fat *etc.* Among all age groups except infancy, an elemental Ca, P ratio of 1:1 should be maintained; both adult man and woman require 600 mg phosphorus/day.

ii) Magnesium

Like calcium, magnesium is closely associated with skeletal system. It is required for cellular metabolism. It is important for maintaining electrical potential in nerves and muscle membranes. Its deficiency leads to neuromuscular dysfunction. About 20-25 g magnesium is present in adult human body and about 60-70 per cent of it occurs in the bone, 25-30 per cent in the muscle, 6-8 per cent in soft tissues and 1 per cent in the extracellular fluid. Plant foods are rich in magnesium and its deficiency is unlikely to occur in the Indian population unless accompanied by malabsorption syndrome (inflammatory bowel disease or sprue) or abnormal disease condition (severe PEM).

Based on the balance studies and tracer turnover studies, the requirement of magnesium has been worked out for adults. It has been observed that absorption among Indian adults is close to 35 per cent. Since, FAO/WHO recommendations are applicable for international use, they can also be used for Indians after adjusting for lower absorption of dietary magnesium. Hence, the ICMR (2011) Committee has recommended the same requirement as suggested by FAO/WHO but corrected for lower absorption *i.e.* 35 per cent as compared to 50 per cent. According to ICMR (2011), RDA for magnesium is. 340 mg/d for adult man and 310 mg/day for adult woman.

II. Electrolyte Minerals

Sodium and Potassium

Sodium and potassium are important constituents of body fluids. Proper concentration of these electrolytes inside and outside the cells is essential for maintenance of osmotic balance. Plant foods contain 10-15 fold higher amount of potassium than sodium. Hence, there is lower content of sodium and higher content of potassium in normal Indian diets based on plant foods. The minimum requirement for a healthy person is 500 mg of sodium for adults while a safe and adequate level is 1100-3300 mg/day.

High intake of sodium in the form of sodium chloride is associated with hypertension and stomach cancer; hence, its intake should be specified. However, there is a lack of information on the body turnover of sodium and potassium in tropical countries like India where there is a significant loss of these electrolyte minerals through sweat. Sodium is also lost through urine. In the absence of such data, specific recommendations for the intake of these electrolyte minerals have not been made.

The diet provides 90 per cent sodium from sodium chloride and only 10-15 per cent originates from natural foods. The daily intake of salt (sodium chloride) in our country ranges from less than 5 to 30 g in different states with almost 40 per cent of families consuming about 10 g. As high sodium intake is related to hypertension, a lower intake (<8 g/day) of salt is advisable. One gram of sodium chloride contains 39 per cent sodium. As the natural diet in India provides less than 1 g of sodium chloride, cooking salt can be reduced from 10 g to 5-6 g/day to avoid long term consequences of excess sodium intake.

Enough potassium should be present in the diet to balance sodium intake. Dietary intake of 140 mmol/day at the upper end is considered to be safe. The recommended intake viewed as safe and adequate is 1875-5625 mg daily (10-28 g of KCl). The ideal desirable sodium: potassium ratio in the diet is 1:1 (in mmol).

III. Trace Elements

Out of 15 trace elements essential for the mammalian species, 9 elements namely, iron, iodine, zinc, copper, chromium, cobalt, manganese, molybdenum and selenium are known to be required for humans. However, deficiency of only a few *e.g.* iron, iodine *etc.* are widespread and known for a long time.

The deficiencies of zinc, copper, chromium and selenium have been reported recently from some geographical areas but requirements of some of these trace elements could not be recommended due to lack of experimental evidences and sufficient data on their availability from food sources. Recently, due to atomic absorption spectrophotometry, it has become possible to accurately estimate the contents of trace elements of foods. ICMR (2011) has suggested daily intake of trace elements for Indian adult (Table 7.10).

Table 7.10: Suggested Daily Intake of Trace Elements by an Indian Adult

Trace Element	*Suggested Intake*
Chromium	33 µg
Copper	1.35 mg
Manganese	2.5 mg

i) Iodine

ICMR (2011) has adopted the recommendations of WHO-UNICEF-ICCIDD for iodine *i.e.* daily allowance of 150 µg iodine per day for both adult man and woman. About 90 per cent of the iodine requirement is met through food, while the rest is obtained through drinking water. Daily intake of 10 g iodized salt having iodine at a minimum level of 15 ppm provides about 150 µg per day, in addition to iodine present in foods consumed. About 30 per cent of it is lost during cooking and 70 per cent of it is absorbed which is comparable to the daily physiological need of the body.

ii) Iron

Iron is essential for the formulation of haemoglobin of red blood cells. It has an important role in transport of oxygen. Its deficiency leads to anaemia which is widely prevalent among Indian population especially adolescent girls, pregnant and lactating women. However, Indian adult men and women both are also prone to this deficiency disorder.

The iron requirement in adult men has been established by determining the body basal iron loss through long term turn over studies is computed factorially as mentioned below:

- **Adult man:** Basal iron loss through GI tract + urine + sweat = 14 µg/kg body weight/day. Therefore, an adult man (60 Kg) requires 0.84 mg of iron/day.
- **Adult woman:** Basal loss (14 µg absorbed iron/kg body weight) + average loss iron through menstrual loss of blood (16 µg absorbed iron/

kg body weight) = 30 μg absorbed iron/kg body weight/day or 1.65 mg iron/day.

Depending upon the quality of diet, iron absorption varies from 5 to 10 per cent. Iron absorption of non-vegetarian diets based on meat/fish is much higher (10 per cent) than from vegetarian plant based diets (2-5 per cent). While computing iron requirement for normal Indian adult males consuming cereal- pulse based mixed diet, an average value of 5 per cent has been used for dietary iron absorption whereas a conservative figure of 8 per cent is considered in women who are expected to have better absorption due to iron deficient stores. Women suffering from anaemia also have higher iron absorption from diet; iron absorption is related to severity of anaemia. Hence, in following conditions iron absorption is enhanced:

- Consumption of diet containing ascorbic acid
- Consumption of meat/fish based diet (iron absorption is 10 per cent)
- Persons suffering from anaemia
- Women having higher iron requirement (iron absorption is 8 per cent)
- During pregnancy (iron absorption is about 8 per cent)

Taking into account the basal iron losses and iron absorption from cereal- pulse based mixed Indian diets, ICMR (2011) has recommended the daily requirement of adults which is given in Table 7.11.

Table 7.11: RDAs for Iron in Indian Adults

Group	Body Weight (kg)	Basal Iron Requirement		Iron Absorption from Mixed Cereal-Pulse Diet (per cent)	RDA (mg/ day)
		μg/ kg/d	Total daily mg/day		
Adult man	60	14	0.84	5	17
Adult woman	55	30	1.65	8	21

iii) Zinc

Zinc is an integral component of many enzymes and is widely distributed in the body and skeletal tissue, and muscle and soft tissues are rich sources. Zinc deficiency causes growth failure, depressed immunity, anorexia, diarrhea, altered skeletal function and reproductive failure. In general, the higher Zn-phytate molar ratio of >15 occurring in Indian diets is supposed to cause low bioavailability (<15 per cent absorption) as against 20-25 per cent seen with low phytate/animal food rich diet. Hence, more than total zinc content, bioavailable zinc is important to maintain adequate Zn status.

Estimates of Zn requirements are mostly obtained from the data using chemical balance methods or from turnover studies employing radio or stable isotopes in adults. It is estimated that the endogenous loss is about 1.67 mg, integumental loss of 0.43 mg and urinary excretion of 0.56 mg, accounting for 2.7 mg of absorbed zinc as the requirement for 51 kg average adult man. Zinc absorption from Indian diet on an average is 25 per cent. For obtaining an absorbed Zn of 2.7 mg from Indian diet, dietary zinc needed is about 10 mg/d. Further correcting for individual variation (2 SC as 25 per cent), the RDA comes out to be 12.5 mg/d or 12 mg/d which is comparable to Zn recommendation values by other countries. Accordingly, ICMR (2011) has recommended 12 mg/day for adult man and 10 mg/day for adult woman.

Vitamins

Vitamins though required in small amounts but are essential for carrying out vital functions of the body. Many of them as coenzymes are needed for the proper utilization of other nutrients. ICMR (2011) has recommended dietary intake of certain vitamins such as vitamin A, vitamin D, vitamin E, vitamin K, thiamine, riboflavin, niacin, pyridoxine, dietary folate and B_{12} as their deficiencies are known to occur in human population. The requirements of these vitamins for Indian adults are discussed below.

Fat Soluble Vitamins

Vitamin A

This fat soluble vitamin is essential for normal vision, maintaining the integrity of epithelial tissues, growth and for a number of metabolic functions. Vitamin A or retinol or preformed vitamin A is present in animal foods such as butter, ghee, whole milk, curd, egg yolk, liver *etc.* only. Plant foods do not contain vitamin A. On the other hand, provitamin A (carotenes) is found in plant foods like fruits and vegetables. In Indian diets, the source of dietary vitamin is mainly β-carotene.

Though provitamin A is converted into vitamin A (retinol) in the small intestine, yet the efficiency of conversion varies depending upon the food source. It is well established that efficiency of β-carotene to retinol is around 50 per cent. However, other carotenes (α and γ) yield even less amount of vitamin A. Further, unlike retinol, β-carotene, is not absorbed completely. Therefore, taking into consideration the physiological conversion of β-carotene to retinol and absorption of β-carotene, ICMR Expert Committee (2011) has modified the extent of conversion efficiency of 1:8 (1μg retinol= 8μg β-carotene) and retained the earlier recommendations of retinal requirements for all groups (ICMR, 1989) except pregnant women. Hence, the requirement of β-carotene is 8 times the requirement of retinol.

The recommended intake of vitamin A or retinol of Indian adult man and woman is 600 μg/day or 4800 μg β-carotene/day.

Vitamin D

The active form of vitamin D *i.e.* 1, 25 $(OH)_2D_3$ formed in kidney is required for absorption of calcium and bone formation. Hence, vitamin D is considered as a prohormone rather than a vitamin. Vitamin D can also be synthesized in the skin due to its exposure to sunlight and also obtained from plant and animal sources. Vitamin D is gaining much more importance now a days due to its involvement in various diseases. Diseases such as osteoporosis, diabetes, psoriasis, hypertension, arthritis, multiple sclerosis and CVD are known to have the involvement of Vitamin D leading to the suggestion that the above spectrum of deficiency diseases be called vitamin D deficiency diseases (VDDD) as in the case of other micronutrients.

The WHO Expert Committees recommended 100 units (2.5μg) per day for adult males in 1988 and increased them later in 2005 to 200 units (5μg) per day. This is due to progressive decrease in the exposure to sun light and the need to obtain the requirements from dietary and supplement sources. ICMR has suggested no recommendation for vitamin D. But in case of women who remain in *purdah* system and are not exposed to sunlight do require vitamin D in their diets. In such cases and under situations of minimal exposure to sunlight, a daily intake of 400 IU (10 μg) vitamin D has been suggested.

Vitamin E and K

These two vitamins are present in abundance in vegetable oils and plant foods. Moreover, there is limited information on the contents of these vitamins in food items. Intake of vitamin E, as an antioxidant is associated with the intake of essential fatty acids (EFA) in the diet *i.e.* 0.8 mg of vitamin E per g of EFA as recommended by ICMR. This roughly works out to be 8-10 mg tocopherol/d, depending upon the edible oil used. Vitamin K deficiency is not common among adults. It is occasionally encountered in premature newborn infants. Hence, ICMR has suggested no recommendations for daily intake of vitamin K for adults. FAO/WHO (2004) has suggested an RDA of 7.5-10 mg α-tocopherol and 55μg of vitamin K/d for adults which can also be accepted by Indians till such recommendations are made by ICMR.

B-complex Vitamins

This group of vitamins includes thiamine, riboflavin, niacin, pyridoxine, folic acid and vitamin B_{12}. Their deficiencies are generally seen in human population. Depletion and repletion study approach has been used to determine the requirements of thiamine, riboflavin and niacin.

Thiamine

Thiamine stores in the body are small and therefore, its regular intake is necessary to maintain its metabolic functions. Single large dose is absorbed poorly. Since, thiamine pyrophosphate, co-enzyme of thiamine has a great role to play in carbohydrate metabolism, therefore, thiamine requirement is related to energy intake, particularly the energy derived from carbohydrates. More the energy derived from carbohydrates, more shall be the requirement for thiamine in daily diet. Due to its water solubility and heat labile nature, some thiamine losses occur during food processing and cooking. Alkaline medium *e.g.* addition of cooking soda causes the destruction of this vitamin. In Indian cooking, almost half of the thiamine present in raw foods is lost depending upon the type of food processing and cooking methods employed. Intake of thiamine above 0.3 mg/1000 kcal are consistent with good health. Keeping this into consideration, an allowance is made for individual variations and cooking losses of this vitamin by almost doubling the minimum requirement figure (0.25-0.35 mg thiamine per 1000 kcals) derived from diet and nutrition surveys, urinary excretion of this vitamin, transketolase activity and depletion-repletion studies. Accordingly, ICMR (2011) has recommended 0.5 mg thiamine intake per 1000 kcals in daily diets of an adult. In the absence of any new information, ICMR (2011) has retained the similar recommendation as made in 1989. However, minimum daily allowance of 1 mg of thiamine has been recommended even if calorie intake is less than 2000 kcal.

Daily energy intake of adult man and woman varies depending upon the type of activity they are engaged in. Hence, accordingly their thiamine intakes have also been computed which are depicted in Table 7.1.

Riboflavin

Riboflavin, an integral part of flavor-proteins has a great role to play in biological oxidation processes in our body. This vitamin is concerned with energy and protein metabolism.

Both the urinary excretion and the concentration of this vitamin in saturated blood cells have been used to determine its requirement. The level of glutathione reductase activation, an enzyme requiring riboflavin as a co-factor in erythrocytes is also used to determine human requirement of riboflavin. Other methods employed for determining riboflavin requirement include diet surveys, clinical signs and depletion-repletion studies. Depending upon the food and the type of cooking, cooking losses of riboflavin vary from 15 to 100 per cent. This vitamin is destroyed when exposed to light. Generally, mean losses to the extent of 20 per cent due to cooking in whole diet are taken into consideration while determining riboflavin requirement. Depending upon the results of various studies and making some allowance for cooking losses, ICMR (2011) has recommended an allowance of 0.6 mg riboflavin per 1000 kcals both for adult men and women. This Committee has also recommended a minimum riboflavin intake of 1.2 mg/day even if the calorie intake is lower than 2000 kcals. Riboflavin requirements for adults engaged in different activities have been depicted in Table 7.1.

Niacin

Niacin, nicotinic acid and nicotinamide are usually used synonymously. This vitamin as a part of co-enzyme NAD, NADP *etc.* is connected with various metabolic reactions such as glycolysis, tissue respiration, synthesis of fat and metabolism of carbohydrate, fats and proteins. Hence, like thiamine and riboflavin, niacin requirement is also dependent upon energy intake.

Niacin has a special relationship with an essential amino acid tryptophan. Tryptophan acts as a precursor of niacin. In well nourished normal individuals, sixty mg of tryptophan can give rise to 1 mg of niacin in the body. Hence, body requirement of this vitamin is met partly from the vitamin present in diet and partly by conversion from tryptophan present in dietary proteins. The contribution of niacin from both the above sources is to be considered in terms of niacin equivalent for assessing the niacin requirement for individuals.

Niacin equivalent (mg) =

$$\text{Niacin content (mg)} + \frac{1}{60} \text{ tryptophan content (mg)}$$

Although some losses of this vitamin occur due to cooking but as compared to other vitamins, niacin is more stable. On an average, 25 per cent loss of niacin due to cooking is assumed in Indian food preparations. Diet surveys and load tests, minimum quantity required for curing pellagra or maintaining urinary excretion of the vitamins have been used for determining niacin requirements. Load tests using niacin have indicated that individuals consuming 6.5 to 7.2 mg niacin equivalents per 1000 kcals have a satisfactory niacin status. The subjects suffering from pellagra have inadequate niacin intake *i.e.* about 3.6 mg per 1000 kcals.

Under certain unusual circumstances where individuals consume about 7.5 mg niacin equivalents per 1000 kcals, still suffer from pellagra. This situation arises when the population group consumes maize or sorghum as the main cereal with no other supplement. Pellagra is commonly seen among maize eaters because maize is a poor source of niacin and tryptophan. Further, a part of nicotinic acid is present in bound form 'niacytin' in maize which is not physiologically available. Pellagra among jowar eaters is attributed to the presence of excessive amounts of the leucine amino acid in protein of jowar and to the resultant imbalance between leucine and isoleucine. Such an imbalance interferes with the conversion of tryptophan to niacin, causing niacin deficiency.

However, under normal circumstances, ICMR (2011) has recommended an allowance of 6.6 mg niacin equivalents per 1000 kcals for adults. Niacin equivalent requirements for adults engaged in different activities have been depicted in Table 7.1.

Pyridoxin (Vitamin B_6)

This vitamin exists in three forms – pyridoxal, pyridoxamine and pyridoxine. Pyridoxine or vitamin B_6 is a co-factor for several enzymes which have a great role to play in metabolism of amino acid and hence, its requirement is affected by dietary protein intake. Pyridoxal phosphate, the co-enzyme form of pyridoxine is required for conversion of tryptophan to niacin. It is also believed to have a role in the production of antibodies.

The human requirement of vitamin B_6 has been established from the intake needed to prevent biochemical changes suggestive of its deficiency. These include urinary excretion of the vitamin and its major metabolic product, 4-pyridoxic acid, excretion of xanthurenic acid in urine following a tryptophan load and determination of the activity of transaminase in red blood cells. Since cooking losses in habitual Indian diets are negligible, hence no allowance for these losses needs to be made. An average daily requirement of this vitamin appears to be about 1.5 mg vitamin B_6 per day. Based on the levels given by the National Research Council, USA, ICMR (2011) has recommended daily adult requirement to be about 2.0 mg B_6 per day in case of both men and women.

Folic Acid

Folic acid is required for multiplication and maturation of red cells. Its deficiency leads to megaloblastic anaemia. The primary biochemical function of folic acid is to transfer single carbon in the synthesis of a number of metabolites in the body. Folic acid and vitamin B_{12} are interrelated and connected as far as synthesis of nucleic acid is concerned.

Folic acid is present either in the free form or in the conjugated form as polyglutamates. The proportion of free and conjugate forms varies considerably in different foods. Free folic acid is well absorbed as compared to polyglutamates. Polyglutamates have to be broken down before absorption into smaller units by the conjugases present in the intestinal lumen. Hence, availability of folic acid depends upon the form in which it is present in a food and it also affects the determination of requirement of this vitamin.

Depending upon the diet surveys, depletion-repletion studies, optimum concentration of folate both in serum and red blood cells, load test and of proven role of folic acid deficiency as an etiological factor for elevated blood homocystiene and CVD and congenital neural tube defect (NTD), ICMR (2011) has recommended a dietary allowance of 200 µg of dietary folate per day for both adult men and women.

Vitamin B_{12}

Like folic acid, vitamin B_{12} is also involved in the maturation of cells and deficiency of this vitamin results in megaloblastic anaemia. It is also needed for proper functioning of central nervous system. As a component of several co-enzymes, this vitamin plays an important role in the synthesis of nucleic acids. Its metabolism is closely associated with that of folic acid.

Vitamin B_{12} is present only in animal foods like liver, eggs, meat *etc.* No plant food contains this vitamin. However, microflora of human large intestine can synthesize vitamin B_{12} but its bioavailability is controversial. Moreover, bacterial contamination due to poor hygiene practice is also believed to provide small quantities of B_{12} to vegetarian population of developing countries like India.

Liver can store considerable amount of this vitamin and it makes the assessment of the daily requirement of B_{12} complicated. In adults, such stores have been estimated to be around 1 mg. Absorption of vitamin B_{12} from the intestine is an active process and dose dependent. For absorption of vitamin B_{12}, presence of 'intrinsic factor' in the small intestine is required.

Depending upon the various methodologies *viz.*, liver stores, dose response, depletion-repletion studies and use of isotope turnover studies, it has been established that the amount of absorbed vitamin B_{12} is only 0.1 µg per day. The daily requirement of vitamin B_{12} for humans has been estimated to be between 0.5 and 1.0 µg per day. After making some allowances for cooking losses and for uncertainty regarding the extent of absorption of vitamin B_{12} from food items, ICMR (2011) has recommended a dietary allowance of 1 µg of vitamin B_{12} per day for both adult men and women.

Vitamin C (Ascorbic Acid)

We cannot synthesize vitamin C in our body as most of the animal species can do. Hence, we have to fulfill the requirement of vitamin C through the foodstuffs we include in our diet. Vitamin C requirement of humans has been determined by depletion and repletion studies. The allowances are derived from the amount of vitamin C that will cure or prevent scurvy, the amount that is metabolized in the body and the amount necessary to maintain adequate body resources. Minimum daily intake of vitamin C to cure scurvy is 6.5 mg. The minimum intake to maintain vitamin C in leukocytes at saturation level has been found between 10 and 20 mg. However, this vitamin present in high concentrations in green leafy vegetables and citrus fruits, is readily destroyed up to 50 per cent extent during processing, cooking and storage. Hence, taking into account these losses, a daily intake of 40 mg of vitamin C has been recommended for adult men and women.

Meal Planning for Adult Man and Woman

Both adult man and woman should take a diet that provides all the essential nutrients in the right amount and proportion as per recommendation by ICMR (Table 7.1) in order to meet body needs adequately. They should take well balanced diets to keep their body in a state of good health. The balanced diets for adult man and woman have been given in Tables 7.12 and 7.13.

Table 7.12: Balanced Diet for an Adult Man (Number of portions) (NIN/ICMR, 2011)

Food Groups	*g/ portion*	*Type of Work*					
		Sedentary		*Moderate*		*Heavy*	
		V	*NV*	*V*	*NV*	*V*	*NV*
Cereals and millets	30	12.5	12.5	15	15	20	20
Pulses	30	2.5	1.5	3	2	4	3
Egg/meat/chicken/ fish	50	-	1	-	1	-	1
Milk	100 ml	3	3	3	3	3	3
Roots and tubers	100	2	2	2	2	2	2
Green leafy vegetables	100	1	1	1	1	1	1
Other vegetables	100	2	2	2	2	2	2
Fruits	100	1	1	1	1	1	1
Sugar	5	4	4	6	6	11	11
Fats and oils (visible)	5	5	5	5	5	8	8

V = Vegetarian; NV = Non-vegetarian.

Table 7.13: Balanced Diet for an Adult Woman (Number of portions) (NIN/ICMR, 2011)

Food Groups	*g/ portion*	*Type of Work*					
		Sedentary		*Moderate*		*Heavy*	
		V	*NV*	*V*	*NV*	*V*	*NV*
Cereals and millets	30	9	9	11	11	16	16
Pulses	30	2	1	2.5	1.5	3	2
Egg/meat/chicken/ fish	50	-	1	-	1	-	1
Milk	100 ml	3	3	3	3	3	3
Roots and tubers	100	2	2	2	2	2	2
Green leafy vegetables	100	1	1	1	1	1	1
Other vegetables	100	2	2	2	2	2	2
Fruits	100	1	1	1	1	1	1
Sugar	5	4	4	6	6	9	9
Fats and oils (visible)	5	4	4	5	5	6	6

V = Vegetarian; NV = Non-vegetarian.

For example, a day's sample diet for Mr. X, not vegetarian, age 40 years, belonging to middle income group who is engaged in sedentary activities (desk work) has been described below.

Step I: Personal Profile

Age	40 years
Sex	Male
Activity	Sedentary
Socio-economic status	HIG
Food habit	Non-vegetarian

Step II: Recommended Dietary Allowances (ICMR, 2011)

Energy	2320 kcals
Protein	60 g
Carbohydrate	335 mg
Fat	70-75 g
(Visble and Invisible)	

Step III: Food Exchange Plan

Exchange	*No.*	*Energy (KCal)*	*Protein (g)*	*CHO (g)*	*Fat (g)*
Milk	2	340	16	24	20
Meat	1	70	7	-	5
Pulse	2	200	14	34	-
Cereal/Starch	16	1120	32	240	-
Vegetable A	2	40	2	7	-
Vegetable B	3	120	6	21	-
Fruit	2	80	-	20	-
Fat	8	360	-	-	40
Sugar	5	100	-	25	-
Total		**2425**	**77**	**373**	**65**

Menu Plan and Distribution of Exchanges

Meal	*Exchange*	*No.*	*Menu*
Early morning	Milk	0.2	Tea
	Sugar	1	
Breakfast	Cereal/Starch	4.0	Stuffed *parantha*
	Meat	1.0	Egg omelet
	Vegetable A	1	Milk
	Vegetable B	1	
	Fat	2	
	Milk	0.5	
	Sugar	1.0	
Lunch	Cereal/Starch	6.0	Chapati
	Vegetable B	1.0	Cauliflower and potato veg.
	Milk	0.5	Cucumber *raita*
	Fat	2.0	Green salad (Radish, onion)
	Fruit	1.0	Orange
Tea	Milk	0.2	
	Sugar	1.5	Tea
	Cereal/Starch	2.0	Biscuits
	Fat	1.0	
Dinner	Pulse	2.0	Tomato soup
	Vegetable A	1.0	Channa dal
	Vegetable B	1.0	Carrot and peas veg.
	Cereal/Starch	6.0	Chapati
	Fat	3.0	Coriander chutney
	Milk	0.6	Fruit custard
	Sugar	1.5	
	Fruit	1.0	

Similarly, a diet plan for an adult woman can be prepared. For the persons belonging to lower income group, the fat, milk and meat exchanges can be reduced whereas cereal, pulse and sugar exchanges can be increased. Instead of expensive ones, locally available inexpensive seasonal vegetables should be judiciously selected.

8

Nutrition during Pregnancy

Adequate nutrition during pregnancy is extremely important to both maternal and foetal health. For women in the deprived conditions as poverty and overwork, pregnancy is amongst the most stressful periods when she sustains a rapidly growing foetus under hostile conditions. The dual role of these women, as mothers and productive workers, is often comprised by poor diets and ill health; not only their own well-being but that of the whole family is affected. A poor diet, frequent acute and some chronic infections, repeated pregnancies, prolonged lactation and a heavy burden of work may all contribute to serious physiological depletion and sometimes to overt malnutrition. Hence, pregnant and lactating women have been widely recognized as a vulnerable group from the health and nutrition point of view.

Pregnancy is a physiologically and nutritionally a high demanding period. A woman prepares herself to meet the nutritional demands by increasing her own body fat deposits during pregnancy. If the nutrition of the mother is inadequate, then her body reserves are drawn upon and depleted to meet the nutrient demands of the growing foetus. As a result, it tells upon the maternal nutritional and health status.

The optimum weight gain for a mother during pregnancy is about 1.5 kg in the first three months. In each subsequent month, the weight gain should be 1.5 kg. At full term, a total gain is about 10-14 kg with an average of 12 kg. In Indian reference woman with pre-pregnancy weight of 55 kg, the body weight gain may be 10 kg. Since the foetal development is accompanied by changes in maternal body composition and metabolism, the nutritional needs of mother increase as pregnancy advances. Increase in the nutritional requirements of a pregnant woman is due to:

- ✰ Rapid growth of foetus
- ✰ Development of placenta
- ✰ Enlargement of maternal tissues
- ✰ Increase in maternal circulating blood volume
- ✰ Formation of amniotic fluid
- ✰ Increased storage deposits
- ✰ Mineralization of the skeletal and bone structure of the foetus

During a 40-week gestational period, the growing foetus accumulates about 400 g of protein, 475 g of fat and 250 g of water containing minerals and other nutrients. These nutrients come from the mother via the placenta. During the same period, mother expands her body fluids by 25 per cent and accumulates 382 g of protein, and 3380 g of fat and minerals in her body tissues. The placenta grows with the foetus and contains 100 g protein and 4 g of fat and minerals.

To meet the above mentioned increased nutritional requirements, a pregnant woman must eat a wide variety of foods to make sure that her own nutritional needs as well as those of her growing foetus are met. There is no need to completely modify the routine dietary pattern. However, the quantity and frequency of usage of various foods should be increased. She should take a well balanced diet as described in Table 8.2.

Nutritional Requirements during Pregnancy

Nearly all the nutrients are required in greater amounts during pregnancy, but the magnitude of the increase is greater for some nutrients than for others. The requirements of various nutrients as recommended by ICMR (2011) during pregnancy (Table 8.1) are discussed as below:

Table 8.1

Nutrients	*Pregnant Woman*		
	Sedentary Worker	*Moderate Worker*	*Heavy Worker*
Energy (Kcal/d)	1900+350	2230+350	2850+350
Protein (g/d)	78	78	78
Fat (g/d)	30	30	30
Calcium (mg/d)	600+600	600+600	600+600
Iron (mg/d)	35	35	35
Retinol (μg/d)	600+200	600+200	600+200
Thiamine (mg/d)	1.0+0.2	1.1+0.2	1.4+0.2
Riboflavin (mg/d)	1.1+0.3	1.3+0.3	1.7+0.3
Niacin (mg/d)	12+2	14+2	16+2
Pyridoxin (mg/d)	2.0+0.5	2.0+0.5	2.0+0.5
Ascorbic acid (mg/d)	40+20	40+20	40+20
Dietary folate (μg/d)	200+300	200+300	200+300
Vit. B_{12} (μg/d)	1+0.2	1+0.2	1+0.2
Magnesium (mg/d)	310	310	310
Zinc (mg/d)	10+2	10+2	10+2

Energy

As per Nutrition Expert Group (ICMR, 2011) for an average Indian reference woman whose body weight is 55 kg, the total energy cost of pregnancy is approximately 77,000 kcals in case if gestational weight gain is 12 kg. Additional energy during pregnancy is required to support the growth of foetus, expansion of maternal tissues and deposition of fat as well as to meet the increased basal metabolic rate (BMR). Increase in BMR relative to pre-pregnancy values can be 5.3, 11.4 and 25.3 per cent during the first second and third trimester, respectively. The energy needs are not evenly distributed during the pregnancy. In the first and second trimesters, the extra energy is mainly needed for growth of maternal tissues like breasts and uterus, increased blood volume and laying down of storage fat. The increased requirements in the third trimester are mainly for the growth of foetus and placenta besides some increase in maternal tissues as well. Various factors like age, pre-pregnancy body weight, pre-pregnancy BMR and physical activity affect the energy requirements of a pregnant woman.

Additional energy requirement of Indian women during pregnancy have been computed on the basis of the reference Indian women and pregnancy weight gain of 10 kg and 12 kg. Additional energy needed for tissue deposition will be 85 kcal, 280 kcal and 470 kcal during the three trimesters, respectively if the body weight gain is 12 kg. As is the practice, if only 2nd and 3rd trimesters are taken into consideration, the average energy requirement would be 375 kcal/day during that period. If the body weight gain is only 10 kg and assuming body weight as well as increase in body weight are assumed to have an influence on PAL, increase in energy requirement during the three trimesters would be 70 kcal, 230 kcal and 390 kcal, respectively or an average of 310 kcal during the 2nd and 3rd trimesters. Computation of additional energy needed during pregnancy for an Indian woman of 55 kg pre-pregnancy body weight is given in Table 8.2.

Table 8.2: Additional Energy Cost of Pregnancy with Gestational Weight Gain of 12 kg (ICMR, 2011)

	I[st] Trimester (g/d)	*2nd Trimester (g/d)*	*3rd Trimester (g/d)*	*Total Deposited (g)*
A. Rate of tissue deposition				
Weight gain	17	6.0	54	12000
Protein deposited	0	1.3	5.1	597
Fat deposited	5.2	18.9	16.9	3741
B. Energy cost of pregnancy (kcal) from energy deposited and increase in BMR				
Protein deposited	0 (0)	7.2 (6.0)	29.0 (24)	3370 (2808)
Fat deposited	48.3 (40.2)	175 (145)	156 (130)	34600 (28832)
Efficiency of energy utilization	4.8 (4.0)	18.2 (15.0)	18.4 (15.3)	3800 (3167)
Basal metabolic rate increase	48 (40)	95 (79)	237 (198)	35130 (29274)
Total energy cost of pregnancy	101 (84)	295 (246)	440 (367)	77100 (64247)
C. Energy cost of pregnancy (kcal) from energy deposited and increase in TEE (assuming same rates of tissue deposition as above)				
TEE increase[a]	20 (17)	84 (70)	311 (259)	38560 (32132)
Total energy cost of pregnancy	69 (57)	266 (222)	496 (413)	76530 (63775)
D. Average of energy cost of pregnancy by the two methods (B and C above)[b]				
Total energy cost of pregnancy	85 (70)	280 (230)	470 (390)	77000 964170)
Average of 2nd and 3rd trimesters	–	**375 (310)**		

a: Includes costs of efficiency of energy utilization; b: Rounded off to nearest 5 kcal.

Figures within parentheses correspond to energy costs if gestational weight gain is 10 kg instead of 12 kg.

Hence, ICMR (2011) has made an average recommendation of 350 kcal/day, as additional requirement during pregnancy for an Indian woman of 55 kg body weight and pregnancy weight gain between 10-12 kg.

Protein

The protein intake must be increased in pregnancy due to its specific contributions to growth. Studies indicate that approximate 900-1000 g protein is deposited in the foetus and accessory tissues of the woman. Protein may be stored in the body at a uniform rate during the entire pregnancy and is made available to the specialized tissues as needed.

Protein requirement during pregnancy has been assessed by the factorial method and nitrogen balance method earlier. The 2007 FAO/WHO/UNU Consultation has also considered newer data on protein deposition during pregnancy, using the total body potassium (TBK) method. These studies showed that the total protein deposited during pregnancy was 686 g, but this was not deposited at a uniform rate during pregnancy. In well-nourished women with a mean gestation weight gain (GWG) of 13.8 kg, this protein deposition was distributed as 1.9 g/day in the 2nd trimester and 7.4 g/day in the 3rd trimester. There was no net accretion or loss of protein during pregnancy, suggesting that the protein deposition during pregnancy was only into foeto-placental tissue.

Therefore, the FAO/WHO/UNU 2007 Report is based on more recent data, although it uses the same factorial approach earlier. If one assumes protein deposition to be proportional to GWG, in reference women (pre-pregnancy wt. 55 kg) with gain 12 kg GWG, the total protein deposited would be 597 g, distributed as 1.6 g/d and 6.5 g/d in the 2nd and 3rd trimesters, respectively. The next question is the efficiency with which dietary protein could meet these deposition needs which has been considered to be 42 per cent. To this were added the maintenance costs, which were based on the mid-trimester body weight of pregnant woman, the adult maintenance value of 0.66 g/kg/d. For the GWG of 12 kg, the mid-trimester weight gain has been assumed to be 0.7, 4.1 and 9.4 kg in the first, second and third trimesters, respectively. Finally the safe level of increased intake by trimester was derived from the estimated average extra requirement, assuming a coefficient of variation of 12.5. Based on these calculations, the extra protein requirement (at safe levels) comes out to be 0.6, 8.1 and 27.0 g/d, during the first, second and third trimesters, respectively to support a 12 kg GWG.

In a woman gaining 10 kg GWG, protein deposition would be 1.4 and 5.4 g/day for the 2nd and 3rd trimesters, respectively. The same calculations as done above indicate that such women require high quality protein at a rate of 0.5, 6.9 and 22.7 g/d during the first, second and third trimesters, respectively (Table 8.3). These values may be lower in Indian woman whose average pre-pregnancy weight is 47 kg and who has a GWG of 7-8 kg. The additional high quality protein requirement in such a pregnant woman gaining 8 kg during pregnancy is 0.4, 5.6 and 18.1 g/d for the first, second and third trimesters, respectively.

These requirements would increase further, if they were to be met only by protein in the cereal-pulse-milk based low cost Indian diet, which has a PDCAAS of 82.5 per cent. After adjusting for the dietary protein quality for a cereal-legume-milk based Indian diet, ICMR Nutrition Expert Group (2011) has recommended 82.2 g for a pregnant woman *i.e.* an additional daily allowance of 27.2 g protein during pregnancy. The extra protein can be obtained by taking more of pulses, milk, eggs, meat and fish. The diet of a vegetarian pregnant woman can be enriched with protein by including more pulses and certain combinations of whole pulses and milk.

Fat and Essential Fatty Acids

Maternal fat is accumulated primarily during the first and second trimester as an energy and essential fatty acid (EFA) reserve for lactation; foetal growth occurs mainly in the last two months of pregnancy. Dietary fat is important in foetal and early infant growth because this is the period of organogenesis where there is a high demand for EFA for the cell structural lipids. The approximate accumulation of EFA during pregnancy is estimated to be 620 g which includes the demand for uterine, placental, mammary gland and foetal growth

Table 8.3: Recommended Additional Protein Intake during Pregnancy for a 10 kg Gestational Weight Gain (ICMR, 2011)

Trimester	*Mid-trimester Weight Gain[a] (kg)*	*Additional Protein for Maintenance (g/d)[b]*	*Protein Deposition (g/d)*	*Dietary Protein Requirement for Deposition[c] (g/d)*	*Mean Extra Protein Requirement (g/d)*	*Safe Intake (g/d)[e]*
1	0.6	0.4	0.0	0.0	0.4	0.5
2	3.5	2.3	1.4	3.3	5.5	6.9
3	8.0	5.3	5.4	12.9	18.2	22.7

a: Women gaining 10 kg during gestation; b: Mid term increase in weight x estimated average requirement for maintenance for adult 0.66 g/kg/d; c: Protein deposited, adjusted for a 42 per cent efficacy of utilization; d: Sum of extra maintenance plus protein deposited; e: Safe intake = Mean extra protein requirement + (1.96 x SD) assuming CV of 12.5$. This requirement refers to high quality protein, with PDCAAS of 100.

and the increased maternal blood volume. To meet these needs, 4.5 per cent of the expected energy intake in the form of EFA is recommended during pregnancy.

The linoleic acid requirement during pregnancy is 4.5 en per cent. The minimum level of total fat should be 20 per cent E. To furnish 20 per cent E, diet of pregnant woman should contain at least 30g visible fat. They should also consume at least 200 mg/d DHA for optimal health and foetal development. Hence, ICMR (2011) has recommended daily visible fat intake to be 30 g during pregnancy to provide higher level of linoleic acid and the necessary calorie density.

Table 8.4: Recommendations for Dietary Fat Intake for Indian Pregnant Woman (ICMR, 2011)

Physiological Group	*Physical Activity*	*Minimum Level of Total Fat (per cent E)*	*Fat from Foods other than Visible Fat[a] Per cent E*	*Visible Fat[b]*	
				Per cent E	*g/p/d*
Adult woman	Sedentary	20	10	10	20
	Moderate				25
	Heavy				30
	Pregnant woman	20	10	10	30

a: If higher than 10 per cent E, visible fat requirement proportionately reduces; b: Cooking oils, butter, ghee and margarine

Carbohydrates

Carbohydrates should comprise 45-64 per cent of daily calories and this includes approximately 6-9 servings of whole grain daily.

Minerals

The need for minerals especially calcium, iron and zinc increases during pregnancy.

Calcium

The pregnant woman must be supplied with sufficient calcium for her own needs and to promote adequate mineralization of the foetal skeleton and teeth. Approximately 27.5 g of calcium are found in the full term infant and 12g of calcium each in the placenta and the maternal fluids and tissues. In this way, the total requirement during pregnancy comes to about 30 g. Most of calcium (300 mg daily) in foetus is deposited during the last trimester, hence it is proposed that an additional amount of 600 mg calcium per day be added in the daily requirement of 600 mg for the non-pregnant state and a total of 1200 mg calcium per day should be given in pregnancy. To meet this requirement, good sources of calcium such as milk, cheese, curd, green leafy vegetables *etc.* should be included in the daily diet of pregnant woman otherwise calcium supplements are needed. Calcium and phosphorus ratio provided through the diet should be 1:1.

Magnesium

There is no additional requirement for magnesium during pregnancy. ICMR (2011) has recommended 310 mg magnesium in the daily diet.

Iron

During pregnancy, iron is needed for the manufacture of haemoglobin in both maternal and red blood cells, iron in the placenta and tissue, expansion of blood volume, blood loss during parturition and foetal growth. Hence, additional iron requirement during pregnancy (12 kg GWG) is computed from iron needs for foetal growth (230 mg), expansion of maternal tissues including the red cell mass (296 mg), the iron content of placenta (100 mg) and obligatory loss (230 mg). These additional requirements are added to the basal requirement (Table 8.5). However, there is a saving (150 mg) of menstrual loss during pregnancy. Based on these considerations, additional iron requirement for entire pregnancy (trimesterwise) of an adult woman (pre-pregnancy weight 55 kg) is estimated 760 mg (10 kg GWG) and 860 mg (12 kg GWG) (Table 8.6). Hence, an average Indian pregnant woman should take 35 mg iron daily (ICMR, 2011).

Table 8.5: Iron Requirement during Pregnancy

Compartments	*Iron Requirement during Entire Pregnancy (mg)*	
	10 kg GWG	*12 kg GWG*
Foetus	190	230
Expansion of maternal red cell mass	250	296
Placenta+ cord	90	100
Obligatory losses[a]	230	234
Total	**760**	**860**

a:@ 14 µg/kg.

Blood lost during parturition considered to be taken care of by the contraction of maternal red cell mass and therefore not accounted for.

Source: ICMR (2009)

Table 8.6: Iron Requirement during Pregnancy (trimester-wise)

Trimester	*Requirement (mg)*	
	10 kg GWG	*12 kg GWG*
1st trimester	130	138
2nd trimester	320	372
3rd trimester	310	351
Total	**760**	**861**

In India, anaemia due to iron and folic acid deficiency aggravates during pregnancy and labour. Repeated succession of rapid pregnancies perpetuates it. Low dietary iron intake (20 mg/day), poor iron absorption (5 to 10 per cent only) from habitually consumed cereal based phytate, fibre and tannin rich diet and parasitic infections have been implicated as major factors responsible for anaemia.

Severe iron deficiency anaemia in pregnancy may lead to premature delivery, low birth weight and increased risk for foetal death in the first few weeks of life. Therefore, a vegetarian woman must take good sources of dietary iron *i.e.* green leafy vegetables including amaranths leaves, mint, spinach, drumstick, green coriander, radish and carrot leaves, jaggery, dates and sesame seeds. Iron from animal sources such as meat, poultry and fish is readily absorbed than iron from plant foods such as cereals and legumes. Those who are vegetarians should take foods rich in vitamin C along with iron as vitamin C increases iron absorption.

It is well recognized that iron requirements of pregnant women from developing countries cannot be met by dietary iron. Therefore, all non-anaemic pregnant women should receive 60 mg of extra iron supplement per day and then increase their intake only when laboratory tests for haemoglobin indicate that more is needed.

Zinc

Zinc is an important mineral for supporting growth and development. The RDA increases from 10 to 12 mg per day during pregnancy. It is based on the calculated additional 0.75 mg required daily for growth of the foetus and placenta. The allowance includes a large safety factor for diet with poor availability of zinc, assuming only 25 per cent absorption of dietary zinc. The extra protein foods in the diet of a pregnant woman should supply most of this zinc as meat, poultry, dry beans, eggs and milk are good sources of this mineral. A poor zinc status in pregnancy increases the risk for having low birth weight infant.

Iodine

Adequate iodine is crucial for thyroid hormone production, which is required for normal neurodevelopment during pregnancy and early childhood. The iodine maternal requirement is increased during pregnancy as a result of an increased requirement for thyroxine (T4) in order to maintain normal metabolism in the mother, a transfer of T4 and iodine from the mother to the fetus and a supposed greater than normal loss of iodine through the kidneys due to an increase in renal clearance.Pregnant woman should take 200 μg of iodine per day (ICMR, 2011).

Sodium

On the basis of average weight gain of 11 kg, of which 70 per cent is water, the sodium requirement in pregnancy is calculated to be about 750 mmol. Therefore, only 3 mmol of sodium/day above the non-pregnant requirement is needed and is easily met by usual diet of pregnant women. In Indian situation, with a gain in weight of about 6.5 kg of which 70 per cent is water, the sodium requirement will be much less and there is no need for extra sodium.

Vitamins

ICMR Nutrition Expert Group (2011) has not recommended additional amounts of vitamin A, vitamin C and vitamin B_{12} to be taken during pregnancy.

Thiamine, Riboflavin, Niacin and Pyridoxine

The requirements of B-complex vitamins such as thiamin, riboflavin and niacin are based upon the energy intake. As we know that additional 350 kcalories per day are needed during pregnancy, hence the additional amounts (mg) of thiamin, riboflavin and niacin as recommended by ICMR (2011) are 0.2, 0.3 and 2.0, respectively. A pregnant woman also needs additional amount of 0.5 mg pyridoxine daily.

Folate

The synthesis of DNA requires folate. This means the growth of foetus and maternal tissues during pregnancy depends on an ample supply of folate. Red blood cells' formation increases in pregnancy and that also requires folate. Serious anaemia can result if folate intake is inadequate. Folate deficiency has been associated with birth defects.

During pregnancy, the RDA for folic acid is increased by an additional 300 micrograms of folate, for a recommended total of 500 micrograms per day. To reach this level, consumption of green leafy vegetables should be encouraged as they are among the best sources of folic acid. Liver, yeast, other green leafy vegetables, legumes, nuts and whole grains also supply folic acid but the amount in most animal foods, fruits, other vegetables and refined grains is poor. Some women have difficulty in consuming sufficient folate from food sources to satisfy the pregnancy needs, hence, additional medicinal folate supplements should be taken especially in the last 12-16 weeks of pregnancy.

Vitamin B_{12}

There is little information on which to recommend the additional needs of vitamin B_{12} during pregnancy. On the basis of B_{12} content of foetus, it has been estimated that foetal demands may be of the order of 0.3 micrograms per day. The recommended intake of vitamin B_{12} for adults is 1 μg per day. An additional intake of 0.2 μg per day during pregnancy would thus, cover the needs for hemopoiesis and liver storage.

The average Indian diet which is largely based on plant foods like cereals and legumes, cannot provide significant amount of vitamin B_{12} as this vitamin is mainly found in foods of animal origin such as milk, egg, meat and other flesh foods. A pregnant woman who is vegetarian or financially poor or due to religious beliefs, does not consume animal foods, should take supplements of vitamin B_{12} along with folic acid so as to prevent megaloblastic anaemia or irreparable damage to nervous system if chronic vitamin B_{12} deficiency is there.

Vitamin D

Vitamin D deficiency in pregnancy has been associated with an increased risk of pre-eclampsia, gestational diabetes mellitus, small for-gestational age infants, impaired fetal skeletal formation causing infant and reduced bone mass, as well as other tissue-specific conditions. Several scientific societies recommend supplementation with 400 U (10 micrograms daily) of vitamin D for all pregnant women. Since there is no evidence from RCTs that supplementation of vitamin D during pregnancy has effects on relevant maternal or neonatal outcomes, WHO guidelines do not recommend routine supplementation, but they do recommend it for women at high risk of deficiency (women of South Asian, African, Caribbean and Middle East origins; women who rarely expose themselves to the sun; women who follow a diet low in vitamin D).

Physiological Adjustments/Changes and Resulting Problems in Pregnancy

Normal pregnancy is accompanied by anatomical and physiological changes that affect almost every function of the body. These changes are necessary to regulate maternal metabolism, promote foetal growth and prepare the mother for labour, birth and lactation.

During pregnancy, the foetus' needs for oxygen, nutrients and excretion increase the burden on the mother's lungs, heart and kidneys. As a result, basal metabolism increases during pregnancy. There is some evidence of reduced gastric tone, motility and secretion. The plasma volume increases by about 50 per cent and the red cell mass by about 20 per cent. The glomerular filtration rate is considerably enhanced. The total body water may increase by as much as seven litres and in late pregnancy, the kidneys may have some difficulty in disposing of the surplus water ingested. Although the mother's digestion and metabolism work very efficiently, some discomforts accompany the changes her body undergoes to accommodate the foetus. Some of the complications/problems associated with pregnancy are discussed below:

Morning Sickness

Nausea and vomiting usually occur during the first trimester of pregnancy. Because these symptoms generally occur on rising, the term 'morning sickness' is often used. Although known as morning sickness, nausea may occur at anytime and persist all day. It is possibly a reaction to pregnancy related hormones circulating in the blood stream. Certain foods that previously have been eaten without difficulty now may cause distress. Fats are a common cause of upset. Fluids taken with meals may also precipitate vomiting. Simple dietary practices will often improve the food tolerance during this time. Small frequent, dry meals consisting chiefly of high carbohydrate foods are more readily tolerated. Other suggestions for the management of nausea are given below.

Suggestions to Alleviate Morning Sickness

Before Going to Bed

- ☆ Be sure to have plenty of fresh air in the room where you sleep.
- ☆ Place some cereal, toast or crackers within reach of your bed.

Before Getting Up in the Morning

- ☆ Eat some of the toast or dry cereals. A little jam on the bread may make its taste better, but do not use butter.

When Getting Up

- ☆ Get up very slowly; take five or six minutes.
- ☆ Avoid sudden movements when getting out of bed.
- ☆ Eat some more toast or cereal shortly after you get up.

Suggestions to Alleviate Nausea during the Day

- ☆ Eat small frequent meals because you are more likely to feel nauseated when your stomach is empty.
- ☆ Do not drink fluids or take soups at mealtime. If you are thirsty, try taking ice cubes.
- ☆ Drink small sips of liquid frequently between meals.
- ☆ Sometime during the day, you will find you can eat a regular meal. Be sure not to overeat at this time.

Foods to be Avoided

- ☆ Fats, fried and greasy foods should be avoided.
- ☆ Highly seasoned foods such as those cooked with onion, garlic, pepper, chilli and other spices should be avoided as they may upset the stomach.
- ☆ Avoid foods that may provide you gas when you are nauseated. Some of these foods are cabbage, onion, turnip, radish, cucumber and dried beans.

Usually nausea stops after the first trimester, but in some cases it can continue throughout the entire pregnancy. Vomiting, if it persists and becomes pernicious, should be treated by a physician.

Constipation and Hemorrhoids

Constipation is common during pregnancy due both to hormonal changes which tend to increase relaxation of the gastrointestinal muscles and to the pressure muscles and to the pressure of the enlarging uterus on the colon. Iron supplementation is also associated with constipation in pregnancy. Straining can lead to hemorrhoids, which are most likely to occur during pregnancy, anyway, because of physiological changes during pregnancy. Consuming more water and use of fibre rich foods such as whole cereals and legumes, fibrous fruits and vegetables, dried fruits *etc.* and exercising can help a pregnant woman avoid constipation and an often accompanying problems, hemorrhoids.

Heartburn

Progestins, the hormones necessary for maintaining pregnancy and lactation, are produced by the placenta and they relax the muscles in both uterus and the intestinal tract. This often causes heartburn as stomach acid slips up through the lower esophageal sphincter into the esophagus. When that happens, the woman should avoid lying down after eating, reduce fat consumption so that foods pass quickly from the stomach into the small intestine, and avoid spicy foods if they are not tolerated. The symptoms of heartburn are usually reduced by eating smaller, more frequent meals. Wearing loose-fitting clothing, eating slowly, avoiding reclining after meals and elevating the head may also help.

Oedema

Hormones *i.e.* estrogens and progestins combine to cause connective tissue to retain fluid during pregnancy. Increase in blood volume during pregnancy also contributes to some oedema. If there is mild oedema, then there is no need to restrict salt or use diuretics. Oedema may limit physical activity late in pregnancy and occasionally require the woman to elevate her feet to control the symptoms. Medical therapy should be taken if fluid retention accompanies hypertension and the appearance of protein in the urine.

Anaemia

During pregnancy, there is slight lowering of the haemoglobin content of the woman's blood due to physiological adjustments. Her blood volume increases significantly to provide for the placental circulation. To supply foetal needs, the mother's blood volume expands upto approximately 150 per cent of normal. The red cell mass expands only 20 to 30 per cent above normal, and this occurs more slowly. This leaves proportionately fewer RBCs in a pregnant woman's bloodstream. The lower ratio of red blood cells is a condition known as physiological anaemia, since it is a normal response to pregnancy rather than the result of poor nutrient intake. The resulting fall in haemoglobin concentration may be about 2 g/100 ml of blood. This physiological change resembles the manifestations of iron deficiency anaemia.

True anaemia occurring during pregnancy is due most often to an iron deficiency. Even healthy women in developed countries also do not have iron stores large enough to meet the demands of pregnancy. Iron and folic acid supplementation can aid greatly in maintaining the haemoglobin at normal level.

Gestational Diabetes Mellitus

Hormones synthesized by the placenta antagonize the action of insulin hormone. This antagonism can precipitate gestational diabetes mellitus, particularly in women with family history of this disease. It is important that the pregnant women should have a regular check for urine or blood glucose levels to detect developing diabetes. Once detected, a special diet or sometimes insulin injections are required. Gestational diabetes often disappears after the birth. Proper control of gestational diabetes mellitus and diabetes mellitus present in the mother prior to pregnancy is extremely important. If not treated, the foetus can grow quite large, often requiring an early delivery, and increasing the risk of birth trauma and malformations.

Pregnancy Induced Hypertension

Another term for this disorder is toxemia. Although the exact cause of toxemia of pregnancy is not known, good nutrition and prenatal care can prevent or limit its severity. It is characterized by an elevation in blood pressure, proteinuria, and rapid weight gain due to oedema. Mild cases are known as preeclampsia; more severe cases are called eclampsia. In the eclamptic stage, there may be convulsions and coma. This disorder resolves once the pregnancy state ends.

Toxemia occurs more frequently in pregnant women on poor diets, and particularly in those on low protein intakes, than in corresponding groups on good diets. Epidemiologic studies show a direct relationship between the incidence of toxemia and an individual per capita income. The poorer the state, the greater the number of cases of toxemia reported.

Hence, good nutrition especially a good calcium intake, may prevent or lessen the disorder. Mild effects can be lessened by bed rest. If not controlled, liver and kidney damage, and even the death of mother and foetus may result. Careful medical attention is needed.

Substance Use and Abuse during Pregnancy

Although a healthy life style is recommended for everyone, it is especially important during pregnancy. The well being of both the mother and foetus is potentially compromised when undesirable substances enter maternal circulation. Since normal foetus growth and development depends directly on what the mother chooses to expose herself to, her awesome responsibility is obvious. The hazardous influences of tobacco, alcohol, caffeine, marijuana, cocaine and certain drugs on the foetus are discussed as below:

Cigarette Smoking

It is widely known that smoking during pregnancy increases the risk of both premature delivery and low birth weight. A recent review of literature on the determinants of low birth weight has, like previous review, concluded that cigarette smoking is by far the single most important modifiable factor responsible for foetal growth retardation in developed countries. Other adverse effects include a moderately increased risk of preterm delivery, perinatal mortality and, possibly spontaneous abortion.

In addition, children of mothers who smoke during pregnancy may have slight but measurable deficits in longterm physical growth, intellectual performance and behavioural development. It is also suspected of increasing the risk of congenital malformations, sudden infant death, and child cancer. These serious problems are believed to be related to the physiological, biochemical, and/or hormonal effects of nicotine cyanides, carbon monoxide, and possibly other compounds present in cigarette smoke on the placental perfusion and oxygen transport to the foetus.

Conclusively, cigarette smoking may affect maternal nutrition and consequently foetal nutrition in two ways. The increased metabolic rate in smokers can lead to the lower availability of calories and exposure to tobacco may increase iron requirements and decrease the availability of certain nutrients such as vitamin B_{12}, amino acids, vitamin C, folate and zinc. In addition, the hypothesized reduction in utero-placental blood flow in smokers could restrict nutrient and oxygen flow to the foetus.

The adverse effects have been found to be proportional to the frequency of smoking and appear to be prevented or reduced if the mother does not smoke during a subsequent pregnancy. The smoking-related effects have been reported to be independent of other factors, such as race, parity, pre-pregnancy weight, maternal weight gain and socio-economic status. These observations support the general accepted conclusion that the adverse effects of smoking represent a cause-and-effect relationship and are not a reflection of different characteristics of smokers and non-smokers.

Alcohol

Women who consume alcohol during pregnancy run high risks of causing foetal alcohol syndrome (FAS) in their babies. Although the foetal alcohol syndrome (FAS) was not described until 1973, alcohol is now considered as a potent teratogenic. Among the awful symptoms resulting from FAS are abnormal facial appearance, delayed development, and mental retardation (Figure 8.1). Other potential alcohol-related birth defects include cardiac and genitourinary abnormalities, but it is difficult to attribute isolated anomalies to alcohol exposure. In addition to FAS, alcohol has been associated with a spectrum of adverse effects that range from spontaneous abortion to subtle behavioural effects in the absence of physical anomalies.

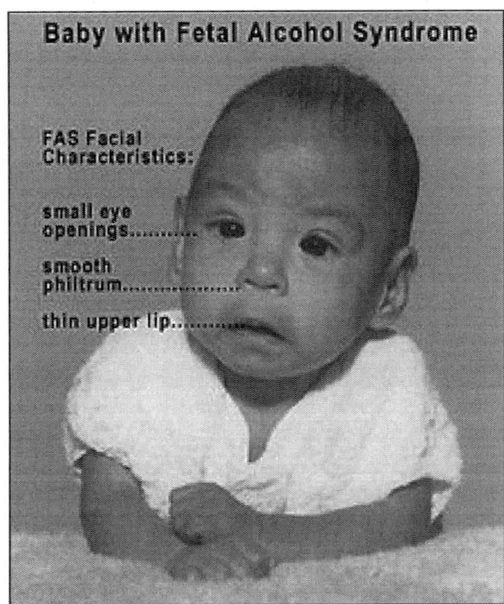

Figure 8.1: Baby having Foetal Alcohol Syndrome.

At present, the exact mechanism by which alcohol adversely affects foetal growth and morphogenesis has not been established. Since alcohol can cross the placenta, the current hypothesis is that high alcohol levels build up in the foetus and produce toxic effects that are most severe in the early phases of pregnancy during blastogenesis and cell differentiation. Another theory is that some of the effects of alcohol may be caused by maternal malnutrition. Women who derive a substantial portion of their daily calorie needs from alcohol may not have an appetite for more nutritious foods. Deficiencies of protein, folic acid, magnesium and zinc are commonly seen in alcoholics. Alcohol intake also impairs the absorption and utilization. Interaction between alcohol and deficiencies of nutrients such as protein, folic acid and zinc may also play a role in the etiology of alcohol related effects on the foetus. Although there is no convincing evidence that nutritional supplementation will counteract the adverse effects of alcohol, multivitamin-mineral supplementation should not replace efforts to encourage women to limit or eliminate alcohol intake during pregnancy.

Caffeine

Caffeine is found in coffee, tea, cola and cocoa beverages. Caffeine is also a common additive in many non-prescription preparations, especially mild analgesics.

The potential danger of caffeine use during pregnancy has been debated for quite sometime. Although evidence so far does not justify recommending abstinence during pregnancy, several recent epidemiological studies have come up with some provocative findings. There is substantial evidence that caffeine is teratogenic in animals but there is no convincing evidence that it is associated with birth defects in humans. Coffee and caffeine consumption have been associated with a reduction in birth weight and an increased risk of low-birth weight infants, especially among full term deliveries, but it is not clear in some of these studies whether the effects were due to caffeine, some other constituent of coffee, or other characteristics of coffee drinkers. Furthermore, most of the studies only assessed coffee consumption, while others included additional sources of caffeine. In addition, the level at which adverse effects have been reported ranges from a total daily caffeine intake greater than 150 mg (equivalent to 1.5 or more cups of coffee daily to seven or more cups of coffee per day). An increased risk of late first and second trimester spontaneous abortions in women consuming more than 150 mg of caffeine daily after adjustment for other risk factors has also been reported. In contrast, some other studies have found no significant association between maternal caffeine or coffee consumption and reduced birth weight.

Hence, information regarding the influence of coffee and caffeine on maternal nutrition is very limited and it is not known whether nutrient supplements would be necessary or beneficial for those who continue to consume caffeine during pregnancy. Even though additional information is needed in this area, it seems justifiable at this point to advocate moderation in caffeine consumption during pregnancy.

Marijuana

Marijuana smoking, like tobacco smoking, is also associated with increased carboxyhaemoglobin levels, which in turn may impair foetal oxygenation and, consequently foetal growth. Marijuana is fat soluble and is excreted at slow rate, thereby, the exposure of foetus to the drug may be prolonged. Its use tends to increase the heart rate and blood pressure, which may lead to reduced uteroplacental blood flow to the foetus.

Animal experiments indicate that marijuana may have a foetotoxic potential, including foetal resorption and reduced birth weight, but some of these effects may have been related to reductions in food and water consumption. Data on the influence of marijuana on pregnancy outcomes in humans are both limited and inconsistent. Adverse effects that have been reported include decreased birth weight and body length, increased frequency of preterm delivery, shortened length of gestation, higher rates of precipitate labour and meconium passage, increased risk of infant compatible with the foetal alcohol syndrome and altered neurobehavioural responses in neonates. Other studies, however, have reported no effect on either birth weight or length of gestation or produced inconsistent results. Hence, no conclusive data are available on the effect of marijuana on the developing foetus but there is however, suggestive evidence that marijuana use during pregnancy may impair foetal growth and thus, should not be taken.

Cocaine

As is true for marijuana, little is known about the nutrition-related effects of cocaine use. Cocaine's vasoconstrictive ability may lead to foetal hypoxia and reduced nutritional supply to the foetus. There is a wide belief that cocaine acts as an appetite suppressant, an inadequate maternal diet may play a role in the growth retardation seen in fetuses of cocaine abusers.

Food Additives

Among the additives receiving much attention in the popular press is aspartame, an artificial sweetener which replaces sugar in foods. Many pregnant women have been led to believe that some component of aspartame may adversely affect the foetus. The phenylalanine component has been of special concern as possible cause of foetal brain damage if maternal exposure is frequent. It is known that high levels of maternal serum phenylalanine, as seen in mothers, suffering from phenylketonuria, seriously interfere with the normal brain development of the foetus. But it is virtually impossible for a woman without phenylketonuria to maintain high serum levels of phenylalanine. Observations of human subjects have generated no data to support that aspartame used by pregnant women adversely affects the course and outcome of pregnancy. Without such data, it is hardly warranted to recommend that pregnant women abstain from using aspartame-sweetened products.

Nutritional Supplements

Nutritional supplements should not be assumed to be benign. Although pregnant women with poor dietary patterns are good candidates for prenatal vitamin/mineral supplements, excessive use of supplements may harm both the mother and the foetus. This is especially true of a high intake of vitamin A by pregnant women which is associated with producing birth defects. Too much vitamin D may cause mental retardation. Women with high serum alpha tocopherol who took mega doses of vitamin E had a higher proportion of spontaneous abortions.

If vitamin A and D in large doses can interfere with normal foetal development, other vitamins may potentially have the same effect. Data supporting this idea, however, are not available. It makes good sense, nevertheless, to be cautious about self-supplementation. The idea that "more of a good thing is better" is foolish.

The pregnant women should try to meet their nutritional needs through sensible food choices. They should delight in knowing that they can provide all the nutrition the foetus needs through wise food choices and generally, without nutritional supplements. If supplements are used, they should be used under medical guidance.

Drugs during Pregnancy

It is well known that many drugs, given in sufficient dosage and at an inappropriate time, can cause foetal damage. It includes extensive list of drugs. The thalidomide caused disaster of the early 1960s. The taking of non-prescribed or excessive number of drugs during the time of rapid foetal development is not desirable. It is now generally accepted that most drugs present in the mother's circulation are capable of reaching the developing embryo or foetus. It is during 15 to 56 days following conception that teratogenic changes induced by some drugs and chemicals can occur. Thus, the nervous system is most sensitive from days 15 to 25, the heart from days 20 to 40, and the limbs from days 24 to 46. After the first trimester, most organs are already formed and are no longer subjected to teratogenic changes. But exception to this are the genital organs, the teeth, and the nervous system, in its further maturation. The list of the drugs known to have teratogenic effect in human is rather small; it includes thalidomide, aminoptorin, busulfan, and androgens; however, antihistaminics, salicylates, and certain antibodies thought to be safe, have caused malformations in animal species. A more rational approach is to evaluate benefits versus risk before a drug is given by a medical doctor to a pregnant woman. The period of sensitivity to teratogens begins 13 to 15 days after conception. Some women are not even aware of being pregnant at this time. Complete restriction of drugs during pregnancy cannot be the answer. Let the final decision be with the physician.

Food Beliefs, Cravings, Avoidances and Aversions during Pregnancy

Most women change their diets during the course of pregnancy and lactation. Some changes are based on medical advice, others on folk medical beliefs, and others on changes in preference and appetite that may be idiosyncratic or culturally patterned. These do exert some influence on diets of mothers.

Many beliefs have been recorded about prenatal diet, such as the idea that the mother can mark her child before birth by eating specific foods. Overuse of a craved food during pregnancy is thought to explain physical or behavioural peculiarities of the infant. More often, unsatiated cravings are thought to explain birthmark that mimic the shape of the desired food (such as strawberry or drumstick shaped marks). Behavioural markings have also been thought to derive from the prenatal diet; that is, the mother's consumption of certain foods has been said to cause the child to like such foods after birth.

Traditional beliefs in food habits also influence profoundly the pattern of foods eaten during pregnancy. For example, in South Pacific Islands, it is believed that certain shell fish eaten during pregnancy will cause the child to be born with scales on its head. In Ethiopia, a pregnant woman must avoid roasted meat as it is thought to cause abortion. In some communities, eggs are thought to cause baldness and sterility and hence, not consumed by pregnant women.

Another important group of beliefs concerns dietary means by which the pregnant mother can ensure an easier delivery. Most important, from the biomedical point of view, are beliefs that a woman to avoid animal protein foods or to avoid 'excessive' weight gain. Some ignorant people think that a smaller weight gain during pregnancy produces a smaller infant since a smaller baby may be 'easier to deliver', low weight gain has been proposed as desirable, especially since it is commonly believed that baby can 'catch up' after birth.

Some people believe that pregnant mothers instinctively know what to eat and that their drive for pickles, tamarind or ice cream is dictated by a natural desire to consume needed nutrients. However, no studies support that notion, and if it were true, one would wonder why women seek out clay, starch, ice and other non-food items during pregnancy. There is a belief that the baby will not be normal if one does not eat pica or starch. There may be a natural instinct to consume the right foods in pregnancy, but we are so far removed from living by instinct that relying on our desires is risky. Good nutritional counseling can focus choices more reliably.

There is also a tendency to avoid certain foods during pregnancy as the woman develops aversions towards those foods. Hence, food avoidances are those foods which the mother consciously chooses not to consume during pregnancy, usually for a reason she can articulate and that seems reasonable to her. The most commonly avoided foods include sources of animal proteins *i.e.* milk, egg, meat, beverages such as tea, coffee *etc.* and curries containing onion and sauce *etc.* Cravings and aversions are powerful urges toward or away from foods, including food about which women experience no unusual attitudes outside of pregnancy. The most commonly craved foods are sweets and dairy products. However, cravings and aversions are not limited to any particular food or food groups and vary from individual to individual too.

Hence, the time of pregnancy when nutritious food is important some women adhere to erroneous beliefs about food. Some common beliefs prevalent among Indian women are as given below:

- ✰ Papaya should be avoided during pregnancy as it causes abortion.
- ✰ Strawberries, pepper, cornstarch, baking soda *etc.* if eaten during pregnancy mark the child.

- ✰ A pregnant woman should indulge in any craving for certain foods because the body needs them.

Pica

Pica is the practice of eating non-food items such as dirt, starch, ice or clay especially while pregnant. The practice of pica is not new nor it is limited to any geographical area, race, creed, culture, sex or status within a culture. This habit has been reported since ancient times from Asia, Africa, Europe and North Central and South America among pregnant women.

Although pica occurs more frequently in population that have poor iron and zinc status; this practice probably results more from habit, culture and superstition than from a need for specific nutrients. The medical implications of pica are not well understood, although several speculations have been put forward. The displacement effect of pica substances could result in reduced intake of nutritious foods, leading to inadequate dietary intake of essential nutrients. Some pica substances may contain toxic compounds or quantities of nutrients not tolerated during pregnancy. Some pica substances interfere with the absorption of certain mineral elements such as iron. Other less commonly reported complications of pica can be as follows:

- ✰ Foetal hemolytic anaemia caused by maternal ingestion of moth balls and toilet air freshness.
- ✰ Parotid enlargement and gastric and small bowel obstructions from ingestion of excessive laundry starch.
- ✰ Parasitic infection from ingestion of contaminated soil or clay.
- ✰ Tender, irritable uterus with dystocia associated with faecal impaction from clay ingestion.
- ✰ Congenital lead poisoning secondary to maternal pica for wall plaster.

The etiology of pica is poorly understood. Several proposals have been put forth. One theory suggests that the ingestion of pica relieves nausea and vomiting. The example of the dog who will eat dirt or grass during illness is often used to illustrate this theory. It has also been hypothesized that a deficiency of an essential nutrient such as calcium or iron results in the eating of non-food substances that contain these nutrients. According to some of the prenatal patients, pica is consumed due to craving for its taste, relieves nervous tension, starch makes the newborn lighter in colour, starch helps the baby to 'slide out' more easily during delivery, clay quiets hunger pains and are pleasant to chew *etc.* Many of these reasons are based upon superstitions, customs, traditions or practices that are passed on from mothers to daughters.

Diet Plan during Pregnancy

The basic meal planning principles are same for the non-pregnant and pregnant women. However, to meet the increased nutritional needs during pregnancy, increased requirement of some of the nutrients is to be met by increasing the quantity of some of the food items namely milk and milk products, green leafy vegetables, fruits and fats and oils.

During the early pregnancy (first trimester), due to physiological and hormonal changes, the expectant woman suffers from nausea and morning sickness. Some carbohydrate rich foods such as dry cereals, toast, a little jam on the bread or crackers, as first food in the morning can prevent nausea of early pregnancy. Highly seasoned foods such as those containing garlic, onion, pepper, chilli and other spices, fried and greasy foods should be avoided. There is no need to increase the food intake during first trimester as nutrient requirement is not increased. Moreover, due to nausea and vomiting, pregnant woman is not able to take extra amount of foodstuffs. She should take small frequent meals instead of three large meals a day.

From the second trimester onwards, extra amount of food should be taken. Frequency of feeding should be increased; she should take 5-6 small meals a day. Foods rich in good quality protein such as egg, milk, meat, fish *etc.* should be taken. Those who are vegetarians, should take a meal containing cereals, pulses, milk and milk products, groundnuts *etc.* Green leafy vegetables, dried fruits, meat *etc.* should be taken to meet the increased iron needs. Whole cereals and pulses, lot of salad, fruits *etc.* should be taken to get good amount of fibre to prevent constipation which is quite common during pregnancy. On the whole, pregnant woman should take a well balanced nutritious diet containing foodstuffs from each food group as described in Table 8.2.

Table 8.7: Balanced Diet for Pregnant Woman (Number of portions) (NIN, 1998)

Food Groups	*g/ portion*	*Type of Work*		
		Seden-tary	*Mode-rate*	*Heavy*
Cereals and millets	30	9	11	16
Pulses	30	2	2.5	3
Milk	100 ml	3+2	3+2	3+2
Roots and tubers	100	2	2	2
Green leafy vegetables	100	1+0.5	1+0.5	1+0.5
Other vegetables	100	2	2	2
Fruits	100	1+1	1+1	1+1
Sugar	5	4	6	9
Fats and oils (visible)	5	4+2	5+2	6+2

For non-vegetarians, substitute one pulse portion (30 g) with one portion (50 g) of egg/meat/fish/chicken

A typical diet plan for Mrs X who is pregnant (2nd trimester), ovo-vegetarian, 30 years old, sedentary worker, belonging to middle income group is as under.

Step I: Personal Profile

Age	30 years
Sex	Female
Activity	Sedentary
Physiological status	Pregnancy (Ist trimester)
Socio-economic status	MIG
Food habits	Ovo-vegetarian

Step II: Recommended Dietary Allowances (ICMR, 2011)

Energy	1900+350 kcals
Protein	82.2 g
Calcium	1200 mg
Iron	35 mg
Dietary folate	500 µg

Step III: Food Exchange Plan

Exchange	No.	Energy (Kcal)	Protein (g)	CHO (g)	Fat (g)
Milk	2	340	16	24	20
Meat	1	70	7	-	5
Pulses	2	200	14	34	-
Cereals	15	1050	30	225	-
Vegetable A	2	24	-	6	-
Vegetable B	2	80	4	14	-
Fruit	2	80	-	20	-
Fat	6	270	-	-	25
Sugar	3	60	-	36	-
Total		2174	71	369	50

Step IV and V

Meal	Exchanges	No.	Menu
Early morning	Milk	0.3	Tea
	Sugar	0.5	Biscuit
	Biscuit	3	
Breakfast	Milk	0.4	Tea
	Sugar	0.5	*Paushtik parantha*
	Pulses	0.5	Curd
	Cereals	4	Apple
	Fruit	0.2	
	Vegetable A	1	
	Fat	2	
Mid morning	Fruit		Pomegranate juice
Lunch	Cereal	4	*Chapati*/Rice Pulav
	Vegetable B	1	Paneer curry
	Vegetable A	1.0	Brinjal *bharta*
	Milk	0.5	*Boondi raita*
	Pulse	0.5	Salad (Radish)
	Fat	2	tomato, cucumber)
Evening tea	Milk	0.3	Tea
	Sugar	0.5	Roasted peanuts
	Fat	1	
Dinner	Vegetable A	0.8	Spinach soup
	Vegetable B	1	Mung *dal*
	Pulse	1	Egg *bhurji*
	Cereal	4	Coriander chutney
	Meat	1	*Chapati*
	Milk	0.5	Carrot *halwa*
	Sugar	1.5	
	Fat	2	

Impact of Malnutrition on the Outcome of Pregnancy

Health and nutritional status of mother during as well as before pregnancy is considered to be strongly associated with pregnancy outcomes. Poor nutritional status and inadequate intake of food in pregnancy not only affect woman's health but also have negative impacts on growth and development of foetus. Maternal health, nutritional status, pre-pregnancy weight, BMI, and gestational weight gain all are strongly related to birth weight of infant. The lack of sufficient calories, of macro- and micronutrients, can lead to deficiencies in building materials for the development and growth of the fetus: moreover there is growing evidence that the maternal nutritional status can alter the epigenetic state of the fetal genome, therefore alterations in nutrition during crucial periods of fetal development may result in developmental adaptations that permanently change the structure, physiology, and metabolism of the offspring, thereby predisposing individuals to metabolic, endocrine, and cardiovascular diseases in adult life. This phenomenon, known as "fetal programming," has led to the recent theory of "fetal origins of adult disease".

Studies have reported that under nutrition during mid or late gestation reduced placental length and area and was associated with smaller babies due to a reduced placental efficiency. Moreover, exposure to under nutrition during any stage of gestation was associated with higher risk of later glucose intolerance. Interestingly, exposure only during early gestation increased risk of obesity, coronary heart disease, schizophrenia and depression, suggesting that early gestation is a pregnancy vulnerable period but also that pre and peri-conceptional nutrition may have negative consequences for the offspring's health, probably based on abnormal placentation occurring early in pregnancy. A recent study in an ovine model found that maternal under nutrition throughout pregnancy may have adverse effects on developing fetal appetitive pathways, increasing orexigenic gene expression in the late gestation fetal hypothalamus, contributing to programmed hyperphagia and obesity in intrauterine growth restricted, low birth weight offspring.

Imbalances or inadequacies of micronutrients in the diet before and during pregnancy can also negatively influence both mother and fetus with significantly high reproductive risks, ranging from infertility to fetal structural defects, abnormal fetal development and growth and long term diseases. On the other side, healthy dietary patterns and micronutrients' supplementation, particularly during the peri-conceptional period, are related to improved birth outcomes, probably through alterations in maternal and fetal metabolism due to role/involvement of micronutrients in enzymes, signal transduction, transcription pathways, oxidative stress and epigenetic modifications. Since different pregnancy stages represent a continuum, from the pre-conception to the post-partum period, an injury acting before conception or in early pregnancy may have long-lasting effects on the well-being of the mother and the fetus, and may further influence the health of the baby at a later age, by programming postnatal pathophysiology. Each stage of fetal development is dependent on and influenced by appropriate maternal nutrient supply; the timing of nutritional insults impacts differently on the nature of adult diseases. Hence, a healthy diet must be balanced not only in terms of macronutrients content (intake of proteins, carbohydrates and fats), but also in terms of micronutrients intake (vitamins and minerals).

Pre-pregnancy obesity has also been linked to increased risk of hypertensive disorders of pregnancy, gestational diabetes mellitus, as well as an entire spectrum of adverse pregnancy outcomes, including increased risk of caesarean delivery, poor lactation practices, obstetric anesthesia related complications, prolonged gestation, maternal infectious morbidity and decreased success with trial of labor. In addition, maternal obesity is a cause for stillbirths, fetal and neonatal death, abnormal fetal growth with large for gestational age infants or fetal growth restriction, preterm birth and increased risk of a range of structural anomalies. Obesity induces a chronic, low-grade intrauterine inflammation with over-expression of maternal C-reactive protein determining a pathological fetal state, compromising placental function and metabolism and altering fetal growth and development. The placental to fetal weight ratio has been shown to be increased in obese pregnancies, suggesting that placentas of obese women are less efficient, with gender specific responses. Similar changes in placental biometric parameters have been associated with increased risks of cardiovascular disease in later life.

Maternal obesity has also been associated with long-term health consequences for the offspring. Maternal obesity perpetuates the obesity epidemic, since children of obese women are more likely to be obese themselves and to suffer from chronic cardiovascular diseases starting from young age. Maternal obesity in early pregnancy has been associated with a four- and two-fold increased risk of childhood overweight and obesity, respectively, likely through genetic as well as pre- and postnatal environmental factors. Childhood overweight is a strong predictor of adult obesity.

Several studies support the evidence that the frequency and severity of pregnancy complications may be reduced through an improvement in the nutritional status of the mother prior to conception and during pregnancy and these are potential strategies to improve birth weight of the infant and to prevent his/her predisposition to adult-onset degenerative diseases later in life.

Nutrition Education, Counseling and Guidance

Spreading the awareness of the importance of maternal nutrition before and during pregnancy and stimulating a cultural change in favor of a balanced healthy diet and high-quality foods consumption, is necessary for improving future global health.

In prenatal care, nutrition education and guidance become major responsibilities of the nutritionist because nutrition plays a primary role in the successful outcome of pregnancy. Perhaps at no other time in the human life cycle is a person more open to nutrition education or more motivated by a sense of responsibility for another human life than during the period of pregnancy. Moreover, nutrition education during this period will go far in carrying over to the young parents' attitudes towards feeding their children as they continue to grow through infancy, childhood, and maturity and in turn contribute to the health habits of another generation. Nutritional assessment and educational considerations for prenatal nutritional counseling are outlined in Table 8.8.

Table 8.8: Considerations in Prenatal Nutritional Counseling

INITIAL VISIT	
Assessment	
1.	Screen for risk factors with Nutrition Questionnaire.
2.	Evaluate adequacy of dietary intake with Pregnancy Diet Intake.
3.	Obtain height and pregravid weight to determine per cent IBW prior to conception and recommendation for weight gain during gestation.
4.	Obtain present weight and week of gestation to begin plot of weight gain on Prenatal Weight Gain Grid.
Education	
1.	Discuss importance of nutrition during pregnancy and lactation; stress key increases in nutrients – protein, iron, folacin, calcium, and energy.
2.	Counsel about nutritional supplements.

3.	Discuss pattern of weight gain during pregnancy.
4.	Describe common nutritional problems associated with pregnancy.
FOLLOW-UP CARE	
Assessment	
1.	Plot weight on Prenatal Weight Gain Grid.
2.	Reevaluate dietary intake with Pregnancy Diet Intake form.
3.	Check *hgb* and *hct* values.
Education	
1.	Discuss diet-related problems.
2.	Give feedback on dietary intake in relation to balanced diet.
3.	Discuss diet and lactation.
4.	Discuss infant feeding.

Positive Reinforcement of Good Food Habits

When counseling a woman regarding her food habits, it is important to provide positive reinforcement. To strengthen food habits, the counselor can give positive feedback immediately for those food behaviours which help to fulfill recommended choices and amounts for pregnancy. By the following actions, the counselor can provide positive reinforcement:

- ✰ Relate the mother's food behaviour to her pregnancy. Guide her to identify for herself how her particular food practices are meeting the nutritional needs of her pregnancy.
- ✰ Review the reasons for these increased needs. Further reinforce her good habits by guiding her to see why it is important that she should get these increased nutrients during her pregnancy. Relate the functions of key nutrients to the changes taking place in her body and to the rapid growth of the baby and its protection.
- ✰ Give verbal recognition for the efforts. Commend her openly and warmly for her good habits. Simple phrases such as 'you are doing a good job with your diet', "Good work' are needed to further reinforce good food habits.

Positive Teaching to Correct Nutritional Deficiencies

If food intake is not optimal, then the counselor should provide positive teaching which can be done by the following actions:

- ✰ Help the mother to relate her food habits to unmet nutrients' needs of her pregnancy. Guide the mother to identify for herself the ways in which particular food intake fails to meet the nutritional demands of her pregnancy.
- ✰ Review the reasons for increased nutrient needs of pregnancy.
- ✰ Identify the reasons for the deficiencies. What are her problems in obtaining needed foods? What limiting factors exist in her personal living situation?
- ✰ Explore possible solutions to problems or alternative practices available.
- ✰ Many religious and cultural food patterns provide marginal levels of the nutrients needed in increased amounts during pregnancy and lactation. Such factors should also be taken into consideration during nutrition counseling.

9

Nutrition during Lactation

Lactation is one of the most complex and nutritionally demanding phases of the human life cycle. Lactation is a physiological process accomplished by females since the origin of mammals. Today, as in past time, the process of breastfeeding is successfully initiated by at least 99 per cent of women who try.

Stages of Lactation

Lactation or more adequately the process of breastfeeding results from the inaction and interplay of the hormones and of instinctive reflexes, and learned behaviour by mother and the newborn. Four overlapping stages may be recognized:

i) Milk Initiation (Lactogenesis)

As a result of stimulation of the breast alveolar cells by placental lactogen, lactogenesis starts *in utero* in the last part of pregnancy. This is continued after birth as an automatic process in all women. Full lactogenesis is under the control of complex interacting hormones, but is closely related to rising levels of the anterior pituitary hormone, prolactin. This hormone is stimulated by the post-delivery rapid fall in inhibitor ovarian and placental steroid, especially progesterone that follows expulsion of the placenta.

ii) Milk Secretion (Maintenance of established lactation)

This is primarily related to sufficient production of the anterior pituitary hormone prolactin and to some extent maternal nutrition.

Full lactation does not begin as soon as the baby is born. During the first 2 and 3 days after birth, a small amount of colostrum is secreted. In subsequent days, a rapid increase in milk secretion occurs, and in usual cases lactation becomes reasonably well established by the end of the first week. In primiparas, however, the establishment of lactation may be delayed until the third week or even later. Generally, therefore, the first 2 or 3 weeks is a period of rapid lactation initiation, and is followed by the longer period of maintenance of established lactation. These two phases are not caused precisely by the same stimuli, but the basic physiological mechanisms that are operative are similar in both cases.

iii) Milk Ejection

In human, as in other higher mammals, movement of milk from the alveoli, where it is secreted, to the lacteal sinuses and, to some extent, the mount of the infant is an active process within the breast, brought about by the let down or milk ejection reflex.

The milk 'ejection' or 'let down reflex', is a neurohormonal mechanism regulated in part by central nervous system factors and it is necessary for breastfeeding (Figure 9.1). The primary stimulus is suckling on the nipple, which releases the hormone oxytocin from the posterior pituitary. Oxytocin is carried in the bloodstream to the myoepithelian cells around the alveoli of the breast tissues to "let-down" or release the milk from storage sites to travel to the nipple area. A tingling sensation signals the let-down reflex shortly before milk flow begins. If the let-down reflex does not operate, little milk is available to the infant. The infant gets frustrated, and this can frustrate the mother also.

The let-down reflex appears to be sensitive to a variety of environmental and emotional factors although the mechanisms involved are poorly understood. The

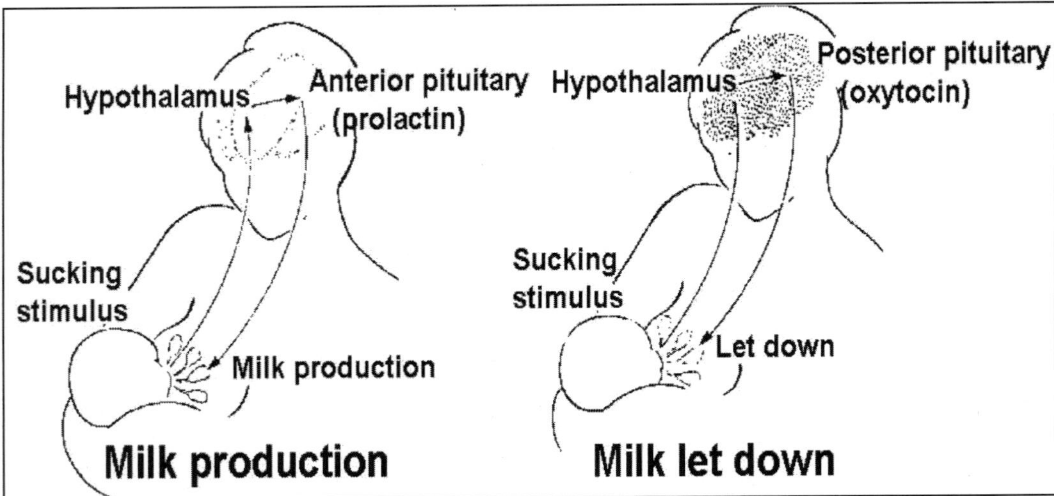

Figure 9.1: Let Down Reflex.

let-down reflex is easily inhibited by nervous tension, a lack of confidence and fatigue. Mothers should be aware of the link between tension and a weak let-down reflex. They need to find a relaxed and supportive environment where they can breast-feed. After a few weeks, the let-down reflex becomes automatic. The mother's response can be triggered just by thoughts about the baby or by seeing or hearing another baby.

iv) Milk Ingestion

It is the process of milk sucking by the baby. This is dependent on interaction between mother and the newborn and of course functioning of the let down reflex on maternal side and on the existence of the neonatal sucking and swallowing reflex.

The hormone prolactin stimulates the synthesis of milk and suckling stimulates prolactin release. Milk synthesis then occurs as an infant nurses. The more the infant suckles, the more milk is produced. Milk production closely parallels infant's demand. Demand is the driving force for milk production. Once milk production and secretion have been accomplished, the baby may obtain this milk by promoting 'ejection' from the alveoli and ducts.

Breastfeeding Reflexes

Successful breastfeeding depends on mutual reinforcement of interacting neonatal and maternal reflexes.

Neonatal Reflexes

The following neonatal reflexes are directly related to milk ingestion:

- Rooting
- Sucking
- Swallowing

These neonatal reflexes are feeble or absent in smaller premature or low birth weight babies or in a variety of general illnesses of the newborn including generalized infection, severe jaundice or cerebral birth damage. Certain congenital anatomical anomalies in the newborn such as severe cleft palate may impair the sucking reflex.

Maternal Reflexes

The three maternal reflexes involved in breast feeding are:

i) Prolactin Reflex (Milk secretion reflex)

Prolactin is believed to be the key lactogeneic hormone involved in the initiation and maintenance of milk secretion. Its production by the anterior pituitary is mainly the result of the prolactin reflex, resulting from the infant sucking at the breast. This reflex may be diminished by severe maternal malnutrition or by the hormonal responses to environmental psychological stress.

ii) Nipple-erection Reflex

The stimulation of the mother's nipple by the baby's mouth leads to nipple erection thereby assisting in the milking of the breast by the sucking reflex.

iii) Let-down Reflex

It is also known as milk ejection reflex. This reflex helps in making the milk available from the alveoli to the terminal lacteal under the influence of oxytocin secreted by the posterior pituitary. With a successful let down reflex, 90 per cent of the available milk can be obtained by the baby in seven minutes or less. Stress due to anxiety or uncertainty can inhibit or block this reflex.

Milk Volume

A mother's milk production is influenced mainly by the demands for her baby, whose sucking stimulates breast milk secretion. The more the baby sucks, the more the mother will produce milk. The amount increases from about 100 to 200 ml on the third day after the baby's birth to 400 to 500 ml by the time the baby is ten days old. Production can continue to increase to as much as 1000 to 1200 ml per day. A healthy, normally growing four month old infant of average weight will, if exclusively breastfed, receives 700 to 850 ml of breast milk in 24 hours. One litre of breast milk gives about 750 Kcal to the baby. Provided the babies can suckle as much as they want, they will always get enough milk.

Although successful lactation can continue as long as adequate suckling stimulation is maintained, a gradual fall in the amount of milk produced generally occurs after 12 months. This reduction in milk output is mainly related to reduction of demand and cessation of stimulation of nipple by the infant. Various other factors affecting milk volume are given in Table 9.1.

Table 9.1: Factors Affecting Milk Volume

Maternal nutrition	Frequency of feeding
Infant size	Duration of feeding
Supplementation	Stress
	Smoking Activity

Lactation Failure

Lactation has been termed a 'confidence trick' and failure in healthy well nourished women with normal full term babies is mostly due to anxiety-nursing-cycle based on emotional interference on the let down reflex. The attitudes of mothers towards breast feeding, whether positive or negative, are powerful factors in achieving successful lactation and in influencing milk production. In most of the cases, lactation failure is mainly due to psychosocial interference with the release of milk. Only rarely true lactation failure has been documented.

Lactation failure can be due to the following factors:

- ☆ Depression or inhibition of anterior pituitary functions which are responsible for synthesis of milk in the alveolar cells.
- ☆ Depression or inhibition of the transport or discharge of the synthesized products from the cells to the alveolar lumen
- ☆ Depression or inhibition of the let down reflex which results ultimately in cessation of milk synthesis

Various other possible causes of lactation failure have been given in Table 9.2.

Table 9.2: Various Causes of Lactation Failure

I. Causes in the infant
☆ Poor sucking reflex – neurological
☆ Structural abnormalities
o Cleft lip
o Pierre Robin's syndrome
o Micrognathia
II. Maternal causes
☆ Poor milk production
o Anatomical
o Hormonal
o Nutritional
o Pharmacological
o Stress due to fatigue or illness
☆ Poor let down
o Hormonal
o Pharmacological
o Stress due to pain or psychological
☆ Infrequent feeds

Management of Lactation Failure

It should involve:

- ☆ Reassurance and psychological support
- ☆ Early initiation of breast feeding
- ☆ Frequent sucking at the breast by the infant
- ☆ Management of breast engorgement, sore nipples *etc.*
- ☆ Adequate rest for the mother with infant
- ☆ Additional food for malnourished mother
- ☆ Treatment with lactogogues

Maintaining Hygiene during Nursing

It is advisable to wash the nipple with plain water once a day. Clean and wash the clothes. A few drops of expressed milk may be allowed to dry on the nipple. This protects it against infections and makes the nipple less likely to crack. The following steps are recommended as a care towards tender or sore nipple before pain, cracking, and abrasions develop:

- ☆ Expose the nipple to dry air.
- ☆ Avoid use of soap.
- ☆ Avoid applying alcohol, and tincture of benzoin.
- ☆ Do not use boric acid.
- ☆ Nurse more frequently; avoid prolonged breastfeeding.
- ☆ Express milk manually.
- ☆ Feed baby in different positions.

- ✰ Keep the breast dry between feedings.
- ✰ Protect from rubbing of clothing.
- ✰ Some mothers may be comfortable by wearing properly fitting bra day and night. Plastic liners should be removed.

Human Milk: Its Nutrient Composition and Digestion

Human milk is a dynamic, multi-faceted fluid containing nutrients and bioactive factors needed for infant health and development. It is the nature's gift to the newborn baby. The value of the human milk in infant feeding has been appreciated for several thousands of years. For this few days after the birth, colostrum is secreted and thereafter, mature milk comes. The composition of colostrum and mature milk is discussed below:

Colostrum

Colostrum is the yellowish or straw coloured thick fluid produced by the breasts for the first few days after the birth of the baby. This type of milk is produced for a few days to a week after birth. It is secreted at a rate of about 40 ml per day for 3 to 4 days. Women who have nursed babies previously, usually demonstrate colostrum output sooner and in greater volume than other women.

The concentration of major milk constituents in colostrum and mature human breast milk are given in Table 9.3. It has deep yellow colour, a feature associated with its relatively high carotene content. It has more proteins, vitamins and minerals than mature milk. The vitamins present in colostrum include vitamin A, thiamine, riboflavin, nicotinic acid and ascorbic acid. The minerals include calcium, phosphorus, sodium, potassium, chloride, magnesium, sulphur, manganese, copper, zinc, iodine, selenium, and iron. Colostrum protects immature newborn against respiratory tract and gastrointestinal infections. The secretory IgA forms a protective coating that prevents the entry of microbes into the body. These immune factors compensate for the infant's immature immune system in its first few months of life. Colostrum has potent laxative properties that help the baby pass meconium, a stool produced during foetal life. One compound in colostrum, the *Lactobacillus bifidus* factor, encourages the growth of *Lactobacillus bifidus* bacteria. These bacteria limit the growth of potentially toxic bacteria in the intestine, such as *Escherichia coli,* and so promote the intestinal health of the breast fed infant.

Table 9.3: Composition of Human Colostrum and Mature Breast Milk (per 100 ml)

Constituent	*Colostrum (1-5 days)*	*Mature Milk (>30 days)*
Energy (Kcal)	58	70
Total solids (g)	12.8	12.0
Total protein (g)	3.3	0.9
Casein (mg)	140	187
α-Lactoalbumin (mg)	218	161
Lactoferrin (mg)	330	167
IgA (mg)	364	142
Fat (g)	2.9	4.2
Lactose (g)	5.7	7.0
Water (g)	87.0	88.0

In developing countries, a number of traditional practices influence the first feed taken by the newborn. There is a tendency to delay it especially among the urban poor and even more so in rural areas. A very small proportion of mothers initiate breast feeding within 12 hours after delivery while the majority do only after 2nd to 4th day. Colostrum is discarded by many mothers both in rural areas and urban slums. This is done due to a number of reasons including the need to wait for certain ceremonies such as cleaning of the breasts by a female relative, or for night-fall to sight a particular star. Some feel that the early milk is unclean as it was in the breast during the latter part of pregnancy. Many feel that they have no milk as their breasts feel 'light'. During the first few days, prelacteal or inaugural feeds consisting of a number of items such as sugar, glucose, honey, *gur*, *Janam ghuttis* and other mixtures of herbs and spices are dissolved in water and given to the infant.

Administration of above mentioned items carry potential risks of infection, aspiration from forced feeding, and even toxic effects in few cases. Mothers, therefore, need to be educated regarding the dangers of the traditional concoctions and also the special beneficial features of colostrum. The mothers should start breastfeeding the newborn as soon as possible. Besides colostrum, early commencement of breastfeeding and frequent suckling on demand will promote milk formation and prevent breast engorgement too. Conclusively, initial feeding of colostrum only is needed and then breast milk for the first 6 months of life.

Mature Milk

After the first few days of milk secretion, colostrum is replaced by a transitional form of milk that gradually becomes the mature breast milk by the 3rd or 4th week. It has more fat and carbohydrate than colostrum (Table 9.4).

Mature human milk is thin, almost watery, in appearance and often has a slight bluish tinge. It has a number of impressive nutritional qualities:

Protein

The lactoalbumin proteins present form a soft, light curd in the infant's stomach, easing digestion. The lactoferrin proteins bind iron, reducing the growth of iron requiring bacteria. Many of these types of bacteria cause diarrhoea. The immunoglobulin proteins offer important immune protection. The amino acid composition of the protein is noted for its low concentration of phenylalanine, for which the infant has a limited ability to metabolize. Human milk also has a high concentration of the amino acid taurine, which is needed for bile synthesis to aid lipid digestion.

Lipid

Lipids in the milk are high in linoleic acid and cholesterol, which are needed for brain growth. Milk delivered near the end of a breastfeeding session, the hind milk, is richer in fat and, therefore, energy dense than the early-fore milk. Lipase enzymes in the milk improve the intestinal digestion of the lipid.

Lactose

The major carbohydrate in breast milk is lactose, which creates an acidic environment in the intestine by stimulating the growth of acid-producing bacteria. It also provides galactose for nerve sheath synthesis.

Iron

While human milk is low in iron, but the amount of iron provided by breast milk appears to be adequate to prevent iron deficiency anemia for at least the first 6 months of life. However, if the same amount of iron is given in any food other than breast milk, this amount is inadequate and iron deficiency anemia does occur. Thus, iron in breast milk seems to be utilized by the infant to a unique extent. In addition, there is some evidence that the high degree of iron utilization in the breast-fed infant has a role in host defense mechanisms. Beyond 6 months of age, the infant needs another source of iron. By that time, the baby's iron stores formed *in utero*, are probably depleted.

Epidermal Growth Factors (EGFs)

Epidermal growth factors play an important role in postnatal development of the gastro-intestinal tract, especially small intestine. It is responsible for functional maturation of gastro-intestinal tract during foetal development and in the newborn.

Host Resistant Factors

Host resistant factors in human milk include immunoglobulins, cellular, enzymic and protein components, lymphoid cells, lysozyme, lactoferrin and antiviral factors *etc.* A list of the antibacterial and antiviral factors present in human milk is as follows:

- Secretory IgA (0.5-1 g/day)
- IgG (100 mg/day)
- Lysozyme (0.1 mg/ml)
- Vitamin B_{12} binding protein
- Bifidus factor
- Oligosaccharide
- High molecular weight (HMW) non-immunoglobulin component
- The component factors HC_3 and HC_4
- A rotavirus neutralizing factor
- Antiviral activity (Lipid fraction *e.g.* monoglyceride of lauric acid)

Protective Functions of Human Milk Constituents

i) Secretory IgA

- Binds bacteria, so they do not enter the body
- Neutralizes toxins and viruses
- Neutralizes poliovirus
- Neutralizes influenza virus
- Binds to *E. coli* 'O' antigen (prevention from diarrhoea)
- Binds to *Salmonella, Shigella,* 'O' antigen (prevention from diarrhoea and bloody dysentery)

  Secretory IgA levels are low in malnourished child.

ii) Lysozyme

Lysozyme enzyme activity is directed to the wall of gram positive bacteria resulting in their death.

iii) Vitamin B_{12} Binding Protein

It binds with vitamin B_{12} and, thus, controls the growth of unwanted gut bacteria.

iv) Bifidus Factor

It favours the growth of lactobacilli which lower pH and, thus, decreases the number of pathogenic *E. coli* in the gut of the breastfed baby.

v) Oligosaccharides

It inhibits the adherence of pneumococci group of bacteria in the oropharynx, thus, prevents infection in the oral cavity and ears.

vi) HMW Nonimmunoglobulin Component (GM Ganglioside)

It blocks the attachment of hemolyticus influenzae in the pharynx, and blocks the activity of *Vibrio cholerae* and *E. Coli* in the gut. This protein prevents influenza, sore throat, diarrhoea, and cholera in the infants.

vii) Antiviral Activity

Antiviral activity is found in a component present in the cream fraction of human milk. It destroys the envelope of many viruses such as herpes simplex, flavoviruses, and onocoviruses (*e.g.* leukemia and tumor virus).

viii) Bacterial Colonization

Bacterial colonization of *Shigella, antaroviruses, Vibrio cholerae* is checked in a breast-fed child. Breastfeeding prevents growth of disease-causing germs, and protects against diarrhoea, and cholera. There are many milk antibodies and defensive factors present in the milk to eliminate the bacteria. Breastfeeding protects against ottis media.

ix) Anti-inflammatory Host Defense Mechanism

Many factors such as IgA, IgG, IgM, lactoferrin, antioxidants, lipids, protease inhibitors, histaminase and arylsulfatase counteract the phases of inflammation.

x) Protection against Specific Bacterial and Viral Infections

Human milk contains antibodies against many enterobacteria, viruses, *Pneumococci, Comebacterium diphtheria, Clostridium tatani, Bordetella pertussis,* rotaviruses *etc.*, to provide:

- ☆ Protection against ottis media
- ☆ Protection against neonatal sepsis
- ☆ Protection against respiratory syncytial virus
- ☆ Protection against salmonella infection
- ☆ Protection against campylobacter diarrhoea

xi) Anti-parasitic Factors in Human Milk

- ☆ antibodies to *Giardia lambia* and *Entamoeba histolytica*
- ☆ lipoprotein lipases (20 types); some kill *Giardia trophozoites*

xii) Cellular Aspects of Immunologic Reactivity in Human Milk

Types of cells present in human milk are:

A. *Macrophages*

- ☆ Destroy herpes simple type I
- ☆ Synthesize lactoferrin, lysozyme, complement components, properdin factor, EGFs, T-cell suppressor factor(s), and B-cell helper factor(s)
- ☆ Regulate T-cell function
- ☆ Destroy *Staphylococcus aureus, E. coli* and *Candida albicans*

B. *Lymphocytes*

C. *Natural Killer Cell (NK)*

So human milk is not only superior in relation to the nutrients to bovine milk (cow's, goat's) and infant formulas, but contains the above mentioned protective factors which could not be present in infant. No scientist or any manufacturer can ever provide these protective factors in any infant formula. Bottle-fed (formula) babies can never be superior to the breast-fed babies.

Digestion of Human Milk Constituents

Protein

Human milk contains about 1.1 per cent proteins. The proteins of human milk are divided into the whey and casein fractions or complexes, with each comprised by a remarkable array of specific proteins and peptides. The most abundant proteins are casein, α-lactalbumin, lactoferrin, secretory immunoglobulin IgA, lysozyme, and serum albumin. Non-protein nitrogen-containing compounds, including urea, uric acid, creatine, creatinine, amino acids, and nucleotides, comprise ~25 per cent of human milk nitrogen. The protein content of milk obtained from mothers who deliver preterm is significantly higher than that of mothers who deliver at term. Protein levels decrease in human milk over the first 4 to 6 weeks or more of life regardless of timing of delivery

If a baby drinks 156 ml milk/day in the beginning, then from mother's milk, baby will receive 1.71 g protein/kg/day. Breast-fed babies take 400 mg nitrogen/kg a day of which 336 mg are obtained from protein and amino acids. Baby absorbs about 80 per cent of nitrogen (as protein) for the purpose of growth. They lose in the urine about 30 per cent of the nitrogen taken in the food, which is within the capacity of the function of their kidney.

The first colostrum secreted has ten times as high a concentration of protein (12 g/100 ml) as mature milk and 96 per cent of this is in the form of immunoglobulin IgA. By day 4, the total protein levels fall to 3 g/100 ml and 25 per cent of it is IgA. Mature milk, with 1.1 per cent protein, has about 10 per cent of its protein in the form of IgA. IgA resists being digested in the intestine through the action of some inhibitor. It also limits the multiplication of bacteria and viruses within the intestine, Mother's milk contains another protein named lactoferrin which binds with iron and makes iron unavailable to *E. coli* in the intestine and so inhibits their growth. Mother's milk also contains other proteins that bind the zinc, vitamin B_{12}, and folate. These proteins help in the absorption of these nutrients and deprive *E. coli* from getting these nutrients for their growth. Breast milk contains another protein called lactoperoxidase, which forms part of a system involved in killing bacteria.

Fat

Mature milk contains 3.6-4.2 per cent fat, the amount of which is known to increase from the beginning to the end of the feed. Palmitic and oleic acids are quantitatively most important fatty acids in milk. Hindmilk, defined as the last milk of a feed, may contain two to three times the concentration of milk fat found in foremilk, defined as the initial milk of a feed.

Newborn babies have a low activity of pancreatic lipase and also relatively small amount of bile acids in their intestine. In the infants, digestion of fats starts in the stomach. The enzyme lingual lipase in infants does not require bile salts for fat digestion and absorption. In human milk fat, 74 per cent of the palmitic acid is esterified and absorbed efficiently.

Breast milk contains lipase (bile salt-stimulated lipase) that remains stable at low pH of the stomach. It digests triglycerides to fatty acid and glycerol. It has to be appreciated that the nutrients of human milk are in right proportion for a human body; there is a lower mineral (solute) load and more unsaturated fatty acids; protein/fats are more easily digested; and carbohydrate content is appropriate for growth.

Carbohydrates

The carbohydrate of human milk is lactose which is present in fairly high amount (about 7 per cent). The concentration of lactose in human milk is the least variable of the macronutrients, but higher concentrations of lactose are found in the milk of mothers producing higher quantities of milk. The other significant carbohydrates of human milk are the oligosaccharides (HMOs), which comprise approximately 1 g/dL in human milk, depending on stage of lactation and maternal genetic factors. The oligosaccharides are among the non-nutritive bioactive factors present in breast milk.

Lactose is split by enzyme lactase into galactose and glucose, which is then actively absorbed. Babies can also digest sucrose readily with the help of enzyme sucrase whose activity exists from birth. Breast milk contains amylase that digests starch given to children in the complementary foods along with breast milk.

HMOs are not digested by the infant but help to make up the intestinal flora. They act as decoy receptors that block the attachment of disease causing pathogens, which may help to prevent infectious diseases. They also alter immune cell responses, which may benefit the infant. To date more than a hundred different HMOs have been identified; both the number and composition vary between women and each HMO may have a distinct functionality.

Composition of Human Milk Vs Animal/ Bovine Milk

Large number of investigations during the past decade have been aimed at defining the biochemical and nutritional properties of different types of mammalian milk (Table 9.4). It is very clear from these efforts that the composition of human milk is very different from that of milk obtained from cow, buffalo and goat.

Cow's milk provides about three times more protein and four times more calcium, but only about 60 per cent of the carbohydrate present in human breast milk. The net protein utilization (NPU) of the human milk protein has been reported to be higher than that of the proteins of cow's milk. Due to lower casein content, human milk forms a soft flocculent, easy to digest curd in the infant's stomach. The increased casein content of fresh cow's milk causes it to form a tough, cheesy, hard to digest curd. So human milk protein is easily digestible than that of cow's milk. The fat contents of cow's and goat's milk are nearly of the same order as that of human milk but buffalo milk contains nearly twice as much fat as milks of human, cow or goat. There are qualitative differences between the fats of human and cow's milk. The fats of each are composed principally of the triglycerides olein, palmitin, and stearin. Human milk, however, contains twice as much of the readily absorbed olein. The volatile fatty acids (butyric, capric, caproic and caprylic) account for only about 1.3 per cent of the fat of human milk, in contrast to about 9 per cent in cow's milk.

Table 9.4: Composition of Human Milk vs Animal Milk

Constituents	*Human*	*Cow*	*Buffalo*	*Goat*
Protein (g)	1.2	3.3	3.8	3.3
Casein (g)	0.4	2.8	3.0	2.5
Lactoalbumin (g)	0.3	0.4	0.4	0.4
Lactoglobulin (g)	0.2	0.2	0.2	0.3
Fat (g)	3.8	3.7	7.5	4.1
Lactose (g)	7.0	4.8	4.4	4.7
Calories (Kcal)	71	69	100	76
Minerals (g)	0.21	0.72	0.8	0.77
Calcium (mg)	33	125	210	130
Phosphorus (mg)	15	96	130	106
Chlorine (mg)	43	103	112	159
Magnesium (mg)	4	12	15	16
Potassium (mg)	55	138	142	181
Sodium (mg)	15	58	65	41
Iron (mg)	0.15	0.10	0.2	0.05
Copper (mg)	0.04	0.03	0.02	0.04
Iodine (mg)	0.007	0.021	0.004	-
Manganese (mg)	0.7	2	-	8
Zinc (mg)	0.53	0.38	-	-
Vitamin A (I.U.)	160	158	200	120
Vitamin D (I.U.)	1.4	2.0	-	2.3
Thiamine (mg)	0.017	0.04	0.05	0.05
Riboflavin (mg)	0.04	0.18	0.10	0.12
Niacin (mg)	0.17	0.08	0.28	0.12
Pantothenic acid (mg)	0.20	0.35	-	-
Vitamin B_6 (μg)	0.001	0.035	-	-
Folic acid (μg)	1.3	5.6	3.3	0.17
Biotin (μg)	0.4	2.0	-	1.5
Vitamin B_{12} (μg)	0.03	0.50	0.30	0.10
Vitamin C (mg)	4.0	2.0	2.5	2.0

Except some trace elements *i.e.* iron and copper, the contents of other minerals in animal milks are about 2 to 5 times those of human milk which can impose strain on the kidneys of young infants. Cow's milk contains inadequate iron; breast milk iron, while low, may be sufficient for the infant because of better absorption. The deficiency is compensated for in the first six months or so of life by iron stored during foetal life.

Human milk from adequately nourished mothers is a reliable source of fat soluble vitamin A and E and the water-soluble vitamins. The concentration of vitamin D in human milk has been reported to be influenced by maternal intake of vitamin D. Cow's milk is a reliable source of vitamin A and B vitamins. It contains less vitamin E and more vitamin K than human milk. Cow's milk contains four times as much vitamin K as human milk. In breast-fed infants, the prothrombin level is 20 to 50 per cent of that of the adults in first few days. Haemorrhagic disease of the newborn may occur if vitamin K is not given. Hence, a prophylactic dose of 0.5 to 1 mg is usually given to breast-fed infants at birth.

Human milk is essentially free from bacterial contamination, pathogenic organisms in significant numbers may gain access to the milk from mastitis. Cow's milk is regularly contaminated, but in most instances, the bacteria are not harmful to man. Milk, however, is a good culture medium for pathogenic bacteria and many infections are human-borne. Furthermore, certain bacteria, which do not affect older children or adults, may cause diarrhoea in infants.

Hence, it is clear from the above facts that each type of milk is unique and is very different in composition from that of human's milk. Most of the mothers consider cow's milk to be an ideal substitute of breast milk but as evident from Table 9.4, it is not so.

For the human infant, breast milk is superior to cow's milk because of:

- ☆ Higher content of vitamin C and iron
- ☆ Higher concentration of PUFA
- ☆ Higher concentration of cystine in proteins
- ☆ Low curd tension due to low casein content and high albumin content
- ☆ Low mineral content and high buffering capacity
- ☆ Less load on kidney

Unless altered, cow's milk should not be used in infant feeding until the infant is 6 to 12 months old. Many authorities have recommended against using cow's milk until the infant is at least 1 year old as it is too high in minerals and protein and does not contain enough carbohydrates to meet infant needs. In addition, the major problem in cow's milk, casein, is harder for an infant to digest than the major protein formed in human milk, called lactalbumin.

Variations in Composition and Output of Human Milk

Human milk composition is not constant; it varies from one human to another, from one period of lactation to the next and even hourly during the day.

The variations in fat content are most significant. Fat concentration is higher towards the end of each feed. Lowest concentration is around 6 a.m. and highest around 10 p.m. These differences can be as much as 15 to 30 g/litre. The fat concentration of human milk increases with lactation up to about 84 days postpartum, providing for the increasing calorie needs of the growing infant.

The protein, carbohydrate, and mineral contents also change with the stage of lactation. To begin with, protein and salt content is high, but their levels start declining as the secretion of mature milk is established and then decrease slowly with advancing lactation. Intake of protein influences the total amount of milk produced rather than the protein content of milk. The levels of lactose increase rapidly during the first week and then relatively slowly, from mean 57 to 70 g/l at the end of the first month, and to 79 g/l at the end of 6 month. The composition of human milk varying at stage, time of day, during each feed suits the need of a growing baby for different nutrients. The major components of mature milk are given in Table 9.4.

The composition of a given human milk sample is related not only to the amount secreted and the stage of lactation but also to the timing of its withdrawal and to individual variations among lactating mothers which may be affected by variables such as mother's age, parity, health and social class.

Milk produced from mothers delivering prematurely has been found to be significantly higher in energy and protein and lower in lactose contents than term milk. This difference persists from 3 to 5 weeks. The protein needs of the preterm infant are higher, so his mother's milk is most important and superior in nutritional value to any other alternative feed. In women who deliver preterm baby, their milk contains essential fatty acids and polyunsaturated fatty acids in adequate quantity to sustain brain growth in preterm infants. However, preterm infants need high amount of calcium and phosphorus because their relative rate of growth is higher than that of full term infants. Supplementation with calcium, phosphorus and protein is recommended for low birth weight infants.

Effect of Maternal Nutritional Status on Composition of Human Milk

Although much data have been recorded on the differences in samples of human milk, the general picture is same throughout the world. The energy yielding nutrients of human milk are not significantly affected by maternal diet or nutritional status. Nutrients in human milk are derived from the diet of the mother and may be subsidized by using nutrient stores. The fat composition of human milk is affected by maternal diet. The fat content of human milk is predominantly oleic acid with substantial amounts of linoleic acid. Maternal consumption of higher proportions of unsaturated fatty acids can increase linoleic acid above the 10 per cent average level. This type of the maternal dietary intervention may be appropriate in the management of the hypercholesterolemic newborn or in infants at high risk for cardiovascular disease. A mother's fluid intake has not been proven to have an impact on milk output. Hence, except for vitamin and fat content, the composition of human milk appears to be largely independent of malnutrition until it becomes severe. Prolonged maternal undernutrition may reduce milk volume but not quality, except for water soluble vitamins, as it is well known that severely undernourished women during the time of famine often manage to feed their babies reasonably well.

Nutrient Requirements during Lactation

The new born infant's food is the mother's milk. Breastfeeding places a greater nutritional demand on the mother than pregnancy, because the woman has to nourish a fully developed and rapidly growing baby whose food need increases day by day. The baby is born with liberal stores of iron, protein, vitamin C and other vitamins drawn from the mother's body.

It must be remembered that whatever nutrients the mother consumes, they are transferred to the infant through breast milk. Even the poorly nourished woman is able to nurse her child satisfactorily at the breast and the composition of breast milk is not seriously affected by maternal diet in the case of protein, carbohydrate and minerals. The vitamin content of breast milk is less when the maternal diet is deficient. In mothers suffering from thiamine deficiency, the milk is also poor in thiamine and the child develops infantile beri beri and dies before the age of 3-4 months. For these reasons, the maternal diet should contain adequate vitamins.

The health of mother is seriously affected if her diet is poor in quality and quantity. The required nutrients are drawn from the body reserves as a consequence of which the woman may, develop certain deficiency diseases. Thus, lactating women are also said to belong to the vulnerable group as their nutritional needs are more demanding. The nutritional requirements of nursing mothers as given in Table 9.5 are briefly discussed.

Energy

The energy requirement during lactation is computed from the volume of milk secreted, its energy content and the efficiency of conversion of food energy into milk energy. There is no evidence for additional demand for lactation besides the energy content of milk secreted for breast feeding. There is no change in total energy expenditure (TEE) during lactation period over that in the non-pregnant period. The efficacy of utilization of milk energy is 80 per cent. Assuming an average milk output of 688 ml/day and conversion efficiency of 80 per cent, the additional intake of energy recommended during lactation by ICMR (2011) is 600 kcal per day for the first 6 months, if the woman is doing exclusive breast feeding. Since Indian women continue to beast-feed beyond 6 months, albeit with reduced milk output, upto one year, an extra allowance of 520 Kcal per day is recommended for periods from 7 months to 12 months. In this computation, the contribution of fat deposited during pregnancy is also taken into account.

Protein

A factorial approach has been adopted to derive the protein requirements during lactation. During lactation, protein requirement has been computed on the basis of secretion of 9.4 g per day of protein in milk during 0.6 months and 6.6 g during 6-24 months. Breast milk contains a relatively high concentration of non-protein nitrogen equivalent to 20-27 per cent of total milk nitrogen, much of it being urea. However, for the purpose of calculating requirements, the protein requirement was assumed to be only for the protein component of the total N in milk. NPN is excluded from the total milk N while computing protein concentration of milk. The factor of 6.25 is used to convert protein nitrogen in milk to protein equivalent. The efficiency of conversion of dietary nitrogen to milk protein equivalent can be assumed to be 47 per cent on Indian diets in case of adults. The safe intake is calculated as mean + (1.96

Table 9.5: Daily Recommended Dietary Allowances for Indian Lactating Women (ICMR, 2011)

Group	*Energy (Kcal)*	*Protein (g)*	*Fat (g)*	*Ca (mg)*	*Fe (mg)*	*Vit. A (μg)*	*Thiamine (mg)*	*Riboflavin (mg)*	*Niacin (mg)*	*Pyridoxin (mg)*	*Vit. C (mg)*	*Dietary Folate (μg)*	*Vit. B_{12} (μg)*	*Mg (mg)*	*Zinc (mg)*
Adult Woman															
Sedentary	1900	50	55	600	21	600	1.0	1.1	12	2.0	40	200	1	310	10
Moderate	2230	50	55	600	21	600	1.1	1.3	14	2.0	40	200	1	310	10
Heavy	2850	50	55	600	21	600	1.4	1.7	16	2.0	40	200	1	310	10
Lactation															
0-6 months	+600	+25	77.9	+600	25	950	+0.3	+0.4	+4	2.5	80	300	1.5	310	12
6-12 months	+520	+18	70.2	+600	25	950	+0.2	+0.3	+3	2.5	80	300	1.5	310	12

Source: Dietary Guidelines for Indians – A Manual (2011), NIN, Hyderabad.

SD) with 12.5 per cent CV. Rounded off figures would be 19g/d for safe allowance for a lactating woman during 1-6 months and 13g from 6 to 12 months. ICMR recommendation (2011) in terms of cereal-pulse-milk based dietary protein with PDCAAS of 82.5 per cent is additional protein intake of 22.9 and 15.2 g/d during 0-6 months and 6-12 months of lactation, respectively.

Fat

The minimum level of total fat should be 220 per cent E and accepted macronutrient dietary range (ADMR) *i.e.* 20-35 per cent E is the same as for general population. Adult lactating females should consume at least 200 mg/d DHA for optimal adult health and infant development. To provide 20 per cent E total fat, diet for lactating woman should provide at least 30 g visible fat (ICMR, 2011; Table 9.6). This level of fat in the diet would provide adequate energy to enable a nursing woman to meet her higher energy needs.

Calcium

Using factorial method, calcium intake needed to maintain calcium adequacy has been obtained for lactating woman. The quantity of calcium present in 850 ml of milk will be 290 mg. To meet this need, a total allowance of 1200 mg daily is recommended for the lactating woman (ICMR, 2011). Calcium and phosphorus should be in 1:1 ratio in the diet.

Magnesium

There is no additional requirement for magnesium during lactation. Magnesium requirement during lactation *i.e.* 310 mg/d remains the same as in case of adult non-pregnant and non-lactating woman.

Iron

Iron requirement during lactation is the sum of the requirement of the mother and that required to make up the iron lost in breast milk. Due to lactational amenorrhoea, the basal iron requirement will be the same as in adult man *i.e.* 14 µg per kg/day (0.77 mg/d). For lactation, the requirement is about 0.5 mg per day or 9 µg per kg/d. So the total iron requirement during lactation is about 1.27 mg/d; hence her iron requirement remains the same as for a normal woman. Amenorrhoea during lactation takes care of the amount of iron secreted in the breast milk. Hence, no additional requirement for iron has been recommended. A lactating woman should take 25 mg iron per day (ICMR, 2011).

Zinc

During lactation, 12 mg zinc per day is required (ICMR, 2011).

Iodine

Lactating women require 200µg iodine per day.

Vitamin A

The additional needs of vitamin A during lactation are calculated on the basis of vitamin A secreted in milk. The FAO/WHO Expert Committee used an average milk secretion of 700 ml per day with a retinol content of 50 µg per 100 ml and recommended an additional daily allowance of 350 µg vitamin A during lactation to meet the above needs. ICMR has also suggested an additional allowance to be about 350 µg of retinol or 7600 µg of β-carotene for Indian lactating women.

Vitamin C

The additional needs during lactation are on the basis of vitamin C secreted in milk. Assuming a daily milk secretion of 700 ml with vitamin C content of 3 mg per 100 ml by well nourished women, the additional requirement during lactation will be 20 mg per day. Taking into consideration of the cooking losses (50 per cent), ICMR has recommended an additional intake of 40 mg vitamin C per day during lactation.

B-complex Vitamins

Thiamine, Riboflavin, Niacin and Pyridoxine

The requirements of thiamine, riboflavin and niacin are related to the calories; hence, the additional amounts recommended during lactation are based on the additional energy intake, their amounts secreted in the milk and energy cost of lactation. Thus, the additional amounts (mg) recommended during first 6 months of lactation are 0.3 thiamine, 0.4 riboflavin and 4 niacin equivalents and during 6-12 months of lactation are 0.2, 0.3 and 3 mg, per day, respectively. During lactation, an additional amount of pyridoxine recommended is 0.5 mg per day.

Table 9.6: Recommendations for Dietary Fat Intake for Lactating Woman (ICMR, 2011)

Physiological Group	*Physical Activity*	*Minimum Level of Total Fat (per cent E)*	*Fat from Foods Other than Visible Fat[a] (per cent E)*	*Visible Fat[b]*	
				Per cent E	*g/p/d*
Adult woman	Sedentary	20	10	10	20
	Moderate				25
	Heavy				30
	Lactating woman	20	10	10	30

a: If higher than 10 per cent E, visible fat requirement proportionately reduces; b: Cooking oils, butter, ghee and margarine

Folic Acid

The folic acid content of breast milk secreted by Indian women is 1.6 μg per 100 ml, a value lower than 3.0 μg reported from the west. At the higher level, the amount of folate lost by mother would be about 25 μg a day. An additional allowance of 100 μg of folate has been provided for lactation by ICMR (2011) for meeting the extra factorial needs. Thus, the RDA of folic acid during lactation would be 150 μg.

Vitamin B_{12}

The vitamin B_{12} of breast milk in normal women is about 300 pg per ml and the amount of the vitamin secreted in milk per day is thus between 0.25 and 0.30 μg. An additional intake of 0.5 μg per day would thus cover the needs of a lactating woman.

Diet during Lactation

After the birth, the baby is nursed on the mother's breast, therefore, in order that the mother may breast feed her infant without any undue strain on her body, she must continue to eat an adequate and balanced diet to meet the requirements of lactation. For proper and trouble free lactation, nursing mother should consume large amounts of body building, protective foods and energy yielding foods. They help in facilitating the information and secretion of breast milk.

The diet during this period should be rich in high quality protein, vitamins, minerals and energy. The deficit in the mother's diet may affect the quantity but not the nutritional quality of the milk except perhaps in case of some vitamins. The quality of milk is maintained even if it means drawing of the nutrients from maternal reserves and tissues like bones. This may be desirable from the infant's point of view, but certainly not so from the mother's health point of view. A weak mother becomes even weaker. Hence, a nursing mother should consume a well balanced nutritious diet containing different food items in right proportion so as to maintain good health and nutritional status of herself as well as of the infant.

While planning balanced meals for nursing mothers, foods from various food groups should be included and the following basic principles should be kept in mind:

- Include as many varieties of foods as possible. Attempt should be made to include foods from each of the five groups so as to ensure adequate supply of energy and nutrients.
- If cereals, millets and pulses form the main stay of diet, protein requirements of individuals are automatically met provided they are consumed in amounts sufficient to meet the energy needs. Hence, adequate quantities of mixed cereals and pulses should be included in the diet.
- Consumption of whole cereals and legumes is recommended so as to get good amount of dietary fibre.
- Regular consumption of fruits and vegetable should be ensured as they are the good sources of vitamins, minerals and fibre.
- Consumption of milk and milk products by vegetarian nursing women is crucial as they besides providing good quality protein, calcium and certain vitamins, are the only source of vitamin B_{12} for them.
- Condiments and spices should be used sparingly. Foods with strong flavour should be avoided, *e.g.* raw onion, as they may impart a peculiar smell to the breast milk, which may not be liked by the infant.
- Judicious use of vegetable oils, rich in essential fatty acids shall help to reduce the dietary bulk as well as provide satiety.
- Plenty of water and other fluids should be taken.

Balanced diet for lactating woman has been given in Table 9.7.

Table 9.6: Balanced Diet for Lactating Women (Number of portions) (ICMR/NIN, 2011)

Food Groups	*g/portion*	*Type of Work*		
		Sedentary	*Moderate*	*Heavy*
Cereals and millets	30	9+1	11+1	16+1
Pulses	30	2+2	2.5+2	3+2
Milk	100 ml	3+2	3+2	3+2
Roots and tubers	100	2	2	2
Green leafy vegetables	100	1+0.5	1+0.5	1+0.5
Other vegetables	100	2	2	2
Fruits	100	1+1	1+1	1+1
Sugar	5	4	6	9
Fats and oils (visible)	5	4+2	5+2	6+2

Between 6-12 months of lactation, diet intake should be gradually brought back to normal

For non-vegetarians, substitute one pulse portion (30 g) with one portion (50 g) of egg/meat/fish/chicken

Menu Planning

While planning meals, nutrient adequacy, density and distribution of calories taken during the day, should be considered. In order to provide the daily requirements of energy and nutrients in right quantity and proportions through meals habitually consumed, appropriate menu planning is essential. The practical tips to be followed in menu planning are:

- About one-third of day's nutrient requirement be ensured in each of the three main meals which are usually eaten.
- Use of liberal amount of raw vegetables *i.e.* salads at meal time should be advocated.
- Use assorted fruits and vegetables in all the areas of the menu, including appetizers, salads, side dishes, and desserts.
- The colour, texture and taste of dishes should be such that they impart a sense of pleasantness and satisfaction.
- Inclusion of variety of foods at each meal is desirable.
- Simple methods of cooking should be used.
- Preparations should not be spicy.
- Be sure to use iodized salt.
- Be sure to have good sources of problem nutrients: folacin, iron, calcium, vitamin B_6 and zinc.
- Socio-cultural and economic factors that influence affordability of food by the individual family, regional customs and traditions and food preferences should be kept in mind.

In this way, a well balanced nutritious diet can be formulated for nursing women, taking into consideration, variety of foods, their nutritive value, nutrient requirements, principles of menu planning and socio-economic and cultural factors.

Points to Remember for Good Milk Supply by a Lactating Woman

- The secretion of milk requires additional energy.
- The quantity of milk produced is related to the food intake of the mother.
- The fat content of milk is dependent on the dietary intake of the mother.
- A liberal protein intake by a nursing woman is recommended.
- Animal proteins favour the production of milk protein.
- Positive calcium and phosphorus balance is difficult to maintain.
- The need for iron is important like during pregnancy.
- Additional iodine is needed.
- The vitamin A requirements are higher than those of pregnancy.
- Vitamin D is important during lactation.
- The requirements for the B-complex vitamins are more during lactation than during pregnancy.
- Marked variations in milk secretions occur.
- Nervous strain and over activity is bad for optimal milk production.
- An optimal diet facilitates adequate milk production and maternal welfare.
- Certain drugs may pass from the mother's blood to her milk and should be avoided by the lactating mother. Drugs that are known to affect the breastfed infant are chloral hydrate, diazepam, caffeine, penicillin, Phenobarbital and most laxatives. Nursing is contraindicated in women taking such drugs as chloramphenicol, metronidazole, lithium and antithyroid drugs. An infant can become addicted to heroin if the nursing mother is taking large doses of it.

Diet Plan

A typical diet plan for Mrs X, engaged in nursing her 4 month old child, 23 years old, vegetarian, engaged in moderate activity and belonging to MIG, is as under:

Step I: Personal Profile

Age	23 years
Sex	Female
Physical activity	Moderate activity
Physiological status	Lactation (0-6 months)
Socio-economic status	MIG
Food habits	Vegetarian

Step II: Recommended Dietary Allowances (ICMR, 2011)

Energy	2230+600 Kcals
Protein	77.9 g
Calcium	1200 mg
Vitamin A	950 µg
Vitamin C	80 mg

Step III: Food Exchange Plan

Exchange	No.	Energy (Kcal)	Protein (g)	CHO (g)	Fat (g)
Milk	3	510	24	36	30
Meat	-	-	-	-	-
Pulses	3	300	21	51	-
Cereals	19	1330	38	285	-
Vegetable A	1.5	18	-	4.5	-
Vegetable B	2	80	4	14	-
Fruit	2	80	-	20	-
Fat	8	100	-	-	40
Sugar	5	360	-	-	-
Total		**2778**	**87**	**359.5**	**70**

Step IV and V: Menu Plan and Distribution of Exchanges

Meal	Exchanges	No.	Menu
Early morning	Milk	0.5	Tea
	Sugar	1	Biscuits
	Cereal	2	
Breakfast	Milk	1	Wheat porridge with
	Sugar	1	Milk
	Cereal	3	Toast
	Fruit	1	Butter
	Fat	1	Papaya
Mid morning	Cereal	2	*Paushtik ladoo*
	Pulse	1	
	Fat	1.0	
	Sugar	2	
Lunch	Vegetable A	1.0	*Sarson ka saag*
	Vegetable B	1	*Makki ki roti*
	Cereal	5	Carrot and pea veg.
	Milk	0.5	Lassi
	Fat	2	Salad
	Fat	2	
Evening tea	Vegetable B	0.5	Vegetable cutlet
	Cereal	3	Orange juice
	Fruit	1	
	Vegetable A	0.5	
	Vegetable B	0.5	
Dinner	Pulse	2	Mixed veg. soup
	Cereal	4	Channa dal with
	Fat	2	Ghia, Cabbage and potato veg., Papad
At bed time	Milk	1	Chapati/Rice
	Sugar	1	Milk

Calculate the amount of various nutrients provided by such a diet and modify the amounts or type of food if a particular nutrient is less or more than the RDA.

Special Foods with Lactogenic Effects

Some of the herbal concoctions or special foods are usually recommended by elderly women to increase milk production. Herbal tea, garlic, tamarind, ginger and cottonseed are some of the foodstuffs used as lactagogues in various countries for lactating women. In our country, dietary surveys have reported the consumption of following special foods during lactation.

- Special foods such as *ajwain, sonth* (dry ginger), *gond* (acacia gum), *jeera* (cumin seeds), jaggery, fenugreek, *battisa*, dry coconut, dry fruits, ghee, garlic, tamarind juice, *dalia*, milk and *panjiri* have been reported to be included in the diet of lactating mothers. The purpose of inclusion is usually to relieve pain, provide extra energy, promote early recovery, provide warmth, easy digestion and induce milk secretion. These foods are given in special preparation of wheat flour or *besan* in the form of laddu.
- Soft diet prepared from rice and pulses, and milk is given for easy digestion.
- Lactating mothers have been reported to increase consumption of special meats and soups, richer proteins and a carbohydrate rich diet of dates, wheat flour and fat.
- Consumption of strong black coffee immediately after delivery has also been reported.
- It is believed that garlic, ginger, cottonseeds and tamarind are given to lactating mothers because these foods possess galactogogic properties.

Although there are no well controlled studies regarding lactagogues but these have had such a long and respected history in the traditional societies that it would be unwise for us to deny categorically that any of them could possibly enhance milk production.

Food Beliefs and Avoidances during Lactation

In some communities, a lactating mother has to observe traditional dietary restrictions in addition to taking any customary lactagogues. These have been reviewed and given as under:

- Milk should be given in restricted amount to lactating mother as it causes gastro-intestinal upsets in mothers and infants. Tubers, chillies, condiments, pickles, citrus fruit, tamarind, spices and legumes are undesirable during early lactation as they lead to indigestion in both mother and infant.
- Fried foods should be avoided by lactating mothers.
- Papaya and pineapple make mother's milk sour, cause vomit and diarrhoea in infants.
- Green leafy vegetables should not be eaten by the lactating mother because they reduce milk secretion and also the colour might show up in mother's milk and the child may develop gastro-intestinal disorders.
- Fish and milk are poisonous if eaten together during first few weeks following childbirth.
- Sweet milk gives milk fever.
- Cabbage should be excluded from diet of lactating mothers as it taints mother's milk.
- Citrus fruits to be avoided by the lactating mother as they cause inflammation of the breastfed baby's throat, flatulence and indigestion.
- Rice, chilli powder, pepper, garlic and ginger are health foods during lactation.

- ✰ Tomato and mango are thought to make milk sour and discolour it. Tomatoes also produce skin rash in infants.
- ✰ Lactating mothers should consume pure ghee and nuts in large quantities.
- ✰ Lactating mothers have been reported to avoid citrus fruits, sweet potato, potato, pumpkin, green leafy vegetables, pulses, curds, coconut, buttermilk, egg and fish, as these foods are believed to cause cold and gas or indigestion.
- ✰ Mango, spicy foods, chillies, tea, tamarind, black gram *dal*, pickles *etc.* are restricted as they are hot foods and believed to cause gastro-intestinal disturbances. During early lactation *illaichi* (cardamom) and dates boiled milk is a recommended food.
- ✰ The foods avoided during three to six months of lactation are brinjal, cluster beans, bengal gram *dal*, black gram *dal*, potato, colocassia, tamarind and fish as they believe to produce 'knee joint' or 'cold and cough', 'vatam'.
- ✰ Hot foods are believed to reduce milk secretion, cause diarrhoea in infant and inflammation in throat of mother and infant.
- ✰ The cold foods as pumpkin, banana, okra, salad leaves and beet root are believed to produce cold, fits, chills, fever and diarrhoea in infant and lower milk secretion.
- ✰ Raw and parboiled rice, jackfruit and yam eaten during lactation could be hot and heavy for infant, difficult to digest.

The nutritional significance of these food related behaviours is difficult to evaluate. Available information has often been collected in an anecdotal or one-sided manner. Thus, there is limited detailed information on dietary alterations that appear to be detrimental. As a result, it is difficult to quantify the nutritional effect of restrictive beliefs and avoidance. The nutritional importance of such practices cannot be assessed without reference to the rest of the woman's diet. Overall, so long as the diet is nutritionally well balanced, beneficial and harmless, it should not be discouraged. Adequate nutrition for the woman especially if she is malnourished, is extremely important. According to old traditions, milk is considered as the best lactagogue. However, there is a need to educate lactating mothers that only taking milk in sufficient quantity cannot fulfill their enhanced nutritional requirements (Table 9.5). They should take a well balanced diet consisting of foods from all the major food groups in adequate amounts to ensure good health of themselves as well as of their infants.

10

Nutrition during Infancy

Infancy, the period from birth to one year of age, is a period of tremendous body growth and development. Due to rapid growth during the first year, infancy is one of the most critical periods in the lifecycle as far as food is concerned. Nutrient needs in relationship to size are high, and optimum nutrition at this time is very important to health and vigour throughout life. Many mothers feel insecure with their first babies and are concerned about what and how to feed the infant. Nutritionists should advise and help the mothers develop skills in this area. Let us explore this critical stage of life and the key role the nutrients play in this stage.

Milestones in Infant Growth and Development

Growth refers to an increase in physical size of the body as a whole or an increase in any of its parts associated with an increase in cell number or cell size. Development is defined as the acquisition of function associated with cell differentiation and maturation of individual organ systems or a change from a lower to a more advanced stage of complexity for greater facility in functioning.

The infant's growth and development refers to:

I. Physiological development
II. Motor development
III. Physical development

I. Physiological Development

Before discussing the nutritional requirements of infancy and childhood, it is important to first know about the pattern of growth in various organ systems during infancy. The systems which change most at the time of birth and during early infancy; these are heart, lungs, brain, the nervous system, and the skeleton. Those systems which could be affected by nutritional deficit are specially discussed below:

The Cardiovascular and Respiratory System

A number of important changes occur in the cardiovascular and respiratory system in the infant, immediately after birth. The heart beats at the rate of about 160 beats per minute at birth gradually declines in rate throughout childhood. By age two, the average heart rate is 110 beats per minute. Within the first week of life, blood pressure is 80-85 (systolic)/45-50 (diastolic) mm Hg and rises to about 100/65 mm Hg by two years. Babies breathe mainly at a rate of 40-46 breaths/minute and with periodic breathing (having short periods of cessation of breathing).

The Digestive System

One has to note that protuberance of the abdomen occurs after feeding. At birth digestive system is fairly mature. There is an increase in the size of the system and functional maturity all through infancy. The amount of trypsin and lipase enzymes is rather low. Digestive enzyme levels increase gradually throughout infancy. Bile acids production by the liver is low resulting into poor tolerance of fats in some infants. Liver is relatively immature as far as some enzymatic activities are concerned. Infants require some solid foods to promote tooth and gastrointestinal development around 6 months of age. Faecal material (called meconium) composed of digestive juices and shed cells from the lumen of the gut, is normally passed by 24 hours in the normal infant. Stools from breast-fed infants are soft, sour, pasty, and yellow. Two to four stools per day are passed by a breast-fed infant. By two years of age, the stool becomes formed and darker.

Hematopoietic System

At birth, blood cells are produced in the bone marrow. Infants have low levels of a number of clotting factors (*e.g.*, Factor VII, IX, and X), yet bleeding tendencies in the newborns are rare. Breast-fed infants derive small amounts of vitamin K from milk and it provides them some protection. Injection of vitamin K, required for formation of some clotting factors has become a routine to prevent bleeding episodes of the newborn.

Urinary-Genital System

The kidney grows with the general body growth, doubling in size by 6 months and tripling by one year. The functional immaturity of the kidney results in two major functional handicaps (1) inability of the kidney to excrete a sodium (salt) load (2) and poor ability to excrete a water load. The infant kidney has a narrow margin of safety to prevent acidosis, dehydration and oedema.

Testes, ovaries, and internal and external genitalia are present at birth but their growth and function starts at the onset of puberty.

The Immune System

The immunologic defense mechanism of the newborn is less efficient and competent to cope up with microbial infections. Every component of the immune system is present but lacks experience. In addition, the infant gets protection in the form of antibodies which cross placenta freely. This acquired immunity from mother will protect the infant for 6-9 months after birth. Immunologic protection is also provided through the breast milk. The infant appears to be particularly susceptible to a number of infections. Some microbes are pathogenic for infants and seldom cause disease thereafter.

Skeletal Growth in Infants

Growth of long bones generally follows an orderly pattern throughout infancy and childhood. The growth is influenced by race, sex, environmental factors including some nutrients. After birth, the hipbone and lower limbs grow more rapidly than the rest of the bones of the body and with walking, these bones become thicker and stronger. The skull is large at birth to accommodate the rapidly growing brain, it continues to grow rapidly during infancy, doubling its volume by age two (from 400 to 950 ml).

The Dentition

The first deciduous teeth erupt in most children between 5 and 9 months. The first to erupt are the lower central incisors, followed by the upper central and then the upper lateral incisors. By the age of one year, most children have 6-8 teeth. Some infants have 2 teeth at one year without evidence of any growth disturbance.

Development of the Nervous System in Infancy

Though growth of the brain occurs during the foetal life, rapid growth and functional maturation continues throughout infancy. By two years after birth, brain completes half of its growth. Most primitive part of the brain completes half of its growth. Most primitive parts of the brain complete their development first and are fully developed functionally at birth (*i.e.* the medulla, pons, and medulla). The cerebellum, which controls posture and equilibrium and coordinates movements, has a spurt of growth after birth. It is known that there are critical periods of brain growth when injury may result in irreversible damage. Such injuries may occur during early life by irradiation or infections (such as rubella) interrupting brain growth. This happens in the first trimester and has no effect during the third or early infancy. But malnutrition during the third or early infancy causes brain damage leading to reduced intelligence but not affecting other primitive brain function. The cerebellum with its late growth after birth would be particularly vulnerable to injury during infancy. Sound is more important to foetus, but after birth, visual function is more rapidly developed. By 3 months of birth, the motor areas and cells controlling the arm and upper trunk develop earlier than those controlling the leg. So the baby can control arms and trunk well, but begins to use the legs later.

Visual System

The baby keeps the eyes closed most of the time. In the first week after birth, because of photophobia and because tearing is not well established until about 2 months of age, most newborns can see objects placed before them at about 4 months. Binocular vision develops by about 6 months.

Auditory Sensation

Hearing ability is present at birth, but proper response to sound develops by about 3 to 4 months of age. The baby can localize sound by about 3 months and by 4 months makes movements in response to a new sound. Listening and language skill develops towards the end of the first year.

Taste

Taste is developed at birth while differentiation is poor. By 2 to 3 months, taste sensation becomes more acute. The infant can now discriminate change in food (formulas) and may reject any formula that contains condiments or medications.

Tough Sensation

Touch sensation is poorly developed at birth. The

baby will respond to painful stimuli by crying or by body movements. Touch sensitivity is well developed by 12-17 months.

II. Motor Development

The changes that occur in the infant's ability to control the movement of the muscles of the body are referred to as motor development. During the first three years of life, the child progresses from a relatively helpless infant state of motor activity to a state of independence and mobility.

The development of motor skills involves coordination between the nervous system, the skeleton, muscles, and sensory mechanisms. Following factors influence the motor development of infants:

- ☆ Genetics
- ☆ Nutrition
- ☆ Maturation and learning
- ☆ Body size
- ☆ Cultural influences

III. Physical Development

Rapid growth characterizes infancy. The major indicators used to assess normal growth during infancy include:

- ☆ Body weight
- ☆ Height/Length
- ☆ Head circumference

Some of the other anthropometric measurements *e.g.* mid-upper arm circumference, skin fold thickness, chest/head circumference ratio, mid-upper arm circumference ratio *etc.* (Table 10.1) are also used for nutritional assessment of infants and children. The

Table 10.1: Some Anthropometric Measurements Applied in Nutritional Assessment

Sl. No.	*Measure-ments*	*Age Groups*	*Nutritional Indication*	*Reprodu-cibility*	*Advantages*	*Disadvantages*	*Observer Error*	*Interpretation*
1.	Weight	All groups	Present nutritional status; Under and over	Good	Common in use	Difficult in field, cannot tell body composition, need accurate age, high related (insensitive)	< 100 g in children <250 g in adults	<60 per cent severe 60-80 moderate 80-90 mild 90-110 normal 110-120 over 120 and over obese
2.	Height	All groups	Chronic nutritional status (under)	Good	Common in use	Other factors play a role	≤0.5 cm <3.0 cm in adults	<80 per cent dwarf 80-93 short 93-105 normal >105 giant
		7 years child	Chronic under-nutrition in early childhood	Good	Simple to do in field			
3.	Head circumference	0-4 years	Intrauterine and childhood nutrition (chronic nunder-nutrition)	Good	Simple	Other factors play a role *e.g.* brain development	<0.5 cm	
4.	Mid-arm circumference	All groups	Present under and over nutrition	Fair	Simple, age dependent, child needs to be stripped, suitable for rapid survey	No limits for overnutriton No standard for adult	<0.5 cm	<75 per cent severe 75-80 moderate 80-85 mild >85 per cent normal
5.	Skinfold thickness, subscapula or triceps	All groups	Present under and over nutrition	Fair	Measures body composition, detects obesity in adults	Needs expensive caliper, difficult with child and in the field, ethnic differences	1.0-1.5 mm	Similar to item (1)
6.	Weight/height/ age ratio	All ages	Present under and over nutrition	Good	Index of body build, age independent 1-4 years and adults	Need proper scales, need trained personnel		<75 per cent severe 75-80 moderate 85-90 mild 90-110 mild 110-120 over >120 obese

Sl. No.	Measurements	Age Groups	Nutritional Indication	Reproducibility	Advantages	Disadvantages	Observer Error	Interpretation
7.	Mid-arm/ head ratio	3 months- 48 months	Present under nutrition	Good	Simple, age independent, sex independent, any person can do it in field	No standards for adults		<0.25 severe 0.25-0.28 moderate 0.28-0.31 mild 0.31-0.35 normal >0.35 obese
8.	Chest/head circumference ratio	1-2 years	Present under nutrition	Fair or poor	Simple, age independent	For limited age No classification		<1 malnourished >1 normal

measurements like skin fold thickness and mid-upper arm circumference should not be taken for infants but should be used for preschoolers and above.

Body Weight

The body weight of the newborn is determined by the mother's pre-pregnancy weight and weight gain during pregnancy. The average weight of a normal full-term newborn is approximately 3500 g. Immediately after birth, there is a weight loss *i.e.* 5 to 10 per cent of birth weight resulting from the loss of fluid and some catabolism of tissues. Thereafter, birth weight is usually recovered by about 10 days of life. Average weight gains for the first 4 months of age are 20-25 g per day whereas weight gain is 15 g per day during the last 8 months. The infant doubles his birth weight by the age of 4-6 months and triples by 12 months of age. Males double their birth weight earlier than do females. Smaller newborns double their birth weights sooner than do heavier infants.

Height/Length

An infant's length increases by 50 per cent in the first year of age, doubles it by 4 years of age and triples it by 13 years of age. The average birth length of 50 cm increases to 74 to 76 cm at 1 year of age, 100 to 105 cm at 4 years of age, and 150 to 155 cm in the preadolescent years. This increase in height stops when the bone ends (epiphyses) fuse together, halting further significant linear growth (Figure 10.1).

Tanner's (1978) relative maturity percentages are sometimes used as a rough estimate of adult height. At 2 years of age, a boy will be 49.4 per cent as tall as he will grow to be as an adult. A girls of the same age has reached 52.7 per cent of her mature height. However, it is not possible to accurately predict the final body size of an infant because there are too many unknown variables that can affect growth.

Head Circumference

During infancy, head circumference is a major index of normal growth. At birth, the head circumference normally ranges from 12 to 15 inches. The head circumference of the typical (59th per centile) 6 month-old infant measures about 17 inches, increasing to around 18 inches at 1 year of age, and around 19 inches at 2 years. Generally, males have a slightly larger head than females during infancy period. If the head circumference measures below the 3rd per centile or above the 97th per centile, the infant should be carefully evaluated for problems such as microcephaly or hydrocephaly.

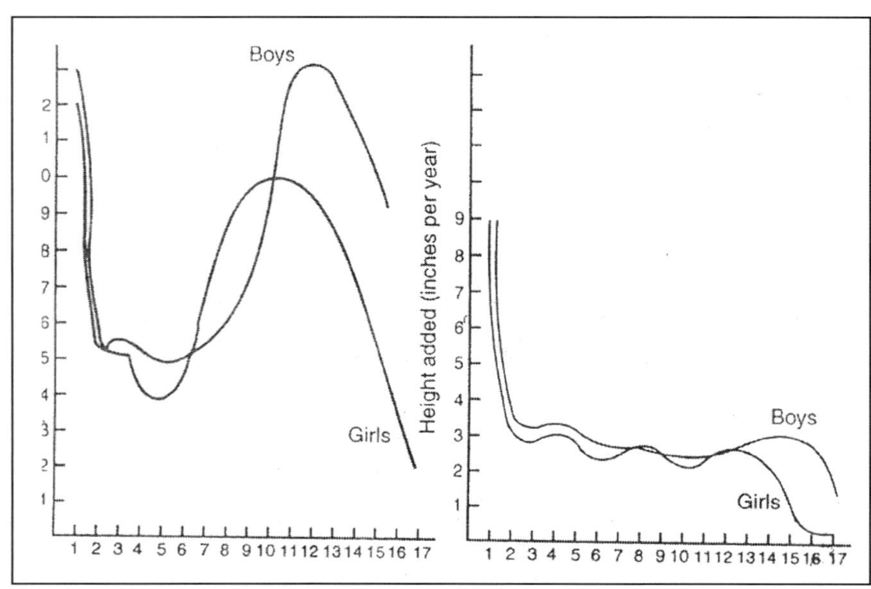

Figure 10.1: Growth Rates.
(A) Average gain in weight for girls and boys. (B) Average additions to height for girls and boys.

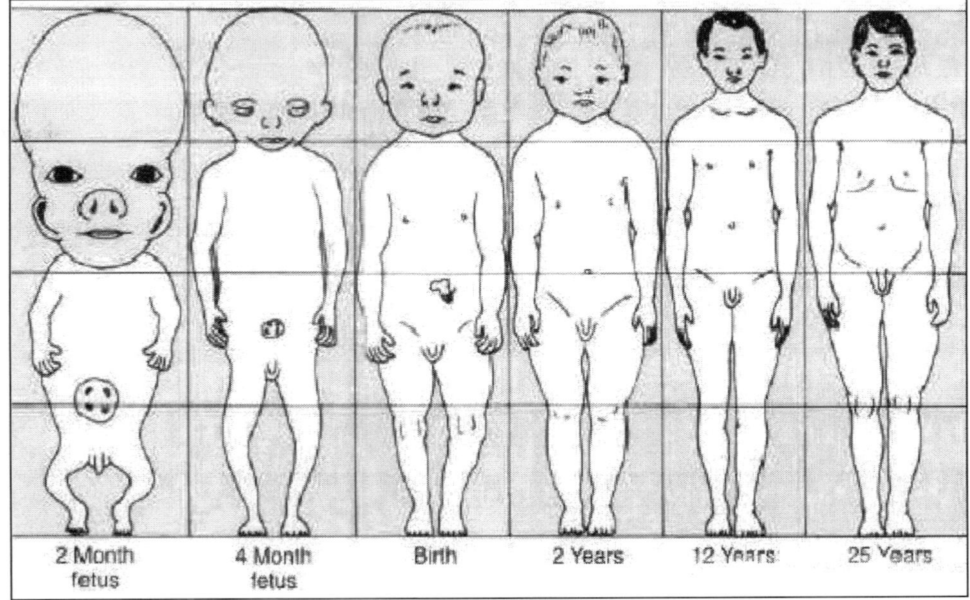

Figure 10.2: Proportionate Changes in Growth in Head, Trunk and Limbs.

At birth, the head accounts for approximately one-fourth of the total body length, however, the head is one-eighth of body length by the time growth has ceased. During infancy, the head grows slower, hence there is a significant change in its proportion to body length. Brain growth is 75 per cent complete by the end of the second year, and the brain is 100 per cent of adult size by age 6 to 10 years. Leg length increases from approximately three-eighth to one-half of total body length between birth and adult (Figure 10.2).

Mid-upper Arm Circumference

The measurement used to estimate muscle and fat bulk is made at the midpoint of the upper part of the left arm. A paper insertion tape is used for this purpose. Arm circumference is measured to the nearest 0.1 cm. Arm circumference may be estimated by the following formula:

$$\text{Arm muscle diameter} = \frac{\text{Arm circumference (mm)} - \text{triceps fat fold (mm)}}{\pi}$$

This is an easy and useful measurement. Normally, the arm circumference increases rapidly from birth to one year, from about 11 to 16 cm. Between the first and fifth birthdays, it generally remains constant at about 16 to 17 cm among well nourished children.

Fat Fold Thickness

For it, triceps is the easiest site of measure and is representative of body fatness. The triceps measurement is taken midway between the acromion and olecranon processes, using special calipers that have been calibrated to provide a constant tension.

Growth requires a generous nutrient intake to support body needs. We know from observing Egyptian mummies that infants were about the same size in 300 BC as they are today. However, adult mummies are much smaller than adults are today. People of the past ate poor diets that did not supply enough nutrients for maximum growth.

In the third world today, about half the population is short and underweight for age. In these countries, when breastfeeding ceases, children are often left to consume a high carbohydrate and a low-protein diet. This diet supports growth but does not allow them to achieve their optimum genetic potential. At this stage adequate protein intake is critical for growth. Weight primarily reflects current nutrient intake, while height is a better measure of long-term nutrient intake.

Development Milestones of Feeding Behaviour

Before discussing the nutritional requirements of infancy, it is important to first know about the developmental milestones of feeding behaviour.

Sequence of Development of Feeding Behaviour

The sequence of development of feeding behaviour is given in Table 10.2 and described below:

0-4 months

Newborn infants can neither focus their eyes nor direct their hands, yet they find nourishment. The rooting reflex helps an infant to turn towards the stimulus so that the mouth comes in contact with it. Stimulus placed on the lip causes involuntary involvements towards it. These reflexes help the infant to suck and receive nourishment. The rooting and suckling reflexes can be

Table 10.2: Sequence of Development of Feeding Beahviour

Age	Reflexes	Oral, Fine, Gross/Motor Development
1-3 months	☆ Rooting, sucking and swallow reflexes are present at birth ☆ Tonic neck reflex present	☆ Head control is poor ☆ Secures milk with a suckling pattern, the tongue projecting during a swallow ☆ By the end of third month, head control is developed
4-6 months	☆ Rooting and bite reflexes fade ☆ Tonic neck reflex fades by 16 weeks	☆ Changes from a suckling pattern to a mature suck with liquids ☆ Suckling strength increases ☆ Munching pattern begins ☆ Grasps with a palmar grasp ☆ Grasps, brings objects to mouth, and bites them
7-9 months	☆ Gag reflex is less strong as chewing of solids begins and normal gas is developing ☆ Choking reflex can be inhibited	☆ Munching movements begin when solid foods are eaten ☆ Rotary chewing begins ☆ Sits alone ☆ Has power of voluntary release and resecural ☆ Holds bottle alone
10-12 months		☆ Develops an inferior princer grasp ☆ Reaches for a spoon ☆ Bites nipples, spoons, and crunchy foods ☆ Grasps bottles and foods and brings them to the mouth ☆ Can drink from a cup that is held ☆ Tongue is used to lick food morsels off the lower lip ☆ Finger feeds with a refined princer grasp

elicited when the infant is hungry but are absent when he is satisfied.

When the infant is 5 weeks old, he can focus the eyes on the faces and can shift the gaze by 12 weeks of age. The infant starts recognizing the breast or bottle as the source of food when he is 10 weeks of age.

4 to 6 months

More mature sucking pattern establishes and tonic neck reflex fades when the infant is 16 weeks of age. He can move the tongue back and forth as opposed to the earlier up and down motions. He can close his hands on the bottle. By 20 weeks of age, the infant can grasp on tactile contact with a palmar squeeze. By 24 weeks of age, he can reach for and grasp an object on sight. He grasps, brings objects to mouth, and bites them.

7-9 Months

The infant can move his jaws up and down and develops an interior pincer grasp. Between 24 and 28 weeks of age, rotary chewing begins. This, combined with the ability to grasp, the hands to mouth route of rasped objects and sitting posture, indicates a readiness of the infant to finger-feed.

At this stage, infant grasp with a palmar grasp. Therefore, the shape of the food presented for finger feeding to the infant is important. Cookies, crackers, teething biscuits *etc.* should be introduced at this age.

The infant starts sitting alone when he is between 28 and 32 weeks of age. He gains control of the trunk. He has greater mobility of his shoulders and arms. Hence, he can reach and grasp any nearby object freely. The infants are able to help themselves to their bottle in sitting postures. The infant is able to transfer items from one hand to another and has the power to voluntarily release and resecure the objects. The tongue shows more maturity in regard to spoon feeding than to drinking. He can suck any liquid from the cup. By 32 weeks of age, infants bring their heads forward to receive the spoon as it is presented to them. They can manipulate food in the mouth with definite chewing movements.

It is a critical stage of development. Soft, mashed foods can be introduced at this age. Custards, puddings and ice cream become favourites of the infants of this age group.

10-12 Months

The infant develops a refined pincer grasp. He can manage the bottle-feeding alone. He can resecure the bottle if it is lost. He can drink from a cup that is held. He can use the tongue to lick the food morsels off the lower lip. He bites nipples, spoons and the crunchy foods.

Infants at this age often imitate the models set for them. By 1 year of age, the patterns of eating change from sucking to beginning of rotary chewing movements. They have voluntary hand to mouth movements. This prepares them to learn self-feeding, a behaviour they learn and refine in the second year.

Critical Stages of Infant Feeding

Suckling

Even though the normal neonate is well prepared to suck and swallow at birth, the physical and neuromotor maturation during the first year alter both the form of the oral structure and the methods by which the infant extracts liquid from a nipple.

At birth the tongue is disproportionately large in comparison with the lower jaw and essentially fills the oral cavity. The mandible is retruded relative to the maxilla. When the mouth is closed, the jaws do not rest on top of each other, but the tip of the tongue lies between the upper and lower jaws. The lips of the neonate also help in suckling and have characteristics appropriate for their function at this age.

The newborn infant sucks reflexively; the young infant (beginning at 2 to 3 weeks of age) suckles, and as the infant grows older, he learns mature sucking. The process of breast suckling and bottle suckling are similar. The nipple of the breast becomes rigid and elongated during breastfeeding so that it resembles a rubber nipple in shape. The infant grasps the nipple in the mouth. The oral cavity is sealed off by pressure from the median portions of the lips assisted by the mucosal folds in the jaws. The nipple is held in the infant's mouth with the tip located close to the junction of the hard and soft palates. The mucosal fold disappears by the third or fourth month, when the lips develop muscular control to seal the oral cavity.

Sucking

Matured sucking is an acquired feature of orofacial muscles. It is not a continuous process. By 16 weeks of age, the more mature sucking pattern becomes evident. As the infant grows older, the oral cavity enlarges so that the tongue no longer oral fills the mouth. The tongue attains motility in the larger oral cavity. Mature sucking is characterized by separate movements of the tip, tongue and gum pads or teeth.

The neonates cannot usually tighten lips around the areola of the breast, so some milk leaks out at the corners of the mouth. The infant swallows air and so has 'wind'. As the infant grows, he is able to completely approximate lips to the areola and so he gets less wind.

Cup-feeding

In the first 4 months of life, infant tends to push food out if food is placed on the front of the tongue. Therefore, food should be placed towards the back. Babies can very well place their lips to the rim of a cup by 4 or 5 months and cup feeding at this time is likely to be successful (Figure 10.3).

Chewing

The child begins to chew at 6 months together with the ability to hold, grab and transfer objects from one hand to another. About this time, teeth begin to appear, and infant begins to handle 'finger foods' with some dexterity. Now a child is ready to feed himself with a biscuit. Dry toast slice or biscuits offer hours of enjoyment. All full-term infants can chew by the age of 7 months.

Figure 10.3: Feeding the Child with Cup and Spoon.

Spoon-Feeding

At any time after 6 months, the infant begins to hold spoon and some can feed very well with the spoon by 9 or 10 months, though the average time for many children for feeding themselves with the spoon is about 15 months. In the early months of self-feeding, the fingers go into the food, and much is split. Feeding time may be messy. Food is seen as something not only to eat but also feel, smear and drop. Children can manage a knife and fork by the age of 2 ½ to 3 years. Food can be used to learn colours, shapes and textures. Independent feeding plays a role in the development of motor skills.

Elimination

Babies usually empty the bowel or bladder immediately after a feed. They can be conditioned at any age (*e.g.* at a month or so) to eliminate when placed on the potty. Voluntary control begins to develop by 15-18 months of age. By the age of two to 2 ½ years, baby pulls his pants down and climbs on to the lavatory seat without help. Bowel control is usually achieved before control of the bladder.

Assessing Infant Growth and Development

The components of assessment of a child before he can be labeled normal are as follows:

- Growth
- Development
- Protection against diseases (immunization)

- ✰ Intake of optimum nutrition
- ✰ Absence of disease

In this chapter, only assessment of growth and development shall be discussed which is as follows:

Growth

The usual measurements for assessing the growth are weight, length or height, head circumference and mid arm circumference. Health professionals monitor a child's increase in height, weight and head circumference, and interpret the records using growth charts. All these parameters are useful indicators of nutritional health. The typical charts contain seven per centile divisions, which represent 90 per cent of children (Figure 10.4). A per centile simply represents the rank of the person among 100 age – and gender-matched peers. If your baby, for example, is at 90th per centile height for age, this description means that if you measure 100 children of the age, 10 children will be taller than your baby, 89 will be shorter. A child at the 50th per centile is considered average. Fifty children will be taller than this child, while 49 will be shorter.

The two growth charts that give the most useful information are height for age – which assesses linear growth and weight for height – which assesses weight status based on height. For children under 3 years of age, head circumference for age is also a useful measure to growth. Since children less than 2 to 3 years of age measured stretched out, lying on their backs, the term length is used instead of height.

Individual growth charts are available for males and females, for ages ranging from 0 to 36 months. Growth of infants and children should be plotted regularly during their health check ups. It takes one to three years for an infant to establish his or her 'genetic' per centile. Once this figure is established such as length- (height) - for-age, the child's measurement should then track along that per centile. If the child's growth does not keep up with his or her per centile, the physician needs to investigate whether a medical or nutritional problem is impeding the predicted growth.

When under nutrition exists, gain in weight is affected earlier and to a greater extent than gain in height. If malnutrition persists, height will eventually be retracted or halted and/or pubertal maturation and epiphyseal closure will be delayed. In general, weight gain is decreased when energy intake is deficient, while linear growth is more delayed when protein intake is inadequate. Over nutrition usually results in taller, heavier, more physically mature children.

Development

Assessment of development involves both a recall of milestones as well as demonstration of the present developmental level. All the four parameters namely

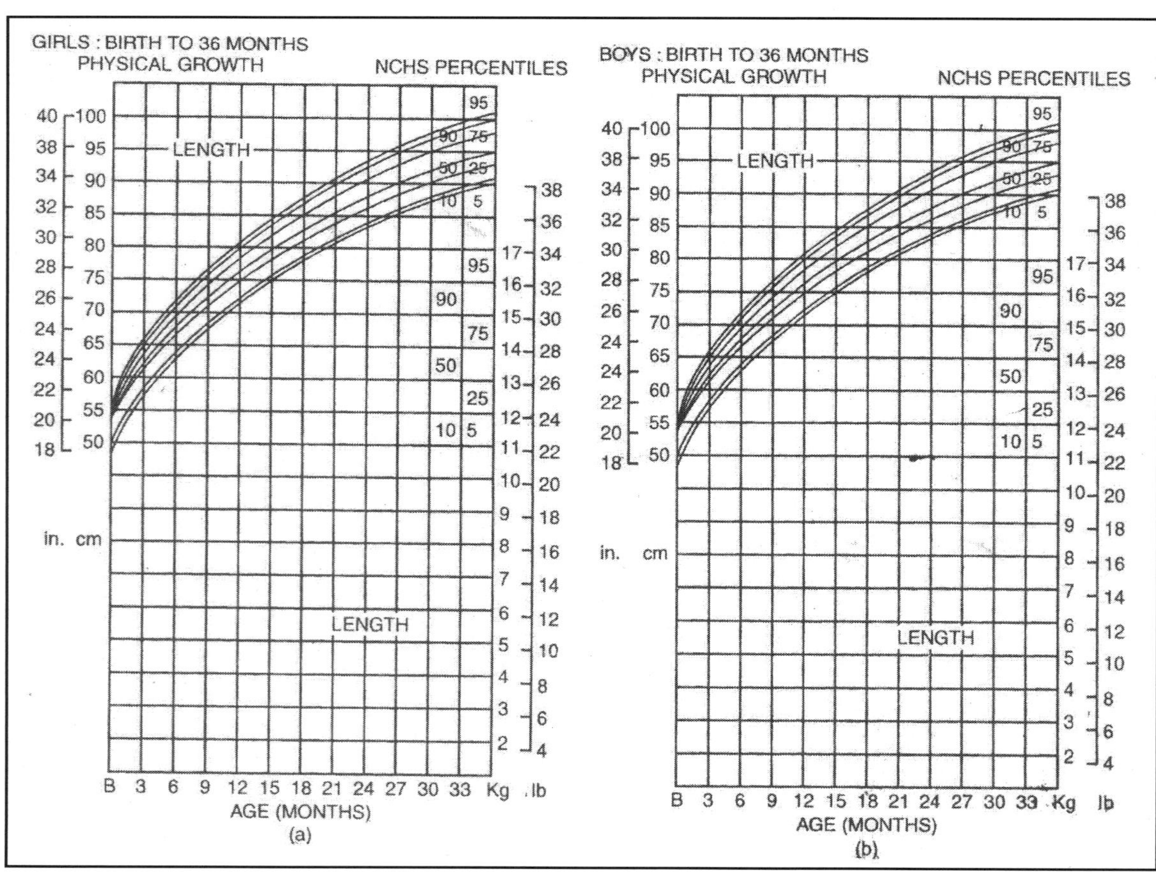

Figure 10.4: Growth Charts Used to Assess Height (Length), Weight and Head Circumference in Young Boys and Girls.

gross, fine motor and adaptive skills, language and social development require evaluation. It must be kept in mind while evaluating that considerably wide range of time of acquisition of developmental milestones exists.

Nutritional Requirements of Infants

Nutrient requirements vary as we progress through different stages of growth. Due to rapid growth during the first year, infancy is one of the most critical periods in the life cycle as far as food is concerned. Nutrient needs in relationship to size are high, and optimum nutrition at this time is very important to health and vigour throughout life. Many mothers feel insecure with their first babies and are concerned about what and how to feed the infant. Nutritionists should advise and help the mother develop skills in this area. Let us explore this critical stage of life and the key role nutrients play in this stage.

Our development is influenced by the genes we are born with and the environment we are exposed to (such as nutrients). The newborn infant is totally dependent on nutrients provided by others. Balance in the amount and the types of nutrients ingested is very important. Too much or too little is not proper for normal growth and development. The very precise amounts of individual nutrients required for infant's optimal growth are not known. Requirements are determined by a number of factors such as age, sex, rate of growth, physical activity, climate, temperature and health, and the extent of deposits or body fat. The requirement for different nutrients required during infancy is discussed as below:

Energy

Energy is required for function of the brain, liver, kidney, and other organs and for the body to perform physical activity. Energy requirements are the greatest in

Table 10.3: Energy Requirements of Infants

Age (Months)	*Weight (kg)*	*Weight gain (g/d)*	*Total Energy Expenditure[a] (kcal/d)*	*Energy Deposition[b] (kcal/d)*	*Daily Energy Requirement[c]*	
					Kcal/d[d]	*Kcal/kg/d[e]*
Boys						
0-1	4.58	35.2	306	211	520	115
1-2	5.50	30.4	388	183	570	105
2-3	6.28	23.2	457	139	600	95
3-4	6.94	19.1	515	53	570	80
4-5	7.48	16.1	565	45	610	80
5-6	7.93	12.8	603	36	640	80
6-7	8.30	11.0	636	17	650	80
7-8	8.62	10.4	664	16	680	80
8-9	8.89	9.0	688	14	700	80
9-10	9.13	7.9	710	21	730	80
10-11	9.37	7.7	731	21	750	80
11-12	9.62	8.2	753	22	780	80
Girls						
0-1	4.35	28.3	286	178	460	105
1-2	5.14	25.5	356	161	520	100
2-3	5.82	21.2	416	134	550	95
3-4	6.41	18.4	469	68	540	85
4-5	6.92	15.5	514	57	570	80
5-6	7.35	12.8	552	47	600	80
6-7	7.71	11.0	584	20	600	80
7-8	8.03	9.2	612	17	630	80
8-9	8.31	8.4	637	15	650	80
9-10	8.55	7.7	658	18	680	80
10-11	8.78	6.6	679	15	690	80
11-12	9.00	6.3	698	14	710	80

Daily energy requirement (kcal/d) is calculated from linear regression analysis of total energy expenditure on weight, plus allowance for energy deposition in tissues during growth.

a: TEE (kcal/d)= -99.4=88.6 kg; b: Weight gain energy accrued in normal growth; c: Requirement = total energy expenditure + energy deposition; d: Rounded off to the nearest 10 kcal; eRounded off to the nearest 5 kcal.

Source: Joint FAO/WHO/UNU. 2004. Joint Expert Report on Energy and protein requirements.

the newborn infant and show a gradual decline during childhood and adult life. Growth of infant depends on the availability of adequate calories for tissue building and energy expenditure. Energy is derived from fats, carbohydrates, and proteins in different amounts.

Daily energy needs in infancy are about 92 to 80 kcals per kilogram per day (Table 10.3). Based on body weight, this is proportionately 2 to 4 times the energy needs of an adult. Infants need an easy way to meet their high-energy needs. Either human milk or formula is ideal for this in the first few months (Table 10.3). Both are high in fat and supply about 700 Kcals per litre. If a baby is breast-fed, it is estimated that 850 ml of mother's milk daily will provide the energy needs of the first few months of life. Later human milk or formula, with solid foods, can provide even more energy.

The high energy needs of the infants are due to rapid growth and a high metabolic rate. Infants need to be provided with calories throughout the day because of their limited intake capacity. If energy needs are not met, infants grow at slower-than-normal rates and may lose weight. Excessive calorie intake may result in an overweight baby.

Protein

The average protein requirement (APR) during infancy and childhood is calculated as the sum of maintenance requirement plus protein deposited.

APR = Maintenance + (Deposition/Efficiency of utilization)

Protein requirement for maintenance in young children is 0.58 g/kg/d which increases at 6 months to the adult value of 0.66 g/kg/d. The efficiency of protein utilization for growth has been assumed to be 66 per cent in case of infants below 6 months and 58 per cent for infants above 6 months of age. The safe level (exceeding the requirement of 97.5 per cent of the population) is then estimated assuming that the requirement follows a long normal distribution *i.e.* safe level is the average level plus 1.96 SD, with total variability of maintenance and deposition calculated from the root mean square of CV of 12 per cent for maintenance (as used in case of adults) and for 24 per cent for the protein deposition rates between 0-2 y and observed values for the older children. Safe intakes of protein by the Indian Infants in terms of good quality protein, proposed by FAO/WHO/UNU are given in Table 10.4 and these are further corrected for Indian cereal-pulse-milk based balanced diets having PDCAAS 77.4 per cent. As per ICMR (2011) recommendations, infants require 1.16 and 1.69 g protein per kg body weight during 0-6 and 6-12 months, respectively.

One litre of formula or human milk contains about 13 grams protein (Table 10.5a). About 40 per cent of this protein should be from essential amino acids. Total protein intake should not exceed 20 per cent of kcal needs. Feeding having less than 1 g/kg/day of protein leads to manifestations of protein deficiency. Severe protein deficiency continuing over an extended time results in the cessation of growth. On the other hand, excess protein supply may often cause high fever and azotemia. Excess nitrogen and minerals supplied by high-protein diets would overtake the infant's kidney functions.

Table 10.4: Safe Level of Protein Intake (g protein/kg body wt/day) for Infants Aged 6-12 Months

Age (Months)	*Maintenance Requirement*[a]	*Growth Requirement*[b]	*Average Requirement*[c]	*Safe Level*[d] *1.96 x SD*
1	0.58	0.83	1.41	1.77
2	0.58	0.65	1.23	1.5
3	0.58	0.55	1.13	1.36
4	0.58	0.49	1.07	1.24
6	0.58	0.40	0.98	1.14
12	0.66	0.29	0.95	1.14

a: Calculated from N balance studies with milk/egg; b: Protein deposition rates taken from Butte *et al.*, 2000. Pediatr. Res. 47:578-585 and adjusted for 66 per cent efficiency of utilization (from N balance studies with milk/egg); c: Sum of maintenance and protein deposition rate (latter adjusted for efficiency of utilization); d: Mean + (1.96 x root mean square of SD values for protein deposition during growth, adjusted for efficiency of utilization and the maintenance).

Table 10.5a: Distribution of Calories in Human Milk

Nutrient	*Calories Derived (Per cent)*
Carbohydrate	45-55
Protein	9-15
Fat	35-45

Carbohydrates

Carbohydrates form bulk of the diet and supply most of the body's energy needs. In its deficit, body uses proteins and fats for energy. It is stored in the body chiefly as glycogen in the liver and muscle. Because the size of infant's liver is 10 per cent of that of the adult's liver and the muscle mass is 2 per cent, the infant's glycogen reserve is much smaller (about 3.5 per cent of that of the adults).

Carbohydrates are consumed in different forms such as glucose, fructose, galactose, lactose, sucrose, starches, dextrin, and glycogen. Pentoses are poorly absorbed. Lactose from mother's milk forms the major source of carbohydrate for the infant with fructose and sucrose introduced later with fruits, and starches taken through cereals and breads. Glucose which is not used immediately on ingestion or stored as glycogen, is converted to fat for storage. A prime advantage of carbohydrate for energy production is that protein is spared for other vital functions. Human milk contains 6.5 to 7 per cent of lactose. About 10 per cent of the carbohydrate in human milk consists of polysaccharides and glycoproteins.

Fat

Fats are an important part of the infant's diet because they are energy-dense. This characteristic helps to resolve the dilemma of high needs but small stomach capacity. Upto 6 months of age, fat should make up about 30 to 50 per cent of an infant's energy intake. More than that may lead to poor fat digestion. The fat content of human milk is relatively constant at 3-4 per cent by weight and provides 50-60 per cent E to infants upto 6 months of age which is sufficient to provide energy density. Approximately 4-5 per cent of total energy in human milk is present as linoleic and linolenic acid, 1 per cent as long chain PUFA derived from these essential fatty acids. Hence, about 6 per cent total energy is supplied by EFA. Essential fatty acids should make up at least 3 per cent of total energy. Upto 6 months of age, breast milk is sufficient to meet the infant's requirements for saturated fatty acids, monounsaturated fatty acids, total PUFAs (n-6 + n-3) and EFAs Human milk substitutes/infant formulas should have fat and individual fatty acid content (including AA and DHA) similar to the levels in human milk.

From age 6 months to 2 years, fat intake should be reduced gradually, depending on the physical activity of the child, to 35 per cent E. The mix of fat from breast milk and complementary foods should provide infants with at least 3-4.5 per cent E from linoleic acid (upper limit of ADMR 10 per cent E) and 0.4 - 0.6 per cent E from ALA (upper limit of AMDR 3 per cent E). Assuming 10 per cent E fat from breast milk and/infant formula plus fat from complementary foods (except visible fat), ICMR (2011) has recommended 19g visible fat per day for infants of 6-12 months of age. This level of visible fat would provide adequate energy density in the child's diet and contribute to prevention of chronic energy deficiency and also ensure adequate supply and absorption of fat soluble vitamins. The provision of dietary sources of DHA to infants should come from human milk/infant formula and complementary foods. *In-situ*ations where breast milk is low, the level of fat, LA and ALA in complementary foods should be increased so as to achieve the recommended intake. (Table 10.5b).

Minerals

Minerals are not mobilized in the body; some of them function as cofactors in many biochemical reactions. Major (macrominerals) elements whose requirements are 100 mg or more include sodium, potassium, calcium, magnesium, phosphorus, and chloride. The trace elements are required only as few milligrams and include iron, iodine, fluoride, zinc, and copper. Most minerals are readily available from a well-balanced diet. Human milk contains considerable amount of these minerals. As many as 80 minerals are used in many diverse functions in the body. In infancy, calcium, phosphate, and magnesium are needed for bone growth.

Calcium and Phosphorus

Calcium is necessary for muscle activity, nerve functions, heart activity, blood coagulation, and production of milk. A large percentage of calcium is reabsorbed in a breast-fed baby. The rate of bone growth of the breast-fed baby is excellent. Infants require 500 mg calcium and 750 mg phosphorus per day up to 1 year of life (Ratio of calcium and phosphorus should be 1:1.5). Deficiency of calcium leads to poor mineralization of teeth and bones, rickets, and impaired growth. Phosphorus is a constituent of bone and teeth; plays a role in energy transfer and in metabolism of cell. Breastfeeding meets the calcium and phosphorus requirements of the infant.

Iron

Iron and copper are needed for haemoglobin formation (in red blood cells) and essential for many enzyme systems. Considering the breast milk intake to be around 600 ml during this period with iron content of 0.78 mg/L, the iron intake per day is about 0.47 mg/d. It is estimated that infants can use greater than 50 per cent of the iron in breast milk and healthy full term infants are born with supply of iron stored in the bone marrow, liver and spleen which can last or 4-6 months. Iron in human breast milk is well absorbed by infants. It is therefore, recommended that infants be exclusively breast fed for the first 6 months of life.

Iron requirements increase markedly, especially in relation to body size and energy intake during later 6 months of life. Infants of 6-12 months of age require

Table 10.5b: Recommendations for dietary fat intake for Indian infants (ICMR, 2011)

Age Group	*Physical Activity*	*Minimum Level of Total Fat (Per cent E)*	*Fat from Foods Other than Visible Fat[a] (Per cent E)*	*Visible Fat[b]*	
				Per cent E	*g/p/d*
Infants	0-6 months	40-60	Human milk[c]		
	7-24 months	35[d]	10[e]	25	25

a: If higher than 10 per cent E, visible fat requirement proportionately reduces; b: Cooking oils, butter, ghee and margarine; c: Infant formulae/milk substitutes should mimic contents of fat and fatty acids in human milk including arachidonic and docosahexaenoic acid; d: Gradually reduce depending upon the physical activity; e: Human milk/infant formula+ complementary foods.

46μg iron/kg/d (ICMR, 2011). Breast-fed infants should be given iron enriched complementary foods to supply extra iron after 6 months of age. After 6 months of age, an excessive milk intake but with little solid food is conducive to cause iron deficiency anaemia in infants. The gradual introduction of egg yolk, green leafy vegetables, meat and cereals at 6 months prevents iron deficiency anaemia in infants. Formula fed infants should use an iron-fortified formula.

Iodide and Zinc

It is important for an infant to receive adequate amount of iodide and zinc to support growth. Human milk and formula supply these needs if the infant's energy needs are met. In human milk, a protein binds with zinc and improves its availability in neonate nutrition. In addition to its role in enzyme system, zinc plays an important part in growth process.

WHO-UNICEF-ICCIDD has recommended 90 μg iodide/d for infants of 6-12 months of age which ICMR (2011) has also adopted. It should be met through breast milk and complementary foods. For zinc, ICMR (2011) has made no recommendations for infants from 0-12 months of age.

Sodium

The main determinants of sodium requirements of infants are rate of growth, composition of faeces and losses through skin. Sodium concentration of fat free tissues after birth is 90 mmol/kg of lean body mass. Obligatory urinary losses and cutaneous and faecal losses are 2 mmol/day and added growth needs to arrive at an estimate of minimal sodium requirements of infants and young children 2.5 mmol/day. Infant needs of sodium and replacement losses could be generously met with intakes of 4-8 mmol/day.

The sodium content of breast milk is higher immediately after delivery (65-75 mmol/L) and falls by day there. Mature human milk contains 7 mmol/L. Thus, breast milk is adequate for infant growth, even if dermal losses are higher. Formula fed infants in general have a high intake (10-50 mmol) from 2 months to 10-12 months. Hence, there is no need for extra sodium intake by infants and ICMR (2011) has given no recommendation for it.

Potassium

Newborn infants contain 50 mmol of potassium per kg fat free body weight. Obligatory urinary and cutaneous losses are the major determinants of potassium needs. The need for growth has led to a minimal potassium requirement of 2.3 mEq/d (90 mg) for infants and children.

Potassium content of human milk is greater in early milk (19 mmol/L) than mature milk (15 mmol/L). The ratio of sodium to potassium is 1:2 in mature milk. Cow's milk contains 38 mmol/L and commercial formulae have around 18 mmol/L. This requirement is fulfilled by breast milk and complementary foods up to 12 months. Therefore, ICMR (2011) has made no recommendations for Na and K for infants.

Vitamins

The requirements for various vitamins during infancy have been given in Table 10.5c.

Table 10.5c: Vitamin and Mineral Requirements during Infancy (ICMR, 2011)

Vitamins	*Infants (months)*	
	0-6	*6-12*
Vitamin A (μg/day)	-	350
Thiamine (μg/kg/day)	55	50
Riboflavin (μg/kg/day)	65	60
Niacin (μg/kg/day)	710	650
Pyridoxin (mg/day)	0.1	0.4
Ascorbic acid (mg/day)	25	25
Dietary folate (μg/day)	25	25
Vitamin B_{12} (μg/day)	0.2	0.2
Calcium (mg/day)	500	500
Iron (μg/kg/day)	-	46
Magnesium (mg/d)	30	45
Zinc (mg/d)	-	-

The vitamin contents of milk may vary with the maternal intake. Breast milk may have adequate amount of vitamin C, if mother consumes vitamin C rich foods, and enough amount of vitamin D if she is well exposed to sunlight. Mother's milk contains adequate amount of vitamin A and the B complex vitamins for the nutritional needs of infants in the first year of life. The vitamins present in the human milk include vitamins A, D, E and K and riboflavin, niacin, pyridoxine, pantothenate, folicin, vitamin B_{12}, and vitamin C. Vitamins are organic compounds required in small amounts to catalyse cellular metabolism. They are usually not formed in the body and must be included in food offered to the infant during weaning to help in their normal growth and body activity.

Water

Water is important for infants. There is heavy water loss through the lungs, and skin, and the kidneys have reduced capacity to concentrate urine and save water. Water is essential for existence; a deficit of it results in death in few days. Water comprises 70-75 per cent of the infants' body weight and it is relatively higher than that of adults. The infant must consume much larger amounts of water per unit of body weight than the adult, but when calculated per unit of calorie intake, the amounts required are identical. An infant requires 1.5 millilitres of water per Kcalorie daily. This corresponds to 150 millilitres per kg of body weight per day. This is needed for growing tissues, and high fluid turnover, and to compensate for low concentrating ability of the kidney. Water requirements of low-birth-weight infants are estimated at 85-170 ml/kg/day. Exclusive breast milk feeding for 6 months usually meets this need of the infant.

The usual complementary foods fed to infants (6-12 months) contain large amount of water; most of the solid foods given to children contain 60-70 per cent water, and many of the fruits and vegetables contain 90 per cent water. In hot climates, supplemental water may be necessary. In addition, diarrhoea, vomiting and fever also often lead to a need for supplemental water. It is important to note that an infant is easily dehydrated. Dehydration can result in a rapid decrease in kidney function and the infant may then require hospitalization for rehydration.

Diet and Feeding Pattern

The famous paediatrician Paul Gyorgy said, "Cow's milk is the best for baby cows and human breast milk is the best for human babies". No one can deny the truth of this statement. Human milk is tailored to meet infant needs for the first 6 months of life. A vast body of research from all over the world sustains the recommendation that only breast milk be fed to infants for the first six months of life.

In India, majority of the mothers initiate breastfeeding but duration of lactation is variable, being longest in the rural areas and far less in the urban areas, especially among the upper social classes. The progressive decline among the latter is a cause of great concern not only for the group itself but because they are often the trend setters for the less privileged who can ill afford to imitate them. Therefore, there is a dire need that the nursing mothers belonging to all strata should be appraised of some of the advantages of colostrum, breastfeeding and provided closer and more continued support in order to sustain lactation.

Advantages of Breastfeeding

The advantages of breastfeeding over bottle-feeding and the reasons why it is so strongly recommended are as follows:

1. Convenience

Breastfeeding is convenient; the food is readily available for the infant, and no special preparation or equipment is needed. Hence, breastfeeding frees the mother from time and expense involved in buying and preparing formula and washing bottles. Breast milk is 'already prepared' and sterile. This allows the mother to spend time with her baby.

Anti-infective Properties

Breastfeeding reduces the general risk of infections to the infants (Table 10.6). This is partly due to the antibodies present both in colostrum and mature breast

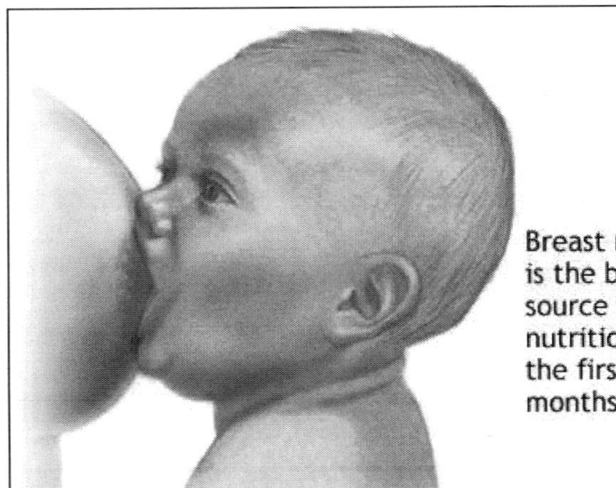

milk that an infant can use. As already mentioned previously, these antibodies reduce the risk of respiratory and intestinal infections. Ear infections are also reduced because infants cannot sleep with a breast in the mouth, as bottle-fed infants often do. While an infant sleeps with a bottle in its mouth, milk pools in the mouth and backs up through the throat into the eustachian tubes and into the ears. This creates a growth media for bacteria. Infant ear infections are a common problem that parents want to avoid to decrease discomfort for the infant and trips to the doctor and to prevent possible hearing loss.

2. The breast milk is fresh and free of bacteria, and thereby reduces the incidence of diarrhoea.

Table 10.6: Nutritive and Anti-infective Features of Human Milk

Nutritive features (compared to cow's milk)
✰ Lower protein with appropriate casein whey ratio of 40:60
✰ Higher NPU
✰ Low curd tension of proteins
✰ High lactose content-promotes growth of probiotic organisms
✰ Presence of medium chain triglycerides
✰ Better availability of iron-adequate for the first few months
✰ High calcium-phosphorus ratio-better bioavailability of calcium
✰ Low solute load and low buffering capacity – less load on kidney
✰ Vitamins adequate for the first few months if maternal diet is well balanced and nutritious
✰ Higher ascorbic acid content
✰ Higher cystine content
✰ Anti-infective factors
✰ Secretory IgA
✰ Lactoferrin
✰ Lysozymes
✰ Complements (C 1-9)
✰ Interferon
✰ Lipids (unsaturated fatty acids)
✰ Lactoperoxidase
✰ Cells (T and B Lymphocytes and macrophages)

3. Economical

Breastfeeding is more economical than bottle feeding, which involves costs for infant formula or cow's milk, the bottles and teats and the fuel necessary for sterilization.

4. Reduced Risk of Allergy and Intolerance

Allergy and intolerance caused in some babies to bovine milk with significant gastrointestinal, skin and respiratory disturbances are not found in the breastfed infants. Colic and atopic eczema are less common in infants getting mother's milk.

Hence, breastfeeding reduces the incidence of allergies, especially in allergy prone infants. Cow's milk contains a number of potentially allergy-causing proteins that are absent from human milk. Breastfeeding also avoids the possibility of infant intolerance to formulas.

5. Normal Psychological Development

Breastfeeding has been referred to as the period of 'exterior gestation' because it provides continuity with the intrauterine environment while providing nourishment and security. Breastfeeding promotes strong emotional ties while meeting the infants' most basic physical need.

Breastfeeding stimulates most of the senses by close, cozy and warm maternal body contact. Close body contact gives opportunity and chance to the baby to recognize his/her mother's unique body, smell and to hear the throbbing of her heartbeat; the visual "en face" position allows eye contact between the mother and child. Hence, there appears to be a certain amount of warmth, nurturing, and skin contact inherent in breastfeeding situations, which facilitates the development of mother-infant bonding. However, claims that there is a relationship between breastfeeding and bonding, have been disputed. There is no scientific data suggesting that maternal-infant bonding is superior in breast-fed infant, or is there 'proof' that psychological development is different between breast-fed and formula-fed infants.

Through the year, researchers have tried to find a relationship between infant feeding methods and later psychological functioning. Strong evidences that breastfeeding results in superior cognitive performance and personality adjustment has not been forthcoming. Moreover, some studies have reported that breastfeeding is not associated with performance on tests of vocabulary and visual-motor coordination at five.

6. No Risk of Formula Contamination and Dilution

Breastfeeding is sterile and is not subject to contamination. While mixing formula for bottle feeding, there is a possibility of using contaminated water or wrong amount of water. These types of problems may not occur very often among well-informed mothers with adequate resources. However, the proper use of bottle milk is a problem of major concern in under-developed countries where bacterial contamination, under-dilution, or over dilution of formula mixture, existence of unhygienic environment, use of unclean utensils, bottles and teats used for preparing formula mix and feeding *etc.*, are more to occur, thereby leading to diarrhoea and poor health of the infant.

7. Optimum Nutrition

Breast milk provides a proper balance and quantity of nutrients ideal for the human infant (Table 10.6). Breast milk proteins and fats are easier for the baby to digest and absorb. In addition, the waste products are more easily processed by the immature kidneys of the infant. Human milk is used as the standard for determining the nutritional needs of the infant for the first six months of life.

8. Maternal Benefits

Breastfeeding causes the uterus to contract after pregnancy, facilitating return to its normal size and reducing the danger of haemorrhage. The stimulation of nipples and resultant secretion of prolactin suppresses ovulation in many women, particularly when supplements and solids are not offered. This is believed to be the major significance in promoting short-term child spacing in the developing countries. When solid foods become a major source of nourishment, nipple stimulation and prolactin levels decrease in the lactating mothers, and ovulation and menstruation usually begin again. It has also been suggested that breast cancer is rarer in countries in which breastfeeding is common. Researchers have found indications that prolonged lactation reduces the risk of breast cancer. However, the evidence is not sufficiently strong at this point to advise women to breastfeed specially to decrease the risk of cancer.

Another prime benefit expressed by most women who breastfeed is the ease with which feeding is managed, particularly at night.

Frequency and Duration of Breastfeeding

Breastfeeding can be commenced as soon as possible, preferably soon after delivery. Give the baby colostrum. Colostrum, the first secretion of the mammary gland, is of great value to the baby. It has superior nutritive value in protein and in some of the minerals and vitamins. Colostrum is also rich in enzymes and immune bodies, all of which play significant roles in the growth and health of newborn infants.

Almost all experts now agree that the infant should be exclusively breastfed for the first 6 months. No other food or fluid, such as cereal, water, butter, honey, glucose water, juices or any other concoction are needed for a baby getting adequate breast milk. A normal infant has nutrient stores *e.g.*, liver glycogen for this period of temporary 'starvation'.

Much research has been focused on the frequency and duration of breastfeeding during the first few days of life. The rigid 3 or 4-hour schedules once in vogue have been clearly shown to be detrimental to the establishment of successful breastfeeding.

Demand feedings have been shown to have a beneficial effect on milk supply. The epic poem "Song of Lawino", by the Ugandan poet Okot p' Bitek, praises breastfeeding on demand, in sickness and in health and satirizes the mostly western practice of regulated feeding, now widely acknowledged as harmful:

When the baby cries
Let him suck
From the breast
There is no fixed time
Breastfeeding.
When the baby cries
It may be he is ill:
The first medicine for a child
Is the breast,
Give him milk
And he will stop crying

Let the baby suckle on-demand. Give the breast when baby cries and feed also at night. If newborn sleeps with mother, he may suckle as often during the night as during the day. If baby sleeps in a separate cot, pick up and feed when he cries. Suckle both breasts for only a few minutes each time during the first few days to prevent nipples sore and cracks. Infants who cannot be fed on-demand should be brought to the mother every 3 hourly during the day and every 4 hour during the night.

Many infants are hungry within 2 hours of a satisfying feed, and about 75 per cent of the breast milk is filled up again for supply. A hungry baby usually empties the first breast in a few minutes. Most mother-babies settle down to a routine in which infant suckles about 5 to 10 times a day. A baby feeding on-demand may suckle between 12 to 15 times in 24 hours in early days.

Infant must be allowed to suckle when hungry, whether or not there is any milk. Early establishment of normal vigorous nursing by letting the infant empty the breast frequently during the time when only colostrum is available is very important. Most potent stimulus to the production of breast milk is regular and complete emptying of the breast; milk secretion is reduced when the secreted milk is not emptied. The first 2 weeks of neonatal period are very important and crucial for establishing successful breastfeeding.

Feeding on demand reduces the incidence of engorgement, results in better weight gain by the infant and reduces the frequency and severity of nipple discomfort. Attention to "demand" is essential if the "supply" is to be increased.

In order for the self-demand to be successful, mother must try to recognize hunger symptoms and she must be definite that the baby is really satisfied at the end of nursing. Most infants want to be fed five or six times in 24 hours and that, on the "self-regulation" schedule, except during the late afternoon and evening when the demand is likely to be more. On such a self-demand schedule most babies double their birth weight in less than 5 months and are neither fat nor flabby. There is remarkable decrease in fussing and crying.

The self-demand plan may not work for premature and sickly infants. The length of the nursing time varies; the baby usually takes 15 to 20 minutes to nurse. If a period of nursing on self-demand exceeds 30 to 40 minutes, or a demand to be fed is more than seven or eight times a day, it is usually an indication that the supply of milk is not sufficient. It is well established that sucking by the infant initiates neuroendocrine reflex that supports lactation and milk ejection. The hormones involved include prolactin for milk secretion and oxytocin for milk ejection. In lactating woman, prolactin appears to be the single most important hormone for initiation and maintenance of milk secretion.

As far as duration of breastfeeding is concerned, it can be continued as long as possible provided that there is a fair amount of milk production and the breast is not acting merely as a 'pacifier'. The tradition in most parts of the world is to breastfeed for 2 to 3 years. Prolonged breastfeeding without additional semisolids is insufficient by itself for the rising needs and declining stores of the rapidly growing infant. Many young children spontaneously give up breastfeeding at about 1 year of age or soon thereafter, especially if they are receiving adequate supplementary foods. There is a data to indicate that the occurrence of malnutrition among infants on prolonged breastfeeding is due to inadequate supplementary semisolid foods, which must be introduced in the diet of infant when he is 5-6 months old. According to WHO recommendation, breastfeeding should be continued till 2 years of age of the child along with complementary feeding.

Establishing the ideal duration of breastfeeding should be carried out by the individual mother-child couple. Following factors should be taken into consideration:

- ☆ The convenience of the mother
- ☆ The physiological as well as psychological needs of the infant
- ☆ The availability of satisfactory alternative or supplemental feeds
- ☆ Customs in the community

Breastfeeding the Premature Infant

There are substantial grounds to prove the nutritional, psychological, and immunologic benefits to using a mother's breast milk to feed her premature infant. Of course very small preterm infants do not gain weight as early on breast milk, because the content of protein is less in breast-feed for the need. For this reason it is advised to supplement breast milk with proteins and some electrolytes as well to match an infant's rapid growth. But the quality of the protein in mother's milk is most suitable for her premature infant. Infants have fewer metabolic problems if they are fed mother's milk (azotemia, hyperaminoacidemia, and metabolic acidosis). Prematures, fed on breast milk have much lower incidence of infections and complications due to the protective and immunologically active components of breast milk (lysozyme, lactoferrin, lactoperoxidase, complement, leukocytes, and specific antibodies). Most important advantage is the psychological benefit for the mother. Mother is upset with such a baby. By giving her own milk, the mother gets actively involved in the infant's care, which gives her pleasure and satisfaction.

The mother of premature infant has to be advised and instructed regarding the various methods of milk expression, collection, and preservation. Hand expression or pumps could be efficiently used in collecting breast milk. Electric pumps could be used, if available. Sometimes decreasing the interval between milk expression and using oxytocin maintains an adequate milk supply. Breast milk may be refrigerated for 24 hours at 1? to 5?C or frozen at -18?C to -23C for up to 3 months.

Preterm babies are easily infected by bacteria. They have thin skin and no fat under it to keep them warm. They grow fast and need high nutrition. But their sucking and swallowing is very weak and they have small stomach. Milk may easily get into their lung, but they cannot cough it out. He will probably survive if he is fed well, kept warm and infections are prevented. Small frequent feedings are essential in order to compensate for the very small stomach capacity, and to prevent hypoglycemia. Vomiting and regurgitation are common as a result of weak cardiac sphincter and the small stomach. They have to be managed in the intensive care nursery.

Early postnatal malnutrition may permanently impair brain growth. Most preterm requires 84 to 142 Kcal/kg/day. Protein intake of 2.25 to 2.75 g/day is adequate but protein must provide sufficient amount of all amino acids. Thus, the quality of protein is most important in feeding low-birth-weight infants because they lack certain enzymes to metabolize certain amino acids. The administration of multivitamin preparation is begun early.

Babies who are able to suck on the nipple should be fed every 3 hour and tube-fed babies every 4 hour. Burping should be done if any distention is noted. The nipple method is usually used for babies over 1500 g who can suck without tiring. Bottle-feeding should not exceed 20 minutes.

In other cases, special feeding problems may prevent uses of human milk or necessitate some supplementation with formula. Sometimes even total parenteral nutrition is the only option. Working as a team, the pediatrician, neonatal nurses and registered dietitian should guide the parents in this decision.

Breastfeeding and AIDS/HIV Infection

The most common route for transmission of the Human Immunodeficiency Virus (HIV) to the infant is from an infected mother across the placenta. Most of the women bearing children who develop AIDS or have evidence of HIV infection, but the woman transmitting the virus to the foetus need not be having symptoms of disease during pregnancy. The exact mechanism of transmission from the mother to the foetus or infant is not known. Transmission could occur *in utero* through passage of the virus across the placenta; around the time of delivery through exposure to vaginal secretions, ingestion of maternal blood or maternal foetal transfusion during labour and delivery; and in infancy through ingestion of the virus in breast milk. This virus has been isolated from breast milk but transmission from mother to infant through breast milk is currently thought to be relatively rare.

In many countries, HIV infection has been reported in 25 to 45 per cent of infants born to HIV mothers. About 80 per cent of women with AIDS are in their prime child bearing age. A pregnant woman with poor vitamin A status is more likely than others to pass the HIV infection to the foetus. According to consultation of the World Health Organization (WHO) and the United Nations Children's Fund (UNICEF), infectious diseases and malnutrition are the main causes of infant deaths and the infant mortality rate is high; breast feeding should be the usual advice to pregnant women, including those who are HIV infected. This is because their baby's risk of HIV infection through breast milk is likely to be lower than the risk of death from other causes if he is not breastfed. Under poor personal hygienic and environmental sanitation conditions, the many advantages of breast feeding far outweigh the risk to the infant of AIDS infection through breast milk from an HIV positive mother. Only where the common causes of morbidity and mortality in infancy are not infectious diseases, should public policy advise the use of bottle-feeding in place of breast-feeding to reduce the possibility of AIDS transmission.

All infants and children with AIDS show enlarged liver and spleen, lung infection (interstitial pneumonitis), and poor health. Parotid gland inflammation and recurrent bacterial infections are more common in such infants and children. Delays in acquiring motor skills are most common deficiencies. Those diagnosed with HIV infections after 1 year of age have better survival than those diagnosed earlier. Skilled nursing, good nutrition, good medical care during sickness, and psychological support are needed in caring for these children. These children should be immunized for poliovirus, diphtheria, pertussis-tetanus (DPT) vaccine and hemophilus type B conjugate according to recommended schedule. Passive immunization with immune globulin should be given to HIV infected children. These children should be breast-fed and given well-balanced weaning diet at appropriate time. Nutritional support, including the use of parenteral feedings, is important. Though AIDS virus is present in the milk of infected mother, breast-feeding is not contraindicated as it protects the baby from many infections and provides a nutritious food during the early month of life.

Contraindications of Breastfeeding

From maternal point of view there are few contraindications to breastfeeding. Inverted nipples, engorged breasts, and mastitis can be corrected and there is no need to stop nursing for these problems. Acute severe infection in the mother may contraindicate breastfeeding if the infant does not have the same infections. When the mother's condition improves, the breast may be emptied and the expressed milk given to the infant. In the following conditions, nursing is contraindicated:

- Septicemia
- Nephritis and eclampsia
- Severe haemorrhage
- Acute tuberculosis
- Typhoid fever and malaria
- Chronic poor maternal nutrition
- Debility and severe neurosis and postpartum psychosis

Weaning and Complementary Feeding

Breast milk alone is the ideal food for the infant for the first 6 months after which some semisolid foods need to be introduced to make up the short-fall because of the decline in the quantity of breast milk and rapid rate of growth of the infant. Introducing solids much earlier than this will provide no nutritional advantage and may do harm. On the other hand, excessive delay may result not only in malnutrition but also in unwillingness on the part of the infant to accept the new type of food.

Weaning

The World Health Organization (WHO/UNICEF, 1987) defines weaning as the process by which milk-fed infants gradually become accustomed to the range of foods characteristic of the society into which they are born. Weaning actually begins the first time the infant receives supplemental or solid nourishment. Many paediatricians and researches believe that the need for sucking and the sucking reflex persist well into the second year of life and the infants weaned from the breast before that time should be given a bottle. For parents, the term most likely means the process of taking away the breast or bottle. Many women wean the child from the breast and bottle when he can take liquid from the cup. Hence, weaning is difficult to be defined with precision because there are a variety of transitions that can be made. The process of weaning may include a partial transition from breast to bottle, from a bottle to cup, from liquids to semi-solids, and solids, and from caretaker – administered feeding to self-feeding.

Weaning does not imply stopping of breastfeeds. Milk will continue to be the major source of protein and fluid for the infant for several months yet. Therefore, breastfeeding should be offered before or at a separate time from solid food.

Age to Begin Weaning and Introducing Complementary Foods

The timing of introduction of semisolid, non-liquid strained and supplementary foods is another important issue for discussion.

Opinions about the time for addition of supplementary foods to the diet of the breast or formula fed infant have undergone marked changes over the last 80 years. At the start of the twentieth century, no solid foods were introduced until the end of the first year. In 1960-1970, the pendulum swung all the way to the opposite extreme of offering the infant semisolids during the first month of life. Concern that this early introduction of semisolid foods predisposed infants to obesity and allergic reactions caused many health care professionals in the paediatric community to reexamine the appropriate age for the introduction of these foods.

Various medical groups have expressed concern about possible harmful effects of introducing semisolid food at 1 or 2 months of age. The practice of feeding infants any food other than breast or formula milk before 6 months of age is strongly discouraged. The reasons for delaying the introduction of semisolids until that time are:

- The extrusion reflex (tendency to push food out to mouth) does not disappear until 3 or 4 months of age. It is about 5 or 6 months before an infant can sit up, turn the head away when full, and control tongue thrusting.
- The neuromuscular co-ordination necessary for the swallowing of semisolids is not fully developed until 16 to 18 weeks of age.
- Semisolid foods impose a higher eliminative load upon immature kidneys. The kidneys are not able to excrete waste products from high levels of dietary protein or minerals.

- The addition of solids may diminish the immune protection conferred by breast milk.
- The addition of solids increases possible sources of allergy reactions. During the first three months of life, absorption of whole proteins is possible but undesirable-through the intestinal tract. Whole protein absorption may set up the child for future allergies, including food allergies. It is best to minimize the number of different types of proteins in a child's diet during the first 3 months to limit chances of developing an allergy.
- Introduction of solids before the infant is able to indicate a desire for food probably represents a type of forced feeding.
- The early introduction of semisolid food may interfere with establishing sound eating habits and may contribute to overfeeding.
- There is no evidence that the addition of semisolid food helps babies sleep through the night. Sleeping through the night is a developmental milestone for the infant and has nothing to do with how much the infant eats before going to bed.
- Iron stores of the infant are exhausted by about 6 months of age. Either solid foods or iron supplements after 6 months are often needed to supply iron if the child is breast-fed or is fed a formula not supplemented with iron. No nutritional needs besides iron require solid foods before 6 months of age.

Keeping these considerations in mind *i.e.* the nutritional need, physiological and physical readiness, and allergy prevention, WHO recommends that solid foods not be introduced before 6 months of age. By that age, the infant is ready for solid food, and feeding is much easier.

In many Indian families, introduction of solid food is associated with a religious ceremony commonly known as "Anna Prashna" and it is performed when the infant is 6 month old. Health workers can take advantage of it to propagate the introduction and continuation of semi-solid food for young infants. However, in some communities, the ceremony is performed only at the stage of one year. In such cases, parents and elders in the family need to be educated and told of the risk of delaying administration of semisolid food. In case, traditions are deep rooted, it may be sensible to find out what is the exact food used in the ceremony. Frequently this is rice. Mothers can often be persuaded to give other food items to the infant and reserve whole rice for the ceremony. Respect for community customs where possible is helpful in winning over the confidence of the family.

Some mothers decide to introduce semisolids when the infant's teeth commence erupting. Again, while some infants cut their first tooth by 6 to 7 months, others may not have a single one until after their first birth day. Hence, it is essential to stress that teeth are not required for eating semisolid food, which is soft and mashed. Mothers may be cited examples of elderly persons without teeth who are able to eat soft foods. Parents should be persuaded to start feeding semisolids to their infants when they are 6 months old. It should neither be started too early nor delayed beyond 6 months.

Balanced Diet for Infants

Balanced diet for infants of 6-12 months is depicted in Table 10.7.

Table 10.7: Balanced Diet for Infants (Number of portions) (NIN, 2011)

Food Groups	*g/portion*	*Infants (6-12 months)*
Cereals and millets	30	0.5
Pulses	30	0.25
Milk (ml)	100	4[a]
Roots and tubers	100	0.5
Green leafy vegetables	100	0.25
Other vegetables	100	0.25
Fruits	100	1
Sugar	5	2
Fats and oils (visible)	5	4

[a]Quantity indicates top milk. For breastfed infants, 200 ml top milk is required

For non-vegetarians, one portion of pulse may be exchanged with one portion (50 g) of egg/meat/chicken/fish

For infants, introduce egg/meat/chicken/fish around 9 months

Order of Introduction of Different Foods

Table 10.8 gives suggested guidelines for the introduction of semi solid foods to normal infants. From a nutritional standpoint, the order of introduction of various foods appears to be relatively unimportant. However, parents or caregivers are advised to introduce one new single food ingredient at no less than weekly intervals. The purpose of the procedure is to identify specific food allergies as they occur.

Table 10.8: Introduction of Semisolid Foods and Table Foods in Onfancy

Foods		*Remarks*
Iron fortified cereals for infants	Add	
Vegetables	Add strained	Gradually delete strained foods, introduce table foods
Fruits	Add strained	Gradually delete strained foods, introduce chopped or well- cooked foods
Finger foods such as biscuits, toast, *chapati*	Add those that can be grasped	Increase the use of small foods as the pincer grasp develops

Foods		Remarks
Well cooked mashed or chopped table foods, prepared without added salt or sugar	Add	
Juice/Soup by cup	Add	

6-7 Months

Milk is the basic food of the infant as it provides a greater part of the required nutritional essentials. Other foods are added to the diet in order to provide dietary essentials lacking in milk and also to develop good food habits as food tastes are formed early in life. These foods include pre-cooked cereals, fruit juices, strained vegetables, fruits and meat, a source of vitamin D *etc.*

Infant cereals, which provide additional energy and iron, are usually introduced as the first semisolid food, when the infant is 6 months. Rice cereal is the best cereal to begin with because it is least likely to cause allergies. One can start with cereal like *suji* also. If *suji* is not available, porridge can be made with broken wheat, ground rice and millet *etc.* Those who can afford to spend more money, commercial pre-cooked cereals like Farex, Balamul and Lactogen can also be given. Fruits like mashed banana, papaya, sapota, mango *etc.* can be given. If apple is to be given, it should be stewed.

The infant can also be given seasonal vegetables like peas, potato, carrots and greens after boiling or steaming. After cooking, the vegetable should be sieved and given in a semisolid form. Little amount of oil should be added to the boiled and strained vegetables. Some mothers give only the water in which the vegetables have been cooked. This is not a good practice, as this water does not provide calories to the infant. Vegetable soups, which contain mashed vegetables or flour and little oil, can be given to the infant. Fruit juices should be given when the child has developed the ability to drink from a cup. Juices fed from a bottle may cause 'nursing bottle syndrome'.

7-8 Months

Combination of rice and legumes like *khichdi* and other foods *viz. dalia, kheer,* curd, bread, a thick *chapati* softened in milk or *dal* can be added in the diet of an infant when he is 7-8 months old. Egg can also be given as boiled, poached or scrambled. At his age, the child starts sitting and teething, hence, he would enjoy if given a piece of toasted bread, crisp *chapati*, biscuit, a piece of radish or carrot *etc.* to nibble on.

8-12 Months

By the age of 9-12 months, all the above mentioned foods for infants of 7-8 months of age, can be given, in increased amounts. Gradually, strained foods should be deleted and chopped well-cooked foods should be introduced. Use of finger foods should be increased. By 9-12 months of age, the infant's desire to explore experience, and play with food hinders feeding. It is best to relax and take this phase of infant development in stride.

12-18 Months

When the child is 1-1 ½ years old, he should eat the routine household diets containing foods which are softened and are without spices. The child should be fed more frequently *i.e.* upto 5 times a day at an interval of about 2-3 hours. During this period, finger feeding becomes more efficient, drinking from a cup improves and chewing is easier as more teeth erupt. The child's eating becomes neater, but experimentation and a lack of predictability still rule the day.

Some Practical Aspects of Complementary Feeding

- Semisolids should be introduced when the infant is 6 months of age.
- The mother should be advised to continue breastfeeding during weaning and complementary feeding.
- In general, attempts should be made to try and feed babies home cooked foods from the 'family pot'. Special recipes should be avoided. This has a number of advantages in that it is economical, saves time and the infant grows up accustomed to foods traditionally eaten by the family. Foods such as *dhals*, rice and vegetables are cooked for the family. Portions of these can be removed before adding spices and seed condiments such as mustard and cumin (*jeera*). They can be mashed and salt added to taste and kept aside in covered containers and given to the infant when required.
- One should respect regional variations in feeding practices *e.g. 'suji kheer'* and *'dalia'* (porridge) and *'khichri'* are used in the North and items such as 'ragi', 'pongal' and *'idli'* are more acceptable in South.
- Start with about half to one teaspoonful of the mashed food. Initially, the infant tends to 'suck' on the spoon and as a result, some of the food may spill out. This does not mean that the infant dislikes the food. When an infant does not like the taste of a particular item, it can be made out of his facial expression and by the fact that he will consistently push it out with the tongue every time.
- If a food is acceptable, give as much as the infant will take at a time from the next day. Feeding the infant from a separate cup is helpful as it gives an idea of how much the infant has taken.
- Infants must receive the semisolids frequently, initially once or twice a day and subsequently

three times. Children around 1 to 2 years or more need to be fed about 4-5 times a day.

- If the infant does not like a particular food, withhold it for a day or two and then try again. If the child consistently indicates his dislike, do not offer it.
- Infants should not be force-fed as it runs the risk of aspiration, vomiting and emotional disturbances.
- Allow infants to feed themselves, where they indicate their wish to do so, even if it means a little mess. It is beneficial in the long run.
- Thin soups and *dhal* water are not nutritious. Mothers should be advised to mash the *dhal* with water used for cooking and feed it to the infants.
- Initially, all the semi-solid foods need to be finely mashed. Later, as the infant grows older, the foods can be coarser, although they still need to be soft.
- To serve a nutritious semi-solid, it is easier to cook rice, *dhal* and vegetables together, mash and feed to the infant.
- Mashed fruits should be preferred over fruit juices.
- Mixtures of cereal and pulses *e.g. khichri* and *pongal* are better than either alone.
- At teething time, finger foods should be given, as infants tend to bite or mouth hard objects.
- Non-vegetarian foods can be given. As far as egg is concerned, it can be introduced any time after 6 months. The inner yellow yolk of a boiled egg can be started first either by itself or mixed with milk or porridge with salt added to taste. The white of an egg, soft boiled or scrambled is generally started later as there is a potential risk of allergic reactions with it. Cooked egg is better than a raw egg.
- Minced meat or fish too is good and can be offered when the infants is 9 months old. Care should be taken to remove small bits of cartilage and bone.
- Feed during illness if the infant accepts the food.
- Salt and sugar should be used sparingly, if at all.
- Keep the foods covered.
- Do not overcook because of the loss of heat-sensitive nutrients.
- Maintain hygienic conditions while preparing, cooking and giving the foods.
- Wash hands before cooking and giving foods.

What Not to Feed an Infant

Several foods and practices to avoid when feeding an infant are:

- **Overly salty and overly sweet foods:** Infants do not need a lot of sugar or salt added to their foods. They enjoy bland foods, much better than do adults.
- **Certain foods that tend to cause choking:** These include candy, nuts, raw carrots, grapes, popcorn *etc.*
- **Honey and corn syrup:** These products may contain *Clostridium botulinum*. The spores can eventually develop into bacteria in the stomach and lead to botulism food poisoning which is often fatal in children under 1 year.
- **Feeding more than 1.2 litre of formula or 1 litre of milk per day:** Solid foods should play a greater role in satisfying an infant's increased appetite after 6 to 8 months of age. About to 1 litre of human milk or formula per day is ideal after 6 months of age, with food supplying the rest of energy needs. This is mainly because foods can contain much iron, whereas human milk, cow's milk and formulas do not.
- **Excessive apple juice:** The sorbitol contained in this juice can lead to diarrhoea because sorbitol is a poorly absorbed sugar alcohol.

11

Nutrition during Childhood

Childhood period covers the following age groups:

1. Toddlers (1-3 years)
2. Preschoolers (4-6 years)
3. School going children (7-12 years)

Before studying nutritional requirements of childhood, it is of paramount importance to know growth and development pattern, changes in body composition and behavioural changes which occur during childhood.

Growth and Development

The rapid growth rate of infancy quickly tapers during the next few years. The time from one year of age until puberty is often referred to as the 'latent' period of growth in contrast to accelerated growth during infancy and adolescence.

Toddlerhood is the transition period between infancy and childhood, extending from 1 to 3 years of age. During second year of life, the growth rate slows down. A toddler gains about 2 to 2.5 kg and 10 cm during the second year. The gain in height during the second year is about one half of that occurs in the first year. The child gains about 2.0 kg and 8 cm during the third year. After 3 years, the gain in height averages 5-6 cm per year and occurs evenly throughout the childhood period (Table 11.1). The child attains about one half of the adult's height at the age of 2.5-3 years. There is little difference in height increment of the boys and girls during childhood. During the second year, gain in weight is about 2.0-3.0 kg per year and afterwards, it averages about 1.5 to 2 years among children of 3-6 years; gains are relatively constant throughout childhood.

The muscles and bones grow larger and stronger during pre-school period. The chubby body appearance gives way to a lean appearance among preschoolers. This is due to proportionately greater increase in height as compared to weight. During the school age, rate of growth is slow. But during this period, reserves are laid down to be used during the rapid growth phase of adolescence period. To monitor growth in children upto six years of age, growth charts can be used. Growth variations are large among school going children due to genetic makeup and environmental influence.

Psychomotor skills develop and improve during toddlerhood. Instead of eating with hands and fingers, the child tries to use utensils for eating. Initially his movements are not well coordinated but slowly become coordinated by the time child reaches pre-school age. The process of teething continues. Due to eruption of molars, the toddler requires chunky foods for chewing. Bones, teeth and muscles grow and develop during childhood. Milk teeth are shed and replaced by permanent ones during the school going age and biting may be difficult. Food which is cut into small pieces or food which is easy to bite should be served to the child during this period. Muscle control continues to develop.

Body Composition

Body composition remains relatively constant in preschool and school age children. The percentage of body water and fat decreases, while the percentage of lean body mass increases. In general, boys have more lean body mass per centimeter of height than that of girls. Girls have a slightly higher per cent of body weight as fat when compared to that of boys of the same age

Table 11.1: 95th Centile Values of Weight (kg), Height (cm) and BMI by Age and Gender: Rural India (16 states)

Males			*Age (years)*	*Females*		
Weight (kg)	*Height (cm)*	*BMI*		*Weight (kg)*	*Height (cm)*	*BMI*
11.2	82.4	16.5	1+	10.7	81.6	16.1
13.0	90.7	15.8	2+	12.6	89.8	15.6
14.8	99.1	15.1	3+	14.4	98.2	14.9
16.5	105.7	14.8	4+	16.0	105.1	14.5
18.2	111.5	14.6	5+	17.7	111.0	14.4
20.4	118.5	14.5	6+	20.0	117.5	14.5
22.7	124.3	14.7	7+	22.3	123.6	14.6
25.2	130.1	14.9	8+	25.0	129.2	15.0
28.0	134.6	15.5	9+	27.6	135.0	15.1
30.8	140.0	15.7	10+	31.2	140.0	15.9
34.1	144.8	16.3	11+	34.8	145.3	16.5
38.0	151.1	16.6	12+	39.0	150.2	17.3
43.3	157.0	17.6	13+	43.4	153.8	18.3
48.0	163.0	18.1	14+	47.1	157.0	19.1
51.5	166.3	18.6	15+	49.4	158.8	19.6
54.3	168.3	19.2	16+	51.3	159.7	20.1
56.5	170.0	19.6	17+	52.8	160.2	20.6
58.4	171.3	19.9	18-19	53.8	161.1	20.7
60.5	172.5	20.3	20-24	54.8	160.7	21.2
62.0	172.3	20.9	25-29	56.1	161.0	21.6

Source: ICMR (2011).

group. Except at the age of 11 to 12 years, boys are taller and heavier than girls.

Behavioural Changes

A toddler starts developing a sense of autonomy. He wants to have his own individuality. It is reflected in his demands to decide what to eat, what not to eat, where to eat, how to eat *etc.* With the development and improvement in verbal expression, he clearly and specifically tells about his food related likes and dislikes. His increased mobility allows him to have access to food and other items. Sometimes, he cannot distinguish between the edible and non-edible items. Therefore, lot of attention is to be paid to prevent the child from ingesting harmful substances.

A preschooler strives for initiative. He has an increasing sense of independence, imagination and inquisitiveness. Boys try to imitate their fathers while the girls try to identify themselves with their mothers. This behaviour is reflected at meal times too. Hence, this is a right time to begin teaching a child simple concepts about food and teach sound food habits.

A school going child is quite independent and he begins to make his own food selections. He spends a lot of time in the school and therefore, school friends and adults other than his parents also influence his food habits. Children like large variety of foods. Some food likes and dislikes established in the early childhood may also persist during this period. School work stress may also have an influence on his appetite.

I. Nutritional Requirements and Recommended Dietary Allowances of Preschoolers (1-6 Years)

The need for various nutrients and their recommended dietary allowances for pre-school children (1-6 years) are discussed as under:

Energy

Energy is required for growth and activity. If energy from food is not given in sufficient amount, it shall result in under nutrition which ultimately hinders growth. Energy requirements of children vary according to age, body size and physiological activity. Recommended energy intakes for children of both sexes of 1-3 years and 4-6 years are 1060 and 1350 kcals per day, respectively (Table 11.2). The energy needs of these children have been computed from energy needs per kg and the body weights of well-to-do Indian children. ICMR has suggested that energy should be provided on the basis of ideal weight for age.

Table 11.2: Recommended Dietary Allowances for Preschool Children

Nutrient	1-3 Years	4-6 Years
Weight (kg)	12.9	18.0
Energy (kcal)	1060	1350
Protein (g)	16.7	20.1
Fat (g)	27	25
Calcium (mg)	600	600
Iron (mg)	9	13
Vitamin A (μg)	400	400
β-carotene (μg)	3200	3200
Thiamine (mg)	0.5	0.7
Riboflavin (mg)	0.6	0.8
Nicotinic acid (mg)	8	11
Pyridoxine (mg)	0.9	0.9
Ascorbic acid (mg)	40	40
Dietary folate (μg)	80	100
Vitamin B_{12} (μg)	0.2 to 1	0.2 to 1
Magnesium (mg)	50	70
Zinc (mg)	5	7

Protein

During the first year of life, body protein increases from 11-14.7 per cent and by 4 years of age, it reaches the adult value of 18-19 per cent. This is accompanied by increase in size and deposit in lean body mass. For increased muscle mass, positive nitrogen balance in the body is required which is met by protein intake of 1.09 to 1.47 g/kg body weight. Due to increase in body size, a larger vascular system is required for transportation of nutrients to the whole body and disposal of waste products away from the tissues. It means there is a great demand for nutrients including protein, iron, folacin etc. which are required for blood formation. Bone growth also requires protein intake.

As in adults, protein requirements for preschoolers have also been computed by the factorial method as adopted by FAO/WHO/UNU Expert Group (2007). The average protein requirement is calculated as the sum of maintenance requirement plus protein deposited. Protein required for maintenance in children above 6 months of age is taken as 0.66 g/kg/d. Efficiency of utilization of protein is assumed as 58 per cent. The safe level is then estimated assuming that the requirement follows a log normal distribution *i.e.* safe level is the average value plus 1.96 SD, with the total variability of maintenance and deposition calculated from the root mean square of CV of 12 per cent for maintenance and observed value for protein deposition (Table 11.3). Daily intake of protein can be derived from the proposed safe intake per kg per day. Maintenance of N requirements of children as given by the FAO/WHO/UNU Expert Group and the N content of body weight increment of normally growing children have been used. The body weight standards of well to do Indian children have been used for computing the total daily allowance of proteins for children at different ages.

Table 11.3: Safe Level of Protein Intake for Children (1-10 years, genders combined)*

Age (Year)	Maintenance [a]	Growth [b]	Total	Safe Level [c]	Safe Level Indian Diet [d]
	g Protein/kg Body Weight/d			1.96xSD	
1	0.66	0.29	0.95	1.14	1.47
1.5	0.66	0.19	0.85	1.03	1.33
2	0.66	0.13	0.79	0.97	1.25
3	0.66	0.07	0.73	0.90	1.16
4	0.66	0.03	0.69	0.86	1.11
5	0.66	0.06	0.69	0.85	1.09
6	0.66	0.04	0.72	0.89	1.15
7	0.66	0.08	0.74	0.91	1.17
8	0.66	0.09	0.75	0.92	1.18
9	0.66	0.09	0.75	0.92	1.18
10	0.66	0.09	0.75	0.92	1.18

* For total daily protein requirement in each age band, values need to be multiplied by the normative attained weight in that age band. For example, the age band of 10 years represents the class interval from 9.1-10.0 years. The weight of a boy in this age band is 28 kg. Then, the total protein requirement will be = 1.18x28=33 g/day.

a: From N balance studies; b: Protein deposition value adjusted for efficiency of utilization of 58 per cent from N balance studies; c: SD calculated; d: Corrected for protein from Indian cereal-pulse-milk based diet having PDCAAS of 77.4 per cent for children upto 10 years.

Children of 1-3 years and 4-6 years require 16.7 and 20.1 g protein per day, respectively (Table 11.2).

Fat

Energy dense foods should be given to so as to reduce the bulk of diet to be fed to children. Hence, fat intake of children should be more. ICMR has recommended linoleic acid requirement to be 3 en per cent which can be satisfied by minimal intake of 10 g/day of visible fat. However, to reduce the bulk of cereal-based diets, more fat should be given to provide optimal energy density and adequate calories to the young preschool children. It is suggested that daily visible fat intake of pre-schoolers should be 27 g for 1-3 years and 25 g for 4-6 years old children (ICMR, 2011).

Calcium

Calcium requirements of pre-school children have been calculated on the basis of the amount of calcium accretion in the body. Calcium is mainly deposited in the bones and teeth. However, calcium deposition is not uniform throughout growing period but would be relatively greater during early childhood. Deficiency of calcium shall result in poor skeletal growth.

The calcium requirements are based on balance studies which measure the intake and output of calcium over a period of time. ICMR (2011) has recommended 600 mg Ca per day for children of 1-6 years.

Magnesium

ICMR (2011) has recommended 50 and 70 mg magnesium per day for 1-3 and 4-6 y old preschoolers.

Iron

Iron requirement of children is computed on the basis of basal iron loss and iron required for growth and increase in haemoglobin concentration. During pre-school years (1-3 y), when growth rate slows down and the body mass increases by 1.9kg/y, there is virtually no reserve store of iron between the ages of 6 months and 2 years. Added to basal losses, the total iron requirement during this period is about 0.434 mg/d.

Table 11.4: Iron Requirement for Pre-school Children (1-3 y)

Body weight	12.9 kg
Basal loss	0.181 mg/d
For growth	0.270 mg/d
Total	0.451 mg/d

Table 11.5: RDA for Iron for Children

Children	*RDA*				
	Body Weight (kg)	*Require-ment (µg/ kg/d)*	*Require-ment (mg/d)*	*Absor-ption Assumed (per cent)*	*RDA (mg/d)*
1-3 y	12.9	35	0.45	5	9
4-6 y	18.0	35	0.63	5	13
7-9 y	25.1	31	0.77	5	16

In childhood, the mean increase in body weight is 2.8 kg/y, which necessitates an iron requirement of 0.633 mg/day. The average iron requirement for growth would be 17µg/kg/day. Total iron requirement is 35 mg/kg/d for 1-6 y old children. The RDA for iron after multiplying with the bioavailability factor of 5 per cent is 9 and 13 mg/day for preschoolers of 1-3 and 4-6 y, respectively (ICMR, 2011).

Iodide

Children require 90µg/d (ICMR, 2011).

Zinc

Children of 1-3 y and 4-6 y require 5 and 7 mg of zinc per day, respectively (ICMR, 2011).

Vitamin A

Vitamin A requirements of children have been computed from the requirement figures for infants (50 µg/kg) and adults (9.3 µg/kg) taking into account growth rate at different ages. On this basis, vitamin A requirement of children will be 300-400 µg retinol/day. Since vitamin A deficiency is very common among pre-schoolers and serum vitamin A levels are generally low among Indian children whose dietary intake is less than 100 µg. ICMR (2011) has recommended intake to be 400 µg vitamin A per day for children of 1-6 years.

Other Fat Soluble Vitamins

ICMR has not given specific recommendations for vitamin D and K for children. The requirement of vitamin E suggested is 0.8 mg/g of essential fatty acids.

Thiamine

There are no studies on the thiamine requirements of preschoolers. The committee accepted the earlier recommendation of 0.5 mg per 1000 Kcal for children also; it is the same as accepted for adults. ICMR (2011) has recommended 0.5 mg and 0.7 mg thiamine per day for children of 1-3 and 4-6 years, respectively.

Riboflavin

There are no data on intakes of riboflavin by normal Indian children. It is therefore, suggested that allowances for preschool children be computed on the basis of 0.6 mg per 1000 kcals as recommended in case of adults. ICMR (2011) has suggested 0.4 and 0.6 mg riboflavin intake per day for 1-3 and 4-6 years old children, respectively.

Niacin

In the absence of any information on the niacin needs of children, Expert Committee has recommended that their dietary allowance should be computed on the basis of 6.6 mg niacin equivalents per 1000 kcals as is done in case of the adults (ICMR, 2011). Children of 1-3 and 4-6 years age groups should take 8 and 11 mg niacin per day in the diet, respectively (ICMR, 2011).

Pyridoxine

Relevant data on which requirements for children can be based are scanty. The human requirement of vitamin B_6 is closely related to the amount of protein intake and is satisfied at the level of 0.02 mg/g of protein. ICMR (2011) has therefore, recommended a higher pyridoxine intake for children than that for infants. The daily recommended pyridoxine intakes for 1-3 and 4-6 years old children are 0.9 and 0.9 mg, respectively (ICMR, 2011).

Folic Acid

For children of 1-6 y, a factorial extra need of 40-50µg has been added to the minimum daily intake to meet the RDA. The daily recommended folic acid values are 80 µg and 100µg for children of 1 to 3 and 4-6 years, respectively (ICMR, 2011).

Vitamin B_{12}

The requirement of vitamin B_{12} for children has been computed from interpolation of values for infants (0.2 µg/day) and adults (1 µg/day). ICMR (2011) has recommended 0.2-1.0 µg vitamin B_{12} in daily diets of children (1-6 years).

Vitamin C

ICMR (2011) has recommended a daily intake of 40 mg vitamin C for pre-schoolers (1-6 years); the value is same as in case of the adults.

How to Establish Sound Food Habits among Children and Choose Nutritious Foods

The preschool years are the best time to start a healthful pattern of living and eating. Self-esteem and successful eating are closely tied. Parents and other caregivers are role models. If parents and other family members eat a variety of foods the children will also eat the same. Children should be pursued to take at least one bite of every food prepared and served at home. The sound food habits should be established right from the beginning of life. These can be established by following the guidelines as mentioned below:

- ✰ One way adults can encourage young children to eat nutritiously well balanced meals is to repeat their exposure to new foods. If a child sees a food over and over again, there is a great chance that it will be accepted eventually, especially if adults and older children are seen eating the food and enjoying it.
- ✰ One needs to feed the preschool child affectionately.
- ✰ This is a time for children to experience new foods. The diet should include a variety of foods. The child should have access to items from all food groups on a regular basis. By constantly being introduced to new foods, children of this age can expand their nutritional choices.
- ✰ It is also a time for them to develop their own likes and dislikes. Vegetables are generally disliked but like sweets very much. Parents should not make an issue of it but should handle this situation carefully.
- ✰ Preschool children are very unpredictable. One should be ready for setbacks. If adults can be patient and persevere, children eventually build good food habits.
- ✰ Food should be served in a colourful, interesting and attractive manner.
- ✰ Research shows that children like certain foods, especially in the crisp textures and mild flavours, and familiar foods. Young children are especially sensitive to and reject hot temperature foods. A balance should be maintained between hot and soft foods and strong and mildly flavoured foods in the meals. Fruits are well liked but vegetables are generally disliked.
- ✰ Child should not be forced to eat more than his capacity. Toddlers have a short attention span and are easily distracted from eating. Therefore, child should be fed in a pleasant and peaceful atmosphere so that he does not get distracted while eating.
- ✰ Children should be served finger foods so that they can eat independently.
- ✰ Good table manners should be taught to the child. Using good table manner leads to harmony and positive environment for learning good nutrition habits.

As a toddler's growth rate tapers, food habits may change and feeding problems can appear. One major cause of potential feeding problems stems from the slower growth and accompanying decrease in appetite that characterizes the preschool years. Preschoolers normally have less hunger than they did as infants.

One nutritional challenge at this age is to adapt food choices to the new growth rate. When little food is consumed, the nutrient density of each food choice should be carefully considered. This usually means watching fat intake while increasing intake of whole cereal grains, pulses, fruits and vegetables. There is no need to decrease fat intake severely, but fatty food choices should not be preferred over the nutritious ones.

At this age, food preferences change very frequently. Children eat what they are exposed to. Children can eat a well-rounded and healthy diet if one is served to them. If offered whole grain breakfast cereals, whole-wheat bread, vegetables, salad and fresh fruits regularly, young children accept most of these foods and eat them. But some preschoolers are difficult eaters at times. Their appetite is erratic. They may enjoy eating one meal but refuse to eat the same when served at some different meal. Many of them tend to throw eating or not eating tantrums to exert pressure on their parents or caretakers to get the things done in their own way.

A preschooler usually imitates his or her elder sibling. If he observes his sibling eating a variety of wholesome foods, he may also copy the same. He does not want his eating behaviour to be compared to that of his older siblings. Rather a preschooler needs appreciation for his own accomplishments.

Television has a great influence on the dietary intake of preschoolers. Colourful advertisements, animated cartoons, catchy rhymes and other age mates have a great impact on a young child. When he goes to a grocery store, he may create a great fuss to force his parents to buy him the same food items which he saw on the TV. Examples are 2-minute maggie noodles, cup noodles, pizza, *kurkure*, Uncle Chips, Lays, Candies, ice creams, chocolates *etc*. If the child next door is eating a particular candy or ice cream, the preschooler wants the same. It is difficult to make him understand that a particular food is not healthy.

Many preschoolers join creche or day care centres if both the parents are working. The kind of food and the manner in which it is served, both should be of significant to the parents. However, in India, mostly food is sent from home to the creche/day care centres. In such a situation, often it is observed that children generally like and want to eat the food brought by his friend than the one packed by his mother. If his friend brings cheese toast or *parantha* or noodles, he also wants the same in his tiffin the next day.

Sometimes, the preschooler wants to eat the same food of his choice all the time in the morning, afternoon and night. During food jags also, parents should continue serving the child a variety of foods and they should not make it an issue if the child does not eat. Usually these food jags are temporary. Within a few days, he starts eating a variety of foods.

Childhood Feeding Problems

Tension between parents or between parents and children often contributes to eating problems. Getting to the root of family problems and create a more harmonious family atmosphere is an important part of resolving many childhood feeding problems. In addition, parents must often be educated to know what to expect of a preschool child and what goals to set. Following are a few typical problems, their causes and suggestions for correcting the problems.

'My child won't eat as much, or as regularly, as did as an infant'

This is to be expected. The growth rate shows from that of infancy, so that child does not need as much food. Parents must know that they should not expect a 3 year-old to eat voraciously as an infant, nor to eat adult-sized portions. In addition, appetite varies depending on a child's activity level and how the child feels whether well or sick.

Parents should also be reminded that food likes and dislikes change rapidly in childhood. They are influenced by food temperature, appearance, texture and taste. Children usually prefer mild flavours but there are exceptions to this rule.

Battling with a child to get him to eat more is rarely worth the effort. There is little reason to beg or bribe a child to eat. Parents should present the food, eat it themselves and let their child decide. In addition, they should recognize that this is an important age for children to explore the world around them. Even 'good' eater are sometimes more interested in exploring than in a meal.

'My child is always snacking, yet never finishes the meal'

Children have small stomachs. Offering many small meals succeeds better than limiting children to 3 meals per day. Actually 3 meals per day is just a social custom. There is nothing nutritionally important about that type of plan. Snacking is fine as long as dental care is practiced. It is really not when one eats, but what one eats, that is available. Nutrition snacks should be readily available. When the child becomes hungry in mid-morning or mid-after noon, there are good foods to offer.

'My child never eats vegetables'

Everyone dislikes certain foods, one-bite policy should be encouraged and guidelines can be set to discourage fussing over unfamiliar foods. Children eventually learn that they can eat foods they do not particularly like without first gagging, choking and yelling. It takes child to become first enthusiastic about a new food, but with continual exposure and positive role model, chances are the child may even grow to like it.

Hunger is still the best sauce. It may work to feed a child vegetables at the start of a meal, when he is hungriest. Offer new foods with familiar ones. When a child reaches the age at which he can safely handle raw vegetables without fear of choking (4-5 years), offer thin carrot or radish or cucumber sticks as an option to the cooked ones. The child may find the taste and texture more pleasing. Vegetable puree can be mixed with wheat flour for making *chapati/parantha/puri*. Vegetables soups can be served.

'How do I know if my child is eating healthfully'

Parents want their children to eat well and be healthy. They should include food items from each food group and serve a balanced diet providing all the necessary nutrients in right amount to the child (Table 11.3.). If a child eats foods from all the food groups and shows a regular increments in height and weight that is evidence of a good balanced diet.

Role of Parents in Helping Children to Eat Well

Parents want their children to eat well and be healthy. They worry that their children eat too much or too little or won't eat the vegetables or drink milk. They feel guilty and responsible when children leave the table without eating and angry when they come back 10 minutes later wanting a snack. Dinner table should not become a battleground between the child and the parents. Generally, parents turn into reluctant food hustlers, insisting that child should eat a regulation "four bites of carrot before you can have ice cream". Children gag. Adults feel pity. On the contrary, parents should respect the child's capability.

- ☆ Children have built-in motivators to eat. They get hungry, they are interested in eating, they have hearty appetites for good food, and they are interested in survival. But the way they operate with eating can fool anybody.
- ☆ Children are wary of new food. They are neophobic about food: if it is new, they do not like it. They work to master new foods and new eating skills in the same way that they work to master other skills. They see new foods, taste them (as many as 15 or 20 times), and eventually learn to like them.
- ☆ While children are learning to like new foods, the adults get anxious and try to hurry them along. It does not work. Bribing backfires. Studies have shown that pre-schoolers who were rewarded for eating a new food were less likely to go back to it later than were pre-schoolers who were allowed to approach the food on their own. In addition, food rewards produce a negative side effect. Children who get dessert for eating the vegetables learn to like the dessert more and the vegetables less.
- ☆ Some parents try to pressure children into trying new foods. Pressure tactics might get children to eat, but these methods will not teach long term eating skills. As soon as the pressure is off, children revert to eating the food they like.
- ☆ Children naturally eat a variety of foods. An internal process called sensory specific satiety ensures that they will tire of even favourite foods and eat something different.
- ☆ Adults have the same tendency but they ignore it. They override their appetites and eat because food is good for them, or because they paid for it, or to keep from getting hungry later.
- ☆ Children know how much they need to eat. They respond to their internal sense of hunger and fullness more strongly than adults do, and they eat the right amount of foods for proper growth. Unlike adults, they stop when they are full rather than when the food is gone.
- ☆ Some children submit to parental controls. Others fight back. Children whose parents attempt to overfeed them may be revolted by food and may tend to under-eat when they get the chance. Children whose parents attempt to underfeed them may become preoccupied with food and then overeat when they get the chance.
- ☆ Parents should provide appropriate support. They should provide a variety of attractive, wholesome foods in pleasant surroundings and approach feeding in a positive way. One cannot force a child to eat.
- ☆ Children eat best and are more likely to learn to like a variety of foods if they have regular meals and snacks at predictable times and are not allowed to panhandle between times. Consistent eating times help children to come to the table hungry and, therefore, more likely to accept the food.
- ☆ Children master their eating when adults provide oppourtunities to learn, give support for exploitation, and limit inappropriate behaviour.

Diet for Preschoolers

Table 11.6 depicts the balanced diet for preschoolers.

Table 11.6: Balanced Diet for Preschool Children (Number of portions) (NIN, 2011)

Food Groups	*g/portion*	*1-3 yr*	*4-6 yr*
Cereals and millets	30	2	4
Pulses	30	1	1.0
Milk (ml)	100	5	5
Roots and tubers	100	0.5	1
Green leafy vegetables	100	0.5	0.5
Other vegetables	100	0.5	1.0
Fruits	100	1	1
Sugar	5	3	4
Fats and oils (visible)	5	5	5

For non-vegetarians, one portion of pulse may be exchanged with one portion (50 g) of egg/meat/chicken/fish

A Day's Sample Plan for 3 Year Old School Going Child

Practical Assignment

Note

Instead of using food exchange tables, we can use Indian Food Composition Tables especially if micronutrients are also to be calculated along with macronutrients as given in the following example.

Aim

Plan a day's sample diet plan for a three year old Saloni, daughter of a clerk (vegetarian) very active and playful. Calculate energy, protein, fat, vitamin A and iron in her diet.

Step I: Personal Profile

Age	3 years
Sex	Female
Socio-economic status:	MIG
Food habits	Vegetarian

Step II: Recommended Dietary Allowances (ICMR, 2011)

Energy	1060 kcals
Protein	16.7 g

Fat 27 g
Vitamin A 400 µg
Iron 9 mg

Step III: Menu plan

Early morning	Milk
Breakfast	French toast Milk
Mid morning (Packed tiffin)	*Paushtik parantha* Mango pickle
Lunch	Rice *pulav* *Channa* curry Curd Green chutney
Evening Tea	Mango shake Fruit cake
Dinner	Tomato soup *Chapati* *Moong dal* *Methi*-potato vegetable
At night	Milk

Step IVa: Deciding the no. of servings of each food item and converting into amounts of raw food ingredients

Meal	Menu	Ingredients Used	Amount (g)
Early morning	Milk	Milk	150
		Sugar	5
Breakfast	French toast	Bread	35
	Milk	Egg	25
		Fat	5
		Milk	150
		Sugar	5
Mid morning	*Paushtik parantha*	Wheat flour	25
	Mango pickle	Besan	5
		Spinach leaves	10
		Mango pickle	5
		Fat	5
Lunch	Rice pulav	Rice	25
	Channa curry	Peas	10
	Curd	Carrot	10
	Green *chutney*	Cauliflower	5
		Potato	10
		Kabuli channa	15
		Onion	10
		Tomato	5
		Fat	5
		Milk	50
		Coriander leaves	10
		Onion	5
Evening tea	Mango shake	Mango	100
	Fruit cake	Milk	50
		Sugar	10
		Maida	10
		Egg	10
		Butter	5
		Tuty-fruity	10
Dinner	Tomato soup	Tomato	35
	Moong dal	Moong dal	10
	Methi-potato vegetable	Methi leaves	40
	Chapati	Potato	10
		Onion	10
		Fat	5
		Wheat flour	25
At bed time	Milk	Milk	100
		Sugar	5

Step IVb. Using the Indian Food Composition Tables, calculate the amount of various nutrients specified in the various food items as mentioned in the planned diet.

Meal	Menu	Ingredients Used	Amount (g)	Energy (kcal)	Protein (g)	Fat (g)	Iron (mg)	Vit. A (µg)
Early morning	Milk	Milk	150					
		Sugar	5					
Total								
Breakfast	French toast	Bread	35					
	Milk	Egg	25					
		Fat	5					
		Milk	150					
		Sugar	5					
Total								

Meal	Menu	Ingredients Used	Amount (g)	Energy (kcal)	Protein (g)	Fat (g)	Iron (mg)	Vit. A (μg)
Mid morning	*Paushtik parantha*	Wheat flour	25					
	Mango pickle	Besan	5					
		Spinach leaves	10					
		Mango pickle	5					
		Fat	5					
Total								
Lunch	Rice *pulav*	Rice	25					
	Channa curry	Peas	10					
	Curd	Carrot	10					
	Green *chutney*	Cauliflower	5					
		Potato	10					
		Kabuli channa	15					
		Onion	10					
		Tomato	5					
		Fat	5					
		Milk	50					
		Coriander leaves	10					
		Onion	5					
Total								
Evening tea	Mango shake	Mango	100					
	Fruit cake	Milk	50					
		Sugar	10					
		Maida	10					
		Egg	10					
		Butter	5					
		Tuty-fruity	10					
Total								
Dinner	Tomato soup	Tomato	35					
	Moong dal	Moong dal	10					
	Methi-potato vegetable	Methi leaves	40					
		Potato	10					
	Chapati	Onion	10					
		Fat	5					
		Wheat flour	25					
Total								
At bed	Milk	Milk	100					
		Sugar	5					
Total								

Step IVc. Compare the given food items vs. RDI and given nutrients vs. their RDAs. Accordingly, make adjustments in the amount of given food items so as to arrive near to the RDAs of specified nutrients.

Food Groups/ Exchanges	Recom-mended (g)	Given (g)	Difference (g)
Cereals			
Pulses			
Other vegetables			

Food Groups/ Exchanges	Recom-mended (g)	Given (g)	Difference (g)
GLVs			
Roots and tubers			
Fruits			
Milk and milk products			
Fats and oils			
Sugar and jaggery			
Fish, meat, chicken *etc.*			

Nutrients	Recom-mended	Given	Difference
Energy (kcals)			
Protein (g)			
Fat (g)			
Iron (mg)			
Vit. A (µg)			

A Day's Sample Plan for a 5-Years Old Child

Practical Assignment

Plan a one day diet for Gaurav who is five year old son a university teacher. He attends a nursery school from 8.30 am to 12.00 noon. He takes packed food in the tiffin. His family is non-vegetarian. Calculate energy, protein, fat, vitamin A and iron in his diet.

Step I: Personal Profile

Age	5 years
Sex	Male
Socio-economic status	MIG
Food habits	Non-Vegetarian

Step II: Recommended Dietary Allowances (ICMR, 2011)

Energy	1350 kcals
Protein	20.1g
Fat	25 g
Vitamin A	400 µg
Iron	13 mg

Step III: Menu plan

Early morning	Milk
Breakfast	Wheat Porridge Milk Apple
Mid-morning (Packed Tiffin)	*Idli* Coconut Chutney
Lunch	Boiled rice *Besan* curry Mint chutney
Evening Tea	Banana shake, Vegetable cutlets
Dinner	Spinach soup *Chapati* Chicken curry Salad
At bed	Milk

Step IV: Deciding the no. of servings of each food item and converting into amounts of raw food ingredients

Meal	Menu	Ingredients Used	Amount (g)
Early morning	Milk	Milk	150
		Sugar	5
Breakfast	Wheat porridge	Broken wheat	50
	Milk	Milk	150
	Apple	Sugar	10
		Apple	50
Mid morning	*Idli*	Rice	30
	Coconut *chutney*	Urd dhal	15
		Coconut	10
		Roasted gram	5
		Lemon	5
Lunch	Boiled rice	Rice	65
	Besan curry	Besan	5
	Milk	Curd	25
	Mint *chutney*	Mint	10
		Onion	25
		Fat	5
Evening tea	Banana shake	Banana	50
	Vegetable cutlets	Milk	75
		Sugar	10
		Potato	30
		Peas	10
		Cabbage	10
		Carrot	15
		Bread	10
		Fat	10
Dinner	Spinach soup	Spinach	40
	Chicken curry	Chicken	25
	Salad	Onion	20
	Chapati	Tomato	10
		Cucumber	15
		Radish	10
		Fat	10
		Wheat flour	65
At bed	Milk	Milk	100
		Sugar	5

Step V: Using the Indian Food Composition Tables (2017), calculate the amount of various nutrients specified in the various food items as mentioned in the planned diet.

Meal	*Menu*	*Ingredients Used*	*Amount (g)*	*Energy (kcal)*	*Protein (g)*	*Fat (g)*	*Iron (mg)*	*Vit. A (μg)*
Early morning	Milk	Milk	150					
		Sugar	5					
Total								
Breakfast	Wheat porridge	Broken wheat	50					
	Milk	Milk	150					
	Apple	Sugar	10					
		Apple	50					
Total								
Mid morning	*Idli*	Rice	30					
	Coconut *Chutney*	Urd dhal	15					
		Coconut	10					
		Roasted gram	5					
		Lemon	5					
Total								
Lunch	Boiled rice	Rice	65					
	Besan curry	Besan	5					
	Mint *chutney*	Curd	25					
		Mint	10					
		Onion	25					
		Fat	5					
Total								
Evening tea	Banana shake	Banana	50					
	Vegetable cutlets	Milk	75					
		Sugar	10					
		Potato	30					
		Peas	10					
		Cabbage	10					
		Carrot	15					
		Bread	10					
		Fat	10					
Total								
Dinner	Spinach soup	Spinach	40					
	Chapati	Chicken	25					
	Chicken curry	Onion	20					
	Salad	Tomato	10					
		Cucumber	15					
		Radish	10					
		Fat	10					
		Wheat flour	65					
Total								
At bed	Milk	Milk	100					
		Sugar	5					
Total								

Step IV: Compare the given food items vs. RDI and given nutrients vs. their RDAs. Accordingly, make adjustments in the amount of given food items so as to arrive near to the RDAs of specified nutrients.

Food Groups/ Exchanges	Recom-mended (g)	Given (g)	Difference (g)
Cereals			
Pulses			
Other vegetables			
GLVs			
Roots and tubers			
Fruits			
Milk and milk products			
Fats and oils			
Sugar and jaggery			
Fish, meat, chicken *etc.*			

Nutrients	Recom-mended	Given	Difference
Energy (kcals)			
Protein (g)			
Fat (g)			
Iron (mg)			
Vit. A (µg)			

Nutritional Requirements and Recommended Dietary Allowances of School-Going Children (7-12 Years)

During this period, due to slow rate of growth, food requirement per unit body weight declines. Boys usually grow slower than the girls during their pre-adolescent years. The recommended dietary allowances of various nutrients for school children aged 7-12 years (Table 11.4) are being discussed below:

Energy

The energy requirements are increased steadily. For children of 7-9 years, energy requirements for boys and girls are the same *i.e.* 1690 kcals per day. However, in the age group of 10-12 years, energy requirements for boys and girls differ as the energy needs of boys are increased than those of girls. ICMR (2011) has recommended a daily intake of 2190 kcals for boys and 2010 kcals for girls of 10-12 years age group.

Protein

Protein requirements of this age group are also computed by the factorial method as adopted by FAO/WHO/UNU Expert Consultation (2007). Children of 7-9 years irrespective of the sex, require 29.5 g protein in their daily diets. Their protein requirements increase and differ sex-wise when they enter 10-12 years age group. Girls have higher protein requirements *i.e.* 40.4g per day than the boys who require 39.9 g per day (ICMR, 2011). Girls require more protein between 10-12 years than boys for approaching menarche.

Table 11.4: Recommended Dietary Allowances of School Children

Nutrient	7-9 Years	10-12 Years	
		Boys	Girls
Weight (kg)	25.1	34.3	35.0
Energy (kcal)	1690	2190	2010
Protein (g)	29.5	39.9	40.4
Fat (g)	30	35	35
Calcium (mg)	600	800	800
Iron (mg)	16	21	27
Vitamin A (µg)	600	600	600
β-carotene (µg)	4800	4800	4800
Thiamine (mg)	0.8	1.1	1.0
Riboflavin (mg)	1.0	1.3	1.2
Nicotinic acid (mg)	13	15	13
Pyridoxine (mg)	1.6	1.6	1.6
Ascorbic acid (mg)	40	40	40
Folic acid (µg)	120	140	140
Vitamin B_{12} (µg)	0.2-1	0.2-1	0.2-1
Zinc (mg)	8	9	9
Magnesium (mg)	100	120	160

Fat

The linoleic acid requirement of young as well as older children is 3 en per cent. In the age group of 7-9 years, more fat is needed in their bulky cereal based diets to provide enough calories to meet their energy needs. Total fat intake below 25 per cent E is considered to affect growth in children. To provide 25 per cent E of total fat calories, the minimum level of visible fat should range from 25-30g/day in the diet of 7-12 y old children. Hence, ICMR (2011) suggested that the daily visible fat intake should be 30g/day in case of both boys and girls of 7-9 years. However, the desirable level of visible fat intake for 7-9y old children should be 30g while for both boys and girls of 10-12 years, it should be 35g/day.

Calcium

The calcium requirements of children in the age group of 7-9 years are the same as for the younger age group (1-6 years) children; they require 600 mg calcium per day. Calcium requirement of children in the age group of 10-12 years is increased to 800 mg per day. Both boys and girls who are 10-12 years require similar amount of calcium *i.e.* 800 mg per day but higher than that of adults. Their calcium requirement is higher than that of the adults so as to meet increased skeletal growth demands, during this period. Phosphorus requirement is similar to that of calcium in the age groups of 7-9 and 10-12 y for girls and boys both *i.e.* 800mg/d.

Iron

Iron requirement is derived by the factorial approach in which the requirements for different physiological functions are added. Daily iron requirement in case of children needs to cover basal iron loss, amount required for growth, expansion of blood volume and that for the improving iron stores. Besides it, body weight and iron absorption are also taken into account while aiming at RDA for iron. On this basis, ICMR (2011) has recommended daily iron intake of 16 mg for children of 7-9 years. In the age group of 10-12 years, different RDAs for iron have been made for boys and girls. Iron requirement of girls in higher than that of boys in the age group of 10-12 years due to slightly higher body weight and expansion of blood volume (Table 11.5). ICMR (2011) has recommended a daily iron intake of 21 mg for boys and 27 mg for girls of 10-12 years.

Table 11.5: Iron Requirement during Childhood

RDA for Iron	*Age Group (10-12y)*	
	Boys	*Girls*
Body wt (kg)	34.3	35.0
Gain in body wt (kg/y)	3.5	3.7
Basal loss (mg/d)	0.49	0.49
Blood volume (mg/d)	0.27	0.27
Muscle mass (mg/d)	0.13	0.13
Store (mg/d)	0.16	0.16
Blood loss (mg/d)	-	0.28
Total requirement (μg/kg/d)	31	38
Total requirement (mg/d)	1.05	1.33
Absorption assumed (per cent)	5	5
RDA for iron (mg/d)	21	27

Magnesium

Based on balance studies and tracer turnover studies, RDAs for magnesium have been suggested as 100 mg for 7-9 y children, 120 mg for boys and 160 mg for girls of 10-12 y (ICMR, 2011).

Zinc

Children of 7-9 y have been recommended 8 mg zinc/d while both boys and girls of 10-12 y should take 9 mg Zn/d (ICMR, 2011).

Iodine

Children of 7-12 y should take 120 μg iodine per day.

Vitamin A

Vitamin A requirements of the children of 7-12 years have been computed from the requirement figures for infants (50 μg/kg) and adults (9.3 μg/kg) taking into account growth rates at different ages. ICMR (2011) has recommended a daily allowance of 600 μg retinol for school age children (7-12 years) irrespective of the sex.

Other Fat Soluble Vitamins

ICMR has not made recommendations for Vitamin D and K. The requirements for Vitamin E are related to essential fatty acid intake.

Thiamine

There are no studies on the thiamine requirements of children. ICMR (2011) accepted the earlier recommendation of 0.5 mg per 1000 kcal for children. The RDAs for thiamine are 0.8 mg for 7-9 years children. In 10-12 years age group, boys have a little higher thiamine requirement (1.1 mg/day) than that of girls (1.0 mg/day) due to higher energy requirement (ICMR, 2011).

Riboflavin

ICMR (2011) has suggested that allowance for children may be computed on the basis of 0.6 mg/1000 kcal as for adults. Riboflavin requirement of children of 7-9 years is 1.0 mg/day. However, boys of 10-12 years (1.3mg/d) require a little higher amount of riboflavin than girls (1.2mg/d) due to higher energy requirement.

Niacin

There is no information on niacin needs of children and therefore, the committee recommends that their dietary allowance be computed on the basis of 6.6 mg niacin equivalents per 1000 kcals as is done for adults. Niacin requirements of children of 7-9 years and girls of 10-12 years are similar *i.e.* 13 mg/day. However, boys who are 10-12 years old, require more amount of niacin than the girls in their daily diets i.e. 15 mg/day as their energy needs are higher.

Pyridoxine

ICMR (2011) has recommended pyridoxine intake of children at a level higher than those for infants. Children in the age group of 7-12 years irrespective of the sex require 1.6 mg pyridoxine per day.

Folic Acid

ICMR (2011) has recommended that children of 7-9 years require 120 μg folic acid per day while both boys and girls of 10-12 years require a little higher *i.e.* 140 μg per day.

Vitamin B_{12}

ICMR (2011) has recommended 0.2-1.0 μg vit. B_{12} per day for children in the age group of 7-12 years.

Ascorbic Acid

ICMR (2011) has recommended 40 mg vitamin C for children (7-12 years); this value is the same as in case of adults.

Diet for the School Child

School children are subject to many stresses which affect their appetite. School work, competition with the classmates, emotional stresses in getting along with other children and illness may adversely affect the appetite. Besides parents, school friends and teachers also have increasing influence on food habits of school children. Group acceptance is very important at this time. At school, he may be exposed to food patterns that may be different from those at home. Within a group he wants to accept or try the unfamiliar foods which otherwise if alone, may not try.

School-age children like large variety of foods but some likes and dislikes established in early childhood may also persist during this period. Appropriate and sound food practices established in early childhood need reinforcement during the school years. Table manners should be taught to the children. Nutrition education should be taught to children by teachers in the school so that they can choose right nutritious foods in their diets. Therefore, children are greatly influenced by TV advertisements and messages. This is also the right time to educate children that what so ever food related advertisements are shown on TV and other media, may not be true in real sense.

While planning the menu for school children, one should keep the following points in mind:

- Breakfast is a very important meal. If a child is in a hurry to go to school, he may skip or shall take the breakfast hurriedly. Some children do not have appetite in the morning or they may like the food served to them in the breakfast late in the morning. Studies have shown that children who go to school without taking breakfast are inattentive in the class. They cannot concentrate well in studies and their class performance is poor. An appealing and nutritious breakfast should be served to the child. One should keep ready-to-eat nourishing foods on hand for occasions when the time is short.
- Children easily get bored if the same menu is repeated every day. Menu planned for the child should have variety in colour, taste, flavour and texture.
- If a child does not like vegetables in the form of salads or curry or dry vegetables these can be given in some other form. Mixed vegetable soup, *parantha* stuffed with grated vegetables etc. can be tried out.
- Children like snacking in between the meals. Nutritious recipes like nutritious *ladoos*, sprouted *dal chat*, vegetable sandwich, mango shake, fruit juice, banana milk shake, baked vegetables, cheese fingers *etc.* can be given.
- Children should be encouraged to take meals on the dining table along with other family members. This shall allow them to have interaction with the family members and learn good food choices and table manners from them.
- In hot weather, lot of liquids should be given.
- Care should be taken that high carbohydrate and high fat foods do not replace the protective foods.
- Food items from each food group should be included in the whole day's diet.

A well balanced diet for the school children has been depicted in Table 11.6

Table 11.6: Balanced Diet for Preschool Children (Number of portions) (NIN, 2011)

Food Groups	*g/ portion*	*7-9 yr*	*10-12 yr*	
			Girls	*Boys*
Cereals and millets	30	6	8	10
Pulses	30	2	2	2
Milk (ml)	100	5	5	5
Roots and tubers	100	1	1	1
Green leafy vegetables	100	1	1	1
Other vegetables	100	1	2	2
Fruits	100	1	1	1
Sugar	5	4	6	6
Fats and oils (visible)	5	6	7	7

For non-vegetarians, one portion of pulse may be exchanged with one portion (50 g) of egg/meat/chicken/fish

A Day's Sample Plan for 11 Year Old School Going Child

Practical Assignment

Aim

Plan a day's diet for Geeta aged 11 years who is vegetarian. She is the daughter of a doctor. Her mother is lecturer in a college. Geeta attends Campus School from 8.30 a.m.–2.30 p.m. and carries mid-morning tiffin. Calculate energy, protein, iron, calcium and vitamin A in her diet.

Step I: Personal Profile

Age	11 years
Sex	Female
Physiological studies	School going child
Socio-economic status	HIG
Food habits	Non-vegetarian

Step II: Recommended Dietary Allowances (ICMR, 2011)

Energy	2010 kcals
Protein	40.4 g
Iron	27 mg

Calcium	800 mg
Vitamin A	600 µg

Step III: Menu plan

Breakfast	*Paushtik parantha*
	Butter
	Curd
Mid-morning (Packed tiffin)	Cheese sandwich
	Tomato ketchup
Lunch	Boiled rice
	Sambhar
	Potato raita
	Salad
Tea	Cold coffee
	Swiss roll
Dinner	*Chapati/tandoori roti*
	Channa dal with ghia
	Karela vegetable
	Green chutney
	Mango *kulfi*

Step IV: Deciding the no. of servings of each food item and converting into amounts of raw food ingredients

Meal	*Menu*	*Ingredients Used*	*Amount (g)*
Breakfast	*Paushtik parantha*	Wheat flour	70
	Curd	Besan	10
	Butter	Methi leaves	50
		Curd	100
		Fat	10
Mid morning	Cheese sandwich	Bread	70
	Tomato ketchup	Cheese slices	25
		Tomato ketchup	5
Lunch	Boiled rice	Rice	100
	Sambhar	Arhar dal	20
	Potato *raita*	Tamarind	5
	Salad	Carrot	25
		Peas	20
		Onion	30
		Egg plant	10
		Potato	25
		Curd	75
		Cucumber	10
		Tomato	20
		Radish	20
		Fat	5
Evening tea	Cold coffee	Coffee powder	2
	Swiss roll	Milk	100
		Sugar	20
		Maida	20
		Egg	10
		Fruit jam	5
Dinner	*Chapati/ tandoori* roti	Wheat flour	70
	Channa dal with ghia	Channa dal	20
	Karela vegetable	Ghia	10
	Green *chutney*	Karela	20
	Mango *kulfi*	Coriander leaves	30
		Mint leaves	20
		Onion	10
		Tomato	5
		Fat	10
		Mango	100
		Milk	200
		Sugar	10

Step V: Using the Nutritive Value of Indian Foods, calculate the amount of various nutrients specified in the various food items as mentioned in the planned diet.

Meal	*Menu*	*Ingredients Used*	*Amount (g)*	*Energy (kcal)*	*Protein (g)*	*Fat (g)*	*Iron (mg)*	*Vit. A (µg)*
Breakfast	*Paushtik parantha*	Wheat flour	70					
	Butter	Besan	10					
	Curd	Methi leaves	50					
		Curd	100					
		Fat	10					
Total								
Mid morning	Cheese sandwich	Bread	70					
	Tomato ketchup	Cheese slices	25					
		Tomato ketchup	5					
Total								

Meal	*Menu*	*Ingredients Used*	*Amount (g)*	*Energy (kcal)*	*Protein (g)*	*Fat (g)*	*Iron (mg)*	*Vit. A (μg)*
Lunch	Boiled rice	Rice	100					
	Sambhar	Arhar dal	20					
	Potato *raita*	Tamarind	5					
	Salad	Carrot	25					
		Peas	20					
		Onion	30					
		Egg plant	10					
		Potato	25					
		Curd	75					
		Cucumber	10					
		Tomato	20					
		Radish	20					
		Fat	5					
Total								
Evening tea	Cold coffee	Coffee powder	2					
	Swiss roll	Milk	100					
		Sugar	20					
		Maida	20					
		Egg	10					
		Fruit jam	5					
Total								
Dinner	*Chapati/tandoori* roti	Wheat flour	70					
	Channa dal with ghia	Channa dal	20					
	Karela vegetable	Ghia	10					
	Green *chutney*	Karela	20					
	Mango kulfi	Coriander leaves	30					
		Mint leaves	20					
		Onion	10					
		Tomato	5					
		Fat	10					
		Mango	100					
		Milk	200					
		Sugar	10					
Total								
Grand total								

Step IV: Compare the given food items vs. RDI and given nutrients vs. their RDAs. Accordingly, make adjustments in the amount of given food items so as to arrive near to the RDAs of specified nutrients.

Food Groups/ Exchanges	*Recom-mended (g)*	*Given (g)*	*Difference (g)*
Cereals			
Pulses			
Other vegetables			
GLVs			
Roots and tubers			
Fruits			
Milk and milk products			

Food Groups/ Exchanges	*Recom-mended (g)*	*Given (g)*	*Difference (g)*
Fats and oils			
Sugar and jaggery			
Fish, meat, chicken *etc.*			

Nutrients	*Recom-mended*	*Given*	*Difference*
Energy (kcals)			
Protein (g)			
Calcium (mg)			
Iron (mg)			
Vit. A (μg)			

12

Nutrition during Adolescence

The transition period in the life cycle, linking childhood to adulthood is called adolescence. Adolescent period is characterized by onset of rapid physical, hormonal and emotional development. This is the final growth spurt of life cycle. Under influence of hormones, many bodily changes occur. However, rate of growth spurt varies in boys and girls. Boys tend to mature later than the girls. A rapid growth spurt begins in most girls between the ages of 10 and 13 years and in most boys between the ages of 12 and 15 years. These growth spurts last for about 3 years. The velocity of growth, the rate at which growth proceeds, is greater in males than in females.

During this growth spurt, girls gain about 10 inches in height and boys gain about 12 inches. Girls also tend to lay down both lean and fat tissues, whereas boys tend to lay down mostly lean tissue. This growth spurt provides about 42 per cent to 51 per cent of ultimate adult weight and 15 per cent to 25 per cent of ultimate adult height. The growth spurt in girls surrounds the onset of menstrual cycle, called menarche and they need more iron in the diet. Little further increase in height occurs beyond 2 years after menarche. There is accumulation of subcutaneous fat in the abdominal area. Bone development around the pelvic region occurs. Boys, although slow in growth have more height and weight than the girls and put on much more muscle mass.

Fortunately, as the growth spurt begins, so does an increase in food intake. Boys need more food than the girls as boys gain more lean body mass. If teens choose nutritious foods, they can take advantage of their increased hunger and easily satisfy their nutritional requirements.

Nutritional Requirements

The recommended dietary allowances for various nutrients are given in Table 12.1.

Table 12.1: Recommended Dietary Allowances of Adolescents (ICMR, 2011)

Nutrient	*13-15 Years*		*16-17 Years*	
	Boys	*Girls*	*Boys*	*Girls*
Body weight (kg)	47.6	46.6	55.4	52.1
Energy (Kcal)	2750	2330	3020	2440
Protein (g)	54.3	51.9	61.5	55.5
Fat (g)	45	40	50	35
Calcium (mg)	800	800	800	800
Iron (mg)	32	27	28	26
Vitamin A (μg)	600	600	600	600
β-carotene (μg)	4800	4800	4800	4800
Thiamine (mg)	1.4	1.2	1.5	1
Riboflavin (mg)	1.6	1.4	1.8	1.2
Niacin (mg)	16	14	17	14
Pyridoxine (mg)	2	2	2	2
Ascorbic acid (mg)	40	40	40	40
Folic acid (μg)	150	150	200	200
Vitamin B12 (μg)	0.2-1	0.2-1	0.2-1	0.2-1
Magnesium (mg)	165	210	195	235
Zinc (mg)	11	11	12	12

Energy

Due to accelerated growth, energy requirements of both boys and girls are increased. Boys (13-15 years) need 2750 kcals a day while girls of this age group require 2330 kcals/day. In the age group of 16-17 years, energy needs of boys and girls are further increased to 3020 kcals and 2440 kcals per day, respectively.

Protein

After the age of 10 years, boys and girls have different growth patterns and their protein deposition rates are also different. Therefore, protein requirements for adolescent boys and girls are different although the principles of calculations remain exactly the same as for children upto the age of 10 years. The protein needs are increased to meet the demands of rapid growth and developing lean body mass (Table 12.2). ICMR (2011) has recommended daily intake of 54.3 g for boys (13-15 years) and 51.9 g for girls (13-15 years). The growth spurt is the highest when boys are 16-17 years; hence they require 61.5 g protein per day. Girls of 16-17 years need 55.5 g protein/day (Table 12.3). The protein intake usually exceeds 1 g protein/kg body weight.

Table 12.2: Safe Level of Protein Intake for Adolescent Boys and Girls

Age (Year)	*Maintenance [a]*	*Growth[b]*	*Total*	*Safe Level[c]*	*Safe Level Indian Diet[d]*
	g Protein/kg Body Weight/d			*1.96xSD*	
Boys					
13	0.66	0.07	0.73	0.90	1.15
14	0.66	0.06	0.72	0.89	1.14
15	0.66	0.06	0.72	0.88	1.13
16	0.66	0.05	0.71	0.87	1.12
17	0.66	0.04	0.70	0.86	1.10
Girls					
13	0.66	0.05	0.71	0.88	1.13
14	0.66	0.04	0.70	0.87	1.12
15	0.66	0.03	0.69	0.85	1.09
16	0.66	0.02	0.68	0.84	1.07
17	0.66	0.01	0.67	0.83	1.06

* For total daily protein requirement in each age band, values need to be multiplied by the normative attained weight in that age band. a: From N balance studies; b: Protein deposition value adjusted for efficiency of utilization of 58 per cent from N balance studies; c: SD calculated; d: Corrected for protein from Indian cereal-pulse-milk based diet having PDCAAS of 78.24 per cent for children from 10-18 years.

Table 12.3: Protein Requirement and Dietary Allowances for Adolescent Boys and Girls

Adolescents	*Requirement[a,b] (g protein/kg/d)*	*Body Weight (kg)*	*Total Daily Requirement (g protein/d)*
Boys			
13-14	1.15	43.3	49.8
14-15	1.14	48.0	54.7
15-16	1.13	51.5	58.2
16-17	1.12	54.3	60.8
Girls			
13-14	1.13	43.4	49.0
14-15	1.12	47.1	52.8
15-16	1.09	49.4	53.8
16-17	1.07	51.3	54.9

a: In terms of mixed Indian vegetarian diet protein, PDCAAS varying from 77.4 to 79.0 for different age groups; b: Requirement for each age band taken as the protein requirement for the lower age limit at that age band.

Fat

Total fat intake below 25 per cent E is considered to affect growth in adolescents. To provide 25 per cent total fat calories, the minimum level of visible fat in adolescent's diet should range between 35-50 g/d. Boys and girls of 13-15 y require 45 and 40 g visible fat per day while boys and girls of 16-17 y require 50 and 35 g/d, respectively.

Table 12.4: Recommendations for Dietary Fat Intake in Indian Adolescents

Adolescents	*Minimum Level of Total Fat (Per cent E)*	*Fat from Foods other than Visible Fat (Per cent E)*	*Visible Fat*	
			Per cent E	*g/p/d*
Boys				
13-15 y	25	10	15	45
16-18 y	25	10	15	50
Girls				
13-15 y	25	10	15	40
16-18 y	25	10	15	35

Calcium

Calcium requirement is calculated on the basis of the amount of calcium accretion during the growth period. During adolescence, calcium is required for increased bone mass. ICMR (2011) has recommended 800 mg calcium per day for both boys and girls in the age group of 13-15 years and 16-17 years. In both the age groups Ca:P ratio should be 1:1.

Iron

During adolescence, there is an increase in body weight, muscle mass and blood volume. In adolescent girls, the added iron loss due to menstruation amounts to about 0.37-0.42 mg iron daily (Table 12.5). The additional iron requirement during adolescence for the growth spurt, expansion of blood volume and increase in Hb concentration is 12 µg/kg/day for boys and 8 µg/kg/day for girls. Girls require an additional 8 µg iron/kg

Table 12.5: RDAs of Iron for Adolescents (ICMR, 2011)

Age Group	Body Wt. (kg)	Total Iron Requirement (mg/d)								Absorption (per cent)	RDA (mg/d)
		Gain in Body Wt. (kg/y)	Basal Loss (mg/d)	Blood Volume (mg/d)	Muscle Mass (mg/d)	Store (mg/d)	Blood Loss (mg/d)	Total Requirement (μg/kg/d)	Total Requirement (mg/d)		
13-15y											
Boys	47.6	4.2	0.66	0.39	0.15	0.40	-	34	1.60	5	32
Girls	46.6	1.7	0.65	0.13	0.06	0.15	0.37	29	1.36	5	27
16-17y											
Boys	55.4	1.5	0.78	0.14	0.05	0.40	-	25	1.37	5	28
Girls	52.1	-	0.73	-	-	0.15	0.42	25	1.30	5	26

to compensate for menstrual blood loss. This needs to be added to the basal losses of 14 μg/kg/day for both boys and girls to arrive at the total iron requirement. Using body weights and iron absorption figure of 5 per cent for a mixed cereal-pulse-milk diet, RDA for iron is computed.

ICMR (2011) has recommended 32 mg iron per day for boys of 13-15 years and 28 mg per day for those who are 16-17 years old. Girls (13-15 years) should daily take 27 mg iron while those in the age group of 16-17 years require 26 mg/day.

Magnesium

ICMR (2011) has recommended 165 mg and 210 mg magnesium for boys and girls of 13-15y and this requirement is further increased to 195 mg and 235 mg/d in case of boys and girls of 16-17y, respectively.

Zinc

Both boys and girls of 13-15y and 16-17y have been recommended 11 mg and 12 mg of zinc per day, respectively.

Iodine

Daily iodine requirement of girls and boys of 13-17 y is similar to that of adults *i.e.* 150μg/d.

Vitamin A

Vitamin A requirement of adolescent boys and girls (13-17 years) is similar to that of the adults *i.e.* 600 μg/day.

Other Fat Soluble Vitamins

ICMR (2011) has not given recommendations for vitamin D and K. Requirement of vitamin E is related to essential fatty acid intake.

Thiamine

Thiamine requirement is computed on the basis of 0.5 mg per 1000 kcals (ICMR, 2011). Boys need more amount of thiamine than the girls as energy requirement of boys is higher. Among the boys, those in the age group of 16-17 years need more thiamine (1.5 mg/day) because of more energy requirement than those who are 13-15 years (1.4 mg/day). The thiamine requirements of girls in both the age groups *i.e.* 13-15 years and 16-17 years are 1.2 mg and 1.0 mg/d, respectively

Riboflavin

The allowances for riboflavin for adolescents are computed on the basis of 0.6 mg/1000 kcals (ICMR, 2011). Due to higher energy requirement, boys require more riboflavin than the girls. Riboflavin requirements of boys in the daily diets are 1.6 mg (13-15 years) and 1.8 mg (16-17 years). Girls of 13-15y and 16-17y need 1.4 mg and 1.2 mg riboflavin per day, respectively.

Niacin

ICMR (2011) has recommended 6.6 mg niacin equivalents per 1000 kcals as is done for adults. Daily niacin requirement of boys is 16 mg (13-15 years) and 17 mg (16-17 years) while those for girls (13-17y) is 14 mg.

Pyridoxine

ICMR (2011) has recommended a daily intake of 2.0 mg for both boys and girls in the age group of 13-15 and 16-17 years.

Folic Acid

The committee recommends that folate requirement for adolescent could be calculated on a kg body weight basis as is done in case of adults. ICMR (2011) has recommended 150 μg folate/day for boys and girls of 13-15 years. Adolescents of 16-17 years irrespective of sex require the same amount of folate as the adults *i.e.* 200 μg per day.

Vitamin B_{12}

ICMR (2011) has recommended 0.2-1.0 μg vitamin B_{12} per day for adolescent boys and girls (13-17 years).

Ascorbic Acid

The vitamin C requirement of adolescent boys and girls (13-18 years) is similar to that of adults *i.e.* 40 mg/day.

Nutritional Problems of Adolescents

Anorexia Nervosa and Nulimia

Sometimes it is also called bulimia nervosa. The term implies loss of appetite; however, denial of appetite is more accurate description. It is characterized by:

- Extreme weight loss
- Rigid dieting
- Helplessness in the presence of food
- Poor body image thinking 'I am too fat', even when emaciated; relentless pursuit of thinness
- Self starvation
- Fear of obesity and weight gain

According to some studies, approximately 1 of every 250 teenage girls between the ages of 12 and 18 years suffers from anorexia nervosa. In this disorder, there is a psychological loss of appetite and self starvation, partly due to distorted body image and various social pressures associated with puberty.

Bulimia means 'ox hunger' or being as hungry as an ox. It is an eating disorder in which large quantities of food are eaten at one time (bingeing) and soon purged from the body by means of vomiting, fasting, use of laxatives, or diuretics. Researchers think that approximately 1 per cent to 4 per cent of college-age women suffer from this disorder. Bulimia is characterized by:

- Secretive binge eating; never overeating in front of others
- Eating when depressed
- Bingeing followed by fasting, laxative abuse, self-induced vomiting or excessive exercise
- Shame, embarrassment, depression, low self-esteem and guilt especially after a binge
- Fluctuating weight resulting from alternate bingeing and fasting
- Erosion of teeth, swollen glands

Nutrition counseling can help correct misconceptions about food and be used to plan a diet within the person's energy needs that still contains a reasonable amount of tasty foods. Nutrition counseling should emphasize setting up regular eating habits, rather than stopping the bingeing and purging. Professionals working with adolescents, young adults, parents and friends should consider the following advice for discouraging future cases of eating disorders:

- Discourage restrictive dieting, meal skipping and fasting
- Provide information about normal changes that occur during puberty
- Correct misconceptions about nutrition, desirable body weight and healthy approaches to weight loss
- Carefully phrase any weight related recommendations and comments.

Less Calcium Intake

A major concern is that many teenage girls stop drinking milk and so they may not consume sufficient calcium to allow bones' maximum mineralization through their early twenties. Many researchers are concerned that adolescent girls who do not drink milk are sowing the seeds for future osteoporosis. Only about 1 in 6 teenage girls consume the recommended amount of calcium.

Less Iron Intake

Iron deficiency anaemia is a common problem among teenage girls after the onset of menses and in boys during their growth spurt. It is important that teenagers choose good sources of iron such as whole grains, enriched cereals, green leafy vegetables, dried fruits, egg and liver. In addition, they should also consume vitamin C rich foods with any plant food or take supplemental source of iron.

If their limited food choices consist of junk/fast foods, little room is left for nutritious food. Iron deficiency anaemia is not a desirable state for a teen. It can produce an increased tendency for fatigue and decreased ability to concentrate and learn. School performance may falter.

More Intake of Saturated Fat

Another problem primarily involves boys during their growth spurt, when food consumption is great. Their great energy intakes can accommodate large intakes of saturated fat and cholesterol, especially if they regularly do snacking. Many adolescents can take some high energy foods such as soft drinks, cookies, ice creams, milk shakes, *samosas*/fried snacks, french fries *etc.* without gaining excess weight. But eating these types of food is not healthy as it may pave the arteries for early coronary heart disease. Heart-healthy foods should be selected, especially during the high energy intake years.

Skipping Breakfast

Under the influence of hormones, the deposition of fat in the abdominal area is normal in case of teenage girls but it becomes an area of great concern to her that she is growing fat. Fear of excess weight gain may cause young girls to limit their energy intake. She eliminates some of the foods which she thinks are fattening such as high carbohydrate foods. Due to energy restriction, she devoids herself of some essential nutrients especially vitamin A, vitamin B_6, vitamin C and zinc.

Due to busy schedule, many of the teenagers skip breakfast and instead of waiting for the regular meals, do in between meals snacking. They do not have appetite at the time when regular nutritious meal is served at home. When away from home, teenagers often take ready to eat foods, inexpensive and acceptable to the peer group. Such foods are often called as fast or junk foods (ice creams, pizza, burgers, *samosa*, patties, *pakoras*, shakes, cold drinks) which are readily available from the school/college canteens, vending machines *etc.* Some of these foods are nutritious too. These foods if taken in moderation do not pose any problem but in excess can seriously affect the nutritional status of adolescents.

Impact of Advertisements

Teenagers are vulnerable to food related advertisements appearing on TV and other media. Most of these advertised fast foods or snacks such as *samosas*, french fries, cutlets, pizza, burgers, potato *bondas*, ice creams, soft drinks *etc.* are rich in calories but lack other essential nutrients. Teenagers often obtain one-fourth of all their energy and major nutrients from snacks, ice creams and cold drinks. They can obtain many nutrients from snacking too, if these are selected judiciously. Even fast food restaurants offer good food choices. Snacks and fast foods are not the problem but poor food choices are. Such foods should be chosen judiciously as a part of well balanced diet and should not be a replacement to the normal meal.

Acne Problem and Diet

Most of the teenagers face acne problem. The sebaceous glands surround hair follicles on the face, ears, back, chest, eyelids and other areas. In these glands are cells that secrete triglycerides, very long chain fatty acids (wakes), and other lipids. Collectively, these substances are known as sebum. If the sebum blocks a duct in the gland, this can lead to an infection and local pressure, resulting in an acne lesion.

Diet appears not to affect acne. Some of the teenagers fear that eating chocolate, nuts, fried foods, and pizza can make acne worse but scientific studies have been unable to show a strong link between any dietary factor and acne. The main culprit promoting acne is over activity by the sebaceous glands in the skin. They respond to testosterone, androgen and use of high doses of birth control pills.

Accutane *i.e.* 13-cis-retinoic acid, a derivative of vitamin A, is known to change the nature of sebaceous gland development, as vitamin A might be expected to do. It decreases the production of sebum and in turn, reduces the number of acne lesions. One should also avoid spicy and oily foods. Lot of fluids should be taken.

Diet for Teenagers

Table 12.6 depicts the balanced diet for teenagers.

Table 12.6: Balanced Diet for Adolescents (Number of portions) (NIN, 2011)

Food Groups	*g/ portion*	*13-15*		*16-18*	
		Girls	*Boys*	*Girls*	*Boys*
Cereals and millets	30	11	14	11	15
Pulses	30	2	2.5	2.5	3
Milk (ml)	100	5	5	5	5
Roots and tubers	100	1	1.5	2	2
Green leafy vegetables	100	1	1	1	1
Other vegetables	100	2	2	2	2
Fruits	100	1	1	1	1
Sugar	5	5	4	5	6
Fats and oils (visible)	5	8	9	7	10

For non-vegetarians, one portion of pulse may be exchanged with one portion (50 g) of egg/meat/chicken/fish

Adolescent girls-Extra cup of milk

Adolescents boys-Diet of sedentary man + extra cup of milk

Sample Diet Plan for Adolescents

Practical Assignment

Plan a day's sample diet plan for an adolescent boy Rajesh (16 years) vegetarian, belonging to middle income group. Calculate energy, protein, iron, calcium and vitamin C in his diet.

Step I: Personal Profile

Age	16 years
Sex	Male
Socio-economic status	Middle income group
Food habits	Vegetarian

Step II: Recommended Dietary Allowances (ICMR, 2011)

Energy	3020 kcals
Protein	61.5 g
Iron	28 mg
Calcium	800 mg
Vitamin C	40 mg

Step III: Menu Plan

Breakfast	Swiss Muesli
	Milk
	Apple
Mid morning	Buttermilk
Lunch	Boiled rice
	Chapati
	Ghia kofta
	Stuffed capsicum

	Cucumber *raita*
	Green *chutney*
Tea	Cold coffee
	Paneer fingers
Dinner	Spinach soup
	Chapati
	Dehusked green gram dhal
	Cauliflower and peas vegetable
	Salad
At night	Milk

Step IV: Deciding the no. of servings of each food item and converting into amounts of raw food ingredients

Meal	*Menu*	*Ingredients Used*	*Amount (g)*
Breakfast	Swiss Muesli	Corn flakes	25
	Milk	Oat flakes	20
	Apple	Barley flakes	20
		Raisins	5
		Almonds	5
		Milk	150
		Sugar	10
		Apple	100
Mid morning	Buttermilk	Curd	50
Lunch	Boiled rice	Rice	75
	Chapati	Wheat flour	50
	Ghia kofta	Bottle gourd	20
	Stuffed capsicum	Besan	5
		Capsicum	20
	Cucumber *raita*	Potato	30
	Green chutney	Cucumber	5
		Tomato	5
		Carrot	10
		Onion	20
		Mint leaves	10
		Coriander leaves	10
		Curd	75
		Fat	10
Evening tea	Cold coffee	Coffee powder	2
	Paneer fingers	Paneer	50
		Corn flour	5
		Maida	10
		Bread crumbs	10
		Milk	100
		Fat	7
		Sugar	10
Dinner	Spinach soup	Spinach	75
	Chapati	Wheat flour	85
	Dehusked green gram dhal	Cauliflower	20
		Peas	20
	Cauliflower and peas vegetable	Ginger	5
		Garlic	5
	Salad	Coriander leaves	5
		Onion	20
		Tomato	10
		Radish	10
		Fat	8
At bed time	Milk	Milk	100
		Sugar	5

Step V: Using the Indian Food Composition Tables (2017), Calculate the amount of various nutrients specified in the various food items as mentioned in the planned diet.

Meal	*Menu*	*Ingredients Used*	*Amount (g)*	*Energy (kcal)*	*Protein (g)*	*Fat (g)*	*Iron (mg)*	*Vit. A (μg)*
Breakfast	Swiss Muesli	Corn flakes	25					
	Milk	Oat flakes	20					
	Apple	Barley flakes	20					
		Raisins	5					
		Almonds	5					
		Milk	150					
		Sugar	10					
		Apple	100					
Total								
Mid morning	Buttermilk	Curd	50					

Meal	***Menu***	***Ingredients Used***	***Amount (g)***	***Energy (kcal)***	***Protein (g)***	***Fat (g)***	***Iron (mg)***	***Vit. A (μg)***
Total								
Lunch	Boiled rice	Rice	75					
	Chapati	Wheat flour	50					
	Ghia kofta	Bottle gourd	20					
	Stuffed capsicum	Besan	5					
	Cucumber *raita*	Capsicum	20					
	Green chutney	Potato	30					
		Cucumber	5					
		Tomato	5					
		Carrot	10					
		Onion	20					
		Mint leaves	10					
		Coriander leaves	10					
		Curd	75					
		Fat	10					
Total								
Evening tea	Cold coffee	Coffee powder	2					
	Paneer fingers	Paneer	50					
		Corn flour	5					
		Maida	10					
		Bread crumbs	10					
		Milk	100					
		Fat	7					
		Sugar	10					
Total								
Dinner	Spinach soup	Spinach	75					
	Chapati	Wheat flour	85					
	Dehusked green gram dhal	Cauliflower	20					
		Peas	20					
	Cauliflower and peas vegetable	Ginger	5					
		Garlic	5					
	Salad	Coriander leaves	5					
		Onion	20					
		Tomato	10					
		Radish	10					
		Fat	8					
Total								
At bed	Milk	Milk	100					
		Sugar	5					
Total								

Step IV. Compare the given food items vs. RDI and given nutrients vs. their RDAs. Accordingly, make adjustments in the amount of given food items so as to arrive near to the RDAs of specified nutrients.

Food Groups/ Exchanges	*Recom-mended (g)*	*Given (g)*	*Difference (g)*
Cereals			
Pulses			
Other vegetables			
GLVs			
Roots and tubers			
Fruits			
Milk and milk products			
Fats and oils			
Sugar and jaggery			
Fish, meat, chicken *etc.*			

Nutrients	*Recom-mended*	*Given*	*Difference*
Energy (kcals)			
Protein (g)			
Fat (g)			
Iron (mg)			
Vit. A (μg)			

13

Nutrition during Old Age

Aging is an irreversible biological phenomenon. The World Health Organization categorizes 'elderly' as all persons over age 60 years, which is half the maximum human survival age of 120 years. In 1980, 8.5 per cent of the world's population was over 60 years, rising to 9.4 per cent in 1990 and 11 per cent in 2000. By 2020, this age group is expected to represent one billion individuals – 13.3 per cent of a projected world population of 7.5 billion. Almost three quarters of those one billion elderly will live in developing nations. The share of India's population aged 60 and older is projected to climb from 8 per cent in 2010 to 19 per cent in 2050, according to the United Nations Population Division (UN 2011); nearly 80 per cent of the world's older persons will live in developing countries *i.e.* with China and India contributing to over one-third that number. Elderly population is the fastest growing section of society in India due to the following factors:

- ✰ Increased life expectancy
- ✰ Better nutrition
- ✰ Advancement in medical/health facilities
- ✰ Gradual fall in mortality rate
- ✰ Low fertility rates
- ✰ Increased awareness

In our country, life expectancy level which was 23 years at the beginning of the century rose to 32 years in 1951. By 2021, an average Indian male and female will live till 69.8 and 72.1 years, respectively according to Union health ministry's life expectation (LE). Better medical care, good nutrition and low fertility have made the elderly the fastest growing section of our society. Improved life expectancy combined with high growth rate of population has contributed to increase in number of persons who are 60+. Their number in India, which was 57 millions in 1991, is expected to reach 198 million in 2030. By 2050, India will be home to one out of every six of the world's older persons. Rising numbers of older people will put new and increasing demands on the health care system.

As India's population ages, the nation will face a shrinking pool of working-age people to support the elderly population. The old-age dependency ratio, the number of people aged 60 and older per person aged 15 to 59 is expected to rise from 12 per 100 to 31 per 100 by 2050. By 2042, the share of Indians 60 and older is projected to exceed children and youth aged 14 and younger. This profound shift in the share of older Indians taking place in the context of changing family relationships and severely limited old-age income support brings with it a variety of social, economic, and health care policy challenges.

This transformation has dramatic implications for society, particularly in the area of nutrition and health, for the first two decades of the new millennium. Hence, lot of interest has been generated recently in 'geriatric nutrition'.

What Causes Ageing

One view of aging describes it as process beginning soon after fertilization and ends with death. Cell aging probably results from both automatic cellular changes and environmental influences. Throughout the animal kingdom, aging is predictable. Eventually, cells lose their ability to regenerate needed components, and they die. As more and more cells in an organ system die, organ function decreases. Why cells age remains a mystery. There are many theories for aging process such as:

- Error in DNA replication
- Aging of the connective tissue
- Build up of toxic products
- Free radical damage
- Poor nutrition
- Changes in hormone function
- Changes in the efficiency of immune system
- Programmed death

In fact, the cause of aging is a combination of all the above mentioned changes.

Ageing and Nutrition

In India, 50 per cent of the elderly population is malnourished. The magnitude of malnutrition among the elderly in India is underreported. Studies have shown that more than 50 per cent of the older population is underweight and more than 90 per cent has an energy intake below the recommended allowance.

Older people are vulnerable to malnutrition for many reasons including physiological and functional changes that occur with age, lack of financial support, health problems and inadequate access to food. Malnutrition leads to decreased independence due to physical weakness and muscle wasting which affects ultimately directly or indirectly their nutritional status. Further, malnutrition compounded with poor health creates a vicious cycle in old age. Hence, good nutrition during old age is as important as during childhood or adolescence and it should be associated with preventing malnutrition, supporting physical function, reducing the risk of chronic disease, supporting mental health and preventing disability.

Changes Associated with Aging which have Impact on Nutritional Status

The role of nutrition increases in importance as the age advances because our health and ability to feel good depends heavily on what and how we eat. Aging individuals are more vulnerable to under nutrition and its associated infectious diseases, as well as to excessive and imbalanced intakes and the concomitant risk of chronic diseases. To predict the nutritional problems of an old person, it is imperative to understand to what extent physiological capabilities have been affected by aging. Let us examine some of the important changes associated with aging which consequently affect the nutritional status of elderly persons.

I. Physiological and Biochemical Changes

Elderly people vary more in health status among themselves than do any other age group. Physiological age is more important than chronological age in the elderly years. Once the body reaches physiological maturity, the rate of degenerative change exceeds the rate of cell regeneration. However, people age at different rates, so the elderly population is not a homogenous group.

Changes in the body composition, systems and organ functions during ageing can affect food choices and ultimately health and nutritional status. The usual physiological and biochemical changes of aging include:

Body Mass and Composition

There is a 5–25 per cent decrease in basal (resting) metabolic rate, leading, most notably, to gain in body weight and body fat, even with the unchanged dietary (energy) intake and exercise habits. For example, for most individuals, body fat starts gradually increasing between 20–25 years of age, until about 65 years. Even more important is the redistribution of fat to the abdominal area and visceral organs, as well as its infiltration into muscle and bone while subcutaneous fat declines in other areas of the body with ageing.

On the contrary, with advancing age, the lean components, such as total body water, skeletal muscle, organ mass, and bone mineral, tend to decrease. Muscle mass peaks at the age of approximately 30 years and then gradually declines. There is about 20–40 per cent decrease in muscle mass by the age of 70 years. In older individuals (70-79 years), skeletal muscle mass declines by about 1 per cent per year.

The loss of muscle mass with aging is increasingly recognized as having important consequences in old age because it may be associated with weakness, disability, and morbidity. In 1989, Rosenberg focused renewed attention on this phenomenon by giving it a name, sarcopenia. Sarco, from the Greek, denotes flesh (muscle), and penia indicates a deficiency; thus, sarcopenia translates loosely as muscle deficiency. It is now used to refer specifically to the gradual loss of skeletal muscle mass and strength that takes place with advancing age. It has been reported that as the muscle ages, some muscle cells shrink and others are lost. Some muscles lose their ability to contract as they accumulate fat and collagen. It is important to distinguish between sarcopenia and dynapenia, the latter being the loss of muscle strength and not necessarily always proportionally accompanied by muscle mass loss. These declines are more pronounced in women than in men.

Recently, a triad encompassing the simultaneous deterioration in bone, muscle and adipose tissues (all the three tissues are interrelated) has been identified and named 'osteosarcopenic obesity syndrome'. Although the original identification of osteosarcopenic obesity syndrome was based on the changes in body composition phenotype in older women, it has recently been shown that such phenotype might exist even in young (18–21 years) overweight adults. Within the osteosarcopenic obesity syndrome, there are two other underlining components that have not been fully recognized until recently: osteopenic/osteoporotic obesity and sarcopenic obesity, of which each one could also exist on its own (Figure 13.1).

Recently, the concept of sarcopenic obesity has emerged whereby elderly persons experience a low muscle mass coupled with a high body fat. Sarcopenic-obesity has been shown to be a particularly strong predictor of functional impairment and physical disability due to increased fat infiltration into muscle with aging. Life style partially determines the rate of muscle mass deterioration. All of these conditions may lead to increased risk of fractures and morbidity and declined functionality. Active life style maintains the muscle mass whereas the inactive one causes loss of muscle mass.

The amount of collagen, found in tendons, ligaments, skin and blood vessels, increases and it becomes rigid. Changes in collagen proteins may decrease flexibility in key body components and so alter organ function. Wrinkled skin is an apparent result. The joints creak and the back becomes bent. Due to decreased lean body mass, energy needs are reduced.

Weight Loss

In some of the old persons, weight loss occurs very rapidly. Such elderly patients with unintentional weight loss are at a higher risk for infection, depression and mortality. Weight loss in the elderly due to voluntary or involuntary (depression, cancer, cardiac disorders and benign GIT diseases) causes has been associated with mortality. A rapid weight loss of ≥ 5 per cent in one month is considered significant and needs to be evaluated immediately by a physician. Maintenance of skeletal muscle should help to attenuate the decline in body weight and its associated consequences.

Bone Health

Bone mineral density, used as a proxy for the assessment of fracture risk, declines with age starting at about 50 years of age. However, equally important is the increase in the bone turnover rate with age, driven by the increased bone resorption, leading to bone loss. Demineralization is much more rapid than mineralization resulting in net bone loss during elderly years.

In women, bone loss due to less production of estrogen especially after menopause occurs, leading to osteoporosis. Women may lose up to 20 per cent of bone mass during the 5–7 years following menopause. Afterward, the loss continues at the rate of 0.5–1 per cent per year (unless there is some adverse underlying condition or immobilization; when the rate is higher).

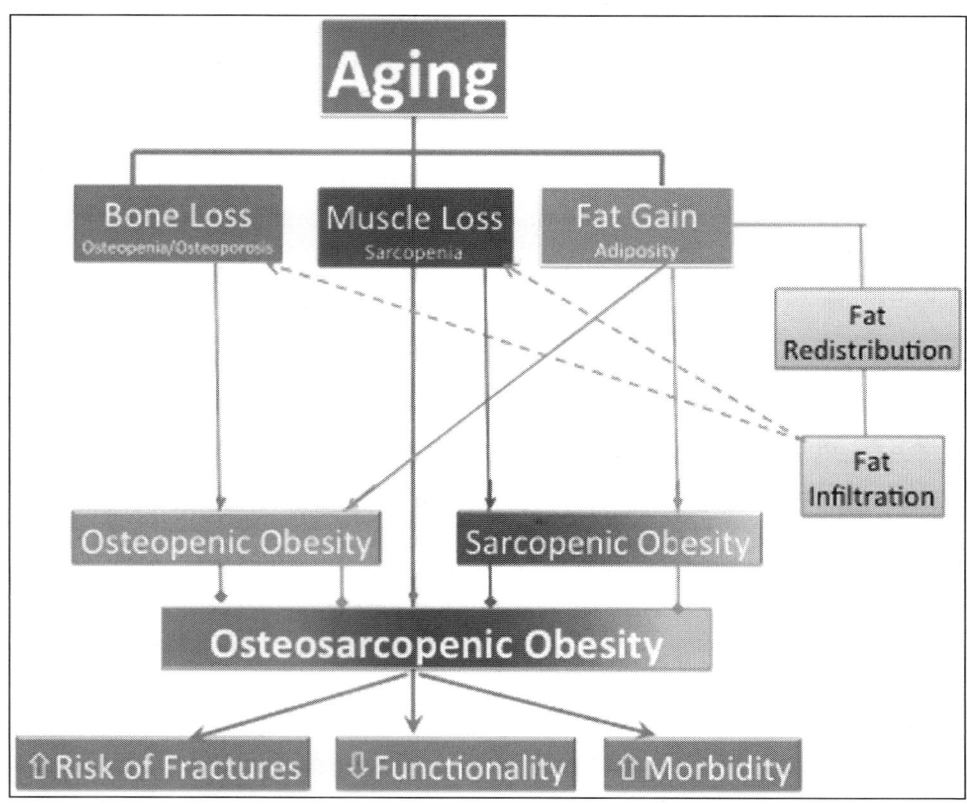

Figure 13.1: A Path of Bone, Muscle and Fat Tissues Deterioration Leading to Osteosarcopenic Obesity and its Consequences.

Men lose bone mass with age too, but the loss starts later in life and persists at about 0.5–1 per cent/year.

Severe osteoporosis may cause the bones in the legs to bow under the weight of the body. This bowing, together with changes of the spine, makes measurement of height unreliable in some elderly people, even in those who are able to stand unaided. If osteoporosis becomes severe, it limits the ability of elderly people to exercise, shop, prepare food and live a normal healthy life. It will lead to a further decrease in nutrient intake.

There is a concern that much hidden osteomalacia may be present in the elderly; it occurs due to less intake of vit. D, possibly poor vit. D synthesis in the skin or lack of sun exposure. If a person is confined inside the home due to physical disability, then a 10 µg of vitamin D per day should be given in the diet. It is important to encourage elderly women to continue consuming the RDA for calcium and to seriously consider estrogen replacement therapy if they are still within 10 years of menopause.

Diseases of the joints and bones such as arthritis occur in many elderly persons. No nutritional therapy has been devised except weight control to minimize burden on joints.

Cellular Changes

Each cell type has a given life span. Each cell can divide only a limited number of times, about 50. Once the number of divisions occurs, the cell automatically succumbs. Some cells *e.g.* those of the gastrointestinal mucosa have a short life span whereas those of the brain have a long life span. Some cells *e.g.* of the gastrointestinal mucosa, hair and skins continue to divide and reproduce throughout life. On the contrary, liver and muscle cells have a programmed death. With a decline in the number of functional cells, performance of the organ is also adversely affected.

Diminished Sense of Taste and Smell

Sensitivity to taste and smell usually declines with aging. The number of taste buds declines with advancing age. By the age of 70 years, about one-third to one-half of a person's taste buds die. Poor diet and possibly zinc deficiency can also contribute to a loss of taste. Some drugs such as diuretics can also cause decreased taste sensitivity. The taste sensitivity to salty foods is usually reduced whereas to that of sweet products is not affected. This problem is further complicated by reduction in olfactory sensitivity due to which flavour of a certain food is not sensed properly.

Such changes are significant considering the importance of smell and taste in contributing to the enjoyment of eating. Food selection, food preparation methods, dietary variety and nutrient intake may be affected. As a result, some of the aged persons lose interest in eating and have diminished appetite. These sensory changes may also lead to difficulty in discriminating between safe and spoiled food.

Oral Health and Dentition

Oral health and dentition can influence, and be influenced by, nutrition. A key component of oral health is the ability to chew food, which relies on the presence of adequate teeth or dentures, and saliva flow. About 50 per cent of people above the age of 60 years have lost all their teeth due to tooth decay and periodontal disease. Those with either few or no natural teeth, or ill-fitting dentures, are more likely to eat a restricted variety of foods, and some older adults without their natural teeth experience pain when chewing. Such experiences may discourage the consumption of foods that are difficult to chew, such as some vegetables, fruit and nuts, and some meats. Chewing difficulties may also affect enjoyment of eating, possibly further limiting food and nutrient intakes. As a result, such people select only soft, pureed, mostly carbohydrate rich foods which do not require much chewing and have deficient intake of protein, minerals and vitamins. Research has shown that micronutrient intakes can be influenced by oral health, including vitamin B_{12}, vitamin C, vitamin E and fibre. Associations among absence of teeth, food selection, nutrient intake and markers of nutritional status have been reported in some studies.

Dry mouth (xerostomia) can result from advancing age and from medication. Xerostomia interferes with the intake of food due to problems ranging from chewing to swallowing difficulties, and can be a significant problem for oral functioning. Dry mouth can also affect taste.

Older adults who retain their natural teeth are prone to dental caries. The development of dental caries is affected by the frequency with which sugars are consumed, which sometimes increases in older people through the use of high-energy snacks or supplements. In addition, older people with functional limitations may find it more difficult to maintain good oral health.

Gastrointestinal Tract

In later years, digestion of foods is affected in a number of ways. There is a decreased secretion of saliva (about 20 per cent), hydrochloric acid (about 20 per cent) and pepsin in the gastric juice and a 30 per cent decrease in trypsin in the pancreatic juice. Fat and protein digestion may be affected by a reduction in pancreatic enzyme secretion if foods containing fat and/or protein are consumed in large amounts. A decrease in stomach acid production in the elderly is usually accompanied by decreased intrinsic factor synthesis. These changes lead to poor absorption of calcium, iron and vitamin B_{12}. Lactase synthesis generally slows with advancing age while the activities of other disaccharidases remain constant throughout adult life. Many elderly people have delay in oesophageal emptying. Hiatus hernia leads to problems of heartburn and intolerance to some foods.

Constipation is another intestinal problem of the old people. It can be due to less intake of fibre rich foods as most of the elderly people have teeth/denture problem. A reduction of the tonus of the musculature of the stomach, small intestine and colon and decline in physical activity lead to less gastric motility. It may cause greater abdominal distention from certain foods and also increase the prevalence of constipation. All age groups should avoid mineral oil as a laxative at mealtimes because it binds fat soluble vitamins.

Liver, Gall Bladder and Pancreas

The liver functions less efficiently in the elderly years, which decreases the ability of liver to detoxify many medications and also increases the possibility for vitamin A toxicity. Elderly people should not take excessive amount of vitamin A in the diet.

Fats are often poorly tolerated because pancreatic production of lipase is inadequate. In old age, chronic biliary impairment may also interfere with the flow of bile to the small intestine. Gall stones, if any, can also hamper fat digestion because of less bile secretion into the small intestine. In such cases, a low fat diet may be necessary.

The digestive function of pancreas may decline in the elderly years, but the pancreas has a large reserve capacity. A sign of failing pancreas is high blood glucose due to decreased insulin secretion in old age. High blood glucose level in the obese elderly can also be caused by insulin resistance or by poor intake or absorption of chromium.

Kidney Function

With the advancing age, the rate of blood flow through the kidney decreases, number of functioning nephrons are reduced which eventually slow down the rate of glomerular filtration. As a result, kidneys filter wastes more slowly and reabsorption of nutrients into the general circulation is also adversely affected. Kidneys deteriorate in aging partly because of excess dietary protein and in some cases, excess calories. The deterioration significantly decreases their ability to excrete the products of protein breakdown. Many elderly persons especially men have elevated blood urea nitrogen and blood uric acid indicating some deterioration of kidney circulation. For adequate kidney functions, elderly persons should have sufficient fluid intake *e.g.* water since the kidneys are not efficient to change the urine concentration. The kidney's ability to concentrate urine diminishes. Foods which increase the renal solute load should be avoided. While an increase in protein intake to 1 gram per kg of desirable body weight has generally been recommended for physically active elderly people that does not apply to cases where decreased kidney function causes elevated blood urea.

Cardiovascular Function

Cardiac output often drops as the age increases. The decline is usually due to inactivity and consequent poor heart conditioning, which allow fatty and connective tissues to infiltrate the heart's muscular wall. If the elderly person remains physically active, cardiac output does not fall, so the decline is not due to aging process per se.

Atherosclerosis and hypertension are the major causative factors of the cardiovascular system problems associated with aging *i.e.* heart attack and stroke. The lumen of the blood vessels gets narrowed down due to the progressive accumulation of atheromatous plaques. Arteries lose elasticity and blood flow is constricted which then elevates blood pressure. With a decline in the blood flow, digestion, absorption and distribution of nutrients is also adversely affected. To limit the buildup of atherosclerotic plaque, serum cholesterol level should be kept below 200 mg per 100 ml. The kind and amount of fat and carbohydrates in the diet, the degree of overweight are other factors which must be given attention as they eventually affect the blood cholesterol level. Restricted amount of sodium (2 g) in the daily diet helps most of the people with hypertension. One can prevent heart attack and stroke just by eating nutritiously, walking briskly or exercising regularly, controlling blood pressure and avoiding smoking.

Thirst

Elderly people often partially lose their sense of thirst. They are then more likely to suffer from dehydration which can lead to confusion. Other symptoms are dry lips, sunken eyes, increased body temperature, decreased blood pressure, constipation, decreased urine output and nausea. They should consume enough fluids *i.e.* 1 milliliter per kcalorie.

Changes in Food Intake

A decline in food intake is common in older adults. Physiological changes in appetite regulation with age include earlier satiety (from slow gastric emptying and a reduced stomach capacity), possible changes in gut peptide hormones that stimulate or inhibit food intake, an increase in the production of cytokines (including serotonin) which are secreted by inflammatory cells in response to stress and have been shown to decrease food intake and body weight and changes to the central nervous system (*i.e.* neurotransmitters). Age-associated changes in dentition, taste and smell also influence food intake. This physiological decline in food intake has been termed 'anorexia of ageing'. Loss of appetite can be associated with physiological changes or due to psychological factors or both.

Some research studies have reported that older people consume less energy (up to 25 to 30 per cent less than younger adults), have smaller meal sizes and slower rates of eating. Reduced energy intake usually leads to weight loss, depending on energy expenditure (which is also often decreased in older adults). Lower energy intakes make it more difficult to achieve adequate intakes of micronutrients, and may lead to deficiencies.

Physical Discomfort

Elderly people are more prone to symptoms of indigestion due to decrease in gastric secretion and loss of teeth. Intake of small, frequent meals, chewing them thoroughly and avoiding foods that are poorly tolerated *i.e.* highly oily and spiced foods, should be avoided. Diet should contain ample amount of foods that are rich in iron, calcium and vit. B_{12} as their absorption is limited in old age.

Immune Function

Immune function operates less efficiently in old persons. Recurrent episodes of sickness, infection and poor wound healing are indicators of poor immune function. To maximize the health of the immune system, adequate intake of protein and zinc is necessary. The deficiency of these nutrients occurs due to limited or insufficient food intake.

Lung Function

In general, lung efficiency declines somewhat but especially in elderly people, who have smoked tobacco products and still continue to smoke. With a decrease in the number of alveoli, breathing becomes shallower, faster and difficult. The decrease in lung deficiency leads to decreased physical activity, endurance and frequently to decreased food intake. These changes ultimately hamper the ability to maintain overall health.

Hearing and Vision

Hearing and vision both decline in the elderly years. People differ as to when or if these losses become disabling. Vision and hearing losses make food shopping difficult. Similarly, this may hamper food preparation and consumption finally.

Metabolism

From age 30 years, there occurs decline (2 per cent) in basal metabolism for each decade. This is due to decreased muscle tension and increased proportion of body fat. The reduction in BMR is less in those elderly persons who remain healthy and physically active in later life. The ability to maintain the normal body temperature is also lessened; hence elderly people have a danger of hypothermia.

Hormonal Changes

Hormonal imbalances generally occur in later life. The activity of various glands such as adrenal cortex and islets of Langerhans is diminished which affects the cell metabolism. A reduction of thyroid activity affects the energy requirement. In women, hormonal imbalances may lead to disturbances of calcium and nitrogen metabolism and thus, resulting in osteoporosis. After menopause, iron requirement of women is decreased.

Neuromuscular Co-ordination

There is decline in neuromuscular coordination with advancing age. It makes the tasks such as holding of utensils and proper eating of food difficult. Those who cook the foods themselves have to reduce the variety in cooking. One avoids taking food especially in the company of other friends and guests so as to avoid embarrassment. These changes affect the types of foods cooked and consumed by the elderly people.

Changes in the nervous system and cognition: With advancing age, risks for cognitive decline increase, affecting the independence and quality of life. Insufficient intake of selected vitamins, or certain metabolic disorders, may affect cognitive processes by disrupting the nutrient-dependent processes within the body that are associated with the management of energy in neurons, which can subsequently affect synaptic plasticity or the ability to encode new memories. The earliest sign of mild cognitive impairment or pre Alzheimer's disease are decreases in the ability to prepare food, forgetting to eat, and inability to access food, which can further impair oral intake. Vitamin deficiencies, particularly vitamin $B_{12,}$ vitamin B6 and folate are associated with cognitive impairment. Other important antioxidants with possibly beneficial outcomes include foods with high levels of phytochemicals and flavonoids. Tomatoes, citrus fruits, blueberries and certain spices are known to reduce oxidative stress and cognitive impairment.

II. Psychosocial and Economic Changes

Established food habits are the result of lifetime social, cultural, psychological and economic influences. It is difficult to change the food habits of an elderly person in the later years of life as they are the part and parcel of his life style for the last so many years. By the time the person reaches 60 or 70 years old, his food habits, dietary patterns, food likings and dislikings are fixed. It is in fact hard to introduce new foods or markedly change his dietary pattern. They do not develop the taste for novel foods and would prefer to stick to previous eating pattern. If he had poor food habits throughout his life time, good diet in later years cannot completely compensate for food inadequacy of the past years or correct irreversible changes. However, the poor state of nutrition can get benefit from the application of good well balanced dietary practices and principles of optimum nutrition. Dietary modifications should be suggested with great care and consideration to elderly people so that these fit in very well within their psycho-social, cultural and economic arena.

Loneliness and Social Isolation

As eating is a social event, loneliness and social isolation can have a negative effect on the nutritional health of elderly people. After retirement, many elderly persons lose their self esteem and sense of worth as they are not consulted in deciding family matters by their own children and it hurts their self-respect. They feel powerless, irrelevant and redundant and it makes them to go into depression which ultimately affects the food selection, preparation and consumption adversely. In the modern Indian society joint family system is disappearing. In the nuclear families, after the settlement of children, parents suffer from empty nest syndrome. If living alone, they do not have motivation to cook and consume regular meals having variety and develop erratic eating patterns leading to malnutrition. Some people eat more to assuage their feelings of loneliness despair and depression. Erratic eating habits further aggravate mental depression. Social isolation can also be increased by reduced communication abilities such as loss of hearing or sight. The loss of a loved one through death, indifference or separation can also impair food intake.

Depression and other Mental Illnesses

Mental illness can result in a decrease food intake and can contribute to social withdrawal. About 12 to 14 per cent of elderly people experience significant depression. That combined with isolation and loneliness as family and friends die, move away, or become less mobile frequently contributes to disinterest in eating and so to weight loss. Depression can descend into poor appetite, weakness and then into an even poorer appetite. Depression is the most common cause of weight loss and anorexia in the older adult. In the elderly, resulting poor nutritional state can lead to further mental confusion and increased isolation and loneliness. Early diagnosis is important to begin antidepressant therapy and nutritional support.

Cognitive function, mental impairment, dementia and Alzheimer's disease may all have a negative impact on the life of an older person, because they affect their autonomy and independence. A person's nutritional status may be affected by their refusal or forgetting of meals, poor or erratic eating habits, and safety issues in cases where they are discouraged from using kitchen equipment in order to avoid harming themselves.

Anxiety

Elderly persons develop emotional stress and anxiety due to one or the other reason. This leads to loss of appetite many a times. Moreover it may also cause changes in attitude towards food, modifications in food habits and changes in digestion, absorption and utilization of nutrients. Anxiety along with hormonal changes may also result in poor utilization of nutrients due to insufficient secretion of digestive juices and secretions.

Economic Factors

Having a low income can make it difficult to obtain the foods required for a healthy diet although healthier foods are not necessarily more expensive, a person may need to cut back on the food budget to accommodate the cost of their medications. The cost of transportation and housing are also factors.

Lack of Nutrition Knowledge

Lack of nutrition knowledge in combination with long held food fads and fallacies often lead to poor selection of food items by the elderly persons. They generally have fixed notions and beliefs regarding certain foods and show resistance in changing their poor food habits. However, with a little support and motivation, it is possible to make some changes in their dietary patterns.

III. Other Factors that Influence Nutritional Status of the Elderly

Multiple Medications

Use of certain medications can profoundly affect nutrient needs in the elderly. Often at this age, several medications must be taken for long periods of time. Drug related nutritional problems can be significant. Drugs have side effects *e.g.* loss of appetite, nausea, vomiting, diarrhoea *etc.* and can affect the body's ability to consume food and to retain nutrients from food. The most notable ones are an increased need for potassium when consuming some types of diuretics in case of hypertension and changes in appetite when taking some types of antidepressants or antibiotics.

Intake of proton pump inhibitors (PPI) and H2 blockers have been reported to have interaction with the absorption of vitamin B_{12}, vitamin C, calcium, magnesium and iron. Similarly, based on the recent review, the long-term use of metformin, a drug widely used by millions for pre-diabetes and diabetes treatment, has also been shown to increase the risk of vitamin B_{12} deficiency.

Aspirin, the most common self prescribed drug for pain can causes microscopic bleeding in the GIT tract leading to loss of iron. The use of some laxatives such as mineral oil is known to interfere with the absorption of fat soluble vitamins. Nutrients most at risk include folic acid, vit. B_{12}, B_6, C, D and K as well as the minerals phosphorus, calcium, magnesium and zinc.

Complications associated with polypharmacy among the elderly include increased falls, functional decline, trouble in performing daily tasks and increased risk of malnutrition. While the benefit of most prescription drugs outweighs the nutritional risk, there is a lack of research reporting on the cumulative nutritional impact of multiple drugs in the elderly.

Chronic Illness

Diseases, especially chronic ones such as chronic obstructive lung disease, diabetes, hypertension, cardiovascular disease, cancer and infections are more common in the elderly and contribute to malnutrition. Many chronic conditions can make it difficult to swallow properly, which interferes with food intake.

Nutritional Requirements of Elderly

Various changes associated with aging as mentioned above do affect the nutritional requirements of elderly persons. Moreover, individual variations in digestion, absorption and utilization of nutrients due to occurrence of degenerative diseases, physical disability, emotional stress, medication *etc.* make the task of generalizing the nutritional needs and requirements of elderly persons, more difficult. As metabolism slows, energy needs decrease but since absorption becomes less efficient, there is a higher need for nutrient-dense foods, typically referred to those which contain more nutrients (minerals, vitamins, protein, complex carbohydrates) but less energy (kcal) per unit of weight

Indian Council of Medical Research has not given any specific separate recommendation for various nutrients except energy for the elderly population. Overall, it is generally considered that nutrient needs for the healthy elderly are similar to those for younger adults but individual modifications are necessary to compensate for specific needs and diseases present.

Table 13.2: Energy Requirements for Indian Elderly Males and Females with different Body Weights (kcal/24 hour)

	Body Weight kg)	*Sedentary (kcal)*	*Moderate (kcal)*
Male	45	1590	1870
	50	1688	1985
	55	1786	2101
	60	1883	2216
	65	1981	2331
	70	2079	2446
	75	2177	2565
Female	40	1477	1737
	45	1553	1627
	50	1630	1917
	55	1706	2007
	60	1782	2097
	65	1860	2200
	70	1936	2277

Source: NIN (ICMR), 2009. Nutrient requirement and recommended dietary allowances for Indians.

Energy

Due to decline in lean body mass, BMR along with reduced physical activity, the energy requirements of elderly are less as compared to younger individuals. Energy intake should be adequate to maintain the body weight. Energy requirements of elderly men and women with different body weights engaged in sedentary activity as recommended by ICMR (1990) have been given in Table 13.2.

Hence, elderly people should make a wise selection of foods that are low in calories but provide adequate amount of protein, vitamins and minerals. The amount of fat and sugar should be decreased in the diet so as to increase the diet's nutrient density.

Carbohydrates

Since energy intake is reduced, hence fewer calories should be derived from carbohydrates; they should provide 55-65 per cent of the daily energy requirement. Instead of simple carbohydrates, foods sources containing complex carbohydrates should be included in the diet.

Protein

Elderly persons are vulnerable to protein-energy malnutrition associated with a progressive decline in the body protein manifested by declining fat-free mass. The reduction in fat-free mass is attributed mainly to the loss of skeletal muscle and is associated with reduced muscle strength as well as predisposition to many metabolic disorders. Protein requirements for elderly are the same as for the younger adults *i.e.* one gram per kg body weight per day.

There have been suggestions that the elderly (56-80 y) require more protein intake (1.4-1.6 g/kg/day), owing to a reduced absorptive capacity, but N balance studies do not provide enough evidence to make a recommendation for an increased protein intake. However, protein and energy are inter-related. In an elderly person whose activity is maintained, there is sufficient energy intake to ensure adequate protein intake. In the sedentary elderly person, with reduced energy requirement, there is a need to consider the protein energy ratio of the diet, since protein intake can reduce if the total diet is reduced.

Fats

The energy requirement is reduced with the advancing age. Hence the requirement of major concentrated energy providing nutrient *i.e.* fats should also be reduced in overweight or obese elderly person. But there is no need to restrict fat intake beyond 30 per cent of the daily energy intake (WHO, 2002). According to NIN (2011), 20 g fat in the daily diet is recommended for old person. However, consumption of saturated fat should be minimized and should not exceed 8 energy per cent. It is desirable to include a variety of fats in the diet, particularly n-3 fatty acids found in fish, soy, linseed, canola seed and oil and green leafy vegetables. Fats and oils such as ghee, butter and refined oils should be taken in moderation. Fried foods, pastries and fatty meat should be avoided.

Recent studies have reported that energy imbalance, lower protein intakes, high level of simple carbohydrates and low omega-3 (n-3) PUFA may contribute to osteosarcopenic obesity syndrome

Minerals

Calcium and Phosphorus

Osteoporosis is very common among the elderly especially women due to increased mobilization of calcium from the bones resulting in fragile bones which fracture easily. Postmenopausal women receiving no estrogen replacement therapy require more calcium. Moreover, calcium absorption also decreases with age. However, ICMR (1990 and 2011) has not recommended any modification for calcium and phosphorus; elderly men and women of all ages require 600 mg calcium in the daily diets. The ratio of Ca and P should also be 1: 1.

Iron

Anemia is another common nutritional problem in old age. Aging does not alter the iron requirement. Hence, iron requirements of elderly men and women (post menopause) are similar to those of adult men *i.e.* 17 mg of dietary iron per day. The iron deficiency in the elderly can be due to less iron intake, poor diet, less food intake, chronic disease *etc.* Chronic ulcers, hemorrhoids and aspirin use may necessitate an increased iron intake over the RDA. Diet should contain iron rich foods such as green leafy vegetables, fresh seasonal foods, jaggery, rice flakes, dates *etc.*

Vitamins

The requirement of vitamins for the elderly is similar to that of adults. However, poor appetite, impaired absorption and decreased storage of fat soluble vitamins due to less fat absorption, chronic disease may show symptoms of some vitamin deficiencies. Therefore, the symptoms of vitamin deficiencies may be due to poor intake rather than increased need for vitamins.

Vitamin A

Vitamin A requirement for elderly of all ages is similar to that of adult man *i.e.* 600 µg of retinol or 4800 µg of β-carotene.

Other Fat Soluble Vitamins

If the elderly people are confined indoors due to physical disability, vitamin D status can be poor due to limited exposure to sunshine. Such elderly persons should be given vitamin D supplements with the consultation of a physician. Otherwise, ICMR has not given any special recommendation for vitamin D, E and K for elderly persons.

Thiamine, Riboflavin and Niacin

The requirements of these B-complex vitamins are related to energy intake for all ages. As the energy requirement of elderly declines with aging, correspondingly the total requirement for thiamine, riboflavin and niacin also decreases and can be calculated on the basis of energy intake of the elderly. In some studies, it has been reported that requirement for riboflavin increases but however, it has concluded now that its requirement should be the same as for adults.

Other B-complex Vitamins

The requirements of elderly for folic acid (100 µg), pyridoxine (2.0 mg) and vit. B_{12} (1 µg) in the daily diets remain the same as for adults.

Vitamin C

Elderly require 40 mg vitamin C per day (ICMR, 1990).

Water

Elderly persons should consume adequate amount of water and other fluids every day even if they do not feel thirsty. Otherwise there can be constipation and decreased urine output.

Fibre

To avoid constipation, a common complaint of old age, an elderly person should take 25-30 g dietary fibre per day.

Vitamin-Mineral Supplements

An elderly person should take a multi-vitamin and mineral supplement if his energy intake is less than 1500 kcals.

Dietary Guidelines

Table 13.3 depicts the balanced diet for the elderly. Compared with the recommendations for young adults, the nutritional requirements in the elderly remain about the same, but their energy needs decrease. This means that older people should consciously select foods with a high nutritional value. One must take special care to reduce the energy sources accordingly, to prevent a positive energy balance which otherwise will result in a slow but steady weight gain. According to dietary guidelines issued for elderly persons, ICMR (2011) has recommended that they should take nutrient dense low fat, micronutrient rich foods. These are foods that are high in vitamins, minerals, secondary plant materials and fibre.

Elderly persons should take 2-3 serving portions of cereals and millets less than those recommended for adults but must take one extra serving portion of fruits than the adults. Energy sources such as oils, butter, whole milk products, sweets, fried foods, baked goods, cold drinks *etc.* should be taken sparingly as they contain a lot of calories in the form of sugar and fat which can contribute to obesity. Just as in younger years, attention must also be given in old age to maintaining a healthy body weight. A BMI between 18.5 and 23 which is within the normal range should be maintained.

Table 13.3: Balanced Diet for Elderly Persons (Number of portions) (NIN, 1998)

Food Groups	g/portion	Type of Work					
		Sedentary		Moderate		Heavy	
		Man	Woman	Man	Woman	Man	Woman
Cereals and millets	30	11	8	13	10	20	14
Pulses	30	2	2	3	2.5	3	3
Milk	100 ml	3	3	3	3	3	3
Roots and tubers	100	2	1	2	1	2	2
Green leafy vegetables	100	1	1	1	1	1	1
Other vegetables	100	1	1	1	1	1	1
Fruits	100	2	2	2	2	2	2
Sugar	5	5	4	8	5	11	9
Fats and oils (visible)	5	4	4	7	6	11	8

For non-vegetarians, substitute one pulse portion with one portion (50 g) of egg/meat/fish/chicken.

Protein is important for the development and regeneration of body cells, especially the muscles. As energy requirements decrease, the protein density of the diet should be greater for both men and women *i.e.* more protein containing foods such as lean meat, low fat milk and dairy foods, eggs and variety of pulses should be eaten. Flesh foods and eggs add to the quality of diet. However, there is no compelling evidence that suggests that protein requirements for the elderly are increased but definitely protein quality of the diet should be good. It is usually helpful to give preference to lean foods and use low fat preparation methods.

Apart from cereals and pulses, they need daily at least 200- 300 ml (low fat) of milk and milk products and 400 g of vegetables and fruits to provide fiber, micro-nutrients and antioxidants. Calcium requirements should be met through adequate intake of milk and milk products and green leafy vegetables. If one suffers from lactose intolerance, consult a specialist about obtaining a general calcium supplement.

Provide a variety of nutrient-dense foods like fruits, vegetables, wholegrain cereals and pulses, low fat milk and dairy products, nuts, soybean products, fish, egg white, lean meat (if non-vegetarian) *etc.* A diet consisting of foods across all food groups in proper amounts and a wide selection within each of these groups will provide all the required nutrients. They will help in prevention of micronutrient malnutrition and certain chronic diseases such as cardiovascular diseases, cataract and cancer. Variety of whole grains, millets and pulses should be included in the daily diet. Include foods rich in alpha-linolenic (ALA) acid such as legumes, green leafy vegetables, fenugreek and mustard seeds. Foods rich in the antioxidant vitamins, vitamin E, carotenoids and vitamin C, continue to receive attention because of their potential to improve immune functions and they should be provided in the daily diets. About 4-5 servings of fruits and vegetables (all colors) should be provided. Seasonal fresh fruits are nutritionally superior to fruit juices. Encourage intake of salads; in case of chewing problem, grate the vegetables in the salad or slightly steam them before consumption.

The daily intake of oil should not exceed 20 g. Foods rich in saturated fats such as ghee, butter, *vanaspati* and coconut oil should be avoided. Limit fatty spreads in cooking or on bread containing saturated fats and *trans* fats. Use a variety of vegetable oil blends (cold pressed or 'extra virgin') containing mono-unsaturated and poly-unsaturated dietary visible fats which can be obtained from mustard, ricebran, soybean, safflower, groundnut, sunflower, til, olive oil *etc.* Added oils may assist in the absorption of fat soluble nutrients and phytochemicals from plant foods. But at the same time, it should be kept in mind that the vegetable oils are also to be taken in restricted amount; not more than 20 per cent of total calories should be taken. Substitute nuts and oilseeds in place of fats and oils.

Limit the salt intake. Higher sodium intake leads to greater calcium excretion which may result in reduction in bone density. Existing evidence reveals a deleterious impact of high salt intake on blood vessels, blood pressure, bones and gastrointestinal tract. Salt intake should not be more than 6 g per day.

Avoid or limit the intake of salty, spicy and fried foods such as sweets, pickles, bakery and other processed goods in the diet. Processed foods being rich in fats, salt, sugar and preservatives may pose a health risk if consumed regularly and therefore, should be better if avoided completely.

Consume adequate amount of water as it is the most important nutrient of all to avoid dehydration, hyponatraemia and constipation and helps in the upkeep of health. Adequate water-intake guidelines are 1.0 ml water/kcal energy consumed (for example, 1.8 L for a 1,800-calorie intake), or 25-30 ml/kg of weight for most individuals.

In case elderly who are confined to bed, supplement of Vitamin D as prescribed by the physician should be given. Bone loss can be slowed down by including calcium and vitamin D rich foods in the diet also helpful is adequate outdoor exercise.

Pattern of Eating and Life Style

Elderly people are more susceptible to certain digestive disorders as well as constipation, diarrhoea, bloating, feeling of fullness or loss of appetite because the capacity of the digestive organs decreases in old age. Other causes lie in the way individuals eat: too little fibre rich foods, lack of fruits, habitually eating too large a meal or eating too fast. The causes may also be related to medications or diseases, as well as personal circumstances involving stress, mental tension or depression.

All of the above mentioned problems can be dealt as given under:

- ☆ Healthy and positive food concepts and cooking practices are foundation for good health. The balanced diet so planned needs to be well cooked in less oil, soft, less salty and spicy.
- ☆ A high fibre rich diet having whole grain products, vegetables and fruits can decrease constipation, but fibre intake should be gradually increased to about 35 g, making sure a glass of water or other fluid follows each dose of fibre.
- ☆ Food should be properly chewed as it promotes digestion. Soft and easily chewable foods should be taken if there is a dental problem.
- ☆ Small quantities of food in divided portions over a day should be consumed at more frequent intervals.
- ☆ Caffeine containing beverages should be limited especially at bed time if there is a problem of gastritis and insomnia.
- ☆ Enjoy food and eating in the company of others. Eating with others brings great pleasure and decreases stress. Solitary elderly people can get together to cook and eat together but avoid the regular use of energy dense (nutrient poor) celebratory foods and beverages which are high in fat and or sugar.
- ☆ Food should be served in pleasant surroundings.
- ☆ Maintain the balance of food intake with physical activity. Body movements in the elderly in the form of exercises like walk, yoga *etc.* regularly can be added to prevent muscle deterioration, improve flexibility, mood, sense of well-being and self esteem, promote mental equilibrium and to ward off depression and the diseases of civilization.
- ☆ Use the stairs instead of elevators, go shopping on foot, take morning or evening walk *etc.* Outdoor exercise builds up vitamin D from the sunshine which, along with calcium, influence bone density.
- ☆ Energy consumption is also raised by regular exercise. People who exercise can also eat more and this makes it easier to meet daily vitamin, mineral and trace element requirements. Exercise and abdominal massage keep the intestines moving.

Sample Diet Plan

Practical Assignment

A sample day's diet plan for Mr. X, 65 years old (retired officer) sedentary worker, vegetarian, 70 kg body weight is as under:

Step I: Personal profile

Age	65 years
Body weight	70 kg
Sex	Male
Physical activity	Sedentary
Socio-economic status	Middle
Food habits	Vegetarian

Step II: RDA (ICMR, 1990)

Energy	2176 Kcal (However, it varies from person to person depending upon his energy expenditure and health status)
Protein	60-65 g
Calcium	600 mg
Vitamin C	40 mg

Step III: Sample Menu

Early morning	Tea
Breakfast	Oat porridge
	Scrambled egg
	Toast with butter
	Milk
Mid morning	Orange
Lunch	*Chapati*
	Stuffed tinda
	Peas and potato vegetable
	Salad
Evening tea	Tea
	Biscuits

Dinner — Mixed vegetable soup, *Chapati*, *Dal Palak*, *Paneer bhurji*, Fruit custard

Step IV: Deciding the no. of servings of each food item, convert into amounts of raw food ingredients

Meal	*Menu*	*Ingredients Used*	*Amount (g)*
Early morning	Tea		
Breakfast	Oat porridge Scrambled egg Toast with butter Milk		
Mid morning	Orange		
Lunch	*Chapati* Stuffed *tinda* Peas and potato vegetable *Salad*		
Evening tea	Tea Biscuits		
Dinner	Mixed vegetable soup *Chapati* *Dal Palak* *Paneer bhurji* Fruit custard		

Step V. Using the Indian Food Composition Tables (2017), Calculate the amount of various nutrients specified in the various food items as mentioned in the planned diet.

Meal	*Menu*	*Ingredients Used*	*Amount (g)*	*Energy (kcal)*	*Protein (g)*	*Fat (g)*	*Iron (mg)*	*Vit. A (µg)*
Breakfast								
Total								
Mid morning								
Total								
Lunch								
Total								
Evening tea								
Total								
Dinner								
Total								
Total								

Step IV: Compare the given food items vs. RDI and given nutrients vs. their RDAs. Accordingly, make adjustments in the amount of given food items so as to arrive near to the RDAs of specified nutrients.

Food Groups/ Exchanges	*Recom- mended (g)*	*Given (g)*	*Difference (g)*
Cereals			
Pulses			
Other vegetables			
GLVs			
Roots and tubers			
Fruits			
Milk and milk products			
Fats and oils			
Sugar and jaggery			
Fish, meat, chicken *etc.*			

Nutrients	*Recom- mended*	*Given*	*Difference*
Energy (kcals)			
Protein (g)			
Calcium (mg)			
Vit. C (mg)			

Unit 2

14

Principles of Diet Therapy

Diet therapy is a broad term for the practical application of nutrition as a preventative or corrective treatment of disease. It has been known to improve the symptoms of many ailments—whether the illness is acute, chronic or critical—and is used to promote better health in general, even if a person is not technically sick. This usually involves the modification of an existing dietary lifestyle to promote optimum health. Diet therapy can be regarded as a change in the diet we normally consume in order to treat an illness or disease in our body. It can be as basic as outlining which foods to eat and avoid, or it may follow a prescribed method,

Diet therapy is often considered a complementary form of alternative medicine. Naturally, when a person is sick, she should consume only food that promotes healing and lessens discomfort. Diet therapy is an effective method of healing, especially when utilized as a complementary method of care. It is used for people who are trying to increase or decrease their body weight, have a sensitive digestion system that may be prone to allergies, who may not be able to digest certain foods, are trying to produce specific effects in their body (like regulating blood sugar), or are trying to overcome vitamin or mineral deficiencies. For instance, an anaemic person will be advised to eat a diet high in iron. A dietician will outline which foods are high in iron and make sure the patient is eating an adequate amount of these foods a day.

Objectives of Diet Therapy

Diet therapy is concerned with the modification of normal diet to meet the requirements of sick individual and its objectives are:

1. To maintain good nutritional status.
2. To correct deficiencies by the addition of food rich in some necessary element (*e.g.*, supplementing the diet with iron in treating macrocytic anemia).
3. To afford rest to the whole body or the body organ that is affected.
4. To provide ease of digestion by omitting irritating substances, such as fiber, spices, or high-fat foods.
5. To maintain body's ability to metabolize the nutrients.
6. To bring about changes in body weight whenever necessary.
7. To adjust the diet to the body's ability to use certain foods.

Principles of Diet Therapy

Following principles are important in diet therapy:

1. **Type of disease:** It is essential to determine the diet in accordance to the disease, otherwise it may have adverse effects *e.g.*, low carbohydrate diet in diabetes and high fibre diet in case of constipation.
2. **Duration of disease:** The diet should be planned after determining the duration of the disease whether it is of long or short duration. In long duration diseases, the diet planning should not be always the same. Make changes from time to time depending upon the condition of the patient as it is very important.

3. **Selection of food products according to the disease:** The amount and type of food should be changed after analyzing the modification required in food products according to the prevailing disease condition.
4. **Case history of food intake of the patient:** Information about the food habits, liking and disliking, meal timings, economic conditions, availability of food, knowledge of cooking methods *etc.*, should be obtained and diet should be planned accordingly.
5. **Psychology of the patient**: Dietician should take care of the psychological condition of the patient. His whole personality should be considered as one unit *i.e.*, emotional, economic status, social status, *etc.*, should be kept in mind. The diet of a patient should be planned in such a way that the patient should not feel it different from that of the others.
6. **Variety and attraction in meals:** The patient is normally disturbed by his disease and if he is given the same diet every day, he will lose interest in food. Therefore, it is important to add variety to the diet of the patient. Food should be attractive, properly cooked, and served in congenial environment. A well planned meal is fruitful only if the patient eats it completely. Taste of the food is also very important.
7. **Selection of appropriate diet for the medical condition:** It is often necessary to cater to a patient's appetite, since many individuals become especially hard to please when sick. In some disease states, such as cancer, patients experience marked taste changes. Because of the importance of the nutritional elements in feeding the sick, try to carry out the patient's wishes whenever possible. A tactful and observant Hospital Corpsman can be of great benefit to the physician and dietitian in carrying out the dietary regimen. You must be aware of what comprises a well-balanced diet and should be able to recognize when dietary adjustments need to be made in special situations. This is important to meet the changing needs of the diseased body's ability to make use of foods. The patient should be made to feel that the utmost cleanliness and care have been observed in the preparation and service of their food. The patient's face and hands should be cleaned before food is served, and the lips and teeth cleaned before and after the meal. If the mouth is dry, it should be moistened periodically. When special or modified diets are ordered, check the contents of the tray with the written orders. An error in serving a special diet may cause discomfort, serious illness, or even death.

Basis of Therapeutic Nutrition

Generally, therapeutic nutrition is simply the modification of normal diet. Hence, the basis of therapeutic nutrition is the normal diet and doing so has certain advantages while planning therapeutic diets which are as under:

1. There is no doubt that there are quantitative and qualitative differences in nutritional requirements of normal and ill persons but psychological and social needs of them are the same. By modifying the normal diet for the sick person, their psychological and social needs are well taken care of.
2. No extra efforts and special preparations are required for preparing the food for the sick person. Cooking of therapeutic food is simplified as the modified therapeutic diet shall also be based upon the food pattern and food habits as of rest of the family members.
3. It will be easier to know from the normal diet plan that if certain food items are added or omitted in the modified diet based on that, what nutrients shall become deficient or shall increase by doing so. For example, if instead of whole cereals and pulses in the normal diet, dehusked cereals and pulses have been used, the modified diet shall contain less dietary fibre. If salt content has been restricted in the normal diet, the modified diet shall be having low sodium content.

What is Therapeutic Diet?

Therapeutic diet is the planned modification of a normal diet in which this modified diet is taken in such a way that it is nutritional and acceptable at the same time to the patient. It is part of the treatment of a medical condition and are normally prescribed by a physician

Therapeutic Diet

- ☆ Modifications of normal diet used to improve specific health conditions
- ☆ Normally prescribed by doctor and planned by dietitian
- ☆ May change nutrients, caloric content and/or texture
- ☆ May seem strange and even unpleasant to the patient
- ☆ Use patience and tact to convince patient to eat food
- ☆ Understand purpose of diet and provide simple explanations to patient

and planned by a dietitian. It is modified or tailored to fit the nutrition needs of a particular person. During times of illness, a person moves from one type of modified diet to the other diet gradually and thus, ultimately leading to one's normal or goal diet. For instance, if a person has under gone a surgery, a liquid diet is very essential and eventually it becomes a stepping stone that leads to everyday base diet. The main intention of providing this kind of diet plan is to help the patient to return to the normal absorption and digestion patterns.

Purpose of Therapeutic Diets

- ✰ Regulate the amount of food
- ✰ Assist body organs to maintain normal function
- ✰ Aid in digestion
- ✰ To maintain nutrition status
- ✰ To correct nutritional status
- ✰ To restore nutritional status
- ✰ To provide extra calories for weight gain or decrease calories for weight control
- ✰ Modify the intervals of feedings
- ✰ To improve specific health conditions
- ✰ To balance the amount of various nutrients in different diseases
- ✰ To restrict sodium in certain diseases or increase protein in some disease conditions
- ✰ To provide texture modifications due to problems with chewing and/or swallowing
- ✰ To exclude foods in food allergies or food intolerances

Various Types of Modified or Therapeutic Diets

Modified or therapeutic diets are modifications of the regular diet and are designed to meet specific patient needs. They are modified for:

- ✰ Method of preparation (*e.g.* Baking, boiling, or broiling)
- ✰ Consistency (*e.g.*, ground or chopped)
- ✰ Texture
- ✰ Flavour (bland foods)
- ✰ Total calories (*e.g.*, high or low calorie)
- ✰ Nutrients (*e.g.*, altering carbohydrate, protein, fat, vitamins, and minerals)
- ✰ Allowing only specific foods (*e.g.*, diabetic diet).
- ✰ Food allergies or food intolerances

Diets used in the treatment of disease are often spoken of by specific names that show a special composition and often indicate the purpose for which the diet is intended such as

- ✰ Regular diet
- ✰ Liquid diet
- ✰ Soft diet
- ✰ Diabetic diet
- ✰ Calorie controlled diet
- ✰ Low cholesterol diet
- ✰ Fat restricted (low-fat) diet
- ✰ Sodium restricted diet
- ✰ Protein diet
- ✰ Bland diet
- ✰ Low residue diet

Recommended dietary allowances are for the normal healthy persons and donot apply to persons who are malnourished or undernourished or those suffering from various infections and diseases. Therefore, planning a diet for any diseased condition shall imply ability of the person to adapt to the principles of normal nutrition as mentioned above for various regimens.

For planning each diet, diet history and assessment of nutritional status of the patient should be done at the first go. Accordingly, the diet should be planned and executed for him.

Factors to be Considered while Planning Therapeutic Diets

The alteration of the normal diet requires on appreciation of:

1. The underlying disease conditions which require a change in the diet.
2. The possible duration of the disease.
3. The factors in the dietary which must be altered to overcome these conditions.
4. The patient's tolerance for food by mouth.
5. In planning meals for the patient, his economic status, his food preferences and his occupation and time of meals should also be considered.

The planning of therapeutic diets implies the ability to adapt the principles of normal nutrition to the various regimens for adequacy, correctness, economy and palatability. It requires recognition of the need for dietary supplements such as vitamins and minerals concentrations when the nature of diet itself improves severe restrictions, the patient's appetite is poor, or absorption and utilization are impaired so that the diet cannot meet the needs of optimum nutrition. Intelligent planning of therapeutic diets necessitates a consideration of food costs, the avoidance of waste and retention of nutritive value so that the diet is economically practicable. The likes and dislikes of patients are respected because food habits are deep-seated and it is not possible to change them overnight. The planning of therapeutic diets requires considerable encouragement and understanding on the part of the doctor-nurse-dietitian team to bring about important changes in the diet.

Diet Prescription

To formulate the most appropriate nutritional therapy for each patient, individual requirements must be considered with initial metabolic, biochemical and anthropometric data obtained by nutritional assessment of the patient. The dietary prescriptions are made on the basis of patient's age, sex, activity, body weight and the type of treatment he is undergoing.

The main steps that are involved in constructing a dietary prescription are as follows:

1. Estimation of body weight and its comparison with desirable body weight.

$$\% \text{ desirable weight} = \frac{\text{Present weight}}{\text{Desirable weight}} \times 100$$

2. Calculation of daily calorie and other nutrient intake

Diet prescription should meet all the nutrient requirements of the patient. To introduce variety and flexibility in diet prescriptions, food exchanges lists should be used. While the likes and dislikes of the patient are taken into account, it becomes easier for the patient to adhere to his recommended food and nutrient intake.

Finally, when prescribing a diet, the philosophy is to stress how few foods are restricted and many foods, patient can choose from. The patient should not be made a prisoner of his dietary prescription.

15

Therapeutic Adaptations of Normal Diet

The regular normal diet is composed of all types of foods, is well balanced, adequate in nutritional value, attractively served to stimulate a possible poor appetite and capable of maintaining a state of good nutrition On the other hand, the regular hospital diet is simple in preparation, character, ease of digestion and calculated to afford maximum nourishment with minimum effort to the body. It is intended for convalescing patients who do not require a therapeutic diet. At times, it has a slightly reduced caloric content. Foods such as rich desserts, cream sauces, salad dressings and fried foods may be decreased or omitted. But diet modifications of normal or hospital diets are done in case of special diseased conditions such as diabetes, coronary heart disease, hypertension *etc.* Modifications of normal diets can be done in several ways depending upon the disease condition. Some of such examples have been mentioned as under:

Table 15.1: Modifications of Diets

Modifications	*Example*
Restrictions of a nutrient	Low Calorie Low Protein Low Sodium Low Fibre Low Fat
Excess of a nutrient	High Calorie High Protein High Fibre Iron Rich
Change in texture	Regular Diet Mechanical Soft Diet Pureed Diet
Change in consistency	Clear Liquid Diet Full Fluid Diet Soft Diet Semi Solid Diet
Rearrangement of meals	Increased/Decreased frequency of meals
Omission of foods	Gluten/Peanuts/Mushrooms/milk *etc.* in case of allergy

1. Modification of Diet According to Consistency

The modification in diet may mean change in several dietary factors. For instance, modification in consistency may mean a liquid diet.

A. Diets without Solids (Liquid Diet)

Liquid diets consist of a variety of foods that are liquid or liquefy at room temperature. It is the recommended diet therapy for patients who cannot tolerate solid foods or are unable to chew and swallow their food. A liquid diet may include many clear liquids, fruit drinks, fruit waters, vegetable juices, milk and possibly specialized sport drinks. For example, water, fruit juices, sugarcane juice, vitamin water and other clear beverages. These liquids provide clear fluids to relieve thirst and prevent dehydration. They maintain the electrolyte balance in the body and provide minimum bowel residue. Such a diet is generally given:

- ✰ After surgery or a heart attack
- ✰ Patients with acute infections or digestive problems
- ✰ To replace fluids lost by vomiting or diarrhea
- ✰ Before some X-rays of digestive tract

The liquid diets are recommended only for shorter periods of time, typically after a surgery. This is because of the fact that a liquid diet contains inadequacies in all types of nutrients and usually causes fast weight loss, however that fact may be a desired result of a liquid diet.

The liquid diet is of two types:

i) Clear liquid diet
ii) Full liquid diet

i) Clear Fluid Diet

It is a temporary diet of clear liquids which is made up of clear liquids that leave no residue. This diet is non-gas forming, non-irritating and non-stimulating to parasitic action. It consists of water, fat free clear soups, whey water, cereal honey water, tender coconut water, clear fruit juice *e.g.* apple or grape juice, fat-free broths, plain gelatin, ginger ale, tea with lemon and sugar, black coffee, carbonated beverages *etc.* Milk and liquids prepared with milk are omitted and also fruit juices which contain fruit pulp.

This diet is given at 1-2 hours intervals. The amount of fluid is usually restricted to 30-60 ml/hour at first and then gradually increased. This diet is recommended in acute illness which causes vomiting, diarrhoea, anorexia and distention, in acute inflammatory condition of intestinal tract, following operations upon the colon or rectum when it is desirable to prevent evacuation from bowel. The clear fluid diet is also given to relieve thirst, to supply the tissue with water, to add removal of gas.

This diet on an average contains 500-600 Kcal, 5-10g protein, minimal fat, 120-130 g of carbohydrates and small amount of sodium and potassium. It is deficient in calories, protein, minerals, vitamins and minerals and should be used for short periods. It is usually continued only for 24-48 hours. After that a more liberal liquid diet is given. As the patient's tolerance improves, the amounts can be increased.

ii) Full Fluid Diet

This diet bridges the gap between the clear fluid and soft diet. It includes everything on clear liquid diet plus strained soups and cereals, fruit juices (lemon/orange/mosumbi/pineapple/apple/sugarcane) and vegetable juices and soups (tomato/lettuce/spinach/cabbage), yogurt, hot cocoa, custard, ice cream, pudding, milk shakes, sherbet, and eggnog *etc.* It is free from cellulose and irritating condiments. The diet is used for patients who are acutely ill, unable to chew, swallow, or digest solid foods. Such diet is prescribed following operations, in acute gastritis, acute infections and in diarrhoea. The diet is given at 2-4 hour intervals. Six or more feedings can be given daily.

The average full liquid diet contains 1000-1500 Kcal with 40-50 g protein, 50-65g fat and 150-170 g of carbohydrates. The protein content of the diet can be increased by incorporating whole egg, egg white, mashed/grated processed cheese, nonfat dry milk in beverages and soups. The caloric value of the diet can be increased by adding butter to cereal gruels and soups,

Table 15.2: Contents of Soft, Full Fluid and Clear Fluid Diets

Type of Food	*Foods Allowed*		
	Soft Diet	*Full Fluid Diet*	*Clear Fluid Diet*
Cereals	Refined, finely ground whole grains	Gruels, porridges, *kanji*, ragi malt	Barley water
Pulses	All *dals*	*Dal* soup, *dal* payasan	*Dal* water
Vegetables and fruits	Juices, pureed, cooked and mashed, baked, ripe banana	Strained juices, cooked and pureed fruits	Clear strained fruit juice
Milk	Milk and milk products, cheese, fine cream	Milk and milk beverages, milk shakes, *lassi*	Whey water
Fats and oils	Butter, oil, cream, margarine	Butter, oil and cream	–
Meat and fish	All except pork, minced fish, poultry	–	–
Eggs	All except fried	Only in beverages	Egg white well beaten in fruit juices
Sugar and jaggery	All	Sugar, jaggery and glucose	Sugar or glucose
Nuts and oilseeds	None	None	None
Beverages	All	Tea, coffee, non-carbonated beverages	Tea, coffee (without milk), carbonated beverages, coconut water
Soups	All	Strained	Fat free broth
Desserts	Custards, *kheer*, puddings	Custard, ice cream, plain gelatin	Plain gelatin

Source: Sri Lakshmi. 2001. Dietetics. New Age International (P) Ltd.

glucose in beverages and using ice creams and desserts. If decreased volume of fluid is desired, nonfat dry milk can be substituted for the part of liquid milk.

iii) Commercial Liquid Formulas

These are used to supplement other external diets. These formulas vary in composition and source of nutrients. Most of formulas are milk based. For persons who cannot tolerate milk, protein source can be meat, soy or casein hydrolysate. Fat and carbohydrate composition and proportions also vary to accommodate persons with different needs and restrictions.

Following should be the criteria for selection of appropriate formulae:

- The protein source
- The composition and proportion of fats and carbohydrates
- The osmolarity
- The palatability
- The cost

B. Diets with Solids

i) Soft Diet

It is similar to regular diet but foods must require little chewing and be easy to digest. It is a transition between a liquid and regular diet. The soft diet is made up of simple, easily digestible foods and contains no harsh fibre and no rich highly seasoned foods. The soft diet is often called a low fiber diet too as it is free from most of the fibers of a normal diet and is easily digestible. This diet is soft in texture and consists of liquids and semi-solid foods such as milk, fruit juices, ripe fruits without skin or seeds, eggs, cheese, custards, puddings, strained soups and vegetables, boiled and mashed potato, rice, ground lamb, fish, wheat, corn, or rice cereals and breads *etc.* Foods may be softened by cooking or mashing. Canned or soft-cooked fruits and vegetables may be used in place of raw or dried varieties. Refined breads and cereals are usually recommended over coarse, whole-grain types. Rich desserts, all fried foods, raw fruits and vegetables, coarse cereals, strong spices, nuts and raisins are avoided. It is nutritionally adequate when planned on the basis of normal diet. Many hospital patients are placed on this diet until a diagnosis is made.

It is indicated in certain postoperative cases, for convalescents who cannot tolerate a regular diet, in acute illnesses, and in some gastrointestinal disorders. It bridges the gap between acute illness and convalescence. It may be used in acute infections, following surgery for patients who are unable to chew. It can help to ease difficulty in chewing and/or swallowing due to dental problems or extreme weakness, and it is sometimes recommended to relieve mild intestinal or stomach discomfort. The soft diet can be especially helpful to patients who are undergoing treatments like chemotherapy, or radiation of the head, neck or abdominal areas, which may cause digestive problems or make the mouth and throat very sore.

The average composition of the soft diet is 1800 to 2000 Kcals. However, energy, protein, carbohydrate levels are based on selection of menu, patient's preferences and other restrictions that may be imposed.

Table 15.3: Sample Menu of Soft Diet

Breakfast	Milk/curd/*lassi*/tea/coffee Bread butter/oats *dalia*/*suji kheer*/poached egg/ scrambled egg
Mid morning	Fresh fruit juice/fruit stew
Lunch/Dinner	Soup/vegetable puree Bread butter/rice gruel Milk/custard/egg custard
Evening tea	Milk/coffee + biscuits

ii) Semisolid Diet

This diet is the softer form of normal balanced diet. It is given to patients on way to recovery from acute illness and to patients having chewing problems. This diet is nutritionally adequate and patient may be kept on this diet for longer period. The foods allowed are beverages like tea and coffee, cornflakes, noodles, rice, whole wheat *dalia*, *suji* porridge *etc.* may be given but hard bran should be avoided. Raw fruits such as banana, grapes, orange, apple, pear and peaches all without skin can be used. Vegetables like potato without skin and carrot, peas and pumpkins all without skin, tender green leafy vegetables, tomato *etc.* can be added. For protein source, egg, *paneer*, minced meat, chicken and well cooked fish can be used.

Table 15.4 Sample menu of semi-solid diet

Breakfast	Milk, Bread butter/toast butter/bread jam/ cornflakes Boiled egg/poached egg/scrambled egg
Mid morning	Fresh fruit juice/fresh fruit without skin and seeds
Lunch/Dinner	Vegetable soup/chicken soup *Khichdi*/*dalia*/*upma*/soft rice Vegetable stew Curd/*lassi* Custard pudding
Evening tea	Tea/Milk/coffee/+ biscuits or sandwich

2. Modification of Diet According to Texture

Such modified diets are used in the treatment of gastro- intestinal tract, oral surgeries or during recovery stage. They can vary from a very low residue diet to a very high fibre diet. Method of feeding is by mouth, unless otherwise indicated. Texture modified diets may be classified as:

Regular Diet

All foods are permitted and there are no restrictions.

Mechanical Soft Diet

The mechanical soft diet is a close cousin of the soft diet. It gets its name from the fact that household tools and machines, like a blender, meat grinder, or knife, are used to make foods easier to chew and swallow.

In contrast to the soft diet, the mechanical soft diet does not restrict fat, fiber, spices, or seasonings. Only the texture and consistency of foods are changed. Fruits and vegetables may be soft-cooked or pureed. Meats, fish, and poultry can be cooked, ground, and moistened with sauce or gravy to make chewing and swallowing more comfortable. Breads and crackers, bread sticks, dried fruits, raw vegetables, pop corn, deep fried crispy foods, potato or snack chips *etc.* are avoided as they can be dry and difficult to swallow. Milk and other dairy products like pudding, custard, and smooth yogurt can be taken as such.

The mechanical soft diet is appropriate for patients who are recovering from head, neck, or mouth surgery, who have dysphagia (difficulty swallowing), narrowing of the esophagus, or who are too ill or weak to chew. The diet also benefits those who have poorly fitting dentures, no teeth, or other dental problems.

Pureed Diet

Food should be of mashed potato consistency, not too runny or liquid. Pureed foods should be properly seasoned in the same way as regular foods. A blender or food processor is used to achieve proper consistency. Food may require additional liquids or thickening during preparation to achieve proper consistency. Commercially prepared pureed foods are available as are formed purees at home.

3. Modifications on the Basis of Energy and Various Nutrient Contents of the Diet

i) Low Calorie Diet

Low calorie diet essentially involves consuming foods that are high on nutritional content and yet have fewer calories. It is defined as a diet which contains less amount of energy than the recommended allowance per day for a particular age group *e.g.* a daily diet containing 1000 or 1200 Kcals instead of 1900 Kcals for an adult sedentary female worker. A low calorie diet is any diet that usually allows 800 to 1000 calories or less in a day. This type of diet is not recommended for children, adolescents, and pregnant women. Individuals who usually go on such a diet are extremely overweight or morbidly obese. This type of diet is also recommended for reducing the incidence of life style diseases. Foods such as butter, cream, whole milk, cream soups or gravies, sweet soft drinks, alcoholic beverages, salad dressings, fatty meats, candy and rich desserts *etc.* are completely avoided or limited.

Table 15.5: Indications/Contraindications for Low Calorie Diets

Indications	*Contraindications*
☆ Obesity/Overweight ☆ Diabetes mellitus ☆ Cardiovascular disease ☆ Hypertension ☆ Gout ☆ Gall bladder disease ☆ Preceding surgery	☆ Unstable cardiac or cerebrovascular disease ☆ Acute and chronic renal failure ☆ Severe of end stage renal disease ☆ Psychiatric disorder that could interfere with compliance
☆ Side effects may be fatigue/dizziness, constipation, diarrhoea, dry skin, hair loss, menstrual changes and cold intolerance	

Variations in Low Calorie Diets

Low calorie diets are varied with regard to either decrease in fat composition or carbohydrate composition, with an increase in fibre and fluid intake *e.g.*

- ☆ Low fat, normal protein, normal carbohydrate diets
- ☆ Low carbohydrate, high protein, low-normal fat diet

ii) High Calorie Diet

A high calorie diet is the one that provides 1000 or more calories a day beyond what is ordinarily recommended to maintain weight. It may be prescribed for nursing mothers, patients with severe weight loss caused by illness, or people with abnormally high metabolic rates or energy requirements, such as certain athletes or outdoor workers, patients who are underweight or who have anorexia nervosa, hyperthyroidism, or cancer. Extra proteins and carbohydrates are included in such a diet. In such a diet, one should avoid high-bulk foods such as green salads, watermelon and fibrous fruits. Avoid high-fat foods such as fried foods, rich pastries, and cheese cake because they are digested slowly and spoil the appetite.

iii) High Carbohydrate Diet

A high carbohydrate diet means a diet that provides more than 70 per cent of the day's total energy from complex nutrients-carbohydrates. High carbohydrate, low fat and keeping the protein constant in diets have been associated with proportional rise in plasma leptin levels which keeps a check on appetite and is used in weight loss therapies also. In sports, physical exertion activities, high carbohydrate diet for a short term is required. For liver regeneration, in certain neurological diseases and before sports even, more amount of carbohydrate is given. Contraindications are in liver failure, uncontrolled diabetes, pregnancy with gestational diabetes mellitus *etc.*

iv) Low Carbohydrate Diet

A low carbohydrate diet is a diet that restricts the consumption of carbohydrates to 20-60 g per day. These diets are used in the treatment of diabetes, obesity, chronic fatigue syndrome, polycystic ovarian disease, epilepsy *etc.*

iii) Low Cholesterol Diet

A low cholesterol diet is the one which is designed to reduce the amount of cholesterol circulating in the blood. Such a diet is indicated for persons with high serum cholesterol levels *i.e.* hypercholesterolemia, hyperlipidemia, or hyperlipoproteinemia, cardiovascular disorders and obesity. Such a diet restricts the amount of animal foods such as beef, liver, pork, lamb, egg yolk, cream, cheese, natural cheeses, shellfish, whole milk and saturated fats *e.g.* coconut and palm oil products. It concentrates instead on poultry, fish, vegetables, fruits, low-fat dairy products, and polyunsaturated fats.

iv) Low Fat Diet

A low fat diet is one where less than 30 per cent of the total calories are derived from fats. A very low fat diet is one where 15 per cent or less of the total daily calories comes from fat. Diets containing low amounts of fat and cholesterol are designed to reduce the risk of heart diseases and, in some cases, cancer. Low fat diets are rich in whole grains, fresh fruits, vegetables and legumes and limit or exclude the consumption of processed oils, dairy products (cream, ice cream, cheese, *khoa*), rich desserts, nuts, oilseeds, fried foods, baked goods like cakes, cookies, pies and doughnuts, regular salad dressings and meat. Low fat diet is indicated in obesity, gallbladder disease, heart disease, malabsorption syndromes, nontropical sprue, celiac disease, cystic fibrosis and hyperlipidemia.

v) Low Sodium Diet

The mineral content of the diet may also be modified. Four levels of sodium restrictions are most often used *i.e.* 250, 500, 1000 and 2400-4500 mgs. Hence, a low sodium diet is a diet that includes no more than 1,500 to 2,400 mgs of sodium per day. Sodium occurs naturally in most foods. The most common form of sodium is sodium chloride, which is table salt. Milk, beets and celery also naturally contain sodium, as does drinking water, although the amount varies depending on the source. Sodium is also added to various food products. Some of these added forms are monosodium glutamate, sodium nitrite, sodium saccharin, sodium bicarbonate and sodium benzoate in various processed foods. Fast foods and pickles are generally very high in sodium. Unprocessed, fresh foods, such as fresh fruits, vegetables, lean meats, poultry, fish and unprocessed grains are low in sodium content. People with certain medical conditions such as high blood pressure, kidney disease, oedema, prolonged treatment with hormones especially sex hormones, ACTH *etc.*, retention of fluid due to severe PEM, ascitis, liver failure and heart problems can benefit from a diet that is low in sodium. A low-sodium diet also can help decrease breathing difficulties caused when the weakened heart has difficulty pumping excess fluid out of the body. Such patients should avoid or limit addition of salt to any food, smoked meats or fish, processed foods, pickles, sauerkraut, olives, and processed cheeses

vi) High Protein Diet

This is a regular diet with added protein rich foods such as pulses, meats, milk, cheese and eggs. Such a diet is generally recommended for children and adolescents who need additional growth, pregnant and lactating women. This type of diet is also given before and/or after surgery and to patients suffering from burns, fevers, or infections. Body builders also recommend it for building muscles. While adequate protein is required for building skeletal muscle and other tissues, there is ongoing debate regarding the use and necessity of high protein diets in weight training and body building. High-protein diets have become a popular way to lose weight because emerging research has hinted that protein may be able to satisfy hunger better than either fats or carbohydrates. But the American Heart Association doesn't recommend high-protein diets for weight loss. Indications for high protein diet are fever, hyperthyroidism, burns, nephritic syndrome, hemorrhage, post operative condition *etc.*

vii) Low Protein Diet

It is a regular diet with limited or decreased protein rich foods. Protein free diet is recommended in cases like hepatic coma, acute aneuria *etc.* Indications for low protein diet are hepatic encephalopathy, acute and chronic glomerulonephritis, acute and chronic renal failure, hepatic precoma and coma, diabetic nephropathy and diabetic retinopathy. This type of diet focuses on obtaining most of a person's daily calories from complex carbohydrates rather than from proteins. Low protein diet (4-8 per cent protein) is used routinely to treat patients with liver disease, kidney failure, and disorders involving the urea cycle, the metabolism, and amino acids. To reduce the nitrogen metabolites and ammonia in individuals with toxic bowel, liver disease or kidney failure, dietary protein must be limited to 0.6 gram/kg of body weight or about 40 to 50 grams per day. In any case, a diet which is especially low in protein should only be undertaken under medical direction.

viii) Bland Diet

This diet is designed to provide adequate nutrition during treatment of inflammatory or ulcerative conditions of the oesophagus, stomach, and intestines and may be helpful for gastritis, hyper-acidity, haemorrhoids, peptic ulcers, and other GI disorders. This diet is often recommended following stomach or intestinal surgery too. It is intended to decrease irritation of the mucosa and aid in physical comfort. A bland diet allows the digestive tract to heal before introducing more difficult to digest foods. This diet consists of foods that are generally soft, low in dietary fibre, cooked rather than raw, and not spicy. It is devoid of known irritants to the gastric mucosa including coarse cereals, alcohol, black pepper, carbonated beverages, coffee, tea, chilli powder, cocoa and certain drugs. Fried and fatty foods, strong cheeses and whole grains are also avoided.

ix) Low Residue/Fibre Diet

Low fibre soft diet includes both liquids and solid foods which contain restricted amount of indigestible carbohydrates and no tough connective tissue. It is recommended for patients with digestive and rectal diseases such as colitis or diarrhea. Such a diet eliminates or limits foods high in bulk or fiber such as raw fruits and vegetables, whole grains and cereals, nuts, seeds, beans and peas, coconut, and fried foods.

Residue includes any food, including fiber that remains in the intestinal tract, is not digested and contributes to stool. The low residue diet is made up of foods which can be completely absorbed, thereby leaving little or no residue for formation of faeces. It limits the amount of roughage intake and material that passes undigested through large intestine. A low residue diet is used to reduce the size and number of stools. It helps to relieve abdominal pain, diarrhea, certain digestive problems like IBD, abdominal surgery, ulcerative colitis in initial stages, diverticulitis *etc.* A low residue diet cannot provide all the nutrients to stay healthy and therefore, should be uses for a short period. This diet is contraindicated in diabetes, cardiac disease, constipation, obesity, cancer *etc.*

Factors to Be Considered While Planning Therapeutic Diets

The alteration of the normal diet requires on appreciation of

1. The underlying disease conditions which require a change in the diet.
2. The possible duration of the disease.
3. The factors in the dietary which must be altered to overcome these conditions.
4. The patient's tolerance for food by mouth.
5. In planning meals for the patient his economic status, his food preferences and his occupation and time of meals should also be considered.

The planning of therapeutic diets implies the ability to adapt the principles of normal nutrition to the various regimens for adequacy, correctness, economy and palatability. It requires recognition of the need for dietary supplements such as vitamins and minerals concentrations when the nature of diet itself improves severe restrictions, the patient's appetite is poor, or absorption and utilization are impaired so that the diet cannot meet the needs of optimum nutrition. Intelligent planning of therapeutic diets necessitates a consideration of food costs, the avoidance of waste and retention of nutritive value so that the diet is economically practicable. The likes and dislikes of patients are respected because food habits are deep-seated and it is not possible to change them overnight. The planning of therapeutic diets requires considerable encouragement and understanding on the part of the doctor-nurse-dietitian team to bring about important changes in the diet.

16

Nutrition Screening and Nutrition Assessment of Hospitalized Patients

The Nutrition Support Service (NSS) available in big multispecialty corporate hospitals, is a multidisciplinary team comprised of physicians, a pharmacist, a nurse and dietitians (clinical dietitians, food service dietitians *etc.*) with certification in nutrition support. The team is headed by the physician and clinical dietitian is a co-director while others are the members of the team. If a nutrition support service is not established, nutrition care should be addressed by a team approach. The membership of this team should include the patient's physician, nurse, registered dietitian, and pharmacist. The NSS provides inpatient consultation for all aspects of nutrition support. The service manages the nutritional therapy for diverse patient populations, including surgery, trauma, critically-ill, medicine, obstetric and bone marrow transplantation patients. Quality assurance of nutrition support is provided by monitoring appropriateness, duration and complications of nutritional therapy.

Goals of Nutrition Support Service

- Nutrition screening and assessment of nutritional status of patient
- Determining appropriate energy and protein needs
- Prescribing enteral and parenteral nutrition regimens
- Addressing drug-nutrient interactions
- Recommending and monitoring parenteral intravenous access sites
- Reviewing fluid and electrolyte challenges
- Placing nasoenteric small bowel feeding tubes

ABCD's of Nutrition Assessment

- Anthropometric data
- Biochemical parameters
- Clinical/physical assessment
- Dietary history/assessment

1. Nutrition Screening of Hospitalized Patient

First step in nutrition care process of hospitalized patients is nutrition screening, nutrition assessment,

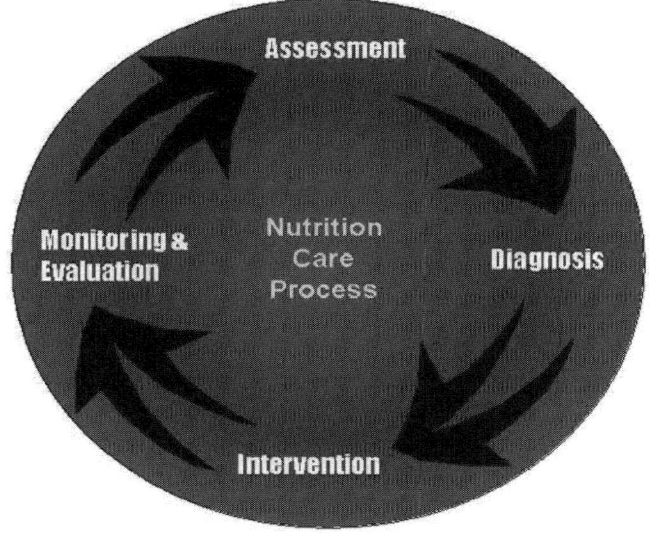

nutrition diagnosis, nutritional intervention and nutritional monitoring and evaluation. All patients admitted to the hospital for a specified length of stay undergo a nutrition screening process using subjective and/or objective criteria by the clinical dietitians. All patients identified as nutritionally-at-risk by the patient screening mechanism then undergo a formal nutrition assessment. Nutrition screening is defined by the American Society for Enteral and Parenteral Nutrition (ASPEAN) as, "a process to identify an individual who is malnourished or who is at risk for malnutrition to determine if a detailed nutrition assessment is indicated".

According to Academy of Nutrition and Dietetics (The Academy) Quality Management Committee, it is defined as "The process of identifying patients, clients, or groups who may have a nutrition diagnosis and benefit from nutrition assessment and intervention by a registered dietitian (RD)."

Screening may be conducted in any practice setting as appropriate but it should be done within 24 hours of admission to an intensive care center.

- Tools for nutrition screening should be quick, easy to use, valid, and reliable for the patient population or setting. Nutrition screening tools that are simple, quick, and easily completed by nonprofessional staff, patients, or family are preferred.

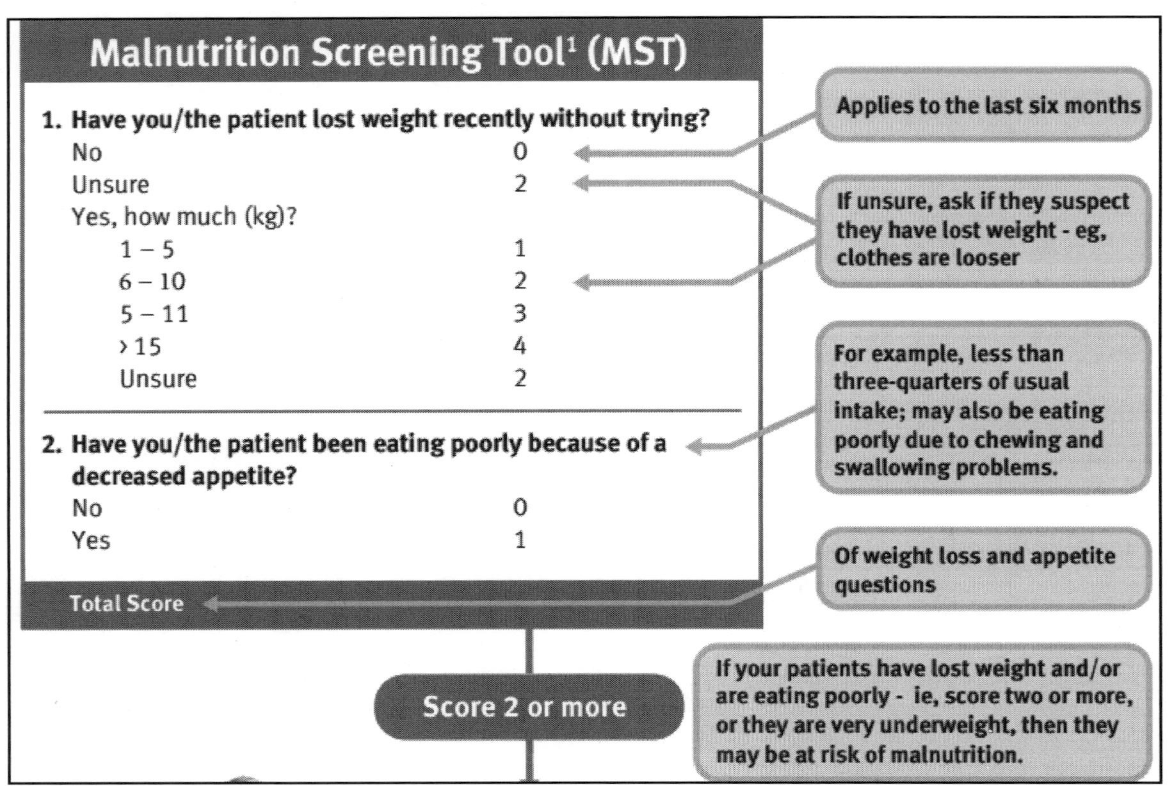

Source: Ferguson M, Capra S, Bauer J, Banks M. (1999). Development of a valid and reliable malnutrition screening tool for adult acute hospital patients. Nutrition, 15, 458-64.

Intervene with nutrition support in case of score is more than 2.

Commonly Used Nutrition Screening Tools: Malnutrition Universal Screening Tool (MUST)

'MUST' is a five-step screening tool to identify adults, who are malnourished, at risk of malnutrition (under nutrition), or obese. It also includes management guidelines which can be used to develop a care plan. It is for use in hospitals, community and other care settings and can be used by all care workers. 'MUST' has not been designed to detect deficiencies or excessive intakes of vitamins and minerals and is of use only in adults. This guide contains:

- A flow chart showing the 5 steps to use for screening and management
- BMI chart
- Weight loss tables
- Alternative measurements when BMI cannot be obtained by measuring weight and height.

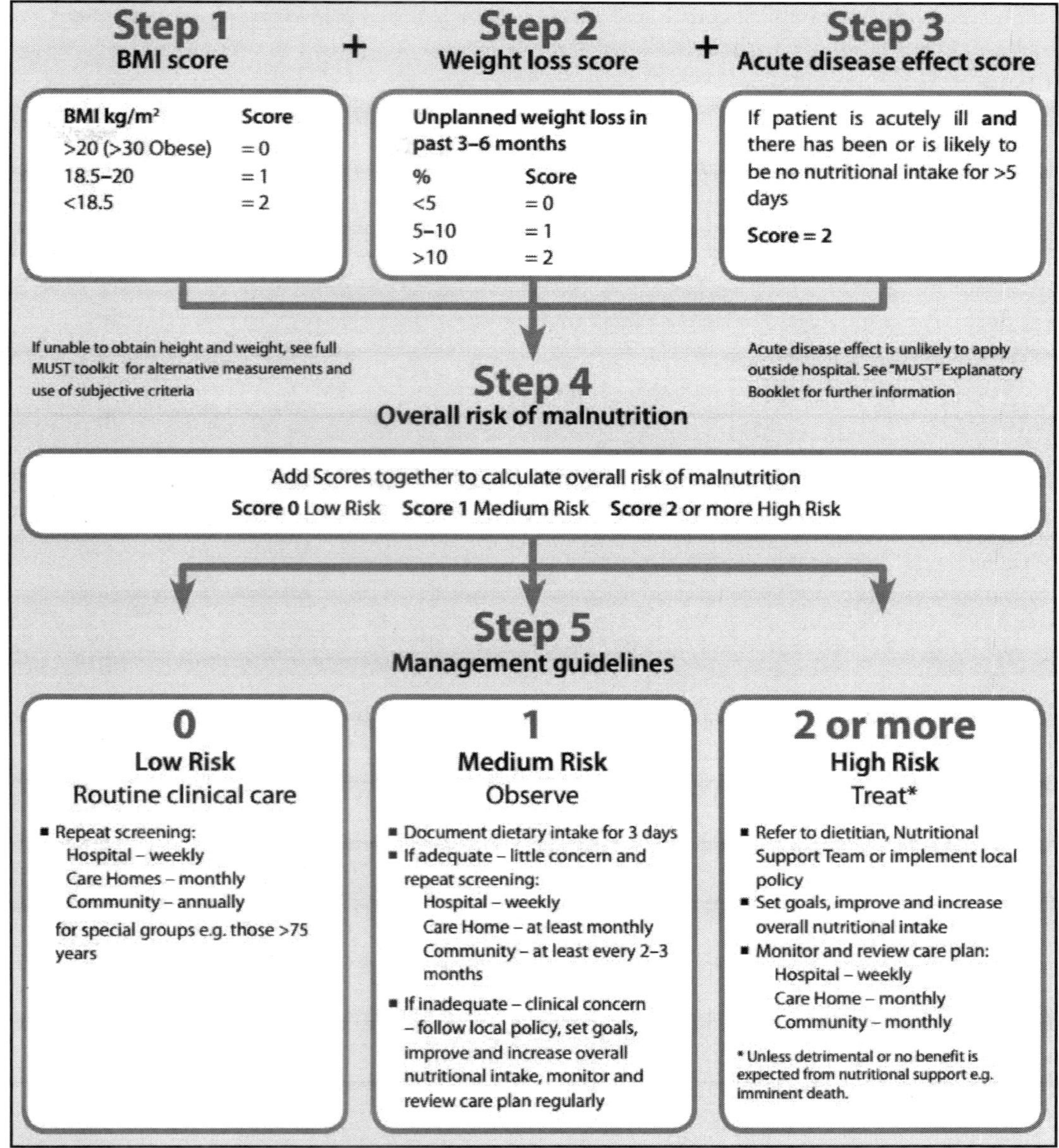

- ☆ Tools and parameters established by RDs should be used, but the screening process may be performed by other personnel, dietetic technicians, registered nurse, and other trained personnel
- ☆ Nutrition screening can be implemented to identify patients or clients with various nutrition diagnoses.
- ☆ Most important function is to identify **risk** for **malnutrition**.
  - ❒ By identifying malnutrition risk early through nutrition screening, nutrition intervention can be implemented immediately.
  - ❒ Increased potential to improve health outcomes as well as length of stay, readmission rates, and hospital costs are all reduced.
  - ❒ In the absence of nutrition screening, malnutrition may be overlooked particularly in patients who fall within the traditional "healthy" weight range or who remain overweight even after losing significant amounts of weight.

2. Assessing Nutritional Assessment of Hospitalized Patients

It provides nutrition foundation for the nutrition care process of the patient. Nutritional assessment is a comprehensive process that combines objective measurements, a focused history and physical examination, and clinical judgment to arrive at a decision regarding a patient's nutritional status. According to JADA (2008), it is defined as a systematic method for obtaining, verifying, and interpreting data needed to identify nutrition related problems and their causes and significance. It is ongoing, non-linear, dynamic process that involves initial data collection as well as continual reassessment and analysis of the patient's status compared with specified criteria.

2.1 Objective

- ☆ Identify any specific nutrition risk(s) or clear existence of malnutrition
- ☆ Identify individuals who require aggressive nutrition support
- ☆ Restore or maintain an individual's nutrition wellness
- ☆ Identify appropriate medical nutrition therapy

2.2 How and What Information to Collect?

Nutrition assessment data can be obtained from a variety of sources, including screening or referral form, patient/client interview, medical or health records, consultation with other caregivers, including family members *etc.*

Following information is to be collected:

- ☆ Food and nutrition-related history
- ☆ Assessment of nutritional adequacy (diet history)
- ☆ Anthropometric data (height, weight, BMI)
- ☆ Biochemical data
- ☆ Medical tests and procedures
- ☆ Nutrition-related physical examination findings
- ☆ Patient/client's history (medical, social)

2.3 Parameters of Data Collection

- ☆ **Medical history:** It usually includes the following information:
- ☆ Chief complaint
- ☆ Present or past illness
- ☆ Current health
- ☆ Allergies
- ☆ Past or recent surgeries
- ☆ Family history of disease
- ☆ Psychosocial data
- ☆ Review of problems by body system from the patient's perspective

These histories shed light on the following nutritional concerns:

- ☆ Alcohol and drug use
- ☆ Increased metabolic needs
- ☆ Increased nutrient losses
- ☆ Chronic disease
- ☆ Recent major surgery or illness
- ☆ Disease or surgery of GI tract
- ☆ Recent significant weight loss

Medication History

It is important because food and drugs can interact in many ways that affect nutrition status and drug therapy effectiveness and is important to any nutrition assessment.

Social History

We can have information on the following:

- ☆ Socio-economic status
- ☆ Ability to purchase food – then prepare it!
- ☆ Living alone
- ☆ Physical or mental handicaps
- ☆ Smoking, drugs, alcohol
- ☆ Lack of socialization at meals
- ☆ Unsuitable or unstable housing

Assessment of Dietary Intake

Balance between nutrient intake and nutrient requirement is equivalent to nutritional status. In patient, food intake can be affected by the disease, SES, eating behavior, emotions, cultural pressures and influences *etc.* Food absorption may also be influenced by disease, physiologic stress, mechanical problems *etc.*

Food intake and its absorption both will influence the nutrient intake. On the other hand, nutrient requirement shall depend upon infection, disease, fever, physiologic stress, growth, psychological stress, body maintenance and well being of the person.

From the diet history, one can get information regarding:

- ☆ **Allergies, intolerances, food avoidances *etc.*:** One should gather information on foods being avoided, why they are being avoided, how long they have been avoided?
- ☆ **Appetite:** Assess if good, poor, any changes. What are the factors that are influencing his appetite?
- ☆ **Attitude toward eating:** Disinterest in eating? Information about irrational ideas about food, eating, body image *etc.*, if any should be collected.
- ☆ **Chronic disease:** Treatments, length of treatment time *etc.*
- ☆ **Dietary modifications:** Self-imposed or physician prescribed, date of modification
- ☆ **Past nutrition and diet education:** Compliance, barriers to compliance
- ☆ **Dental and oral health:** Problems with chewing, dry mouth, pain with eating
- ☆ **Gastrointestinal factors:** Frequency of problems, avoidance of specific foods, swallowing difficulty, choking or food sticking
- ☆ Heartburn, gas, bloating

- **Problems with nausea, vomiting, diarrhea, constipation:** Certain foods make it worse?
- **Home life and meal patterns:** Who does the shopping, cooking; Ability to prepare foods; Socialization during meals *etc.*

Method of Data Collection for Dietary Intake

Broadly, there are two methods:

Retrospective

- 24- hour recall
- Food Frequency Questionnaire (FFQ)

Prospective/concurrent

- Food Diary
- Calorie count (Nutrient Intake Analysis)

From the above information, review the following concerns:

- Dietary intake for factors that affect health conditions and nutrition risk.
- Evaluate health and disease condition for nutrition-related consequences.
- Evaluate psychological, functional, and behavioral factors related to food access, selection, preparation, physical activity, and understanding of health condition.
- Evaluate patient's knowledge, readiness to learn, and potential for changing behaviors.
- Identify the standards by which data will be compared.
- Identify possible problem areas for making nutrition diagnoses.
- Observing for nonverbal and verbal cues that can guide and prompt effect interviewing methods.
- Distinguishing between important and unimportant data.

Conclusively, on the basis of anthropometric data, biochemical parameters, clinical/physical, assessment and dietary history/assessment, clinical dietitian shall assess the nutritional status of the patient and decide further course of nutrition intervention, if required.

2.4 Tools for Assessment of Nutritional Status

i) Subjective Global Assessment (SGA)

SGA is a simple, reliable and well proven validated nutrition assessment tool that correlates well with nutrition risk indices and other assessment data in hospitalized patients. SGA fulfills the requirements of a desirable system of nutritional assessment by identifying malnutrition, distinguishing malnutrition from a disease state, predicting outcome and identifying patients in whom nutritional therapy can alter outcome.

Subjective Global Assessment Form for Infants

PATIENT DATA		Birth Date	
DATE		AGE	
LAST NAME		SEX	
FIRST NAME		HEIGHT (cm)	
ROOM		WEIGHT (kg)	
PIN		PREV WT (kg)	
DATE ADMIT		RECALLED WT	
Diagnosis			
Nutrition support indication			

SUBJECTIVE GLOBAL ASSESSMENT			
Weight Change	No weight loss or gain	<10% loss	>10% loss
Dietary Intake	No change	Sub-optimal	Poor
Oral Motor Skill Development (<2YR)	At par	Premature	Significantly delayed
Feeding Tolerance	None	Nausea, vomiting Diarrhea, anorexia <2 months	Nausea, vomiting, Diarrhea, anorexia > 2 months
Functional Capacity	No dysfunction	Dysfunction - weeks	Dysfunction > 2 months
Muscle Wasting (" MAC)	0 or +1 >15%ile	2+ 5-15%ile	3+ <5%ile
Ankle Edema	None	None	+1 or +2
SGA Grade	A 0	B 1	C 3
Weight for Height (Wasting)	11-89%ile 0	5-10%ile 90-95%ile 1	<5%ile >95%ile 2
Length/Height for Age (Stunting)	> 11%ile 0	5-10%ile 1	<5%ile 2
Head Circumference 0-3yrs	> 10%ile/>3yr 0	5-10%ile 1	<5%ile 2
NUTRITIONAL STATUS	NORMAL/MILD 0	MODERATE 1	SEVERE 2
RISK STRATIFICATION			
BMI	NORMAL 0	OW 1	UW or OB 2
SGA	A 0	B 1	C 2
Albumin	> 3 gm 0	2.1 - 3 gm 1	< 2 gm 2
TLC	> 1500 0	1000 - 1500 1	< 1000 2
NUTRITION RISK	MILD 0	MODERATE 1-3	SEVERE >=4

PERFORMED BY: ____________________ DATE PERFORMED ____________

Subjective Global Assessment Form for Adults

MEDICAL HISTORY

Patient name: ____________________ **Date:** ______/______/______

NUTRIENT INTAKE

1. ❑ No change; adequate
2. Inadequate; duration of inadequate intake ________
   ❑ Suboptimal solid diet ❑ Full fluids or only oral nutrition supplements
   ❑ Minimal intake, clear fluids or starvation
3. Nutrient Intake in past 2 weeks*
   ❑ Adequate _______ ❑ Improved but not adequate _______ ❑ No improvement or inadequate _______

WEIGHT

Usual weight ____________ Current weight ____________

1. Non fluid weight change past 6 months Weight loss (kg)
   <5 per cent loss or weight stability 5-10 per cent loss without stabilization or increase
   >10 per cent loss and ongoing
   ❑ If above not known, has there been a subjective loss of weight during the past six months
   ❑ None or mild ❑ Moderate ❑ Severe
2. Weight change past 2 weeks* Amount (if known)
   ❑ Increased ❑ No change ❑ Decreased

SYMPTOMS (Experiencing symptoms affecting oral intake)

1. ❑ Pain on eating ❑ Anorexia ❑ Vomiting ❑ Nausea ❑ Dysphagia ❑ Diarrhea
   ❑ Dental problems ❑ Feels full quickly ❑ Constipation
2. ❑ None ❑ Intermittent/mild/few ❑ Constant/severe/multiple
3. Symptoms in the past 2 weeks*
   ❑ Resolution of symptoms ❑ Improving ❑ No change or worsened

FUNCTIONAL CAPACITY (Fatigue and progressive loss of function)

1. No dysfunction ________
2. Reduced capacity; duration of change
   ❑ Difficulty with ambulation/normal activities ❑ Bed/chair-ridden
3. Functional Capacity in the past 2 weeks*
   ❑ Improved ❑ No change ❑ Decrease

METABOLIC REQUIREMENT

High metabolic requirement ❑ No ❑ Yes

PHYSICAL EXAMINATION

1. Loss of body fat ❑ No ❑ Mild/Moderate ❑ Severe
2. Loss of muscle mass ❑ No ❑ Mild/Moderate ❑ Severe
3. Presence of edema/ascites ❑ No ❑ Mild/Moderate ❑ Severe

SGA RATING

A. ❑ Well-nourished
Normal

B. ❑ Mildly/moderately malnourished
Some progressive nutritional loss

C. ❑ Severely malnourished
Evidence of wasting and progressive symptoms

CONTRIBUTING FACTOR

1. ❑ CACHEXIA - (fat and muscle wasting due to disease and inflammation)
2. ❑ SARCOPENIA - (reduced muscle mass and strength)

Subjective Global Assessment Guidance for Body Composition

Subcutaneous Fat			
Physical Examination	***Normal***	***Mild/Moderate***	***Severe***
Under the eyes	Slightly bulging area	Somewhat hollow look, Slightly dark circles.	Hollowed look, depression, dark circles.
Triceps	Large space between fingers	Some depth to fat tissue, but not ample, Loose fitting skin.	Very little space between fingers, or fingers touch
Ribs, lower back, sides of trunk	Chest is full; ribs donot show. Slight to no protrusion of the iliac crest	Ribs obvious, but indentations are not marked. Iliac Crest somewhat prominent.	Indentation between ribs very obvious. Iliac crest very prominent.

Muscle Wasting			
Physical Examination	***Normal***	***Mild/Moderate***	***Severe***
Temple	Well-defined muscle	Slight depression	Hollowing, depression
Clavicle	Not visible in males; may be visible but not prominent in females	Some protrusion; may not be all the way along	Protruding/prominent bone
Shoulder	Rounded	No square look; acromion process may protrude slightly	Square look; bones prominent
Scapula/ribs	Bones not prominent; no significant depressions	Mild depressions or bone may show slightly; not all areas	Bones prominent; significant depressions
Quadriceps	Well defined	Depression/atrophy medially	Prominent knee, Severe depression medially
Interosseous muscle between thumb and forefinger (back of hand)**	Muscle protrudes; could be flat in females	Slightly depressed	Flat or depressed area

Fluid Retention			
Physical Examination	***Normal***	***Mild/Moderate***	***Severe***
Edema	None	Pitting edema of extremities/pitting to knees, possible sacral edema if bedridden	Pitting beyond knees, sacral edema if bedridden, may also have generalized edema
Ascites	Absent	Presnt (may only be present on imaging)	

A: Well-nourished no decrease in food/nutrient intake; < 5 per cent weight loss; no/minimal symptoms affecting food intake; no deficit in function; no deficit in fat or muscle mass **OR** *an individual with criteria for SGA B or C but with recent adequate food intake; non-fluid weight gain; significant recent improvement in symptoms allowing adequate oral intake; significant recent improvement in function; and chronic deficit in fat and muscle mass but with recent clinical improvement in function.

B: Mildly/moderately malnourished definite decrease in food/nutrient intake; 5 per cent - 10 per cent weight loss without stabilization or gain; mild/some symptoms affecting food intake; moderate functional deficit or recent deterioration; mild/moderate loss of fat and/or muscle mass **OR** *an individual meeting criteria for SGA C but with improvement (but not adequate) of oral intake, recent stabilization of weight, decrease in symptoms affecting oral intake, and stabilization of functional status.

C: Severely malnourished severe deficit in food/nutrient intake; > 10 per cent weight loss which is ongoing; significant symptoms affecting food/ nutrient intake;severe functional deficit **OR** *recent significant deterioration obvious signs of fat and/or muscle loss.

Cachexia – If there is an underlying predisposing disorder (*e.g.* malignancy) and there is evidence of reduced muscle and fat and no or limited improvement with optimal nutrient intake, this is consistent with cachexia.

Sarcopenia – If there is an underlying disorder (*e.g.* aging) and there is evidence of reduced muscle and strength and no or limited improvement with optimal nutrient intake.

**In the elderly prominent tendons and hollowing is the result of aging and may not reflect malnutrition.

ii) Mini Nutritional Assessment (MNA)

Nutritional assessment in older people to detect malnutrition or risk of malnutrition is essential to avoid adverse nutrition-related outcomes. Poor nutritional status appears to be a major contributing factor for poor prognosis in malnourished individuals. After 20 years of clinical practice and research, the Mini Nutritional Assessment (MNA) seems to be the tool most widely accepted by health carers and patients for the assessment of nutritional impairment in comprehensive geriatric assessment (CGA). It is a reliable and quick method for evaluating nutritional status in older adults, >65 years old.

The MNA is able to classify older people as well nourished, at risk for malnutrition or malnourished. The full MNA consists of 18 self reported questions derived from four parameters of assessment: anthropometric assessment, general assessment, dietary assessment and self- assessment (see Figure given below) administered in two steps *i.e.* Step I and Step 2. When a quick screening is all that is needed, just the first six questions, also known as MNA Short Form (MNA-SF) can be completed.

Step I

The first step, the MNA-SF, takes only few minutes to complete. It is a screening tool composed of the first six items (of the 18 questions) that permits detection of a decline in ingestion over the past three months (loss of appetite, decline of food intake, digestive problems, chewing or swallowing difficulties), weight loss in the past three months, current mobility impairment, an acute illness or major stress in the past three months, a neuropsychological problem (dementia or depression) and a decrease in body mass index (BMI).

The maximum score for this part is 14; a score of 12 points or greater indicate that the patient has an acceptable nutritional status and that it is not necessary to complete the full MNA. However, a score of 11 points or below is an indication of possible malnutrition and it requires proceeding further with the complete version of the MNA *i.e.* to Step 2.

Step 2

If indicated after Step 1 (MNA-SF), the risk of malnutrition should be assessed using the full MNA. The full MNA can be performed in 10 to 15 minutes. It evaluates living arrangements, the presence of polypharmacy or pressure ulcers, the number of full meals eaten daily, the amount and frequency of specific foods and fluids, and the mode of feeding. The patient reports nutritional and health status, and the practitioner determines weight and height (to calculate BMI), and mid-arm and mid-calf circumferences.

The maximum score for the second part is 16. Scores from the two parts must be added to obtain the malnutrition indicator score. Each answer in the two parts has a numerical value and contributes to the final score, which has a maximum of 30; 24 and higher indicates a well nourished patient; 17 to 23.5 indicates a risk of malnutrition; lower than 17 indicates malnutrition. Assessment of the cognitive status of the patient should be obtained by the caregiver, or by the medical or nursing staff. If a cognitive decline is present, answers should be checked by proxies (caregiver, or medical or nursing staff). If the patient suffers from severe cognitive decline, the caregiver or nursing staff can complete the MNA instead. In fact, the physician, dietician, or nurse in charge of an institutionalized patient can easily complete the MNA with a minimal risk of bias.

MINI NUTRITIONAL ASSESSMENT
MNA®

ID# ________________

Last Name: ______________ First Name: ______________ M.I. ____ Sex: ____ Date: ________

Age: ____ Weight, kg: ________ Height, cm: ________ Knee Height, cm: ________

Complete the form by writing the numbers in the boxes. Add the numbers in the boxes and compare the total assessment to the Malnutrition Indicator Score.

ANTHROPOMETRIC ASSESSMENT

	Points
1. Body Mass Index (BMI) (weight in kg) / (height in m)^2 a. BMI < 19 = 0 points b. BMI 19 to < 21 = 1 point c. BMI 21 to < 23 = 2 points d. BMI ≥ 23 = 3 points	☐
2. Mid-arm circumference (MAC) in cm a. MAC < 21 = 0.0 points b. MAC 21 ≤ 22 = 0.5 points c. MAC > 22 = 1.0 points	☐.☐
3. Calf circumference (CC) in cm a. CC < 31 = 0 points b. CC ≥ 31 = 1 point	☐
4. Weight loss during last 3 months a. weight loss greater than 3kg (6.6 lbs) = 0 points b. does not know = 1 point c. weight loss between 1 and 3 kg (2.2 and 6.6 lbs) = 2 points d. no weight loss = 3 points	☐

GENERAL ASSESSMENT

5. Lives independently (not in a nursing home or hospital) a. no = 0 points b. yes = 1 point	☐
6. Takes more than 3 prescription drugs per day a. yes = 0 points b. no = 1 point	☐
7. Has suffered psychological stress or acute disease in the past 3 months a. yes = 0 points b. no = 2 points	☐
8. Mobility a. bed or chair bound = 0 points b. able to get out of bed/chair but does not go out = 1 point c. goes out = 2 points	☐
9. Neuropsychological problems a. severe dementia or depression = 0 points b. mild dementia = 1 point c. no psychological problems = 2 points	☐
10. Pressure sores or skin ulcers a. yes = 0 points b. no = 1 point	☐

DIETARY ASSESSMENT

	Points
11. How many full meals does the patient eat daily? a. 1 meal = 0 points b. 2 meals = 1 point c. 3 meals = 2 points	☐
12. Selected consumption markers for protein intake • At least one serving of dairy products (milk, cheese, yogurt) per day? yes ☐ no ☐ • Two or more servings of legumes or eggs per week? yes ☐ no ☐ • Meat, fish or poultry every day? yes ☐ no ☐ a. if 0 or 1 yes = 0.0 points b. if 2 yes = 0.5 points c. if 3 yes = 1.0 points	☐.☐
13. Consumes two or more servings of fruits or vegetables per day? a. no = 0 points b. yes = 1 point	☐
14. Has food intake declined over the past three months due to loss of appetite, digestive problems, chewing or swallowing difficulties? a. severe loss of appetite = 0 points b. moderate loss of appetite = 1 point c. no loss of appetite = 2 points	☐
15. How much fluid (water, juice, coffee, tea, milk,...) is consumed per day? (1 cup = 8 oz.) a. less than 3 cups = 0.0 points b. 3 to 5 cups = 0.5 points c. more than 5 cups = 1.0 points	☐.☐
16. Mode of feeding a. Unable to eat without assistance = 0 points b. self-fed with some difficulty = 1 point c. self-fed without any problem = 2 points	☐

SELF ASSESSMENT

17. Do they view themselves as having nutritional problems? a. major malnutrition = 0 points b. does not know or moderate malnutrition = 1 point c. no nutritional problem = 2 points	☐
18. In comparison with other people of the same age, how do they consider their health status? a. not as good = 0.0 points b. does not know = 0.5 points c. as good = 1.0 points d. better = 2.0 points	☐.☐

ASSESSMENT TOTAL (max. 30 points): ☐☐.☐

MALNUTRITION INDICATOR SCORE		
≥ 24 points	well-nourished	☐
17 to 23.5 points	at risk of malnutrition	☐
< 17 points	malnourished	☐

Ref.: Guigoz Y, Vellas B and Garry PJ. 1994. Mini Nutritional Assessment: A practical assessment tool for grading the nutritional state of elderly patients. *Facts and Research in Gerontology*. Supplement #2: 15-59.

17

Specialized Feeding Methods for Hospitalized Patients-Enteral Feeding

Introduction

Adequate nutrition support is very important as 20-50 per cent of the adult patients hospitalized in our country are affected by malnourishment. Patients with chronic or debilitating disease processes are often already malnourished on admission to hospital. Without appropriate nutrition support, malnourished patients continue to deteriorate in their nutritional status during their hospitalization, especially if repeatedly placed 'nil by mouth' for investigations and surgical interventions. If uncorrected, malnutrition may result in prolonged and complicated recovery from illness or surgery with increased risk of infection and compromised wound healing and persisting functional deficits. This in turn, leads to a longer hospital stay with an associated increase in cost and a negative impact on quality of life. Hence, healthcare professionals should consider specialized methods of feeding in hospitalized patients who are malnourished or at risk of malnutrition. The special feeding methods depend on the type of disease, the patient's conditions and his tolerance to food. The different modes of feeding patients are:

1. Enteral
2. Parenteral

The foods are administered via a tube and hence, enteral feeding is also called tube feeding. Enteral tube feeding should be considered in patients who are malnourished or at risk of malnutrition and have:

- ☆ An inadequate or unsafe oral intake

AND

- ☆ A functional, accessible gastrointestinal tract

Enteral nutrition has been shown to be safe, cost effective and compatible with the body's normal processes. Decision on route, content and management of nutritional support are best made by the Nutrition Support Service group.

Enteral Feeding

By definition, enteral means 'within or by the way of gastrointestinal tract'. Enteral feeding is most commonly used to deliver feed into a patient's stomach. When necessary, it is possible to feed directly into the duodenum or jejunum. The route and type of tube used will depend on the individual circumstance of the patient. As far as possible, the patient should be encouraged to ingest food through the oral route. Supplements may be added whenever necessary.

Indications of Enteral Feeding

Whenever possible, oral food intake is always preferred. Enteral feeding is only indicated when a patient's nutritional needs cannot be met orally. Artificial nutrition support is needed when oral intake is absent or likely to be absent for a period of >5-7 days. Support may also be needed in patients with inadequate oral intakes over long periods.

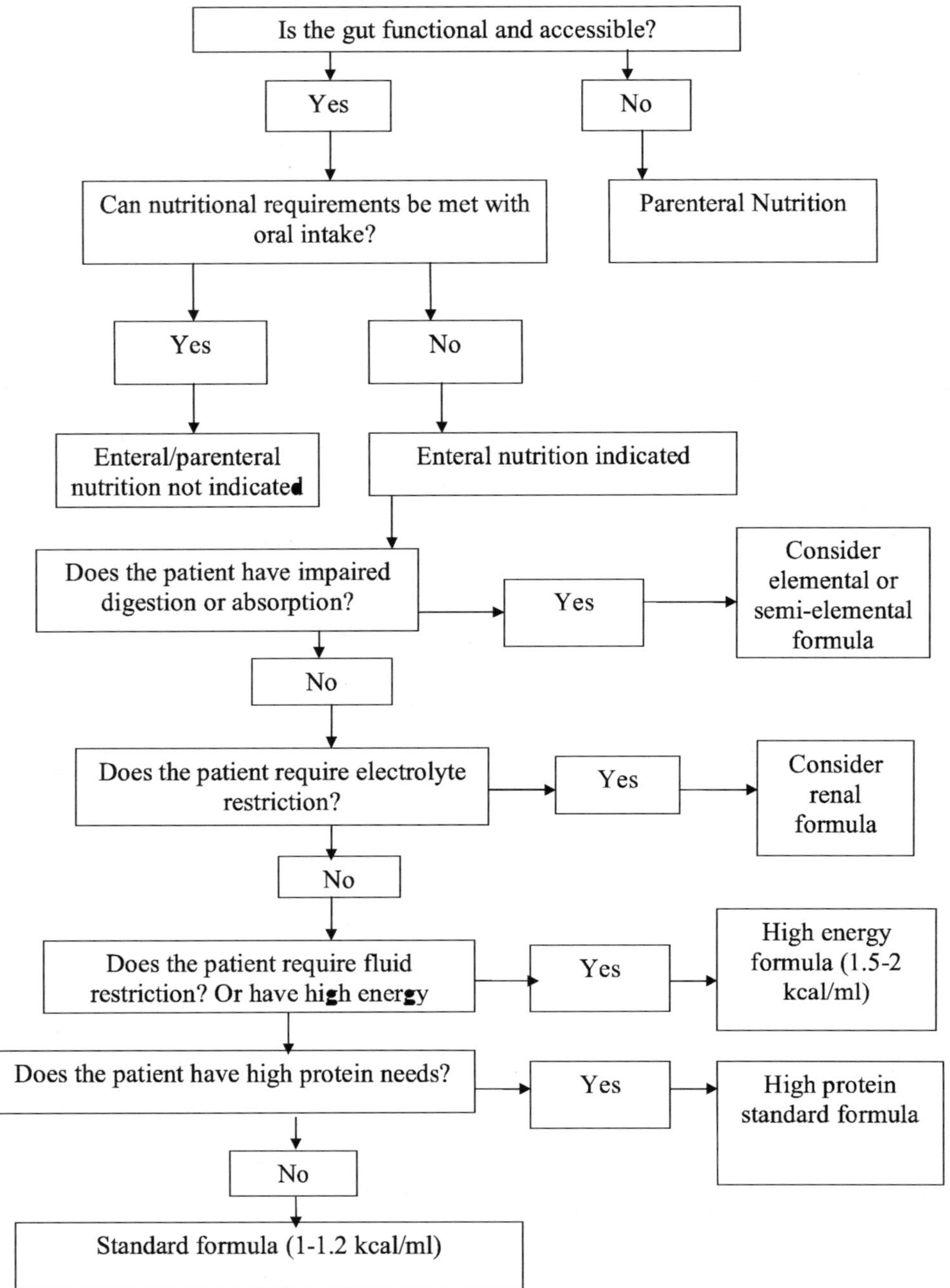

Flow Sheet for Nutrition Support Planning.

Common indications for enteral feeding include the following (Table 17.1):

- Swallowing disorder (*i.e.* motor neurone disease, multiple sclerosis)
- Partial intestinal failure
- Upper GI obstructions
- Short Bowel Syndrome
- Cystic fibrosis
- GI dysfunction or malabsorption, severe diarrhoea
- Neurological and psychiatric disorders: Cerbrovascular accidents, trauma, inflammation, demyelinating diseases, anorexia, failure to thrive, severe depression *etc.*
- Head and neck cancer (due to side effects of chemo/radiotherapy)
- Stroke or head injury
- Oropharyngeal and esophageal: Neoplasm's, inflammation, trauma, oral and maxilla facial surgery *etc.*
- Unconscious or semi-conscious patient
  - Low birth weight babies
  - Post surgery- Early ETF after major gastrointestinal surgery reduces infections and shortens length of stay (grade A)

- Early post pyloric ETF is generally safe and effective in postoperative patients, even if there is apparent ileus (grade A).
- In all post surgical patients not tolerating oral intake, ETF should be considered within 1–2 days of surgery in the severely malnourished, 3–5 days of surgery in the moderately malnourished, and within seven days of surgery in the normally or over nourished (grade C).
- ETF can be used for the support of patients with uncomplicated pancreatitis (grade A).

☆ Miscellaneous: Burns, AIDS, organ transplant *etc.*

Table 17.1: Indications for Enteral Feeding

Physiological Problem	*Clinical Problem*
Inability to ingest food	Carcinoma of esophagus or stomach, dental or oral surgery, inflammatory disease of esophagus, coma
Inability to digest food	Pancreatitis, biliary tract disease
Decreased ability to handle colonic residue	Radiation therapy, sprue, inflammatory bowel disease
Inability to handle colonic residue	Inflammatory bowel disease, pre-surgical preparation, ileostomy, colostomy, draining fistula
Inability to meet nutritional requirement fully with normal food	Major surgery, burns, trauma, extended fever, anorexia of chronic illness, anorexia nervosa

When is Enteral Nutrition Not Used?

In some clinical conditions, such as gut failure, intestinal obstruction, or an inability to gain enteral access, prolonged ileus, severe diarrhea, high output GI tract (>500 ml/day), severe enterocolitis, severe GI haemorrhage *etc.*, it may be necessary to bypass the gut and deliver nutrients directly into the blood stream.

Aggressive nutrition support is not always appropriate in the care of elderly patients. Careful consideration should be given to patient and family wishes concerning the appropriateness of initiating or continuing enteral feeding. Potential benefits, including quality of life; possible complications and expected outcomes should be considered.

Types of Enteral Feeds

I. Homemade/Blenderized Enteral Feedings

Blenderized tube feeding (BTF) is whole homemade or hospitalized foods liquefied using a food blender and administered as a bolus through a gastrostomy tube. Before the introduction of commercial enteral formulas in the late 1960s, most tube feedings used in hospitals were made from blended hospital or homemade foods. Over the last few decades, commercial formulas have been the mainstay of tube feedings in western world. However, recently, BTFs have made a comeback as some people believe that they may be more natural, fresh and unprocessed forms of food without synthetic ingredients or preservatives, better tolerated and cost-effective than commercial formulas. Increasingly, RDs are being asked to develop nutrition plans for BTFs.

In India, various nutritionists and dietitians have developed Low Cost Enteral Formula (LCEF) to be blenderized for feeding to the patients. Some of the examples are given as under:

Low Cost Enteral Formula (LCEF) designed originally by the dietician in a tertiary care cancer center in exchange to other commercial supplements is a high calorie high protein enteral formula (Tables 17.2 and 17.3).

Table 17.2: Development of Low Cost Enteral Formula to be Blenderized

Ingredients	*Quantity*
Whole Wheat	300g
Rice milled	100g
Ragi	100g
Green gram whole	300g
Soya beans	100g
Til	100g
Flax seed	100g
Sugar	200g
Skim milk powder	100g
Total	**1400g**

Choosing and Preparing the Healthy and Well Balanced Blenderized Food Formula at Home

Foods that are popular for blending include various grains (rice, bread), legumes (green gram *dal*, chickpea), vegetables (spinach, pumpkin, carrots, potatoes, sweet potatoes *etc.*), nuts, fruits (banana, mango *etc.*), milk (cow's, soy, almond, coconut, skim milk powder *etc.*), curd, refined oil, sugar *etc.* for preparing blenderized feeding. Other liquids include water, broths, and juices. Foods such as cereals and legumes have to be cooked. Vegetables should be cut into chunks and boiled. All the food items are placed in a blender along with enough liquid to allow blending to a smooth consistency. You may need to strain the blend to remove small chunks or seeds. To increase the calorie density, refined oil is added. Depending upon the patient' nutritional demands, many types of blenderized recipes can be prepared. One way to make a balanced meal is to portion out foods in amounts that would otherwise have been eaten by mouth, add liquid, and blenderized. Another way is to blend a specific number of portions each day from the different food groups: carbohydrates (grains and starchy vegetables), protein, vegetables, fruits, dairy (or dairy substitutes), and fats and oils (Table 17. 4). RDs will calculate calorie, protein, water, and micronutrient

Table 17.3: Nutritive Value of Low Cost Enteral Formula Developed at Home and Cost Analysis

Ingredients	*Wheat (Whole)*	*Rice (Milled)*	*Ragi*	*Mung (Whole)*	*Soya Bean*	*Til*	*Flax Seed*	*Sugar*	*Skim Milk Powder*	*Total*	*Per 100g*
Amount (g)	300	100	100	300	100	100	100	200	100	1400g	
Energy (kcal)	1038	345	328	1002	432	563	530	800	496	5534 Kcal	395 Kcal
Protein (g)	35.4	6.8	7.3	72	43.2	18.3	20.3	-	25.8	229.1g	16.4g
CHO(g)	213.6	78.2	72	170.1	26.9	25	28.9	200	38	846.7g	60.5g
Fats (g)	4.5	.5	1.3	3.9	19.5	43.3	37.1	-	26.7	136.8g	10g
Na (mg)	51.3	-	11	84	-	-	-	-	-	136.3mg	9.8mg
K (mg)	852	-	408	2529	-	-	-	-	-	3789mg	270mg
Ca (mg)	123	10	344	372	240	1450	170	-	95-	3659mg	261mg
P (mg)	918	160	283	978	690	570	370	-	730	4699mg	335mg
Iron (mg)	15.9	0.7	3.9	13.2	104	9.3	7.9	-	.6	61.6mg	4.42mg
Fiber (mg)	3.6	0.2	3.6	12.3	3.7	2.9	4.8	-	-	31.1mg	2.2mg
Zinc (mg)	8.1	1.4	2.3	9	3.4	12.2	-	-	-	36.4mg	2.6mg
Mg (mg)	414	90	137	381	170	-	-	-	-	119.7mg	85.5mg

Cost Analysis					
Name of Company	***Products***	***Package/Quantity***	***Cost (Rs)***	***Dilution***	***Cost per Feed Rs.***
Nestle	Resource (plain)	400gms in 1 tin	342	50gms/200ml water	42.75
	Resource HP	200gms in 1 tin	168	15gms/200ml water	12.6
	Resource (Diabetics)	200gms in 1 tin	220	22.4gms/200ml water	24.6
	Resource (Renal)	200gms in 1 tin	160	22gms/200ml water	17.6
	Resource (Hepatic)	200gms in 1 tin	215	26gms/200ml water	28

Source: Nair, A.B., Madan, J. and Dhaaki, N. 2015. Formulation and Nutritional Appraisal of Low Cost Enteral Formula. International Journal of Scientific and Research Publications, Volume 5, Issue9, September 2015.1 ISSN 2250-3153.

Table 17.4: Blenderized Tube Feeding (Each recipe is for the whole day)

Nutrients	*Calorie Levels[3]*										
	DRIs[1]	*800*	*1000*	*1200*	*1500*	*1800*	*2000*	*2200*	*2400*	*2600*	*3000*
Kcals	–	799	989	1205	1478	1784	1986	2216	2408	2600	3002
Protein (g)	–	40	48	58	63	71	79	93	96	99	112
Total Fat (g)	–	35	47	60	72	89	102	118	126	133	160
Saturated Fat (g)	–	16	21	25	32	42	48	51	57	61	68
Monounsaturated (g)	–	11	15	19	22	27	31	36	38	40	48
Polyunsaturated (g)	–	5	8	12	13	14	15	23	23	24	34
Carbohydrate (g)	–	84	95	112	151	181	197	202	234	263	289
Sugar (g)	–	35	46	57	79	82	83	91	114	124	137
Fiber (g)	–	4	4	4	7	7	9	9	9	9	9
Calcium (mg)	1200	673	965	1032	1195	1636	1729	1889	1965	2150	2391
Iron (mg)	10.5	17	17	18	19	33	34	35	35	42	50
Magnesium (mg)	370	154	187	199	250	329	344	374	394	430	488
Sodium (mg)	15004	400	520	586	656	744	833	955	1006	1060	1124
Potassium (mg)	4700	1472	1842	1969	2516	2874	3096	3451	3767	3901	4370
Phosphorus (mg)	700	703	930	1019	1152	1491	1643	1815	1887	2029	2252
Zinc (mg)	9.5	6.1	7.0	8.0	8.8	10.1	11	13	14	14	16
Vitamin A (RE)	800	1565	1640	1673	1842	1991	2142	2148	2223	2288	2374
Vitamin C (mg)	82	97	100	101	149	151	195	198	239	240	283
Thiamin (mg)	1.1	1.1	1.2	1.2	1.4	2.2	2.3	2.4	2.5	2.9	3.5
Riboflavin (mg)	1.2	1.7	2.1	2.2	2.5	3.4	3.8	4.1	4.2	4.6	5.2

Nutrients	***Calorie Levels***[3]										
	DRIs[1]	***800***	***1000***	***1200***	***1500***	***1800***	***2000***	***2200***	***2400***	***2600***	***3000***
Niacin (mg)	15	14	14	17	17	26	27	29	29	34	41
Pantothenic Acid (mg)	5	2.8	3.5	4.1	4.8	5.3	6.3	7.0	7.4	7.6	8.5
Folate (mcg)	400	92	104	112	176	189	215	227	251	256	290
Vitamin B_6 (mg)	1.5	0.7	0.8	1.0	1.1	1.3	1.4	1.6	1.7	1.8	2.1
Vitamin B_{12} (mcg)	2.4	3.6	4.5	4.9	5.2	5.8	6.6	8.0	8.2	8.4	8.9
Vitamin D (mcg)	10	133	230	234	250	294	330	394	403	413	423
Vitamin E (mg)	15	6.8	10.8	15	16	16	17	29	29	30	42
Vitamin K (mcg)	105	39	49	49	52	54	80	87	91	91	93
Water per cent[2]	–	64	62	64	64	64	64	64	64	64	64

☆ The average recommended value for a healthy male or female adult. For more information: Refer to RDAs for Indians by ICMR, 2010.

☆ Water may need to be added to thin down the formula; furthermore, separate water bolus' will be needed to meet hydration needs.

☆ Numbers shaded and in bold print highlight those nutrients that fall below the average DRI's for adults – a Centrum vitamin/mineral supplement (or equivalent) can be crushed and flushed 4-7 days per week as needed to ensure nutrient adequacy of tube feeding.

☆ In some circumstances, additional sodium may need to be added to these mixtures.

Source: *Theresa A. Fessler, 2014.* Home Tube Feeding with Blenderized Foods. Nov.-Dec., 2014. *Lifeline letter,* Oley Foundation, USA.

needs based on age, height, gender, laboratory information, medical diagnoses, post-surgical status (if applicable), activity levels, weight, and, for children, growth goals and select the quality and quantity of foods to be included in the blenderized formula.

Food Safety Issues of Enteral Feeds

General food safety principles are very important, especially if the person using tube feeding has compromised immune function. As with any food preparation, hand washing is essential, and fresh vegetables and fruits and dried legumes should be carefully cleaned to remove any soil, microorganisms, and other contaminants. Dried legumes should be inspected for small pebbles prior to cooking.

Raw foods can be used, but foods that normally require cooking (such as meats, legumes, brown rice, potatoes) should be cooked just as they would if they were being eaten by mouth. Blenders and syringes need to be thoroughly cleaned right after each use. Unused blends should be promptly covered and refrigerated for up to twenty-four hours.

Feeding the Blenderized Formula

Blenderized foods can be administered the same ways as regular canned formula–by gravity, bolus, or pump. However, most often they are infused into G-tubes (including PEG-tubes and low-profile devices) slowly with a 60 ml syringe, by using gentle pressure on the syringe plunger (called bolus feeding). All methods work well with tubes of 14 French size or larger. Although very uncommon, some people use nasogastric (NG-) tubes in the home setting; using blenderized food is difficult or not possible with small-bore NG-tubes.

As with any tube feeding, the person being fed should be seated upright or have their upper body elevated at least 30 degrees to help prevent reflux. Tubes need to be flushed with warm water before and after each feeding and whenever medications are administered. Water flushes are critically important to clean out the tube and prevent clogging. If any lumps are seen in the blend, pour it through a strainer.

With pump feeding, the blend should be infused within two hours to avoid spoilage. If the blend is left in feeding bags too long, ingredients may separate out in layers, increasing risk for clogging and inconsistent intake of nutrients. Some manufacturers of feeding pumps have specifically stated that their pumps are not to be used with anything but commercial formula.

Advantages of Blenderized Low Cost Enteral Formula

The potential benefits of LCEF include the following:

☆ The commercially available enteral formulae are usually polymeric or elemental defined formula diets. They have high osmolarity and can lead to partial gastric and ileac discomfort because none or very little digestion is required. Moreover, since nutrients are mostly in pure form, later adaptation of normal food intake takes time.

☆ Commercial formulas are expensive; it may not be feasible to the patients to continue on commercial formulas. Low cost enteral feed (LCEF) is cheaper, easy to prepare and provides comfort to the patient. As reported in the study conducted by Nair *et al.* (2015), the outcomes of LCEF formula were similar to those obtained from commercial feeds, but at a cheaper rate (Tables 17.2 and 17.3).

☆ There is flexibility and variety in choosing ingredients for BTF recipe. Commercial formula also provide with a fixed profile of ingredients and nutrients which are not modifiable, whereas formulated feeds can be modified. Formulated feeds can be modified by addition or removal of certain ingredients. Hence, it has the ability to add specific foods to target a nutrition or health concern, such as healthy, unprocessed natural foods for a diet lower in sugar and with phytonutrients and fibers. Higher amount of dietary fiber helps to relieve constipation unlike commercial formula.

- Some patients have allergy or intolerance to certain ingredients that are present in most standard EN products, such as corn, cow's milk, or soy proteins and it can be avoided in LCEF.
- The uses of familiar food bring about emotional comfort to the patient.
- It can be easily prepared at home with nutrients in natural form. So that later adaptation to normal foods becomes easy. It has a consistency that would easily pass through the feeding tube. The overall acceptability and tolerance is good.
- It has been observed that enteral food akin to normal diet but of improved digestibility and texture perform better nutritional support to patients with than elemental diet.
- Psychosocial considerations for clients and their family members, such as relationship bonding between family members when preparing and sharing foods at mealtime.
- Although anecdotal reports, decreased gastrointestinal intolerances and improved tolerance to feeding volumes have been reported with LCEF.
- Some who are concerned about the environment like blenderized foods because there are fewer cans and cartons to throw out.

Disadvantages

Most nutrition support clinicians discourage the use of homemade formulas for several reasons.

- Blenderized formulas increase the chance of food borne illness, a heightened concern in immuno-compromised patients.
- With advancing technology, commercial enteral formulas have addressed many concerns associated with BTFs such as high microbial loads, risk of bacterial contamination, tube clogging due to high viscosity, and inadequate macro- and micronutrients delivery.
- In addition, there is an increased work burden on the patient or caregiver as blenderized formulas can be very time consuming.
- Perhaps most important, blenderized formulas must be carefully made to ensure nutritional adequacy, a challenging task for the care-giver.

II. Commercial Enteral Nutrition Formulae

Commercially prepared formulas are produced by industry. They are delivered as liquids of various viscosities or in a powder form. Commercially prepared formulas are always sterile. They usually fall under one of the following categories:

Commercially prepared enteral nutrition formulas may be classified as elemental and semi-elemental, standard or polymeric, specialized/disease specific and immune enhancing. Many formulas are available within each category, often containing significant differences in nutrient composition and are discussed as under.

Table 17.5: Commercial Enteral Formulas Used at One of the Centers (Prepared by Dietetic Department)

Type of Formula	*Contents*		*Nutritional Value/100ml*	
Elemental	Protinex	50g	Calories	110
	Glucose	100g	Proteins	2.6g
	Refined oil	30g	CHO	19.5g
	Water to make	1000 cc	Fats	3g
Polymeric	Milk	500g	Calories	150
(Milk based)	Sugar	50g	Proteins	4g
(*Suji kheer*)	*Suji*	20g	CHO	7.2g
	Oil	20g	Fats	7.5g
Polymeric	Rice	50g	Calories	66
(Lactose free)	Sugar	45g	Proteins	3g
	Oil	30g	CHO	8.4g
	Water to make	1000 cc	Fats	3.7g
Special formulas milk (Renal formula)	Skimmed Milk			
	Powder	25g	Calories	344
	Fats	125g	Proteins	3g
	Sugar	150g	CHO	33g
	Tonned Milk	250cc	Fats	22g
Special formulas	Milk	250cc	Calories	100
(Hepatic Rice formula)	Sugar	50g	CHO	20g
	Rice	150g	Proteins	2g
	Corn flour	25g	Fats	1.5g
	Oil	5g		
	Water to make	1000 cc		

Source: Ainsley M. Malone, 2005. Enteral Formula Selection: A Review of Selected Product Categories, Mt. Carmel West Hospital, Department of Pharmacy, Columbus, OH.

Table 17.6: Examples of Commercially Available Enteral Formulas

Formula	Company	Calories
Simy1 MCT	FDC	459 Kcal/100 g
Prosoyal	FDC	506 Kcal/100 g
Nusobee	Wockhardt	480 Kcal/100 g
Pedisure	Ross	100 Kcal/100 g
Tocal	Galen Medical Limited	500 Kcal/100 g
Ten-O-Lip	Criticare	300 Kcal/40 g (To be reconstituted in 225 ml of milk)

1. Elemental Diets or Formulas

Elemental formulas (also known as monomeric, hydrolysed, or chemically defined formulas) are monomeric formulas of low molecular weight which require minimal digestive effort and are easily absorbed without leaving much residue. Carbohydrates are present in these diets as oligosaccharides, sucrose and glucose. Proteins are present in form of short chain peptides and fat is usually present in the form of medium chain triglycerides along with small amounts of essential fatty acids. Elemental formulas are nutritionally complete, which means they contain all of the nutrients needed to maintain nutritional sustenance.

They generally have caloric density of 1 kcal/1 mL and nitrogen concentration of 6-8 g/1 L.

Because of the presence of multiple small particles, they are highly osmotic (500-900 mOsm/L) and may therefore, cause osmotic diarrhea. They cannot be given orally because of their taste. Optimental, Pivot, Tolerex and Vivonex are all examples of elemental formulas. As these diets are predigested, they are easily absorbed and are ideal for patients with pancreatic insufficiency, acute pancreatitis and malabsorptive states.

Elemental formulas are different from standard formulas because their macronutrients have already been partly or fully broken down for those patients who have compromised digestive systems or nutrient absorption problems. One of the major differences between chemically defined 'elemental diets' and standard polymeric liquid diets used for enteral feeding is that elemental diets contain partially or completely hydrolysed protein as their nitrogen source whereas polymeric diets contain undigested protein. Despite the fact that elemental diets are more hypertonic and expensive than polymeric diets, they have been widely used in the clinical setting, not only in patients with impaired gastrointestinal function, but also in those with normal gastrointestinal function, all of whom would appear to be perfectly capable of assimilating whole protein.

These elemental formulas are not commonly needed by oral cancer patients. J tube patients usually eat elemental formulas, and some G tube patients may be prescribed elemental formulas in certain instances.

Here are common ingredients/nutrients in elemental formulas:

Carbohydrates

Carbohydrates may provide between 40 and 60 per cent of the energy making them the major energy contributor. Maltodextrins are usually the main source of carbohydrates in polymeric formulas, the advantage being that they are more soluble than starch, have a low osmotic load, and are rapidly hydrolyzed in the gut. Sucrose in small amounts may be added, increasing the osmotic concentration but improving palatability (useful for oral intake). Some formulas may also contain starch. Sources are:

- ✰ Cornstarch
- ✰ Hydrolysed cornstarch
- ✰ Fructose
- ✰ Maltodextrin

Proteins

Protein provides between 15 and 25 per cent of total energy in polymeric formulas. Protein content ranges from 30 to 80 g l^{-1}, with non-protein calorie: nitrogen ratio between 75:1 and 200:1 kcal g^{-1} N. Sources separated from the various original foods include:

- ✰ Soy protein isolate
- ✰ Crystalline L-amino acids
- ✰ Hydrolysed casein
- ✰ Hydrolysed whey protein
- ✰ Hydrolysed lactalbumin

Fats

Lipids provide an isotonic, calorie dense source and contribute significantly to the non-protein energy content in polymeric formulas. Corn and soybean are commonly used lipid sources in enteral formulas. Safflower and canola oil (monounsaturated fatty acids) may also be found. These vegetable oils provide mostly long-chain triglycerides and besides contributing with essential fatty acids, they help limit osmolality. Lipid content of polymeric formulas can vary between 25 and 40 per cent of total calories.

Medium chain triglycerides (MCT), may replace all or part of the lipid content. MCT's do not require bile salts or pancreatic lipase, bypassing the lymphatic system and being absorbed directly into the portal circulation, which renders them particularly useful in cases of malabsorption or chylothorax. However, MCT oil does not contain essential fatty acids and may delay gastric emptying, causing intolerance. Generally, polymeric formulas include the following:

- ✰ Fatty acid esters
- ✰ MCT oil

- Soybean oil
- Safflower oil
- Fish oil
- Sardine oil
- Structured lipids
- Soy lecithin

Electrolytes and Micronutrients

Nutritionally complete enteral formulas when provided in adequate volume, usually meet 100 per cent of the RDA for vitamins, minerals and trace elements. However, situations of increased requirements or loss of specific nutrients must be taken into account and enteral or parenteral supplementation instituted and adequately monitored.

Water

Energy density of enteral formulas is dictated by their water content. Formulas providing 1 kcal ml[1] contain 85 per cent of water, while more energy dense diets (2 kcal ml[1]) provide only 70 per cent.

2. Semi-elemental (Oligomeric)

Chemically defined oligomeric and monomeric enteral formulas contain macronutrients that have been enzymatically hydrolyzed to varying degrees, thus requiring minimal digestion and being almost completely absorbed. Both types of enteral formulas are lactose and gluten free, as well as being low in residue. In enteral diets, osmolality is inversely related to the molecular size of nutrients in solution. Amino acids and small peptides, because of their reduced particle size, have a major impact on the osmolality of hydrolyzed formulas. An example is Peptamen AF®.

These are designed for use in patients suffering from maldigestion and malabsorption such as Crohn's disease, short bowel syndrome, exocrine pancreatic insufficiency, intestinal obstruction and/or fistulas and radiation enteritis in cancer patients.

Formula Composition

Semi-elemental formulas contain dipeptides, tripeptides and some free amino acids as nitrogen, simple sugars, glucose polymers or starch and variable doses of fat, primarily as medium chain triglycerides (MCT), minerals, vitamins and trace elements. The di- and tripeptides of semi-elemental formulas have specific uptake transport mechanisms and are thought to be absorbed more efficiently than individual amino acids or whole proteins, the nitrogen sources in elemental and polymeric formulas, respectively. Individual and free amino acid residues, as found in elemental formulas, were poorly absorbed while amino acids provided as dipeptides and tripeptides were better absorbed. The semi-elemental formulas containing casein and lactalbumin hydorlysates, but not the fish protein hydrolysates, also stimulated jejunal absorption of water and electrolytes. These formulas are nutritionally complete and have a lower osmolality than elemental diets and are easier to absorb.

Disadvantages

Chemically defined formulas have some disadvantages namely:

- Complications (osmotic diarrhoea) due to high osmolality;
- Poor taste (caused by amino acids) usually compromises oral ingestion; and
- Higher cost when compared to polymeric formulas.

3. Standard or Polymeric Formulas

Regarded as the "standard" approach to enteral nutrition, polymeric formulas are nutritionally complete and made up of mostly intact nutrients, which implies a functional digestive system, and are suitable for use both in hospitals and home care. Polymeric formulas contain whole proteins, complex carbohydrates and long chain triglycerides – meaning, the nutrients are intact/have not been broken down, which requires additional work by the digestive system. For individuals who have full digestive function, there is no issue digesting a polymeric formula. They may also include fiber, and depending on the formula, may be consumed orally or via a tube. Polymeric formulas are often used for longer periods of time (6+ weeks) and are less expensive than elemental or semi-elemental formulas. Boost® and Ensure® are examples of high calorie polymeric formulas.

Standard formula is the largest group and includes concentrated formulas, high protein formulas, and calorie dense formulas. Standard formulas comprise the enteral product category most often used in patients requiring tube feedings. Their nutrient composition is meant to match that recommended for healthy individuals.

Formula Composition

- Whole protein as the nitrogen source
- Carbohydrates from oligosaccharides, maltodextrins or starch
- Lipids from vegetable oils
- Minerals, vitamins and trace elements

Carbohydrates

- Corn syrup solids
- Maltodextrin
- Hydrolysed corn starch
- Glucose
- Fructose

- Sucrose
- Sugar alcohols

Proteins

- Casein
- Whey protein concentrate
- Milk protein concentrate
- Soy protein isolate
- Lactalbumin
- Sodium, calcium, potassium, and magnesium caseinates

Fats

- Soybean oil
- Corn oil
- Canola oil
- Palm kernel oil
- Medium chain triglycerides (MCT) oil
- Safflower oil
- High-oleic sunflower oil
- Borage oil
- Fish oil
- Monoglycerides
- Diglycerides
- Soy lecithin

They contain proteins that haven't been broken down (intact proteins) and polysaccharides. The proteins are normally extracted from cow's milk, soybeans, or protein isolates derived from other food sources. The carbohydrate component is typically glucose polymers, such as maltodextrin, modified starches, and all varieties of sugar. The fats for standard formulas are typically heat- or chemically-extracted from plant sources. They could also include fish oils or modified oils such as MCT oils taken from coconut and palm oils.

Their goal is to provide the entire spectrum of macro- and micro-nutrients necessary for human nutrition. The differences among standard formula brands and products come from proportions and sources of ingredients.

Polymeric formulas are lactose-free and most of them contain no gluten. Since nutrients are not hydrolyzed, the osmolality is reasonably close to the physiological levels (300 mOsmol l), which is a plus for enhanced tolerance. Caloric density may vary from 0.5 to 2 kcal ml allowing for adaptation to individual patient needs:

There are several standard formulas available in pharmacies, grocery stores, or online in the form of flavored oral supplements like Ensure, Boost, or Carnation Instant Breakfast. Also, some stores have generic, store brands of standard formulas at a cheaper price *e.g.* Equate, Market Pantry Nutrition Shakes, *etc.* They can be ready-to-go out of the can or in the form of powder, requiring reconstitution with milk or water.

These diets consist of intact proteins, complex carbohydrates, fat and residue. The various types of polymeric formulas include milk based diets and lactose free diets which are preferred in critically ill patients and in lactose-intolerant patients. Polymeric diets are of high molecular weight and are of relatively low osmolality. Therefore, they are preferred for duodenal and jejunal administration.

Calorie Dense Standard Formula Products

Standard formulas are intended for patients who are able to absorb and digest nutrients without difficulty. They are most commonly used by G-tube patients who have no specific medical conditions or higher nutrient needs. The formulas commonly come in different calorie strengths for patients with higher calorie needs or when volume needs to be taken into consideration. For example, there are 1.0 Kcal/ml (1,000 Kcalories per 1,000 ml), 1.2 Kcal/ml (1,200 Kcalories per 1,000 ml), 1.5 Kcal/ml (1,500 Kcalories per 1,000 ml), and 2.0 Kcal/ml (2,000 Kcalories per 1,000 ml). 0.5–1 Kcal ml [1]are useful for initiating enteral feeding; Whereas 1.5–2 Kcal/ml may provide increased needs as well as respond to the need for fluid restriction.

The products may or may not contain fiber. Calorie dense formulas may be used with volume sensitive patients or patients needing fluid restriction. Such conditions may include congestive heart failure, renal failure or syndrome of inappropriate diuretic hormone (SIADH). However, this intervention may not always be clinically significant. For example, if a patient requires 1800 Kcal/day, changing a 1.0 Kcalorie/mL to a 2.0-Kcalorie/mL product would reduce the water content by 900 mL, but to change a patient from a 1.5 to a 2.0 Kcal/mL product represents a mere 300 mL difference per 24 hour period. Calorically dense formulas are most practical for use in patients requiring nocturnal and/or bolus feeding.

Fiber Supplemented Standard Formulas

The advantage of adding fibre to enteral formulas is still controversial, in spite of undisputed evidence of benefits in a normal diet. Commercially available blenderized enteral feeds contain around 5 g l [1] of natural sources of fibre. Several fibre enriched enteral formulas are currently available, with amounts ranging from 5 to 15 g of TDF per litter.

Sources of fiber in enteral formulas include soluble and insoluble. A recent fiber addition to selected formulas (Ross products) is fructo-oligosaccharides (FOS). FOS are defined as short-chain oligosaccharides and, similar to other dietary fibers, are rapidly fermented by the colonic bacteria to short-chain fatty acids (SCFA).

Table 17.7: Dietary Fibre Content of some Enteral Formulas

Product	*Total Dietary Fiber (g/L)*	*Per cent Insoluble Fiber*	*Per cent Soluble Fiber*	*Cost/1000 Kcal ($)**	*Manufacturer*
Compleat	4.3	74.0	26.0	10.9**	Novartis
Fibersource Std	10.0	75.0	25.0	5.83	Novartis
Fibersource HN	10.0	75.0	25.0	6.13	Novartis
Isosource 1.5	8.0	48.0	52.0	4.40	Novartis
Isosource VHN	10.0	48.0	52.0	8.80	Novartis
Jevity 1.0	14.4	100.0	0.0	6.60	Ross
Jevity 1.2	22.0	75.0	25.0	6.50	Ross
Jevity 1.5	22.0	75.0	25.0	6.37	Ross
Nutren 1.0 w/Fiber	14.0	95.0	5.0	5.98	Nestle
NutriFocus	20.8	75.0	25.0	4.44	Nestle
Novasource Pulmonary	8.0	48.0	52.0	6.72**	Novartis
Peptamen w/FOS	4.0	0	100.0	23.76	Nestle
Probalance	10.0	75.0	25.0	6.83	Nestle
Promote w/Fiber	14.4	94.0	6.0	6.60	Ross
Protain XL	9.1	94.0	6.0	5.86**	Novartis
Replete w/Fiber	14.0	95.0	5.0	8.45	Nestle
Ultracal	14.4	70.0	30.0	7.70	Novartis
Ultracal Plus HN	10.0	73.0	27	7.23	Novartis

Source: Ainsley M. Malone, 2005. Enteral Formula Selection: A Review of Selected Product Categories, Mt. Carmel West Hospital, Department of Pharmacy, Columbus, OH.

Table 17.7 provides a listing of enteral formulas and their fiber content.

Based on the proven importance and advantages of dietary fibre in the normal diet, it is tempting to extrapolate to enteral formulas and propose that unless fibre restriction is required, polymeric formulas should resemble normal feeding in terms of fibre content. However there is still no clear cut evidence proving efficacy or optimum amounts of fibre in enteral formulas. Further research and guidelines are needed.

Comparison of Elemental and Polymeric Formulas

Standard formulas tend to be less expensive than the elemental types of formulas. Although elemental and semi-elemental formulas cost more than polymeric formulas; they are widely used because of being better absorbed, less allergenic, better tolerated in patients with malabsorptive states and cause less exocrine pancreatic stimulation in patients with pancreatitis.

Table 17.8 provides a comparison of nutrient sources in polymeric and hydrolyzed products.

Table 17.8 Macronutrient Sources in Enteral Formulas

Enteral Formula	*Carbohydrate*	*Protein*	*Fat*
Polymeric/ standard	Corn syrup solids	Casein	Borage oil
	Hydrolyzed cornstarch	Sodium, calcium, magnesium and	Canola oil
	Maltodextrin	Potassium caseinates	Corn oil
	Sucrose	Soy protein isolate	Fish oil
	Fructose	Whey protein concentrate	High oleic sunflower oil
	Sugar alcohols	Lactalbumin	Medium chain triglycerides
		Milk protein concentrate	Menhaden oil
			Mono- and diglycerides
			Palm kernel oil
			Safflower oil
			Soybean oil
			Soy lecithin
Hydrolyzed	Cornstarch	Hydrolyzed casein	Fatty acid esters
	Hydrolyzed cornstarch	Hydrolyzed whey protein	Fish oil
	Maltodextrin	Crystalline L-amino acids	Medium chain triglycerides
	Fructose	Hydrolyzed lactalbumin	Safflower oil
		Soy protein isolate	Sardine oil
			Soybean oil
			Soy lecithin
			Structured lipids

Historically, soluble fiber has been difficult to add to enteral formulas due to its viscous nature. Many early fiber supplemented enteral formulas, therefore, contained soy polysaccharide as their primary fiber source. Subsequent technological advances have enabled the inclusion of soluble fiber sources to enteral formulas and many now contain a combination of both soluble and insoluble fibers.

Table 17.9: Water Content of Various Enteral formula Densities

Calorie Density	*Per cent Water*	*Volume/1800 Kcal (ml)*	*Water by Density for 1800 Kcal (ml)*
1.0 kcal/mL	84	1800	1530
1.2 kcal/mL	82	1500	1230
1.5 kcal/mL	76	1200	930
2.0 kcal/mL	70	900	630

Source: Ainsley M. Malone, 2005. Enteral Formula Selection: A Review of Selected Product Categories, Mt. Carmel West Hospital, Department of Pharmacy, Columbus, OH.

4. Modular Diets/Disease Specific Formulas

The availability of specialized enteral formulas has burgeoned in the last 20 years, many touting pharmacologic effects in addition to standard nutrient delivery. Well over 100 Specialty formulations are available for patients with unique "disease specific" or "organ specific" nutritional requirements; making formula selection rather challenging. This is a growing area of enteral nutrition, where improved knowledge of disease processes has led to the development of a multitude of specialty products. Currently specially designed formulas exist for liver disease, renal disease, diabetes, pulmonary insufficiency, heart failure, GI dysfunction as well as situations of metabolic stress such as trauma or sepsis. Elemental and fiber supplemented formulas are also frequently recommended for use in those with certain types of gastrointestinal dysfunction. These formulas are available as individual nutrients. Various nutrients can be mixed according to the need of the patient. They contain biologically active substances or nutrients such as glutamine, arginine, nucleotides, essential fatty acids *etc.* These products are more expensive than standard enteral nutrition and may result in complications when used inappropriately.

Examples of some of the special formulas (disease specific) are given as under:

i) Liver Formulas

The specialized formulas for hepatic failure and hepatic encephalopathy have a higher proportion of branched-chain amino acids (BCAA): valine, leucine, and isoleucine; and reduced amounts of aromatic amino acids (AAA): phenylalanine, tyrosine and tryptophan, and methionine, compared to standard products in order to assist in correcting the abnormal plasma ratio of these amino acids and increase the Fischer ratio (BCAA/AAA). These alterations have been purported to promote a reduced uptake of AAA at the blood brain barrier, reducing the synthesis of false neurotransmitters and thereby, ameliorating the neurological symptoms that occur with hepatic encephalopathy (HE).

Two enteral formulas with increased BCAA are available (Table 17.10). Evidence supporting the use of hepatic formulas is very limited. The routine use of BCAA enriched enteral formulas in patients with advanced liver disease and/or HE is not recommended at this time. Standard enteral formulas can successfully be used with most patients at a much lower cost. However, in those patients who are refractory to routine drug therapy for HE and are unable to tolerate standard protein intakes without precipitation of HE, the use of BCAA enriched enteral formulas may be worth a short trial.

Table 17.10: Enteral Formulas Designed for Hepatic Disease

Product	*Manufacturer*	*Kcals/ mL*	*Per cent CHO Kcals*	*Per cent Fat Kcals*	*Per cent Protein Kcals*	*Comments*	*Cost/1000 Kcal**
Hepatic-Aid II	Hormel Healthlabs	1.2	57.3	27.7	15.0	☆ Increased levels of leucine, isoleucine and valine	$41.56
						☆ Minimal phenylalanine tryptophan and tyrosine content	
						☆ Contains negligible amounts of vitamins and minerals	
NutriHep	Nestle	1.5	77.0	11.0	12.0	☆ Contains standard amounts of vitamins and minerals	$35.55
						☆ 50 per cent BCAA and 50 per cent AAA	
						☆ 66 per cent of fat is MCT	

* Based on 1-800 Company Home Delivery Numbers.

Source: Ainsley M. Malone, 2005. Enteral Formula Selection: A Review of Selected Product Categories, Mt. Carmel West Hospital, Department of Pharmacy, Columbus, OH.

ii) Renal Formulas

Acute renal failure patients are usually hypercatabolic and hypermetabolic. The aim of enteral nutrition in these patients is to minimize blood urea nitrogen, reduce accumulation of toxic products, while maintaining water and electrolyte balance as well as nutritional status. Stable pre-dialysis patients require low protein energy dense formulas enriched in essential amino acids. Dialysis patients need a high protein energy dense formula. Since water and electrolyte balance need to be carefully monitored, renal enteral diets are calorically dense to facilitate fluid management.

Formulas designed for patients with renal disease vary in protein, electrolyte, vitamin and mineral content (Table 17.11). Generally, renal formulas are lower in protein, calorically dense and have lower levels of potassium, magnesium and phosphorus when compared to standard formulas.

iii) Gastrointestinal (GI) Dysfunction Formulas

Patients with GI dysfunction *i.e.* pancreatic insufficiency, short bowel, inflammatory bowel disease (IBD), diverticulosis, ischemia *etc.*, may benefit from hydrolyzed or peptide containing formulas. Gut recovery may be enhanced by supplementation of glutamine and soluble (fermentable) fibre, which is a precursor of short-chain fatty acids (SCFA). Glutamine promotes enterocyte proliferation and differentiation, thereby improving the gut barrier. SCFA resulting from the fermentation of soluble fibre also maintain and support colonic mucosal function. Patients with diverticulosis, diverticulitis and constipation, may benefit from enteral diets enriched with insoluble (less fermentable) fibre.

iv) Diabetes/Hyperglycemia

Several formulas have been developed for use in patients with diabetes mellitus (DM) (Table 17.12). These formulas offer a lower amount of total carbohydrate and a higher amount of fat than standard formulas as well as a variation in type of carbohydrate. The carbohydrates in diabetic formulas have a lower glycemic index so that the sugars are converted more slowly before they're released into the bloodstream. These formulas also contain more fiber, to slow the breakdown of carbohydrates even more. Carbohydrate sources generally consist of oligosaccharides, fructose, corn starch and fiber. In normal subjects, the use of more complex carbohydrates, such as fructose, corn starch and fiber has been shown to improve glycemic control as a result of delayed gastric emptying and reduced intestinal transit. Formulas designed for patients with DM are

Table 17.11 Enteral Products Designed for Renal Disease

Product	Manufacturer	Kcals/mL	Protein (gm)**	K (mEq)**	P (mg)**	Mg (mg)**	Cost/1000 Kcals ($)**
Renal Formulas							
Magnacal Renal	Novartis	2.0	37.5	16	400	100	3.47
Nepro	Ross	2.0	35.0	14	343	108	6.08
NovaSource Renal	Novartis	2.0	37.0	14	325	100	5.64
Suplena	Ross	2.0	15.0	14	365	108	3.73
Nutri-Renal	Nestle	2.0	17.0	Negligible	Negligible	Negligible	4.17
Standard Concentrated Formulas							
Deliver 2.0	Novartis	2.0	37.5	21.5	555	200	4.30
NovaSource 2.0	Novartis	2.0	45.0	19	550	210	3.81
Nutren 2.0	Nestle	2.0	40.0	25	670	268	2.98
Two-Cal HN	Ross	2.0	42.0	31	538	213	3.21

* Per 1000 kcals; ** Based on 1-800 Company Home Delivery Numbers (see Table 17.17).

Source: Ainsley M. Malone, 2005.

Table 17.12: Enteral Formulas Designed for Diabetes Mellitus

Product	Manufacturer	Kcals/mL	Per cent CHO Kcals	Per cent Protein Kcals	Per cent Fat Kcals	Fiber (g/1000 mL)	Cost/1000 Kcal ($)*
Choice DM	Novartis	1.06	40.0	17.0	43.0	14.4	10.48
DiabetiSource AC	Novartis	1.0	36.0	20.0	44.0	4.3	8.33
Glucerna Select	Ross	1.0	22.8	20.0	49.0	21.1	**
Glytrol	Nestle	1.0	40.0	18.0	420	15.0	8.20
Resource Diabetic	Novartis	1.06	36.0	24.0	40.0	12.8	6.22

* Based on 1-800 Company Home Delivery Numbers; ** Ross Products was unable to provide this information.

based on this premise. Due to the inherent viscosity of soluble fiber, most enteral formulas for DM contain a combination of soluble and insoluble fiber. Two prime examples of specialized formulas for diabetic patients are Glucerna and Boost Glucose Control.

v) Pulmonary Formulas

The deterioration of nutritional status in critically ill patients with respiratory insufficiency is associated with reduction of respiratory muscle mass, muscle weakness and problems with weaning from ventilating support. Patients with pulmonary insufficiency retain CO_2 and experience O_2 depletion. Nutrition therapy, particularly carbohydrate rich formulas, may precipitate respiratory failure in these patients due to increased O_2 consumption and CO_2 production. Therefore, in patients with pulmonary compromise, a formula containing decreased carbohydrate calories should be selected.

To prevent this problem, there are two approaches: reduce or withhold feeding, or change to a formula with higher proportion of fat vs. carbohydrate. Before deciding on which solution, it is important to determine if overfeeding is not the under-lying cause of excess CO_2 production and ventilator dependency. Specialized enteral formulas have been developed for two types of pulmonary disease: chronic obstructive pulmonary disease (COPD) and acute respiratory distress syndrome (ARDS). While there are similarities with these products, distinct differences do exist.

vi) Allergies

It is important to be aware of the composition of enteral feeding products for patients with suspected or documented food allergies. Nuts, fruits and milk are the most common triggers. Epidemiological data indicate that these reactions are caused by different mechanisms, with only about a third of the reactions in children and 10 per cent of those in adults due to actual food allergy. The majority of adverse reactions to food are non-immunologic in origin with lactose intolerance being the most common type of adverse reaction worldwide. However, true food allergies are thought to affect up to 6 per cent to 8 per cent of children under the age of ten and between 1 per cent – 4 per cent of the adult population, which makes it likely that a sub-set of patients receiving enteral feeding will have food allergies. Allergy to cow's milk, eggs, wheat and soy is more common in infants and young children while seafood, peanuts and tree nuts are the more common causes of food allergy in adult life. Although not an allergy, but an autoimmune process, patients with celiac disease need to avoid gluten-containing foods, including enteral formulas should they be necessary. Tables 17.13–17.16 provide a listing of enteral products that may be considered for use in patients with allergy to corn, casein, soy, whey, egg, and gluten intolerance.

vii) Sepsis, Trauma and Burns

Stress caused by sepsis, burns and trauma, results in release of various hormones in the body. These hormones cause skeletal muscle proteolysis and hydrolysis of branched chain amino acids (BCAA). Therefore, in such stress conditions, formulas rich in BCAAs are preferred.

viii) Wound Management Formulas

Wound management formulas can provide higher amounts of protein (approximately 25 per cent of total calories vs. 15 per cent from standard formulas) and may contain added nutrients, such as arginine and glutamine, which have been shown to assist with the healing process.

Type	*Subtype*	*Description*	*Indications*
Polymeric	Standard	Nutrient distribution as a normal diet	Normal GI function
	High protein	Protein >15 per cent of total energy	Catabolic states
			Wound healing
	Energy dense	2 kcal ml [1]	Fluid restriction
			Electrolyte imbalance
	Fibre enriched	5–15 g l [1] (TDF)	Bowel function disorders
Oligomeric	Partially hydrolyzed	Composition varies; one or more	Impaired digestion and absorption
	Peptide based	nutrients are hydrolyzed.	
Monomeric	Free amino acids		
Disease specific	Renal	Less protein, low electrolyte load	Renal failure
	Hepatic	High BCAA, low AA, low electrolyte	Hepatic encephalopathy
	Pulmonary	Higher fat content	ARDS
	Diabetic	Low CHO load	Diabetes mellitus
	Immune enhancing	Arginine, glutamine	Metabolic stress
		ω-3 fatty acids, nucleotides, antioxidants	Impaired immunity

Source: Zdenek. Zada´k, Luiza.2009. Kent-Smith/e-SPEN, the European e-Journal of Clinical Nutrition and Metabolism. 4 (2009) e212-e215.

Table 17.13: Formulas/Modulars that do not Contain Casein in Product Formulation

	Ross (1)	***Novartis***	***Nestle***
Adult Products			
Tube Feeding Formulas	EleCare	Diabetisource AC	f.a.a.
	EquaLYTE	Fibersource, Fibersource HN	Peptamen, VHP, PreBio 1, 1.5
		Isosource, Isosource HN	
		Subdue Plus	
		Tolerex	
		Vivonex Plus, RTF, TEN	
Oral Supplements	Juven	Boost Breeze	None
		Impact Recover	
		Peptinex	
		Resource Arginaid	
		Resource Arginaid Extra	
Modulars	Polycose	Benefiber	None
		Beneprotein	
Pediatric Products			
Tube Feeding Formulas	EleCare	Vivonex Pediatric	Peptamen Junior
			Peptamen Junior Powder
			Peptamen Junior with PreBio1
Infant Formulas	EleCare	None	Goodstart Essentials
			Goodstart Supreme
			Goodstart Supreme with DHA and ARA
			Goodstart 2 Essentials
			Goodstart 2 Supreme with DHA and ARA
			Goodstart Supreme Soy with DHA and ARA
			Goodstart 2 Essentials Soy

Table 17.14: Formulas/Modulars that do not Contain Soy in Product Formulation

	Ross (1)	***Novartis***	***Nestle***
Adult Products			
Tube Feeding Formulas	EleCare	Compleat (3)	Crucial (4)
	EquaLYTE	Comply (2)	f.a.a. (4)
		Deliver 2.0 (4)	Glytrol (2)
		Impact (2), Impact 1.5 (3)	Modulen (2)
		Isocal HN Plus (2)	Nutren 1.0 (2), 1.5 (2), 2.0 (2)
		Lipisorb (4)	Nutren Fiber (2)
		Magnacal Renal (2)	NutriHep (2)
		Novasource 2.0 (2)	NutriRenal (2)
		Novasource Renal (2)	NutriVent (2)
		Peptinex DT (4)	Peptamen (4), VHP (4), with PreBio1(4), 1.5 (4)
		Respalor (2)	Renalcal (2)
		Subdue, Subdue Plus	Replete (2)
		Tolerex	
		Traumacal (4)	
		Vivonex Plus (4), TEN, RTF (4)	
Oral Supplements	Enlive!	Boost (2)	Carnation Instant Breakfast (2)
	Ensure Pudding	Boost Plus (2)	Carnation Instant Breakfast for the Carb Conscious (2)
	Juven	Boost Breeze	Carnation Instant Breakfast Juice Drink (2)
		Impact Recover	Carnation Instant Breakfast Lactose Free (2)

	Ross (1)	***Novartis***	***Nestle***
		Peptinex (4)	Carnation Instant Breakfast Lactose Free Plus (2)
		Resource Arginaid (2)	NutriHeal (2)
		Extra 2.0 (2)	
Modulars	Polycose	Benefiber	Additions (2)
		Beneprotein (2)	
Pediatric Products			
Tube Feeding Formulas	EleCare	Compleat Pediatric (4)	Nutren Junior (4), Nutren Junior with Fiber (4)
		Pediatric Peptinex DT (4)	Peptamen Junior (liquid and powder) (4)
		Vivonex Pediatric (4)	Peptamen Junior with PreBio1 (4)
Oral Supplements	None	Resource Just for Kids (4)	Carnation Instant Breakfast Junior (2,4)
Infant Formulas	EleCare	None	Goodstart Essentials (2,4)
			Goodstart Supreme (2,4)
			Goodstart Supreme with DHA and ARA (2,4)
			Goodstart 2 Essentials (2,4)
			Goodstart 2 Supreme with DHA and ARA (2,4)

Table 17.15: Formulas/Modulars that do not Contain Gluten in Product Formulation

	Ross (1)	***Novartis***	***Nestle***
Adult Products			
Tube Feeding Formulas	AlitraQ	All tube feeding formulas are	Crucial
	EleCare	gluten free EXCEPT Boost	f.a.a.
	EquaLYTE (3)	chocolate malt flavor.	Glytrol
	Glucerna		Modulen
	Glucerna Select (3)		Nutren 1.0, 1.5, 2.0
	Jevity 1 Cal		Nutren Fiber
	Jevity 1.2, 1.5 Cal (1,2,3)		NutriHep
	Nepro (3)		NutriRenal
	Optimental (3)		NutriVent
	Osmolite, 1, 1.2, 1.5 Cal		Peptamen, VHP, with PreBio1, 1.5
	Oxepa		ProBalance
	Perative (3)		Renalcal
	Pivot 1.5 Cal (3)		Replete
	Promote		Replete with Fiber
	Promote with Fiber (2)		
	Pulmocare		
	Suplena		
	TwoCal HN (3)		
	Vital HN		
Oral Supplements	Enlive!	All liquid oral supplements are	Carnation Instant Breakfast Juice Drink
	Ensure	gluten free EXCEPT Boost	Carnation Instant Breakfast Lactose Free
	Ensure Fiber with FOS (2,3),	chocolate malt flavor.	Carnation Instant Breakfast Lactose Free Plus
	Healthy Mom Shake, High Calcium,		Carnation Instant Breakfast Lactose Free VHC
	High Protein, Plus, Plus HN, Powder,		NutriHeal
	Pudding (3)		
	Glucerna Shake (3), Weight Loss Shake (3)		
	Hi-Cal		
	Juven		

	Ross (1)	*Novartis*	*Nestle*
	NutriFocus (1,2,3)		
	ProSure Shake (3)		
Modulars	Polycose	Benefiber (EXCEPT tablet form)	Additions
	ProMod		
Pediatric Products			
Tube Feeding Formulas	EleCare	All tube feeding formulas	Nutren Junior
	PediaSure Enteral Formula	are gluten free.	Nutren Junior with Fiber
	PediaSure Enteral Formula with Fiber		Peptamen Junior (liquid and powder)
	(1,2,3)		Peptamen Junior with PreBio1
Oral Supplements	PediaSure	All liquid oral supplements	None
	PediaSure with Fiber	are gluten free.	
Infant Formulas	EleCare	None	Goodstart Essentials
			Goodstart Supreme
			Goodstart Supreme with DHA and ARA
			Goodstart 2 Essentials
			Goodstart 2 Supreme with DHA and ARA
			Goodstart Supreme Soy with DHA and ARA
			Goodstart 2 Essentials Soy

Table 17.16: Formulas/Modulars that do not Contain Egg in Product Formulation

	Ross (1)	*Novartis*	*Nestle*
Adult Products			
Tube Feeding Formulas	AlitraQ	All tube feedings are	Crucial
	EleCare	egg free.	f.a.a.
	EquaLYTE		Glytrol
	Glucerna		Modulen
	Glucerna Select		Nutren 1.0, 1.5, 2.0, Fiber
	Jevity 1 Cal, 1.2 Cal, 1.5 Cal		NutriHep
	Nepro		NutriRenal
	Optimental		NutriVent
	Osmolite, 1 Cal, 1.2 Cal, 1.5 Cal		Peptamen, Peptamen with PreBio1, 1.5, VHP
	Oxepa		ProBalance
	Perative		Renalcal
	Pivot 1.5 Cal		Replete, Replete with Fiber
	Promote, Promote with Fiber		
	Pulmocare		
	Suplena		
	TwoCal HN		
	Vital HN		
Oral Supplements	Enlive!	All liquid oral supplements	Carnation Instant Breakfast
	Ensure	are egg free.	Carnation Instant Breakfast for the Carb Conscious
	Ensure Fiber with FOS, Healthy Mom		Carnation Instant Breakfast Juice Drink
	Shake, High Calcium, High Protein,		Carnation Instant Breakfast Lactose Free
	Plus, Plus HN, Powder, Pudding		Carnation Instant Breakfast Lactose Free Plus
	Glucerna Shake		Carnation Instant Breakfast Lactose Free VHC

	Ross (1)	Novartis	Nestle
	Glucerna Weight Loss Shake		NutriHeal
	Hi-Cal		
	Juven		
	NutriFocus		
	ProSure Shake		
Modulars	Polycose	None	Additions
	ProMod		
Pediatric Products			
Tube Feeding Formulas	EleCare	All tube feeding formulas	Nutren Junior, Nutren Junior with Fiber
	PediaSure Enteral Formula	are egg free.	Peptamen Junior
	PediaSure Enteral Formula with Fiber		Peptamen Junior Powder
			Peptamen Junior with PreBio1
Oral Supplements	PediaSure	All oral liquid supplements	Carnation Instant Breakfast Junior
	PediaSure with Fiber	are egg free.	Goodstart Essentials
Infant Formulas	EleCare	None	
			Goodstart Supreme
			Goodstart Supreme with DHA and ARA
			Goodstart 2 Essentials
			Goodstart 2 Supreme with DHA and ARA
			Goodstart Supreme Soy with DHA and ARA
			Goodstart 2 Essentials Soy

ix) Stress and Immunomodulatory Formulas

These specialized formulas are theoretically designed to affect the inflammatory response and to enhance resistance to infection by reducing gut bacterial translocation and promoting the gut associated lymphatic tissue. Specific nutrients are found in these so-called "immune enhancing diets" – glutamine, arginine, omega 3-fatty acids, nucleotides and BCAA – either individually or in combination. These diets have shown some promising results in critically ill patients, for which they seem particularly suited.

x) Immunonutrition

Septic, injured and other critically ill patients have immunosuppression, manifesting as infections. It has been seen that early administration of enteral feeds may be associated with decreased frequency of hospital-acquired infections. Certain nutrients have the potential to alter cellular responses to inflammatory mediators. To obtain this effect, these nutrients are required in doses higher than those required for general nutritional support. Use of these nutrients is called immunonutrition. Various nutrients known to enhance immune function are dietary peptides, arginine, glutamine, Vitamin C, Vitamin A, Vitamin E and nucleic acids. Most of the trials on immunonutrition have used formula containing intact proteins and have added arginine, RNA and mixture of omega-3 and omega-6 polyunsaturated fatty acids. In a meta-analysis it was concluded that immunonutrition causes decrease in duration of hospital stay, rate of infections and ventilatory days.

Some Factors to Consider When Choosing a Formula with Your Doctor or Nutritionist

- ✰ Formulas commonly come in different strengths or caloric densities, classified with numbers from 0.5 to 2. This refers to the number of calories per milliliter. A 'regular' strength formula is a 1, which is about 30 Kcal/oz. One should be prescribed a Low-Cal formula (0.5), a 1.2, a Hi-Cal formula (1.5) or a Two-Cal (2.0). For the most part, the lower the caloric density, the easier the formula will be to tolerate. Higher caloric densities could be prescribed when patients are unable to take in enough volume to meet their nutritional needs, or if they have fluid restrictions or above average energy needs.
- ✰ It's important to remember that formulas with high-caloric density have lower water content. Therefore, unless the patient is fluid-restricted, he/she will be required to drink more **free water** with higher strength formulas. Lower strength formulas, like Ensure or Boost, are around 85 per cent water, while 2 Kcal/ml formulas (Two Cal or Boost VHC) are about 70 per cent. It's important to factor this in when determining the patient's total daily water intake.

- Free water is water by itself, without being mixed in with foods, supplements, electrolytes, or medicines. It is vital for all tube-fed patients to have free water because the body reacts differently to water on its own than it does to anything else. When we eat or drink anything with calories, our bodies can discern that the food needs digestion and act accordingly. The body secretes acid, tailors enzymes according to the nutrients, and activates paristalsis (muscle contractions in the digestive tract). Our systems can even assess the fat content of the food we've eaten and slow down gastric emptying as a result. This happens every time we consume something needing digestion. However, it doesn't happen with pure water. When we drink water on its own, the water is absorbed quickly through the gut walls to be used for vital bodily functions.
- There are so many reasons why free water is necessary for tube-fed patients. One obvious reason is that water cleans feeding tubes before and after meals. It also aids stomach emptying between meals. Drinking free water 30 minutes to an hour before meals can help volume tolerance because it 'primes' the stomach to begin digesting. Water assists with general digestion health, fixing many constipation, diarrhea, and reflux issues. Finally, water helps to flush out our kidneys.
- Another factor to consider when choosing a formula is that regular-strength formulas, high-caloric-density formulas and elemental formulas are absorbed by the gut at different rates. Elemental and high calorie formulas are described as **hypertonic**, and are not absorbed as readily as regular formulas **(isotonic** formulas). This could cause diarrhea, or some other type of digestive issue, and it's another reason to put more free water through the tube.
- Formulas are more than likely to contain some type of **fiber**, whether it's soluble or insoluble. The soluble fiber-of-choice for most formulas is FOS (fructo-oligo-saccharides). FOS acts as fuel for good bacteria in stomach, helping them perform necessary dietary functions. Soluble fiber also strengthens the lining of GI tract and aids water and sodium absorption. Insoluble fibers give bulk to fecal matter and help it pass through the intestines more quickly. Insoluble fiber is another vital part of diets. It helps to manage diarrhea and constipation, and supports the regulation of blood glucose levels. However, certain patients with intestinal or digestive problems should be aware of fiber levels in their formula as fiber may exacerbate their issues.
- Another factor is formula intolerance. Commercial formulas can be difficult to tolerate, especially over the long-term. This intolerance presents itself like any other type of food intolerance or allergy. The patient could see various symptoms like reflux, constipation, diarrhea, nausea, retching, vomiting, or any other allergic reaction. This could be from a specific ingredient in the formula, or it could simply be a matter or drinking the same product multiple times a day for months or even years. Formula manufacturers claim that their products provide patients with all their dietary needs. This is a dubious claim. New compounds in our food are discovered with amazing health-giving properties on a weekly basis. Commercial companies could never replicate the nutrients contained in a varied diet also, as stated earlier, formulas contain a lot of sugar–usually in the form of corn syrup–which is not ideal long-term nutrition.
- Therefore, while commercial formulas are perfectly fine to consume as a sole source of nutrition over a short period of time (6-8 months), the Oral Cancer Foundation recommends that oral cancer patients with long-term feeding tubes look into alternate ideas, such as blending meals, for a complete and healthy diet. Of course, some patients have no choice in this matter and cannot deviate from a commercial formula diet. The patients need to consult their doctor or nutritionist before making dietary changes.

Routes of Enteral Feeding

Enteral feeding tubes may enter the body at a number of different sites such as through nose to stomach which is called nasogastric tube feeding or stomach or duodenum directly by surgical operation known as gastrostomy and jejunostomy feeding. The choice of enteral feeding route depends on several factors, such as the intended duration of nutrition support, the patient's condition, and any limitations to access (such as trauma or obstructions).

Duration of Tube Feeding

1. Short Term Feeding (upto 4 weeks)

i) Naso-Gastric Tube (NG) Feeding

A fine bore feeding tube (French gauge 5-8) is inserted via the nose into the stomach. Bolus or pump feeding can be used with an NG tube. Long term fine bore tubes should be replaced every 4-6 weeks, swapping them to the other nostril. Regular checks of NG tube placement are imperative as there is a risk that tubes can be misplaced into the lungs on insertion or move from the stomach at a later stage.

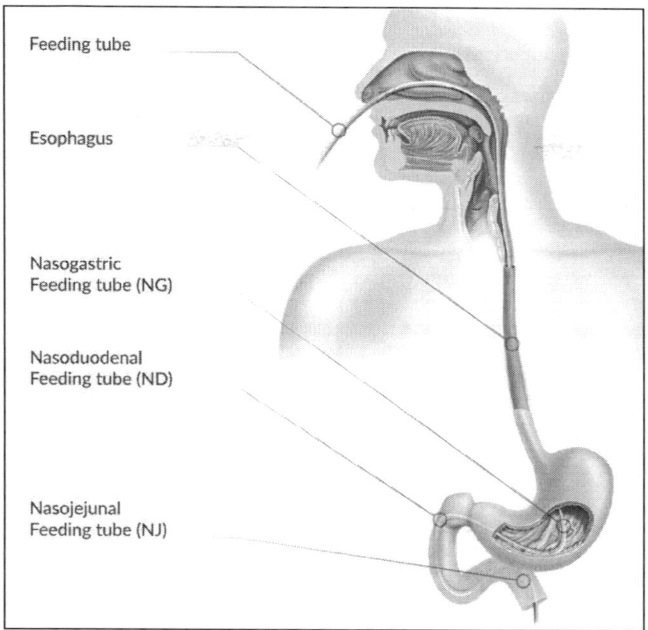

ii) Oro-Gastric Feeding

A feeding tube (French gauge 5-8) is inserted via the mouth into the stomach. As with NG tube, placement must be checked regularly and tube should be replaced every 4-6 weeks.

iii) Naso-Jejunal Feeding

A fine bore feeding tube (French gauge 6-10) is inserted via the nose into the jejunum. This tube may also have a lumen of which the distal end allows deflation of the stomach. It is indicated for patients with gastric reflux or delayed gastric emptying.

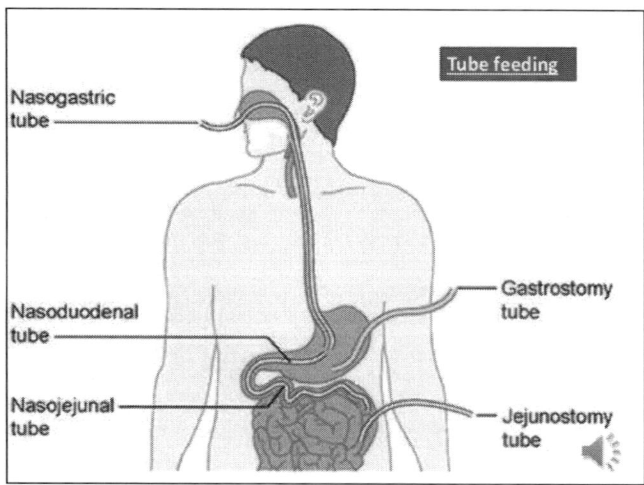

2. Long Term Feeding (longer than 4 weeks)

i) Percutaneous Endoscopic Gastrostomy (PEG)

A tract is made into the stomach via endoscopy under local anesthesia and a feeding tube is inserted. It is held in place by an external fixation device and a soft plastic bumper internally.

ii) Surgically Placed Gastrostomy

A gastrostomy feeding tube is inserted surgically under general anesthesia. This is often used when a patient is unable to tolerate an endoscopy or an endoscope cannot be passed. These tubes may or may not be held in place by sutures so check before removing any sutures. A balloon gastrostomy can replace these feeding tubes once the stoma tract is formed.

iii) Radiologically Inserted Gastrostomy (RIG)

A gastrostomy tube is inserted under X-ray (fluoroscopy or ultrasound) guidance and is usually indicated if an endoscopic procedure cannot be performed. It can be replaced by balloon gastrostomies once the stoma site has healed.

iv) Balloon Gastrostomy

The gastrostomy is held in place by a balloon filled with sterile water. The volume of the balloon should be checked weekly to ensure it is inflated sufficiently to prevent tube displacement.

v) Low Profile Gastrostomy Device (LPGD)

LPGD also known as button gastrostomy is a small device that sits close to the skin and is usually held in by a balloon. It has the same function as a PEG but is less cumbersome, easier to conceal, less obtrusive and may be useful for those patients that pull at their gastrostomy. Extension sets are connected onto the "button" part to enable water, feed, or medications to be administered. Once completed, the extension set is removed. LPGDs can be inserted into most patients once a stoma tract is established but are usually used more frequently in children than adults.

vi) Jejunostomy

A feeding tube is inserted directly into the jejunum during surgery or endoscopically.

vii) Gastrojejunostomy (PEG – J)

This is an endoscopically placed extension of a PEG. The extension is passed through the PEG into the stomach and down past the pylorus into the jejunum. These are used if there is a problem with the stomach or gastric emptying.

Equipment Required for Tube Feeding

Equipment involved in enteral feeding may include the following:

- Enteral feeding tube
- Administration reservoir (Feed bag, bottle *etc.*)
- Giving set
- Syringes
- Enteral pump

Enteral Feeding Pump

Use of a pump for enteral feeding allows precise delivery of formula, at an even rate, which should maximize feeding tolerance and reduce the likelihood that the tube will block. High-fibre and energy-dense formulae may be easier to deliver with a pump, as the viscosity of these feeds can make gravity feeding problematic. Pumps can be used to deliver bolus and intermittent feeds, as well as for continuous feeding.

Pump feeding is the preferred method of feeding directly into the small bowel, as the small bowel may not tolerate large changes in volume infusion. Rapid administration of formula into the jejunum may result in dumping syndrome, or signs of feeding intolerance such as abdominal distension, cramps and diarrhoea.

Many different types of enteral pumps are available, including lightweight, compact pumps for ambulatory patients and pumps with the capacity for programming automated water flushes. A power source is required for regular use of enteral pumps. Useful features of the pump include battery back-up, portability, simplicity of use and ease of cleaning, alarm sound, quiet operation, excess pressure cut-out, a hold facility for administration of medications and a method of attaching the pump to the bed/IV pole/other location.

Gravity Drip Feeding

If an enteral pump is unavailable, gravity feeding can be used. By adjusting the roller clamp of the giving set to deliver a specified number of drops of formula per minute, a consistent feeding rate can be achieved. This method is not precise and the drip rate should be checked regularly to ensure the desired rate of formula is being given.

Table 17.17: Approximate Drip Counts to Achieve a Desired Feed Rate

mls/hr	*No. of Drops/Minute*
25	7
50	13
75	20
100	27
125	33
150	40
175	47
200	53

Tube Feeding Rate

The Table 17.18 provides a guide to common practice in starting and advancing feed rates.

Advantages of Enteral Nutrition

- Enteral feeds lead to decrease in infection rate in critically ill patients. It maintains integrity of gastrointestinal mucosal structure and function. In experimental animals, enteral nutrition is associated with higher secretion of secretary IgA and decreased secretion of catabolic hormones. In addition, enteral feeding also causes decrease in bacterial leak through the intestinal mucosa. All these factors result in decreased mortality rate after a septic or hypotensive insult.
- Experimental studies have also shown that enteral nutrition results in better wound healing after abdominal surgery and there is reduced hyper-metabolic as well as catabolic phase after burn injury in experimental animals.

Table 17.18: Guide to Common Practice in Starting and Advancing Feed Rates

Mode of Delivery	*Start Rate*	*Increase by*	*Comments*
Nasogastric	40-50ml/h	10-50ml/h every 4 hours	Stomach can usually tolerate high rates of feeding (as in bolus or intermittent feeding). This varies between individual patients and regimens. To estimate maximum tolerable feed rates, it is useful to relate it to the volume of nutrient-dense fluid that the individual patient could drink: maximum bolus might be the volume of a small glass of milk (150 ml) for a smaller/older person or bigger/younger person. Maximum continuous rate might be up to one cup of milk (250 ml) every hour.
Nasoduodenal or nasojejunal	20-30 ml/h	10-30 ml/h every 4-8 hours	Continuous feeding at a controlled rate is usually needed due to lack of reservoir capacity in small bowel. Feeds may be tolerated at rates as high as 100-120 ml/h. Some patients may tolerate higher rates; others may experience abdominal distension or discomfort. Sterile water should be used for tube.
Gastrostomy/PEG	As for nasogastric	As for nasogastric	After the insertion of a gastrostomy or jejunostomy feeding tube, a water trial is sometimes used to check for leaks before feed is given. Feeds are usually commenced 12-24 hours after insertion of the tube, but there is no evidence to support delaying feeds for more than 3-4 hours after PEG placement.
Jejunostomy	As for nasojejunal	As for nasojejunal	As for nasojejunal. Surgeon will usually advise when feeds can commence (usually within 24-48 hours of insertion); this depends on the surgical procedure.

Contraindications of Enteral Feeding

Some of the reported contraindications include diffused peritonitis, intestinal obstruction that prohibits use of bowel, intractable vomiting, paralytic ileus, severe diarrhea, severe pancreatitis, enterocutaneous fistulae, gastrointestinal ischemia *etc.*

Factors to be Considered when Choosing a Feeding Formula

Feeding formulas should be evaluated using following characteristics

- Osmolarity
- Viscosity
- Appropriate macronutrient ratio to meet assessed needs
- Suitability for contribution of fluid and electrolyte
- Cost effectiveness

Factors to be Considered for Choosing Feeding Regimen

The dietitian, in consultation with doctors and other health professionals, will decide on method of delivery or choosing the feeding regimen. The choice of enteral feeding regimen is based on assessment of the individual needs of the patient. The goal is to provide safe enteral nutrition and hydration appropriate to the clinical status of the patient, taking quality of life issues into consideration.

Factors to be considered are:

- Patient's current and past nutritional status
- Feeding environment
- Reason for feeding enterally proposed
- Any complications
- Present time period of enteral feeding
- Patient wishes

A written regimen specifying feed, rate of feeding, and additional water flushes is provided by the dietitian.

Methods of Feed Delivery

Feeds are either given by continuous drip, intermittent infusions, and bolus feed.

i) Continuous Drip

Continuous feeding at low volume is often used as the first step to commencing a patient on an enteral feeding regimen. It is the preferred method of use if the patient has rapid intestinal transit and may be more suitable for critically ill patients than other regimens. Tube feedings are administered at constant, steady rate usually for 24 hours continuously including a break of at least 4 hours in 24 hours to allow the stomach to re-acidify. 24 hour feeding is used in critical care or patients on sliding scale insulin. Feeding is given either by gravity drip or feeding pump. In most of the cases, enteral feeding is administered by gravity drip.

Advantages

- Allows the lowest possible hourly feed rate to meet nutrient requirements
- Better gastrointestinal tolerance due to the lower feed rate
- Better control of blood glucose levels due to continuous carbohydrate input

Disadvantages

- Physical attachment to the feeding apparatus (may affect quality of life)
- Expense of equipment (pump and giving sets)

ii) Cyclic/Intermittent Infusion

Tube feeds are administered at specific intervals throughout the day. The volume of desired feeding is divided into equal portion and been fed 4-6 hourly per day. The feedings are usually given by gravity drip over period of 30 min. to 1 hour.

Advantages

- Allows greater patient mobility (may improve quality of life).
- Allows breaks for physical activity, for administration of medications that are incompatible with feeds, and to encourage oral intake if applicable.
- Useful in the transition from continuous to bolus feeding, or from tube feeding to oral intake.
- Daytime feeds may reduce aspiration risk if it is difficult to maintain a 30° elevation overnight.
- Feeding during daytime only is more physiological and may therefore, have benefits such as helping to re-establish the diurnal cycle; promoting normal gastrointestinal motility and promoting re-acidification of the stomach (which protects against bacteria). There is no evidence for a benefit of gut rest, however.

Disadvantages

- Compared with continuous feeding, a higher infusion rate is required to provide the same volume of feed. This may be less well-tolerated, with a higher risk of problems such as reflux, aspiration, abdominal distension, diarrhoea and nausea.

iii) Bolus Feeding (without pump)

It is defined as rapid administration of a bolus feed/water by syringe (usually by gravity, without the plunger – speed of delivery is controlled by holding the syringe higher or lower). Most of the patients tolerate this method of feed. If there are any signs of intolerance then another feeding method should be sought.

Bolus feeding is usually into the stomach, which has the reservoir capacity to tolerate a large volume of feed. Bolus feeding involves the delivery of 100mls to 300mls of feed over a period of 10-30 minutes and can be given 4-6 times a day depending on patient's individual feeding regime.at regular intervals. The patient must have a competent oesophageal sphincter and be able to protect his/her airway adequately to minimize aspiration risk, as this risk is higher with larger feed volumes and administration rates.

Advantages

- Physiologically similar to a typical eating pattern
- Allows greater patient mobility
- Convenient for gastrostomy feeding
- Can be used to supplement oral intake
- Can be flexible to suit the patient's lifestyle and improve quality of life
- May facilitate transition to oral intake
- Avoids the use of expensive equipment

Disadvantages

- Large boluses may be poorly tolerated, especially in small bowel feeding
- Requires more nursing time compared with pump-controlled feeding
- Highest risk of aspiration, reflux, abdominal distension, diarrhoea and nausea

Complications of Enteral Nutrition

1. Gastro-intestinal Complications

- Diarrhoea
- Gastric bloating and abdominal distension
- Abdominal pain
- Nausea and vomiting
- Regurgitation-inhalation pneumonia

2. Mechanical Complications

- Misplacement of feeding tube
- Occlusion of feeding tube
- Inflammation, bleeding, perforation of nose, pharynx, oesophagus, stomach

3. Metabolic Complications

- Hyperglycemia
- Electrolyte alterations

4. Feeding Related Complications

- Tube blockage; seen with blenderized feeding in hospital feeding, because of high viscous and large sized food granules/particles
- Osmotic diarrhea: this is common due to high osmolar feeds especially between 650-1000 momls/litre and high calorie density feeds between 1.5 to 2 kcals/ml.
- Sudden increase in feed quantity/strength.

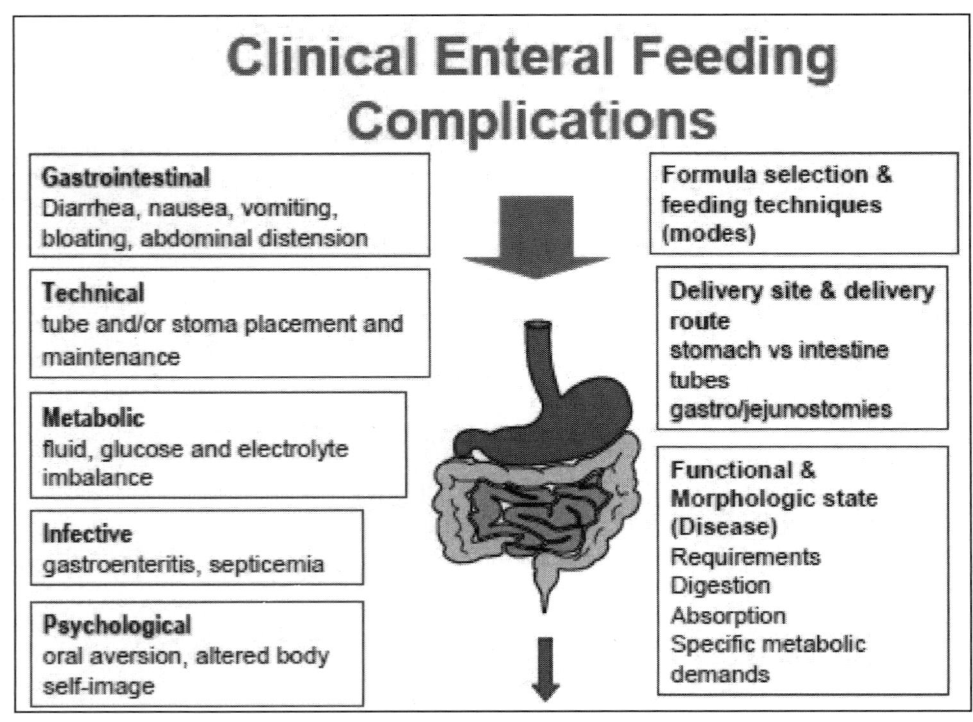

Risk of Infections, their Prevention and Control

There are potential hazards associated with enteral feeding which can make it a source for the growth of microorganisms. Liquid nutrients provide an ideal medium for bacteria and can cause cross contamination to the feeding system during setting up and handling the equipment.

It is important to always wash hands thoroughly before handling feeds and components of the feeding system. Preparing the equipment and opening of feed must be done in a clean environment and a no-touch technique should be adopted when preparing the feed during priming and connecting to the administration set/feeding tube.

Enteral feeding tubes should be flushed regularly with at least 30 ml tap water using a 50 ml syringe and flushing should be documented. For immuno-compromised patients or those fed directly into the jejunum, sterile bottled water should be used. Recommendations for flushes are as follow:

- Every 4 to 6 hours during the day
- Before and after feeding
- Before and after drug administration.

Feed, feed reservoirs and giving sets must not be reused and should be discarded after 24 hours. Single patient use syringes should be discarded after 24 hours.

Commercially produced, pre-filled ready to hang feeds must be used wherever possible as these are least likely to become contaminated during preparation and use. Feed that has been decanted or reconstituted from powder and hung must be discarded after 4 hours.

Monitoring

Monitoring of patients on enteral feeds should be done very carefully. The frequency of monitoring of patients on enteral feeds depends on nutritional and clinical condition. The suggested monitoring schedule is shown in Table 17.19.

Table 17.19: Suggested Monitoring Protocol for Enteral Nutrition

Tolerance	☆ Nausea ☆ Vomiting ☆ Diarrhea ☆ Constipation ☆ Abdominal distension
Nutrition and metabolic	☆ Bodyweight ☆ Serum Na+, K+, osmolality ☆ Acid base balance ☆ Blood glucose, urea, nitrogen ☆ Serum Mg++, Ca++, PO4++ ☆ Urine examination ☆ Liver function tests
Mechanical	☆ Confirm tube patency and location before each use ☆ Irrigate feeding tube for intermittent feeds, crushed medication administration.

18

Parenteral Feeding

Introduction

All people need food to live. Sometimes a person cannot eat any or enough food because of an illness. The stomach or bowel may not be working normally, or a person may have had surgery to remove part or all of these organs. When this occurs, and one is unable to eat, nutrition must be supplied in a different way to the patient. No doubt, enteral nutrition (EN) should always be the first option when feeding a patient who is not able to consume adequate food orally. Parenteral nutrition (PN) should be used when the gastrointestinal tract (GIT) is not available or not able to absorb nutrients supplied on a short or long term basis. PN (intravenous nutrition) can be used either alone or in combination with enteral or oral foods to meet patients' estimated nutrition requirements.

People of all ages can receive parenteral nutrition. It may be given to infants and children as well as to adults. People can live very well on parenteral nutrition for as long as it is needed. Many times, parenteral nutrition is used for a short time; then it is removed when the person can begin to eat normally again. Normal digestion occurs when food is broken down in the stomach and bowel, then absorbed in the bowel. These absorbed products are carried by the blood to all parts of the body.

Parenteral nutrition bypasses the normal digestion in the stomach and bowel. It is a special sterile chemical liquid food mixture given directly into the blood through an intravenous (IV) catheter (needle in the vein). The mixture contains proteins, carbohydrates (sugars), fats, vitamins and minerals (such as calcium). This special mixture may be called parenteral nutrition and was once called total parenteral nutrition (TPN), or hyperalimentation.

Some Terms Related to Parenteral Nutrition

Parenteral nutrition: Parenteral nutrition (PN) is intravenous administration of nutrition, which may include amino-acid (protein) solutions, hypertonic glucose, lipid emulsions, electrolytes, vitamins, minerals and trace elements into a central vein of patients who cannot eat or absorb enough food through tube feeding formula so as to maintain their good nutrition status. Achieving the right nutritional intake in a timely manner can help combat complications and be an important part of a patient's recovery. Parenteral nutrition may, however, under specified conditions, be administered peripherally.

Total parenteral nutrition (TPN): Parenteral nutrition is sometimes called Total Parenteral Nutrition (TPN) as it supplies all daily nutritional requirements. It serves as the sole source of nutrition for the individual. TPN can be used in the hospital or at home. Because TPN solutions are concentrated and can cause thrombosis of peripheral veins, a central venous catheter is usually required.

Partial parenteral nutrition: It supplies only part of daily nutritional requirements, supplementing oral intake. Many hospitalized patients are given dextrose or amino acid solutions by this method.

Supplemental parenteral nutrition: It is used as an adjunct therapy to enteral nutrition in patients.

Aims of Parneteral Nutrition

- Preservation of nutritional status
- Prevention of complications of protein malnutrition
- Decrease in postoperative complications of surgery

Indications and Contraindications of Total Parenteral Nutrition

A clear indication for the need for TPN, and where possible a small amount of enteral or oral feeding, should be considered in order to maintain the integrity of the gastrointestinal tract and reduce the risk of hepatobiliary complications by stimulating gall bladder contractility. TPN may be the only feasible option for patients who do not have a functioning GI tract or who have disorders requiring complete bowel rest. Patients may need TPN for any variety of diseases or conditions that impair food intake, nutrient digestion or absorption. Some diseases and conditions where PN is indicated include:

- GI fistulas
- Bowel obstruction
- Prolonged bowel rest
- Some stages of ulcerative colitis
- Certain pediatric GI disorders (e.g, congenital GI anomalies, prolonged diarrhea regardless of its cause)
- Inadequate absorption resulting from short bowel syndrome.
- Short bowel syndrome due to surgery
- Critically ill patients
- Severe acute pancreatitis
- Severe malnutrition, significant weight loss and/or hypoproteinaemia when enteral therapy is not possible.

Some patients may require this therapy for a short time and there are other patients who have received PN at home for a lifetime.

Indications and contraindications for PN are summarized in Table 18.1.

Table 18.1: Indications and Contraindications of TPN

Usual Indications	*Relative Indications*	*Contraindications*
☆ High dose chemotherapy ☆ Severe malnutrition with a non-functional GIT ☆ Severe catabolism in the presence of malnutrition and a non-functional GIT for 3-5 days ☆ Malnutrition (intensive surgery) ☆ GI route not available for 3-5 days ☆ Documented inability to absorb adequate nutrients from the GI tract: ❒ Massive small bowel resections/short bowel syndrome. (PN should be indicated in the initial period at least) ❒ Radiation enteritis ❒ Severe diarrhoea ❒ Untreatable steatorrhea/malabsorption – if pancreatic insufficiency, small bowel bacterial overgrowth and celiac disease have been ruled out ☆ Complete bowel obstruction ☆ Severe catabolism with or without pre-existing malnutrition where enteral feeding is not possible for 5 – 7 days ☆ Inability to obtain enteral access ☆ Inability to provide sufficient nutrients enterally ☆ Persistent GI haemorrhage ☆ Acute abdomen/prolonged ileus ☆ Work-up requiring nil per os (NPO) status for several days in a patient with pre-existing malnutrition ☆ High output enterocutaneous fistula (> 500ml per 24hrs) with inability to gain enteral access distal to fistula site or if enteral feeding worsens fistula output ☆ Trauma requiring repeat surgical procedures and lengthy periods of NPO status	☆ High dose chemotherapy ☆ Enterocutaneous fistula (<500 ml output per 24h) ☆ Hyperemesis gravidarium when nausea and vomiting persists for 5 – 7 days and enteral feeding is not possible ☆ Partial small bowel obstruction ☆ Severe mucositis ☆ Intractable vomiting when small bowel feeding is not possible. ☆ Chylous ascitis or chylothorax when low fat or fat free enteral nutrition does not reduce output adequately. ☆ Abdominal compartment syndrome ☆ Moderate to severe acute pancreatitis or pancreatitis accompanied by abdominal pain with jejuna feeding	☆ Functional GIT ☆ Need for emergency surgery ☆ Patient refusal ☆ Prognosis that does not warrant aggressive nutrition support intervention ☆ PN risk exceeds potential benefits ☆ Inability to obtain venous access

Beginning TPN Administration

Artificial nutrition is regarded as medical nutrition therapy and thus, subject to all the ethical dilemmas and debates regarding treatment. Patient consent needs to be obtained before commencement of nutrition therapy, in this case TPN. Patients with uncontrolled cardiac failure and electrolyte imbalances should be stabilized prior to the commencement of TPN.

Nutrition Assessment of Patient Requiring TPN

It is must that prior to start TPN, knowledge about patient's nutritional assessment; information regarding his baseline weight data, his venous access evaluation and baseline lab investigations should be there.

i) Anthropometric Data

- ✰ Current weight
- ✰ Current height or substitute measure *e.g.* arm span
- ✰ Recent changes in weight
- ✰ Level of oedema

Nutritional status in patients requiring TPN will vary and it is therefore important to assess their metabolic and nutrition profile with care. Information about appetite and weight change is important and should be elicited from the patient where possible.

ii) Biochemical Data

- ✰ Full blood count
- ✰ Na, K, Cl, urea, creatinine
- ✰ Calcium, magnesium, phosphate
- ✰ Serum triglycerides
- ✰ Serum glucose
- ✰ Liver function tests, including international normalized ratio (INR)
- ✰ C-reactive protein
- ✰ Albumin –Should not be used as an independent criterion

Baseline laboratory data is vital and should be reviewed prior to the commencement of TPN. In the severely malnourished or stressed patient, particular importance should be placed on the need for information regarding potassium, magnesium, calcium, phosphate, haemoglobin and coagulation. In those patients requiring long term PN, information about other trace elements will be required, as well as liver function tests.

iii) Clinical

A thorough history of the medical condition and anticipated treatment plan should be reviewed before the TPN prescription can be compiled. Special attention should be given to the following:

- ✰ Diagnosis
- ✰ Anatomy – resections, ostomies or previous surgery
- ✰ Pre-existing conditions – *e.g.* diabetes mellitus, renal failure, liver disease
- ✰ Organ function
- ✰ Treatment plan

iv) Dietary History

v) Prior Medication History

An initial assessment of all patients should be done by an appropriate qualified nutrition expert *e.g.* dietitian before commencement of any TPN in order to determine goals of nutrition therapy. Within this assessment a number of key questions should be asked *e.g.*

- ✰ Is the nutrition therapy intended to maintain or replete body composition?
- ✰ Is PN required for short (less than two weeks) or long term?
- ✰ Are there any pre-existing medical conditions which may impact on substrate utilization?

Nutritional status in patients requiring PN will vary and it is therefore important to assess their metabolic and nutrition profile with care. Information about appetite and weight change is important and should be elicited from the patient where possible. Baseline laboratory data is vital and should be reviewed prior to the commencement of PN. In the severely malnourished or stressed patient, particular importance should be placed on the need for information regarding potassium, magnesium, calcium, phosphate, haemoglobin and coagulation. In those patients requiring long term PN, information about other trace elements will be required, as well as liver function tests. A thorough history of the medical condition and anticipated treatment plan should be reviewed before the TPN prescription can be compiled.

Dietary and medicine intake history regarding the following should be recorded:

- ✰ Food/drug allergies or intolerances
- ✰ Dietary intake prior to hospital admission
- ✰ Period nil per mouth within hospital/when was last meal taken
- ✰ Special dietary needs interventions
- ✰ Drug-nutrient interactions
- ✰ Herbal/supplement use
- ✰ Home and current medication
- ✰ Assess for any drug interactions or drug induced deficiencies

Nutritional Recommendations

For making nutritional recommendations, on the basis of nutritional assessment, dietitian and other members of the Nutrition Support Service Team will have to identify the patients who are at risk of developing refeeding syndrome. The following guidelines have been recommended by National Institute for Health and Care Excellence (NICE) for identifying patients at high risk of developing refeeding syndrome. Guidelines are as under:

The dietitian is responsible for the calculation of the individualized nutritional requirements of all patients. Disease-specific guidelines are used and daily adaptations are made to the prescription (Table 18.3).

Table 18.2: Guidelines for Identifying Patients at High Risk for Refeeding Syndrome

Patient presents one or more of the following:
✰ Body mass index <16
✰ Unintentional weight loss>15 per cent in the past 3-6 months
✰ Little or no nutritional intake for >10 days
✰ Low levels of potassium, phosphate or magnesium before initiating feeds
Or, the patient presents with TWO or more of the following:
✰ Body mass index 18.5
✰ Unintentional weight loss>10 per cent in the past 3-6 months
✰ Little or no nutritional intake for >5 days
✰ History of alcohol misuse or drugs, including insulin, chemotherapy, antacids or diuretics

A. Macronutrients

See the Table 18.3.

B. Micronutrients

Pre-existing micronutrient (vitamins and trace elements) deficiencies are often present in hospitalized patients. Deficiencies occur due to inadequate or inappropriate administration, increased or altered requirements, and increased losses, affecting various biochemical processes and resulting in organ dysfunction, poor wound healing, and altered immune status with deleterious sequelae.

There are commercially available preparations, which provide daily requirements for fat and water-soluble vitamins in addition to trace elements. Micronutrient supplementation should begin as soon as parenteral nutrition is started and continued daily as its role is crucial. All parenteral nutrition prescriptions should include a daily dose of multivitamins and trace elements to prevent and/or correct nutrient deficiencies. Any additional supplementation should be disease-specific.

Suggested composition of parenteral multivitamin and trace trace-elements are indicated in Table 18.4.

Nutrition requirements with respect to the patients' anthropometric measurements and clinical condition especially with regards to sepsis, severe malnutrition and or other factors should be determined. If there is an evidence of muscle wasting, depletion of subcutaneous fat, presence of peripheral oedema, ascites, features of vitamin and mineral deficiencies, echymosis and easy bruising, BMI is <19 or >10 per cent decrease, albumin is <30 mg/dl, pre-albumin <12 mg/dl, transferring <150mmol/l, total lymphocyte count is <1800/mm^3, clotting time is less, trace minerals are low in the serum *etc.* all indicate the start of nutrition support to the patient.

Each patient should have a nutrition prescription calculated by an appropriately qualified nutrition expert *e.g.* each nutrition prescription should include at least the following information:

- ✰ Patient name
- ✰ Hospital number
- ✰ Ward
- ✰ Date
- ✰ PN code prescribed
- ✰ Total volume/administration rate
- ✰ Route of administration (centrally or peripherally)
- ✰ There are two main forms of parenteral nutrition:
- ✰ Peripheral Parenteral Nutrition
- ✰ Central (Total Parenteral Nutrition)
- ✰ Both differ in composition of feed, primary calorie source, potential complications, method of administration *etc.*

Key Features of Peripheral Parenteral Nutrition

These are:

- ✰ Given through peripheral vein
- ✰ Short term use
- ✰ Given to mildly stressed patients
- ✰ Patients having low calorie requirement
- ✰ Patients who require large amounts of fluid
- ✰ Patients having contraindications to central TPN

Requirements for Central Venous Access for TPN

Patient needs venous access to a 'large' central line with fast flow to avoid thrombophlebitis. A special intravenous (IV) catheter is placed in a large vein in the chest or arm. It can stay in place for as long as needed. The nurse cares for the catheter. Proper care is required to avoid infection and clotting. Different kinds of catheters may be used. Common names of these catheters are **Hickman, Broviac, PICC, triple lumen, double lumen or single lumen catheters, and ports**. Nutrition is given through this large vein.

Table 18.3: General and Disease Specific Requirements for Macronutrients and Micronutrients

I. General Information and Intensive Care

	Energy	Protein	Carbohydrate	Fat	Fluid	Other
General information						
	☆ Energy requirements are indicated as non-protein energy (NPE) or total energy (TE) where stated in the literature ☆ Consider energy contribution from dextrose containing IV fluids and lipid based sedatives.	$1gN_2$ = 6,25g protein	☆ Basal carbohydrate requirements = 2g/kg ☆ Maximum glucose oxidation rate = 4 – 7 mg/kg/min (aim not to exceed 5mg/kg/min) ☆ Hyper-glycaemia should be avoided and carbohydrate provision adjusted if hyperglycaemia occurs	☆ Essential fatty acids include linoleic acid (omega 6 FA) ☆ Daily lipid requirements: 0.7-1.5 g/kg/day ☆ Essential fatty acids should be provided at 7-10g/day, equating to 14-20 g LCT from soy oil or 30-40g LCT from olive/soy oil mix. ☆ Fish oil administration should be 0.1-0.2g/kg/day ☆ In case of hyper-triglyceridemia or poor fat clearance, the amount of lipid infused should be reduced and/or the type of fat should be changed. ☆ Fish oil containing lipid emulsions have been shown to be anti-inflammatory and contain less hepatotoxic phytosterols.	☆ Maintenance at 20-40ml/kg/day ☆ Assess for and replace ongoing losses on a daily bases *e.g.* diarrhoes, losses due to fever, wound and fistula losses	☆ Parenteral nutrition prescriptions should include a daily dose of multivitamin and trace elements

	Energy	Protein	Carbohydrate	Fat	Fluid	Other
			Intensive care			
	20-25kCal/kg/d TE (acute phase)) 25-30 Kcal/Kg/d TE (recovery phase) **Obese ICU patient:** **(BMI>30)** **Hypo-calorie feeding** 11-14Kcal/kg Actual Body Wt. **Eucaloric feeding** 21Kcal/kg Actual Body Wt.	1.3-1.5g/kg/d **Obese ICU patient:** (BMI>30) Hypo-calorie feeding BMI 30-39.9≥ 2g/kg/IBW BMI 40≥ 2.5g/kg/IBW **Eucaloric feeding** Following disease specific guidelines	Minimum 2 g/kg Maximum glucose oxidation rate in the stressed patient is 4-7 mg/kg/min. Aim not to exceed 5 mg/kg/min Hyperglycemia (>10 mmol/l) should be avoided.	0.7-1.5g/kg/day Fish oil containing lipid emulsions probably decrease length of stay.		All parenteral nutrition prescriptions should include a daily dose of multivitamin and trace elements. Electrolyte requirements are highly variable and should be determined by electrolyte monitoring

II. Liver Diseases

	Energy	Protein	Carbohydrate	Fat	Fluid	Other
Alcoholic steato-hepatitis **Weight to be used:** ☆ Without ascitis: Actual body weight ☆ With ascitis: Ideal body weight	1.3x BMR	Well nourished/moderately malnourished: 1.2 g/kg Severely malnourished: 1.5g/kg	50-60 per cent NPE	40-50 per cent NPE Use a lipid emulsion with lower omega 6 content than traditional pure soybean lipid emulsion		**Water soluble vitamins:** ☆ All water soluble vitamins according to RDA ☆ Thiamine prior to commencement due to high risk for Wernicke encephalopathy ☆ Prophylaxis 250 mg IM daily for 3-4 days ☆ Treatment 500 mg IVper day for 2-3 days **Fat soluble vitamins** ☆ All fat soluble vitamins according to RDA ☆ IV vitamin K if jaundice with fat malabsorption **Minerals and trace elements** ☆ Daily ☆ Routine administration of twice the RDA

	Energy	Protein	Carbohydrate	Fat	Fluid	Other
Liver cirrhosis **Weight to be used:** ☆ Without ascitis: Actual body weight ☆ With ascitis: Ideal body weight	1.3xresting energy expenditure (REE) 30-35 Kcal/kg dry body weight	1.2-1.5 g/kg in encephalopathy grade III or IV consider use of solutions high in branched chain amino acids (AAA), methionine and tryptophan	50-60 per cent NPE given as glucose in case of hyperglycemia: 2-3 g/kg/d + IV insulin infusion	40-50 per cent NPE Lower in omega 6		Water, electrolytes, water- and fat soluble vitamins, trace elements according to RDA If ascites: Na<2000mg Administer thiamine prior to starting glucose infusion in patients with alcoholic disease to reduce the risk of Wernicke's encephanlopathy. Prophylaxis: 250 mg IM daily for 3-5 days Treatment: 500 mg IV t.i.d. for 2-3 days
Liver transplant and surgery	Pre-operative follow guidelines for liver cirrhosis 1.3xREE (NPE) (until 6-12 months post-operative)	PN: 1-1.5g/kg/d Consider BCAA containing solutions in encephalopathic patients	Patients might be more prone to hyperglycemica due to immunosuppressant therapy	Lipid emulsion with lower omega 6 content like MCT/Olive oil or MCT/fish oil containing emulsions were less immunosuppressive and pro-inflammatory		Water, electrolytes: water- and fat soluble vitamins, trace elements according to RDA Close monitoring of Mg levels in post-transplanted patients to detect cyclosporine or Tacrolimus induced hypomagnesaemia
Acute liver failure	1.2-1.3xREE	0.8-1.2g/kg/d	2-3g/kg/d to avoid hypoglycemia	0.8-1.2g/kg/d		Strict control of glucose, lactate and ammonia levels

III. Renal Diseases and Burns

	Energy	Protein	Carbohydrate	Fat	Fluid	Other
	Renal					
Acute kidney injury (AKI) **Weight to be used:** ☆ **Normal BMI:** Actual body weight ☆ **Over weight and critically ill:** Ideal body weight	20-30 Kcal/kg/day increase to 35 Kcal/kg/day if undergoing continuous renal replacement therapy (CRRT) 1.1-1.2xREE	**Conservative therapy, minimal metabolism:** 0.8-1.0g/kg/d **Dialysis:** **CRRT:** 1.8-2.5g/kg/d **Intermittent hemodialysis (IHD):** 1.5-2.0 g/kg/d	5-7 g/kg/day	1.2-1.5g/kg/day		In patients managed conservatively minimize exposure to renally regulated electrolytes and minerals *e.g.* potassium, calcium, phosphorus, magnesium, vitamin A, B, C as well as zinc, selenium, copper and aluminum AKI patients on RRT might have increase needs for water soluble vitamins. Inappropriate vitamin C supplementation might result in secondary oxalosis but patients might need more than the normal 50mg/day. Losses can be as high as 100mg/day of vitamin C and 600nmol/day of thiamine in the ultra-filtrate. Restrict vitamin C to <100 mg/day. Increase to maximum of 200 mg in patients on CRRT. In patients receiving CRRT provide: ☆ Folate 1 mg/daily ☆ Pyridoxine **10** mg/day ☆ Thiamine 25-100 mg/day Increased losses of zn, Se, Cu and Al is uncommon and can be replaced with a standard trace element preparation.
Chronic Kidney Disease	**Non-dialysed:** ≥ 30-35 Kcal/kg/day TE **Dialysed:** <60 years: 35Kcal/kg/day TE >60Years: 30-35 Kcal/kg/day TE	**Non-dialysed:** **Glomerular filtration rate (GFR)=25-70ml/min** **0.55-0.6g/kg/d** **GFR<25 ml/min: 0.6-0.75g/kg/d** **Dialysed:** Haemodialysis (HD): 1.2-1.4 g/kg Continuous ambulatory peritoneal dialysis (CAPD): 1.2-1.5 g/kg	50-60 per cent of TE	20-25 per cent of TE	**Dialysed:** 1000 ml+urine volume	**Non-dialysed:** Phosphate: 600-1000 mg/d Potassium: 1500-2000 mg/d Sodium: 1.8-2.5 g/d **Dialysed:** Phosphate: 800-1000 mg/d Potassium: 2000-2500 mg/d Sodium: 1.8-2.5 g/d
	Burns					
Burns	**Toronto Formula**	Adults 1.5-2.0g/kg/d	55-60 per cent of TEMax 5 mg/kg/min infusion rate corresponding to 7g/kg/day in an adult patient. Keep blood glucose levels between 4.5-8.0 mmol/l using insulin therapy	<20 per cent TE		Zn, Cu, Se B vitamins, Vit. C, folic acid, Vit. D, Vit. E, Vit. A Duration of supplementation is related to total body surface area burned. ☆ 2 weeks for burns 40-60 per cent ☆ 30 days for burns>60 per cent

IV. Acute Pancreatitis and Gastroenterology

	Energy	Protein	Carbohydrate	Fat	Fluid	Other
Acute Pancreatitis						
	25 Kcal/kg NPE increasing to maximum 30 Kcal/kg NPE Restrict to 15-20 Kcal/kg NPE in case of systemic inflammatory response syndrome (SIRS) or multiple organ dysfunction syndrome (MODS) or when risk for refeeding syndrome	1.2-1.5 g protein/kg/d	50-70 per cent TE 4-7 mg/kg/min	Temporarily discontinue lipid infusion if hypertriglyceridemia persist for more than 72 hours (>12 mmol/l)		RDA for vitamins and trace elements. Insufficient data to recommend supranormal doses
Gastroenterology						
	Crohn's Disease: 25-30 Kcal/kg/d TE **Short bowel syndrome** 32 Kcal/kg/TE **Open abdomen:** 25-35 Kcal/kg/NPE **Fistula:** **Low output:** 25Kcal/kg TE **High output:** At least 30 kcal/kg TE	**Crohn's Disease:** Depending on disease complication **Short bowel syndrome** 1-1.5 g/kg/day Should take increased digestive losses into account 2g N_2 per liter of gastric losses **Open abdomen:** 1.5-2.5 g/kg/PLUS 29g/l of abdominal losses **Fistula:** **Low output** 1-1.5g/kg **High output** 1.5-2g/kg PLUS 2g/l effluent	**Short bowel syndrome** 60 per cent NPE	**Short bowel syndrome** 40 per cent of NPE Restrict to <1g/kg lipid		**Crohn's Disease:** Yearly monitoring of serum Vit. B_{12} and folate **Short bowel syndrome** Total micronutrient requirement should be given IV. Some should be given in increased amounts due to increased losses *e.g.* zinc and Mg. Special consideration should be given to electrolyte monitoring especially sodium and magnesium **High output fistulae:** 2x dietary reference intakes (DRIs) for vitamins and trace elements. Up to 5x DRIs for vitamin C and zinc. At high risk of vitamin B_{12}, Zn, Mg and Se deficiency

V. Non-surgical Ooncology

	Energy	Protein	Carbohydrate	Fat	Fluid	Other
			Non Surgical Oncology			
	Ambulate patients: 30-35Kcal/kg/day TE **Bedridden patients:** 20-25 Kcal/kg/day TE **Weight gain:** 30-40 Kcal/Kg/day TE **Hypermetabolic/stressed:** 35 Kcal/kg/day of TE **Haemopoietic cell transplant:** 30-35 Kcal/kg/day TE	**Non stressed:** 1-1.2 g/kg/day **Severe stressed:** 1.5-2.5 g/kg/day **Hypercatabolic:** 1.2-1.6 g/kg/day **Haemopoietic cell transplant:** 1.5-2g/kg/day	50 per cent NPE	50 per cent NPE		Micronutrients-100 per cent DRI
			Surgery			
	25 Kcal/kg ideal body weight (IBW) TE Up to 30 Kcal/kg TE in severe stress	1.5 g/kg or 20 per cent of TE	30 per cent of TE 50-70 per cent of NPE	30-50 per cent of TE Lipid emulsion should probably include omega 3 fatty acids		100 per cent of RDA for micronutrients

VI. Home Parenteral Nutrition

	Energy	Protein	Carbohydrate	Fat	Fluid	Other
	20-35Kcal/kg TE MAX 40 Kcal/kg	Unstressed patient: 0.8-1 g/kg NPE: N_2= 100-150: 1	60 per cent of NPE Administration of > 7 mg/kg/min is associated with parenteral nutrition associated liver disease	Long term patients: <1g/kg/d Fish oil containing lipid emulsions		☆ Restrict the amount of sodium to losses, to avoid sodium induced hypercalciuria ☆ The calcium, magnesium and phosphate content of the home parenteral nutrition (HPN) should maintain normal serum concentrations and 24-h urinary excretion ☆ The recommended ratio is 1mmol of calcium to 1mmol of phosphate ☆ The amount of amino acids prescribed should not be greater than losses, in order to limit hypercalciuria ☆ The recommended intravenous dose of vitamin D is 200 IU/day. Consider vitamin D withdrawal in patients with low bone mineral density (BMD), low serum parathyroid hormone, and 1, 25-dihydroxyvitamin D concentrations associated with normal 25-hydroxivitamin D. Reducing infusion rates may decrease hypercalciuria • bisphosphonates (such as clodronate 1500 mg IV or pamidronate 20 mg iv every 3 months), may maintain BMD in patients with osteopenia ☆ Biochemistry and anthropometry at every visit ☆ Monitoring of trace elements and vitamins are recommended at intervals of 6 months ☆ Bone mineral density assessment by dual energy X-ray absorptiometry(DEXA) scanning is recommended at yearly intervals

Source: National Total Parenteral Nutrition Practice: Guidelines for Adults. 2016. Deptt. of Health, Republic of South Africa.

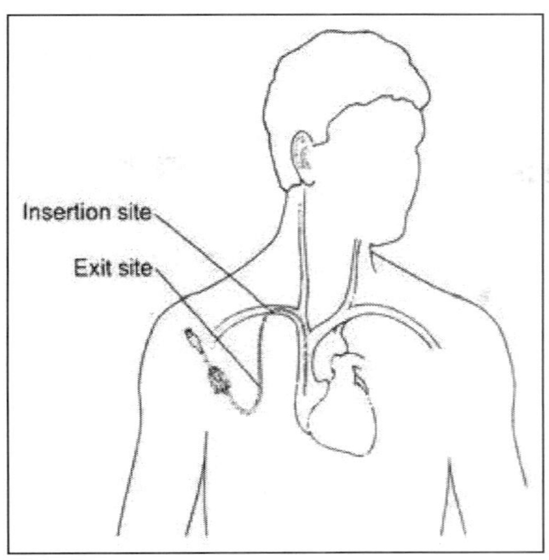

Table 18.4: Suggested Composition of Parenteral Multivitamin and Trace Elements

Nutrient	*Suggested Composition*
Fat soluble vitamins	
Vit A	1mg
Vit D	5 μg (200 IU)
Vit E	10 mg
Vit K	150 μg
Water soluble vitamins	
Vit C	200 mg
Folic Acid	600 μg
Niacin	40 mg
Riboflavin	3.6 mg
Thiamine	6.0 mg
Pyridoxine	6.0 mg
B_{12}	5 μg
Pantothenic acid	15 mg
Biotin	60 μg
Trace elements	
Zinc	2.5–5.0 mg
Selenium	20–60 μg
Copper	0.3–0.5 mg
Chromium	10–15 μg
Manganese	60 – 100 μg
Iron	-
Fluoride	-
Iodine	-
Zinc	2.5 – 5.0 mg
Selenium	20 – 60 μg
Copper	0.3–0.5 mg
Chromium	10 – 15 μg
Manganese	60 – 100 μg
Iron	-
Fluoride	-
Iodine	-

Source: National Total Parenteral Nutrition Practice: Guidelines for Adults. 2016. Deptt. of Health, Republic of South Africa.

Because the central venous catheter needs to remain in place for a long time, strict sterile technique must be used during insertion and maintenance. The TPN line should not be used for any other purpose. External tubing should be changed every 24 h with the first bag of the day. In-line filters have not been shown to decrease complications. Dressings should be kept sterile and are usually changed every 48 h using strict sterile techniques.

If TPN is given outside the hospital, patients must be taught to recognize symptoms of infection, and qualified home nursing must be arranged.

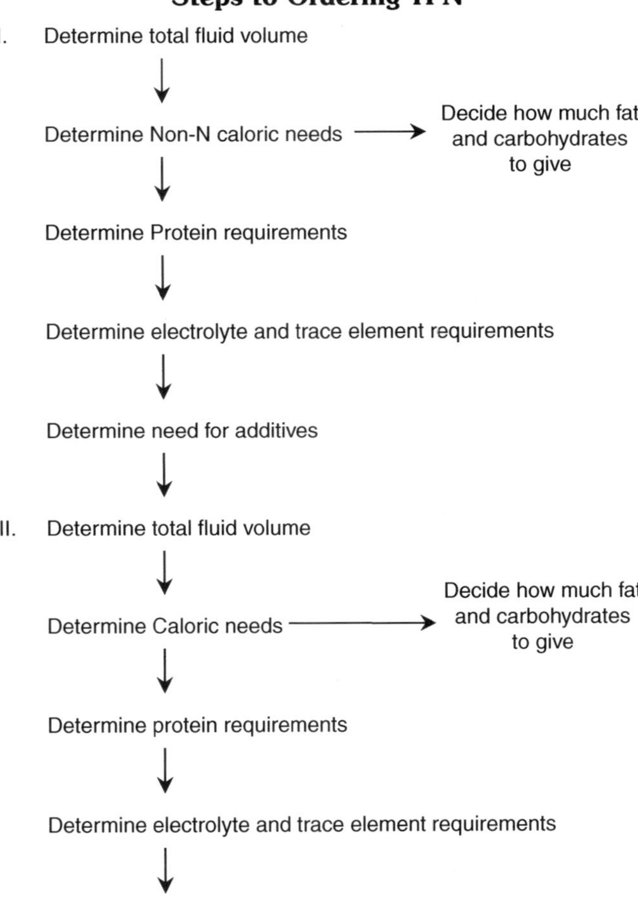

How much volume is to be given depends and must be able to cater for maintenance and on- going losses, insensible fluid losses also *e.g.* add 10 per cent for every °C rise in temperature, add on- going losses based on I/O chart; sufficient for normal maintenance requirements which is calculated on the basis of body weight or alternatively, 30-50 ml/kg/day.

Fulfillment of Nutritional Requirements through Constituents of Commercial TPN Formulas

Caloric Requirements

They are calculated either as per recommendations

given in Table 18.3 for a patient suffering from particular disease or it can also be calculated otherwise as per the formula mentioned below. Calorie requirement shall be based on total energy expenditure which can be estimated using predictive equations.

TEE=REE + Stress Factor + Activity Factor

It can also be measured using metabolic chart.

Values for stress factors have been given in Table 18.5a.

Table 18.5a: Stress Factors for different Injuries and Infections for Calculating Caloric Expenditure

Stress Factor	
Malnutrition	+30 per cent
Peritonitis	+15 per cent
Soft tissue trauma	+15 per cent
Fracture	+20 per cent
Fever (per °C rise)	+13 per cent
Moderate infection	+20 per cent
Severe infection	+40 per cent
<20 per cent BSA burns	+50 per cent
20-40 per cent BSA burns	+80 per cent

Similarly, values for activity factor for calculating the energy expenditure are given in Table 18.5b.

Table 18.5: Activity Factor for Calculating Energy Expenditure

Activity Factor	
Bed bound	+20 per cent
Ambulant	+30 per cent
Active	+50 per cent

Carbohydrates

Carbohydrate in the form of dextrose (glucose) is given as it has nitrogen sparing effect. Dextrose is least expensive and most commonly used intravenous energy source. Commercially made formulas are available in concentrations ranging from 5 per cent to 70 per cent. Central Parenteral Nutrition (CPN) formulations use 50 per cent to 70 per cent which is diluted to a final concentration of between 15 per cent and 30 per cent. The dextrose in intravenous solution is hydrated, so each gram of dextrose monohydrate provides 3.4 Kcal. Dextrose should not be given more than 4 mg/kg/min *i.e.* less than 6g per kg per day as excess of everything is bad. Carbohydrates usually form 70-75 per cent of calories.

Fats

Commercially available lipid emulsions are often added to supply essential fatty acids and triglycerides; 20 to 30 per cent of total calories are usually supplied as lipids. However, withholding lipids and their calories may help obese patients mobilize endogenous fat stores, increasing insulin sensitivity.

Lipid emulsions provide an intravenous source of fat calories including the essential fatty acids *i.e.* linoleic and linolenic acids. These emulsions are available as a 10 per cent (1.1 Kcal/ml) or a 20 per cent (2 Kcal/ml) solution and are approximately 6 times more expensive than the equivalent glucose calories. Lipid emulsions contain soybean oil or a combination of soybean and safflower oil triglycerides, egg yolk phospholipids as emulsifying agent and glycerin to achieve isotonicity with plasma. The osmolarity of current formulas range from 270 to 340 M OsM per liter. The emulsified particles are approximately the same size and structure as chylomicrons. Commercially made lipid emulsions are distinct from chylomicrons in that the emulsions contain mostly essential fatty acids and are devoid of cholesterol or protein. Currently available emulsions contain approximately half to two thirds of their fatty acids as linoleic acid and approximately 5 per cent to 10 per cent as linolenic acid.

Fats usually form 25-30 per cent of calories. In any case, fat should not provide more than 40-50 per cent calories. In severe stress, fat requirement is increased. Fat intake should maintain the serum TG levels <350 mg/dl or 3.95 mmol/l. According to Moore and Cerra (1991), lipid intake should not be more than 0.7mg/kg/min *i.e.* less than 1g/kg/day.

Protein

Protein to be given should be based on calorie: nitrogen ratio, degree of stress and body weight and nitrogen balance. Normal calorie: nitrogen ratio is 150 Kcal: 1g nitrogen. In critically ill patients, it should be 85 to 100Kcal: 1g nitrogen. In non-stress patients, protein intake should be 0.8g/kg/day while with increase in the stress level, protein requirement is also increased (Table 18.6).

Depending on the nitrogen balance, aim is to maintain positive balance and for it, protein requirement would be 1.5 to 2 g/kg/day in most of the patients.

Table 18.6: Protein Requirement during Stress Time

Degree of Stress and Protein Requirement	
Non stress patients	0.8 g/kg/day
Mild stress	1.0-1.2 g/kg/day
Moderate stress	1.3-1.75 g/kg/day
Severe stress	2.0-2.5 g/kg/day

Standard commercially made solutions are composed of crystalline amino acids in concentrations of between 2.75 per cent and 15 per cent. These solutions usually contain 40 per cent to 50 per cent essential and 50 per cent to 60 per cent non-essential amino acids. The proportion of nitrogen consisting of essential amino acids is more than twice that required by normal person and increases amino acid profile and is its electrolyte and trace mineral, but it is not known whether these variations are clinically significant. Most formulas completely lack, or have very little of, the non-essential amino acids glutamine, glutamate, aspartate, asparagines, tyrosine and cyseine.

Electrolyte Requirements

Electrolyte requirement is calculated on the basis of maintenance and replacement needs which has been given as under:

Table 18.7: Electrolyte Requirement

Trace Element	*Requirement*
Na^+	1 to 2 mmol/kg/d (or 60-120 meq/d)
K^+	0.5 to 1 mmol/kg/d (or 30-60 meq/d)
Mg^+	0.35 to 0.45 mmol/kg/d (or 10-20 meq/d)
Ca^{++}	0.2 to 0.3 mmol/kg/d (or 10-15 meq/d)
PO_4^{2-}	20 to 30 mmol/kg/d (or 60-120 meq/d)

Total requirements are not well established. Commercial preparations exist to provide RDA *e.g.*

- Zinc 2-4 mg/day
- Cr 10-15µg/day
- Cu 0.3-0.5 mg/day
- Mn 0.4-0.8 mg/day

Other Additives

Give one ampoule multivitamin per bag of TPN. Give 2-3x of vitamin that recommended for oral intake. Multivitamin does not include vitamin K which can be given 1 mg/day or 5-10 mg per week.

Medications are other additives. In case of insulin, one give initial SI based on sliding scale according to glucose q6h (keep <11 mmol/l). Once stable, give 2/3 total requirements in TPN and review daily. There are other alternate regimes which should be followed on the advice of the physician.

Feed Preparation and Administration

Generally, TPN requires water (30 to 40 mL/kg/day), energy (30 to 45 kcal/kg/day, depending on energy expenditure), amino acids (1.0 to 2.0 g/kg/day, depending on the degree of catabolism), essential fatty acids, vitamins, and minerals. Children who need TPN may have different fluid requirements and need more energy (up to 120 kcal/kg/day) and amino acids (up to 2.5 or 3.5 g/kg/day).

Basic TPN solutions are prepared using sterile techniques, usually in liter batches according to standard formulas. Normally, 2 L/day of the standard solution is needed. Solutions may be modified based on laboratory results, underlying disorders, hypermetabolism, or other factors. Solutions may vary also depending on other disorders present and patient age, as for the following:

- **For renal insufficiency not being treated with dialysis or for liver failure**: Reduced protein content and a high percentage of essential amino acids
- **For heart or kidney failure:** Limited volume (liquid) intake
- **For respiratory failure:** A lipid emulsion that provides most of non-protein calories to minimize carbon dioxide production by carbohydrate metabolism
- **For neonates:** Lower dextrose concentrations (17 to 18 per cent)

The solution is started slowly at 50 per cent of the calculated requirements, using 5 per cent dextrose to make up the balance of fluid requirements. Energy and nitrogen should be given simultaneously. The amount of regular insulin given (added directly to the TPN solution) depends on the plasma glucose level; if the level is normal and the final solution contains 25 per cent dextrose, the usual starting dose is 5 to 10 units of regular insulin/L of TPN fluid. Vitamins like folic acid are infused with the solution, but vitamin B_{12} must be prescribed separately.

Other important guidelines for TPN are:

- PN should be administered at the rate prescribed on the patient fluid balance chart.
- Standard PN solutions should be used as far as possible to prevent errors in prescription and compounding and to achieve better nutrient delivery.
- Individualized bags should only be used for specific patients and checks should be in place to ensure that prescriptions are correct.
- Do not exceed 24 hours of hang time for all PN bags. After 24 hours the infusion of the remaining PN in the bag should be discontinued and discarded and a new bag should be started without delay to prevent hypoglycemia.
- Medication should be added to the PN at ward level. If co-infusion (same line, but different port) is the only option (due to limited line access), medication administration has to be discussed with the pharmacist/physician in charge.

Home Parenteral Therapy

Home parenteral nutrition support should be used in patients who cannot meet their nutritional requirement by enteral intake, and who are able to receive therapy outside an acute care setting. Home parenteral nutrition (HPN) was introduced as a treatment modality in the early 1970s primarily for the treatment of chronic intestinal failure in patients with benign disease. The relatively low morbidity and mortality associated with HPN has encouraged its widespread use in western countries. Demand for home parenteral nutrition (HPN)-to facilitate hospital discharge- is rising but access to local services may be limited. The aims of HPN include preventing or treating undernutrition and/or dehydration, helping a patient become fit for surgery, reducing diarrhoea/vomiting, reducing abdominal pain caused by oral/enteral feeding and improving the quality of life.

Patient must receive training and information on HPN prior to discharge. There should be a formal teaching program for the patient and/or carer. The teaching program should include catheter care, pump use, and preventing, recognizing and managing complications. Experienced nurses are usually best placed to take responsibility for the teaching program. The patient must be supported by a skilled team, which includes specialist nutritionist nurses, dieticians and district nurses. Patient must be competent in the management of feeding systems and aware of common complications.

To make the parenteral nutrition therapy at home as easy and safe as possible, one can use so-called 'all-in-one' systems that get mixed up just before use, that are contained in one bag, and that contain all nutritional elements (amino acids, carbohydrates, lipids, electrolytes, trace elements and vitamins) as per prescription of the physician and dietitian.

Using only one infusion line, one can receive everything that the body needs on a daily basis in terms of nutrition. This system, with its multi-chambers (2 or 3-chamber bags) and an individual mixing bag component (compounding), offers a wide spectrum of all-in-one systems that is ideal for parenteral nutrition therapy at home.

Three-chamber bags contain amino acids with/without electrolytes, glucose and fat all in individual chambers. The individual chambers are mixed by applying gentle pressure. An additional lipid emulsion gets added via the use of a special transfer set in the case of two-chamber bags. Individually produced mixing bag systems contain all the necessary macro- nutrients in exactly defined gram dosages and come in a mixed state.

Trace elements and vitamins can be injected prior to the application of the nutritional solution. Preparation of a nutritional regime can be carried out in a pharmacy or directly at home provided that in the latter case, hygienic measures are adhered to.

The multi-chamber bags can be stored in a cardboard box at room temperature in an unmixed state for up to two years. Individually produced mixing bags must be stored in a refrigerator and must be used within 4 weeks. These bags must be taken out of the refrigerator 8 hours before the infusion.

Complications of Parenteral Feeding

Parenteral nutrition is a life-saving but complex therapy, which is not without risk of complications. Some of these complications include infection, metabolic, and fluid issues. Management by an interdisciplinary Nutrition Support Team can optimize patient outcomes associated with this therapy.

Mainly the complications are classified as under:

- ☆ Metabolic
- ☆ Physiological
- ☆ Mechanical
- ☆ Infectious

i) Metabolic Complications

Hyper/Hypoglycemia

Glucose abnormalities are common. Upto 30 per cent of the patients receiving nutritional support are hyperglycaemic. Hyperglycemia can be avoided by monitoring plasma glucose often, adjusting the insulin dose in the TPN solution, and giving subcutaneous insulin as needed. Hypoglycemia can be precipitated by suddenly stopping constant concentrated dextrose infusions. Treatment depends on the degree of hypoglycemia. Short-term hypoglycemia may be reversed with 50 per cent dextrose IV; more prolonged hypoglycemia may require infusion of 5 or 10 per cent dextrose for 24 h before resuming TPN via the central venous catheter. To manage this complication, avoid exceeding upper limit of dextrose provision. Avoid administration of other IV dextrose fluid if possible. Adjust the insulin dose, if needed.

Fluid Overload

Volume overload (suggested by > 1 kg/day weight gain) may occur when patients have high daily energy requirements and thus, require large fluid volumes and there is in appropriate fluid retention. To manage it, decrease the volume provision if possible. Follow the medical therapy for fluid retention, if indicated.

Abnormalities of Serum Electrolytes and Minerals

During starvation, intracellular electrolyte stores, particularly phosphate, are depleted despite normal serum concentrations; Feeding stimulates the cellular uptake of electrolytes and can lead to electrolyte disturbances with profound hypo-phophataemia. Clinical features usually develop within 4 days of re-feeding, but are often non-specific. Later manifestations include cardiac failure, hypotension, respiratory failure, seizures and coma. Recommendations for electrolyte replacements as mentioned in Table 18.7 should be followed.

This complication should be corrected by modifying subsequent infusions or, if correction is urgently required, by beginning appropriate peripheral vein infusions. Vitamin and mineral deficiencies are rare when solutions are given correctly. Elevated BUN may reflect dehydration, which can be corrected by giving free water as 5 per cent dextrose via a peripheral vein.

Metabolic Bone Disease

Or bone demineralization (osteoporosis or osteomalacia), develops in some patients given TPN for > 3 mo. The mechanism is unknown. Advanced disease can cause severe periarticular, lower-extremity, and back pain.

ii) Physiological

Steatosis

Liver dysfunction may be transient, evidenced by increased transaminases, bilirubin, and alkaline phosphatase; it commonly occurs when TPN is started. Fat gets accumulated in the liver, AST and ALT are increased which generally occur within 2 weeks of starting PN due to overfeeding with dextrose and/or total calories or lipids. Delayed or persistent elevations may result from excess amino acids. It can also happen due to deficiency of carnitine or choline in the formula. Pathogenesis is unknown, but cholestasis and inflammation may contribute. Progressive fibrosis occasionally develops. Reducing protein delivery may help.

For its management, avoid overfeeding, start enteral feeding as soon as possible. Give antibiotics if bacterial over growth is a possibility. Add taurine to PN.

Gallbladder Complications

Inclue gall bladder stones (cholelithiasis), gallbladder sludge, and cholecystitis. These complications can be caused or worsened by prolonged gallbladder stasis. Stimulating contraction by providing about 20 to 30 per cent of calories as fat and stopping glucose infusion several hours a day is helpful. Oral or enteral intake also helps. Treatment with metronidazole, ursodeoxycholic acid, phenobarbital, or cholecystokinin helps some patients with cholestasis.

Painful Gepatomegaly

Suggests fat accumulation; carbohydrate delivery should be reduced.

Hyperammonemia

It can develop in infants, causing lethargy, twitching, and generalized seizures. Arginine supplementation at 0.5 to 1.0 mmol/kg/day can correct it.

If infants develop any hepatic complication, limiting amino acids to 1.0 g/kg/day may be necessary.

Gastro-Intestinal Atrophy, Enteric Bacteremia and Sepsis

Atrophy of villi due to lack of enteral nutrients may take place while enteric bacteremia and sepsis can be without clear source. For its management, enteral feeding should be started as soon as possible.

Adverse Reactions to Lipid Emulsions

E.g, dyspnea, cutaneous allergic reactions, nausea, headache, back pain, sweating and dizziness are uncommon but may occur early, particularly if lipids are given at > 1.0 kcal/kg/h. Temporary hyperlipidemia may occur, particularly in patients with kidney or liver failure; treatment is usually not required. Delayed adverse reactions to lipid emulsions include hepatomegaly, mild elevation of liver enzymes, splenomegaly, thrombocytopenia, leukopenia, and, especially in premature infants with respiratory distress syndrome, pulmonary function abnormalities. Temporarily or permanently slowing or stopping lipid emulsion infusion may prevent or minimize these adverse reactions.

iii) Mechanical

Catheter Related Complications

Immediate complications are mainly related to insertion and include hemorrhage, pneumothorax, haemothorax, arrhythmias *etc.* Long term catheter complications include thrombosism, nerve injury and pulmonary embolism, pleural *etc.* along chylothorax and venopulmonary fistula. Catheter-related sepsis occurs in probably ≥ 50 per cent of patients. Catheter should be removed at the earliest possible. Medical review and management of the complications is required.

iv) Infectious

Infections can be catheter or non-catheter related too. This occurs in 1.3 per cent to 26.2 per cent of patients with central venous catheters used to administer parenteral nutrition. Higher rates are found across the world, particularly in high-risk groups, *e.g.* intravenous drug users.

Infections with staphylococcal and enterococcal species are common. *Canida* spp., *Klebsiella pneumoniae* and *Pseudomonas aeruginosa* are also found. In the long term TPN population, approx. 60 per cent of infections are caused by coagulase-negative staphylococcus.

There must be strict adherence to asepsis and solution bags and giving sets must be discarded after 24h use.

Monitoring

An interdisciplinary nutrition team, if available, should monitor patients. Progress should be followed on a flowchart. Weight, CBC, electrolytes, and BUN should be monitored often (*e.g.* daily for in-patients). BMI calculations and anthropometric measurements should be repeated at two week intervals. Plasma glucose should be monitored every 6 h until patients and glucose levels become stable. Fluid intake and output should be monitored continuously. When patients become stable, blood tests can be done much less often.

Liver function tests should be done. Plasma proteins (e.g, serum albumin, possibly transthyretin or retinol-binding protein), prothrombin time, plasma and urine osmolality, and calcium, magnesium, and phosphate should be measured twice/wk. Changes in transthyretin and retinol-binding protein reflect overall clinical status rather than nutritional status alone. If possible, blood tests should not be done during glucose infusion.

The frequency of most tests can be reduced once the patient's condition is stable. In addition, there should be daily attention to peripheral lines for signs of thrombophlebitis and centrally sited lines for signs of infections or inflammation.

19

Diet in Infections and Fevers

Interrelationship between Infections, Nutrition and Fever

Infection as defined by the medical dictionary is the successful invasion, establishment and growth of microorganisms in the tissues of the host. Hence, infection occurs when viruses, bacteria, or other microbes enter the body and begin to multiply. Disease occurs when the cells in the body are damaged as a result of infection and signs and symptoms of an illness appear. The incidence of disease among those infected varies greatly depending on the particular pathogen and individual susceptibility. Pathogenic microbes challenge the immune system in many ways. Viruses make us sick by killing cells or disrupting cell function. Our bodies often respond with fever (heat inactivates many viruses), with the secretion of a chemical called interferon (which blocks viruses from reproducing), or by marshaling the immune system's antibodies and other cells to target the invader. Many bacteria make us sick in the same way that viruses do, but they also have other strategies at their disposal. Sometimes bacteria multiply so rapidly they crowd out host tissues and disrupt normal function. Sometimes they kill cells and tissues outright. Sometimes they make toxins that can paralyze, destroy cells' metabolic machinery, or precipitate a massive immune reaction that is itself toxic.

In response to infection, the immune system springs into action. White blood cells, antibodies, and other mechanisms go to work to get rid the body of the foreign invader. An infection weakens the immune response of a person. If a person is already in poor nutritional status, it will further result in lowered resistance to infections. Because of the multiple effects on nutritional status, infections are likely to precipitate frank nutritional disease in individuals whose nutritional status is borderline. Kwashiorkor may follow measles, chickenpox, German measles, and whooping cough. Diarrhea is reported from many countries to be the most common contributory cause of kwashiorkor. This is the case even when not necessarily severe, because it may be only the latest in a long series of diarrheal, respiratory and other infections which progressively deplete the young child, when there is neither sufficient food or time for complete recovery between episodes.

Fever often is a response of humans and many animal species to infection, inflammation and trauma. However, this view of fever is merely an oversimplification as a growing body of evidence now suggests that fever represents a complex adaptive response of the host to various immune challenges whether infectious or non-infectious. Many of the symptoms that make a person suffer during an infection *i.e.* fever, malaise, headache, rash *etc.* result from the activities of the immune system trying to eliminate the infection from the body. Fever is one of the oldest clinical indicators of disease in the mammalian host as well as one of the most common reasons for medical consultations worldwide. Fever often occurs in response to infection, inflammation and trauma. However, this view of fever is merely an oversimplification as a growing body of evidence now suggests that fever represents a complex adaptive response of the host to various immune challenges whether infectious or non-infectious. Although elevated body temperature is an indispensable component of the febrile response, it is not synonymous with fever. It is generally agreed that fever is a regulated rise in body temperature above normal daily fluctuations occurring in conjunction with an elevated thermoregulatory set point.

Infection affects the nutritional status of an individual adversely and the extent depends upon the following:

1. Nature of the infection or infectious disease
2. Severity of the disease
3. Period for which it persists
4. Presence or absence of fever
5. Nutritional status of the person before the infectious disease.

The effects of infection are seen as follows (Figure 19.1):

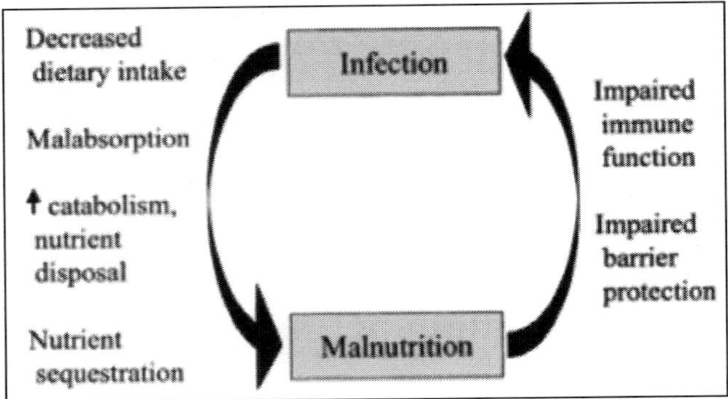

Figure 19.1: Relationship between Nutrition and Infection.

- ☆ Micronutrient deficiencies and infections are intricately linked, with deficiencies being related to poor growth, impaired intellectual performance, and predisposition to infections (mainly bacterial, viral, and parasitic). There are increased losses of nitrogen, some electrolytes like potassium and phosphates and other elements such as zinc and magnesium during infection.
- ☆ If there is fever along with an infection, the energy needs are increased since metabolic rate also increases.
- ☆ A catabolic response occurs with all infections even when they are subclinical and not accompanied by fever.
- ☆ There is decrease in food intake due to nausea and loss of appetite.
- ☆ Loss of nutrients is increased due to increased perspiration, vomiting and diarrhoea.
- ☆ There is decrease in the absorption of nutrients.
- ☆ In all kinds of infections, nutritional requirements are increased.

What is Fever?

Fever, pyrexia, controlled hyperthermia and febrile *etc.* are all synonymous terms. Pyrexia is from the Greek *pyr* meaning *fire*. Febrile is from the Latin word *febris*, meaning *fever*, and archaically known as *ague*. Fever is when a human's body temperature goes above the normal range of 36–37° Centigrade (98–100° Fahrenheit). It is a common medical sign. People's normal body temperatures may vary and are affected by factors such as eating, exercise, sleeping, and what time of the day it is. Our body temperature is usually at its highest at around 6 p.m. and at its lowest at about 3 a.m.

Although elevated body temperature is an indispensable component of the febrile response, it is not synonymous with fever. It is generally agreed that fever is a regulated rise in body temperature above normal daily fluctuations occurring in conjunction with an elevated thermoregulatory set point. To highlight the adaptive nature of the febrile response, the International Union of Physiological Sciences Commission for Thermal Physiology in 2001 defined fever as a state of elevated core temperature, which is often, but not necessarily, part of the defensive responses of multicellular organisms (host) to the invasion of live (microorganisms) or inanimate matter recognized as pathogenic or alien by the host.

Hence, fever is an elevation of body temperature that exceeds the normal daily variation and occurs in conjunction with an increase in the hypothalamic set point. When the hypothalamic set point is raised or as the body temperature goes up, the body is perceived to be cooler than the new set point. Shivering is initiated to generate heat. Blood is shunted from the periphery to the core to conserve heat and sweating is diminished. The generated heat will raise the body temperature to match the elevated set point. When the hypothalamic set point is lowered, either as part of the normal diurnal fluctuations that occur during an infection or in response to antipyretic agents, heat is lost by evaporation (sweating) and radiation (cutaneous vasodilation).

A high body temperature, or fever, is one of the ways our immune system attempts to combat an infection. Usually, the rise in body temperature helps the individual resolve an infection. However, sometimes it may rise too high, in which case, the fever can be serious and lead to complications. Doctors say that as long as the fever is mild, there is no need to bring it down; if the fever is not severe, it is probably helping to neutralize the bacterium or virus that is causing the infection. Medications to bring down a fever are called antipyretics. If the fever is causing undue discomfort, an antipyretic may be recommended. When a fever reaches or exceeds 38° Centigrade (100.4° Fahrenheit), it is no longer mild and should be checked every couple of hours.

Causes of Fever

Fever may be caused by:

- Endogenous factors (Antigen-antibody reaction, malignancy and graft rejections)
- Exogenous factors (Bacteria, fungi, virus *etc.*)
- Certain inflammatory conditions such as rheumatoid arthritis.
- Some medications, such as antibiotics and drugs used to treat high blood pressure or seizures.
- Some immunizations, such as the diphtheria, tetanus and acellular pertussis (DTaP) or pneumococcal vaccines
- Heat exhaustion, extreme sunburn

Overall, main causes of fever include infections caused by parasites, viruses, bacteria, richetsia, Chlamydia, immune reactions (including the defects in collagen, immunological abnormalities and acquired immunodeficiency. Other causes of fever are destruction of tissues, such as trauma, local necrosis (infarction), and inflammatory reaction in tissues and vessels (flebitis, arthritis), pulmonary infarction, and rhabdomyolysis. Specific inflammations (sarcoidosis, granulomatous hepatitis), inflammation of intestine and intra-abdominal inflammatory processes with the participation of lympho-endothelial system and hemopoetic system, solid tumours (Grawitz tumour of the kidney, carcinoma of the pancreas, pulmonary and skeletal tumours, hepatoma) can also cause fever. Acute metabolic failures such as arthritis urica, porfyria, addison's crisis, thyreotoxic crisis, and feochromocytoma, administration of some drugs, dehydration and administration of salts (that is why fever occurs with diarrhea), administration of foreign proteins may be the reason of fever's origin.

Mechanism of Fever

Any imbalance in which heat production exceeds heat loss will lead to the storage of heat and a rise of body temperature. Temperature is ultimately regulated in the hypothalamus. In response to an infection, illness, or some other cause, the hypothalamus may reset the body to a higher temperature. Although the most common causes of fever are common infections such as colds and gastroenteritis, other causes include infections of the ear, lung, skin, throat, bladder, or kidney. A trigger of the fever, called a pyrogen, causes release of prostaglandin E2 (PGE2). PGE2 in turn acts on the hypothalamus, which creates a systemic response in the body, causing heat-generating effects to match a new higher temperature set point.

In many respects, the hypothalamus works like a thermostat. When the set point is raised, the body increases its temperature through both active generation of heat and retention of heat. Peripheral vasoconstriction both reduces heat loss through the skin and causes the person to feel cold. Norepinephrine increases thermogenesis in brown adipose tissue, and acetylcholine stimulates muscle to raise metabolic rate. If these measures are insufficient to make the blood temperature in the brain match the new set point in the hypothalamus, then shivering begins in order to use muscle movements to produce more heat. When the hypothalamic set point moves back to baseline either spontaneously or with medication, the reverse of these processes (vasodilation, end of shivering and non-shivering heat production) and sweating are used to cool the body to the new, lower setting.

This contrasts with hyperthermia, in which the normal setting remains, and the body overheats through undesirable retention of excess heat or over-production of heat. Hyperthermia is usually the result of an excessively hot environment (heat stroke) or an adverse reaction to drugs. Fever can be differentiated from hyperthermia by the circumstances surrounding it and its response to anti-pyretic medications.

Following infection, inflammation or trauma, immune cells activate the release of pyrogenic cytokines as well as cryogens. Fever signals are transmitted via neural and humoral pathways to the brain to reset the thermal balance point. Vagal afferents may not necessarily be involved in fever generation. Cryogens prevent excessive elevation of the balance point. Heat gain mechanisms and acute phase responses are activated whereas heat loss is minimized. Ultimately, body temperature rises to new thermal balance point producing fever.

General Symptoms

When somebody has a fever, signs and symptoms are linked to what is known as sickness behavior, and may include:

- Rise in temperature of body heat
- Perspiration or Shivering
- Restlessness and agitated temper
- Pain and soreness all over the body but some limbs may be extra painful and sore
- Thirst
- Loss of Appetite
- Dehydration – preventable if the person drinks plenty of fluids
- Hyperalgesia, or increased sensitivity to pain
- Lethargy

If the fever is high, there may also be extreme irritability, confusion, delirium, and seizures.

Types and Patterns of Fever

A number of types of fever were known as early as 460 BC to 370 BC when Hippocrates was practicing medicine including that due to malaria (tertian or every 2 days and quartan or every 3 days). It also became clear

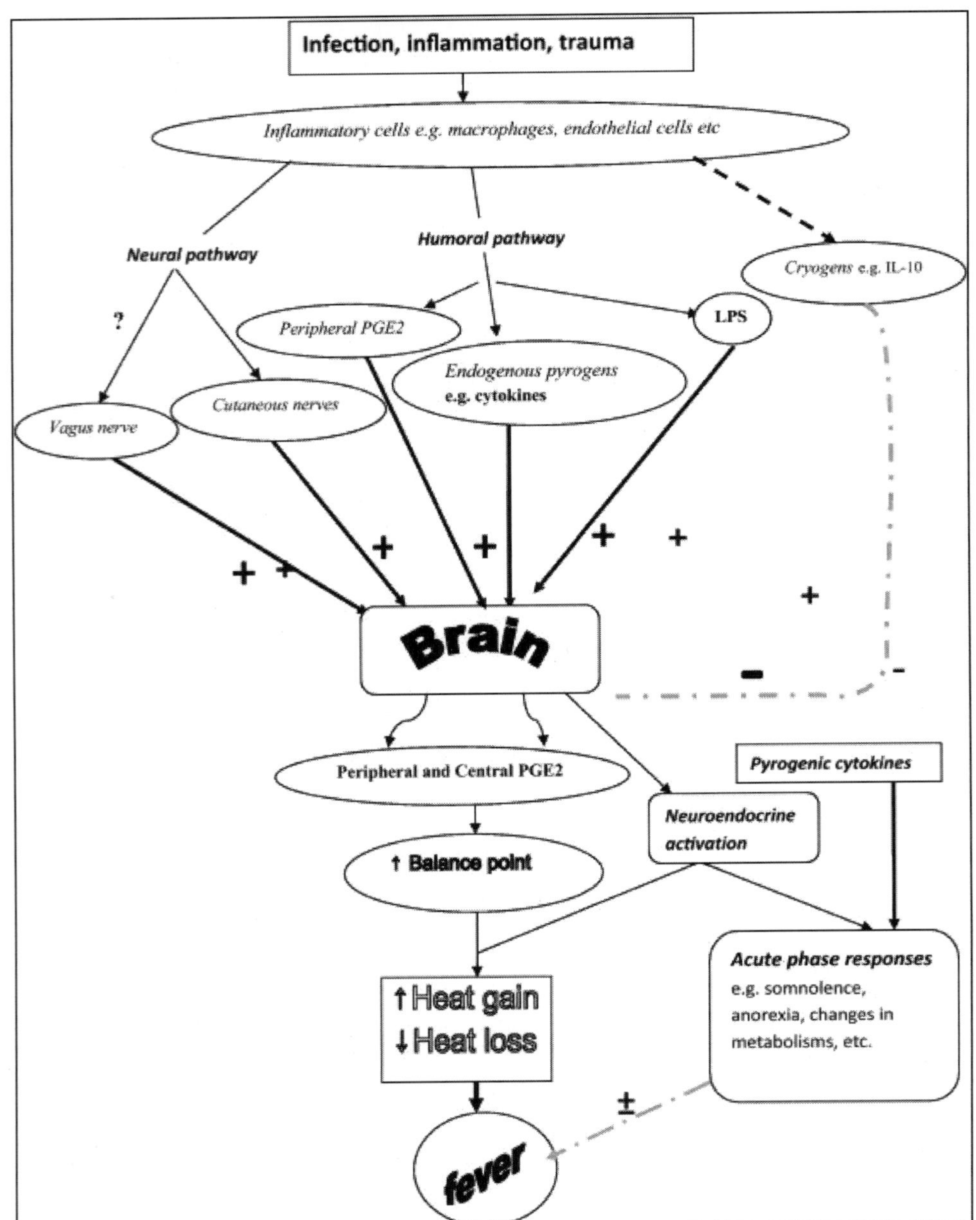

around this time that fever was a symptom of disease rather than a disease in and of itself.

A) Based on Duration of Fever

Fevers can be arbitrary classified into acute, sub-acute and chronic fevers based on duration.

i) **Acute fevers** (<7 days in duration) are characteristics of infectious diseases such as malaria and viral-related upper respiratory tract infection.

ii) **Sub-acute fevers** (usually not more than 2 weeks in duration) may be seen in cases of typhoid fever and intra-abdominal abscess, among others.

iii) **Chronic or persistent fevers** (>2 weeks duration) are typical of chronic bacterial infections such as tuberculosis, viral infections like HIV, cancers and connective tissue diseases. However, any cause of acute fever can become persistent or chronic if untreated.

B) Based on Height of Fever

Based on the height of body temperature, fever can also be classified into low grade, moderate grade, high grade and hyperpyrexia (Table 19.1).

C) Based on Pattern of Fever

Three major fever patterns have been described including sustained/continuous fever, intermittent fever and remittent fever.

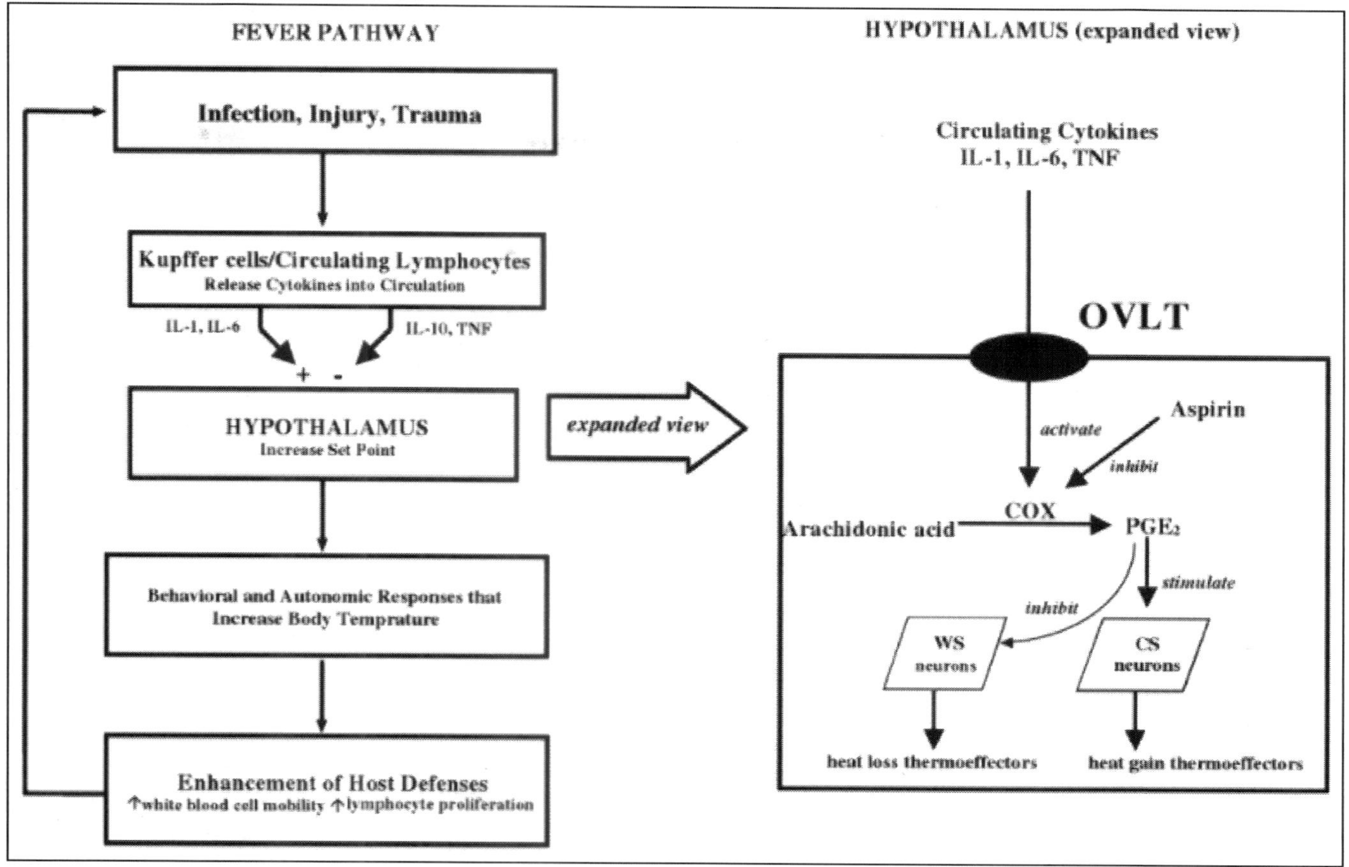

Table 19.1: Normal and Febrile Body Temperature Ranges (Rectal temperatures)

Body Temperature	*°C*	*°F*
Normal	37–38	98.6–100.4
Mild/low grade fever	38.1–39	100.5–102.2
Moderate grade fever	39.1–40	102.2–104.0
High grade fever	40.1–41.1	104.1–106.0
Hyperpyrexia*	>41.1	>106.0

**Note*: The difference between fever and hyperthermia is the underlying mechanism. Different sources have different cut-offs for fever, hyperthermia and hyperpyrexia.

The conversion formula used is:

T°F = 9/5 (T°C) + 32

T°C = 5/9 (T°F) – 32

Hyperpyrexia in severe malaria is defined as rectal temperature above 40 °C.

C, centigrade; F, Fahrenheit.

Continuous or Sustained Fever

It is defined as fever that does not fluctuate more than about 1 °C (1.5 °F) during 24 h, but at no time touches normal. Continuous fevers are characteristics of lobar and gram-negative pneumonia, typhoid, acute bacterial meningitis, urinary tract infection, drug fever among others. Fever characterized by slow stepwise temperature rise and a high plateau are classical of typhoid fever.

Intermittent Fever

Intermittent fever is defined as fever present only for several hours during the day and falls to normal each day. This pattern of fever can be seen in malaria, pyogenic infections, tuberculosis (TB), schistosomiasis, lymphomas, leptospira, borrelia, kala-azar, or septicemia. In malaria, depending on the species of parasite, fever can occur with a periodicity of 24 h, 48 h or 72 h.

Remittent Fevers

Remittent fever is defined as fever with daily fluctuations exceeding 2 °C but at no time touches normal. They are often associated with infectious diseases such as infective endocarditis, rickettsiae infections, brucellosis, among others.

Relapsing Fevers

They refer to those that are recurring and separated by periods with low-grade fever or no fever. Periodic or relapsing fevers are seen in malaria, lymphoma, borrelia, cyclic neutropenia, and rat-bite fever.

Metabolic Changes during Fever

- Increase in BMR
- Decrease in body fluids due to increased sweating
- Diminution of glycogen in the liver and muscles in neurogenic fever and in the fever of infection
- Increase in protein turn over. There is an increased breakdown of proteins, especially in typhoid fever, malaria, poliomyelitis *etc.*
- Greater increase in urinary nitrogen in infectious fever and its excretion through kidney.
- The stores of adipose tissue are decreased since they are burnt up to meet the increase in the calorie needs.
- Electrolyte balance may disturb due to increased excretion of sodium and potassium.

Nutrient Requirements and Dietary Considerations in General

Overall, all infections no matter how mild, decrease nutrient intakes and increase nutrient losses even when subclinical. The losses include decreased intestinal absorption, direct loss of nutrients in the gut, internal diversion for metabolic responses to infection and increased BMR when fever is present. Infection influences in this way not only protein and energy status but also that of most other nutrients. The clinical importance of these consequences of infection depends on the prior state of the individual, the nature and duration of the infection and the diet of the individual during the infection, particularly dietary intake during the convalescent period and whether full recovery takes place before another infection occurs. Particular attention must be paid to the nutrition of hospitalized patients since they are frequently debilitated by their primary disease, morbidity, and their nutritional status. Morbidity and mortality are increased by nosocomial infections to which the poorly nourished individual is more susceptible. Hence, adequate nutrition is must to be given to a person suffering from any kind of infection whether it is acute or chronic.

Energy

It has been known since early in the 20th century that a rise in temperature is associated with an increase in metabolic rate. For each degree Celsius rise in body temperature, metabolism rate increases by 10 to 13 per cent which means one has to expend an extra 100 to 130 calories per day if the individual is normally taking 2,000 calories daily because body temperature accounts for 50 per cent of the energy expenditure. In general, the needs for energy are increased by 10 to 50 per cent during and after acute infections; 30 to 40 Kcal/kg/day for adults and 100 to 150 Kcal/kg/day for children are needed. Energy intake is increased by 50 per cent if the temperature is high and tissue damage is high. For most respiratory disease, the energy cost would be trivial, but for diarrheal disease and the common communicable diseases of childhood, it could add significantly to either a chronic energy deficit or to increased dietary energy need depending on dietary adequacy during the infection.

There are "savings" of dietary energy during the acute phase of an infection that are due to anorexia and reduced physical activity. Patient may be able to ingest only 600-1200 kcal daily during high fever. During the recovery period, however, this must be made up together with the losses due to the deficits arising from the malabsorption, increased catabolism, and internal synthesis associated with infections. Thus, with each infectious episode there is a small decrease in de facto dietary requirement at the time of the infection and an increase during the convalescent period that is difficult to quantitate because it is so variable. Obviously, the severity of the caloric deficit is related to the duration and magnitude of the morbidity in each individual situation, and rarely poses a major threat to nutrition unless the infection is chronic or recurrent.

High calorie diet (energy value upto 50 per cent more than the normal maintenance intake for a person in bed) with frequent feeding should be given as soon as the fever is controlled.

Carbohydrate

The catabolic responses during fever have as one principle function *i.e.* the provision of amino acid substrate for gluconeogenesis. Thus, a continual conversion of alanine carbon to glucose carbon occurs with acute infection, even when exogenous carbohydrate is adequate. It appears to be the rate of release of glycogenic amino acid substrates from peripheral tissues that determines the rate of hepatic gluconeogenesis. All of the hormones that regulate carbohydrate metabolism participate in host responses to infection. A number of groups have documented an increased fasting concentration of both glucagon and insulin in serum. Despite the initial stimulation of gluconeogenesis, the body may eventually become severely hypoglycemic. Lethal hypoglycemia may result in septic neonates and in severe viral infections of the liver such as fulminating hepatitis) or as shown in yellow fever.

The protein sparing property of carbohydrate diet is also of practical importance. To avoid loss of muscle protein, a large proportion of the food ingested should be in the form of carbohydrate. To replenish the glycogen, liberal intake of carbohydrates especially glucose which is readily absorbable is essential. This prevents ketosis, which may arise if patient is burning his own body fat in increased amounts owing to the raised metabolism. In acute infections, glucose is given by IV drip to the individual if he is anorexic or does not feel like eating the food.

Protein

The protein catabolism increases during fever and infections. Powanda (1977) had summarized data from a wide variety of acute infectious diseases by adding the total nitrogen losses and dividing them by the number of days over which these losses occurred. For all infections, the average loss of 0.6 g of protein lost per kg per day, was equal to the mean estimated total protein requirement for adults. Diseases associated with diarrhea or dysentery produced an average loss of 0.9 g of protein/kg per day. Higher losses were observed with typhoid fever and other severe infections, reaching 1.2 g or protein/kg per day. Another approach to the magnitude of the catabolic response in infection is to measure urinary 3-methylhistidine as a measure of muscle protein catabolism in the septic patient. Such calculations are underestimates because they do not include energy expended for the multiple anabolic responses.

Hence, a liberal amount of protein is needed to make good the loss of body protein. During chronic prolonged fevers, a high protein diet supplying 1.25-1.5g protein/kg body wt (about 100 g protein daily) should be fed during the recovery period to replenish the body protein stores. Protein supplements can be incorporated in the beverages besides the dietary protein at regular meals.

Fat

Fat metabolism is not much affected in acute infections. However, in chronic infections and prolonged fevers, adipose stores are used to provide energy. For increasing the energy intake in the diet, amount of fat can be increased to certain extent depending upon the food intake during and after the infection episode. But fried and junk foods rich in *trans* fats should be avoided.

Vitamins

Vitamins requirements are increased if infection is accompanied by swelling and necrosis. The vitamin intake should be increased by 2 to 3 times of minimum daily requirements. Vitamin A blood levels are appreciably reduced in pneumonia, rheumatoid arthritis, acute tonsillitis, infectious hepatitis *etc.* Lower serum carotene and vitamin A levels have been reported with hookworm disease. The capacity of infections to precipitate xerophthalmia and keratomalacia in individuals already marginally deficient is well established and the effect is particularly severe with measles and also noted for chicken pox. Vitamin A malabsorption occurs during systemic febrile illnesses. Enteric infections of bacterial and viral origin also have this effect on the absorption of vitamin A. Requirement of B-vitamins is also increased during infections.

Foods rich in β-carotene or Vitamin A, Vitamin C and B-vitamins in the form of soups, smoothies, vegetables *etc.* should be included in the diet.

Minerals

Requirement of zinc and iron increases since both are involved in the immune process. Zinc excretion in urine also increases. Salt depletion occurs from sweating and this should be made by adding sodium chloride in drinks. Fruits or fruit juices rich in potassium like coconut water, banana, lime *etc.* should be included in the diet.

One multivitamin-mineral tablet should be added in the diet during chronic fever or they should be given with glucose IV drip in chronic infections and fever if the person is suffering from malabsorption.

Fluid Intake

Fluids should be given in abundance. Owing to the loss of fluids in sweat and the disinclination of patients to take fluids and foods, dehydration may develop in fever. Fluids are required to be given in greater amounts than are usually ingested but excessive amounts should not be forced on the patient. A daily total fluid intake of about 2 ½ to 3 litres is ample. If vomiting and diarrhoea persist, parenteral treatment may be required.

As the patient improves and can take solid food, this diet should be modified by addition of bread with butter, milk puddings, custards, stewed or fresh fruits, pureed vegetables, eggs, fish.

Meal Pattern

Foods should be served initially in liquid or semi-liquid and soft texture forms which must be easily digestible, well cooked and appetizing. Serving size should be small but frequent. Large meals should be avoided and small foods should be given every two or three hours. During recovery period, consistency and texture of food can be changed.

Convalescence

Diet has an important part to play in determining the speed with which a patient recovers from a surgical operation, fever or any serious illness. Ample amount of protein are needed to make good the drain on body's reserves of protein and if the patient is under-weight, plenty of energy to build up tissues. Protective foods *e.g.* fruits and vegetables should be added liberally in the diet to provide vitamins and minerals.

Typhoid Fever

Typhoid fever is an acute illness associated with fever caused by the *Salmonella enterica serotype Typhi* bacteria. It can also be caused by *Salmonella paratyphi*, a related bacterium that usually causes a less severe illness.

Sources of Infection and Transmission

The infection is often passed on through contaminated food and drinking water, and it is more prevalent in places where hand washing is less frequent.

It is spread to other people in the area through the faeces and urine of the patients or the asymptomatic carriers who do not know they carry the bacteria or by house flies. No animals carry this disease, so transmission is always human to human.

S. typhi enters through the mouth by taking contaminated food or drink with the faeces of an infected person and spends 1 to 3 weeks in the human intestine. After this, it makes its way through the intestinal wall and into the bloodstream. From the bloodstream, it spreads into other tissues and organs. The immune system of the host can do little to fight back because *S. typhi* can live within the host's cells, safe from the immune system.

Typhoid is diagnosed by detecting the presence of *S. typhi* via blood, stool, urine, or bone marrow sample.

Symptoms and Signs

Symptoms normally begin between 6 and 30 days after exposure to the bacteria. There are two major symptoms of typhoid *i.e.* fever and rash. Typhoid fever is particularly high, gradually increasing over several days up to 104 degrees Fahrenheit, or 39 to 40 degrees Celsius. The rash, which does not affect every patient, consists of rose-colored spots, particularly on the neck and abdomen. Acute illness is characterized by:

- Prolonged fever
- Nausea
- Loss of appetite
- Weakness
- Abdominal pain
- Constipation or sometimes diarrhoea
- Headaches

Symptoms are often non-specific and clinically non-distinguishable from other febrile illnesses. However, clinical severity varies and severe cases may lead to serious complications or even death. In serious, untreated cases, the bowel can become perforated. This can lead to peritonitis, an infection of the tissue that lines the inside of the abdomen, which has been reported as fatal in between 5 and 62 per cent of cases. It occurs predominantly in association with poor sanitation and lack of clean drinking water.

Principles and Suggestions for Dietary Management

A high energy, high good quality protein, high carbohydrate, more liquid, low fat, low fibre and soft and bland diet is required for typhoid patients.

Nutritional requirements and general dietary principles as mentioned earlier for fever hold true in typhoid too.

- Typhoid fever is a common condition that causes abdominal pain and high fever, among other symptoms. A typhoid diet plan plays an important role in treatment as it places less stress on the digestive system of the person who has typhoid. A diet for typhoid patients should be one that is light and easy to digest.
- The diet should be light, soft, bland and easy to digest and at the same time supply enough energy and protein to the patient. Keep in mind that a patient who has typhoid has a low appetite. It is vital therefore, to ensure that such a patient is given a high calorie and high protein diet. This is essential as it aids the patient's recovery. A patient's energy and protein reserves are depleted at this time due to the lack of appetite and subsequent low intake.
- There should be adequate nutrition. Though bland, the diet should be nutritionally adequate to meet daily calorie requirements of the body. The body's basal metabolic rate is 10 per cent high during fever, which means that more tissue is being broken down to give energy. So the diet given during typhoid should provide sufficient proteins, meet the nutrient requirement of the body and maintain a water and electrolyte balance. This is mainly to avoid irritation to the mucus membrane, to reduce gastric acidity, which would further inflame the gastrointestinal tract

Dietary Management

In acute cases, to meet the nutritional requirements, clear liquid should be given to the patient orally or through IV drip depending upon the severity of illness. It should be followed by semi-liquid, semi-solid and soft bland diet devoid of fibre once the patient becomes capable of taking food through oral route. Meals should be easily digestible, soft and not hard on the intestine. Milk, fruit juices, glucose water, buttermilk, coconut water and barley water are good options in the beginning and they should be consumed until body temperature comes back to normal. As the symptoms subside and patient's appetite gets better, semi solids can be given. When the patient has no fever, when appetite of patient improves and there is no feeling of nausea and vomiting, bland semi-solid foods for instance, custards, vegetable/chicken/lentil soups, soft boiled or poached egg, baked apple, white bread, porridge, boiled rice with yogurt, soft fruit smoothies without fibre and seeds and so on can be incorporated to provide high energy and high protein diet for recovering quickly. A typhoid recovery diet plan should include yogurt and eggs, as they are easier to digest than meat and provide adequate protein. For vegetarians, lentils, legumes and cottage cheese, should meet the protein needs.

The main motive behind a typhoid patient's diet is to supply proper nutrition, and to balance the water and electrolyte proportion in the patient's body. An ideal typhoid diet is one which has small and frequent meals in it. Such a diet should contain foods that are soft, bland, digestible and low in fiber as well. It is important to remember that such a diet should not irritate the patient's weak intestinal system.

Foods to be Allowed

Following foods are allowed in the diet of typhoid patient:

- ✰ Liberal intake of carbohydrates that are well cooked and easily digestible (starches, glucose, cane and honey) should be included in the diet for easy digestion and absorption. Simple carbohydrates like jam, honey, biscuits, *etc.* can be included in the diet.
- ✰ Proteins of high biological value like eggs, soft cheese *etc.* should be included in the diet. Proteins can also be derived from tender meat, organ meats, fish, and poultry but these foods will also be high in fat and difficult to digest; so for this reason these should be avoided for a few weeks till the symptoms subside.
- ✰ Fever can cause a lot of fluid loss from the body in the form of sweating (if diarrhea and vomiting are also present, then fluid losses are more). So to recoup these water and electrolyte losses, it is important to provide sufficient fluids at short intervals. Drink at least six to eight glasses of water every day. Increase intake of liquid foods such as clear or vegetable broth, fresh strained fruit juice, buttermilk, glucose water, coconut water, sugarcane juice and other vegetable soups and juices when given can provide some calories along with fluids and electrolytes to maintain their balance. These foods help to replace electrolytes that are lost in the diarrhea. If someone becomes severely dehydrated, he may need to receive fluids intravenously in the hospital.
- ✰ Soft fruits like papaya, banana, sapota *etc.* should be given.
- ✰ Low fiber foods such as refined cereals and their products like white bread, pasta, macaroni *etc.*, dehusked pulses, well cooked fruits, tender vegetables in soft and puree form, potatoes *etc.*
- ✰ Milk and milk based beverages like shakes, custard, buttermilk, yogurt *etc.*

Foods from each food group should be included but they should be soft, tender, bland, digestible and low in fibre so as not to irritate the intestinal mucosa.

Foods to be Avoided

Certain foods can cause dietary problems and load the already compromised liver in such a patient. Foods that should be avoided while being treated for typhoid include the following:

- ✰ High fiber foods like whole grain cereals and their products (broken wheat, wheat flour, brown bread, oats), raw fruits and vegetables, whole pulses and pulses with husk *etc.*
- ✰ Excessive milk, flavoured milk and other milk based products.
- ✰ Fried fatty food such as *samosas, pakoras, ladoos, halwa, pastries* and other fried foods *etc.*
- ✰ Strongly flavored fermented foods, vegetables like cabbage, capsicum, turnip, radish, garlic, beetroot *etc.*
- ✰ Avoid vegetables like cabbage, capsicum and turnip as these can cause bloating and gas.
- ✰ Stay away from oily foods, butter, ghee, rich pastries, desserts, thick cream soups, pickles, sauces, chutneys spices and seasonings like pepper, cayenne and chilli powder to make sure that the digestive tract does not inflame all the more.
- ✰ Avoid solid food unless the person feels hungry.

Post-typhoid Diet Plan

Typhoid is a bacterial infection characterized by intestinal disorder which leaves the patient weak and lagging in energy once he gets rid of typhoid. The infection adversely affects the immunity and digestive system. So, one need to follow a post-typhoid diet plan to restore the immunity, digestive power, and balance the levels of electrolytes and fluids in the body. Following are the post-typhoid dietary tips:

1. **Eat meals at short intervals:** Typhoid drains all the energy and affects the appetite adversely. Even after recovering from typhoid, one may experience low appetite for a few weeks. However, one should eat food so as to provide energy to the body. Post typhoid, the patient may not be able to eat big meals at once. Make sure the individual includes nutritious foods in his small meals and have them more often. This will help increase his calorie intake, nourish his body, and restore the lost energy soon.
2. **Include easily digestible foods:** Digestion power is bound to get affected adversely due to typhoid, and it may take some time for the digestion and intestinal system to get back to normal post typhoid. So, after recovering from typhoid, make sure that the diet comprises of easily digestible foods. Include well-cooked, soft, yet nutritious foods in the post-typhoid diet also, include healthy fluids, such as vegetable soups in the diet.
3. **Have protein-rich diet:** The body's protein needs increase significantly post typhoid, as typhoid fever and low food intake tend to affect the body adversely. So, one needs to include foods

rich in proteins following typhoid. One should take eggs, *paneer*, *dals*, fruits, such as dates, jack fruit, dried apricot, avocados, and prunes, curd rice, yogurt, fresh buttermilk, dry fruits, such as almonds, and other foods that have high amounts of proteins. Consuming protein rich foods helps to quickly recover from weakness caused by typhoid.

4. **Drink healthy fluids:** Fluid loss from the body is obvious due to typhoid fever. Vomiting and sweating drain a lot of fluid and electrolytes out of the body during typhoid. So, post typhoid, one needs to recoup all electrolyte and fluid loses by including healthy fluids in the diet. Drink plenty of water to keep the body hydrated. Also, drink healthy fluids, such as coconut water, fresh fruit juices, vegetable soups, lime juice, sugarcane juice, and glucose water, regularly after recovering from typhoid. Have these healthy fluids in adequate quantities at short intervals.
5. **Add Vitamin A, B, and C rich foods:** Adding plenty of vitamin A, vitamin B, and vitamin C rich foods to the regular diet post typhoid is quite essential. Including these crucial vitamins to the diet helps to recoup the loss of vitality that patients suffered due to typhoid. So, one should eat cooked carrots, pumpkin, sweet potato and mango for vitamin A. Citrus fruit juices, such as juices of orange, grapefruit, aonla and other vitamin C rich fruits should be taken to fulfill the body's vitamin C needs. Eat cooked vitamin B rich vegetables, soya beans, and other foods.
6. **Do not gobble spicy, fried foods:** As the symptoms of typhoid subside, while following a healthy diet, the patient also needs to avoid the temptation for eating junk food. Completely avoid eating spicy foods, junk and fried foods, and other unhealthy oily and fatty foods post typhoid at least for two complete weeks. Eating unhealthy foods can hamper the digestion process. Add simple fatty foods, such as butter and cream, and spices to the diet gradually.
7. **Avoid insoluble fiber:** The digestive tract might suffer an injury or damage due to typhoid. So, it is best to avoid eating foods containing insoluble fiber post typhoid. Avoid eating foods, such as seeds and skin of fruits and vegetables. Also, foods, such as fermented foods, raw vegetables, and semi-cooked pulses, tend to cause dietary disturbances and load the compromised liver post typhoid. So, avoid eating them after typhoid.

Tuberculosis Fever

Tuberculosis has been plaguing humankind for thousands of years; it has been and still is one of the most deadly infectious diseases in the world. It is a bacterial disease caused by the bacterium *Mycobacterium tuberculosis* that infects the lungs and may spread through blood stream to any organ in the body like brain, kidney, bones, joints, lymph nodes, spine *etc.*

Symptoms

The symptoms of tuberculosis are varied and often include:

- Weight loss
- Night sweats
- Muscle weakness and fatigue
- Persistent cough (lasting at least three weeks)
- Phlegm or blood in cough
- Chest pain
- Wheezing sound when breathing
- Breathing problems

Types and Diagnosis of Tuberculosis

Latent TB: Latent TB, also called inactive TB is the condition in which the bacteria remain in the body in an inactive state, cause no symptoms and it does not spread. People with latent tuberculosis infection have the bacteria in their lungs and test positive for the illness, but have not yet experienced any symptoms. People with latent tuberculosis infection may not necessarily develop active tuberculosis, but some will.

Active TB: People with active tuberculosis do experience symptoms and can spread this infection to others. It can occur in the first few weeks after infection with the TB bacteria, or it might occur years later

The most commonly used diagnostic tools for tuberculosis are skin tests, blood tests, sputum tests and chest X-rays. Medications are available to help keep tuberculosis from developing into a serious illness.

How Tuberculosis is Spread

The TB bacteria are picked up in the air by breathing air that an infected person sneezed or coughed into, or even just spoke into. If one breathes contaminated air, he or she may contract TB. One cannot get TB through:

- Drinking or eating after someone with TB
- Kissing someone with TB
- Using the same toothbrush
- Physical contact, like shaking a hand
- Coming into contact with clothing, sheets, towels, or toilet seats of the infected person

People at Risk for Tuberculosis

Anyone can get tuberculosis. It does not discriminate between young and old, men or women. But there are some people who are at a greater risk of contracting tuberculosis than others *e.g.*:

- Infants
- Elderly population
- Anyone with a compromised immune system, such as people on chemotherapy or transplant medications and those with HIV or AIDS
- People with a poor diet
- People living in an unsanitary environment
- People who work or live with someone infected with tuberculosis

Food and Nutritional Factors that Increase TB Risk

In India, TB occurs more often in the poor who are likely to have chronic undernutrition, food insecurity and more likely to encounter delays in initiation of treatment. In such patients, TB worsens the already poor nutritional status. The link between tuberculosis and malnutrition has long been recognized; malnutrition may predispose people to the development of clinical disease and tuberculosis can contribute to malnutrition.

Nutritional deficiencies are generally associated with an increased risk for contracting TB and have an effect on the severity of the disease. Micronutrients deficiency is considered to be the most frequent cause of secondary immunodeficiency and infection related morbidity including tuberculosis. Patients with active TB also suffer from micronutrient deficiency, especially iron and folate, vitamin A, zinc, vitamin D. These micronutrient deficiencies are important as they can impair the cell mediated immune responses (*e.g.* zinc, vitamin D), anemia can impair quality of life and physical function.

Hence, the poorer the diet, the more likely it is that a person will develop complications associated with this disease. In addition, the nutritional status of people (whether they get all the nutrients their bodies need) and utilization of nutrients (how their bodies use these nutrients) are also adversely affected after they have been infected.

What's more is that malnutrition could diminish the effectiveness of the drug regimens, which must be taken by TB patients. A weakened immune system caused by malnutrition can be quickly and easily corrected by just giving the body the nutrients it needs. Thus, nutritional intervention in combination with appropriate pharmaceutical therapy, as is the case in HIV-infected patients, could improve the outcome in malnourished TB patients.

Nutritional Response to Infection

Once infected with TB, the body starts using more and more energy trying to fight the infection, and in addition, the infection often causes a loss of appetite in the patient, resulting in the patient losing weight.

An increase in protein breakdown leads to muscle wasting in these patients. The breakdown of protein and other reserves due to fever may also worsen under nutrition and further impair resistance against the infection.

The increased energy expenditure and tissue breakdown associated with infection are thought to increase the requirements of micronutrients such as vitamin A, E, B_6, C, D and folate. It is also known that a decrease in trace elements such as iron, zinc and selenium occur during the infection.

Nutritional Treatment of TB

The risk and morbidity of infections are influenced by the nutritional status of the individual. Likewise, the nutritional status and the intake and utilization of foodstuffs are profoundly altered during the body's response to infection. Factors that affect food intake, such as food availability, appetite, eating patterns, medication side effects, traditional food taboos, lifestyles (smoking, alcohol, physical activity, caffeine intake, use of social drugs), psychological factors (stress and depression), stigma, and economic factors are also very important to be considered.

Nutritional Needs in Tuberculosis

Energy

Energy needs of TB patients are increased because of the disease itself. The energy requirements of patients with active TB, varies with their age, gender, their activity levels as in persons without active TB. The requirements of energy in a patient with active TB can be divided into the energy requirements appropriate to their age, gender and activity levels, the additional requirements in view of the active TB disease, and the energy requirements for recovery of nutritional deficit. It has been suggested that a desirable rate of weight gain should be in the 2.5-5 per cent per month in those with moderate undernutrition and about 5 per cent per month in those with severe undernutrition. Usually, an additional approx. 325 -625 kilocalories per day are required over the usual recommended intake in different patients depending upon the physical activity and extent of infection to regain ideal body weight in about four to six months in TB patients who are undernourished. In the case of Aids patients having TB infection, energy requirements increase by 20 to 30 per cent to maintain body weight.

Protein

The requirements of protein would be 1.2-1.5 g/kg ideal body weight per day for adults and 3-3.5 g/kg for children. The higher requirement of protein is in view of the metabolic stress related to the active infectious disease. Protein is important to prevent the wasting of body stores (for example muscle tissues).

Micronutrients

WHO states that TB patients have low circulating concentrations of micronutrients, such as vitamins A, E and D, and the minerals iron, zinc and selenium. Moreover, anti-TB drugs can also result in micronutrient deficiencies. Hence, a good multivitamin and mineral supplement, providing 50 to 150 per cent of the recommended daily allowance, is advisable since it will be most unlikely that a person with TB will be able to meet the increased requirements for vitamins and minerals with diet alone (due to a poor appetite).

Dietary Management

While the treatment of active tuberculosis is very long term *i.e.* from 9 months up to a year of daily antibiotics which have significant impact on nutrient absorption including vitamin B_6 or pyridoxine, vitamin D which in turn impacts calcium absorption, and other minerals. Hence, the body needs healthy and balanced amount of nutrients now more than ever. Moreover, people who are malnourished or underweight are more likely to get tuberculosis and are also more susceptible to re-infection or relapse of TB even after treatment. Malnutrition leads to decreased immunity, and body needs to be as strong as possible to defend itself against those tough tuberculosis bacteria. Hence, it is a vicious cycle. Poor nutrition can actually encourage the persistence of active tuberculosis disease, and active tuberculosis leads to worsening malnutrition. So to keep the body primed to fight TB, one has to take healthy, nutritious and well balanced diet for treatment of tuberculoses and prevention of its relapse. A patient suffering from tuberculosis should take care of the following while planning his diet:

- ✰ Nutrition and dietary habits play an important role in the prevention and treatment of TB. Therefore, the patient should take a well balanced diet containing foods from each food group in optimum amount so as to get all the nutrients in right amount to fight active tuberculosis and regain strength and stamina. The diet has to be energy dense and rich in good quality protein to meet up rising metabolic demands and to prevent further weight loss.
- ✰ Carbohydrates are major sources of energy and patients can consume them during meals as well in snacks between meals to increase their energy intake. It is important to emphasize that the intake of complex carbohydrates and adequate dietary fibre are essential.
- ✰ Eat foods high in B-vitamins and iron, such as whole grains (in case of no allergy). Plenty of mixed whole grains like whole wheat *dalia*, pastas, breads, and cereals, rice, oat, barley bajra, maize *etc.* should be taken daily in one or the other form and according to the season. Energy dense foods include cereal porridge or *kheer*, rava or sooji halwa, peanut *chikki*, paushtik *laddoo*, wheat and ragi sprouted malt porridge or drink, *khichdi* and so on.
- ✰ For taking good quality protein, include milk and milk products in daily diet.
- ✰ Vegetarian patients should take preparations of cereal, pulses and milk products together at one meal as good quality protein is required for the patient. Non-vegetarian patient can replace one serving of pulses by meat, fish, poultry or egg *etc.* Eat fewer red meats and more lean meats, egg, cold water fish, tofu (soy, if no allergy), or beans for protein.
- ✰ Make a choice of healthy foods in the daily diet such as include leafy, dark-colored greens like mustard leaves, spinach, fenugreek leaves, mint and coriander leaves *etc.* for their high iron, β-carotene, B-vitamins and mineral contents.
- ✰ Seasonal vegetables and fruits (3-4 servings each) should be included in the daily diets Antioxidant-rich, brightly-colored seasonal vegetables such as carrots, peppers, pumpkin, ladyfinger and squash, and fruits, like tomatoes, lemons, guava, ber, aonla, lime, guava, mango, banana, blueberries, and cherries *etc.* should be taken in the diet daily. Dark green leafy vegetables, legumes and starchy vegetables including bitter gourd, drumsticks, spinach and broccoli can help the body in its fight against the tuberculosis. Bottle gourd is the best vegetable as it is easily digestible.
- ✰ Strawberries are high in potassium, vitamins, and minerals that help boost the immune system and help fight the disease. Banana and coconut water also contain good amount of potassium. Pineapple is considered to be very useful in treating TB. Drinking the juice of the fruit once a day will help in clearing mucus and aid recovery.
- ✰ One should prefer to take whole fruit instead of fruit juice. But during loss of appetite, it is better to take fruit juice. Orange juice is rich in vitamin C, a powerful antioxidant that can help the body to fight tuberculosis bacteria. Other juices that are beneficial include carrot, tomato, gooseberry and pineapple juice. They should be taken. Fresh juice of Indian gooseberry and honey every morning is highly effective in curing tuberculosis.
- ✰ Take handful of mixed nuts daily if they are affordable. Most nuts and seeds including sunflower seeds, chia seeds, pumpkin seeds, sesame, flax are good sources of both selenium and zinc. If not able to eat due to poor appetite, the dry fruits and nuts can be powdered finely and mixed into a milkshake or added to the *roti* to give more energy and protein.

- Take prescribed amount of mixed fats and oils (about 30 per cent of total calorie need daily). Avoid foods rich in *trans* fats such as bakery products.
- Lot of fluids should be taken daily.
- To prevent side effects of medication and nutrient deficiencies, on the recommendation of the doctor, include one multivitamin-mineral tablet daily. Vitamin A and zinc containing supplementation should be taken.
- In acute phase, a high calorie and protein diet in fluid and soft diet form should be given.
- Once the appetite is regained, absorption is well, normal high calorie, high protein, rich in multivitamins and minerals should be continued. As the patient is on antibiotics for a long time, it is beneficial to provide probiotic curd daily in the diet.

Foods to be Avoided

- Donot take spicy and fried foods.
- Avoid carbonated drinks
- Avoid excess of tea and coffee or their intake with food
- Limit the intake of refined products, like sugar, white breads, and white rice.
- Skip tobacco in all forms.
- Don't drink alcohol as it can add to the risk of liver damage from some of the drugs used to treat TB.
- Avoid high-fat, high-cholesterol red meat and instead load up on leaner protein sources like poultry, beans, tofu, and fish.
- One should strictly avoid canned foods, pies, puddings, sauces, caffeinated beverages, pickles, condiments and processed foods.
- Reduce or eliminate trans-fatty acids, found in commercially baked goods such as cookies, crackers, cakes, French fries, onion rings, donuts *etc.*

Human Immunodeficiency Virus (HIV) and Acquired Immunodeficiency Syndrome (AIDS)

HIV and AIDS are a global catastrophe. HIV (Human Immunodeficiency Virus) is the virus that causes HIV infection. The abbreviation "HIV" can refer to the virus or to HIV infection. AIDS stands for Acquired Immunodeficiency Syndrome and is the most advanced stage of HIV infection. The human immunodeficiency virus (HIV) was identified in 1983 as the biological agent responsible for the acquired immunodeficiency syndrome (AIDS). HIV attacks and destroys the infection-fighting CD4 cells of the immune system. The loss of CD4 cells makes it difficult for the body to fight infections and certain cancers. Without treatment, HIV can gradually destroy the immune system and advance to AIDS.

HIV is a slow-acting virus that may take years to produce illness in a person. During this period, an HIV-infected person's defense system is impaired, and other viruses, bacteria and parasites take advantage of this "opportunity" to further weaken the body and cause various illnesses, such as pneumonia, tuberculosis and oral thrush. This is why the infections and cancers seen in HIV-infected individuals are called "opportunistic". When a person starts having opportunistic infections, he/she has AIDS. The amount of time it takes from HIV infection to become full-blown AIDS depends on the general health and nutritional status before and during the time of HIV infection. The average time for an adult is approximately ten years.

Global HIV/AIDS Statistics

- In 2017, there were 19.6 million people living with HIV (53 per cent) in eastern and southern Africa, 6.1 million (16 per cent) in western and central Africa, 5.2 million (14 per cent) in Asia and the Pacific, and 2.2 million (6 per cent) in Western and Central Europe and North America.
- 36.9 million [31.1 million–43.9 million] people globally were living with HIV in 2017.
- 35.1 million [29.6 million–41.7 million] adults
- 17.8 million [15.4 million–20.3 million] women (15+ years)
- 1.8 million [1.3 million–2.4 million] children (<15 years)
- 21.7 million [19.1million-22.6 million] people living with HIV (59 per cent) were accessing antiretroviral therapy globally in June 2017, an increase of 2.3 million since 2016 and up from 8 million in 2010.
- 1.8 million [1.4 million–2.4 million] people became newly infected with HIV in 2017.
- 940 000 [670 000–1.3 million] people died from AIDS-related illnesses in 2017.
- 77.3 million [59.9 million–100 million] people have become infected with HIV since the start of the epidemic.
- 35.4 million [25.0 million–49.9 million] people have died from AIDS-related illnesses since the start of the epidemic.
- Worldwide 1.8 million [1.4 million–2.4 million] people became newly infected with HIV in 2017 compared to 3.4 million [2.6 million-4.4 million] in 1996. New HIV infections have been reduced by 47 per cent since the peak in 1996.
- AIDS-related deaths have fallen by 51 per cent since the peak in 2004.

- ✰ Tuberculosis remains the leading cause of death among people living with HIV, accounting for around one in three AIDS-related deaths.
- ✰ In 2016, there were an estimated 10.4 million cases of tuberculosis disease globally, including 1.2 million [11 per cent] among people living with HIV.
- ✰ Tuberculosis-related deaths among people living with HIV fell by 33 per cent between 2005 and 2015.

Source: UNAIDS, 2017

How HIV is Transmitted

The spread of HIV from person to person is called HIV transmission. The spread of HIV from a woman with HIV to her child during pregnancy, childbirth, or breastfeeding is called mother-to-child transmission of HIV. HIV is transmitted through three primary routes:

- Having unprotected sex with a person already carrying the HIV virus;
- Transfusions of contaminated blood and its by-products; or use of non-sterilized instruments such as sharing non-sterilized needles, razors and other instruments for surgical procedures;
- From infected mother to her child (mother-to-child transmission, MTCT) during pregnancy, childbirth or breastfeeding.

One can't get HIV by shaking hands or hugging a person who has HIV. One also can't get HIV from contact with objects such as dishes, toilet seats, or doorknobs used by a person with HIV. HIV does not spread through the air or through mosquito, tick, or other insect bites.

In most of the developing world, HIV is mostly transmitted through sexual contact with an infected person. Women are more likely to be infected with HIV than men. **HIV transmission is possible at any stage of HIV infection even if a person with HIV has no symptoms of HIV.** Many people do not know that they are infected with the virus. They may appear healthy, but still be capable of transmitting the virus through unprotected sexual intercourse or by reusing contaminated needles, razors or other devices, or in the case of a pregnant or lactating mother, during labor and delivery or through breastfeeding her child.

Symptoms

Most of the recently infected persons with HIV will not have any symptoms. However, within 2 to 4 weeks after a person becomes infected with HIV, they may have flu-like symptoms, such as fever, chills, or rash, headache, tiredness, and enlarged lymph nodes. The symptoms may last for a few weeks after they become infected and then usually disappear. Hence, these symptoms are often mistaken for those of common viral infection. During this period, HIV is present in large quantities in body fluids like blood, semen, vaginal fluids *etc.* and people are very infectious to transmit the disease.

After this earliest stage of HIV infection, HIV continues to multiply but at very low levels. More severe symptoms of HIV infection, such as signs (fatigue, rapid weight loss, frequent fevers *etc.*) of opportunistic infections (TB, Cancer *etc.*), generally don't appear for many years. Without treatment with HIV medicines, HIV infection usually advances to AIDS in 10 years or longer, though it may take less time for some people.

Treatment

There is no cure for HIV / AIDS. Some antiretroviral (ARV) drugs have been developed recently. People on antiretroviral therapy (ART) take a combination of HIV medicines (called an HIV regimen) every day which prevents HIV from multiplying and reduces the amount of HIV in the body. Having less HIV in the body protects the immune system and prevents HIV infection from advancing to AIDS. ART can not cure HIV, but it can help people with HIV to live longer, healthier lives. ART also reduces the risk of HIV transmission. However, ART is very expensive and may not be within the reach of many patients.

Nutritional Challenges of HIV/AIDS

People living with HIV / AIDS face increased challenges in maintaining proper nutrition. Despite developments in medical treatment, nutrition remains a key component in managing this condition. The challenges that those living with HIV / AIDS face can be the result of the viral infection itself or from the effects of anti-HIV therapy (ART).

Infection with HIV causes reduction in the body's natural defence system against diseases and infections leading to the AIDS. As the immune system weakens, the body becomes less able to fight infections. Other germs take advantage of this opportunity and cause various illnesses such as pneumonia, tuberculosis, oral thrush and other opportunistic infections further weakening the body. When a person starts developing these opportunistic infections he/she has AIDS. The amount of time it takes for the person to advance from HIV infection to AIDS and his / her overall quality of life depends on the general health and nutritional status of the person. Good nutrition is important to everyone. A person who is well-nourished may have a better quality of life. He / she is stronger and better able to fight infections. This is true for all people but is especially important for people with HIV who are more susceptible to recurring infections. On the other hand, people living with HIV are at a higher risk of malnutrition as HIV increases nutritional needs even during early stages of HIV infection, when no symptoms are apparent. The demand increases significantly during the course of the infection posing additional challenges to people living with HIV and their care providers. People living with HIV and their families need care and support to enable them to be better nourished.

Relationship between Nutrition and HIV/AIDS

Nutrition and HIV are strongly related to each other:

- Any immune impairment as a result of HIV/AIDS leads to malnutrition, and
- Malnutrition leads to immune impairment, worsens the effect of HIV and contributes to more rapid progression to AIDS.

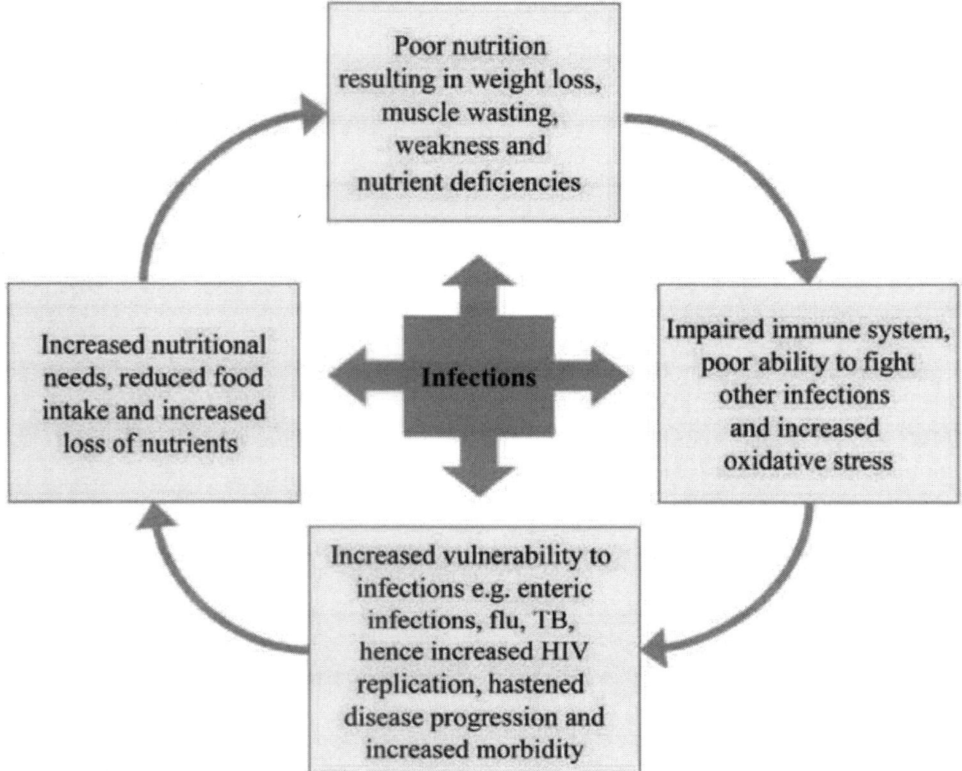

Thus, malnutrition can both contribute to and result from the progression of HIV. A person who is malnourished and then acquires HIV is more likely to progress faster to AIDS, because his/her body is already weak and cannot fight infection. A well-nourished person has a stronger body for coping with HIV and fighting illness. While people with HIV and AIDS have special nutritional needs, it is important to note that all people will benefit from adequate nutrition. Good nutrition increases resistance to infection and disease by improving his nutritional status.

Impact of HIV/AIDS on Nutritional Status

Many of the conditions associated with HIV/AIDS affect food intake, digestion and absorption, while others influence the functions of the body. Many of the symptoms of these conditions (*e.g.* diarrhoea, weight loss, sore mouth and throat, nausea or vomiting) are manageable with appropriate nutrition. Good nutrition will complement and reinforce the effect of any medication taken.

The three main ways in which HIV can affect the nutritional status of a person are by:

- Increasing nutritional requirements
- Lowering food absorption
- Reducing food intake

Therefore, an HIV-infected person is more at risk for malnutrition due to the combination of the following factors which lead to poor nutrition in people with HIV/AIDS:

i) Increased Nutritional Needs

When a person has an infectious illness, his body's immune response to the virus uses up more energy and nutrients than normal. When opportunistic infections are present, the body needs even more nutrients. People with HIV/AIDS often need to make up for protein losses, which may result from malabsorption due to diarrhoea. Protein loss leads to muscle-tissue breakdown. Concerns about health can also lead to high stress levels, which affect the immune system negatively. The infected person needs higher amounts of certain nutrients during stressful periods to keep the immune system strong.

ii) Poor Food Intake

There are several reasons for infected persons to reduce their food intake. Adults with HIV/AIDS suffer from anorexia and may have difficulty in eating or swallowing, may have altered taste, nausea and vomiting. Physical symptoms such as mouth and throat soreness can interfere with eating. Repeated infections and fever often result in poor appetite. Medications used to treat an illness may have side effects such as nausea, vomiting, metallic taste in the mouth, diarrhea and abdominal cramps which may cause reduced appetite. Depression and anxiety from dealing with a fatal disease and possible social stigma can also cause people to lose their appetite and reduce their food intake. Fatigue can make food preparation, and even eating, difficult, particularly in the late stage of AIDS. Treatment costs and reduced work output due to illness may leave the person with less money to spend on food. Overall, in case of reduced food intake, nutrient intake shall also be less than the recommended allowances, ultimately leading to various nutritional deficiency diseases in the long run which shall further worsen the health condition.

iii) Poor Nutrient Absorption

HIV/AIDS affects how the body uses the foods that are consumed, resulting in poor absorption of various nutrients. Poor absorption of nutrients accompanies diarrhea, which is common with HIV infection. Due to diarrhea, the food passes too quickly through the gut, reducing the time for digestion and absorption. Parasites like giardia and other infections caused by bacteria and viruses in HIV/AIDS can also cause diarrhea and reduce absorption especially if the environmental conditions of the HIV infected person are not clean and is not following the food safety and hygienic norms. HIV may also damage intestinal cells that affect the absorption of fats and carbohydrates. Poor fat absorption also affects absorption of micronutrients like Vitamins A and E, which are important for the proper functioning of the immune system. Poor absorption can cause difficulty in controlling blood sugar levels, which may lead to diabetes.

iv) Disruption of Nutrient Metabolism

If food intake and absorption are not sufficient for meeting increased energy needs, muscles are broken down ('muscle wasting'). In turn, excessive muscle breakdown leads to weight loss, swelling of the feet and/or other parts of the body, reduced production of saliva and other digestive fluids needed to break down foods into nutrients, which further reduces the absorption of food. With poor nutrient absorption, individuals may not be able to utilize the nutrients properly, particularly fats, carbohydrates and proteins.

v) Muscle wasting or loss in lean body tissue: Food intake is already less and what so ever nutrients are taken through food, they are not properly absorbed and utilized in HIV/AIDS affected persons. It shall lead to loss in lean body tissue, weight loss and increase damage to immune system which shall result in causation of associated chronic infection and illnesses.

vi) Chronic Infections and Illnesses having Nutritional Impacts

The vicious cycle sets in with the poor food intake in HIV. The poor intake of food leads to loss of weight, body weakness, and malnutrition, which further weakens the body's natural defense mechanism, thus aggravating this 'bad cycle' In the early stages of infection, a person shows no visible signs of illness. But later many of the signs of AIDS will become apparent, including weight loss, fever, diarrhoea and other opportunistic infections such as, sore throat and tuberculosis (TB). Poor nutrition may accelerate the onset of these infections. Fevers and infections that accompany an HIV infection lead to greater nutrient requirements and poor use of the nutrients by the body. Furthermore, people who are chronically ill may also have a reduced appetite, which leads to reduced food intake and weight loss. The results of these factors include weight loss, loss in lean muscle tissue and increased damage to the immune system. These factors are most common for adults but they are also prevalent in children infected with HIV. A number of other symptoms and illnesses commonly caused by HIV infection have nutritional consequences that can lead to malnutrition which are explained as under:

a) **Anorexia** or loss of appetite, occurs during many different infections and when fever is present. It leads to general weight loss, and is common when individuals are depressed or are living in socially and emotionally unfavorable environments.

b) **Diarrhea** occurs when a person has several watery or loose bowel movements in a day. There are several causes for diarrhea: unclean drinking water, infections, parasites or even some medical treatments. It results in losses of water, nutrients and minerals and leaves a person at greater risk of dehydration. Diarrhea also reduces appetite and leads to poor nutrient absorption. If diarrhea continues for a prolonged period of time, severe malnutrition results.

c) **Fever** means that the body feels warmer than normal. People with fever may have the chills, sweat more, have muscle and joint aches or be fatigued. Fever is common in people living with HIV/AIDS, and does not necessarily indicate a serious illness. There are many reasons for fever, and it is often hard to determine whether fever is due to HIV or to other illnesses such as malaria. From a nutritional point of view, fever may result in increased nutrient requirements, since the body utilizes nutrients poorly when fever is present.

d) **Nausea/Frequent Vomiting** can result from the drugs used to treat HIV/AIDS or from the opportunistic infections. Nausea also leads to reduced appetite and poor utilization of the food consumed.

e) **Thrush** is a fungal (Candida yeast) infection common in HIV-infected people who have damaged immune systems. Thrush refers to whitish spots on the inside of the mouth, tongue, vagina or anus. Although these sores are uncomfortable, they are not life threatening. The danger is that these sores can result in difficulty eating foods, loss of appetite and reduced food intake and thus, lead to weight loss.

f) **Anemia** results from an inadequate number or quality of red blood cells in the body. This is most often due to a lack of iron in the diet and to infections such as malaria and hookworm, which destroy red blood cells and may lead to bleeding and/or loss of blood. Women and children are especially prone to anemia. Persons who are anemic often experience tiredness and weakness and may have pallor in the eyes, tongue, palms and nail-beds. The relationship between HIV and anemia is not clearly understood, although it is known that HIV-infected persons who are anemic generally progress faster to AIDS than those who are not.

Other nutritional impacts of HIV: In households with HIV-infected members, others may also experience nutritional impacts, particularly young children and orphans. Sometimes when adults become infected with the virus, they are too sick to work and cannot continue to provide for themselves and their families. This can result in reduced income for the household and hence, fewer resources (labor and money) to obtain food. In addition, other family members suddenly find themselves working more, either to produce income or to provide care for the HIV-infected individual, which can also lead to deterioration in nutritional status for themselves and their children.

Which Nutrients are Deficient among HIV Infected Patients?

It is common for HIV-positive people to have multiple nutrient deficiencies, even early on in the course of infection. Researchers have reported that in both children and adults, deficiencies of zinc, selenium, copper, B_6, and B_{12} (all of which are important for an intact immune response) are common in HIV-positive people, even before their immune system becomes damaged.

In people with AIDS, nearly every specific nutrient is deficient. Researchers have also reported that these deficiencies appear to speed disease progression and that replenishing these nutrients (including B_6, B_{12}, and zinc) can actually help boost CD4 (T4 cell) counts.

Many other researchers have reported that deficiencies of glutathione and other important antioxidants (including vitamins C and E and the mineral selenium) are common. This is most likely because one of the body's responses to infection is the creation of unstable molecules, usually referred to as "free radicals," which are intended to destroy viruses and other disease-causing germs. These free radicals serve an important immune purpose but after their mission is carried out, they need to be countered by antioxidant nutrients in order to stop a chain reaction that could otherwise damage the body. So-called "oxidative stress" results when insufficient antioxidants are present to counter oxidative damage to cells and tissues in the body. Researchers have shown that oxidative stress is very common in both HIV disease and hepatitis C, and is a factor in progression of both diseases. Keeping optimal levels of antioxidants in the body is crucial to lessen oxidative stress and prevent body damage.

The amount of this oxidative damage increases early in HIV disease, and tends to worsen over time. For example, studies have shown that decreased levels of glutathione, the most important antioxidant found in cells, occur within weeks of HIV infection. The lowered levels of glutathione lead to immune cell dysfunction in multiple ways, and allow body cells and tissues to become damaged from the oxidative stress. In fact, researchers have shown that lowered levels of glutathione are strongly tied to an increased risk for disease progression. Insufficient glutathione also means that the liver is less able to properly break down drugs and other toxins, increasing the potential for liver damage from medicines. Thus, boosting glutathione levels is important for anyone living with HIV.

HIV-infected patients may be at nutritional risk at any point in their illness. Severe malnutrition and weight loss, particularly loss of lean tissue, and delayed weight gain and height velocity in children, can affect morbidity and mortality. Fear of developing fat-redistribution syndrome, with central obesity and loss of subcutaneous fat, may prevent patients from beginning or continuing potent antiretroviral therapies. Development of hyperglycemia and lipid abnormalities may increase the risk of diabetes, heart disease, and stroke. The appearance of these metabolic changes may induce patients to cease HAART or to change to less effective antiretroviral regimens.

Impact of Good Nutrition for People Living with HIV/AIDS

Good nutrition cannot cure AIDS or prevent HIV infection, but it can help to maintain and improve the nutritional status of a person with HIV/AIDS and delay the progression from HIV to AIDS-related diseases. It can therefore, improve the quality of life of people living with HIV/AIDS. Nutritional care and support are important from the early stages of the infection to prevent the development of nutritional deficiencies. A healthy and balanced diet will help to maintain body weight and fitness. Eating well helps to maintain and improve the performance of the immune system, the body's protection against infection and therefore, helps a person to stay healthy.

Adequate nutrition, which is best achieved through consumption of a balanced healthy diet, is vital for health and survival for all individuals regardless of HIV status. If people living with HIV are well-nourished, they are likely to have improved quality of life because a stronger body can fight infection and cope with HIV. A good diet is vital to keep the body working. Due to better nutritional care and support, HIV infected persons shall be able to work and contribute to the family's income, have prolonged good health, remain active and able to care for themselves and help with the care of children and other dependants, have reduced illnesses and recover more quickly from infections, thereby reducing costs for health care. They shall maintain a good appetite and stable weight. If children with HIV are well-nourished, they can go to school regularly which shall result in better education and development; they will have more energy to play and have fun. Therefore, the provision of good nutrition care and support to people living with HIV and their families is very essential to promote their well-being, self-esteem and a positive attitude towards life.

Positive effects of adequate nutrition for People living with HIV and AIDS (PLWHA)

- Prevents malnutrition and wasting
- Achieves and maintains optimal body weight and strength
- Enhances the body's ability to fight opportunistic infections
- May help delay the progression of HIV
- Improves the effectiveness of drug treatments
- Improves the quality of life

How to Maintain Good Nutrition

Nutritional management is integral to the care of all patients infected with human immunodeficiency virus (HIV). HIV infection results in complicated nutritional issues for patients, and there is growing evidence that nutritional interventions influence health outcomes in HIV-infected patients.

Good nutrition for all individuals, but especially people living with HIV/AIDs (PLWHAs), require the consumption of an adequate amount of all the essential nutrients *viz.* macronutrients (Carbohydrates, Proteins, and Fats) and micronutrients (Vitamins and Minerals) through energy giving, body building and protective foods.

I. Macronutrients

A deficiency in macronutrients, also known as "protein-energy malnutrition," manifests itself in the weight loss and wasting that is typical of AIDS patients. These symptoms occur as a result of reduced food intake; poor absorption of nutrients; and changes in metabolism that affect cell growth, enzymatic processes and immune system reactions. Because of these effects, an HIV-infected person has additional nutrient requirements.

i) Energy

Weight loss in adults and growth failure in children are common in HIV/AIDS infected children and adults. Regardless of the individual variants, the total daily food intake is somewhat increased by this chronic viral infection since the energy demanded by an ongoing immune response to the virus is constantly burning up calories. In addition, any other infections including not only opportunistic infections, but any others which may develop will further increase this need for additional calories. Resting energy expenditure is increased by around 10 per cent in asymptomatic HIV infected adults and children. An additional 20–50 per cent increase in energy needs occurs during the convalescent catch-up period after a severe infection in both adults and children. Therefore, there are high nutrition costs associated with HIV infection and the opportunistic infections associated with HIV as energy needs increase as given under (WHO, 2003; USAID, 2007):

Adults

- Increased energy intake by 10 per cent to maintain body weight in asymptomatic HIV-infected adults.
- Increased energy intake by 20 per cent to 30 per cent for adults during periods of symptomatic disease or opportunistic infection.

  NOTE: This increase may not be feasible during periods of acute infections or illness and might not be necessary due to reduced physical activity.
- Increased energy intake by 30 per cent during the recovery period.

  NOTE: Efforts should be made to aim at the maximum achievable to compensate for losses during periods of infections and illness.
- When possible, people living with HIV/AIDS should meet their additional energy needs by increasing consumption of foods with high nutrient densities, rather than increasing consumption of high-energy foods that are low in protein and micronutrients (such as high fat and high sugar foods).

Children

Energy requirements in children can vary according to the type and duration of HIV-related infections and whether there is weight loss along with acute infection.

- ☆ During the asymptomatic phase, energy requirements increase by 10 per cent.
- ☆ During the symptomatic phase with no weight loss, energy requirements increase by 20 to 30 per cent.
- ☆ During the symptomatic phase with weight loss, energy requirements increase by 50 to 100 per cent.
- ☆ Practically, it is often difficult for children experiencing opportunistic infections and weight loss to consume 50 to 100 per cent more energy than normal levels. Therefore, it is also important to encourage children to consume additional food following bouts of illness and weight loss (WHO, 2003; USAID, 2007).

Pregnant and Lactating Women

At present, there are no specific data on the impact of HIV/AIDS and related conditions on energy needs during pregnancy and lactation over and above those requirements already identified for non-infected women. Pregnant and lactating women living with HIV/AIDS should follow the same recommendations given above for adults and adolescents living with HIV/AIDS. In addition to the extra energy requirements due to HIV infection, it is important to keep in mind that pregnant and lactating women need to consume extra energy, protein and micronutrients required by pregnancy or lactation.

ii) Protein

There is insufficient data at present to support an increase in protein intake for PLWHA above normal requirements for health *i.e.* 12 per cent to 15 per cent of total energy intake. Based on current evidence, protein requirements for all groups *i.e.* adults, adolescents, children, pregnant and lactating women – of people living with HIV/AIDS are the same as for healthy non-HIV-infected individuals of the same age, sex, and physical activity level (WHO, 2003; USAID, 2007). Some of the studies were published which had suggested increased protein intake during HIV infection, but they concluded that these recommendations were not based on rigorously conducted studies.

iii) Fat

There is no evidence that total fat needs are increased beyond normal requirements as a consequence of HIV infection in various age groups. However, special advice regarding fat intake might be required for individuals undergoing antiretroviral therapy or experiencing persistent diarrhea (WHO, 2003).

II. Micronutrients

Micronutrient is a critical component of health for people infected and affected by HIV/AIDS. The impact of taking micronutrient supplements on HIV infection is not well known. The role of micronutrients in immune function and infectious disease is well established. However, the specific role of individual and multiple micronutrients in the prevention, care and treatment of HIV infection and related conditions merits further attention. Several studies on micronutrients and HIV are under way. One review had confirmed the potential role of micronutrients in HIV infection and vice versa. Observational studies indicate that low blood levels and decreased dietary intakes of some micronutrients are associated with faster HIV disease progression and mortality, and increased risk of HIV transmission. The most promising results are for Vitamin A and the B-group vitamins. Vitamin B_6 and B_{12} supplementation have been shown to improve survival and reduce disease progression, although only in healthy populations where micronutrient deficiencies are not common and when intakes were high. Vitamin B_6 (10 mg daily) supplementation is also recommended for TB patients being treated with isoniazid. However, these studies' methodological limitations preclude definitive conclusions about the relationship between micronutrient intake and blood levels, and HIV infection.

Some studies show that there is evidence that supplements of, for example, B-complex vitamins, and vitamins C and E, can improve immune status, prevent childhood diarrhoea and enhance pregnancy outcomes, including better maternal prenatal weight gain and a reduction of fetal death, preterm birth and low birth weight. The effect of these micronutrients on HIV disease progression and mortality is under study.

Micronutrients

i) To ensure micronutrient intakes at RDA levels, HIV-infected adults and children are encouraged to consume healthy diets.

ii) Nevertheless, dietary intake of micronutrients at RDA levels may not be sufficient to correct nutritional deficiencies in HIV-infected individuals.

iii) There is evidence that some micronutrient supplements, *e.g.* vitamin A, zinc and iron, can produce adverse outcomes in HIV-infected populations.

WHO (2003)

But current evidence is inconclusive about the effects of micronutrient supplementation on transmission and progression of HIV infection. Interventions to increase the intake of micronutrients may reduce or even increase the risks of transmission or progression of HIV infection, and it is still not possible to give evidence-based recommendations. Dr Henrik Friis, Professor of International Nutrition and Health, University of Copenhagen, Denmark, discussed the role of nutrition interventions in HIV. He said that infections impair nutritional status and, in turn, malnutrition may impair immune functions. Since HIV infection per se impairs status with respect to nutrients of importance to growth and development, pregnancy outcomes, physical activity and working capacity, and resistance to infections, nutrition interventions are likely to improve all these general outcomes. In addition, there is biological plausibility and some evidence to suggest that nutrition interventions may reduce risk of HIV transmission, as well as progression of HIV infection, and risk of opportunistic infections. Micronutrients that have produced positive health outcomes in HIV-uninfected populations include zinc supplementation for reducing diarrhoea and pneumonia morbidity in children. The safety and effectiveness of zinc supplements in HIV-infected adults and children are now being studied.

Efforts to maintain adequate recommended intake of all essential vitamins and minerals must remain a major emphasis of the public health programmes irrespective of HIV status and particularly in areas where both malnutrition and HIV infection are endemic.

Adults

HIV-infected adults and children should consume diets that ensure micronutrient intakes at RDA levels. However, this may not be sufficient to correct nutritional deficiencies in HIV infected individuals.

Results from several studies raise concerns that some micronutrient supplements, *e.g.* vitamin A, zinc and iron, can produce adverse outcomes in HIV-infected populations. Safe upper limits for daily micronutrient intakes for PLWHA still need to be established.

Children

Periodic vitamin A supplementation has been shown to reduce all-cause mortality and diarrhoea morbidity in vitamin A deficient children, including HIV-infected children. Vitamin A supplementation for children of HIV-infected mothers has been shown to reduce illness, particularly diarrhea. Vitamin A supplements are also recommended for mothers immediately after delivery and for infants 6 months and older.

In keeping with WHO recommendations, 6 to 59-month-old children born to HIV-infected mothers living in resource-limited settings should receive vitamin A supplements every six months (100,000 IU for infants 6 to 12 months and 200,000 IU for children >12 months). There is insufficient evidence at present to recommend an increased dose or frequency of vitamin A in HIV-infected children.

No data are available on the efficacy of other micronutrient supplements for HIV-infected children.

Pregnant and Lactating Women

i) Iron-Folate Supplementation

It is a standard component of antenatal care for preventing anaemia and improving fetal iron stores. WHO recommends daily iron-folate supplementation (400 µg of folate and 60 mg of iron) during six months of pregnancy to prevent anaemia, and twice-daily supplements to treat severe anaemia.

As with other chronic infections, HIV causes disturbances of iron metabolism and anaemia. In view of iron's potential adverse effects, for example due to its pro-oxidant activity, which might accelerate disease progression, research on the safety of iron supplementation in adults and children with HIV infection is recommended. Based on available evidence, however, the approach to caring for HIV-infected women is the same as that for uninfected women.

ii) Vitamin A

Daily vitamin A intake by HIV-infected women during pregnancy and lactation should not exceed the RDA. According to published reports, daily antenatal and postnatal vitamin A supplementation for HIV-infected women in well-designed randomized controlled trials not only did not reduce mother-to-child HIV transmission; in some settings it actually increased the risk. Thus, daily vitamin A intake by HIV-infected women during pregnancy and lactation should not exceed the RDA (WHO, 2003).

In keeping with WHO recommendations for areas of endemic vitamin A deficiency, HIV-infected women should receive a single high-dose of vitamin A (200,000 IU) as soon as possible after delivery, but no later than six weeks after delivery.

WHO Guidelines for High Dose Vitamin A Supplementation (in International Units (IU))

Time	Mother	Infant
At delivery	200,000 IU at delivery or during the safe infertile postpartum period*,	**
Post-Partum period	Low dose does not exceeding 10,000 IU per day or 25,000 IU per week at any time postpartum	
0-5 months		Three doses of 50,000 IU with one month interval. Doses should be given at any health or immunization contact, in particular at each DTP (6,10 and 14 weeks)

Time	Mother	Infant
6-11 months		A single dose of 100,000 IU given at any health or immunization contact, in particular at measles vaccination (9 months)
12-59 months		200,000 IU every six months at any health or immunization contact

* The safe infertile period for breastfeeding mothers to receive a high-dose Vitamin A is within 8 weeks of delivery; for non-breastfeeding women within 6 weeks of delivery.

** WHO recommends that infants born to HIV infected women should receive 50,000 IU dose of Vitamin A at birth, in addition to the three subsequent 50,000 IU doses of Vitamin A administered over the first six months of life.

Source: USAID (2001). HIV/AIDS: A Guide for Nutrition, Care and Support.

Research is currently under way to further assess the effect of single dose, postpartum vitamin A supplementation on HIV-infected women.

iv) Multiple Micronutrient Supplements

Micronutrient deficiencies are common in resource-limited settings where HIV infection is prevalent. Some studies show that different multiple micronutrient supplements may have produced a broad range of beneficial outcomes.

Adults

For HIV-infected adults, the WHO recommended micronutrient intake comes from a good diet at RDA levels; higher intake of vitamin A, zinc, and iron can produce adverse effects in HIV positive adults, and these were not recommended unless there is documented deficiency. A generally healthy diet is to be promoted.

Pregnancy and Lactation

During pregnancy, daily multivitamin supplementation with multiple RDA levels of B-complex vitamins, and vitamins C and E, improved birth outcomes in infants born to HIV-infected women, and increased maternal weight gain during pregnancy, haemoglobin concentration and CD4 cell counts. Daily use of this multivitamin supplement during lactation reduced postnatal HIV transmission and mortality in infants born to nutritionally vulnerable women and to women with immune deficiency. The supplements also reduced the risk of diarrhoea and improved infants' immune status.

Another micronutrient supplement formulation, with single RDA nutrient levels, improved birth weights among infants born to HIV-infected women. The impact of single RDA multivitamin supplements on HIV disease progression and transmission, which was not assessed, requires further study.

Some studies have looked into the use of implementing daily multivitamins into the diet regimens of HIV/AIDS patients. One study done in Tanzania involved a trial group with one thousand HIV positive pregnant women. Findings showed that daily multivitamins benefited both the mothers and their babies. After four years, the multivitamins were found to reduce the women's risk of AIDS and death by approximately 30 per cent. Another trial in Thailand revealed that the use of multivitamins led to fewer deaths, but only among people in advanced stages of HIV. However, not all studies have provided a positive correlation. A small trial done in Zambia found no benefits from multivitamins after one month of use.

According to WHO (2003) recommendations, adequate micronutrient intake is best achieved through an adequate diet. However, in settings where these intakes and status cannot be achieved, multiple micronutrient supplements may be needed in pregnancy and lactation. Pending additional information, micronutrient intakes at the RDA level are recommended for HIV-infected women during pregnancy and lactation.

The optimal micronutrient supplement composition that will be safe, ensure nutritional adequacy, and potentially produce the greatest benefits in HIV-infected pregnant and lactating women in different settings has not yet been defined. Additional research is required to determine the safety of nutrient supplements such as zinc, iron and vitamin A, and to determine whether different multiple micronutrient supplements are needed for HIV-infected women compared with uninfected women.

Limitations in Research Studies Regarding Role of Nutrition Interventions in HIV Infection

The term "nutrition intervention" includes both food-based approaches and micronutrient supplementation.

Examples of limitations in study design that have presented obstacles to understanding the role of micronutrients in HIV infection included lack of baseline assessments of the nutritional status in intervention trials; lack of attention to confounding conditions, including active infections and the consequent acute phase response affecting measurement and interpretation of biochemical assessments; lack of appropriate control groups (*e.g.* HIV-negative samples in areas of endemic malnutrition); and reliance on non-specific outcomes (*e.g.* weight gain, global measures of immune function). The combination of the latter two issues limits the ability to generalize research results or to ascribe findings to an HIV-specific effect rather than a generalized amelioration of a nutritional deficiency. Consequently, it becomes difficult to advocate for an equitable intervention in settings where HIV and chronic widespread malnutrition coexist. With regard to the impact of micronutrients on HIV transmission, results from one randomized clinical trial indicated that supplementation with preformed vitamin A and β-carotene increased mother-to-child HIV transmission, but two other vitamin A trials found no evidence of adverse effects. Because vitamin A is an essential nutrient and vitamin A deficiency is widespread in resource-limited settings where HIV is most prevalent, there is a need to evaluate the safety of vitamin A supplementation for HIV infected adults and children. This information is essential to ensure that the global commitment to combat vitamin A deficiency may be continued without the risk of increasing mother-to-child HIV transmission. Few studies showed that a daily supplementation may reduce HIV disease progression and mortality among adults. However, studies had methodological flaws and were subject to different interpretations. Furthermore, the protocol was different between the studies.

Paramount among the concerns about this trial is that the regimen used consisted of an array of vitamins given at doses well beyond standard recommendations for pregnant women. Baseline data on the status of these women for many of the nutrients evaluated were lacking. The lack of an HIV-negative control group did not allow for the determination of an HIV specific effect because the outcomes–CD4+ and CD8+ counts, morbidity and mortality are all similarly affected by micronutrients status in HIV-uninfected populations. New knowledge is urgently needed to provide the scientific evidence base required for making nutrition recommendations for rapid implementation.

Nutrition and Antiretroviral Therapy (ART)

ART is an essential component of care for PLWHA; it is essential to save lives and efforts to achieve universal access must continue. Improved attention to diet and nutrition may enhance ART acceptability, adherence and effectiveness. Additional research has reinforced several points including:

- Some types of ART are associated with increased energy requirements.
- Body Mass Index (BMI) at time of entry for treatment is an independent and significant predictor of mortality in PLWHA receiving ART.
- Metabolic complications of ARVs are seen in adults and children and have nutrition implications. These complications include "ART Metabolic Syndrome": derangements in lipid storage resulting in changes in body composition and dyslipidemias, (indicative of high CVD risk), insulin resistance/impaired glucose tolerance (possible role for dietary management); Bone related problems: Derangements include osteopenia and osteoporosis (possible role for Vitamin D); Lactic acidemia; generally associated with the class of ARV (NRTI- Nucleoside Reverse Transcriptase Inhibitor (*NRTI*), a *class of antiretroviral drugs*) (possible role of specific micronutrients in either etiology and/or treatment).

Nutrition and Antiretroviral Therapy

- Nutritional interventions should be an integral part of all HIV treatment programmes.
- Improved attention to diet and nutrition may enhance ART acceptability, adherence and effectiveness.
- National health authorities should prepare for ART access by training relevant personnel on counseling and managing ART's long-term nutritional aspects. More research is needed in this regard in resource-limited settings.

WHO (2003)

With specific reference to conditions in the South-East Asia Region, it was noted that approximately 20-25 per cent of urban HIV-uninfected South Asians have evidence of the "metabolic syndrome" (combination of obesity, hypertension, dyslipidemias, insulin insensitivity) indicating a potential predisposition to these complications of HIV and its treatment.

One of the reported study has concluded by noting that while many knowledge gaps remain as challenges to research and care, the core conclusions from the initial review remain the same *i.e.*:

- Highly Active Antiretroviral therapy (HAART) saves lives, but its use may cause metabolic changes, including insulin resistance, dyslipidemia and fat redistribution. These conditions may place patients at risk for other diseases.
- Switching drugs is one response to addressing these concerns. However, the role of dietary and/or nutritional management should also be considered.
- Knowledge of these effects is critical in the implementation of appropriate nutritional policies in conjunction with HAART use.

Assessment of Nutritional Status of HIV Patients

Nutritional status can be assessed time to time in many ways by some general methods, including:

- Weight and other measurements of body fat and muscle mass
- Hemoglobin or hematocrit counts, which measure iron in the blood
- Other blood tests to check levels of important fats (cholesterol and triglycerides), proteins (such as

albumin), vitamins (B_{12}, vitamin D), and minerals (sodium, potassium)

- Hemoglobin A1c levels to diagnose pre-diabetes or to monitor diabetes

Diet Plan for HIV Infected Person

No single diet can be recommended for HIV infected person as nutritional recommendations shall vary from individual to individual, his body weight, muscle mass, his activity level, degree of infection, presence or absence of oppourtunistic infections, his income level, type of medications, his disease symptoms, his nutritional status *etc.* However what so ever is the condition, a healthy diet is a key part of any HIV treatment plan. A diet is simply any food and drink that he or she has to consume regularly to maintain his body weight and muscle mass, keep energy levels high, minimize the negative impacts of HIV drugs, maintain good nutritional status and capacity to build up immune system to fight infections. Diet to be given to asymptomatic HIV patient, persons with HIV and TB, HIV and diabetes, HIV and diarrhea *etc.* shall vary depending upon the associated illness.

For that matter, dietary considerations and management in general for HIV/AIDs patients have been mentioned below:

Dietary Considerations and Management

An HIV infected person already has a weakened immune system. A nutritious well balanced diet can help to maintain the proper functioning of the immune system and provide needed energy, protein and micronutrients to help in maintaining weight and avoiding loss of muscle mass. There is no specific eating plan for people with HIV, but an overall healthy diet can help health a lot. Well-chosen dietary practices can help manage illnesses associated with HIV/AIDS. The following suggestions are provided as general principles in responding to the nutritional needs of HIV affected populations to maintain health:

Eat Small but Frequent Meals

Persons with HIV may experience difficulties with food intake due to illnesses such as thrush (mouth sores) or to poor nutrient absorption. To ensure that PLWHAs receive sufficient nutrients, it is important that they eat smaller portions (so that the stomach can digest foods) and more frequently throughout the day. Eat 6 to 8 small meals each day instead of 3 large meals. Encourage snacking throughout the day to increase intake of food. Eat soft, easy to digest foods with a mild flavor as they may be tolerated better during periods of illness. Sit and relax for 30 minutes after eating.

Eat a Variety of Foods

HIV-infected people have increased nutrient requirements because of the infection. Consuming a variety of foods from each of the main food groups on a daily basis is important for maintaining good nutritional status to help fight illness. In addition, good nutritional status may ensure that a person is able to survive an opportunistic infection such as tuberculosis or pneumonia without further compromising their health and nutritional status.

No single food or group of foods contains all the nutrients that the body needs in the right quantities and combinations, except breast milk for infants up to six months. Eating a variety of different foods will provide the nutrients that are essential for the body. It is important to include different food and food groups in one's diet as well as to alter methods of preparation.

PLWHAs should eat a "balanced" diet consisting of foods from each food group *i.e.* energy giving, body building and protective food groups. By taking care to choose foods that are in season and locally available, eating can be enjoyable, healthy and affordable. It might not be possible for all people to enjoy a variety of foods at each meal. However, it is important that, at least, the meals consumed throughout the day are varied.

In addition, the same food could be cooked differently. For example, instead of always frying food in fat or oil, one can try steaming, baking, sautéing, roasting and grilling.

Eat Wisely

'Eating wisely' means eating a variety of foods in the right quantity, combination and frequency to provide the body with the required nutrients on a daily basis. Eating a diet high in vegetables, fruits, whole grains, and legumes with lots of fluids is good for HIV infected person.

i) Choose Carbohydrates Wisely

Carbohydrates, the main source of energy should form main part of the meal to keep the body weight stable but should be chosen wisely. Carbohydrates come in different forms. Simple carbohydrates found in sugar (as found in sweets, soft drinks), white rice, white flour, fruits and milk are more easily digested, but can cause blood sugar to rise sharply. Complex carbohydrates found in whole grains, beans (legumes), starchy vegetables like corn and potatoes, and brown rice, often contain more fibre and other nutrients than simple carbohydrates; they take longer to digest and hence, do not cause blood sugar to rise as sharply as simple carbohydrates.

For people with HIV and diabetes, it is recommended to make staples or starchy foods having complex carbohydrates, the largest or central or main part of the meal *e.g.* certain breads, cereals porridge, rice preparations, and pastas *etc.* and the rest of the meal should be planned around this food. Such people should avoid eating refined cereals such as white bread, refined wheat flour products, refined, milled and polished rice *etc.*

ii) Eat a wide variety of vegetables and fruits on a daily basis to get vitamins, minerals, fibre and antioxidants

Vegetables and fruits also known as 'protective foods' are especially important for people living with HIV to fight infections as these foods are high in nutrients called antioxidants, which protect the immune system. Try to take more of fruits and vegetables. Include 3-5 servings of all types of vegetables and 2-4 servings of fruits in daily diet. An easy way to meet that goal is to fill half of plate with fruits and vegetables at each meal.

One should try to eat fresh seasonal vegetables and fruits that are dark-green, yellow, orange or red in colour in the diet; examples include green leafy vegetables (spinach, mustard leaves, mint, fenugreek leaves *etc.*), pumpkin, green pepper, apricot, papaya, mango, carrot *etc.* Include other vegetables and fruits rich in vitamin C to help fight infections *e.g.* cabbage, tomato, aonla, ber, guava, lemon, orange and pineapple. Vegetables lose some of their nutrients if soaked or boiled for a long time. Cook them for as short a time as possible and reuse the cooking water in soups and other foods. Some of the vegetables which can be taken in raw form as salad, should be taken daily in the diet. It is important to clean vegetables thoroughly if they are to be consumed raw.

iii) Remember Variety and Color while Selecting Food while Shopping

Each food has its particular strengths and weaknesses in terms of nutrient content so choosing from a wide variety of foods at each pyramid level will help ensure intake of all the nutrients which nature can provide, rather than the more limited number that might result from repeating the same foods over and over. And emphasizing color when one selects that wide variety of foods is in fact providing nutritional insurance. This is also called as the 'Rainbow Theory of Shopping'. Cereal products chosen should be from whole-grain varieties instead of those prepared from refined cereals. While selecting fruits and vegetables, purchase red, purple, green, orange, yellow and blue coloured ones because they are rich in antioxidants. More the shopping basket looks like a rainbow of color, the better off the diet will be.

iv) Get Plenty of Protein

Protein is essential for building healthy body, maintaining muscles and a stronger immune system. During times of infection, protein stored in muscles can get burned as a fuel source. This can lead to loss of muscle, also called muscle wasting. Therefore, it is important that HIV infected person must take at least 3 servings of good quality protein to fight muscle loss.

One can choose from a wide variety of foods that will contribute to total protein intake, including lean meat, poultry, fish, beans, eggs, nuts, seeds, milk, curd, *paneer* and cheese. Some animal sources of protein can be high in saturated fats, and should be used in moderation especially if one is having elevated cholesterol or is at risk for heart disease. It is better to choose lean, low-fat sources of protein.

Vegetarians should include legumes and oilseeds such as beans, peas, lentils, groundnuts and soybeans in daily diet. The vegetarians can take a combination of cereal, pulse and milk product such as curd/buttermilk/ paneer at one meal so as to have good quality protein which is almost as good as that from animal source. Fermented milk *i.e.* curd or yogurt is good for controlling diarrhea. Besides it, curd is easily digested, may help in digestion and absorption of other foods and is a good source of energy.

In case of HIV infected person having serious liver disease, a too-high level of protein may be dangerous. When protein is broken down in the body, one of the byproducts is ammonia. A damaged liver cannot process ammonia. The result can be an overload that results in encephalopathy, a brain condition that can result in a state of mental confusion and, in advanced stages, a coma.

v) Watch Daily Intake of Fats and Sugar and Sugary Foods

To improve energy intake in underweight person, use oils and fats as well as sugar and sugary foods because they are rich in energy. They also add flavour to food, thereby stimulating appetite. During illness, a person with a poor appetite may eat better if small amounts of sugar or fats are added. Adding extra quantities of oil/fat and sugar may help one to gain body weight. But one must know that although fats and sugars are good sources of energy, they are not rich in other nutrients. Thus, they should be eaten in addition to other foods and not in place of them.

People living with HIV who are not experiencing weight loss should be encouraged to regulate their fat and sugar intake. Because fats/oils, sugar and sugary products enhance the flavour of food and boost one's appetite, people tend to overuse them. It can then become difficult to reduce their intake. Hence, in case of excess weight gain, sugar and fat intake should be reduced gradually and taken in moderation, while closely monitoring one's weight.

Care should be taken in the case of patients diagnosed with Candida (oral thrush) as sugar can make this condition worse. In such cases, patients should avoid eating sweet foods such as sugar, honey, sweet fruits, fizzy drinks, cakes, sweets, chocolate and commercial juices. It is important to know that juices often contain large amounts of sugar. Sometimes the sugar is natural (from the fruit), and sometimes there is sugar added to the juice (*e.g.*, cane sugar, corn syrup). Either way, juices tend to have more sugar and less fiber than fresh fruit. As a result, juices can make diarrhea worse and HIV

Table 19.1: Caring for Symptoms and Illnesses Associated with HIV in Adults

Illness	*Food*	*Care and Nutrition*
		Practices
Anorexia (appetite loss)	Try to stimulate appetite by eating favorite foods.	If loss of appetite is due to illness, seek medical attention for treatment.
	Eat small amounts of food more frequently.	
	Select foods that are more energy dense.	
	Avoid strong smelling foods.	
Diarrhea	Drink lots of fluids to avoid dehydration (soups, diluted fruit juices, boiled water).	*Prevention*
	Avoid drinking strong citrus (orange, lemon) juices because it may irritate the stomach.	Drink plenty of clean boiled water.
	Consume foods rich in fiber to help in retaining fluids (millet, banana, peas and lentils).	Wash hands with soap and water before handling, preparing, serving or storing foods.
	Eat starchy foods like wheat, rice, maize, sorghum, Potato and blended cereal-legume foods-khichdi	Wash hands with soap and water after using a toilet or latrine or cleaning a child after defecation.
	For protein, eat eggs, chicken or fish or beans.	
	Drink light teas (herbal), boiled water.	*Treatment*
	Boil or steam foods, avoid fried foods.	Drink more fluids to prevent dehydration.
	Consume fermented foods like porridges, yogurt;	Prepare rehydration solutions using oral rehydration salt packets or a home- made solution of 2 cups of boiled water, 1 cup of sugar and 1 teaspoon of iodized salt (1 cup roughly equals one handful; 1 teaspoon roughly equals one pinch).
	Consume easily digestible foods high in carbohydrates, like rice, bread, millet, maize porridge, potato, biscuits.	Eat small amounts of food frequently and continue to eat following illness to recuperate weight and nutrient loss.
	Eat soft fruits and vegetables like bananas, squash, banana, mashed carrots.	Go to a health center if symptoms such as severe dehydration (low or no urine output, fainting, dizziness, shortness of breath, bloody stools, high fever, vomiting, severe abdominal pain or diarrhea) persist for more than 3 days.
	Drink nonfat milk if no problem with lactose.	
	Foods to avoid/reduce intake:	
	Some dairy products, such as milk.	
	Avoid caffeine (coffee and teas) and alcohol.	
	Reduce intake of fatty foods.	
	Limit intake of gas-forming food such as cabbage, onions, carbonated soft drinks (sodas).	
	Avoid excessively fried foods and extra oil, lard or butter.	
Fever	Eat soups that are rich in foods that give energy and nutrients	Drink fluids to prevent dehydration, particularly clean boiled water.
	Drink plenty of liquids.	Bathe in cool water.
	Drink weak tea, herbal tea *etc.*	Rest.
	Drink more than usual beyond thirst.	Continue to eat small frequent meals
		Go to the health center in case of: ☆ Fever that last several days and is not relieved with aspirin ☆ Loss of consciousness ☆ Severe body pain ☆ Yellow eyes ☆ Severe diarrhea ☆ Fits.
Nausea and Vomiting	Eat foods like soups, unsweetened porridge and fruits like bananas.	Eat small frequent meals. Nausea is worse if nothing is in the stomach.
	Eat lightly salty and dry foods like crackers to calm the stomach.	Avoid lying down immediately after eating; wait at least 20 minutes to avoid vomiting.
	Drink herbal teas and lemon juice in hot water.	Rest between meals.
	If available, drink ginger root: crush ginger in coldwater, boil in water for 10 minutes; place in a covered container; strain ginger and drink liquid	

Illness	*Food*	*Care and Nutrition*
		Treatment
	Avoid spicy and fatty foods.	
	Avoid caffeine (coffee and tea) and alcohol	
	Drink liquids, such as clean boiled water.	
Thrush	Eat soft mashed foods, such as carrot, scrambled eggs, mashed potatoes, bananas, soups, porridge.	Seek medical attention for treatment.
	Eat cold or room temperature foods.	If available, use a spoon or cup to eat small amounts of foods.
	Avoid spicy, salty or sticky foods; these may irritate mouth sores.	Tilt head back when eating to help with swallowing.
	Avoid sugary foods; these cause yeast to grow.	Rinse mouth with boiled warm salt water after eating to reduce irritation and keep infected areas clean so yeast cannot grow.
	Avoid strong citrus fruits and juices which may irritate mouth sores.	Drink liquids.
	Avoid alcohol.	
Anemia	Eat more iron-rich foods, such as animal products (eggs, fish, meat and liver) green leafy vegetables, fruits and vegetables, legumes (beans, lentils, groundnuts), nuts, oil seeds and fortified cereals.	If available, adults take one iron tablet once a day with some food. Best if taken with a source of Vitamin C such as tomatoes or orange juice to help with absorption.
	Take iron supplements.	Treat malaria and hookworm.
Muscle Wasting	Increase food intake by increasing quantity of food and frequency of consumption.	Eat small frequent meals.
	Improve quality and quantity of foods by providing a variety of foods.	Eat soft liquid food if mouth sores present.
		Increase protein in diet.
		Slowly introduce fat in the diet.
		Increase intake of starchy foods in cereals and other staples.
		Use fortified foods.
Constipation	Eat more foods that are high in fiber content	Avoid using cleansing practices, such as enemas and medications.
	Drink plenty of liquids.	Drink plenty of fluids including boiled water.
	Avoid processed or refined foods.	
Bloatedness/ Heartburn	Eat small frequent meals.	Eat long enough before sleeping so food can digest.
	Avoid gas-forming foods (cabbage, soda).	
	Drink fluids.	
Tuberculosis	Consume foods high in protein, energy, iron and vitamins.	Seek medical attention immediately.
		Consult medical personnel about taking food with medications.
		If taking isoniazid for treatment, take a Vitamin B_6 supplement to avoid deficiency of this micronutrient.
Loss of Taste and/ or Abnormal Taste	Use flour enhancers, *e.g.*, salt, spices, herbs and lemon.	
	Chew food well and move around mouth to stimulate receptors.	

Source: USAID (2001). HIV/AIDS: A Guide for Nutrition, Care and Support.

infected people who have diabetes or pre-diabetes are advised to drink juices sparingly.

Researchers have reported that around one-quarter of people in early disease stages have already developed fat malabsorption, and that by later disease stages, this problem may occur in more than half of all HIV-positive people. Thus, many people may find it necessary to keep the fat content of their diet moderately low since high fat intake can be immunosuppressive; it can cause diarrhea in some HIV-positive people, and is tied to an increased risk of progression to cirrhosis in those also living with chronic hepatitis C. If diarrhea is a problem, one may need to cut back on animal fat especially dairy, fried foods, and sugary foods. For those with taking HIV medications causing elevated blood fats, a high-fat diet could also be risky to cardiovascular health. One should avoid products containing *trans* fat.

vi) Use Salt and Food High in Salt Sparingly

People living with HIV are prone to heart disease, especially when taking ARV's that have a side-effect of increasing blood cholesterol levels. Thus, it is important to avoid processed foods and foods that are high in salt.

In HIV affected people having damaged liver and developed ascites, a complication of cirrhosis, too much salt intake can worsen the problem. It may also be important to limit iron intake, since diets higher in iron have been tied to liver damage in those with hepatitis C.

vii) Take Foods Rich in Fibre

Instead of refined foods, one should take whole foods rich in fibre. Foods high in fiber such as oats, barley, whole grains, whole pulses, fruits and vegetables *etc.* can help keep the bowel movements regular and support gut health.

If diarrhea is there, avoid above mentioned high-fiber foods. Instead, eat bland foods such as bread, rice, and soft fruits.

viii) In Case of South Sores

Avoid citrus fruits like oranges and grapefruit. Avoid very hot or cold foods. Don't eat spicy foods. Try not to eat hard foods like chips and hard nuts. Use a straw to drink liquids.

ix) If Nausea and Vomiting is there

One should avoid drinking any liquid with meals. Eat foods at a medium temperature, not too hot or cold. Avoid high fat, spicy or overly sweet foods.

x) If Diarrhoea is there

- ✰ Drink plenty of clear fluids such as water, broth, soups, clear juices.
- ✰ Decrease intake of high-fibre foods like nuts, raw fruits and vegetables, as well as whole-grain breads, cereals and pulses.
- ✰ Consume bananas, white rice or pasta, which are easy to digest and help to bulk up the stool.

Eating During and after Illness

Illnesses such as fever and diarrhea cause the body to underutilize nutrients, leading to weight loss. It is important to continue to eat during illness and to eat more when recuperating to make up for lost nutrients and weight.

Maintain Healthy Weight

With a chronic infection like HIV, body burns more energy (calories). If one is spending more energy than its intake, one may lose weight. It is also possible to eat more calories than one is spending, and thus, gain weight. Either way, if one is not eating healthy foods, one can suffer from malnutrition and hurt his or her health.

If one is underweight or chronically ill, one may need to eat more. Be sure to eat a diet that is high in energy. Eat more when recovering from an illness to make up lost weight. If one is overweight, one may need to lose some weight to be healthier.

Maintaining Appetite

HIV infected persons have poor appetite. Here are some tips for increasing the appetite:

- ✰ Eat smaller meals more often. For instance, try to snack between meals.
- ✰ Make food as nutrient-dense as possible to ensure that even though small amounts of food are eaten, one still manages to consume a fair amount of nutrients. Drink high-calorie protein shakes with meals or between meals.
- ✰ Try to eat as much as possible when feeling best and eat in a pleasant atmosphere with family or friends. Choose favorite foods, and make meals as attractive as possible.
- ✰ Avoid drinking too much right before or during meals. This can make one feel full. It can cause depression, interfere with medications and lead to reduced food intake and hence, weight loss.
- ✰ Avoid carbonated (fizzy) drinks and foods such as cabbage, broccoli, and flatulence producing beans. These foods and drinks can create gas in the stomach and make one feel full and bloated.
- ✰ Try a little exercise, like walking or doing yoga. This can often stimulate the appetite and make one feel like eating more.
- ✰ Avoid smoking, because it reduces appetite and leads to decreased food intake.

Drink Plenty of Clean and Safe Water

Drinking enough liquids is very important when one is suffering from HIV infection. As HIV patient at any time is running a fever or having diarrhea, or is suffering from nausea or vomiting or daytime or night sweats, he may run the risk of dehydrating and hence, should put plenty of fluids at the top of the list of his priorities. Drinking plenty of good liquids is just as critical as eating well.

A person needs about two to three litres or eight large cups of fluids each day. When it is very hot, while working, sweating or suffering from diarrhoea, vomiting or fever, a person needs to drink even more to replace the water that has been lost. Extra water can reduce the side effects of medications, help flush out the medicines that have already been used by the body, help some medications to work well if the person is dehydrated and help to avoid dehydration (fluid loss), dry mouth, and constipation.

Here are some tips on getting the extra fluids one needs:

- ☆ Drink more water than usual. Try other fluids, too, like non-caffeinated teas, flavored waters, or fruit juice mixed with water, bottled fruit or vegetable juices, soups, herbal teas and fresh juices.
- ☆ Avoid colas, coffee, tea, and cocoa. These may contain caffeine and can actually dehydrate. Read the labels on drinks to see if they have caffeine in them.
- ☆ Avoid alcohol.
- ☆ Begin and end each day by drinking a glass of water.
- ☆ Suck on ice cubes.

The risk of water-borne infections is too high among HIV patients. This is important for anyone since there are organisms (like cryptosporidia) that can cause infections even in the HIV-negative, but it is particularly crucial for those who have lower T4 counts (especially under 100), since they are at the highest risk of developing serious waterborne infections. Hence, it is very important that water taken should be clean and safe. Eliminate disease-causing organisms from water by either boiling it or using one of the water purifiers designed to kill or filter out bacteria, protozoa, and other disease-causing organisms.

Physical Exercise

For PLWHAs, being active plays a very important role in maintaining health. Activity improves appetite, develops muscle, reduces stress, increases energy and helps maintain overall physical and emotional health. PLWHAs should be encouraged to be active and continue with their daily routine as long as they are physically able to do so.

Food Safety

Properly handling food and water is especially important as PLWHAs having weakened immunity are more vulnerable to infections and diseases due to foodborne pathogens. They should practice food safety which includes food handling, food preparation and food storage, all to be dealt with carefully to ensure safety from food-borne bacteria. Practice safe hygienic practices. Listed below are guidelines for handling water, protein rich products, fruits and vegetables and general food storage for general food safety.

Water

- ☆ Advise individuals and families to always use safe water such from a safe source Water that is not safe should be boiled (at least 5-10 minutes) before drinking or use for food preparation.
- ☆ Use clean containers to collect and store water with a lid or covered with a clean cloth.
- ☆ Cool drinks and ice cubes should also be made with water that is safe.
- ☆ Always wash hands with soap before and after touching water and foods.

Protein Rich Food Products

- ☆ Buy fresh foods especially such as, meat and fish. Foods with any signs of poor quality are likely to have been contaminated.
- ☆ Cook all animal products (meat, chicken, pork, fish and eggs) at high temperatures until thoroughly cooked.
- ☆ Do not eat soft-boiled eggs or meat that still has red juice.
- ☆ Wash hands with soap and water after touching raw fish, chicken, or meat to help prevent infection.
- ☆ Donot take unpasteurized milk, cheeses, and fruit juices
- ☆ Donot eat raw seed sprouts, such as mung bean sprouts

Fruits and Vegetables

- ☆ Wash all fruits and vegetables carefully. If it is not possible to wash fruits and vegetables properly, remove the skin to avoid contamination.
- ☆ Remove the bruised parts of fruits and vegetables to remove any molds and bacteria that are growing.
- ☆ Use clean water to thoroughly wash all fruits and vegetables that are to be eaten raw to avoid contamination.

General Food Storage and Handling

- ☆ Keep the kitchen clean, wash foods, and be careful about food preparation and storage.
- ☆ Make sure that the areas where one prepares and eats food are free of flies.
- ☆ Clean cooking counters and utensils often. Clean all cutting boards and knives (especially those that touch chicken and meat) with soap and hot water before using them again.
- ☆ Wash hands before preparing or eating food.
- ☆ Prepare food in small amounts to reduce leftovers. Cover food that is not eaten to avoid contamination.
- ☆ Thaw frozen meats and other frozen foods in the refrigerator or in a microwave. Never thaw foods at room temperature.
- ☆ Keep hot food hot and cold food cold.
- ☆ If food products have expiration labels, do not eat after the "best before" date has expired.

- ☆ Avoid storing leftovers for more than a few hours (unless refrigerated). Store fresh foods (especially foods from animals), and cooked food in a cool place, or a refrigerator if available.

Nutrition Counseling

Levels of nutritional risk for HIV/AIDS are categorized as low, moderate, and high risk for nutritional compromise. The risk reflects consideration of multiple factors that may lead to nutritional compromise. These categories are based on those published by the American Dietetic Association (Nerad *et al.*, 2003) and the Los Angeles County Commission on HIV Health Services (1999).

I. High Risk

- ☆ Poorly controlled diabetes mellitus
- ☆ Pregnancy (mother's nutrition; infant: artificial infant formula)
- ☆ Poor growth, lack of weight gain, or failure to thrive in pediatric patients
- ☆ >10 per cent unintentional weight loss over 4–6 months
- ☆ >5 per cent unintentional weight loss within 4 weeks or in conjunction with chronic oral (or esophageal) thrush, dental problems, dysphagia, chronic nausea or vomiting, chronic diarrhea, CNS disease, intercurrent illness or active opportunistic infection.
- ☆ Severe dysphagia
- ☆ Enteral or parenteral feedings
- ☆ Two or more medical comorbidities, or dialysis
- ☆ Complicated food-drug-nutrient interactions
- ☆ Severely dysfunctional psychosocial situation (especially in children)

II. Moderate Risk

- ☆ Obesity
- ☆ Evidence for body fat redistribution.
- ☆ Elevated cholesterol (>200 mg/dL) or triglycerides (>250 mg/dL), or cholesterol <100mg/dL
- ☆ Osteoporosis
- ☆ Diabetes mellitus, controlled or new diagnosis
- ☆ Hypertension
- ☆ Evidence for hypervitaminoses or excessive supplement intake
- ☆ Inappropriate use of diet pills, laxatives, or other over-the-counter medications
- ☆ Substance abuse in the recovery phase
- ☆ Possible food-drug-nutrient interactions
- ☆ Food allergies and intolerance
- ☆ Single medical comorbidity
- ☆ Oral thrush
- ☆ Dental problems
- ☆ Chronic nausea or vomiting
- ☆ Chronic diarrhea
- ☆ CNS disease resulting in a decrease in functional capacity
- ☆ Chronic pain other than oral/gastrointestinal tract source
- ☆ Eating disorder
- ☆ Evidence for sedentary lifestyle or excessive exercise regimen
- ☆ Unstable psychosocial situation (especially in children)

III. Low Risk

- ☆ Stable weight
- ☆ Appropriate weight gain, growth, and weight-for-height in pediatric patients
- ☆ Adequate and balanced diet
- ☆ Normal levels of cholesterol, triglycerides, albumin, and glucose
- ☆ Stable HIV disease (with no active intercurrent infections)
- ☆ Regular exercise regimen
- ☆ Normal hepatic and renal function
- ☆ Psychosocial issues stable (especially in children)

All patients of all levels of risk should be educated about healthy, balanced diets for their given lifestyle and physiologic requirements. Medical nutrition therapy should take into account the patient's age, sex, and physiological state, with special attention paid to pediatric growth and development, pregnancy, obesity, quality of dentition, and exercise practices. During nutrition assessment, family and medical history should be considered, particularly regarding diabetes, coronary artery disease, hypertension, and other cardiac risk factors. Recommendations (individualized care plans) should be adapted to stage of HIV progression, from asymptomatic to advanced stages with active secondary infections. Socio-economic, cultural, and ethnic background should be considered, including a history of mental health disorders or substance abuse as well as literacy level and financial status. Nutritional counseling or medical nutrition therapy should also be tailored to make the most of available access to care.

Healthy and balanced nutrition should be one of the goals of counseling and care for people at all stages of HIV infection. An effective programme of nutritional care and support will improve the quality of life of people living with HIV/AIDS, by: maintaining body weight and strength; replacing lost vitamins and minerals; improving the function of the immune system and the body's ability to fight infection; extending the period from infection to the development of the AIDS disease; improving response to treatment; reducing time and money spent on health care; keeping HIV-infected people active, allowing them to take care of themselves, their family and children; and keeping HIV-infected people productive, able to work, grow food and contribute to the income of their families.

Nutritional counseling has been shown to be effective; it has also been shown to influence health outcomes in HIV infection. When dietary counseling is combined with oral nutritional supplements, there is additional evidence for its value. Nutritional counseling also facilitates access to adequate dietary intake.

20

Medical Nutrition Therapy and Diet in Diabetes Mellitus

What is Diabetes Mellitus?

Diabetes mellitus is a chronic metabolic disorder of multiple etiologies that prevents the body to utilize glucose completely or partially. It is characterized by chronic hyperglycemia (raised blood glucose) with alterations in carbohydrate, protein and fat metabolism. This can be due to defects in insulin secretion, insulin action, or both.

WHO defines diabetes as "a chronic disease that occurs either when the pancreas does not produce enough insulin or when the body cannot effectively use the insulin it produces".

Prevalence of Diabetes in the World and India

Diabetes mellitus, a major lifestyle disease is undoubtedly the most challenging public health problem of 21st century with a worldwide prevalence of 387 million (8.3 per cent) and predicted to be 592 million by 2035. Seventy seven per cent of people with diabetes live in low- and middle-income countries.

According to statistics from the International Diabetes Federation (IDF), India has more diabetics than any other nation of the world. Current estimates peg the number of diabetics in the country at about 62 million – an increase of over 10 million from 2011 when estimates suggested that about 50.8 million people in the country were suffering from the disease. India is second only to China which is home to 92.3 million diabetics. By the year 2030, over 101.2 million people in India are likely to suffer from diabetes, say researchers (Gupta *et al.*, 2015. *Int J Sci Rep.* 1(1):1-2).

Why are Indians Highly Susceptible to Diabetes?

A number of factors in combination make Indians highly susceptible to diabetes.

- ☆ **Genetic**: It is the greatest contributor to the rapid spread of this disease. On an average, Indians are 4 times more likely to develop diabetes than Europeans, based solely on genetic outlook.
- ☆ **Cultural and social factors:** The Indian diet is rich in carbohydrates and saturated fats. A typical Indian diet has more calories and sugar than required by the body. This is the cause of obesity, which in turn leads to diabetes.
- ☆ **Urban migration and change in lifestyle:** The younger generations are increasingly choosing a sedentary lifestyle and moving to urban areas from rural settings. With rising standard of living comes the tendency to consume processed sugary foods.
- ☆ **Socio-economic perspective:** Of all the states, Maharashtra and Tamil Nadu seem to have the highest prevalence of diabetes.
- ☆ A recent report confirmed that increasing obesity in South Asians is primarily driven by nutrition, lifestyle and demographic transitions, increasingly faulty diets and physical inactivity, in the background of genetic predisposition.

Obesity appears to spreading across India in part at least as a result of an invasion of processed Western food.

- India's economic boom has been accompanied by a meteoric increase in the number of people with diabetes and those at risk for the disease.

Types of Diabetes

- Type 1 Diabetes Mellitus
- Type 2 Diabetes Mellitus
- Gestational Diabetes
- Other types:
  - LADA (Latent autoimmune diabetes of adults)
  - MODY (maturity-onset diabetes of youth)
- Secondary Diabetes Mellitus

Type 1 Diabetes

It was previously called insulin-dependent diabetes mellitus (IDDM) or juvenile-onset diabetes. Type 1 diabetes develops when the body's immune system destroys pancreatic beta cells, the only cells in the body that make the hormone insulin that regulates blood glucose. This form of diabetes usually strikes children and young adults, although disease onset can occur at any age. Type 1 diabetes may account for 5 per cent to 10 per cent of all diagnosed cases of diabetes. Risk factors for type 1 diabetes may include autoimmune, genetic, and environmental factors.

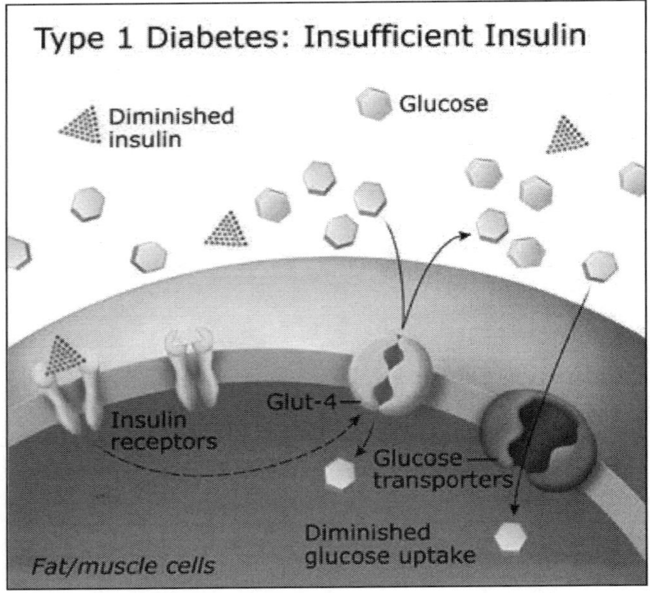

Type 2 Diabetes

It was previously called non-insulin-dependent diabetes mellitus (NIDDM) or adult-onset diabetes. Type 2 diabetes may account for about 90 per cent to 95 per cent of all diagnosed cases of diabetes. It usually begins as insulin resistance, a disorder in which the cells do not use insulin properly. As the need for insulin rises, the pancreas gradually loses its ability to produce insulin. It is associated with older age, obesity, family history of diabetes, history of gestational diabetes, impaired glucose metabolism, physical inactivity, and race/ethnicity. Type 2 diabetes is now increasingly being diagnosed in children and adolescents.

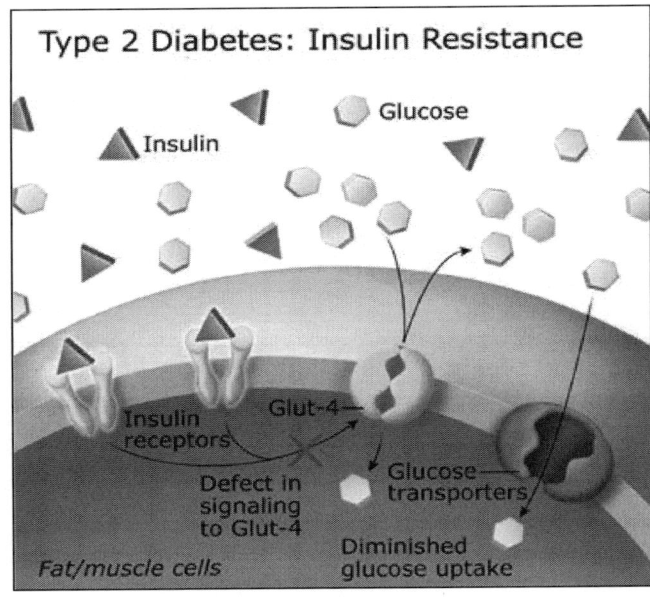

Gestational Diabetes

A form of glucose intolerance that is diagnosed in some women during pregnancy. Gestational diabetes occurs more frequently among African Americans, Hispanic/Latino Americans, and American Indians. It is also more common among obese women and women with a family history of diabetes. During pregnancy, gestational diabetes requires treatment to normalize maternal blood glucose levels to avoid complications in the infant. After pregnancy, 5 per cent to 10 per cent of women with gestational diabetes are found to have type 2 diabetes. Women who have had gestational diabetes have a 20 per cent to 50 per cent chance of developing diabetes in the next 5-10 years.

Risk Factors for Abnormal Glucose Metabolism

- Overweight or obesity
- High percentage of abdominal fat
- Physical inactivity
- Smoking
- Hyperlipidemia
- Hypertension
- Abnormal glucose metabolism is a risk factor for CVD and may progress to type 2 diabetes (Ann. Intern.Med. 2015. 163 (11) 861-868.)

Risk Factors for Diabetes

Some of the important diabetes risk factors include:

- Physical inactivity

- First degree relative with diabetes
- High risk race/ethnicity
- Woman who delivered a baby > 9lb or were diagnosed with GDM
- HDL cholesterol <35 mg/dl ± TG> 250 mg/dl
- Hypertension (≥140/90 or on therapy)
- A1C ≥5.7 per cent, IGT or IFG on previous testing
- Conditions associated with insulin resistance: severe obesity, acanthosis nigricans, PCOS
- CVD history

Persons with these risk factors must go for testing for type 2 diabetes and prediabetes as per tests given under (ADA Diabetes Care 2015. 38 (Suppl. 1): S1-S 93).

Symptoms of Diabetes

Symptoms are:

- High blood levels of glucose
- Blurry vision
- Fatigue
- Frequent thirst
- Painful urination
- Frequent urination
- Sores that donot heal
- Nausea
- Vomiting
- Weight loss
- Frequent hunger

In some cases, there are no symptoms at all. Diabetes type 1 and type 2 symptoms are closely similar, although type 1 is often worse in severity.

Diagnosis of Diabetes

The following tests are used by the doctors to diagnose diabetes and pre-diabetes:

1. **Glycosuria:** The presence of an abnormal amount of sugar in the urine, should be regarded as evidence of diabetes until proved otherwise.
2. **Ketonuria:** High amount of ketone bodies excreted in urine can be detected by nitroprusside reaction which is carried out by using Acetest tables or ketosis paper sticks.
3. **FPG (Fasting Plasma Glucose Test):** Patient must fast for eight hours. Fasted blood sample is drawn by the technician and blood sugar or glucose is tested by the standard prescribed method. If blood glucose is in the range of 100-125 mg/dl, it means the person is pre-diabetic. If blood glucose results in the range of 126 mg/dl and above, it means the person is diabetic. These levels can increase the risk of heart disease and stroke.
4. **OGTT (Oral Glucose Tolerance Test):** By this test, how well the body reacts to average dose of sugar, is measured. Blood is drawn from the overnight fasted patient for determination of blood sugar/glucose. Immediately after that he is asked to dissolve 75 g of glucose in water and drink. If he wants to take breakfast also, the person can take. Exactly after 2 hours, again his blood sample is drawn and blood sugar/glucose is measured. Blood glucose will range from 140-199 mg/dl for prediabetics. For those with type 2 diabetes, the range will start at 200 mg/dl and continue to peak.
5. **A1C testing:** According to American Diabetes Association (2015) guidelines, perform A1C testing at least 2 times each year in individuals who are meeting treatment targets and have stable glycemic control; and quarterly in individuals whose therapy has changed or who are not meeting glycemic targets.

Correlation of A1C with Average Glucose

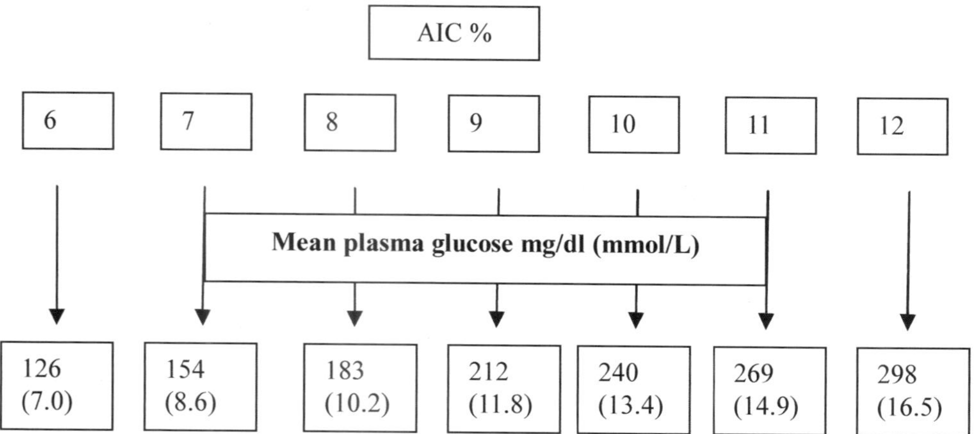

Criteria for Screening for T2D and Prediabetes in Asymptomatic Adults

According to AACE/ACE 2015 guidelines, the criteria are given as under:

✰ Age ≥45 years without other risk factors	✰ Dyslipidemia
✰ Family history of T2D	✰ HDL-C <35 mg/dL
✰ CVD	✰ Triglycerides >250 mg/dL
✰ Overweight	✰ IGT, IFG, and/or metabolic syndrome
✰ BMI ≥30 kg/m^2	✰ PCOS, acanthosis nigricans, NAFLD
✰ BMI 25-29.9 kg/m^2 plus other risk factors*	✰ Hypertension (BP >140/90 mm Hg or therapy for hypertension)
✰ Sedentary lifestyle	✰ History of gestational diabetes or delivery of a baby weighing more than 4 kg (9 lb)
✰ Member of an at-risk racial or ethnic group: Asian, African American, Hispanic, Native American, and Pacific Islander	✰ Antipsychotic therapy for schizophrenia and/or severe bipolar disease
	✰ Chronic glucocorticoid exposure
	✰ Sleep disorders† in the presence of glucose intolerance
✰ Screen at- risk individuals with glucose values in the normal range every 3 years. ✰ Consider annual screening for patients with 2 or more risk factors.	

* At risk BMI may be lower in some ethnic groups; consider using waist circumference.

† Obstructive sleep apnea, chronic sleep deprivation, and night shift occupations.

BMI: Body mass index; BP: Blood pressure; CVD=cardiovascular disease; HDL-C: high density lipoprotein cholesterol; IFG: impaired fasting glucose; IGT: impaired glucose tolerance; NAFLD: nonalcoholic fatty liver disease; PCOS: polycystic ovary syndrome; T2D, type 2 diabetes.

Source: Handelsman, Y. *et al.*, 2015. Endocrine Practice, 21 (supplement) 1: 87.

Criteria for Diagnosing Pre-diabetes

Screen the person for the following tests:

- ✰ IFG
- ✰ 2- h OGTT
- ✰ AIC

According to AACE/ACE (2015) guidelines, the person is prediabetic if:

Impaired fasting glucose (IFG): FPG≥ 100 to 125 mg/dl

or impaired glucose tolerance: 2h PG≥ 140 to 199mg/dl

or AIC: 5.5.-6.4 per cent

- ✰ AIC is a screening tool only.
- ✰ Use FPG or OGTT for definitive diagnosis.

Testing For Type 2 Diabetes and Prediabetes in Asymptomatic Individuals

Type 2 Diabetes Testing

- ✰ All adults who are overweight or obese (BMI ≥25 or ≥ 23 in Asian Americans) who have ≥ 1 diabetes risk factor.
- ✰ Test starting at age 45, especially if overweight or obese.
- ✰ If normal results: repeat testing in ≥ 3-yr intervals.

Prediabetes Testing

- ✰ A1C, FPG or 2-h PG after 75 g OGTT
- ✰ Identify and treat (if appropriate) other CVD risk factors.
- ✰ Consider testing in children and adolescents who are overweight/obese and have ≥ 2 diabetes risk factors.

Table 20.1: Blood Glucose Values Indicating Pre-diabetes or Type 2 Diabetes

Screening Test	*IFG (Impaired fasting glucose) or IGT (impaired glucose tolerance)*	*Type 2 Diabetes*	*Normal*
AIC	5.7-6.4 per cent	≥6.5 per cent	<5.7 per cent
FPG (Fasting plasma glucose)	100 to 125 mg/dl (5.6-6.9 mmol/L)	≥126 mg/dl (≥7.0 mmol/L)	<100 mg/dl (5.6 mmol/L)
2-h OGTT (Oral glucose tolerance test)	140-199 mg/dl (7.8-11.0 mmol/L)	≥200 mg/dl (≥11.1 mmol/L)	(7.8 mmol/L)

Source: Ann. Intern. Med., 2015. 163 (11): 861-868.

Criteria for Diagnosing Type 2 Diabetes

As per AACE/ACE 2015 Guidelines, the person is Type 2 diabetic if:

- FPG(Fasting Plasma Glucose) ≥ 126 mg/dl (After ≥ 8 h fasting)

  OR

- 2-hr PG (Plasma glucose) ≥ 200 mg/dl (2 hrs after ingesting a 75 g oral glucose load in the morning after ≥ 8 hr overnight fast)

  OR

- Random PG (Plasma glucose) ≥ 200 mg/dl + hyperglycemia symptoms (Polyuria, polyphagia, polydipsia)

  OR

- A1C ≥ 6.5 per cent
- Same test should be repeated on different days to confirm diagnosis.
- No confirmation required for glucose level ≥ 200 mg/dl in the presence of symptoms.

Criteria for Diagnosing Type 1 Diabetes

As per ADA Type 1 Diabetes Position Statement 2014, the person is Type 1 diabetic if:

- FPG (Fasting Plasma Glucose) ≥ 126 mg/dl (7.0 mmol/L)

  (Fasting defined as no calorie intake for at least 8 hrs)

  OR

- 2-hr PG (Plasma glucose) ≥ 200 mg/dl (11.1 mmol/L)

  (2 hrs after ingesting a 75 g oral glucose load in the morning after ≥ 8 hr overnight fast)

  OR

- Random PG (Plasma glucose) ≥ 200 mg/dl (11.1 mmol/L)

  (in persons with classic hyperglycemia symptoms or hyperglycemic crisis)

  OR

- A1C ≥ 6.5 per cent (perform in lab. using NGSP-certified and standardized to DCCT assay – (diabetes control and complications trial assay)
- Consider measurement of pancreatic autoantibodies to confirm diagnosis.
- Recommended relatives be tested for type 1 risk.

Diagnostic Criteria for Gestational Diabetes

Test	*Screen at 24-28 Weeks Gestation*
FPG, mg/dL	> 92
1-h PG*, mg/dL	≥ 180
2-h PG*, mg/dL	≥ 153

* Measured with an OGTT performed 2 hours after 75-g oral glucose load.

Complications Associated with Diabetes

Diabetes can affect the heart, kidneys (direct result from large amounts of protein being deposited into the urine), eyes, nervous system, skin and whole body. Long term complications include stroke, peripheral vascular disease, hypertension, urinary infection and coronary artery disease. The most frequent health risk from diabetes is known as cardiovascular disease. Diabetes can cause high blood pressure that then causes an even stronger resistance to insulin. Diabetes raises bad cholesterol and lowers good cholesterol leading to an increase of heart disease and stroke. This can cause the arteries to become clogged with fat. According to (NIDDK, National Diabetes Statistics fact sheet. HHS, NIH (2007),

- 2 in 3 people with diabetes die of heart disease or stroke. At least 65 per cent of people with diabetes die from heart disease or stroke.
- Diabetes is the #1 cause of adult blindness.
- Diabetes is the #1 cause of kidney failure.
- Diabetes causes more than 60 per cent of non-traumatic lower-limb amputations each year.

Preventing Diabetes Complications

- Blood glucose (sugar) control
- Blood pressure control
- Blood lipid (fat) control
- Preventive care practices for eyes, kidneys, feet, teeth and gums
- Low-dose aspirin as directed by a doctor

Management of Diabetes Mellitus

The major components for treatment of diabetes mellitus are:

1. Diet
2. Physical activity
3. Oral hypoglycemic therapy
4. Insulin therapy

Goal of diabetes treatment is to keep blood glucose in control through:

Medical Nutrition Therapy in Diabetes

American Diabetes Association in 1994 coined the term "medical nutrition therapy" constituting 2 phases, namely, adjudging the nutritional requirement of a person and treatment through counselling and nutrition therapy, respectively. Medical nutrition therapy is recommended for all patients with pre-diabetes, type 1, type 2 and gestational diabetes as part of an overall treatment plan, preferably provided by a registered dietician skilled in diabetes MNT. Its main objective is to regulate optimal level of lipids in blood, ideal body weight, and blood glucose level in normal range.

Factors affecting Nutrition Therapy

Nutrition therapy as a therapy for diabetes depends on certain factors such as:

- Patient's age-based nutritive requirements and
- Food preferences as well as
- Other medical conditions together with
- An exercise regime and
- Recommended nutritional requirements depending upon the patient's abilities and health conditions.

Nutrition Therapy Recommendations

There are many types of nutrition interventions that are effective such as:

- Decreased calorie and fat intake
- Carbohydrate counting
- Use of insulin to carbohydrate ratios
- Healthy food choices
- Individualized meal planning
- Behavioural strategies

It must be remembered that there should be 'no one size fits all' eating approach for diabetic persons. The aim of chosen eating approach should be to improve blood glucose, blood pressure and lipids.

Type 1 Diabetes

Insulin doses should match to planned carbohydrate intake. It is important for the management of diabetes.

Type 2 Diabetes

Attention to food intake and patterns of eating are important for the management of diabetes, even if on medications, including insulin. The focus should be on reducing or maintaining a reduced energy intake.

Guidelines Regarding Nutritional Requirements in Diabetes

Macronutrients

According to ADA (2013) diabetes guidelines, there is no ideal percentage of calories from carbohydrates, protein or fats for individuals with diabetes. Macronutrient distribution is to be based upon individual assessment of current eating patterns, preferences and metabolic goals. However, DRI acceptable macronutrient distribution is

- Carbohydrate: 45-65 per cent
- Protein: 20-35 per cent
- Fat: 10-35 per cent

In diabetes, recommendation is that regardless of the macronutrient distribution, total energy intake must be appropriate for weight management. The mix of macronutrients is adjusted to meet metabolic goals and individual preferences of the diabetic person. Total energy intake and a healthy eating pattern should take precedence over the distribution of macronutrients (Higher vs lower CHO, low fat or high-MUFA –No differences in A1C, LDL cholesterol or TG levels, BP or wt). Available evidence does not show an adverse effect on insulin sensitivity from carbohydrate intake; a higher carbohydrate intake may instead improve insulin resistance.

For overweight (BMI 25-29.9 kg/m^2) and obese (BMI ≥30 kg/m^2) individuals with type 2 diabetes, reduce energy intake while maintaining healthy eating pattern. For such individuals, optimal macronutrient intake to reduce weight has not been established. Modest weight loss may improve glycemia, BP and ± lipids particularly early in the disease process. Reduce energy intake by 500 to 1000 Kcal per day for a weight loss of 450-900 g per week. Exercise moderately *e.g.* brisk walking that expends 150 to 225 Kcal per session for 30-45 minutes, 3 to 5 days a week.

Carbohydrates

According to ADA, RDA for carbohydrates is at least 130g/day for adults and children. A high carbohydrate intake is >55 per cent of total energy, low CHO intake is <25 per cent of total energy or <130g/day while very low CHO intake <20 per cent of total energy intake. It is difficult to report the effect of CHO on insulin action. If CHO intake is increased, fat is generally decreased and vice-versa (Franz, M.J. And Evert, A.B. 2013. ADA Guide to Nutrition Therapy for Diabetes). Therefore, it is unknown if the benefit on insulin sensitivity is due to the higher CHO intake or low fat intake. Higher CHO diet did not elevate TGs when energy restriction was prescribed. Therefore, energy intake is a factor when determining the effect of CHOs on TG levels. Ideal amount of carbohydrate intake has not been established. While developing eating plan, it should be kept in mind that amount of carabohydrates and available insulin are the important factors which influence the postprandial glycemic response. Monitor the carbohydrate intake to achieve glycemic control. Include moderate amounts of carbohydrates (45-50 per cent of total energy intake) and that should include fruits, vegetables, whole grains, and low fat dairy foods; all carbohydrate sources should be in the appropriate amounts and portion sizes.

As far as quality of carbohydrates is considered, substitute foods with lower glycemic load for those with higher load and they should bring about modest improvement in glycemic control. Women should consume at least 25 g while men should consume at least 38 g dietary fibre per day (14g fibre/1000 Kcals/ day). More than 50 per cent of all grains should be whole grains.

Protein

ADA (2013) recommendations are:

i) Diabetes without diabetic kidney disease

- ✰ No ideal intake to improve glycemic control of CVD risk
- ✰ Individualize goal

ii) Diabetes and macro- or-microalbuminuria

- ✰ Reduction below usual intake not recommended

iii) Type 2 diabetes

- ✰ Donot use carbohydrate sources high in protein to treat or prevent hypoglycemia

Fat

ADA (2013) recommendations are:

i) Total Fat

- ✰ No ideal intake, individualize goal
- ✰ Fat quality is more important than the quantity

ii) MUFAs

- ✰ Mediterranean style, MUFA-rich eating pattern recommended as alternative to low-fat, high carbohydrate in patients with type 2 diabetes.
- ✰ It may improve glycemic control, CVD risk factors.

iii) PUFAs

- ✰ Limited evidence on effect in people with type 2 diabetes

iv) Saturated Fat

- ✰ < 10 per cent of total calories

v) Cholesterol

- ✰ < 300 mg dietary cholesterol per day

vi) Trans Fat

- ✰ Limit as much as possible

vii) Dietary Omega-3 Fatty Acids Intake

- ✰ Omega-3 fatty acids (EPA, DHA) supplements not recommended for CVD prevention or treatment in diabetic persons.
- ✰ Omega-3 supplementation in persons with type 2 diabetes lowers TG levels, but may raise LDL cholesterol and have no effect on glycemic control or fasting insulin.
- ✰ Increase intake of foods rich in EPA, DHA, ALA as they have benefits on lipoproteins, CVD prevention and health outcomes.
- ✰ Eat fish (particularly fatty fish) ≥ 2 times a week.

Micronutrients and Herbal Supplements

As per ADA (2013) recommendations:

- ✰ Vitamin and mineral supplementations have no benefit in patients without underlying deficiencies.
- ✰ **Antioxidants (Vitamins A, C, E and carotenes):** Not advised due to lack of evidence of efficacy and concern in long term safety studies.
- ✰ Although animal studies have suggested that deficiencies in many of the trace elements-including Zn, Cr, Mg, Cu, Mn and Vit. B_6 may lead to glucose intolerance, evidence for their role in the pathogenesis of glucose intolerance in humans is not definitive.
- ✰ **Chromium and magnesium:** Insufficient evidence to recommend routine use to improve glycemic control in diabetic patients.
- ✰ **Vitamin D:** Epidemiological research indicates that people with low vit. D levels have a significantly higher risk of type 2 diabetes than individuals with higher levels. National Health and Nutrition Examination Survey showed an inverse relationship between serum 25-hydroxy vitamin D and the incidence of type 2 diabetes and insulin resistance. Ultimately, current evidence does not definitively show that daily vit. D supplementation is effective in the treatment of type 1 or type 2 diabetes.
- ✰ **Cinnamon, other herbs and supplements:** Insufficient evidence to support use for diabetes treatment.
- ✰ **Sodium:** Patients with diabetes should reduce salt intake. Reduce sodium to < 2300 mg per day on individual basis.
- ✰ Currently, there is insufficient evidence to support the routine use of supplements for the treatment of diabetes.
- ✰ **Individualize meal planning:** It is better to optimize food choices to meet recommended micronutrient dietary allowance/intake.

Meal Planning for Diabetic Patients

Goals

- ✰ To achieve and maintain healthy and productive life

- To maintain optimal nutrition for adequate growth, development and maintenance.
- To maintain near normal blood sugar level
- To achieve and maintain a desirable body weight
- To provide relief from symptoms
- To prevent, delay or minimise the onset of chronic degenerative complications.
- Moderate amounts of carbohydrates (45-50 per cent of total energy intake) and that includes fruits, vegetables, whole grains, and low fat dairy foods-all carbohydrate sources in appropriate amounts and portion sizes.

The ADA acknowledges that there is no one-size-fits-all eating pattern for individuals with type 2 diabetes

A diabetic meal plan focuses on providing a healthy diet that:

- Controls blood glucose and
- Prevents diabetic complications
- There is no one specific diabetic meal plan, a diabetic meal plan is developed to meet one's own special needs:
  - Individual food likes
  - Daily schedule
  - Medications used
  - Other health issues:
    - Glucose control
    - Weight
    - Blood lipids
    - Blood pressure
- A healthy diet includes a variety of foods from all the food groups. In general, a diabetic meal plan is:
  - Low in fat
  - Moderate in protein
  - High in complex carbohydrates like beans, vegetables and grains
- Total calorie intake matters a lot rather than its proportion from different macronutrients. Actual amounts of each nutrient depend on the number of calories you need.
- Calorie needs depend on gender, size, age, and activity level and should not exceed the RDA. The overweight or obese diabetic patient should reduce 500-1000 Kcal a day depending upon the body weight reduction to attain the normal optimum body weight.
- Meal planning with diabetes is very individualized. Get a personalized meal plan from a Registered Dietitian or Certified Diabetes Educator.
- A diabetic meal plan provides a consistent amount of carbohydrate from day to day, spaced evenly throughout the day to prevent wide changes in blood glucose. Various methods are used to provide the consistency of a diabetic meal plan.
- Methods of meal plans for diabetes include the following:
  - Diabetes Pyramid/My Pyramid Exchange list
  - Plate Method
  - Carbohydrate Counting
  - Food/CHO Exchange List
  - Use of insulin to carbohydrate ratios
  - Healthy food choices
  - Individualized meal planning

I. Diabetes Pyramid

Similar to My Pyramid, the diabetes pyramid divides foods into six groups. The base of the pyramid is the grains, beans and starchy vegetables group while at the top is the fats, oils, sweets and alcohol group. The pyramid gives a range of servings to eat and the range is based on 1600-2800 calories. It is important to note that the exact number of servings and calories one needs to eat will depend on diabetes goals, nutrition needs and lifestyle factors. The diabetes pyramid's food groups are based on the amount of carbohydrate/protein content of the food.

II. Plate Method

A 8 inch dinner plate is the recommended plate to be used by diabetics than the typical dinner plate of 10 inch diameter. It should include the following amounts of foods:

- **¼ plate:** Grains or starchy foods such as wheat, rice, pasta, potatoes, corn, or peas
- **¼ plate:** Protein foods such as pulses, meat, fish, poultry, or tofu
- **½ plate:** Non-starchy vegetables such as carrots, cauliflower, salad and tomatoes
- Add a glass of non-fat milk and a small roll or piece of fruit

III. Food Exchange List Method

This is the oldest method for meal planning. It is based on dietary guidelines and My Pyramid. It includes a variety of foods. Before understanding Food Exchange List/CHO Exchange list, it is important to know about carbohydrates as they are the most important nutrient to be understood in meal planning as they affect the blood sugar the most.

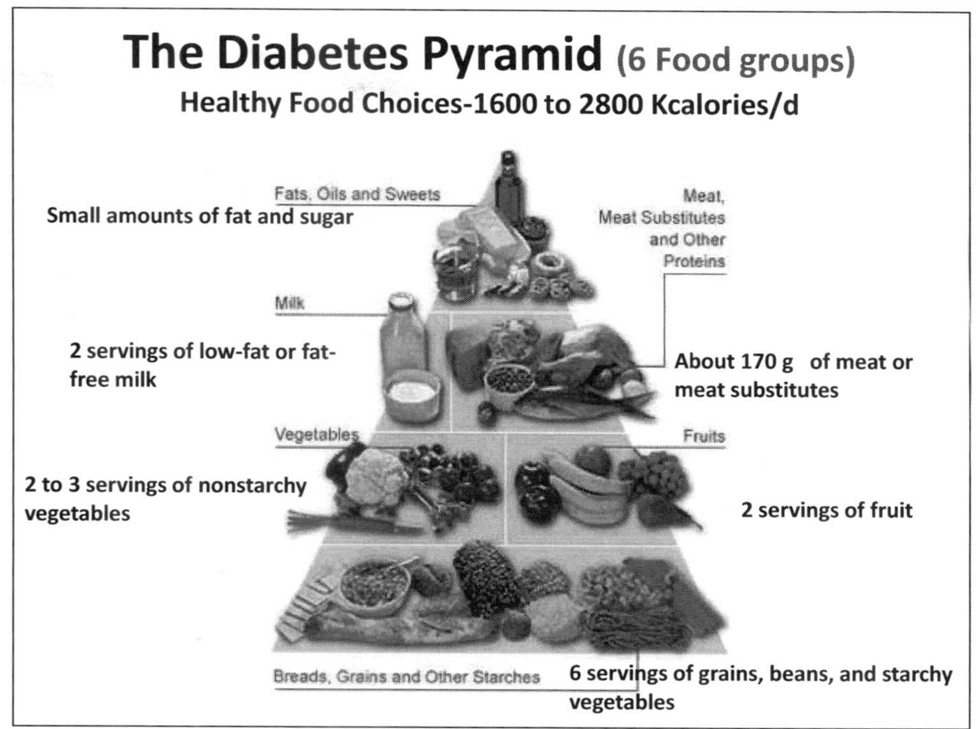

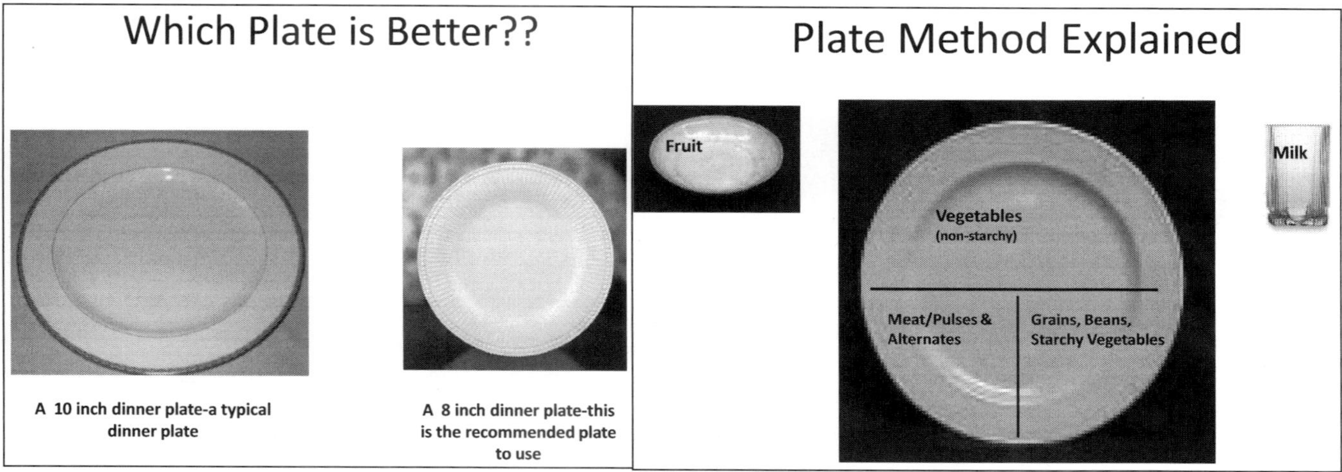

1. Selection of Carbohydrate Foods from Food Exchange/CHO Exchange List

There are two types of carbohydrates- sugars and starches. Total carbohydrate eaten determines how much sugar reaches the blood. Grains, beans and starchy vegetables (potatoes, sweet potato, colocasia, peas and corn), fruits (mango, grapes, sapota and banana), milk and milk products and sweets (sodas, juice drinks, cake, cookies, candy, and chips) are the major sources of carbohydrates. Vegetables (non-starchy such as lettuce, cucumbers, broccoli, bottle gourd, ash gourd, bittergourd and cauliflower have a little bit of carbohydrate but in general are very low); meat and meat alternates and fats and oils donot contain carbohydrates.

The daily amount of carbohydrate, protein, and fat for people with diabetes has not been defined–what is best for one person may not be best for another. Everyone needs to get enough carbohydrate to meet the body's needs for energy, vitamins and minerals, and fiber. It is better to take moderate amounts of carbohydrates (45-50 per cent of total energy intake) and that includes fruits, vegetables, whole grains, and low fat dairy foods-all carbohydrate sources in appropriate amounts and portion sizes (Franz, M.J. And Evert, A.B. 2013. ADA Guide to Nutrition Therapy for Diabetes). Approximately, for 2000 Kcal diet, CHO intake should be 250g/day (@50 per cent calories from CHOs). Carbohydrate intake should not fall below 130 grams/day.

In the past, people with diabetes were told to avoid sugar. Now it is known that complex carbohydrates and sugars have a similar effect on blood glucose levels. The total amount of carbohydrate is the issue, not just sugar. People with diabetes can have sugar, but the carbohydrate from foods containing sugar must be worked into the meal plan. There are some problems with foods high in sugar *i.e.* foods high in sugar often don't provide other important nutrients needed every day. Foods high in sugar are also often high in fat. Therefore, foods high in sugar should be avoided in the diet plan.

Eat carbohydrates rather than drinking. Take whole fresh fruits and vegetables which are minimally processed and free from food additives rather than drinking their juices. Select the foods which contain more of dietary fibre. Patients should try to consume a minimum of 20 - 35 grams of fiber daily (ideally 50 grams/day), from vegetables, fruits, whole grain cereals, breads, nuts, and seeds. Processed foods, such as packaged cookies, crackers and canned fruit, usually contain added salt, sugar, carbohydrates, fat, or preservatives and should be avoided. Eating more whole foods and less processed foods will also make the body work harder to digest them, which is a good thing for weight management.

The low-carbohydrate diets as in Atkin's Diet that are popular today would make one think whether they are ideal for diabetics? The fact is that current scientific knowledge does not support the long term use of low-carbohydrate diets especially in type 2 diabetics.

Carbohydrate Serving

In the CHO Exchange List, a carbohydrate choice is a serving of food which contains 15 gms of carbohydrate. There are three carbohydrate groups *viz.* Bread/starch, Fruits and Vegetables (Roots and tubers) and Milk. One serving from each group should contain 12 - 15 carbohydrate grams. Sometimes, if one wants an extra starch during a meal, one just needs to omit a fruit or milk from that meal.

For example there is about 15 grams of carbohydrate in:

- ☆ 1 small piece of fresh fruit
- ☆ 1/2 cup of canned or frozen fruit
- ☆ 1 slice of bread
- ☆ Chapatti, small 1 (6 inches across)
- ☆ 1/2 cup of oatmeal
- ☆ 1/3 cup of pasta or rice
- ☆ 4-6 crackers
- ☆ 1/2 English muffin
- ☆ 1/2 cup of black beans or starchy vegetables such as mashed potatoes, cooked corn, peas, or lima beans
- ☆ 1/4 of a large baked potato
- ☆ 2/3 cup of plain fat-free yogurt or sweetened with sugar substitutes
- ☆ 2 small cookies
- ☆ 2 inch square brownie or cake without frosting
- ☆ 1/2 cup ice cream or sherbet
- ☆ 1 Tbsp syrup, jam, jelly, sugar or honey
- ☆ 1/4 serving of a medium French fry

Portion control is the key to managing the diabetes. For example, which will have the greater effect on the blood sugar, 1 tsp of sugar or ½ cup potatoes? Potatoes have 15 g of carbohydrate, while 1 tsp. of sugar has only 4 g of carbohydrate. Therefore, potatoes have 3 times the effect on blood sugar than table sugar would.

- ☆ ½ cup cooked vegetable = 1 serving
- ☆ ½ cup vegetable juice = 1 serving
- ☆ 1 cup raw vegetable = 1 serving

i) **Each cereal/starch exchange** supplies 15 g carbohydrate, 2-3.5 g proteins, negligible fat and 80 Kcals.

☆ Barley, cooked	1/3 cup
☆ Oat bran	1/4 cup
☆ Wheat bran	1/2 cup
☆ Bulgur, cooked	1/2 cup
☆ Cereals Bran	1/2 cup
☆ Oats, oatmeal, cooked	1/2 cup
☆ Puffed oats	1 1/2 cups
☆ Shredded wheat	1/2 cup
☆ Sugar-coated cereals	1/2 cup
☆ Unsweetened, ready-to-eat cereals	3/4 cup
☆ Granola, regular or low-fat	1/4 cup
☆ Grits, cooked	1/2 cup
☆ Millet, cooked	1/3 cup
☆ Muesli	1/4 cup
☆ Pasta, cooked	1/3 cup
☆ Rice, white/brown, cooked	1/3 cup
☆ Wheat germ, dry	3 Tbsp
☆ Rice, cooked	1/2 cup

ii) **Carbohydrate exchanges not containing protein** include sugar (sugar, jaggery, syrups, honey, jams, jellies, marmalades, canned fruits, squashes, *etc.*) exchanges and each exchange provides 5 g carbohydrate and 20 Kcal. One teaspoon (5 g) of sugar is the basis of this exchange. In case of other products such as jams, jellies *etc.*, 7 to 8 g (1 ½ tsp) is taken as one sugar exchange, as this much amount provides 5 g carbohydrates.

iii) **A milk exchange** is usually 1 cup or 215 ml. All people with diabetes should avoid artificially sweetened milks. 1 cup milk = 1 serving

- ☆ 170 g (no sugar added) yogurt = 1 serving
- ☆ The difference between whole milk, 2 per cent milk, 1 per cent milk, and skim milk is the amount of fat that they contain.

iv) **Serving size of meats/pulses** in the exchange list is usually 55-85 g at meals.

- ☆ 30 g meat = 1 egg
- ☆ 30g meat = 30g cheese
- ☆ 30g meat = cup cottage cheese or tuna
- ☆ 30g meat = 2 tablespoons of peanut butter
- ☆ ½ cup dried beans count as both 30g of meat and a starch

v) **Serving size of fats and oils** is

- ☆ 1 tsp of margarine or butter = 1 serving
- ☆ 1 tsp of oil = 1 serving
- ☆ 1 tsp of mayonnaise = 1 serving
- ☆ 1 Tbs of salad dressing = 1 serving

vi) **Each fruit exchange** contains about 15 grams of carbohydrates for a total of 60 calories. Vegetables also contain some CHO. Three servings of vegetables also contain 15 grams thus one would need to count these as 1 CHO choice. One or 2 servings of vegetables have very little carbohydrates in them and should not be counted.

- ☆ 1 small piece of raw fruit = 1 serving
- ☆ 1 cup of raw fruit = rule of thumb for 1 serving
- ☆ ½ cup juice, canned fruit, or cooked fruit = 1 serving
- ☆ 1/3 cup cranberry or grape juice = 1 serving

2. Proteins and Fats

Foods in the fat and protein category do not affect blood glucose directly. Foods in this category include meat, poultry, eggs, beans, tofu, peanut butter, most cheeses and fats such as butter, margarine and oils. These foods contain negligible amounts of carbohydrate, thus, one should not need to worry about them. That does not mean that one can eat as much fat or protein as one likes. Their intake should also be as per their RDAs and should not exceed the total daily energy intake recommended.

High-protein diets are not recommended as a method for weight loss at this time. The long-term effects of protein intake >20 per cent of calories on diabetes management and its complications are unknown. Although such diets may produce short-term weight loss and improve glycemia, it has not been established that these benefits are maintained long term, and long-term effects on kidney function for persons with diabetes are unknown (Nutrition recommendations and interventions for diabetes. Diabetes Care 31:S61-S78, 2008).

3. Free Foods

A free food is any food or drink that contains less than 20 calories or less than 5 gms of carbohydrate per serving. Foods like sugar free sodas and beverages, artificial sweeteners, spices and seasonings fall into this category. As no serving size is specified, amount used may be as desired. Items that have a specific serving size should not exceed 3 servings per day.

How to Use Food Exchange/CHO Exchange List Method in Meal Planning

Step I

Write down the specifications such as age, sex, physical activity, socio-economic status, food habits, disease *etc.*

Step II

Using ICMR 2010 tables, write down the recommended Dietary Allowances (RDA) for the nutrients (Energy, protein, carbohydrate, fat, dietary fibre *etc.*) for diabetic patient to be calculated for the above mentioned specifications.

Step III

Write down the amount of different foods/CHO exchanges required in a balanced diet for the diabetic person so as to provide the desired amount of above mentioned nutrients. Her/His food preferences should also be taken into consideration. For example:

Food Exchange	*No. of Exchanges*	*Energy (Kcal)*	*Protein (g)*	*Carbohydrate (g)*	*Fat (g)*
Milk	2	340	16	24	20
Meat	-	-	-	-	-
Pulse	2	200	14	34	-
Cereal/ Starch	14	980	28	210	-
Vegetable A	2	30	2	6	-
Vegetable B	1	40	2	7	-
Fruit	1	40	-	10	-
Fat	4	180	-	-	20
Sugar	3¼	65	-	16.25	-
Total		**1875**	**62**	**307.25**	**40**

Step IV

Distribute the food exchanges in different meals as per example given below:

Meals	*Carbohydrate (g)*	*Carbohydrate Choices*
Total meals (5)	225 g	15
Breakfast	45 g	3
Lunch	60 g	4
Mid morning Snack 15 g	1	
Dinner	75 g	5
Evening Snack	30 g	2

Step V

Keeping in view the principles of meal planning, plan the menu. Depending upon the menu, select the food from different food exchanges. Often start with 3-4 CHO servings per meal (includes fruits, starches, milk, sweets) for women and 4-5 for men plus 1-2 for snack if desired. Evaluate feasibility of meal plan with patient. Trial meal plan and evaluate blood glucose records. Adjust plan as necessary.

Number and Timings of Meals

Small frequent meals should be planned and timings of their intake are also very important. Space the food intake throughout the day. Eat at regular times. Donot skip meals.

IV. Carbohydrate Counting

There are several different ways people with diabetes can manage their food intake to keep their blood glucose (sugar) within their target range and one such method is 'carbohydrate counting'. Carbohydrate counting is a method of calculating grams of carbohydrate consumed at meals and snacks. Counting carbohydrate servings provides an accurate "guess" of how your blood sugar will rise after a meal or snack. The more carbohydrates one eats, the higher will be the spike in blood glucose levels and the more insulin that individual will need to cover that meal. For example a South Indian meal of *idli* and *dosa* will raise blood glucose levels higher than a meal comprising of grilled *paneer* and salad. Hence, one will need more insulin for the South Indian meal.

This method is a meal planning approach to manage blood sugar levels based on the following ideas:

- Foods that contain carbohydrates have the greatest effect on blood glucose compared to foods that contain protein or fat.
- Total amount of carbohydrates is more important than source of carbohydrates.

Reasons behind popularity of this method are (Using Carbohydrate Counting in Clinical Practice, JADA Practical Carbohydrate Counting, American Diabetes Association, 2001):

- Effectiveness
- Flexibility
- Ease of implementation

CHO counting is the newest method in diabetes meal planning. It has been around for almost 10 years and is most widely used for patients who are type 1 or 2 diabetics. Most kids and some adults use the CHO counting method for meal planning because it is easy to use in a variety of settings. Research has shown that CHO counting is more accurate at preventing roller coaster blood sugar levels. It is easier when eating out or away from home because one is concerned with the grams of carbohydrates in that food.

Carbohydrate Counting empowers you to manage your diabetes more effectively matching your own lifestyle, while offering more variety to your meal plan

It is important to know that carbohydrates are only found in starches, fruits, milk, sweets and some vegetables. CHOs are not present in meat, fish and poultry or fats. A meal plan outlines the number of CHO choices a person can eat for meals and snacks. One example is a 45 g CHO for Breakfast, Lunch, and Dinner and 15 g CHO for snacks or 60 g/meal and 30 g/snack. Your doctor or dietitian will work the meal plan out with you. It is important to note that the CHO counting method requires diligence and constant monitoring of blood sugars. Diabetics who require insulin will base the amount of insulin that needs to be given on how many carbohydrates they ate.

How to Count Carbohydrates

Any diabetic whether type 1 or type 2 should know the carbohydrate count for 1 serving of every food group. The ways to count carbohydrates are:

- Carbohydrate measured in grams
- Serving or carbohydrate choices
- Exchanges

A food that contains 15g of carbohydrate is called one carbohydrate exchange or one carbohydrate choice is 15 grams carbohydrate. When carbohydrate counting, it is possible to substitute one food item for another for a similar impact on blood glucose. Example: Exchange one small apple (120g) for two small cookies for a similar effect on blood glucose

1 Exchange = 1 Choice = 1 Carbohydrate = 1 Portion = 1 CP (Carbohydrate Portion) = 1 CU (Carbohydrate Unit) =15 g carbohydrate

But most foods are a mixture of nutrients. So how do you know how much of carbohydrates are there in any given food? One can check serving sizes with measuring cups, spoons or a food scale to calculate the amount of carbohydrates in the food.

1 tablespoon sugar =1 apple = 1slice of bread = 1/3 cup pasta = ½ cup oatmeal = 15 g of carbohydrates

For packaged foods, the easiest way is to read the nutrition facts section on the package. When reading the food label one can go by the total amount of carbohydrates per serving of product.

Carbohydrate Counting

Exchange Groups	*Carbohydrate (g)*	*Carbohydrate Choices*
Starch	15	1
Fruit	15	1
Milk	12	1
Vegetable	5	
Other carbohydrate	15	1

How many Carbohydrates per Day are Allowed

According to latest ADA guidelines, there is no set proportion of calories from carbohydrates, proteins and fat. The important thing is that total calorie intake should not exceed the amount recommended. In general, the recommended number of servings of carbohydrates is based on weight, activity level, diabetes medications, and goals for blood glucose levels of an individual. Hence, the general guideline for most adults with diabetes is 45 to 60 grams of carbohydrate per meal, which are three to four carbohydrate choices or servings. A snack would be around 15 to 30 grams of carbohydrates, or one to two choices or servings. That's just a starting point, however. Ultimately, the total carbohydrate allowance should meet the energy needs, blood glucose targets, and weight management goals.

Time to time all patients with diabetes should test their blood glucose before and 2 hours after the first bite of a meal. This is the only way to tell how the choices and amount of carbohydrates consumed affect their blood sugar. When carbohydrate counting, it is possible to substitute one food item for another for a similar impact on blood glucose. Example: Exchange one small apple (120g) for two small cookies for a similar effect on blood glucose

How to Follow a Consistent Carbohydrate Meal Plan for Keeping Blood Glucose to Target Levels

Example 1

If we have to plan 1300 kcal/day diet for a diabetic patient, follow the steps as given below:

- ✰ 1300 Calories 2 = 650 (½ calories from carbohydrates)
- ✰ 1 gram carbohydrate = 4 calories
- ✰ 650 4 = 162 g carbohydrates/day
- ✰ Divide 162 g carbohydrates into number of carbohydrate choices at each meal and snack to be consumed over the day. Then select the right kind of carbohydrate at each meal and snack and plan the whole day menu.

Meal	*Grams of Carbohydrates*	*Carbohydrate Choices*
Breakfast	30	2
Snack	15	1
Lunch	45	3
Snack	15	1
Dinner	45	3
Snack	15	1
Total (11x 15)	165	11

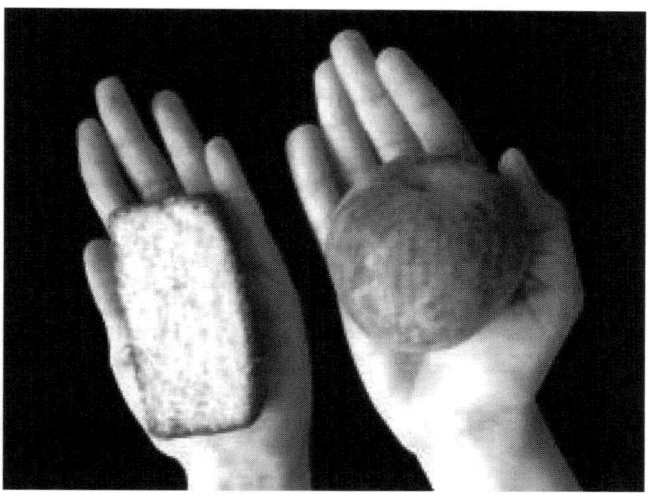

If you have a choice between a cake and an apple.

What do you choose?

Example 2 of Carbohydrate Counting

How many carbohydrates are in this meal?

Meal	*Answers*
Breakfast 2 bread slices=? Small banana (115g) =? 235ml 1 per cent milk=? 1 egg=?	2 Bread slices=30 g Small banana (115g) =15 g 225ml 1 per cent milk=12 g 1 egg=0 g **TOTAL= 57 g**
Lunch 2 slices of bread=? 17 grapes=? 1 Cup raw carrots=? 85g tuna fish=? 1 tsp mayonnaise=?	2 slices of bread=30 g as 1 slice = 15g 17 grapes=15 g (1 small piece of raw fruit = 1 serving =15g) 1 Cup raw carrots=5 g (1 cup raw vegetable = 1 serving= 5g) 85g tuna fish=0 g 1 tsp mayonnaise=0 g **TOTAL= 50 g**
Dinner 1½ C pasta=? 1 oz (28g) of bread=? 1 Cup salad =? 1 tsp olive oil=?	1½ C pasta=68 g (1/3 cup cooked pasta = 1 serving=15 g CHO) 28 g of bread=15 g 1 Cup salad =5 g 1 tsp olive oil= 0 g **TOTAL=88 g**

It is more important to know CHO allowance for each meal and snack than it is to know the total for the day. The amount of CHO eaten at each meal should remain consistent. Here is an example to show how carbohydrate counting can make meal planning easier. Let's say dinner meal plan contains 5 CHO servings or 75 grams of CHO. This is based on a meal plan of 3 starch servings, 4 protein, 1 vegetable, 1 fruit, 1 milk and 3 fat. The label on a frozen dinner says it contains 62 grams of CHO. All you would need to do is figure out how many grams of CHO you would need to make 75.

Example 3 of 1800 calorie carbohydrate counting meal plan for the whole day

Plan the whole day menu like this:

- **Breakfast:** 2 servings of starch, 1 fruit serving, 1 milk serving, 1 meat/pulse serving, 1 fat serving.
- **Lunch:** 2 starch servings, 1 fruit servings, ½ milk serving, 2 vegetable servings, 2 meat/pulse servings, 2 fats.
- **Dinner:** 2 starch servings, 1 fruit serving, 2 vegetable servings, 3 meat/pulse servings, 2 fats.
- **Snack:** 1 starch serving, 1 fruit serving, ½ milk serving.

This meal plan is approximately 60 g of carbohydrates per meal and 30 g of carbohydrates per snack.

Starch list includes cereals such as wheat, rice, maize, sorghum, pearl millet, ragi, breakfast cereals, bread, biscuits, oats, noodles, pasta, *maida*, starchy vegetables, beans, peas and lentils. Avoid refined grains like *maida*, polished rice and white bread and choose healthier options such as whole wheat flour, barley, quinoa, brown rice, multigrain bread, durum wheat pasta *etc.*

> **In general one starch exchange contains**
>
> 1 serving =15 g carbohydrate, 3 g protein, 1 g of fat and 80 calories

Bread (29g)

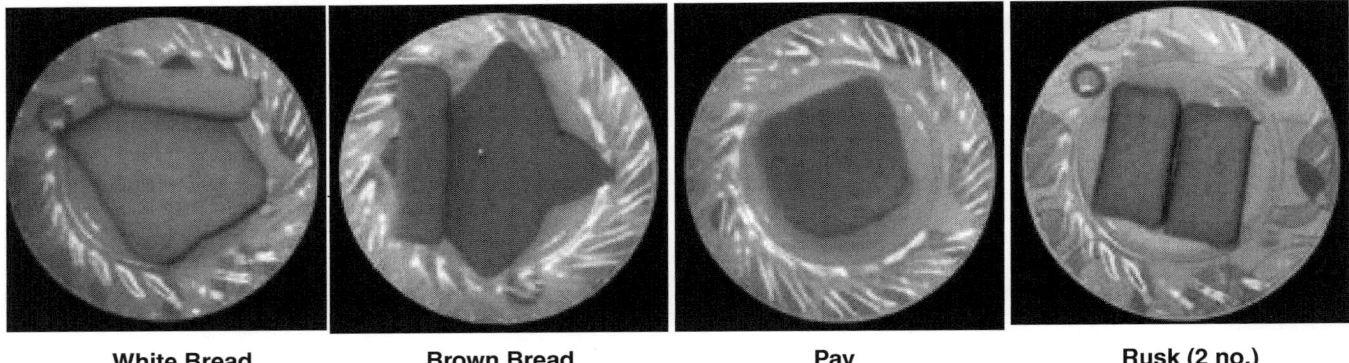

White Bread | Brown Bread | Pav | Rusk (2 no.)

Each of the following food represents one carbohydrate choice or 15 g carbohydrate exchange.

Breakfast Cereals

Muesli (19.2 g) | Oats (24 g)

Chocos (18 g) | Cornflakes (16 g)

Chapati and Paratha

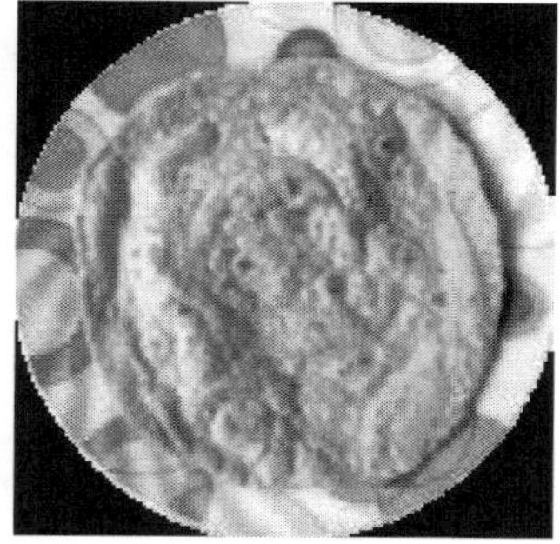

Chapati (Raw weight of atta / flour=22 g). The size of the chapati, paratha and bhakri will vary depending on its thickness.

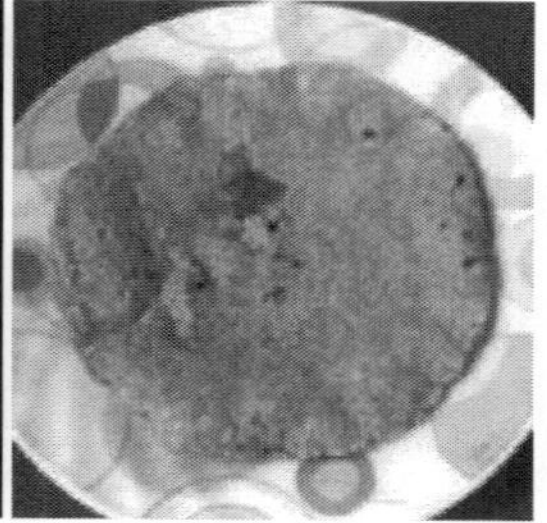

Bhakri (Raw weight = 22 g)

Aloo Paratha (15 g flour and 21 g potato)

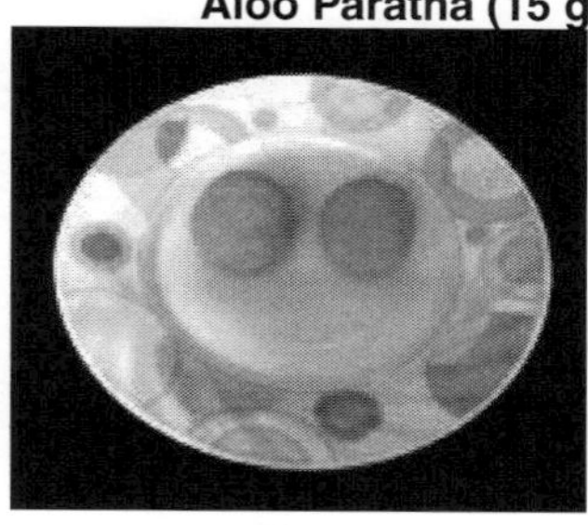

Raw

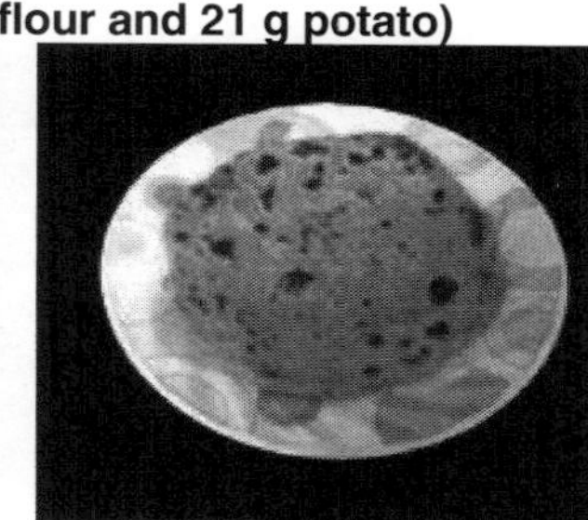

Cooked

Rice

Raw Rice (Raw weight)= 19 g

Cooked Rice (Cooked weight)= 65 g

Noodles (Raw weight = 25 g)

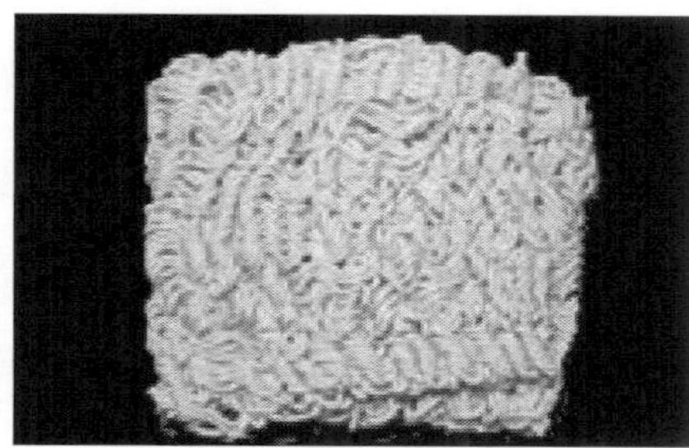

Raw

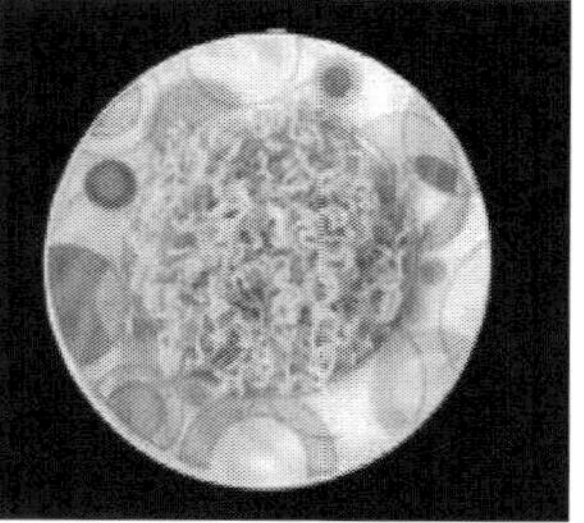

Cooked

Poha (Raw weight = 19 g)

Raw

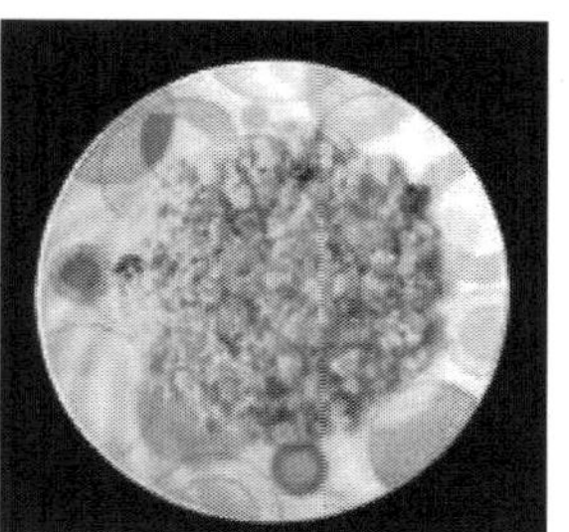

Cooked

Upma (Raw weight =20 g)

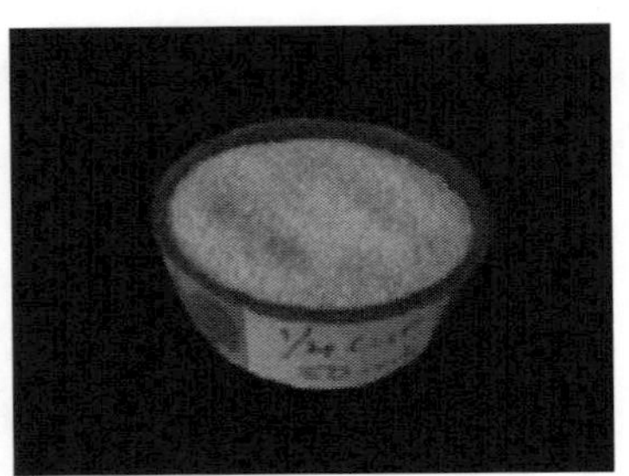

Raw

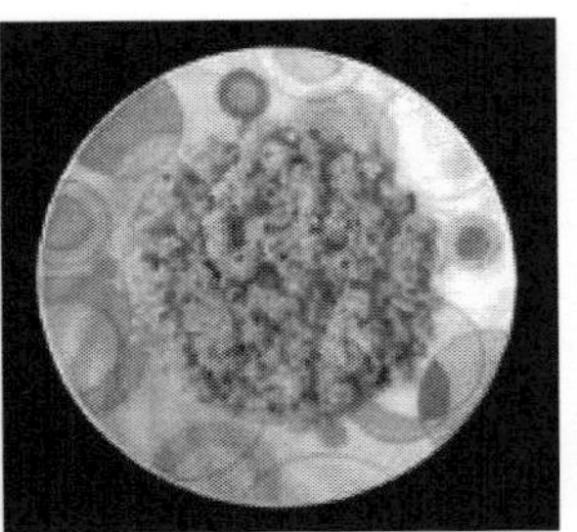

Cooked

Sabudana (Raw weight = 17 g)

Raw

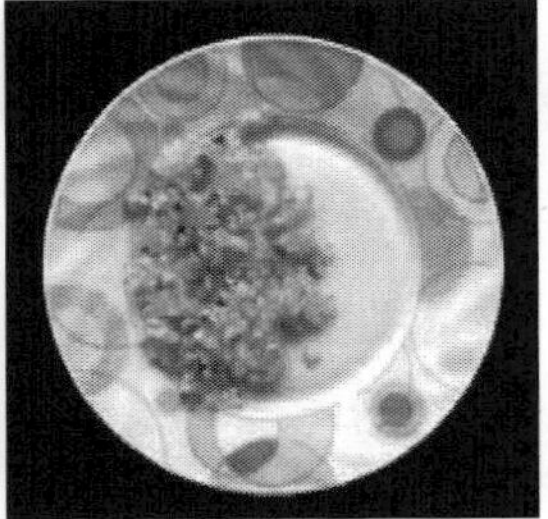

Cooked

Idli & Dosa (Raw weight = 21 g)

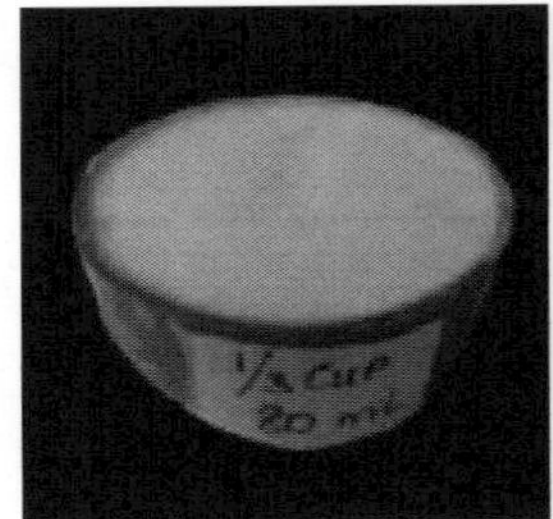

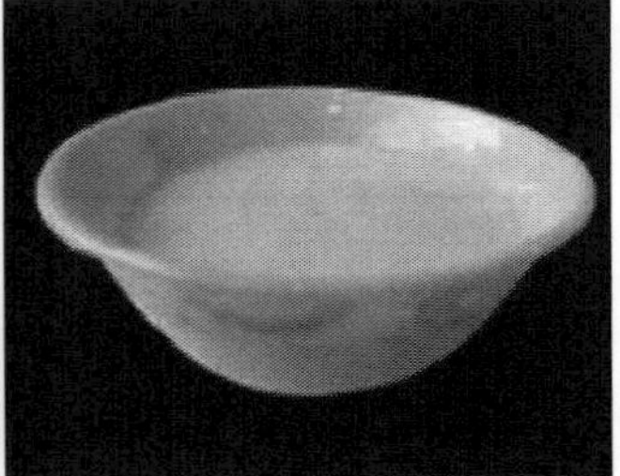

Batter

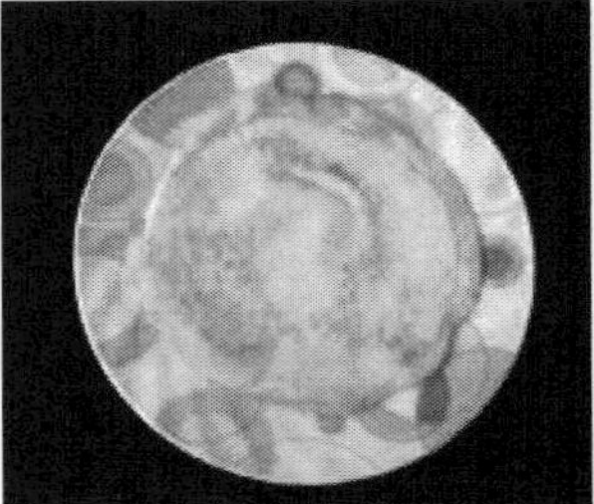

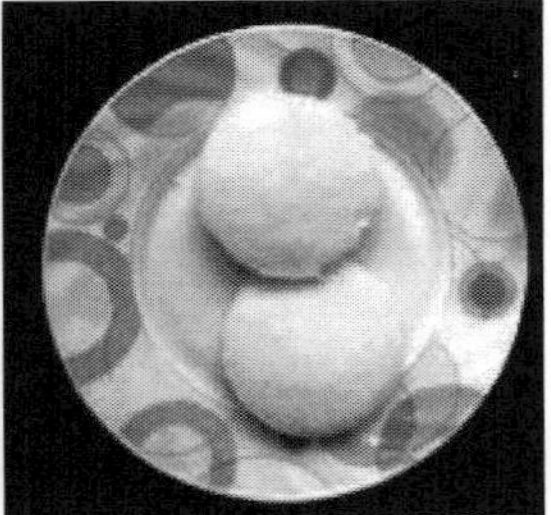

Dosa and idli

Homemade Idli (7.5 g carbs each)

Restaurant Idli (15 g carbs each)

Idli size will vary from restaurant to restaurant

Starchy Vegetables

Corn 1/2 cup

Corn on the cob 1/2 cob (15 cm)

Peas 1 cup

Plantain, ripe 1/3 cup

Potato, boiled, sweet potato 1/2 cup (90 g)

Potato, mashed 1/2 cup

Popcorn 3 cups

Each of the foods shown below represent 15 g carbohydrate exchange

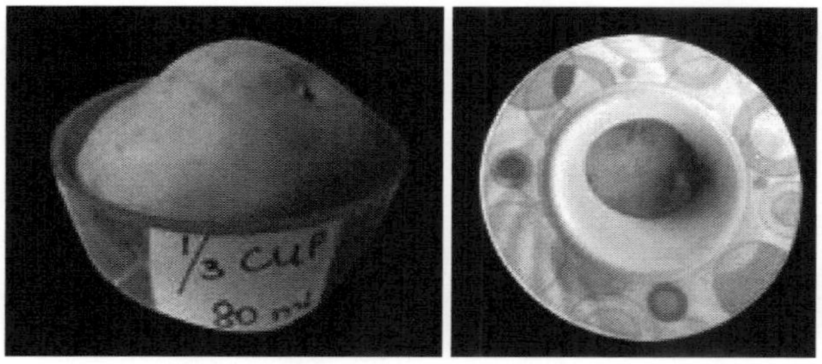

Raw  | Cooked

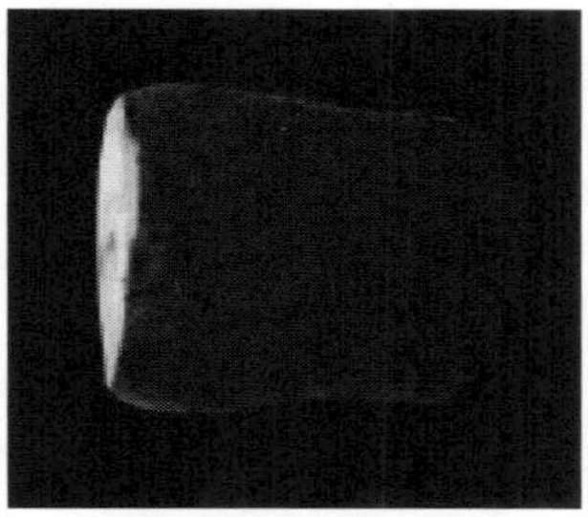

Sweet Potato (53 g)  | Green Peas (94 g)

Sweet Corn (61 g)  | Popcorn (24.6 g)

Pulses

On an average, 25 g of pulses give 15 g of carbs I 1 serving =15 g carbohydrate

Each of the foods shown below represent 15 g carbohydrate exchange

Dal (Raw weight = 25 g)

Raw | Cooked

Moong Chilla (Moong dal raw weight = 25 g)

Batter

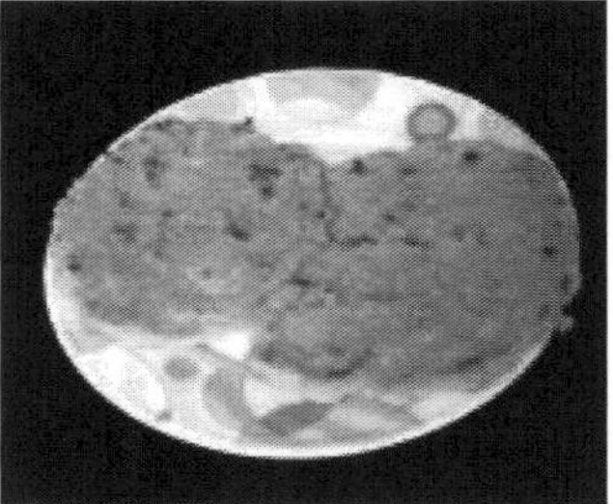

Cooked

Cowpea/ kidney beans (Raw weight=25 g)

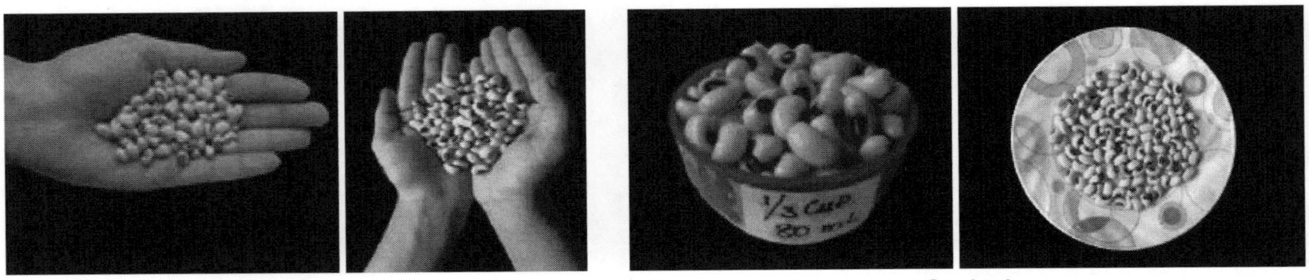

Raw **Soaked**

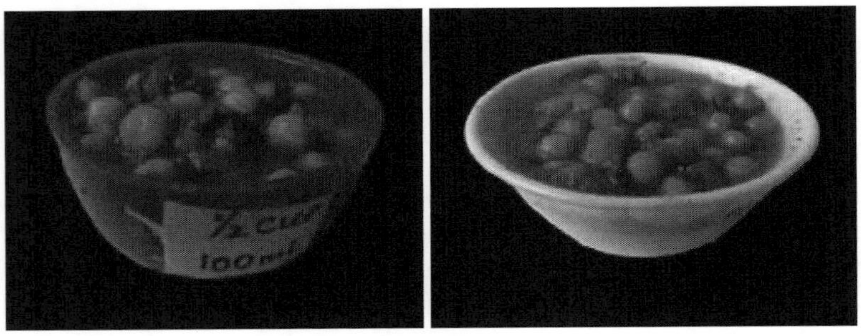

Cooked

Moong (Raw weight = 25 g)

Raw

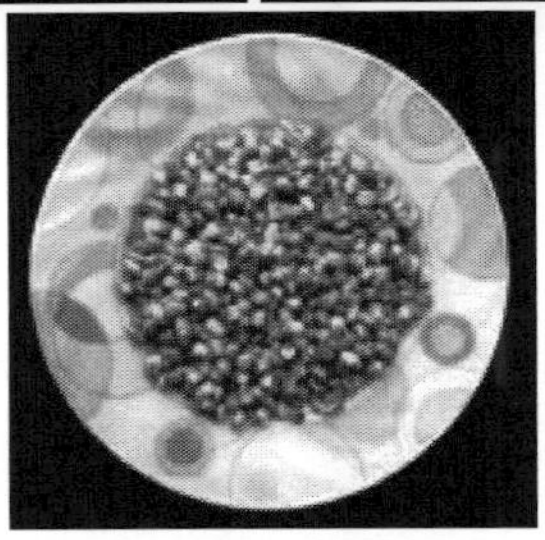

Sprouted (quantity doubles as well as the fiber content increases)

Roasted Chana (Raw weight= 25 g)

Fruit Exchange

All fruits are not the same. Some fruits are higher in calories than the others of the same weight. In such cases, there is a need to keep a check on the portion of fruit one eats. Raw or partially ripe fruits are preferable for people with diabetes.

One serving (exchange) =15 grams of carbohydrate, no protein or fat, and 60 calories

Fruits in the amounts listed below equal one exchange (15 g). Weight includes skin, core, seeds and rind.

Apricots, fresh 4 whole (~ 150 g)	Melon, 1 slice (300 g)
Banana, small 1 (120 g)	Orange, small 1 (180 g)
Blackberries 3/4 cup	Strawberries (medium-7no.)
Cherries, fresh 12 (90 g)	Pineapple, canned 1/2 cup = 2 slices
Figs, fresh 2 medium (100 g)	Plums, small 2 (150 g)
Fruit cocktail 1/2 cup	Plums, dried/prunes 3
Grapefruit, large 1/2 fruit (300 g)	Raisins 2 Tbsp
Grapes, medium 12 (90 g)	Tangerine, small 2
Guava, medium 1 (120 g)	Peach, canned 1/2 cup
Kiwi 1 (100 g)	Pineapple, fresh 3/4 cup = 3 slices
Nectarine, small 1 (150 g)	
Fruit Juices (unsweetened)	
Apple juice 1/2 cup	
Grapefruit juice 1/2 cup	
Grape juice 1/3 cup	
Orange juice 1/2 cup	
Pineapple juice 1/2 cup	

Each picture shown below represents edible portion of fruit containing 15 g of carbohydrate

Mango (90 g,small ½ cup cubed)

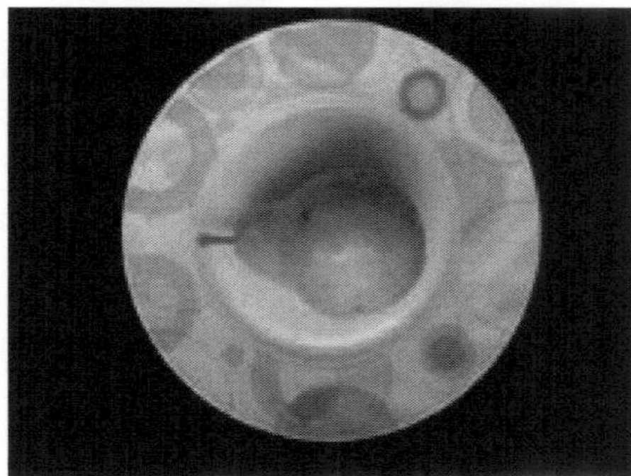

Pear (126 g,1 medium)

Pomegranate (103 g,1small)

Sweet lime (161 g, 1 big)

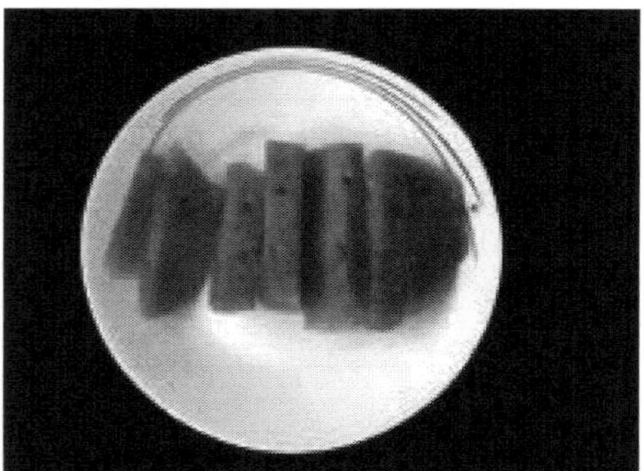

Watermelon (454 g, dinner plate, 10" diameter)

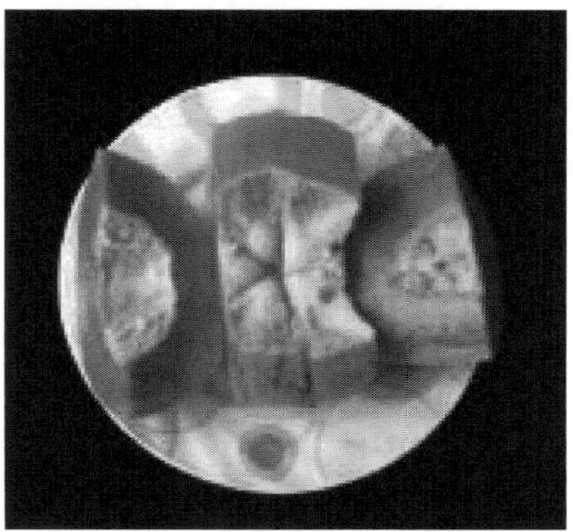

Papaya (208 g, quarter plate, 7" diameter)

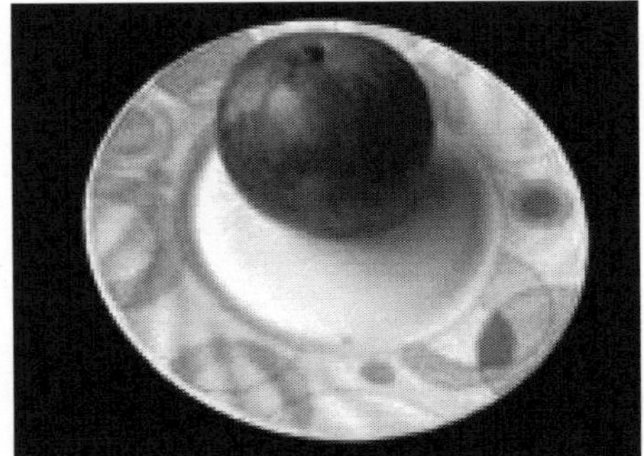

Apple (120 g, 1 medium)

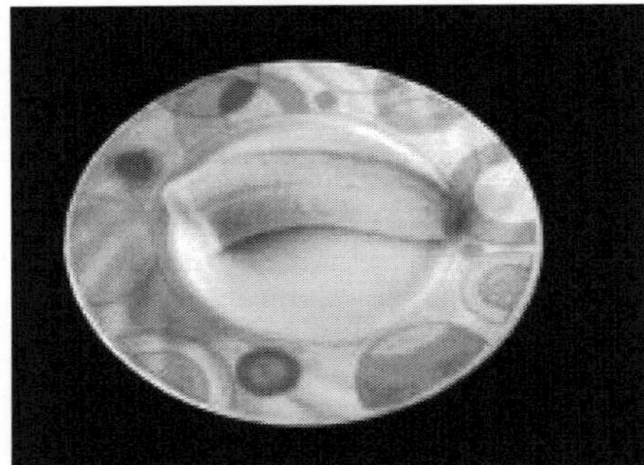

Banana (55 g)

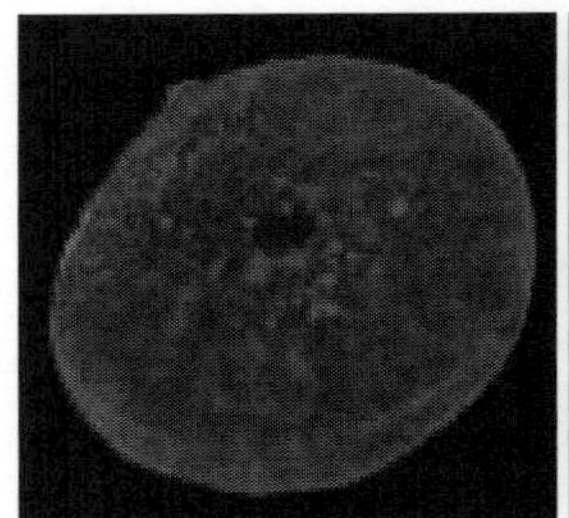

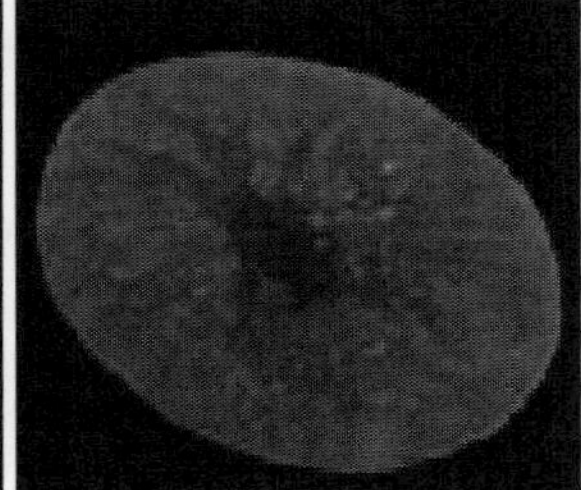

Dry Figs (2 nos)

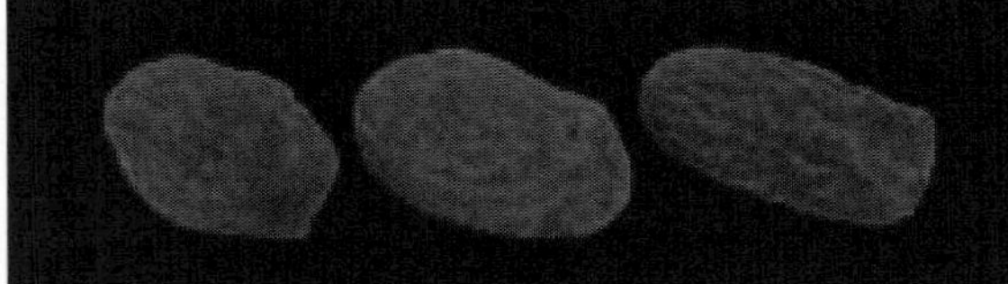

Dried Dates (3 nos)

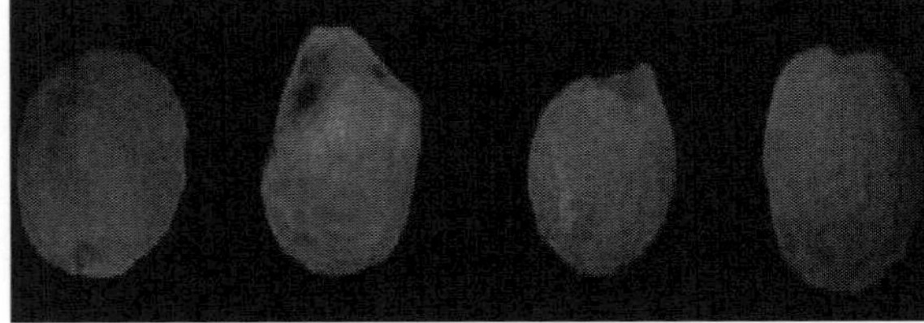

Apricots (4 nos)

Milk Exchange

Milk and milk products like curd, buttermilk, low fat cheese, whey water and low fat *paneer* from the dairy group are excellent sources of high quality protein and calcium. They must be included in the meal plan to ensure strong bones and teeth. Though mik is low in carbohydrates but it contains saturated fat, therefore, it is desirable to include low fat milk and milk products.

Each of the foods shown below represent 5 g carbohydrate exchange

Curd -166 g
5 g carbohydrates

Milk -100 ml
5 g carbohydrates

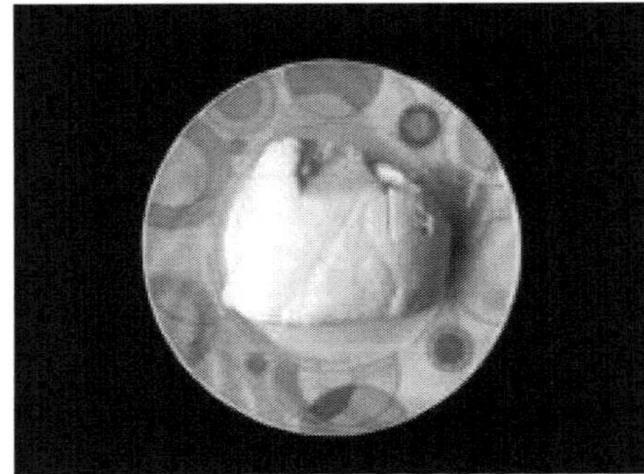

Paneer - 250 g
5 g carbohydrates

Non-Starchy Vegetables

Non starchy vegetables contain small amounts of carbohydrates and calories, but they are rich sources of fiber and micronutrients such as vitamin C, folic acid, B complex vitamins, iron and calcium. One should try to include raw in the form of salad or cooked vegetalbe, two to three times a day in the meals. One should choose more dark green, yellow and red colored vegetables such as leafy vegetables, bell peppers, broccoli and others as they are rich in antioxidants. Fill up with vegetable juices, clear vegetable soups, salads, stir fry or some sautéed vegetables in the whole day meal.

One serving (exchange) = 5 grams of carbohydrate, 2 grams of protein, no fat and only 25 calories.

Examples of such vegetables are:

Amaranth	Brussels sprouts	Peppers, all varieties	Water chestnuts
Asparagus	Bottle gourd, ash gourd, bitter gourd	Soybean sprouts	Eggplant
Baby corn	Cabbage (green)	Spinach, mustard leaves	Green onions
Bamboo shoots	Carrots	Mint, Coriander leaves, bathua	Scallions
Green beans	Cauliflower	Summer squash	Greens, collard, Leeks

Bean sprouts	Celery	Tomato	Mixed vegetables without corn, peas or pasta
Beets	Onions	Vegetable juice cocktail	Mushrooms
Broccoli	Radish	Cucumber	Okra (Lady Finger)

If 3 cups or more of raw vegetables; or one and a half cup or more of cooked vegetables are included in the meal, one should count them as one carbohydrate exchange or 15 g of carbohydrate.

Sweets, Desserts and other Snacks

Try cutting down on "unhealthy junk food" such as chips, chocolates, bakery foods, candy, soft drinks, fried foods, refined foods like white breads, parathas, naan, doughnuts, khari, nankatais, biscuits, brownies, cakes, *etc.* as much as possible because they contribute large amount of calories from fat and sugar without really providing any nutritional value.

If you choose to eat sweets or desserts, do so in moderation and make sure that you do not exceed your calorie and carbohydrate allowance for the day. Balance and moderation is the key here especially if diabetic patient is watching his weight too. This will also prevent blood glucose from rising.

Each of the foods shown below represent 15 g carbohydrate exchange

Note: Remember to include the exchanges in these products as part of your daily allowance.

Sugar

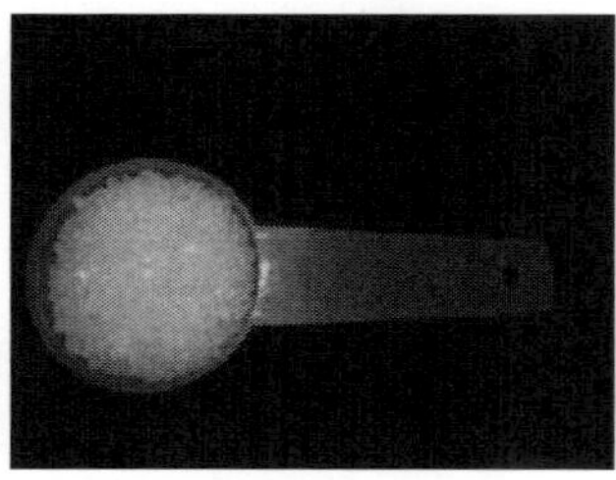

1 tablespoon (15 g)

Honey

1 tablespoon (15 g)

Jaggery

1 tablespoon (15 g)

Gulab Jamun

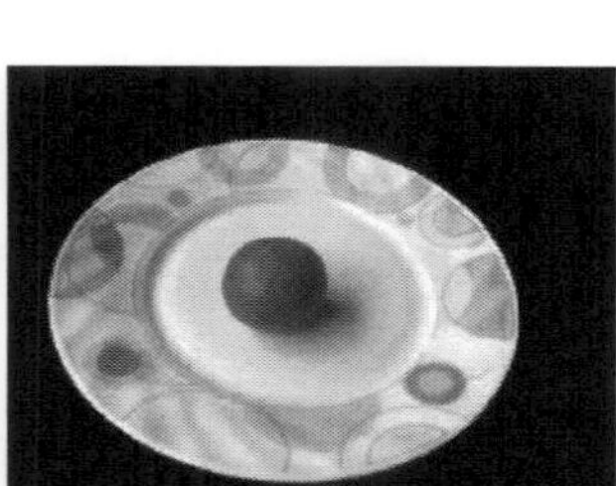

(1 no., no sugar syrup, 40 g)

Seviyan kheer

Motichoor ladoo (1/2)

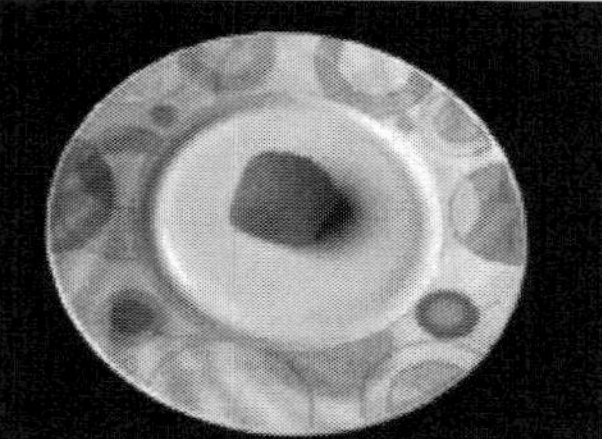

Cup cake

(1 no., edges shaved off)

Rasmalai

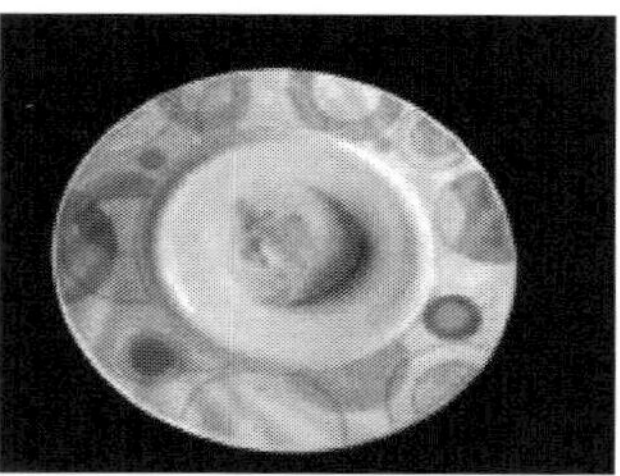

(1 no., no sugar syrup)

Pudding

(1 small pc)

Puran Poli

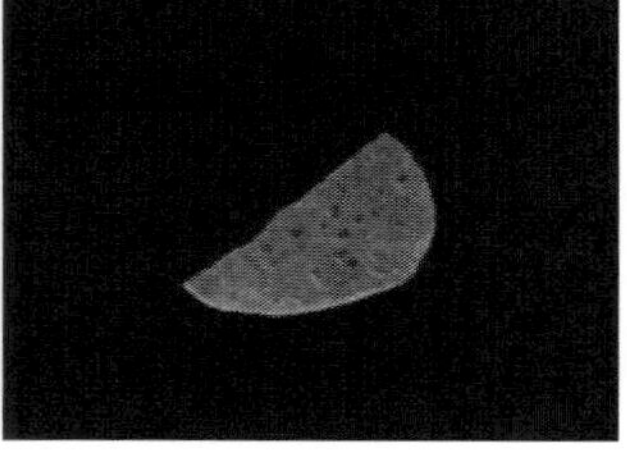

(1/2 no)

Other snacks

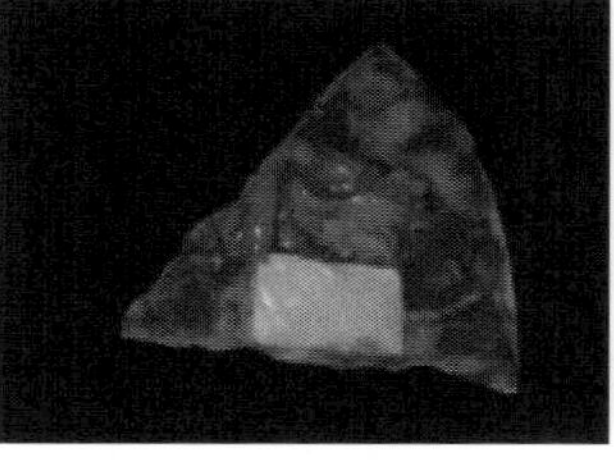

Veg Pizza (6" ,1 slice)

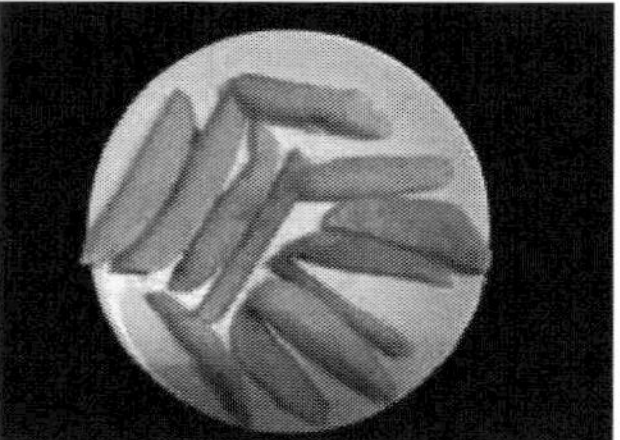

French Fries (12 nos, 66 g potato)

Dhokla (2 nos)

Khandvi (12 medium pcs)

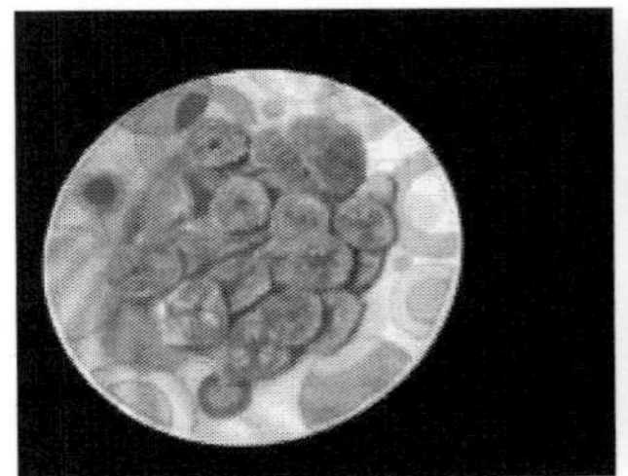

Banana chips (20 chips)

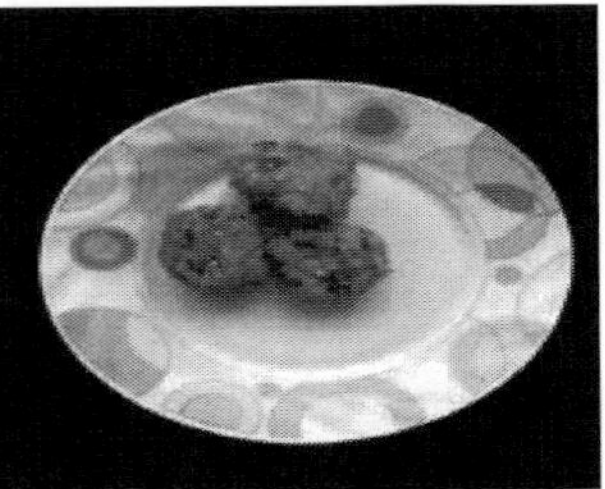

Muthia (3 small)

125 g (1 stick) 17 g

Carbohydrate content of some other common eaten snacks is as given below:

Menu Item	***Serving Size***	***Total Carbohydrates (g)***
Paani Puri	5 puris	20
Dahi papadi	5 papadi	21
Aloo tikki	1 (50g potato)	11
Bhel puri	1 cup	26
Dahi wada	1 wada	22
Sev papadi	5 papadi	33
Bread pakora	1	24
Onion pakoda	1 no. (big)	10
Masala peanuts	30g	10
Khari	30g (3 nos. small)	15
Cheese sandwich	1 no. (2 bread slices)	22
Grilled sandwich (small) with potato filling	1 no. (2 bread slices)	26
Vegetable Frankie or vegetable wrap	1 no.	34
Chicken Frankie or chicken wrap or chicken roti	1 no.	28
Vegetable cutlet	1 no.	21
Samosa	1 no.	35
Kakori kabab	3 nos. (102 g)	3
Shammi kaebab	1 no. (50g)	5
Naan, bhatura, makki ki roti *etc.*		
Wada	1 no.	20
Naan	1 no. (90g)	47
Bhature	1 no. (85g)	35
Makki ki roti	1 no. (24g)	50g
Biscuits (The following numbers of biscuits contain 15 g carbohydrate)		
Parle G 3 biscuits	Marie Gold 4 biscuits	Good day (cashew pista badam) 3 biscuits
Good day (butter) 2 biscuits	Priya Gold classic cream 3 biscuits	50-50 maska chaska 7 biscuits
Chocopie 1 pie	Nutri Choice (ragi/oats) 3 biscuits	Bourbon 1.5 biscuits
Hide and seek 4 biscuits	Oreo 3 biscuits	Mc Vities Digestive 2.5 biscuits
Dark fantasy 2 biscuits	Cream crackers 3 biscuits	Monaco biscuits 5 biscuits
Britannia 5 grain 1 biscuit	Digestive high fiber 3 biscuits	Jim Jam 2 biscuits
Oat meal cookies 3 biscuits	Nice 2 biscuits	
Chocolates		
65 g (10 pcs) of Cadburys Dairy Milk Silk has 37.9 g of carbs. This means 1 pc has 3.8 g of carbs. One can calculate the carb Content depending on the amount one eats.		
Chips		
100 g of Frito lays (India's Magic Masala) has 51.4 g of carbs. A medium sized packet of Frito lays is 26 g. 26 g (17 chips) has 13.4 g of carbs. One can calculate the carbohydrate content of different chips available in a similar manner.		

Juices and Aerated Drinks

From the food label, one can calculate the amount of carbohydrate present. For example, the Réal fruit juice 100 ml has 13.5 g of carbohydrate. If one drinks a glass of Réal juice (200 ml), the carbohydrate content in the juice is 2 x 13.5 = 27 g which is almost equivalent to 2 carbohydrate choices. One should avoid taking juices and aerated drinks. These should be consumed only when blood sugar is too low and there is a need to increase it rapidly.

Nuts and Dry Fruits Exchange

Nuts and dried fruits are rich in calories, hence one should not consume these in excess.

Each of the following is equivalent to 15 g carbohydrate or one carbohydrate exchange.		
Dates (fresh) 6 nos	Dry dates 3 nos	Raisins 35 nos.
Groundnuts 30 nos	Cashewnuts 45 nos	Walnuts 65 nos
Almonds 110 nos	Pistachios 88 nos	

Advantages of Carbohydrate Counting

It places importance in keeping the carbohydrate content of meals and snacks consistent from day to day. Eating the same amount of carbohydrate for the meals and snacks everyday ensures better blood glucose control. Carbohydrate counting ensures that one can have variety and flexibility in the diet and most importantly can follow the traditional diet. This method is also useful for people who are using more intensive methods of adjusting insulin to control diabetes. The amount of meal and snack carbohydrate is adjusted based on the pre-meal blood glucose reading. Depending on the reading, more or less carbohydrate may be eaten. Likewise, insulin may be adjusted based on what the person wants to eat. For example, if one wants to eat a much larger meal than usual, carbohydrate counting can help to determine how much extra insulin to take.

V. Insulin to Carbohydrate Ratio

A commonly used ratio is one unit of rapid-acting insulin per 10g of carbohydrate, or 1:10. This can vary from person to person, and your ratio might end up being 1:15, 2:10, or something else. So, if your ratio was 1:15, eating 45g of carbohydrate with a meal would require you to inject three units of insulin. You can also have different ratios for different times of day. The following methods are used for this purpose:

i) Method #1: Food diary, insulin dose, and SMBG (self monitoring blood glucose) information.

Insulin: Carbohydrate ratio = Grams of carbohydrate at a given meal number of units of insulin taken at that meal

For example: Patient needs 4 units of insulin to cover 45 grams of carbohydrates.

Insulin: Carbohydrate ratio = 45 4 = 11.25

Ratio = 1:11

ii) Method#2: The rule of 450 or 500

For example: Insulin: Carbohydrate ratio = 450 or 500 total daily insulin dose (TDD)

- TDD = 36 units
- Glucose levels are within target range
- Insulin:Carbohydrate ratio = 500 36 = 13.8 (round up to 14)
- Ratio = 1:14

iii) Method#3: Using the insulin sensitivity factor (ISF)

- ISF = 1500 or 1800 TDD (Total daily insulin dose)
- Insulin: Carbohydrate ratio = ISF 0.33

For example:

- TDD = 25 units
- ISF (Insulin sensitivity factor) = 1800 25 = 72 mg/dl
- Insulin: Carbohydrate ratio = 72 0.33= 23.8 (round up to 24)

Ratio = 1:24

Amount of Insulin Needed to Cover a Certain Amount of Carbohydrate

Example: 1 unit of insulin for every 15 grams of carbohydrate is taken. So for something 30 grams one would give 2 units of insulin. Apply correction factor to it.

Correction factor: A specified amount of insulin for specified mg./dL above target blood sugar.

Example: 1 unit of insulin for every 50 mg/dL above 150mg/dL

Example 1: Amount of insulin required for the following meal consisting of:

- Cheese sandwich
- 1 cup raw carrots salad
- Small bag chips
- One cup 2 per cent milk
- Small apple

By carbohydrate counting method, we shall come to know about the total carbohydrates in this meal:

Most sandwiches are made with 2 slices of bread (each slice is 15 grams), carrots are an example of a raw vegetable serving for 5 grams, small bag of chips has a label of 15 grams total carbohydrates, 1 cup milk is 12 grams, 1 small apple 15 grams. Adding all of the carbohydrates equals 77 grams.

For calculating amount of insulin for this meal, we have the following information:

i) Carbohydrates 77 g
ii) Blood Sugar 209 mg/dl
iii) So in this sandwich example, the carbohydrates would be 77g, and if we divide that by 15 we would get 5.13, and then the correction factor 209-150=59, then divide 59 by 50, you get 1.18., and if you add 1.18 and 5.13, you get 6.3units.

Insulin to Carbohydrate ratio = 1 unit for every 15grams= 77/15=5.13 units

Corrective Factor 1 unit for every 50mg/dL above 150mg/dL = 209-150=59/50= 1.18 units

Total amount of insulin to carbohydrate ratio = 5.13+ 1.18= 6.3 units of insulin is required for this meal.

VI. Healthy Eating Practices

Table 20.2: Healthful Eating Recommendations in Diabetes

Carbohydrates	Specify healthful carbohydrates (fresh fruits and vegetables, legumes, whole grains); target 7-10 servings per day Preferentially consume lower-glycemic index foods (glycemic index score <55 out of 100: multigrain bread, whole oats, legumes, apple, lentils, chickpeas, yams, brown rice)
Fat	Specify healthful fats (low mercury/contaminant-containing nuts, avocado, certain plant oils, fish) Limit saturated fats (butter, fatty red meats, tropical plant oils, fast foods) and trans fat; choose fat-free or low-fat dairy products
Proteins	Consume protein in foods with low saturated fats (fish, egg whites, beans); there is no need to avoid animal protein. Avoid or limit processed meats.
Micronutrients	Routine supplementation is not necessary; a healthful eating meal plan can generally provide sufficient micronutrients. Chromium; vanadium; magnesium; vitamins A, C, and E; and CoQ10 are not recommended for glycemic control. Vitamin supplements should be recommended to patients at risk of insufficiency or deficiency.

Basic Strategies for Type 2 Diabetes

- Encourage weight loss (5-10 per cent) in overweight and obese persons with diabetes.
- Moderate calorie restriction (250–500 kcal/day less) is associated with improved control independent of weight loss in overweight/obese individuals.
- Spread nutrient intake, especially carbohydrate (CHO) throughout the day.
- High intake of vegetables, fruits and whole grains.
- Decrease fat intake.
- Low fat dairy, poultry, fish, legumes, non tropical vegetable oils and nuts.
- Limited sweets, sugar sweetened beverages and red meat.
- Reduce per cent calories from saturated and *trans* fat.
- Encourage physical activity; 3-4 sessions per week; 40 min (average) per session; moderate to vigorous intensity.
- Monitor blood glucose, and add medications if needed.
- Stop smoking.
- Diabetic persons with hypertension should limit sodium intake. Diabetic persons with hypertension should maintain 130/80 mm mercury blood pressure.
- Majority of patients with T2D have a high cardiovascular risk. People with T1D are at elevated cardiovascular risk. LDL-C target should be <70 mg/dl for the majority of patients with diabetes who are determined to have a high risk.
- Use a statin regardless of LDL-C level in patients with diabetes who meet the following criteria:
  - >40 years of age
  - ≥1 major ASCVD risk factor
  - Hypertension
  - Family history of CVD
  - Low HDL-C
  - Smoking

Summary of Some Important Tips for Dietary Management

- No standard meal plan or eating pattern works universally for all persons with diabetes.
- Nutrition therapy should be individualized based on health goals, personal and cultural preferences, access of healthful choices, readiness, ability and willingness to change lifestyle pattern.
- Nutrition intervention should emphasize variety of minimally processed nutrient dense foods in appropriate portion sizes as part of healthful eating pattern.
- Eat a variety of foods. Try to include 5 servings of fruits and vegetables, 6 servings of grains (3

whole grains) and 3 servings of low fat dairy daily.

- Substitute brown rice for white rice. Generally people with diabetes are told to avoid rice altogether. This is a misconception. One can eat rice in reasonable quantities as long as the total carbohydrate for that meal does not exceed the limit.
- Instead of eating only rice, try other grains such as cracked wheat, barley, oats, ragi *etc.* Use them like one would use rice in *pulaos etc.*
- Minimize sucrose intake when substituting for starch.
- Avoid displacing nutrient dense foods.
- Consume plant-based diet (high in fiber, low calories/glycemic index, and high in phytochemicals/antioxidants).
- Eat regular meals and snacks; avoid fasting to lose weight.
- Avoid starchy vegetables. When using potatoes, yams or other starchy vegetable, always remember to cut down on the amount of rice/*roti* eaten at that meal. Better still, select green vegetables in place of starchy ones.
- Use green vegetables freely and prepare them in a small amount of oil.
- Avoid frying as a cooking method; learn to use other methods such as dry roasting, baking *etc.*
- Try to cook with a minimum amount of oil. Use olive or canola oils as they are high in monounsaturated fats which are good for heart. Avoid ghee, butter and cream in cooking.
- Switch to skim or 1 per cent milk and yogurt. When making desserts like *kheer,* use 1 per cent milk or evaporated skim milk and use artificial sweeteners.
- Whole milk and yogurt and products made with them like *paneer* are high in saturated fats, which raise cholesterol in blood. Try making own *paneer, chenna* with low fat milk at home.
- Tofu is a good source of high-quality protein especially for vegetarians. Try using it in place of *paneer* for making dishes like *palak paneer.*
- Avoid coconut milk/cream or use less than called for in the recipe. Coconut milk is high in saturated fat.
- Use 100 per cent whole wheat chapatti flour for making *rotis/phulkas.* Avoid putting any ghee/oil when making *rotis.*
- Glycemic load is more important than Glycemic Index itself for the diabetic population. Half a spoon of sugar in tea will increase postprandial sugar level to a much lower level than a bowl of rice. But definitely if one takes rice instead of equal calorific quantity of sugar, their stomach will be fuller.
- Amounts less than 100g of CHO is not advised as it will lead to ketosis. Beside total amount of CHO, type of CHO and its distribution between meals is extremely important.
- Alcohol contributes to calories and upset the dietary regulations and should be avoided.

Physical Activity

Physical exercise plays an important role in facilitating a good control of diabetes. In addition to strengthening and toning up the muscle, it consumes more energy derived from blood glucose and sensitizes the tissue to the action of insulin, which helps in keeping the blood glucose in lower range. For most people, this means finding ways to move around for at least 30 minutes more nearly every day; 150 min/week of moderate-intensity exercise (*e.g.,* brisk walking) plus flexibility and strength training is sufficient.

Walking, bicycling, stationary cycling, swimming and jumping rope are good exercises. Sudden initiation of a vigorous exercise programme in person of sedentary habits is not desirable.

Acute Complications of Diabetes Mellitus

Hypoglycemia

It is defined as mild condition in which there is blood glucose concentration of less than 70 mg/dl. Sweating, trembling, hunger, confusion, drowsiness, incordination and nausea are some of the symptoms of hypoglycemia. Moderate to severe hypoglycaemia can be a medical emergency in diabetic patients which can be life threatening. Hence, the patient and the care giver should also be aware of two of the most common types of hyperglycemic attack as Somoyogi Phenomenon and Dawn Phenomenon.

It is also very common that excessive control of diet, higher dose of hypoglycemic medicine or insulin, excessive exercise to facilitate blood sugar control, irregular insulin secretion or uncoordinated drug and food intake may result in more dangerous effect of occasional hypoglycemia (low blood sugar level) in hyperglycemic diabetic patients. This should be avoided.

In type I diabetes, dose of insulin is the main reason for hypoglycemic reaction. When this hypoglycemic effect (blood sugar 50-70 mg/dl, moderate hypoglycaemia; <50 mg/dl is severe hypoglycemia) is taking place at night (nocturnal hypoglycemia) it may get unnoticed with severe consequences. The patient has to identify the signs and symptoms of this hypoglycemic phase to be able to handle the blood sugar level. Bedtime snacking is the suggested approach to prevent nocturnal hypoglycemia. Any time when there is a hypoglycemic attack one must immediately give glucose/sugar water as drink for faster absorption. If the patient is unable to drink, then one can put sugar powder in the mouth before one gets medical help. Here one point to remember is that symptoms are similar for both acute hypo and hyperglycemic patient. If the patient is hyperglycemic, little more sugar will not affect much and can be controlled later by medicine. But if severely hypoglycemic, little sugar can save the life. Therefore, one must treat with glucose water to a collapsing diabetic person even though one has a doubt about the specific condition. If the patient is alert and conscious, one can be given glucose-containing foods (fruit juice, soft drink, crackers, milk, and glucose tablets); avoid foods also containing fat. Repeat glucose intake if SMBG (self-monitoring of blood glucose) result remains low after 15 minutes. The patient should consume meal or snack after SMBG has returned to normal to avoid recurrence.

If the patient is severely confused or unconscious (requires help), glucagon injection should be delivered by another person. Patient should be taken to hospital for evaluation and treatment after any severe episode.

Ketoacidosis

Any form of stress, acute infection or neglect of treatment due to carelessness can lead to ketoacidosis. There is intense thirst, polyuria, constipation, muscle cramps, and altered vision. Acetone may be smelt in the breath. Blood glucose levels may be as high as 360-720 mg/dl and low plasma bicarbonate and blood pH. The conditions should be treated with utmost urgency in hospital.

Long Term Complications

Diabetes affects the blood vessels, the blood and heart. Diabetic patients are at great risk of developing atherosclerosis. The long term complications of diabetes include angina, cardiac failure, intermittent claudication and gangrene due to atherosclerosis; blindness due to diabetic retinopathy; difficulty in walking with chronic ulceration of feet due to diabetic neuropathy and also diabetic microangiopathy and diabetic neuropathy due to small blood vessel disease. Wound healing is delayed in diabetics because of neuropathy and hyperglycaemia and ischaemia.

21

Diet in Cardiovascular Diseases

What is 'Heart Disease' or 'Cardiovascular Disease'?

The term "heart disease" is often used interchangeably with the term "cardiovascular disease." Heart disease or cardiovascular disease (CVD) is a collective term for diseases of the heart, blood vessels or both. Other heart conditions, such as those that affect heart's muscle, valves or rhythm, also are considered forms of heart disease.

Heart disease refers to more than 30 cardiovascular conditions but their treatment, symptoms, and prevention often overlap. Cardiovascular diseases commonly include diseases of the heart and blood vessels *e.g.* atherosclerosis leading to angina pectoris, coronary artery disease, coronary heart disease (CHD), hypertension, heart failure, cardiomyopathy, congenital heart disease, peripheral vascular disease and cerebrovascular disease (stroke) *etc.*

The term 'Ischaemic Heart Disease' (IHD) (Coronary Heart Disease) implies cardiac damage as a result of inadequate coronary blood supply. In majority of cases, this is due to narrowing of coronary artery due to atherosclerosis.

The heart muscle is living and needs food and oxygen to survive. There are arteries within the heart muscle which supply the food and oxygen to the heart muscle. These arteries are known as coronary arteries. When one or more of these arteries in the heart get narrowed, or partially blocked, then it leads to Coronary Artery disease. When one or more of these arteries get blocked then some part of the heart muscle does not get food and oxygen. This condition is called a **heart attack or infarction.** The term, coronary artery disease is used for problems which include angina, heart attack, and heart failure. Among cardiovascular diseases, those resulting from thromboembolic complications such as coronary occlusion, infarction, and stroke are the main cause of death and disability.

Prevalence

Cardiovascular diseases (CVD) are the major cause of mortality globally, as well as in India. Cardiovascular diseases, including heart disease and stroke, are the world's number one killers, claiming 17.9 million lives a year globally. Heart attack is the single greatest killer and takes more lives than the next seven leading causes of death put together. In 2015, there were more than 400 million individuals living with CVD and nearly 18 million CVD deaths worldwide. It is expected that by 2030, more than 22.2 million people will die annually from CVDs. Populations in low and middle income countries now contribute 75 per cent of the CVD deaths, which leads to 7 per cent reduction of gross domestic product (GDP) in these countries. Heart diseases or disorders are surprisingly high in newborns and then it usually strikes when we have spent two thirds of our lives. However, heart disease often begins in childhood.

In India, chronic non-communicable diseases have emerged as leading causes of mortality and morbidity only at the turn of the present century. The prevalence of Coronary Heart Disease (CHD)/"Heart attack" is rapidly increasing in India. It has become an "EPIDEMIC". It is a major contributor for mortality and morbidity. Registrar General of India first reported this transition while enumerating causes of deaths in the country for years 2001–2003. Non-communicable diseases (NCDs), also known as chronic diseases and better as socially transmitted diseases, caused more than 50 per cent of deaths in these years and cardiovascular diseases (CVDs) emerged as the most important cause. Serial data from the Census of India have also reported increasing proportionate mortality from CVD in India as the number one killer in the Indian population.

The Global Burden of Disease has been periodically estimating mortality and morbidity burden from various diseases and risk factors in almost all countries of the world for the last 20 years. Data for India have been obtained from GBD website based on Global Burden of Disease (GBD) 2015 study. As per the report of Global Burden of Disease (GBD) Study, in the year 1990, there were 5.2 million deaths from CVD in economically developed countries and 9.1 million deaths from the same cause in developing countries. It has been predicted that by the year 2020, there will be an increase by almost 75 per cent in global CVD prevalence, and almost all of this increase will occur in developing countries. It was reported that mortality from CVD was projected to decline in developed countries from 1970 to 2015, whereas it was projected to almost double in the developing countries.

In terms of absolute numbers, there has been a steady increase in persons dying from cardiovascular diseases (Tables 21.1 and 21.2). The increase is due to increasing mortality from IHD and strokes. The Global Burden of Disease (GBD) study estimate of age-standardized CVD death rate of 272 per 100 000 population in India is higher than the global average of 235 per 100 000 population. Age-adjusted mortality from these conditions has increased by 31 per cent in last 25 years in contrast to declining mortality from these diseases in developed countries. The Registrar General of India reported that CHD led to 17 per cent of total deaths and 26 per cent of adult deaths in 2001-2003, which increased to 23 per cent of total and 32 per cent of adult deaths in 2010-2013. The World Health Organization (WHO) and Global Burden of Disease Study also have highlighted increasing trends in years of life lost (YLLs) and disability-adjusted life years (DALYs) from CHD in India.

Some aspects of the CVD epidemic in India are particular causes of concern, including its accelerated buildup, the early age of disease onset in the population, and the high case fatality rate. It is amply clear from the above data that in India, the epidemiological transition from predominantly infectious disease conditions to non-communicable diseases has occurred over a rather brief period of time.

In India, studies have reported increasing CHD prevalence over the last 60 years, from 1 per cent to 9 per cent -10 per cent in urban populations and <1 per cent to

Table 21.1: Global Burden of Diseases Study Data on Absolute Number of Mortality Due to Cardiovascular Disease, Ischemic Heart Disease and Stroke (in thousands) in Males, Females and Overall from 1990 to 2015 in India (Available at: www.ghdx.healthdata.org/gbd-results-tool).

Year	*Cardiovascular Disease*			*Ischemic Heart Disease*			*Stroke*		
	Male	*Female*	*Total*	*Male*	*Female*	*Total*	*Male*	*Female*	*Total*
1990	828.9	673.1	1501.9	476.2	321.0	797.2	247.4	248.9	496.3
1995	953.5	760.1	1713.5	555.5	370.4	925.9	283.0	278.1	561.1
2000	1095.8	863.2	1959.0	651.2	429.8	1081.0	317.2	302.4	619.6
2005	1211.0	936.7	2148.5	729.2	474.0	1203.2	337.7	314.5	652.2
2010	1451.0	1065.8	2516.8	881.4	553.3	1434.7	371.3	343.3	734.6
2015	1634.9	1146.0	2780.9	1001.8	608.3	1601.1	438.0	364.0	802.0

Note: Numbers are in thousands.

Table 21.2: Global Burden of Diseases Study Data on Age-adjusted Mortality Rates/100,000 Due to Cardiovascular Disease, Ischemic Heart Disease and Stroke in Males, Females and Overall from 1990 to 2015 in India (Available at: www.ghdx.healthdata.org/gbd-results-tool).

Year	*Cardiovascular Disease*			*Ischemic Heart Disease*			*Stroke*		
	Male	*Female*	*Total*	*Male*	*Female*	*Total*	*Male*	*Female*	*Total*
1990	183.9	160.3	172.5	105.6	76.4	91.6	54.9	59.2	57.0
1995	191.5	164.0	178.2	111.6	79.9	96.3	56.8	60.0	58.4
2000	200.8	169.9	185.9	119.3	84.6	102.6	58.1	59.4	58.8
2005	204.4	169.9	187.8	123.0	86.0	105.2	57.0	57.0	57.0
2010	227.5	179.9	204.6	138.2	93.4	116.6	61.3	58.0	59.7
2015	240.5	181.4	212.0	147.4	96.3	122.8	64.4	57.6	61.2

Note: Death rates are in numbers per 100,000.

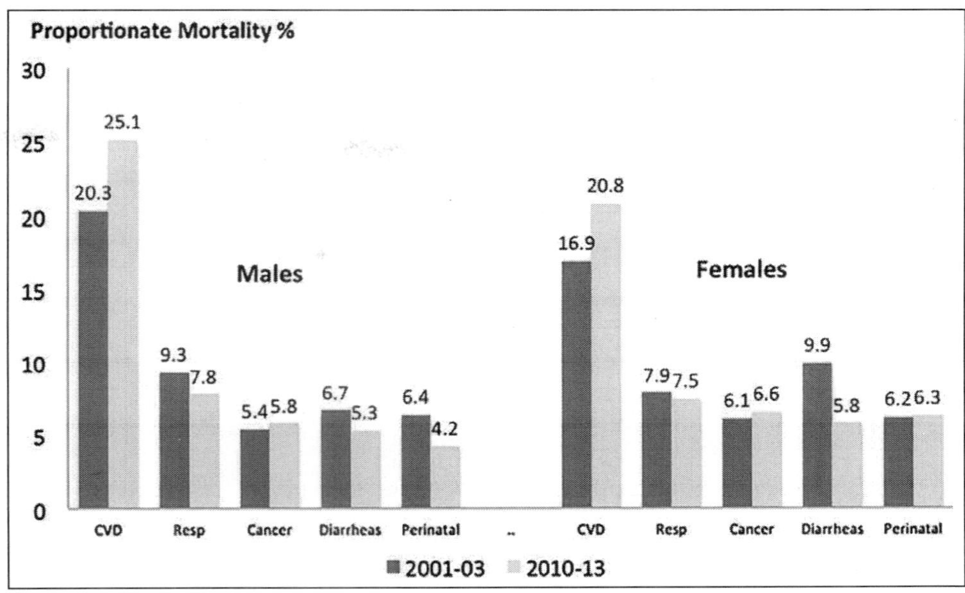

Figure 21.1: Change in Proportionate Mortality from the Top 5 Causes of Deaths in India According to Registrar General of India Reports in Years 2001–03 and 2010–13.

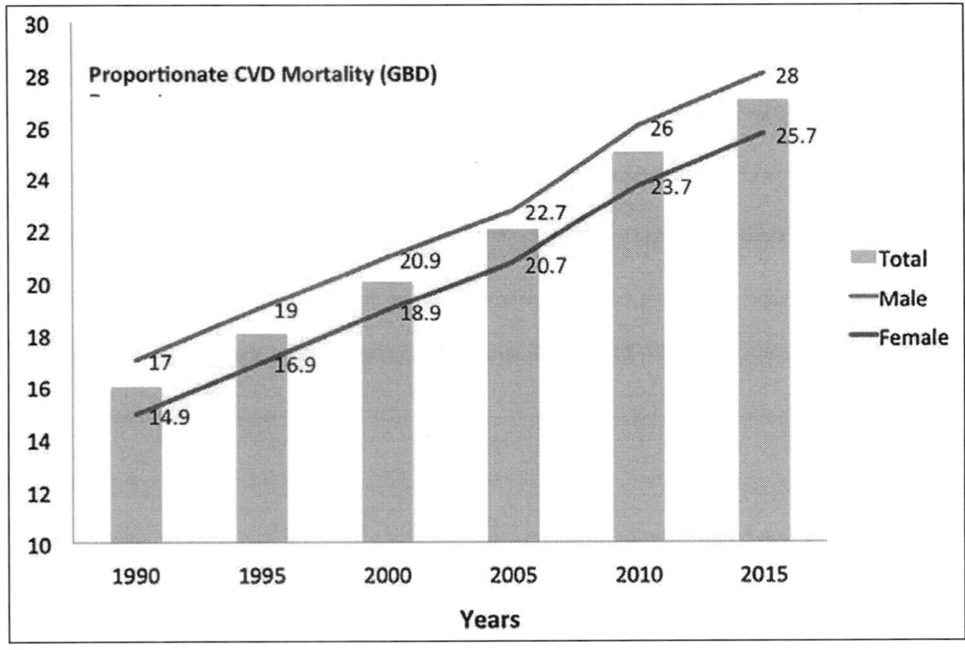

Figure 21.2: Increasing Proportionate Mortality from Cardiovascular Diseases in India. [*Source*: Data are from Global Burden of Disease (GBD) Study 2015].

4 per cent -6 per cent in rural populations. Using more stringent criteria (clinical ± Q waves), the prevalence varies from 1 per cent -2 per cent in rural populations and 2 per cent -4 per cent in urban populations. This may be a more realistic prevalence of CHD in India. Case-control studies have reported that important risk factors for CHD in India are dyslipidemias, smoking, diabetes, hypertension, abdominal obesity, psychosocial stress, unhealthy diet, and physical inactivity (Gupta and Xavier, 2018).

Suitable preventive strategies are required to combat this epidemic. It is high time to make cardiac health a top priority. If the heart is not working, nothing else works. Life to some extent can remain even without a fully functioning brain, but one cannot survive without a heart.

Risk Factors

Knowing risks is the first step to avoiding a heart attack or stroke. Risk factors are conditions that increase the risk of developing a disease. One will not necessarily develop cardiovascular disease if one has a risk factor. But the more risk factors one has, the greater the likelihood that he or she will, unless one takes action to modify the risk factors and work to prevent them compromising heart health.

Risk factors for heart disease are either modifiable, meaning one can take measures to change them, or non-modifiable, which means they cannot be changed. However, many non-modifiable risk factors can be controlled and their effect reduced by making changes to the lifestyle pattern. There are many risk factors associated with coronary heart disease and stroke which are classified and discussed as under:

I. Modifiable and Non-Modifiable Risk Factors

Modifiable Risk Factors

Modifiable risk factors include:

- Diabetes
- Physical inactivity
- Overweight/obesity
- Raised or altered blood cholesterol
- Raised triglycerides with low HDL-cholesterol
- Hypertension
- Smoking
- Excessive alcohol
- Excessive stress

Non-modifiable Risk Factors

- Age
- Gender
- Ethnic background
- Family history

II. Other Probable Risk Factors/Markers

- Other lipid related factors (not cholesterol)
- Vascular endothelial dysfunction
- Oxidative stress
- Coagulation factors *e.g.* fibrinogen
- Inflammatory markers *e.g.* C-reactive protein
- Blood homocysteine
- Maternal/fetal undernutrition
- Vitamin D deficiency

III. There is another way by which cardiovascular risk factors can be broadly divided into the following four groups (Table 21.3):

i) Social determinants
ii) Lifestyle factors
iii) Biochemical factors
iv) Genetic factors

Hundreds of risk factors within these four classes have been identified. There are no prospective epidemiological studies from India that have evaluated relative importance of various CVD risk factors in cardiovascular diseases. Determinants of ischemic coronary heart disease and stroke have been evaluated in large case-control studies in India. Global studies such as INTERHEART (Joshi *et al.*, 2007) and INTERSTROKE (O'Donnell *et al.*, 2016) have reported substantial contribution of standard cardiovascular risk factors in their pathogenesis. Accordingly eight common risk factors are responsible for incident acute coronary disease in South Asians (Gupta and Xavier, 2018):

- High apolipoprotein B/apolipoprotein A1 ratio
- Hypertension
- Diabetes
- Abdominal obesity
- Smoking
- Sedentary lifestyle

Table 21.3: General Classification of Cardiovascular Risk Factors

Group 1 *Social Determinants*	*Group 2* *Lifestyle Factors*	*Group 3* *Biological Factors*	*Group 4* *Genetic Factors*
Urbanization (rapid, low quality)	Sedentary lifestyle	High blood pressure	Genes related to specific risk factors, *e.g.*, lipid abnormalities, hypertension, diabetes, *etc.*
Industrialization (unplanned)	Dietary factors	High LDL cholesterol	
Illiteracy	Smoking and tobacco	Low HDL cholesterol	
Poverty	Alcohol abuse	Triglyceride rich lipoproteins	Inflammatory and thrombogenic genes
Unemployment and working conditions	Indoor air pollution	Lipoprotein (a)	
		Impaired glucose tolerance	Epigenetic factors
Housing		Diabetes	
Outdoor ambient air pollution		Metabolic syndrome	
Maternal nutrition			
Quality of health system			
Political manifesto			

Source: Gupta and Xavier (2018). Hypertension: The most important non communicable disease risk factor in India. Indian Heart J (2018), https://doi.org/10.1016/j.ihj.2018.02.003.

- Low fruit and vegetable intake
- Psychosocial stress

IV. Classification of Risk Factors for CVD Based on GBD (2015) Study for Indians

The GBD studies have estimated importance of various risk factors for disease incidence in India. Estimates in the GBD (2015) study reported the following risk factors as important for disease burden and mortality (Table 21.4). Category 2 risk factors include groups of risk factors and show that air pollution (ambient air pollution and indoor air pollution) and dietary risks (multiple dietary factors) are important. However, when we further sub-classify the risk factors into Category 3 or specific risk factors, elevated blood pressure is by far the most important risk factor.

Table 21.4: Top Ten Category 2 (Broad categories) and Category 3 (Specific factors) Risk Factors for Disease Burden (Disability adjusted life years) and mortality in India: Global Burden of Disease Study (2015)

Ranking	*Category 2 Risk Factors*	*Category 3 Risk Factors*
1	Air pollution	High blood pressure
2	Dietary risks	High fasting blood glucose
3	Child and maternal malnutrition	Ambient particulate matter
4	High systolic blood pressure	Household air pollution
5	High fasting blood glucose	Smoking
6	Unsafe water, sanitation	High total cholesterol
7	Tobacco smoke	Low whole grains intake
8	High total cholesterol	Low glomerular filtration
9	Alcohol and drug use	Unsafe water
10	High body mass index	Low fruit intake

The study also shows that of the top ten Category 2 and 3 risk factors, 7 are related to cardiovascular diseases. These risk factors are:

- High blood pressure
- Air pollution (ambient and household)
- High total cholesterol
- High fasting glucose
- Tobacco
- High body mass index
- Low glomerular filtration rate
- Low physical activity

Some of the important risk factors for CVD as mentioned in various classifications are discussed as under:

Age

As one gets older, the risk of heart disease increases. Percentage of CVD is found to be increasing age wise *i.e.* at the age between 20 and 39, the average percentage of CVD is 11.8 per cent, between 40 and 59, the average percentage of CVD is 38.55 per cent, between 60 and 79, the average percentage of CVD is 73.3 per cent and at the age above 80, the average percentage of CVD is 81.75 per cent. Also age wise death due to CVD is high as compared to cancer.

Gender

Men are at higher risk of heart disease than pre-menopausal women. Women's risk grows and may be equal to men after menopause. Men generally develop CVD at a younger age and have a higher propensity of developing coronary heart disease (CHD) than women. Women, in contrast, are at a higher risk of stroke, which often occurs at older age. Hence, stroke and CHD mortality rates in middle-aged men are typically higher than in middle-aged women, a difference that may persist throughout most of the lifetime. Compared to men, women have less obstructive coronary artery disease and warning signs of CHD in women are often regarded as 'atypical' and tend to go ignored, unrecognized or misdiagnosed.

Ethnic Background

People of some origin *e.g.* from the Indian sub-continent have higher risk.

Family History

If someone in the family has had a heart attack, one should adopt the healthy life style and discuss with the medical doctor.

Homocysteine

It seems clear that high homocysteine levels have potentially deleterious effects on endothelial cells, platelets, and smooth-muscle cells, which could promote a prothrombotic, atherogenic state. Clinical studies have shown those cohorts with high levels of homocysteine levels have increased rates of atherosclerosis involving coronary, cerebral, or peripheral arterial vessels. Among patients with hyperlipidemia, elevated levels of homocysteine levels are independent predictors of atherosclerotic events. High homocysteine levels are seen with renal insufficiency and renal failure.

Lipoprotein a

Genetic studies and numerous epidemiologic studies have identified that high Lp (a) in blood is a risk factor for coronary heart disease, cardiovascular disease, atherosclerosis, thrombosis and stroke.

C-reactive Protein

C-reactive protein (CRP), a plasma protein synthesized by the liver, is a sensitive and dynamic systemic marker of inflammation.

Hypertension

Hypertension is the medical term for high blood

pressure. Blood pressure is the force exerted as the blood pumps through the arteries from the heart and the resistance to the flow of the blood. The optimal BP should be less than 120/80 while high BP is 140/90 or greater. Anything in between is "pre-hypertension". High blood pressure is harmful to the arteries and increases the risk of heart attack, heart failure and stroke. Uncontrolled high blood pressure can lead to a heart attack or stroke. It may also affect kidneys too.

The exact causes of high blood pressure are often not clear but it may be strongly influenced by family history, ageing, dietary pattern, body weight, alcohol intake, physical activity, smoking *etc.* The condition tends to run in families, but blood pressure is also influenced by lifestyle. Hypertension is associated with aging and the lifetime risk for high BP is greater than 90 per cent. The risk for heart disease and stroke *doubles* with every 20 point increase in systolic pressure.

Hypertension is a silent killer for South Asians and the strongest risk factor for stroke. INTERSTROKE study diagnosed hypertension based on measured blood pressure as well as known hypertension and reported that it is the most important risk factor for stroke (both hemorrhagic and ischemic) in India as well as globally. Thus, hypertension is the most important stroke risk factor and among top-3 of coronary risk factors in India and given its high prevalence and high contribution of stroke to CVD mortality in India (40–45 per cent), it is the most important CVD risk factor in India (Gupta, 2018).

To prevent blood pressure from rising, it is important to achieve and maintain a healthy body weight, keep alcohol intake moderate, reduce salt intake, manage and reduce stress and be physically active. Systolic blood pressure can be lowered by as much as 30 points by lifestyle changes. If these measures fail, there are drugs that are effective in reducing elevated blood pressure. Excess alcohol intake seems to be an important contributor to high blood pressure and therefore, one should take alcohol in moderation.

There are effective medications (beta blockers, diuretics, calcium channel blockers, and ACE inhibitors) available for high BP if lifestyle modifications are not enough and one should take medical advice.

Raised Triglycerides with Low HDL-Cholesterol

This combination of having too much of blood TG with low HDL-cholesterol is often linked with premature coronary heart disease. It can be inherited but also occurs in people who are overweight especially when excess fat is carried around the waist. Weight reduction and regular exercise may help to reduce triglyceride levels and increase HDL levels.

Raised or Altered Levels of Blood Cholesterol

Cholesterol is vital for healthy cells. It is so important that the body does not rely on a dietary source, it makes its own. If, however, the body accumulates too much, cholesterol can become deposited in the walls of arteries, which become damaged and may become blocked. If this happens, a heart attack could result. Many people synthesize too much cholesterol when their diet is rich in saturated fats. High blood cholesterol may also be inherited, as in Familial Hypercholesterolaemia (FH) and Familial Combined Hyperlipidaemia (FCH). In these conditions, there are genetic alterations that lead to over-production or accumulation of cholesterol in the blood. To combat this, a rigorous cholesterol lowering diet, usually combined with drug treatment, is required to reduce and control blood cholesterol and the risk of CVD.

Overweight/Obesity

Being overweight increases the chance of having a heart attack. This is in part because people who are overweight are more likely to have high blood pressure, diabetes and high blood fats. In accord with the data on mortality from CHD, the National Health and Nutrition Evaluation Survey (NHANES) III has demonstrated that obesity is associated with an increased prevalence of CHD both in men and in women. Most studies have shown that abdominal obesity (waist to hip ratio) is a better predictor of CVD risk.

Impact of Visceral Obesity on Lipoprotein Physiology

- ☆ Increased sensitivity of visceral adipocytes to lipolytic stimuli
- ☆ Reduced sensitivity of visceral adipocytes to antilipolytic effects of insulin
- ☆ Increased hepatic exposure to free fatty acids with lipolysis in visceral location
- ☆ Increased hepatic triglyceride lipase activity

The increased risk associated with obesity appears to be due not only to changes in lipoprotein metabolism. Weight gain is associated with a clustering of risk factors termed "the metabolic syndrome" (present in Indians), these risk factors include the following:

- ☆ Abdominal obesity
- ☆ Hypertriglyceridemia
- ☆ Low HDL-cholesterol
- ☆ Hypertension
- ☆ Glucose intolerance
- ☆ Small, dense LDL
- ☆ Hypercoagulable state
- ☆ Insulin resistance

Much of the adverse impact of obesity on risk of CHD can be accounted for by the effects of obesity on lipoprotein metabolism. Obesity is associated with undesirable changes in VLDL, LDL, and HDL. Obesity may have a striking impact on serum levels of VLDL and, therefore, on fasting TGs. Marked hypertriglyceridemia in an obese individual can often be effectively treated by loss of weight. Furthermore, body weight is a major factor in the expression of type III hyperlipoproteinemia (dysbetalipoproteinemia).

Effects of Obesity on Plasma Lipoprotein Levels

- ✰ Very low-density lipoprotein (fasting triglyceride) increased
- ✰ Low-density lipoprotein (LDL) cholesterol increased
- ✰ High-density lipoprotein cholesterol reduced
- ✰ Atherogenic pattern of small, dense LDL fostered

Overweight appears to confer coronary risk independent of its association with hypertension, hypercholesterolemia, and glucose intolerance, as demonstrated in the 26-year follow-up in the Framingham Study. Additionally, gaining weight after early adulthood (born as low birth weight baby) appears to contribute independently to CHD risk.

Modest weight loss can reduce risk but weight loss must be maintained. Reducing fat (especially saturated fat), sugar and alcohol, reducing portion sizes, being more physically active and adequate support to enable behavioural change are recognized ways of losing weight.

Insulin Resistance, Dyslipidemia, and Coronary Heart Disease

Hyperinsulinemia and insulin resistance have been reported to be associated with CHD and with other metabolic disorders that themselves predispose to development of CHD. In addition to any independent effect on CHD risk, it is clear that hyperinsulinism is associated with other abnormalities, which themselves increase risk of CHD. Hyperinsulinism is associated with increased prevalence of established risk factors, including reduced HDL-c, hypertension, and obesity. Insulin resistance and hyperinsulinism are felt to be central to the metabolic syndrome, in which there is a clustering of several risk factors. In addition to these very well-established independent risk factors for CHD, hyperinsulinism is also associated with a predominance of small, dense LDL particles, elevated TG levels, impaired hemostasis, and a heightened inflammatory state. It is not clear whether insulin resistance has an impact on LDL-c levels; reports in the literature have not been consistent.

Coronary Heart Disease Risk Factors Associated with Hyperinsulinism

- ✰ Low high-density lipoprotein cholesterol concentrations
- ✰ High low-density lipoprotein (LDL) cholesterol concentrations
- ✰ Small, dense LDL particles
- ✰ Hypertension
- ✰ Hyperglyceridemia
- ✰ Abdominal obesity

Diabetes Mellitus and Coronary Risk

People with diabetes develop atherosclerosis at a younger age and more severely than people without diabetes. Hypertension is more than twice as common in people with diabetes as in people with normal blood glucose levels. People with diabetes are more likely to have a heart attack or stroke, than people who do not, and their prognosis is worse. Diabetes can damage nerves as well as blood vessels so a heart attack can be 'silent', which is lacking the typical chest pain. If one is diabetic, he has a two- to three-fold greater risk of heart failure compared to people without diabetes. Premenopausal women who have diabetes have an increased risk of heart disease because diabetes cancels out the protective effects estrogen.

Some ethnic groups have a much higher rate of diabetes particularly people of South Asian and African Caribbean origin. These people with diabetes have a greater risk of developing coronary heart disease than people who don't have diabetes. People with diabetes are at a much higher risk of CVD because diabetes carries with it a great burden of atherosclerotic risk. Even mild, asymptomatic hyperglycemia leads to increased coronary risk. Uncontrolled diabetes causes damage to body's blood vessels making them more prone to damage from atherosclerosis and hypertension.

Atherogenic disorders of plasma lipoproteins have been associated with asymptomatic hyperglycemia, diagnosed type 1 diabetes, and type 2 diabetes. The major lipoprotein abnormalities of type 2 diabetes are elevated very low density lipoproteins (VLDL), low HDL concentrations, high TG concentrations, and prevalence of small, dense LDL. This combination is commonly termed as diabetic dyslipidaemia and is particularly atherogenic. Type 2 diabetes is strongly associated with a predominance of atherogenic small, dense LDL particles (LDL subclass phenotype B). This association is at least partially a result of hypertriglyceridemia in type 2 diabetes, in men, but not in women, the association of type 2 diabetes with small, dense LDL could be entirely accounted for on the basis of hypertriglyceridemia.

These abnormalities and their pathophysiology are very similar to what is seen in non-diabetics with insulin-resistant states. Additionally, glycosylation of lipoproteins in diabetics whether type 1 or type 2 may enhance the atherogenicity of lipoproteins.

Lipoprotein Abnormalities Characteristic of Type 2 Diabetes

- Hypertriglyceridemia (high very low-density lipoprotein)
- Low high-density lipoprotein cholesterol concentration
- Small, dense low-density lipoprotein (LDL)
- Glycosylated LDL

People with diabetes and a raised cholesterol level experience an even greater risk of heart disease than people without diabetes with the same level of cholesterol. It is possible to have undiagnosed diabetes or impaired glucose tolerance without knowing it, which also increases CHD risk.

Diabetes eliminates the female advantage in CHD rates. The increased relative risk imposed by diabetes in women appears to be due at least in part to increased risk in women with low HDL-c (≤50 mg/dL) and/or high VLDL-c (≥20 mg/dL) levels.

If someone is diabetic, he or she should control blood glucose, blood pressure and cholesterol levels to reduce the risk of coronary heart disease and other cardiovascular diseases. Keeping diabetes under control will help protect the heart health as much as possible. For doing so, one should make life style changes; do more physical activity; eat a healthy well balanced diet; control body weight; give up smoking and alcohol.

Physical Inactivity

Sedentary life style is an important contributor to coronary heart disease. The increasing urbanization and mechanization of the world has reduced our levels of physical activity. The World Health Organization believes that more than 60 per cent of the global population is not sufficiently active. Physical activity, at any age, protects against a multitude of chronic health problems including many forms of cardiovascular disease. Even if you have existing risk factors for cardiovascular disease, if you keep active, the evidence suggests that this may lower your risk of premature death compared to inactive people with no risk factors for cardiovascular disease. But if you do not keep active, the risk to cardiovascular health is similar to that from hypertension, abnormal blood lipids and obesity.

Being active regularly is beneficial for blood pressure, weight control, controlled waist circumference, blood lipid levels, blood glucose levels, blood clotting factors, the health of blood vessels and reducing inflammation, which is powerful promoter of cardiovascular disease. If you are physically active you will increase your life span, regardless of any adverse inherited factors. Studies show that doing more than 150 minutes (2 hours and 30 minutes) of moderate physical activity every week or an hour of vigorous physical activity every day will reduce the risk of coronary heart disease by about 30 per cent.

Smoking

Cigarette smoking, even a few a day, increases the risk of heart disease. Smoking is a major cause of heart disease. It is estimated that smoking increases the risk of stroke, coronary heart disease and impotence by 100 per cent. Smoking increases the risk of death from undiagnosed coronary heart disease by 300 per cent. **Smoking increases the risk of:**

- Heart attack: 2x higher
- Stroke: 3x higher
- Angina: 20x higher
- Peripheral Arterial Disease: 5x higher

Smoking harms the cardiovascular system in many ways:

- It damages the endothelium lining of arteries.
- It reduces the level of HDL.
- It raises the levels of triglycerides and LDL.
- Nicotine present in it increases blood pressure and heart rate.
- The carbon mono-oxide in cigarette smoke displaces oxygen in the blood, reducing the availability of oxygen to the body and the heart has to pump harder to supply oxygen to the body.
- It causes platelets that stick together in the blood stream, leading to clotting.
- It speeds up the development of fatty deposits in the arteries.

Somebody who smokes is four times more likely to die of heart disease (heart attack and stroke) and three times more likely to die from sudden cardiac death. Women who smoke are at a higher risk of heart attack than men who smoke.

Although smoking causes a great deal of damage, quitting smoking effectively reduces cardiovascular risk to close to that of a person who has never smoked over a period of time. After five years, risk of having a heart attack falls to about half that of a smoker and overall risk of coronary heart disease cut by half 1 year after quitting. Second-hand smoke can also cause heart disease in non-smokers.

Excessive Alcohol

Consuming too much alcohol places health at risk in a number of ways. Excess alcohol is harmful to the heart and other organs. It can directly damage the heart muscle and cause irregular beating of the heart. Drinking too much alcohol raises blood pressure and can increase the risk of stroke and heart failure. Alcohol also contributes to weight gain, high triglycerides, strokes and cancer, not to mention accidents and violence. Alcohol in moderation may reduce the risk of heart disease.

Psychological and Social Factors

Chronic stress, chronic hostility and anger, lack of social support, and other psychological and social factors that influence other areas of wellness are also important risk factors of CVD. The cardiovascular system is affected by both sudden, acute episodes of mental stress and the more chronic, underlying emotions of anger, anxiety, and depression.

Stress can exacerbate symptoms in people with pre-existing heart disease, and can contribute to high blood pressure. Stress, particularly emotional upset including anger, as well as depression increase inflammatory hormones such as cortisol and epinephrine in the blood. These contribute to cardiovascular disease by promoting thrombosis (blood clot) in the context of atherosclerosis. Chronic stress is associated with an almost three-fold increase in heart disease risk. In addition, depression is associated with diabetes, obesity, and metabolic syndrome. For those with high cholesterol, stress, anxiety and depression can lead to neglect of a healthy lifestyle, such as poor eating habits or not taking prescribed medication.

Stress can be managed through lifestyle modifications. It is important to find time for relaxation. Simple breathing exercises, sports, music, meditation, reading, engaging in hobbies, taking a walk *etc.* are many ways to reduce stress.

Why Indians are more Prone to CVD?

The burden of atherosclerotic cardiovascular disease (ASCVD) in India is alarmingly high and is a cause of concern. Indians are not only at high risk of developing ASCVD, they usually get the disease at an early age, have a more severe form of the disease and have poorer outcome as compared to the western populations. In comparison with the people of European ancestry, CVD affects Indians at least a decade earlier and in their most productive midlife years. For example, in Western populations only 23 per cent of CVD deaths occur before the age of 70 years; in India, this number is 52 per cent. Reports on CAD in Indians from different parts of the world have shown that Asian Indians are at 3–4 times higher risk of CAD than white Americans, 6 times higher than Chinese, and 20 times higher than Japanese due to the following reasons:

- ☆ Malignant coronary artery disease
- ☆ Multivessel disease and left main coronary artery disease
- ☆ Smaller coronary arteries?

Although prevalence of conventional risk factors such as smoking, hypertension, and hypercholesterolemia is not higher in South Asians than in other ethnic groups, yet, it is seen that some risk factors for atherosclerosis are particularly prevalent among them, including high triglyceride concentration, increased level of total cholesterol to high-density lipoprotein cholesterol ratio, type 2 diabetes mellitus (T2DM) and central or visceral obesity. Throughout the Asia–Pacific region, there are differences in obesity prevalence as well as in body fat distribution.

Some of the important causes for higher prevalence and different pattern of CVD among South Asians especially Indians are as under:

i) Increase Longevity

Life expectancy has improved over the years due to improved health care in India. As a result, the average life expectancy has increased from 49.7 years in 1970-75 to 68.8 in 2018, resulting in a growing aging population which faces an increased risk of heart diseases.

ii) Rapid Industrialization and Globalization

Globally, the cardiovascular disease epidemic has followed socio-economic development and industrialization. The first phase of industrial revolution in the developed countries was characterized with replacement of human effort in creating industrial goods by machines. These machines led to a decline in human effort and resulted in less physical activity and affluence which translated into consumption of unhealthy foods. Creation of industrial cities also led to a profound socio-economic change and environmental pollution. Although it also led to better maternal nutrition and gain in average height of population, in the long-term, the adverse consequences of physical inactivity led to obesity and an exponential increase in multiple cardiovascular risk factors. Similar epidemiological transition is now happening in India.

Higher income levels and globalization have induced a nutritional shift resulting in the rise of unhealthy food and decreased intake of fruits and vegetables. The nutritional shift has moved a number of people to unhealthy eating habits. Between 1983 and 2004, while the per capita consumption of protein went down, the amount of fats' intake increased by more than 25 per cent, both in urban and rural areas. This, coupled with reduced physical activity, gives rise to intermediate risk factors such as hypertension and obesity.

iii) Rapid Urbanization

Most of the rural people are migrating to urban areas for employment and changing their life style pattern under the influence of urbanization and are adopting unhealthy life pattern of urban areas. India's rapid urbanization, with 31.8 per cent of Indians (Census data suggests that 31.8 per cent of India is urban but satellite images suggest that India's urban population is more than twice that figure) living in urban areas and the decadal growth rate for urban areas almost three times the rate in rural areas, has led to a number of issues like reduced physical activity, unhygienic and overcrowded living conditions, growing levels of stress, and higher exposure to pollution which are risk factors for causation of heart diseases.

iv) Life Style Changes

People with changing lifestyles due to growing urbanization, industrialization and globalization are associated with adverse CVD risk factors irrespective of their habitat. Earlier man was taking whole grains, legumes and other foods rich in dietary fiber. Now a days, due to western influence, people have replaced their traditional wholesome foods by intake of more refined, processed ready to prepare and eat junk foods rich in empty calories, having more amounts of sugar and fat but devoid of fibre. Due to consumption of these processed junk foods, energy consumption mainly due to fat and sugars has increased, leading to obesity and its related health problems. Due to availability of energy saving methods of transport and labour saving devices, physical activity has reduced. Stress and strain has increased in the new world. All these life style changes are leading to more occurrence degenerative diseases like CVD.

v) Better Control of Communicable Diseases

Due to availability of better medical facilities, incidence of communicable diseases like diarrhoea, tuberculosis and other respiratory problems has decreased.

vi) Genetic Predisposition

The metabolic syndrome, which can be defined as the constellation of CVD risk factors, is one of the growing public health burdens in the Asia–Pacific region, although they are no more overweight than Europeans and Americans. Indians have genetic/ constitutional predisposition to X- Metabolic Syndrome, a metabolic condition associated with doubling the risk for cardiovascular diseases. Socially we are people of the 21st century but genetically we remain citizens of the Paleolithic era (marked by the agricultural revolution 10,000 years ago). Indians are more prone to central obesity and related heart diseases due to obesity and related insulin resistance syndrome.

India is not just the diabetes capital of the world with more than 50 million patients, it also has the highest prevalence of metabolic syndrome and obesity – 20 million Indians are obese today with 70 million projected by 2025; 20 per cent of Indians suffer from hypertension. These factors together have paced the growth of cardiovascular diseases in India as well as the mortality levels from these diseases.

Table 21.5: Clinical Identification of the Metabolic Syndrome*

Risk Factor	*Defining Level*
Abdominal obesity †‡	Men ≥102 cm (≥40 in)
	Women ≥88 cm (≥35 in)
Triglycerides §	≥150 mg/dL
HDL-c §	Men <40 mg/dL
	Women <50 mg/dL
Blood pressure §	≥130 or ≥85 mm Hg
Fasting glucose	≥100 mg/dL

* Any three of the five risk factors constitutes a diagnosis of the metabolic syndrome† Measure waist circumference on the right at the top of the iliac crest perpendicular to the floor at the end of a nominal expiration; tape should be snug, but not compress the skin. ‡ Some US adults of non-Asian origin (*e.g.* White, Black, Hispanic) with marginally increased waist circumference (*e.g.* 94-101 cm [37-39 inches] in men and 80-87 cm [31-34 inches] in women) may have strong genetic contribution to insulin resistance and should benefit from changes in lifestyle habits, similar to men with categorical increases in waist circumference. Lower waist circumference cutpoint (*e.g.* ≥90 cm [35 inches] in men and ≥80 cm [31 inches] in women) appears to be appropriate for Asian Americans. § Having risk factor or on therapy for this abnormality.

Source: Adapted from: Grundy S.M. *et al.*, Circulation, 2005; 112:2735-2752.

vii) Abdominal Obesity (Thrifty gene hypothesis)

It is noteworthy to mention that in Asian Indians, morbidity and mortality from CVD is occurring in people with lower body mass index (BMI) and smaller waist circumference (WC). Thus, intra-abdominal visceral fat is accumulated in Asian Indians without developing generalized obesity. South Asians (*e.g.* Indians) have a more centralized distribution of body fat and remarkably higher mean waist-hip ratio (WHR) for a given level of BMI compared with Europeans or Americans.

Intra-abdominal fat affects blood pressure, blood lipid levels and interferes with the ability to use insulin effectively. If one cannot use insulin properly, it may lead to development of diabetes, a risk factor of cardiovascular disease.

ix) Stress and Strain

It is a competitive world now. There is lack of social and emotional support due to changing family patterns *i.e.* more of nuclear than joint family system. Otherwise also, socialization has decreased due to coming of digital era. All these have led to physical isolation, stress, strain, anxiety, depression *etc.* which have an effect on heart health directly or indirectly.

Women and CHD

Approximately 50 per cent of all deaths from CHD occur in women. The onset of CHD in women occurs, on average, about 10 years later than in men. As women age, however, the rates of CHD increase; in the postmenopausal woman, the risk from CHD and stroke far exceeds that from cancer. Risk factors in women are similar to those seen in men, although there are some important differences.

HDL-c is a major risk factor in women, and low HDL-c levels are important predictors of CHD mortality in older women. Lp (a) levels in some, but not all studies are a risk factor for premature CHD. Certainly Lp (a) levels rise after the menopause. After menarche, women have higher HDL-c levels than men of similar ages. When women enter the menopause, they often have a rise in weight, higher levels of LDL-c, and lower levels of HDL-c.

Signs and Symptoms of CHD

- Chest discomfort (angina) or chest pain or chest tightness, chest pressure. It is the most common sign.
- Nausea, indigestion, heartburn, or stomach pain
- Pain in throat or jaw or left arm or upper abdomen or back
- Shortness of breath
- Heart palpitation
- Exhaustion or dizziness
- Sweating

Major Forms of Coronary Vascular Diseases

Following are the major forms of CVDs:

- Atherosclerosis
- Heart disease and heart attack (Heart attack, angina pectoris, arrhythmia, sudden cardiac death)
- Stroke
- Congestive heart failure

Atherosclerosis

A form of CVD in which the inner layers of artery walls are made thick and irregular by plaque deposits; arteries become narrowed, and blood supply is reduced.

Coronary Heart Disease (CHD)

It is a heart disease caused by atherosclerosis in the arteries that supply blood to the heart muscle; also called coronary artery disease. Arteries become clogged with cholesterol and other deposits, reducing blood flow to heart. This can make it harder for heart to conduct electrical impulses smoothly.

Angina Pectoris

It is the medical term for chest pain or discomfort due to coronary heart disease. It occurs when the heart muscle does not get as much blood as it needs especially during physical exertion when heart has to work hard. This usually happens because one or more of the heart's arteries is narrowed or blocked, also called ischemia. Angina usually causes uncomfortable pressure, fullness, squeezing or pain in the center of the chest. One may also feel the discomfort in neck, jaw, shoulder, back or arm.

Arrhythmia

A change in the normal pattern of heart beat.

Heart Attack and Sudden Cardiac Failure

Sudden cardiac arrest is different from that of heart attack. A heart attack is when blood flow to the heart is blocked. Sudden cardiac arrest is when the heart malfunctions and suddenly stops beating unexpectedly. A heart attack is a "circulation" problem while sudden cardiac arrest is an "electrical" problem. Sudden cardiac arrest can occur after a heart attack, or during recovery.

Stroke

An impeded blood supply to some part of the brain resulting in destruction of the brain cells, also called CVA (cerebrovascular accident)

Congestive Heart Failure

When sudden cardiac arrest occurs in children or adolescents, it may be due to a heart condition that was present at birth (congenital heart disease). Even adults who have had corrective surgery for a congenital heart defect still have a higher risk of sudden cardiac arrest.

Atherosclerosis

The root cause of most cases of coronary artery disease or cardiovascular disease and stroke is atherosclerosis *i.e.* a buildup of atheroma - a fatty deposit within the inside lining of arteries of heart and brain. A blockage caused by the development of complex atheromatous plaques, composed of inflamed tissues and fatty deposits on the inner surface of the artery. These plaques develop slowly over time and may cause chronic restriction of blood flow leading to pain on exertion (angina) or it may lead to as acute blockage. This most often occurs when the irregular, inflamed surface of the plaque leads to the formation of a blood clot. When this occurs, a region of heart muscle is suddenly deprived of blood and gets damaged. This is what is known as a heart attack. If the damage is relatively limited, the heart can recover but major damage can lead to death. Similar degeneration of arteries in the brain leads to loss of blood supply and stroke or death.

Formation of Atheroma

Three distinct layers make up arterial walls.

i) The intima, or innermost layer bordering the lumen, consists of flattened endothelial cells.

ii) A fenestrated membrane composed largely of elastin separates the intima from the middle layer. This is formed of concentric wrappings of smooth muscle cells, elastin, and fibrous material supported in a matrix of glycoproteins, proteoglycans, and glucosaminoglycans.

iii) The adventitia, or outermost layer, contains fibroblasts and smooth muscle cells loosely arranged among bundles of collagen fibres.

The outer half of the arterial wall is supplied by the capillary network of the vasa vasorum while the inner half relies on diffusion from the lumen for its nutriment.

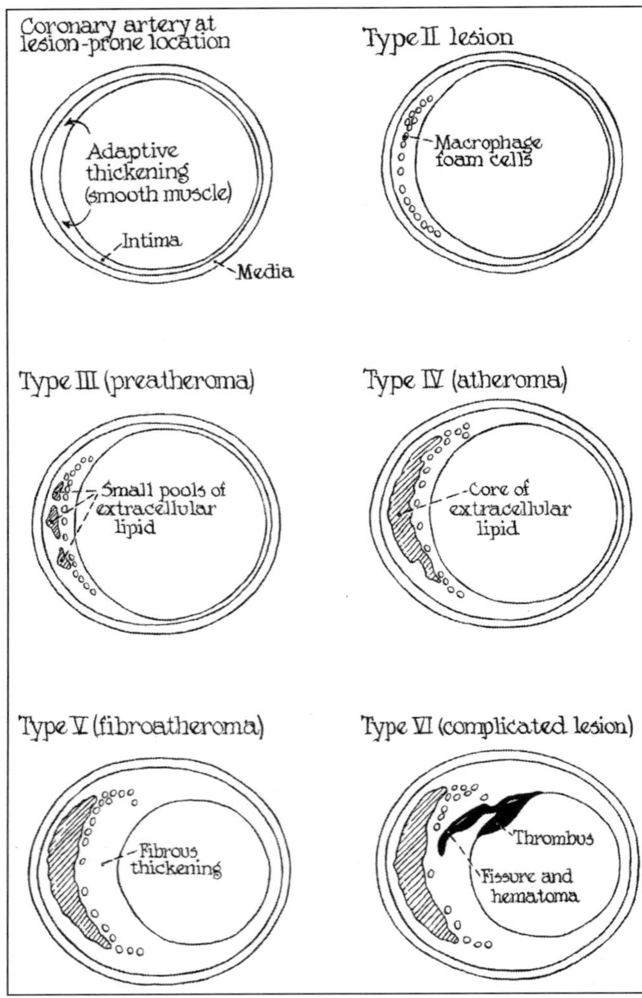

Stages of Atherosclerosis.

Injury to the endothelial cells of the arterial wall increases their permeability and initiates atherosclerosis. This allows the entry of cholesterol from circulating lipoproteins into the deeper layers of the arterial wall. Lipoproteins, which are macromolecular complexes synthesized by the liver for the transport of cholesterol and triglycerides, are also composed of phospholipids and specific proteins termed apoproteins. The various classes of lipoproteins, which include very low density (VLDL), low density (LDL), and high density (HDL) lipoproteins, differ in size and relative proportion of constituent components (triglycerides, cholesterol, phospholipids). The source of cholesterol in atherosclerotic lesions is believed to be the low density lipoproteins (LDLs). Lipid droplets are deposited and can form a dense coating on elastin and collagen fibres. Fat-filled cells derived from smooth muscle cells or circulating monocytes also appear.

Localized thickenings or plaques of fatty and fibrous material on arterial walls cause narrowing of the vessel lumen. Clumps of aggregated platelets called thrombi may form at the sites of atherosclerotic lesions, and impaired circulation and ischemia occur when atherosclerotic lesions and thrombi obstruct the flow of blood.

Various Theories Regarding Atherosclerosis

The exact molecular sequence of events in atherosclerosis is still debated. Three different theories (not mutually exclusive) include the following (Boren *et al.*, 2000).

Response-to-injury hypothesis, emphasizing injury to endothelium and smooth muscle cells by:

- ✰ Oxidized low-density lipoprotein (LDL)
- ✰ Cigarette smoking
- ✰ Hypertension

Oxidation hypothesis, emphasizing oxidative modification of LDL-c and its involvement in the recruitment of macrophages and other inflammatory cells to lesions.

Response-to-retention hypothesis, emphasizing that the initial event in cholesterol-induced atherosclerosis is the retention and accumulation of LDL and other atherogenic (apolipoprotein [apo] B-containing) lipoproteins in the arteries.

The latter theory holds that subendothelial retention of apo B-100-containing lipoproteins is the initiating event in atherogenesis. The events described above lead to a series of biologic responses that promote growth of the atherosclerotic plaque and lead to the formation of "vulnerable" plaques.

Thus, emerging risk factors, such as ultrasensitive C-reactive protein, carotid ultrasound, ankle-brachial index, or calcium quantification by computed tomography (CT) scan, might prove useful in certa*in-situ*ations to indicate those who need more intensive therapy. However, it remains to be proven whether any of these novel risk factors provide clinically useful information beyond that provided by the well-established risk factors, including:

- ✰ Age
- ✰ Gender
- ✰ High LDL-c
- ✰ Low HDL-c

- ☆ Cigarette use
- ☆ Diabetes mellitus
- ☆ Hypertension

Burke looked at CHD risk factors and plaque anatomy and found the following significant correlations:

- ☆ Cigarette smoking and plaque erosion
- ☆ Glycosylated hemoglobin with stable plaque and healed infarct.
- ☆ Hypertension with stable plaque with healed infarct
- ☆ Women >50 years of age and plaque rupture

The focus of therapy then is to prevent the plaque from rupturing and to decrease the thrombogenicity of the plaque.

Atherosclerosis, Lipids and Acute Coronary Syndromes

Coronary atherosclerosis begins in youth and develops over a lifetime. Progression of atherosclerosis varies according to:

- ☆ Heredity
- ☆ Gender
- ☆ Well-defined coronary heart disease (CHD) risk factors
- ☆ Complications of atheroma in advanced stages may occur suddenly, but not necessarily expectedly, as acute events are more likely to occur during:
  - ❒ Early morning hours (6 to 11 AM) due to circadian variation and a host of physiologic variables
  - ❒ Winter months
  - ❒ Heavy physical exertion
  - ❒ Emotional stress
  - ❒ Sexual activity
- ☆ Yet the most concerning and frustrating manifestation of CHD is acute, out of-hospital sudden death. This emphasizes the need to find susceptible but asymptomatic persons in the population and bring them to treatment before sudden death can occur.

Various research studies support the concept that hypercholesterolemia, even when it is present in early life, contributes to the risk of developing CHD. Thus, there is a need to work at improving life-habit risk factors such as:

- ☆ Atherogenic diet (High calorie diet, high saturated fat and cholesterol intake)
- ☆ Increased level of cholesterol in blood
- ☆ Sedentary lifestyle
- ☆ Weight gain, leading to overweight and obesity
- ☆ Stress and strain
- ☆ Drinking soft water

The dynamic relationship between changes as people's diet and health is generally seen in the fast changing patterns of certain diseases. With rapid socio-economic development, decreasing trends in infectious diseases due to better health care and consequently longer life expectancy, chronic degenerative diseases are increasing even in developing countries. The disorders that lead to the processes of atherosclerosis and thrombosis (coronary artery disease, stroke and peripheral vascular diseases) are generally associated with affluence and consequent alterations in lifestyle such as faulty dietary habits, smoking, alcoholism, tobacco chewing and sedentary living.

Major Risk Factors

- ☆ Cigarette smoking (in the past month)
- ☆ Hypertension (BP ≥140/90 mm Hg or taking antihypertensive medication)
- ☆ Low HDL-c (<40 mg/dL)*
- ☆ Family history of premature CHD (CHD in male, first-degree relative <55 years or in female, first-degree relative < 65 years)
- ☆ Age (men ≥45 years, women ≥55 years)

Abbreviations BP- blood pressure; CHD- coronary heart disease; HDL-c -high-density lipoprotein cholesterol.

* HDL-c >60 mg/dL, counts as a "negative" risk factor, its presence reduces the tally of risk factors by one.

Emerging Risk Factors: Lack of Definite Evidence of an Independent Contribution

- ☆ Lp (a)
- ☆ Homocysteine
- ☆ Prothrombotic risk factors (plasminogen-activator inhibitor-1 [PAI-1], fibrinogen)
- ☆ Inflammatory markers
- ☆ Impaired fasting glucose
- ☆ Subclinical atherosclerosis (noninvasive assessment):
  - ❒ Carotid imaging with intimal-medial thickness
  - ❒ Computed tomography (CT) scan of coronary arteries for calcium
- ☆ Ankle-brachial index

Source: Smith SC, *et al.*, Circulation, 2000: 101:111-116.

Types of Fats and their Role in Causation of Atherosclerosis

Many factors have been implicated in the pathogenesis of this disease. Prominent among these factors is the amount and kind of fat in the diet.

Main two types of fats *i.e.* triglycerides and cholesterol are carried in the blood stream. These fats are contributed by the endogenous (*de novo* synthesis in the liver) as well as the exogenous sources (dietary). These fats are transported by lipoproteins in the blood. There are 4 types of lipoproteins.

a) **VLDL or pre β-lipoproteins:** These are synthesized in the liver and transport endogenously produced TG which is derived from circulating CH_2Os and FFAs generally mobilized from the tissues.

b) **LDL or β-lipoprotein:** This accounts for most of the cholesterol in the plasma. It has a short half life of three days and mostly a metabolic product of VLDL after the dissociation of the TG from the VLDL aggregate. It has the highest atherogenic potential. It is also called bad cholesterol.

c) **HDL or alpha-LP**: Carry the remaining cholesterol in the plasma, besides phospholipids. However, its major component is apoprotein. HDL probably has the least disease producing potential. It is also called good cholesterol.

d) **Chylomicrons:** Chylomicrons are the lipoproteins of intestinal origin. These consist almost entirely of dietary TGs with small amounts of cholesterol, phospholipids and proteins. Chylomicrons are present in the plasma for 1-4 h after a meal. The enzyme LP lipase by its action dissociates TG from chylomicron with the resulting FFA being transported to adipose tissue, heart and skeletal muscles for utilization. The glycerol is removed by the liver.

Besides TG and cholesterol, these lipoproteins contain phospholipid and protein. The protein part of the lipoprotein is named as apo-lipoprotein. At least three different apolipoproteins (A, B and C) have been demonstrated in different lipoproteins. There is strong evidence that lipoproteins play a fundamental role in atherosclerosis and their interaction with the arterial wall appears to initiate the cascade of events that leads to atherosclerosis which is a preventable disorder.

Role of Cholesterol in Atherosclerosis

Of all the lipoproteins, it is the LDL cholesterol which plays a central role not only in the initiation of atherosclerosis but also in the progression of atherosclerosis ending in clinical cardiovascular events. Lipoproteins that promote atherosclerosis are termed atherogenic. Apolipoprotein B (apoB) is an important component of atherogenic lipoproteins. Multiple lines of evidence have established that cholesterol-rich LDL and other apolipoprotein B (apoB)-containing lipoproteins, including very low-density lipoproteins (VLDL) and their remnants, intermediate density lipoproteins (IDL), and lipoprotein(a) [Lp(a)], are directly implicated in the development of atherosclerotic cardiovascular disease (ASCVD). Apo-B is the key atherogenic lipoprotein which is a more sensitive measure of risk than LDLc. It provides information on LDL particle size which is difficult to measure directly.

On the other hand, HDL appears to play a protective role, and high levels of HDL particles are associated with less risk of coronary artery disease. HDL facilitates reverse cholesterol transport and delivers cholesterol from the vasculature to the liver where it is excreted in the biliary system or excreted out from the body for example. In doing so, it counteracts the harmful pro-atherogenic lipid particles such as LDL. Besides it, however, HDL exhibits multiple other physiologic effects that may play a role in the reduced risk for atherosclerosis. Genetically, Indians have low HDL level.

Most robust evidence for the role played by LDLc comes from RCT's which had used statins (drug that reduces blood cholesterol). The evidence clearly shows that by reducing LDLc, substantial reduction in CV morbidity and mortality occurs. Many secondary prevention trials, primary prevention trials, and trials conducted in high risk groups, clearly demonstrated the important role of LDLc reduction and the potency of statins in reducing the atherosclerotic cardiovascular risk (Narasingan, 2017. Newer lipid guidelines-interpretations and applications for Indians). It means that amount of dietary fat from sources containing different fatty acids *i.e.* saturated, PUFA, MUFA, omega-3 and 6 fatty acids and dietary cholesterol should be as recommended by the ICMR so as to keep the lipid profile within the recommended range.

Recommendations for Blood Lipid Profile

Lipid levels vary by age, sex, genetic makeup, body weight and life style choices. Although a line between safe and dangerous levels is not easy to draw, there are recent recommendations which are as under (2013 ACC/AHA guidelines):

Cholesterol

Adults

Healthy levels of cholesterol do not vary much for typical adults. Variation of recommended levels tends to change due to other health conditions and considerations. ACC/AHA, 2013 recommendations suggest the following targets:

- ✰ **Total cholesterol levels** less than 200 milligrams per deciliter (mg/dL) are considered desirable for adults. A reading between 200 and 239 mg/

dL is considered borderline high and a reading of 240 mg/dL and above is considered high.

- **LDL cholesterol levels** should be less than 100 mg/dL. Levels of 100 to 129 mg/dL are acceptable for people with no health issues but may be of more concern for those with heart disease or heart disease risk factors. A reading of 130 to 159 mg/dL is borderline high and 160 to 189 mg/dL is high. A reading of 190 mg/dL or higher is considered very high. Indian Heart Association believes that South Asians should have a goal LDL of less than 70 due to their markedly elevated risk for heart disease.
- **HDL cholesterol levels** should be kept higher. A reading of less than 40 mg/dL is considered a major risk factor for heart disease. A reading from 41 mg/dL to 59 mg/dL is considered borderline low. The optimal reading for HDL cholesterol levels is of 60 mg/dL or higher. According to Indian Heart Association, for South Asians, the goal HDL should be 50-60, given their elevated risk.
- **Serum triglycerides** should be 150mg/dl or less while over 200 is considered high. In between is considered borderline risk.

Children

The American Academy of Pediatrics (AAP) recommends all children between 9 and 11 years old should be screened for high blood cholesterol levels due to the growing epidemic of obesity in children. In addition, the AAP recommends cholesterol testing for the following groups of children:

- Those whose parents or grandparents have had heart attacks or have been diagnosed with blocked arteries or disease affecting the blood vessels, such as stroke, at age 55 or earlier in men, or 65 or earlier in women.
- Those whose parents or grandparents have total blood cholesterol levels of 240 mg/dL or higher
- Those whose family health background is not known (*e.g.* many adopted children), or those who have characteristics associated with heart disease, such as high blood pressure, diabetes, smoking, or obesity

Table 21.6: Cholesterol Levels of Children and Adolescents

Classification	*Total Cholesterol**	*Low Density Lipoprotein (LDL)**
Acceptable	<170	<110
Borderline	170-199	110-129
High	>200	>130

* mg/100ml of blood.

- For children in these categories, their first cholesterol test should be after 2 years but no later than 10 years of age.

A child may have high cholesterol for a variety of reasons such as obesity, diabetes, liver disease, kidney disease, or an underactive thyroid. If an initial test shows high cholesterol, the pediatrician will check the child's blood again at least 2 weeks later to confirm the results. If it is still high, the doctor will also determine if the child has an underlying condition. A recent report has indicated that there is good evidence that children with cholesterol problems become adults with high cholesterol. So it is important to monitor the cholesterol of children who may have an increased risk of elevated cholesterol.

Dietary Fat and Cholesterol

If diet is rich in saturated fat, trans fat and cholesterol or there is more production of cholesterol in the liver, it will lead to higher levels of LDL-cholesterol in the blood which is atherogenic. If more amount of fat than the recommended allowance is taken daily, then also LDL-c and triglycerides shall increase in the blood, leading to hypercholesterolemia, hypertriglyceridemia and ultimately atherosclerosis and CHD.

Composition of Dietary Fat

Different types of dietary fats are made up of basic units called fatty acids (Table 21.7). Each type of fat or oil is a mixture of different fatty acids and they have different atherogenic potential and hence, have different effects on cardiac health. These fatty acids are discussed as under:

Saturated Fatty Acids (SFA)

A saturated fat is a type of fat in which the fatty acid chains have all or predominantly single bonds *e.g.* lauric, myristic, palmitic fatty acids *etc.* Saturated fats are usually solid at room temperature. Oils rich in SFA are known as saturated oils. Sources of saturated fat in the diet include:

- Fatty meats
- High-fat dairy products
- Vegetables sources, such as coconut and palm kernel oil.

Table 21.7: Sources of Fatty Acids in the Diet

Name	*Structure**	*Common Sources*
Saturated		
Lauric	C12.0	Coconut oil, palm kernel oil
Myristic	C14.0	Coconut oil
Palmitic	C16.0	Palm oil, beef
Stearic	C18.0	Cocoa butter, beef
Monounsaturated (n-9) †-oleic	C18.1	Olive oil, canola oil, beef, peanut oil, avocado

Name	*Structure**	*Common Sources*
Polyunsaturated (n-6) † - linoleic	C18.2	Corn, cottonseed, safflower, sunflower, soybean oils
Polyunsaturated (n-3) † - linolenic	C18.3	Flaxseed oil, canola oil, soybean oil, nuts (English walnuts)
Eicosapentaenoic	C20.5	Fatty fish such as mackerel, sardines, salmon, and blue fish
Docosahexaenoic	C22.6	Fatty fish such as mackerel, sardines, salmon, and blue fish

* Carbon chain length; double bonds. † By convention, named for the location of the first double bond from the methyl position on the carbon chain.

Saturated fatty acids are known to increase serum total and LDL-cholesterol levels, reduce insulin sensitivity and enhance thrombogenicity and increase CVD risk. Therefore, SFA intake should not exceed 8-10 per cent of total energy (ICMR, 2011).

Saturated fats appear to raise blood cholesterol and LDL-c through their influence on the LDL receptor. Among saturated fats, lauric, myristic, and palmitic are the most cholesterol raising (Table 21.7). Stearic acid has not been found to be cholesterol raising and may, in some circumstances, lower blood cholesterol. Certainly, at the population level, dietary saturated fatty acids are strongly associated with CHD; this was best seen in the 16 cohorts from 14 countries followed in the landmark, Seven Countries study.

Polyunsaturated Fats

Polyunsaturated fats (n-6) are essential fatty acids, thus a diet completely devoid of these fats is not healthy. Linoleic (n-6) and α -linolenic (n-3) acids are the simple PUFA, which are present only in plant foods. These fatty acids contain more than one double bond in their backbone. These fats are represented by the seed oils and often are components of soft margarines. Sources of PUFAs are mentioned in Table 21.8. Chemically, PUFAs are omega-3 [α–linolenic acid, n-3; eicosapentaenoic acid (EPA); and docosahexaenoic acid (DHA)], omega-6 (linoleic acid, n-6) and omega-9 (Erucic acid) fatty acids. Sources of PUFA (linoleic acid) are all vegetable oils except coconut oil *e.g.* safflower, sunflower, cottonseed, corn and soybean oils. Soyabean, rapeseed and mustard oils are the only vegetable oils, which contribute significant proportion of α–linolenic (n-3) acid. Legumes/pulses, mustard and fenugreek seeds and green leafy vegetables are also good sources of α -linolenic (n-3) acid (Table 21.9). On the other hand, fish and fish oils provide long chain n-3 fatty acids *i.e.* eicosapentaenoic acid (EPA) and docosahexaenoic acid (DHA) are present which are biologically more active than alpha-linolenic (n-3) acid present in plant foods. They lower blood cholesterol and LDL-c in relation to their intake in the diet.

PUFAs are essential components of cell membranes. While n-6 PUFAs are predominant in all cells, the nerve tissue has high levels of long chain n-3 PUFA. An appropriate balance of the these two classes of PUFAs, namely, linoleic and α– linolenic acids in the diets is essential for the functioning of vascular, immune, nervous and renal systems and for early human development. PUFA (n-6) decrease plasma cholesterol as well as HDL cholesterol level (only at high intake) and these are also pro-inflammatory in nature and have adverse effects at higher intakes. The lipid lowering and other physiological effects of individual members of the PUFAs vary widely. As compared to linoleic acid, alpha-linolenic (n-3) acid is more beneficial for prevention of inflammation and accumulation of fatty material in blood vessels (atherosclerosis) and clotting of blood (thrombosis); n-3 PUFA increase insulin sensitivity, increase peripheral glucose utilization and decrease adiposity and hence, are anti-atherogenic.

Table 21.8: Major Types of Fatty Acids in Fats and Oils

Saturated	*Monosaturated*	*Polyunsaturated*		
		Linoleic (n-6)		*α-Linoleic (n-3)*
		Low	Red palm oil Palmolein	Rapeseed, Mustard Soyabean
		Moderate	Groundnut, Ricebran Sesame	
Coconut Palm kernel Ghee/butter Vanaspati	Red palm oil Palmolein Groundnut Ricebran Sesame	**High**	Safflower, Sunflower, Cottonseed, Corn, Soybean	

Source: Dietary guidelines for Indians (2011), ICMR.

Table 21.9: Quantities of Foods Required to Furnish 0.1 g ALA

Foods	*Gram*
Cereal/Millet	70
Wheat and Pearl millet (*bajra*)	
Pulses	
Blackgram (*kala chana*), kidney beans (*rajmah*) and cowpea (*lobia*)	20
Other pulses	60
Vegetables	
Green leafy vegetables	60
Other vegetables	400
Fruits	400
Spices	
Fenugreek seed (*methi*)	5
Mustard (*sarson*)	1
Unconventional	
Flaxseed (*alsi*)	0.5
Perilla seeds (*Bhanjira*)	0.3

Source: Dietary guidelines for Indians, ICMR (2011).

The minimum intake levels for essential fatty acids to prevent deficiency symptoms are estimated at a convincing level to be 2.5 per cent E LA plus 0.5 per cent E ALA. Based on epidemiologic studies and randomized controlled trials of CHD events, the minimum recommended level of total PUFA consumption for lowering LDL and total cholesterol concentrations, increasing HDL cholesterol concentrations and decreasing the risk of CHD events is 6 per cent E. Based on experimental studies, risk of lipid peroxidation may increase with high (>11 per cent E) PUFA consumption, particularly when tocopherol intake is low. Therefore, the resulting acceptable range for total PUFA (n-6 and n-3 fatty acids) can range between 6 and 11 per cent E. The adequate intake to prevent deficiency is 2.5 – 3.5 per cent E. Thus, the recommended range (ADMR) for PUFA (n-6 and N-3 fatty acids) according to WHO (2008) is 6 – 11 per cent E. The adequate intake to prevent deficiency in 2.5-3.5 per cent.

As per ICMR (2011), the intake of PUFA should be 8-10 per cent of energy intake. No population has a long-term experience with high amounts of polyunsaturated oils, so their usage is initiated to <10 per cent of energy. Potential side effects from consuming excessive amounts of polyunsaturated oils include:

- ✰ Obesity (fats have 9 kcal/g)
- ✰ Gallstones
- ✰ Lower HDL-c levels

Although excess polyunsaturated fat intake is associated with tumor growth in animals, this has not been seen in the many clinical studies of humans.

Long Chain Omega-3 Fatty Acids and Atherosclerosis

The most common long chain Omega-3 PUFA fatty acids known as EPA (eicosapentaenoic acid) and DHA (docosahexanoic acid) have multiple effects which are very relevant for the prevention and treatment of atherosclerosis.

The long chain n-3 PUFA of fish oils and micro algae have greater antiatherogenic, antithrombotic and anti-inflammatory effects than alpha-linolenic (n-3) acid of plant foods. EPA is required for the production of a special group of substances in the body called prostaglandins, which control blood clotting and other arterial functions. EPA also provides a natural approach in lower blood cholesterol and triglycerides. Thus, EPA is a direct source of an important substance called prostaglandin E_3. Prostaglandin E_3 is directly responsible for making blood platelets less sticky thus, leading to an easier flow of blood throughout our bodies. This means that EPA is intimately involved in bodily process that inhibits blood clots, particularly in the small capillaries of the heart.

Omega-3 also reduces the amount of adhesion molecules measured as soluble ICAM-1 and E-selectin. This is attributed to a reduction of the presentation of these molecules at the transcription level. A reduction of leukotriene B_4 which is regularly induced by Omega-3 could add to this effect. A reduction of oxidized LDL could also indirectly down-regulate the production of adhesion molecules.

Regarding the build-up of atherosclerotic thrombic, omega-3 fatty acids interfere with the aggregation phase of early thrombus formation preventing arterial occlusion with aspirin-like mechanisms. Even short time supplementation with large amounts or a combination of EPA and DHA has shown to have long-lasting effects on the human platelet aggregation, an effect suggested by inhibition of TXA_2/PGH_2 receptor by EPA and or DHA-sensitive mechanisms.

By the same mechanism, Omega-3 reduces the concentration of thromboxane thereby, counteracting the vasoconstriction related to the formation of acute-thromboses. They have influence on platelet aggregability at rather low doses, whereas significant effects on blood lipids and blood pressure can be achieved at higher doses.

Omega-3 reduces morbidity and mortality associated with cardiovascular disorders, as numerous studies show that increased long-term intake of marine oils, rich in EPA and DHA reduces the morbidity and mortality associated with cardiovascular disorders in middle-aged men. It is generally agreed that omega-3 fatty acids reduce the triglyceride levels in the blood by a reduced-synthesis and secretion of VLDL particles.

Omega-3 Fatty Acids-Bare Facts

- PUFAs are made up of chains of fatty acids that differ in length and the number and position of double bonds. These differences determine PUFAs biological properties.
- Molecules with the first double bond between the third and fourth carbon atom from the methyl end are called Omega-3 or n-3 fatty acids. Those with a double bond between the sixth and seventh carbon atoms are known as Omega-6 or n-6 fatty acids.
- The most common long chain Omega-3 PUFA fatty acids are known as Eicosapentaenoic acid and DHA (Docosahexaenoic acid).
- We need to obtain two key PUFAs – Linoleic acid and alpha-linolenic acid – from our diets. These are the 'essential' fatty acids that the body converts into other PUFAs.
- Eating 200 to 300 g of oily fish weekly fulfills the dietary requirements of n-3 long chain fatty acids. Some patients may need to increase their intake through supplements.
- Babies, pre-term infants and pregnant women need PUFAs to ensure normal development of the brain and other nerve tissues.
- People with a personal or family history or heart disease or circulatory problems may also benefit from a higher Omega-3 PUFA intake.
- People suffering from rheumatoid arthritis often find increasing consumption of Omega-3 PUFA reduces pain and stiffness.
- Omega-3 PUFAs may alleviate infantile seborrhoeic dermatitis and diaper dermatitis, also protect smokers from chronic obstructive pulmonary disease and reduce the likelihood of relapse in Crohn's disease.
- Still to be confirmed roles for PUFAs include treating pre-menstrual syndrome, diabetes, multiple sclerosis, migraine, depression and cancer.

Hence, the most potent effect is related to triglyceride reduction which will induce a shift from small, dense LDL-particle to normal LDL. This may offer a way of reducing oxidized LDL and reduce the inflammatory stimuli in the arterial wall. Recent data show that DHA produce a 30-40 per cent greater reduction in triglyceride levels in plasma than a corresponding amount of EPA.

They increase insulin sensitivity, increase peripheral glucose utilization and decrease adiposity and hence, are anti-atherogenic. Furthermore, Omega-3 increases the concentration of HDL which is responsible for the transportation of surplus cholesterol from the periphery back to the liver. A low level of HDL-C regularly coincides with hypertriglyceride and is related to increased coronary heart disease risk.

Recent data also show that DHA has more pronounced inhibitory effect on the expression of cytokines in endothelial cells, which clearly down regulate the inflammatory process and may inhibit the progression of atherosclerosis. Epidemiological and clinical researches have shown that omega-3 fatty acids intervene in the atherosclerotic process at all steps, and that there probably are synergistic effects of EPA and DHA at many levels. Hence, omega-3 with multiple effects on causes and features relevant to atherosclerosis is an anti-atherosclerotic drug more than only a triglyceride-lowering drug.

According to WHO (2008), the available evidence indicates that 0.5 to 0.6 per cent E alpha-linolenic acid (ALA) per day corresponds to the prevention of deficiency symptoms. The total n-3 fatty acid intake can range between 0.5 - 2 per cent E whereas the minimum dietary requirement of ALA (> 0.5 per cent E) for adults prevents deficiency symptoms. The higher value 2 per cent E (ALA) plus n-3 long-chain polyunsaturated fatty acids (LCPUFA) eicosapentaenoic acid (EPA) and docosahexaenoic acid (DHA) (Acceptable macronutrient distribution range *i.e.* ADMR 0.250 g – 2.0 g) can be part of a healthy diet. Whilst ALA may have individual properties in its own right, there is evidence that the n-3 LCPUFA may contribute to the prevention of CHD and possibly other degenerative diseases of aging. For adult males and non-pregnant/non-lactating adult females, 0.250 g/day of EPA plus DHA is recommended, with insufficient evidence to set a specific minimum intake of either EPA or DHA alone; both should be consumed. For adult pregnant and lactating females, the minimum intake for optimal adult health and fetal and infant development is 0.3 g/d EPA+DHA, of which at least 0.2 g/d should be DHA.

Recommended ratio of n-6/n-3 is <10. In India, where the consumption of LC n-3 PUFA is negligible, the n-6/n-3 ratio gains additional importance. Available dietary information suggests that Indians consume, at most, about 20-50 mg n-3 PUFA per day. In such a situation, the feasibility as well as the necessity of consuming at least 200 mg n-3 PUFA (as per the RDA) needs to be questioned. Recent data suggested that even low intakes of LC n-3 PUFA (20- 40 mg/d) during pregnancy resulted in increased birth weight of infants and a reduced incidence of small for gestational age infants. To generate evidence on the requirement of n-3 PUFA or LC n-3 PUFA, well-conducted epidemiological studies as well as randomized controlled trials need to be carried out to better understand the role of not only n-3 PUFA, but also the n-6/n-3 PUFA ratio in Indian diets.

Monounsaturated Fats

Monounsaturated fatty acids (MUFA) are fatty acids that have one double bond in the fatty acid chain with all of the remainder carbon atoms being single-bonded. It is called omega-9 monounsaturated fatty acids (ω–9 fatty acids or *n*–9 fatty acids) if the double bond is present in the omega -9 position; that is, the ninth bond from the methyl end of the fatty acid *e.g.* erucic acid (rapeseed, mustard seed) and oleic acid (main component of olive oil, macadamia oil *etc.*); they are found in plant and animal fats. Unlike omega -3 fatty acids and omega-6 fatty acids, omega–9 fatty acids are not classed as essential fatty acids (EFA). This is both because they can be created by the human body from unsaturated fat, and are therefore, not essential in the diet, But omega-9 monounsaturated fatty acids (MUFAs) are good choices for replacing saturated fatty acids in the diet. The chief monounsaturated fatty acid is oleic acid. Common sources include:

- Olive oil
- Canola (rapeseed) oil
- Peanut oil
- Avocados
- Almonds

In the Mediterranean-type diet, the source of fats is usually olive oil which is very healthy. One potential difficulty is for the patient who overdoes the use of olive oil and gains weight.

Canola oil is low in saturated fatty acids and is a source of omega-3 fatty acid as well (alpha linolenic acid). This makes it a good choice for home use. In addition to lowering blood cholesterol and LDL-c by replacing saturated fats in the diet, MUFAs appear to make LDL less susceptible to oxidation. This tendency may be particularly advantageous in those with combined hyperlipidemia and diabetes where small dense LDL particles are more likely to be present.

MUFA reduces LDL and raises the HDL cholesterol. This fat also helps in developing and maintaining body cells. It is also high in vitamin E, necessary for healthy vision, a healthy immune system, and other benefits.

According to WHO, there is convincing evidence that replacing carbohydrates with MUFA increases HDL cholesterol concentrations and possible evidence that replacing carbohydrates with MUFA improves insulin sensitivity. There is convincing evidence that replacing SFA (C12:0 – C16:0) with MUFA reduces LDL cholesterol concentration and total/HDL cholesterol ratio. There is insufficient evidence for relationships of MUFA consumption with chronic disease end points such as CHD or cancer. There is insufficient evidence for relationships of MUFA consumption and body weight and per cent adiposity and relationship between MUFA intake and risk of diabetes.

The determination of intake of MUFA is unique in that it is calculated by difference, *i.e.* Total fat intake (per cent E) – SFA (E per cent) – PUFA (E per cent) – TFA (per cent E). It can be up to 15-20 per cent E, according to total fat intake. Therefore, the MUFA intake resulting may cover a wide range depending on the total fat intake and dietary fatty acid pattern. Excessive use of highly unsaturated fats should be avoided. Further, to get a good proportion of all the classes of fatty acids, it is advisable to consume more than one type of vegetable oils.

Trans Fatty Acids

During hydrogenation, the liquid oils become solid because the mono and polyunsaturated fatty acids are converted into saturated fatty acids and isomers called *trans* fatty acids. The most common TFA is the trans isomer of oleic acid *i.e.* elaidic acid. Its structure is such that it is straighter than the usual cis isomer and thus, can be more tightly packed and hence solid at room temperature. Common sources of trans fat include:

- Stick margarines and shortenings
- Dairy items, such as milk, butter and cheese
- Commercially processed baked goods

Like saturated fatty acids, trans fatty acids (TFAs) also raise cholesterol. In terms of the potential of raising blood cholesterol, butter, which contains both dietary cholesterol and saturated fat, is the most harmful, followed closely by stick margarine with its higher content of TFAs. On the other hand, small amounts of soft tub or liquid margarine should impact the lipid profile in minimal fashion.

In increased amounts, TFAs raise LDL-c and lower HDL-c. The current recommendation is to keep consumption of TFAs low. The intake of *trans* fatty acids should be less than 1 per cent of energy intake.

Dietary Cholesterol

Cholesterol is a chemical compound that is found in all body cells. It plays a key role in the formation of brain, nerve tissue and some steroid hormones. It is a precursor for vitamin D too. It is synthesized in the body and hence, it is not an essential dietary component. The liver produces about 80 per cent of the body's cholesterol and the rest comes from dietary sources like meat, poultry, eggs, fish, organ meats such as liver, kidney and brain and dairy products. Foods derived from plants including vegetable oils contain no cholesterol.

Daily dietary cholesterol intake should be maintained below 200 mg/day. Higher dietary cholesterol intake than this amount shall increase the blood cholesterol which shall raise LDL-c and atherogenesis. If the cholesterol consumption is high, the blood cholesterol elevating effect of dietary saturated fats shall increase.

Prevention and Control

Certain types of congenital heart diseases, such as heart defects, cannot be prevented. However, many other types of heart disease can be prevented and controlled by making the same lifestyle changes that can improve the heart disease. Such life style changes are as under:

1. Cessation of smoking
2. Life style Modification
3. Healthy dietary Habits
4. Regular Exercise
5. Control of Obesity
6. Hypertension – Detection and effective control of blood pressure
7. Early detection and control of diabetes
8. Cholesterol reduction
   TC <200 mg/dl
   LDLc < 100 mg/dl
   HDLc > 60 mg/dl
   TG < 150 mg/dl
9. Stress control
10. Yoga
11. Meditation
12. Relaxation Techniques

Principles of Treatment

1. Rest and drugs as prescribed by physician.
2. Intake of modified diet

The role of diet is crucial in the development and prevention of cardiovascular disease. Diet is one of the key things one can change that will impact all other cardiovascular risk factors. A diet high in saturated fat increases the risk of heart disease and stroke. It is estimated to cause about 31 per cent of coronary heart disease and 11 per cent of stroke worldwide.

Comparisons between a diet low in saturated fats, with plenty of fresh fruit and vegetables, and the typical diet of someone living in the high-income countries show that in the former there is a 73 per cent reduction in the risk of new major cardiac events.

Nutritional Management

The importance of nutrition in modifying the risk of CVD has been repeatedly emphasized. Over the past few years, increasing attention has been given to dietary patterns and their relationship to health outcomes such as CVD. Historically, the role of dietary components has been the predominant focus; however, foods are typically consumed in combinations rather than individually. The diet recommended should be basically a normal diet which is low in saturated fat, cholesterol, trans fat, high in omega-3 and omega-9, normal protein, low carbohydrate, optimum amount of micronutrients, high fibre and antioxidants.

Energy

The calorie intake should be adequate to meet the requirement. The calorie intake should be in balance with energy expenditure. To maintain desirable body weight is one of the successful approaches for preventing heart diseases. Obese and overweight heart patients should reduce the body weight to normal and maintain it. Therefore, they will have to take less energy per day depending upon the amount of weight loss per week. By taking 500 kcal less daily, one can reduce 500g body weight in a week.

Proteins

The diet should contain adequate amounts of proteins (20 per cent of total calories according to TLC of NCEP ATP III for cholesterol reduction) to meet the recommended allowances. CVD effects of dietary protein have been relatively understudied. Few prospective cohorts have reported on total protein intake and CHD events, with generally null results. In some studies, plant but not animal protein sources in diet associate with lower CHD risk, suggesting that types of foods consumed or overall diet patterns may be more relevant than protein per se.

Fats

In prevention, the fatty acid composition of the diet is more important than the total fat content. The major physio-pathological factors for coronary artery disease are high blood lipids (cholesterol, triglycerides) and alterations in blood clotting mechanisms. Dietary fat strongly influences these processes and therefore, it is important to ensure the optimum quantity and right quality of fat in the total diet.

For cholesterol reduction, the National Heart, Lung and Blood Institute's National Cholesterol Education Program (NCEP) Adult Treatment Panel (ATP III) of USA had recommended TLC dietary guidelines. According to the revised TLC dietary recommendations, limit total fat intake to no more than 25-35 per cent of total daily calories and avoid trans fat. If you eat 2,000 calories per day, then fat intake should be no more than 50-75 grams per day. Substituting polyunsaturated fats for saturated fats will reduce both total and LDL cholesterol but do have the potential to also lower HDL cholesterol levels when consumed in large amounts. That is why they should be consumed to no more than 10 per cent of total calories each day. MUFA should provide upto 20 per cent of total calories. Dietary cholesterol should be less than 200 mg a day in the diet.

Animal foods like eggs, meat, organ meats, milk, butter, ghee and cheese are good sources of saturated fatty acids and cholesterol. According to 2013 AHA/ACC Guideline on Lifestyle Management to reduce cardiovascular risk, reduce per cent of calories from saturated fat. Aim for a dietary pattern that achieves 5 per cent –6 per cent of calories from saturated fat so as to reduce LDLc. Rest of the calories should come from PUFA and MUFA. Fish oils and vegetable oils (*e.g.*, olive and canola) are each major sources of MUFAs, largely oleic acid (18:1n-9). Compared with carbohydrates, MUFA intake lowers LDL-c and TG, raises HDL-C, and lowers BP. Monounsaturated fatty acids have a favourable effect on HDL cholesterol levels when they replace saturated fatty acids or carbohydrates in the diet. Fewer studies have compared MUFAs and PUFAs; as a replacement for carbohydrates, MUFAs may raise HDL-c slightly more and lower LDL-c and TG slightly less than PUFAs, with a similar overall improved TC/HDL-c ratio.

An ideal quality fat for good health is the one which maintains a balance, so as to give a ratio of PUFA/SFA of 0.8-1.0, and linoleic/α -linolenic (n-6/n-3) of 5-10 in the total diet. For ensuring this appropriate balance of fatty acids in cereal-based diets, it is necessary to increase the α -linolenic (n-3) acid intake and reduce the quantity of linoleic (n-6) acid obtained from the cooking oil.

Food rich in trans fat *e.g.* butter, hydrogenated fat, baked goods, processed foods *etc.* should not provide more than 1 per cent of energy intake. Meta-analysis of prospective cohort studies has shown that, on average, a higher trans fatty acid intake of 2 per cent of energy increases the risk of CHD by 23 per cent. Reducing intake of *trans* fatty acids lowers LDL-c, with little or no effect on HDL-c or triglycerides levels. The direction of the relationship between *trans* fatty acids and LDL-c is consistent, regardless of whether the *trans* fatty acids are replaced by carbohydrates, monounsaturated fatty acids, or polyunsaturated fatty acids (AHA/ACC Guideline on Lifestyle Management to reduce cardiovascular risk).

Carbohydrates

Carbohydrates mainly provide the energy as the diet is low in fat content, hence should provide 50-60 per cent of the total energy intake. Total carbohydrate quantity consumed does not associate strongly with CHD risk but the types *i.e.* simple (*e.g.*, glucose, fructose, galactose, sucrose, lactose, lactulose) or complex (*e.g.*, starch, cellulose, hemicellulose, glycogen) and quantity of carbohydrate consumed are important determinants of health effects. Recent evidence indicates that this classification scheme has little relevance to health effects. Specific factors that determine quality and health effects of high-carbohydrate foods include dietary fiber content, glycemic index (GI) and glycemic load (GL), and the extent of processing (*i.e.*, refined grains versus whole grains).

Carbohydrates contribute significantly to the endogenous synthesis of triglycerides and are, therefore, important dietary determinants of circulating triglycerides levels. While the simple sugars such as sucrose, fructose and glucose increase endogenous synthesis of triglycerides (VLDL), the complex carbohydrates reduce blood lipids and improve the glucose tolerance. Foods high in complex carbohydrates having low glycemic index are usually low in calorie density, contain wide variety of vitamins and minerals and are good sources of soluble fibres. Complex carbohydrates yield a low glycemic index and therefore, have better control of blood glucose levels. A higher intake of complex carbohydrates with low glycemic index and glycemix load at the expense of total fat intake is beneficial in endogenous hypertriglyceridaemias and diabetes.

According to 2013 AHA/ACC Guideline on Lifestyle Management to reduce cardiovascular risk, reduce per cent of calories from saturated fat. Aim for a dietary pattern that achieves 5 per cent –6 per cent of calories from saturated fat so as to reduce LDLc. There is strong evidence that reductions in LDL-c were achieved when dietary fat is replaced by carbohydrate, monounsaturated fatty acids, or polyunsaturated fatty acids. Favorable effects on lipid profiles are greater when saturated fat is replaced by polyunsaturated fatty acids, followed by monounsaturated fatty acids, and then carbohydrates. It is important to note that there are various types and degrees of refinement of carbohydrates. Substitution of saturated fat with whole grains is preferable to refined carbohydrates.

Dietary Fibre

Dietary fiber is comprised of non-digestible polysaccharides, resistant starch and oligosaccharides, and lignins in plants. Important sources of fibre are wholegrain products, legumes, fruits, and vegetables. The American Institute of Medicine recommends an intake of 30–45 g/day for adults. This intake is assumed to be the optimal preventive level. Half of this amount or more should be viscous fibre.

Vitamins

The diet should provide daily requirement of all vitamins.

Vitamin A and E

Many case–control and prospective observational studies have observed inverse associations between levels of vitamin A and E and risk of CVDs. This protective effect was attributed to their antioxidant properties.

As animal protein and fat product like whole milk, cream, organ meats, poultry *etc.* are avoided; the diet may become deficient in Vitamin A. Hence, red, orange and yellow colored fruits and vegetables and nuts should be incorporated in the diet.

Vitamin D

Vitamin D deficiency may be an important risk factor in the pathogenesis of CAD (Aggarwal *et al.*, J. Midlife Heathl. 2016. Apr-Jun; 7(2): 56–60.). Vitamin D inhibits the uptake of cholesterol by macrophages and in case of Vitamin D deficiency, cholesterol uptake by macrophages is promoted and these cholesterol-laden macrophages, also known as foam cells, deposit in the endothelium forming atheromatous plaque and promote atherosclerosis. Vitamin D deficiency has also been associated with decreased levels of high-density lipoprotein and apolipoprotein A-1, which promotes atherosclerosis. Vitamin D deficiency has been associated with diabetes mellitus and metabolic syndrome due to its receptor-mediated effects leading to increased insulin resistance and pancreatic beta cell dysfunction. These are the independent risk factors for CAD.

Overall, some epidemiological studies have shown associations between vitamin D deficiency and cardiovascular disease. Conclusive evidence showing that vitamin D supplementation improves cardiovascular prognosis is however lacking, but trials are underway. However, in case of Vitamin D deficiency, its supplements should be taken as per suggestions of the physician.

Vitamin C

The hypothesis that vitamin C may play a role in CVD prevention also draws support from the vitamin's antioxidant capabilities. The epidemiological evidence relating fruit and vegetable intake to reduced risk of CVD may be explained, at least in part, by antioxidant content, and especially the role of these antioxidants in preventing oxidative changes to LDL. Oxidized LDL is a target for scavenger receptors, which incorporate it into plaque. Therefore, the prevention of LDL oxidation by vitamin C may prevent atherosclerosis, thereby mediating a potential role in CVD risk reduction.

The lack of consistency within the body of research on this topic has called into question the roles of antioxidants in the human body, and has even cast doubt on the LDL oxidative hypothesis. A key limitation in understanding the relationship between vitamin C and CVD is the lack of mechanistic studies in humans. Most evidence supporting this link is based on animal studies and observational studies in humans. Overall, current research suggests that vitamin C deficiency is associated with a higher risk of mortality from CVD and that vitamin C may slightly improve endothelial function and lipid profiles in some groups, especially those with low plasma vitamin C levels. However, the current literature provides little support for the widespread use of vitamin C supplementation to reduce CVD risk or mortality (Moser and Chun, Int. J. Mo.l Sci. 2016 Aug; 17(8): 132).

Niacin

Niacin has been used for primary and secondary coronary heart disease prevention for over 40 years. Niacin raises high-density lipoprotein cholesterol (HDL-c) by as much as 30-35 per cent. This effect is caused by a reduction of cholesterol transfer from HDL to VLDL and delayed clearance of HDL. It also lowers total cholesterol, low-density lipoprotein cholesterol (LDL-c), triglycerides, and lipoprotein (a). Hence, there is no doubt that niacin is a very effective lipid-lowering agent (Axel F. Sigurdsson MD. Doc's opinion, June 10, 2018). Niacin has a greater effect on reducing atherogenic lipoproteins in the small, dense LDL region than other pharmacologic treatments including statins. Niacin has other metabolic effects including cellular effects on cholesterol efflux and inflammation too that may benefit patients with atherosclerosis (Superko *et al.*, 2017. Niacin and heart disease prevention: Engraving its tombstone is a mistake J.Clin. Lipidology. Nov.-Dec.11 (6): 1309-1317).

B-vitamins (B_6, folic acid, and B_{12}) and Homocysteine

The B-vitamins B_6, B_{12}, and folic acid have been studied for their potential to lower homocysteine levels, which has been postulated as a risk factor for CVD.

The Cochrane Collaboration concluded in a recent meta-analysis of eight RCTs that homocysteine-lowering interventions did not reduce the risk of fatal/non-fatal myocardial infarction stroke or death by any cause; 3 large secondary prevention trials (SEARCH),(VITATOPS), (SU.FOL.OM3) also concluded that supplementation with folic acid and vitamin B_6 and/or B_{12} offers no protection against the development of CVD. Thus, B-vitamin supplementation to lower homocysteine levels does not lower risk of CVD.

Minerals

Intake of all minerals should be adequate to meet the daily requirements.

Strategies for Dietary Management

Diet is the hallmark of therapy for hyperlipidemia, hypercholesterolemia and cardiovascular diseases. Its usefulness extends beyond simply providing additional low-density lipoprotein cholesterol (LDL-c) lowering for those with hypercholesterolemia. It also can help improve the abnormal high-density lipoprotein cholesterol (HDL-c) and triglyceride (TG) values seen in combined hyperlipidemia. In fact, patients with combined hyperlipidemia are often quite sensitive to weight reduction and exercise. Diet may play an important role in those with low HDL-c and premature coronary heart disease (CHD). Knowledge of the effects of dietary fat, kind of dietary fat, effect of dietary carbohydrate and effect of alcohol may prove useful in the overall care of such individuals. In those with severe hypertriglyceridemia who are prone to pancreatitis, reduction of all dietary fats and cessation of alcohol is crucial to avoiding exacerbations that could lead to the chylomicronemia syndrome and acute pancreatitis. Also, dietary antioxidants may reduce risk of CHD through mechanisms independent of lipid effects. Finally, antithrombotic and antiarrhythmic effects of diet may prove useful.

Also, excess body weight is a factor that must be considered. It is a strong contributor to elevated cholesterol levels in men. Weight gain in adult life can lead not only to abnormal lipid values, but to increased risks for diabetes mellitus and CHD also. Genetic influences on dietary responsiveness are considerable. Those persons with familial hypercholesterolemia cannot usually hope to achieve goal LDL-c values with diet alone. They will have to take medication as per advice of the physician along with the modified diet.

I. Dietary Pattern

Various dietary patterns including the Prudent, DASH-type and Mediterranean type diets have been reported to reduce CVD risk factors in controlled trials.

Prudent Dietary Pattern

It is characterized by higher intake of vegetables, fruits, legumes, whole grains, fish and poultry.

Mediterranean Dietary Pattern

The Mediterranean diet is a diet inspired by the eating habits of Greece, Southern Italy and Spain in the 1940s and 1950s. It is characterized by high consumption of olive oil, legumes, unrefined cereals, fruits, and vegetables, moderate to high consumption of fish, moderate consumption of dairy products (mostly as cheese and yogurt), moderate wine consumption, and low consumption of meat and meat products. Hence, this diet emphasizes:

- Eating primarily plant-based foods, such as fruits and vegetables, whole grains, legumes and nuts
- Replacing butter with healthy fats such as olive oil and canola oil
- Using herbs and spices instead of salt to flavor foods
- Limiting red meat to no more than a few times a month
- Eating fish and poultry at least twice a week
- Snack on fruit, dried fruit and unsalted nuts rather than cakes, crisps and biscuits
- Drinking red wine in moderation (optional) *i.e.* no more than three small glasses per day for men and no more than two small glasses for women.
- Getting plenty of exercise

DASH Dietary Pattern (Dietary Approaches to Stop Hypertension)

This is a diet promoted by USA and National Heart, Lung, and Blood Institute (NHLBI), an agency of United States Department of Health and Human Services.

This diet is a plant-focused diet, rich in fruits, vegetables, whole grains and low –fat dairy foods; includes meat, fish, poultry, nuts and beans; is limited in sugar-sweetened foods and beverages, red meat, and added fats. Total fat in a diet with roughly this composition is 25 per cent to 35 per cent of calories, with saturated fat at 8 per cent or less of calories (Table 21.10). This dietary pattern has provided the highest-quality evidence for a dietary pattern causing improvements in BP and lipid profiles. The LDL-C–lowering effect has been demonstrated in men and women, African Americans and non–African Americans, and in adults of all ages. The evidence suggests that the effects of the recommended dietary pattern persist as long as the pattern is consumed.

It is now recommended by the United States Department of Agriculture (USDA) as an ideal eating plan for all Americans.

Table 21.10: Approximate Nutrient Composition of DASH Dietary Pattern

Nutrient	*Daily Quantity*
Total fat	27 per cent of total calories
SFA	6 per cent of total calories
Carbohydrates	55 per cent of total calories
Protein	18 per cent of total calories
Cholesterol	150 mg
Fiber	31 g
Potassium	4700 mg
Magnesium	500 mg
Calcium	1240 mg

On the basis of above dietary patterns, the following dietary pattern has been recommended as per 2013 AHA/ACC Guideline on Lifestyle Management to Reduce Cardiovascular Risk:

A. Advise adults the following diet to get benefit from LDL–c lowering

i) Reduce percentage of calories from saturated fat. Aim for a dietary pattern that achieves 5 per cent to 6 per cent of calories from saturated fat

There is strong evidence that reductions in LDL-c were achieved when dietary patterns were followed in which saturated fat intake was reduced from 14 per cent to 15 per cent of calories to 5 per cent to 6 per cent.

Reducing saturated fat intake lowers both LDL-c and HDL-c. Because the absolute effect tends to be greater for LDL-c than HDL-c, reducing saturated fat intake has a beneficial effect on the lipid profile.

Given that reducing saturated fat intake lowers LDL-c regardless of whether the saturated fat is replaced by carbohydrate, monounsaturated fatty acids, or polyunsaturated fatty acids, the Work Group did not specify which of these 3 macronutrients should be substituted in place of saturated fat. However, favorable effects on lipid profiles are greater when saturated fat is replaced by polyunsaturated fatty acids, followed by monounsaturated fatty acids, and then carbohydrates. Substitution of saturated fat with whole grains is preferable to refined carbohydrates.

ii) Reduce percentage of calories from trans fat

Reducing intake of *trans* fatty acids lowers LDL-c, with little or no effect on HDL-c or triglycerides levels. The direction of the relationship between *trans* fatty acids and LDL-c is consistent, regardless of whether the *trans* fatty acids are replaced by carbohydrates, monounsaturated fatty acids, or polyunsaturated fatty acids.

B. Dietary Pattern for Lowering LDL-c and Risk of CVD

1. Consume a diet which is rich in vegetables, fruits and whole grains; includes low-fat dairy products, poultry, fish, legumes, non-tropical vegetable oils and nuts; limited intake of sweets, sugar-sweetened beverages, and red meats.
   a) Adapt this dietary pattern to appropriate calorie requirements (e.g, restricted for those attempting weight loss, personal, social and cultural food preferences, and nutrition therapy for other medical conditions *e.g.* diabetes if present in co-existence with CVD.
   b) Achieve this pattern by following plans such as the DASH dietary pattern.
2. Reduce per cent of calories from saturated fat. Aim for a dietary pattern that achieves 5 per cent to 6 per cent of calories from saturated fat.
3. Reduce per cent of calories from *trans fat.*

C. Dietary Components

Select Whole Cereal Grains

Whole grains are good sources of fiber and other nutrients that play a role in regulating blood pressure and heart health. One can increase the amount of whole grains in a heart-healthy diet by making simple substitutions for refined grain products.

Grain Products to Choose	*Grain Products to Limit or Avoid*
☆ Whole-wheat flour	☆ White, refined flour
☆ Whole-grain bread, preferably 100 per cent whole-wheat bread or 100 per cent whole-grain bread	☆ White bread
☆ Brown rice	☆ Muffins
☆ High-fiber cereal with 5 g or more fiber in a serving	☆ Doughnuts
☆ Millets such as barley, maize, pearl millet, oat, barley and sorghum	☆ Biscuits
☆ Whole-grain pasta	☆ Quick breads
	☆ Cakes
	☆ Pies
	☆ Egg noodles
	☆ Buttered popcorn
	☆ High-fat snack crackers

Eat more Vegetables and Fruits

Vegetables and fruits are good sources of vitamins and minerals. They are also low in calories, low in sodium, rich in phytochemicals and dietary fiber. Eating more fruits and vegetables may help to cut back on higher calorie foods and provide flavnols which protect against heart disease. One should try to eat lots of seasonal and fresh fruits and vegetables. Raw and fresh vegetables like green leafy vegetables, carrots, fresh fruits including citrus and tomatoes are good sources of antioxidants (free radical scavengers). Different colored vegetables provide different antioxidants like orange colored provide beta-carotene, red provide lycopene, deep red provide betalines and blue and purple provide anthocynins. All types of green leafy vegetables and other vegetables should be liberally included in the diet in the form of fresh salad without creamy sauces, soups, *raitas*, chutneys and curry or dry seasoned vegetables. Among the roots and tubers, amount of potato, sweet potato and colocassia should be limited while carrots, radish, onion *etc.* can be taken liberally. One should eat 5-6 servings of vegetables and fruits daily.

Take Plenty of Dietary Fiber Rich Foods

Practically speaking, there are two kinds of fibers or non-digestible carbohydrate. The first, which is insoluble, aids in bowel function. An example is wheat bran. The second is soluble fiber, now referred to as viscous fiber, which has an additional cholesterol-lowering effect. Examples include:

- ☆ Dried beans
- ☆ Grains (Oat, barley)
- ☆ Certain fruits
- ☆ Vegetables

Psyllium is a source of soluble fiber used as a dietary supplement or in dry cereals. Psyllium can be a useful adjunct to bile acid sequestrants and augment the LDL-c lowering response. Oat products have the most soluble fiber of any grain. A simple way to improve the diet is that oatmeal, oat bran, or any of the single or multigrain dry cereals be substituted for higher-fat breakfast items. Barley is another cereal grain which is rich in soluble fibre and lowers the blood cholesterol level which should be included in the diet. Many fruits such as avocados, pears, apples, guavas, plums, gooseberries, and oranges are rich source of pectin and should be taken in the diet. Vegetables especially broccoli, brussel sprouts, peas *etc.* are rich in soluble fibre and should be used in food preparations. The DRI recommends consumption of 14 g dietary fiber per 1000 kcal, or 25 g for adult women and 38 g for adult men

Choose Low-Fat Protein Sources

Legumes, lean meat, poultry and fish, low-fat dairy products, and eggs are some of the best sources of protein. But be careful to choose lower fat options and prefer plant protein to animal protein sources. For the vegetarian population, beans, peas and lentils are good protein sources as they contain less fat and no cholesterol making them good substitutes for meat. Substituting plant protein for animal protein for example, a soy or bean burger for a hamburger will reduce the fat and cholesterol intake while increasing fiber intake.

Instead of whole milk, choose skim milk and skinless chicken breasts rather than fried chicken patties. Avoid red meat. Fish is another good alternative to high-fat meats. And certain types of cold water fatty fish such as salmon, mackerel and herring are rich in omega-3 fatty acids, which can lower blood triglycerides.

Proteins to Choose	*Proteins to Limit or Avoid*
✰ Low-fat dairy products, such as skim or low-fat (1 per cent) milk, yogurt and cheese	✰ Full-fat milk and other dairy products
✰ Eggs	✰ Red meat
✰ Fish, especially fatty, cold-water fish, such as salmon	✰ Organ meats, such as liver
✰ Skinless poultry	✰ Fatty and marbled meats
✰ Legumes	✰ Hot dogs and sausages
✰ Soybeans and soy products, such as soy burgers and tofu	✰ Fried meats
✰ Lean ground meats	

Limit Unhealthy Fats

Limit the amount of foods rich in saturated fat. Avoid trans fat rich foods such as cookies, cakes, pastries, patties, ice creams, frostings, crackers and chips. Most of the fried snacks and sweets available in the market such as *pakoras, smosas, kachori, mathi, bhujia,* potato chips, potato bondas, sweets *etc.* and foods available in the sweet shops, *dhabas* and restaurants are fried or prepared in the hydrogenated fat (*vanaspati*) which is a trans fat and therefore, their intake should be limited to bare minimum. Check the food label of ready to eat and processed foods; if they contain butter, margarine, partially hydrogenated or fully hydrogenated fat, avoid those foods.

High-fat animal foods like ghee, meat, egg and organ meats not only contain saturated fat but are rich in cholesterol which is a menacing factor for atherosclerosis. Hence, their intake should also be limited.

Select Heart Healthy Fats and Oils

All the plant oilseed oils contain unsaturated fatty acids which are good for health. The plant oils in addition contain certain useful substances such as lignans (sesame oil), sterols, tocopherols (vitamin E) oryzanole (rice bran oil), carotenoids - all of which reduce cholesterol and reduce oxidant damage due to ageing, inflammation which occur in chronic diseases.

Most of the vegetable oils such as soybean, safflower, sunflower, groundnut *etc.* are rich in PUFA while olive, ricebran, mustard, rapeseed and canola oils are rich in MUFA. Polyunsaturated fats (omega-3 fatty acids) found in avocados, pulses, nuts (walnuts, almonds), and seeds (flaxseeds, fenugreek seeds, mustard seeds), also are good choices for a heart-healthy diet. For the non-vegetarian population, fatty fish are very good choices as they are rich sources of omega-3 fatty acids. The recommendation, therefore, is to eat fish at least twice a week, of which once oily fish.

When used in place of saturated fat, monounsaturated and polyunsaturated fats may help lower the total and LDL cholesterol in the blood. But moderation is essential as all types of fat are high in calories.

Fats to Choose	*Fats to Limit*
✰ Olive oil	✰ Butter
✰ Canola oil	✰ Lard
✰ Vegetable and nut oils	✰ Cream sauce
✰ Nuts and oilseeds (walnut, almond, flax seeds, fenugreek seeds, mustard seeds *etc.*)	✰ Nondairy creamers
✰ Avocados	✰ Hydrogenated margarine and shortening
	✰ Cocoa butter, found in chocolate
	✰ Coconut, palm, cottonseed and palm-kernel oils

It is better to take blend of oils rather than taking single source of fats and oils. ICMR (2011) has recommended the following oil blends:

- ✰ Groundnut or Sesame or Rice bran + Mustard
- ✰ Groundnut or Sesame or Rice bran + Canola
- ✰ Groundnut or Sesame or Rice bran + Soybean
- ✰ Palmolein + Soybean
- ✰ Safflower or Sunflower + Palmolein + Mustard

Source: ICMR (2011)

Nuts

Nuts (tree nuts and peanuts) are nutrient-dense foods with complex matrices rich in unsaturated fatty acids and other bioactive compounds: high-quality vegetable protein (L-arginine), fiber, minerals, tocopherols, phytosterols, phenolic and in mono- and polyunsaturated fatty acids which help to keep the blood vessels healthy. Epidemiological data show a consistent negative association between nuts consumption and CHD risk. Some of the studies found a dose-response pattern of association. An analysis of four studies from the United States concluded that high nut intake is associated with a 35 per cent risk reduction for CVD. In cross-sectional observational studies and controlled trials, nut intake lowers total and LDL-C and variably improves oxidative, inflammatory, and endothelial biomarkers. Nut intake associates with lower body mass index (BMI) in observational studies and similar or greater weight loss in intervention trials. Effects of different types of nuts require further study, but benefits in short-term trials and the magnitude and consistency of lower risk in observational studies support the importance of modest nut consumption for lowering CHD risk. Eating about a handful a day of most nuts such as walnuts, almonds, macadamia nuts, hazelnuts, peanuts, pecans, some pine nuts, cashewnuts, Brazilnuts and pistachio nuts, may reduce the risk of heart disease.

Dairy Products

Despite the contribution of dairy products to the saturated fatty acid composition of the diet, and given the diversity of dairy foods of widely differing fat composition, there is no clear evidence that dairy food consumption is consistently associated with a higher risk of CVD and some evidence that low-fat products may have beneficial effects on blood pressure. The general health recommendation is to prefer low-fat products in order to reduce SFA intake. It is better to take probiotic fermented curd or yoghurt instead of milk for health.

Restrict Sodium Intake

Eating a lot of sodium can contribute to high blood pressure, a risk factor for cardiovascular disease. Consume no more than 2,400 mg of sodium/day (6g salt). Reducing sodium intake to no more than 1,500 mg of sodium a day is an important part of a heart-healthy diet especially if someone is hypertensive too. It is recommended to substitute salt with other spices and herbs.

Low-Salt Items to Choose	High-Salt Items to Limit or Avoid
☆ Herbs and spices ☆ Salt-free seasoning blends ☆ Reduced-salt canned soups or prepared meals ☆ Reduced-salt versions of condiments, such as reduced-salt soy sauce and reduced-salt ketchup	☆ Table salt ☆ Canned soups and prepared foods, such as frozen dinners ☆ Tomato juice ☆ Condiments such as ketchup, mayonnaise and soy sauce ☆ Restaurant meals ☆ *Papad*, pickle, baked goods

Reduce the amount of salt to be added to food at the table or while cooking as it is a good first step. Restrict intake of preserved and processed foods such as *papads*, pickles, sauces, ketch up, salted biscuits, chips, cheese, canned soups, salted fish and baked goods. Eat plenty of fresh fruits and vegetables as they are good sources of potassium which is good for blood pressure and heart.

Coffee and Tea

Coffee consumption has long been suspected of being a contributing factor in the development of CVD, based mainly on case-control studies. Excess consumption of caffeine may increase blood pressure and cause abnormalities in heart beat. On the contrary, as per research data of last few years, consumption of moderate amounts of caffeine is associated with a reduced risk of CVD and may even be protective against CVD. Populations at risk for hypertension, or already with hypertension, may be more sensitive to some effects of caffeine; for example, pre/hypertensive populations experienced acute increases in blood pressure following caffeine consumption ranging from 100 to <400 mg per day compared to their normotensive counterparts (Turnbull *et.al*, 2017. Caffeine and cardiovascular health. https://doi.org/10.1016/j.yrtph.2017.07.025. In: Regulatory Toxicology and Pharmacology. Elsevier Publication. Vol. 89, October 2017, Pages 165-185). The existing literature suggests moderate (400–600 mg/d) caffeine intake is not associated with increased risks of total cardiovascular disease; arrhythmia; heart failure; blood pressure changes among regular caffeine consumers; and hypertension in baseline healthy populations. Lately, the risk for developing type 2 diabetes was found to be lower in individuals who consumed four or more cups of coffee per day compared with those who drank less than two cups per day. Several mechanisms have been proposed to explain the harmful as well as protective effects that certain components of coffee may have on the development of CHD. These include the effects of coffee on blood pressure, serum cholesterol and homocysteine levels, oxidation, and inflammation.

In addition, an association between boiled coffee consumption and elevated levels of total and LDL cholesterol, triglycerides and heart disease has been demonstrated in few studies whereas usual amounts of filtered caffeinated coffee do not. There is disagreement as to whether decaffeinated coffee raises LDL-c. Moderate amounts of filtered coffee or decaffeinated coffee do not appear to increase the risk of CHD. Therefore, individuals with heart disease need to restrict coffee consumption. Also, those who experience adverse effects from caffeine should stop drinking coffee.

Besides caffeine, tea contains theobromine and theophylline. These are known to relax coronary arteries and thereby promote circulation. Tea also contains flavonoids and other antioxidant polyphenols, which are known to reduce the risk for coronary heart disease and stomach cancer. However, as a result of its caffeine content, excess tea consumption is deleterious to health. Green tea consumption may reduce LDL-C; however, the effect-size and the dose needed are still unknown. Prefer tea over coffee. Take decaffeinated coffee and tea in moderation.

Sugar-Sweetened Beverages

Short-term trials have suggested that calories in liquid form may be less satiating and thus, increase the total quantity of calories consumed, compared with solid foods. Observational studies have shown a positive association between sugar-sweetened beverages intake and adiposity or weight gain. One should avoid their intake and prefer to take simple plain water over sugar-sweetened beverages.

Eggs

During the past 40 years, the public had been warned against frequent egg consumption due to the high cholesterol content in eggs and the potential association with CVD. This was based on the assumption that high dietary cholesterol consumption is associated with high blood cholesterol levels and CVD. However, subsequent research suggests that, in contrast to SFA and TFA, dietary cholesterol in general and cholesterol in eggs in particular has limited effects on the blood cholesterol level and on CVD. There is a great variation in the response of blood cholesterol levels to dietary cholesterol, possibly related to the large variability in intestinal absorption of cholesterol. It is also possible that the fat composition of eggs (high MUFA- about 50 per cent of the fat in the egg is MUFA and lower SFA) restrains the blood LDL-C elevation. Eggs are also a source for high biological value protein, as well as vitamins and minerals such as folic acid, vitamin B_{12}, vitamins E and D, selenium, choline, zinc *etc.* It is recommended to take 3–4 eggs per week, including eggs contained in other foods.

Take Alcohol in Moderation

Numerous mechanisms have been proposed to explain the benefit of light-to-moderate alcohol intake on the heart, including an increase in HDL-C, reduction in plasma viscosity and fibrinogen concentration, increase in fibrinolysis, decrease in platelet aggregation, improvement in endothelial function, reduction in inflammation, and promotion of antioxidant effects. However, despite the biological plausibility and observational data in this regard, these are still insufficient to prove causality. Despite the evidence from cohort studies on the inverse association between moderate alcohol drinking and CVD, current guidelines do not recommend to begin consuming alcohol for preventing CVD.

Excessive alcohol intake weakens the heart muscle (cardiomyopathy) and also damages the liver (cirrhosis), brain and peripheral nerves. It also increases serum triglycerides. In patients with familial lipid disorders in which TG abnormalities are prominent, alcohol consumption can greatly increase TG values. In fact, excess alcohol intake is a common cause of secondary hyperlipidemias. In addition to concerns regarding alcohol and substance abuse, chronic, heavy drinking can exacerbate hypertension and lead to increased rates of:

- ✫ Stroke
- ✫ Cardiac dysrhythmia, especially atrial fibrillation
- ✫ Left ventricular hypertrophy
- ✫ Cardiomyopathy
- ✫ Liver disease

Daily intake of more than moderate amounts of alcoholic beverages can also be a risk factor for the development of hypertension, increased plasma TG levels, can serve as a source of excess calories, as well as increased risk for breast and other cancers.

Avoid alcohol. Those who drink should limit its intake. People who regularly consume more than two alcoholic drinks (one equals about 30 ml of ethanol) are at a higher risk for hypertension and stroke. Thus, there can be no public health recommendation for alcohol usage to reduce CHD risk.

D. General Dietary Counselling

A healthy diet and lifestyle management are the best weapons to fight cardiovascular disease.

Control Portion Size

How much you eat is just as important as what you eat. Overloading the plate, taking seconds and eating until one feels stuffed can lead to eating more calories than one should.

Use a small plate or bowl to help control the portions. Keep track of the servings of foods from each food group. Eat larger portions of low-calorie, nutrient-rich foods, such as fruits and vegetables, and smaller portions of high-calorie, high- fat and high-sodium foods, such as refined, processed or fast foods. This strategy can shape up the diet as well as your heart and waistline.

Eat a Variety of Nutritious Foods from all the Food Groups

One may be eating plenty of food, but body may not be getting the right type of nutrients it needs to be healthy. Nutrient-rich foods have minerals, vitamins, protein, carbohydrates, fat and other nutrients but are lower in calories. They may help to control body weight, cholesterol and blood pressure. Hence, choose foods from all the food groups but have a right choice as per calorie requirement.

Eat Less of the Nutrient-Poor Foods

Donot take empty calories and junk foods. The right number of calories to eat each day is to attain the required body weight is important. Limit foods and beverages high in calories but low in nutrients. Also limit the amount of saturated fat, *trans fat* and sodium in the diet. Take healthy fat in optimum amount.

Choose Appropriate Cooking Method

Prefer pressure/steam cooking to deep frying/roasting.

Use up at Least as many Calories as you take in

Try to maintain the energy balance every day. Burn as many calories as you have taken in daily to maintain ideal body weight.

Maintenance of Body Weight

Maintenance of ideal body weight is important for CVD patients. In case of overweight or obesity, weight reduction should be done as it would prove good for heart health.

Physical Exercise

Physical activity is essential to maintain ideal body weight by burning excess calories and is of vital significance for health and prevention of non-communicable diseases. It is recommended to carry out at least 45 minutes of aerobic physical activity of moderate-intensity for at least 5 days in a week to lower BP and reduce LDL-C and non–HDL-C.

Stop Smoking Cigarettes

Smoking decreases HDL-C. Smoking is believed to change LDL cholesterol to a form that promotes the buildup of deposits in the walls of coronary arteries. Smoking significantly increases the overall chances of developing heart disease, because it damages heart and blood vessels.

Reduce Stress

Try to minimize stressful situations as much as possible at work, at home, and elsewhere. Practice yoga and meditation to relieve stress, strain and anxiety.

Hypertension

Hypertension, also known as high blood pressure, is a long term medical condition in which the blood pressure in the arteries is persistently elevated. The higher the pressure in blood vessels, the harder the heart has to work in order to pump blood. If left uncontrolled, hypertension can be a major risk factor for cardiovascular disease (especially stroke, angina, myocardial infarction, heart failure, and left ventricular failure), renal disease, and retinopathy.

Classification of Blood Pressure

Normal adult blood pressure is defined as a systolic blood pressure of 120 mm Hg and a diastolic blood pressure of 80 mm Hg. Normal levels of both systolic and diastolic blood pressure are particularly important for the efficient function of vital organs such as the heart, brain and kidneys and for overall health and wellbeing. In Nov., 2017, the American Heart Association updated the hypertension guidelines. The major change was that systolic blood pressure of 130 mmHg or more or diastolic of 80 mmHg or more is considered hypertension and needs to be controlled. The association lowered the standards from the reading of 140/90 and above to facilitate early management of high BP and reduce disability and deaths related to complications from hypertension. Hence, persons with systolic blood pressure of 130–140 mmHg and/or diastolic 80–90 mmHg are now classified as hypertensive, while they were earlier classified as pre-hypertensive by Joint National Committee of USA for the Detection, Evaluation and Treatment of High Blood Pressure 7 criteria (JNC7, 2004). This relabels a large number of people as having elevated blood pressure and warranting lifestyle changes. The 2017 guideline also emphasizes individualized cardiovascular risk assessment. They use 10 year cardiovascular disease (CVD) risk calculation to decide on treatment threshold. This will also mean that most elderly patients would get treated with dietary modifications and medications earlier.

Table 21.11 New ACC/AHA Hypertension Guidelines

Systolic, Diastolic Blood Pressure (mm Hg)	*JNC7*	*2017 ACC/AHA*
<120 and <80	Normal BP	Normal BP
120–129 and <80	Pre-hypertension	Elevated BP
130–139 or 80–89	Pre-hypertension	Stage 1 hypertension
140–159 or 90–99	Stage 1 hypertension	Stage 2 hypertension

Prevalence

Globally, hypertension has been emerging as a serious threat to public health. The high and increasing worldwide burden of hypertension is a major global health challenge because it increases morbidity and mortality from cardiovascular and kidney diseases and financial costs to society. Researchers have estimated that raised blood pressure currently kills 9.4 million people every year (WHO, 2013). About 54 per cent of stroke and 47 per cent of ischemic heart disease worldwide were attributable to high blood pressure (Jugal Kishor *et al.*, 2016. Prevalence of hypertension and determination of its risk factors in rural Delhi. Int. J. Hypertension. http://dx.doi.rog/10.1155/2016/7962595). In 2010, the worldwide prevalence of hypertension was 1.39 billion persons, representing 31 per cent of all adults. This represented a 5.2 per cent increase in the global prevalence between 2000 and 2010 (Michael J. Bloch. 2016. J. Am. Soc. Hypertension 10 (10): 753–754); from 2000 to 2010, the prevalence of hypertension decreased by 2.6 per cent in high-income countries, but increased by 7.7 per cent in low/middle–income countries. Almost three times as many people with hypertension live in low/middle–income countries (1.04 billion) than in high-income countries (694 million).

Hypertension is the third most important risk factor for burden of diseases in South Asia. It is the leading preventable risk factor for premature death and disability worldwide. Without effective intervention, the increasing burden of hypertension will only exacerbate the global epidemic of cardiovascular and kidney diseases, especially in low- and middle-income countries (Mills *et al.,* 2016. Global disparities of hypertension prevalence and control. Circulation. 134: 441-450).

Hypertension is most important risk factor for CVD in India. Global Burden Disease (GBD) Study has estimated that hypertension led to 1.6 million deaths and 33.9 million disability-adjusted life years in 2015 and is most important cause of mortality as well as disease burden in India. Hypertension is widely prevalent in India with significant regional variations as reported by the NFHS-4 study (Gupta and Xavier, 2018. Hypertension: The most important non- communicable disease risk factor in India. Ind. Heart J. 70 (4): 565-572. doi: 10.1016/j.ihj.2018.02.003. Epub 2018 Feb 12.). The prevalence of hypertension in adult Indians is estimated to be 30 per cent (34 per cent in urban areas and 28 per cent in rural areas). The number of individuals with hypertension is expected to double from 118 million in 2000 to 213.5 million by 2025. In India, the average blood pressure has increased in the past 2 decades, whereas in most Western nations it has declined.

Hypertension is also directly responsible for 57 per cent of all stroke and 24 per cent of all coronary heart disease (CHD) related deaths in India (Gupta *et al.,* 2018. Int. J. Community Med. Public Health. 2018. Mar; 5 (3): 957-962). It is clear from the above mentioned facts that the burden of hypertension causes significant morbidity and mortality. In India, hypertension among young adults is higher than previously estimated and more than in Central and Eastern Europe, the region previously estimated to have the highest rates for this group, says a study led by researchers at Harvard University recently. The prevalence of hypertension was 20 per cent among women and 24.5 per cent among men, the research showed. The study also found unexpectedly high rates of hypertension among young adults. The prevalence of hypertension in the age group 18-25 years was 12.1 per cent, according to the study.

Risk Factors Contributing to Hypertension

Different factors are responsible for increasing the prevalence of hypertension in developed and developing countries and are attributable to rapid transition of lifestyle practices in developing countries including India, as well as increased elderly population due to an increase in life expectancy. Different risk factors leading to hypertension are:

i) Age and Sex

Among the key determinants, age and sex are the important non-modifiable risk factors for hypertension. The risk of hypertension increases with age due to stiffening of blood vessels, although ageing of blood vessels can be slowed through healthy living, including healthy eating and reducing the salt intake in the diet. Until age 64, men are more likely to get high blood pressure than women are. At 65 and older, women are more likely to get high blood pressure.

ii) Behavioral Risk Factors

There are many modifiable behavioral risk factors for the development of hypertension including:

- ✰ **Consumption of food containing too much salt and fat, and not eating enough fruit and vegetables:** Good nutrition from a variety of

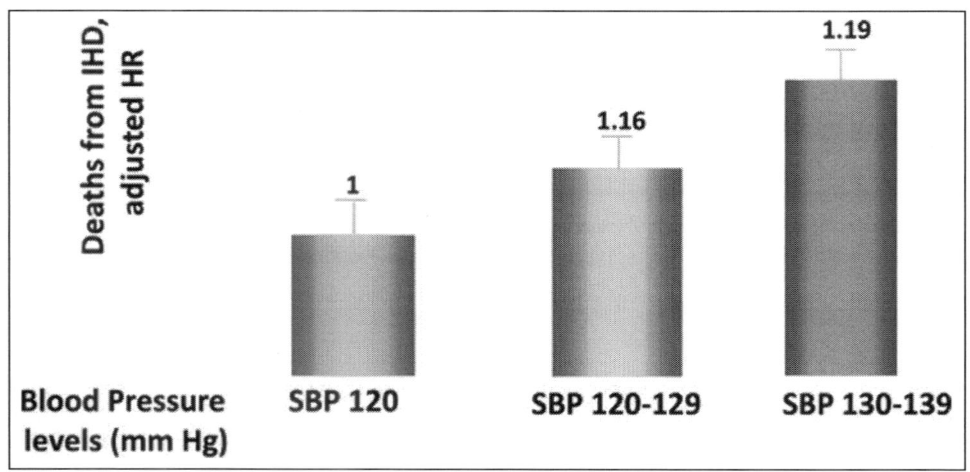

Figure 21.3: Deaths from Ischemic Heart Disease and Blood Pressure Levels from Mumbai/India Cohort Study. SBP Indicates Systolic Blood Fressure.

***Source*: Chopra and Ram, C.V.S. 2019. Recent Guidelines for Hypertension: A clarion call for blood pressure control in India. Circulation Res. 124 (7): 984-986, DOI: (10.1161/CIRCRESAHA.119.314789).**

sources is critical for health. A diet that is too high in salt consumption, as well as calories, saturated and *trans* fat and sugar, carries an additional risk of high blood pressure. Too much sodium in diet can cause body to retain fluid, which increases blood pressure. On the other hand, making healthy food choices can actually help lower blood pressure.

- **High blood lipid profile especially cholesterol:** More than half of people with high blood pressure also have high cholesterol.
- **Harmful levels of alcohol use especially if regular heavy or binge drinking:** Regular, heavy use of alcohol can cause many health problems, including heart failure, stroke and an irregular heartbeat. It can cause blood pressure to increase dramatically and can also increase the risk of obesity and related diseases.
- **Smoking:** Using tobacco can cause blood pressure to temporarily increase and can contribute to damage arteries. Not only does smoking or chewing tobacco immediately raise blood pressure temporarily, but the chemicals in tobacco can damage the lining of artery walls. This can cause arteries to narrow and increase risk of heart disease. Second hand smoke, exposure to other people's smoke, also increases the risk of heart disease for non-smokers.
- **Physical inactivity and lack of exercise:** Obesity especially central adiposity is linked to hypertension. BMI>25 is significantly associated with hypertension. The more one weighs, the more blood is needed to supply oxygen and nutrients to the tissues. As the volume of blood circulated through blood vessels increases, so does the pressure on the artery walls. Hence, carrying too much weight puts an extra strain on the heart and circulatory system that can cause serious health problems. It also increases risk of cardiovascular disease, diabetes and high blood pressure. Indeed in overweight individuals, the influence of hypertension on cardiovascular disease is more substantial than in normal weight individuals.
- **Insulin resistance due to obesity:** Most people with diabetes also develop high blood pressure.
- **Sleep apnea: Obstructive sleep apnea** may increase risk of developing hypertension and is common in people with resistant hypertension.
- **Poor stress management:** Too much stress may contribute to increased blood pressure. Also, too much stress can encourage behaviors that increase blood pressure, such as poor diet, physical inactivity, and using tobacco or drinking alcohol more than usual. Socio-economic status and psycho-social stress can affect access to basic living necessities, medication, healthcare providers, and the ability to adopt healthy lifestyle changes and therefore, cause high blood pressure.

A higher level of BMI and alcohol consumption, improper unhealthy dietary practices, sedentary lifestyle and harmful use of alcohol and tobacco are the major contributing factors to the higher prevalence of hypertension in India. These behavioural risk factors are highly influenced by people's working and living conditions. Addressing these behavioural risk factors can prevent hypertension. If no action is taken to reduce exposure to these factors, cardiovascular disease incidence, including hypertension, will increase.

iii) Metabolic Factors

In addition, there are several metabolic factors that increase the risk of heart disease, stroke, kidney failure and other complications of hypertension, including diabetes, high cholesterol and being overweight or obese. Tobacco and hypertension interact to further raise the likelihood of cardiovascular disease.

iv) Socio-economic Factors

Social determinants of health, *e.g.* income, education and housing, have an adverse impact on behavioral risk factors and in this way, influence the development of hypertension. For example, unemployment or fear of unemployment may have an impact on stress levels that in turn influences high blood pressure. Living and working conditions can also delay timely detection and treatment due to lack of access to diagnostics and treatment and may also impede prevention of complications.

Rapid unplanned urbanization also tends to promote the development of hypertension as a result of unhealthy environments that encourage consumption of fast food, sedentary behavior, tobacco use and the harmful use of alcohol.

v) Other Factors

In some cases, there is no known specific cause for hypertension. Genetic factors may play a role. Family history plays a signification role in the causation of BP as high blood pressure tends to run in families. If parents or other close blood relatives have high blood pressure, there is an increased chance that one will get it, too.

When hypertension develops in people below the age of 40 years, it is important to exclude a secondary cause such as kidney disease, endocrine disease and malformations of blood vessels. Preeclampsia is hypertension that occurs in some women during pregnancy. It usually resolves after the birth but it can sometimes linger, and women who experience preeclampsia are more likely to have hypertension in later life.

Symptoms

Hypertension is a silent, invisible killer that rarely causes symptoms in the early stages and many people go undiagnosed. There is a common misconception that people with hypertension always experience symptoms, but the reality is that most hypertensive people have no symptoms at all. Sometimes hypertension causes symptoms such as headache, shortness of breath, dizziness, chest pain, palpitations of the heart and nose bleeds. It can be dangerous to ignore such symptoms, but neither can they be relied upon to signify hypertension. Hypertension is a serious warning sign that significant lifestyle changes are required. As the condition can be a silent killer, therefore, it is important for everybody to know their blood pressure reading after middle age.

Prevention and Control

- ☆ Promoting a healthy lifestyle with emphasis on proper nutrition and healthy diet
- ☆ Avoiding alcohol
- ☆ Stop smoking
- ☆ Maintaining a normal body weight
- ☆ Regular physical activity
- ☆ Yoga and meditation
- ☆ Health check up
- ☆ Monitor blood pressure
- ☆ Checking blood sugar, blood cholesterol and urine albumin from time to time
- ☆ Regularly taking any prescribed medications for lowering blood pressure.

Strategies for Dietary Management

A. Recommended Dietary Pattern for Hypertensive Individuals to Lower BP

According to AHA/ACC (2013) Guideline on Lifestyle Management to Reduce Cardiovascular Risk, the following dietary pattern has been advised to adults to get benefit from lowering blood pressure:

1 Consume a dietary pattern that emphasizes intake of vegetables, fruits, and whole grains; include low-fat dairy products, poultry, fish, legumes, non-tropical vegetable oils and nuts; and limit intake of sweets, sugar-sweetened beverages and red meats.

   a. Adapt this dietary pattern to appropriate calorie requirements, personal and cultural food preferences, and nutrition therapy for other medical conditions (including diabetes mellitus).

   b. Achieve this pattern by following plans such as the DASH dietary pattern with lower sodium intake

2. Lower sodium intake.

   a. Consume no more than 2,400 mg of sodium/day (6g salt);

   b. Further reduction of sodium intake to 1,500 mg/day is desirable since it is associated with even greater reduction in BP; and

   c. Reduce intake by at least 1,000 mg/day since that will lower BP, even if the desired daily sodium intake is not yet achieved.

B. Dietary Components of DASH Dietary Pattern with Lower Sodium Intake

Diet is the most important environmental factor influencing hypertension, with sodium and potassium being the most important dietary components. The high consumption of dietary sodium is a relatively recent phenomenon. In isolated societies where the consumption of sodium is less than 1 g per day, the prevalence of hypertension is only 1 per cent of that in industrialized communities. Similarly, eating a diet high in sodium, calories, saturated fat, and sugar of diet can increase risk of developing high blood pressure.

According to recent guidelines of AHA/ACC (2013) guidelines, hypertensive patient should adopt DASH dietary pattern with low sodium intake. The DASH diet is an acronym for Dietary Approaches to Stop Hypertension. The landmark Dietary Approaches to Stop Hypertension (DASH) trial demonstrated that modification of diet significantly lowered BP in patients with stage 1 hypertension and high normal BP. The DASH diet, which emphasizes the consumption of fruits, vegetables, high fiber, low saturated fat and low fat dairy products, also lowers BP in persons with isolated systolic hypertension (ISH). Following this diet can help lower blood pressure, coronary disease risk and risk of stroke by 5 to 8 per cent in individuals with hypertension.

I. Nutritional Goals of DASH Dietary Pattern

Macronutrients

Gaining weight can increase blood pressure while losing weight can reduce blood pressure. Hence, calorie intake should be such so as to maintain optimum body weight. Obese or overweight hypertensive patients should take low calorie diet to reduce body weight.

The DASH diet is based on a 2,100-calorie diet. According to the DASH program, person should aim to consume 27 per cent of their total calories from fat, 18 per cent from protein and 55 per cent from carbohydrates. Saturated fat should account for no more than 6 per cent of total caloric intake and rest should be from PUFA and MUFA while cholesterol intake should be restricted to 150 milligrams or less. Opt for heart-healthy monounsaturated and polyunsaturated fats — olive oil, fatty fish, nuts and seeds instead of saturated and trans fats, which are detrimental to cholesterol levels.

Following a 2,100-calorie diet means that one needs 260 to 300 grams of carbohydrates each day. Select whole foods that are rich in carbohydrates as well as fiber, vitamins and minerals. Fruits, starchy vegetables, beans and other plant foods are healthy carbohydrate sources. One will also get carbohydrates and protein from milk, but opt for low-fat milk to avoid consuming a high amount of calories. DASH dieters should also aim to consume at least 30 grams of fiber each day.

Micronutrients

While the DASH diet does not provide specific guidelines for all vitamins and minerals, it provides recommendations and limitations for sodium, potassium, calcium and magnesium. According to the DASH program, dieters should limit sodium intake to 2,400 milligrams a day, obtain 4,700 milligrams of potassium, 1,250 milligrams of calcium and 500 milligrams of magnesium. A sodium intake of 1,500 milligrams or less is recommended for dieters with high blood pressure.

Sodium

The effect of sodium intake on BP is well established. Reductions in dietary salt lessen the amount of sodium the kidney has to excrete to restore normal blood volume. Aortic and carotid artery compliance is significantly improved in older patients with hypertension when sodium intake is reduced. Reduction in sodium intake also improves arterial vasodilation. Sodium content is high in processed foods, such as bread (approximately 250 mg/100 g), processed meats like bacon (approximately 1500 mg/100 g), snack foods such as salty biscuits, cheese puffs and popcorn (approximately 1500 mg/100 g), as well as in condiments such as soy sauce (approximately 7000 mg/100 g) and should be avoided. A meta-analysis estimated that even a modest reduction in sodium intake of 1 g/day reduces systolic blood pressure by 3.1 mmHg in hypertensive patients and 1.6 mmHg in normo-tensive patients. The DASH trial showed a dose–response relationship between sodium reduction and BP reduction.

Potassium

It is another mineral that affects blood pressure. A higher potassium intake has been shown to reduce blood pressure. Potassium relaxes the walls of the blood vessels, lowering blood pressure and protecting against muscle cramping. A number of studies have shown an association between low potassium intake and increased blood pressure and higher risk of stroke. On the flip side, people who already have high blood pressure can significantly lower their systolic blood pressure by increasing their potassium intake when they choose to eat healthy foods. Risk of stroke varies greatly with potassium intake; the relative risk of stroke in the highest quintile of potassium intake (average of 110 mmol/day) is almost 40 per cent lower than that in the lowest quintile of intake (average intake of 61 mmol/day). Potassium-rich foods help to reduce blood pressure. A meta-analysis of 33 potassium supplement trials demonstrated modest reductions in systolic blood pressure and diastolic blood pressure (–3.1 and –2.0 mm Hg), with effects appearing strongest when dietary sodium intake was high.

The recommended potassium intake for an average adult is 4,700 milligrams (mg) per day. Many of the elements of the DASH diet *viz.*, fruits, vegetables, fat-free or low-fat (1 per cent) dairy foods and fish are good natural sources of potassium. For example, a medium banana has about 420 mg of potassium and half a cup of plain mashed sweet potatoes has 475 mg. Since people with high blood pressure may also be trying to lose weight, consider potassium rich foods that are low in calories and carbohydrates. Good examples include broccoli, water chestnuts, spinach, and other leafy greens. Coconut water and citrus fruits are rich sources of potassium. Also good although slightly higher in carbohydrates and calories are butternut squash, banana and sweet potatoes, and fruits such as cantaloupe and kiwi.

Calcium

Calcium is a major mineral essential for human health. In addition to playing an important role in building and maintaining strong bones and teeth, calcium aids in muscle contraction and regulates blood pressure. In fact, consuming a diet rich in calcium, along with other minerals, such as potassium and magnesium, is recommended for reducing blood pressure. Trials of calcium supplements that often included vitamin D found small reductions in systolic and diastolic blood pressure (–1.9 and –1.0 mm Hg). Thirteen trials using calcium alone in hypertensive patients demonstrated similar small reductions in systolic and diastolic blood pressure (–2.5 and –0.8 mm Hg).

Per DASH diet guidelines, one should strive to consume approximately 1,250 mg of calcium daily. This may come from food sources alone, dietary supplements, or a combination of food sources and dietary supplements. Calcium-rich foods include dairy products, such as yogurt and milk. Cheese is calcium-rich but also high in sodium. The calcium in milk and yogurt is typically well absorbed and used efficiently by the body. About 230 g of milk provides approximately 300 mg of calcium.

Magnesium

Magnesium helps to regulate hundreds of body systems, including blood pressure, blood sugar, and muscle and nerve function. We need magnesium to help blood vessels relax, and for energy production, bone development, and transporting calcium and potassium. Just like potassium, too much magnesium can be lost in urine due to diuretic use, leading to low magnesium levels. In some clinical studies, magnesium intake has shown a wide range of BP reduction, with some showing no change in BP. The researchers found that taking 368 mg of magnesium supplements daily for three months reduced people's systolic blood pressure by an average of 2 millimeters of mercury (mm Hg), and reduced their diastolic blood pressure by an average of 1.8 mm Hg. Twelve trials of magnesium supplements in hypertensive patients showed modestly lower diastolic blood pressure (–2.2 mm Hg), Overall, the evidence indicates that potassium modestly lowers BP, more so in hypertensive patients or when dietary sodium is high. The combination of increased intake of magnesium and potassium coupled with reduced sodium intake is more effective in reducing BP than single mineral intake and is often as effective as one antihypertensive drug in treating hypertension.

It is best to get this mineral from food, especially dark, leafy green vegetables, unrefined grains, and legumes. Good sources are wheat germ, buckwheat, almond, cashew nuts, green leafy vegetables, chickpeas, coco *etc.* The RDA of magnesium is 420 milligrams (mg) per day for men aged 50 and older; 320 mg/day for women aged 50 and older.

II. Recommended Food Intake as Per DASH Diet with Lower Sodium Intake

In addition to the standard DASH diet, there is also a lower sodium version of the diet.

- ✰ **Standard DASH diet: One** can consume up to 2,300 milligrams (mg) of sodium a day.
- ✰ **Lower sodium DASH diet**: Ideally, a hypertensive should have a goal to consume up to 1,500 mg of sodium a day.

Both versions of the DASH diet aim to reduce the amount of sodium in the diet. The standard DASH diet meets the recommendation from the Dietary Guidelines for Americans to keep daily sodium intake to less than 2,300 mg a day. The American Heart Association recommends 1,500 mg a day of sodium as an upper limit for all adults as it further lowers the blood pressure.

Both versions of the DASH diet include lots of whole grains, fruits, vegetables and low-fat dairy products. The DASH diet also includes some fish, poultry and legumes, and encourages a small amount of nuts and seeds a few times a week. One can eat red meat, sweets and fats in small amounts. The DASH diet is low in saturated fat, cholesterol and total fat.

Here's a look at the recommended servings from each food group for the 2,100-calorie-a-day DASH diet.

Grains and Grain Products

Grains include bread, cereal, rice and pasta. Examples of one serving of grains include 1 slice whole-wheat bread, 30g dry raw cereal, or 1/2 cup cooked cereal, rice or pasta. The DASH diet recommends that individuals who consume a 2,100-calorie diet should include 7 to 8 servings of grains and grain products each day, and at least three from whole grain foods as they contain more fibre and nutrients than do refined grains. For instance, use brown rice instead of white rice, whole-wheat pasta instead of regular pasta and whole-grain bread instead of white bread. The DASH diet emphasizes consuming the majority of calories from these complex carbohydrates to minimize excess fat in the diet.

Fruits and Vegetables

The DASH diet recommends consuming 4 to 5 servings of fruits and 4 to 5 servings of vegetables each day, respectively. Fruits and vegetables can contribute to daily carbohydrate needs, while also providing an abundance of vitamins and minerals that cannot be met through other dietary sources.

Examples of one serving of vegetables include 1 cup raw leafy green vegetables or 1/2 cup cut-up raw or cooked vegetables. Tomatoes, carrots, broccoli, sweet potatoes, greens and other vegetables are full of fiber, vitamins, and such minerals as potassium and magnesium. They should be served as the main dish for a meal. Choose vegetables which are low in sodium content.

Many fruits need little preparation to become a healthy part of a meal or snack. Like vegetables, they are packed with fiber, potassium and magnesium and are typically low in fat; coconuts are an exception. Examples of one serving include one medium fruit, 1/2 cup fresh, frozen or canned fruit, or 115 ml of juice. Have a piece of fruit with meals and one as a snack, then round out day with a dessert of fresh fruits topped with a dollop of low-fat yogurt. Leave on edible peels whenever possible. The peels of apples, pears and most fruits with pits add interesting texture to recipes and contain healthy nutrients and fiber. In case of canned fruit or juice, make sure that no sugar is added.

Dairy Products

Dairy products are important to the diet because of their vitamin, mineral and protein content, although they tend to be high in fat and cholesterol. DASH dieters should consume only low-fat or fat-free dairy products, and limit their dairy servings to just 2 to 3 each day. Examples of one serving include 1 cup skim or 1 per cent milk, 1 cup low fat yogurt, or 40 g skim cheese. *Paneer* from low fat milk should be preferred over cheese as even regular fat-free cheeses are typically high in sodium.

Meat, Fish and Poultry

The DASH diet plan recommends consuming only lean sources of meat, fish and poultry to meet protein needs, and limiting servings to just two or less each day. Trim away skin and fat from poultry and meat and then bake, broil, grill or roast instead of frying in fat. Eat heart-healthy fish, such as salmon, herring and tuna.

These types of fish are high in omega-3 fatty acids, which can help lower total cholesterol in the blood. While animal products are high in protein, they can be high in fat, cholesterol and sodium, all of which contribute to hypertension. Preferably, one should meet much of protein needs through plant foods.

Nuts, Seeds and Legumes

Nuts, seeds and legumes are healthy sources of lean and cholesterol-free protein, unsaturated fat and vitamins and minerals. Almonds, sunflower seeds, kidney beans, peas, lentils and other foods in this family are good sources of magnesium, potassium and protein. They are also full of fiber and phytochemicals, which are plant compounds that may protect against some cancers and cardiovascular disease. The DASH diet plan recommends consuming 4 to 5 servings of nuts, seeds and legumes each week. Examples of one serving include 1/3 cup nuts, 2 tablespoons seeds, or 1/2 cup cooked beans or peas. Since sodium intake is a concern, aim to consume unsalted nuts and seeds.

Fats and Oils

The DASH diet strives for a healthy balance by limiting total fat to less than 30 per cent of daily calories from fat (2 servings of fat per day), with a focus on the healthier monounsaturated fats. Example of one serving includes 1 teaspoon fat or oil. Saturated fat and trans fat are the main dietary culprits in increasing risk of coronary artery disease. DASH helps to keep daily saturated fat to less than 6 per cent of total calories by limiting use of meat, butter, cheese, whole milk, cream and eggs in the diet, along with foods made from lard, solid shortenings, and palm and coconut oils. Avoid trans fat, commonly found in such processed foods as baked goods and fried items. Dietary sources of MUFA and PUFA should provide rest of the calories from fat. MUFA is found in olive and rapeseed oil as well as walnut oil and avocados. PUFA is found in sunflower, corn and safflower oils *etc.* Excess of these PUFA and MUFA is also not good because they are concentrated sources of energy. For example, one tablespoon of olive oil has the same calories as four apples or two slices of bread. Fats and oils should be taken in limited amounts.

Sweets

According to the DASH diet, five servings or fewer per week of sweets are allowed. Examples of one serving include 1 tablespoon sugar, jelly or jam, 1/2 cup sorbet, or 1 cup lemonade. Choose those sweets that are fat-free or low-fat, such as sorbets, fruit ices, jelly beans, hard candy or low-fat cookies. Cut back on added sugar which has no nutritional value but can pack on calories. Artificial sweeteners such as aspartame (NutraSweet, Equal) and sucralose (Splenda) may help to satisfy sweet tooth while sparing the sugar. But remember that these are to be used sensibly too.

Alcohol

Finally, the guidelines recommend no more than two alcoholic drinks per day for men and one for women and people of lighter weight. Drinking too much alcohol can increase blood pressure.

Caffeine

The DASH diet does not address caffeine consumption. The influence of caffeine on blood pressure remains unclear. But caffeine can cause blood pressure to rise at least temporarily.

III. Physical Activity

According to recent guidelines of AHA/ACC (2013) guidelines, in general, adults should be advised to engage in aerobic physical activity to lower BP; 3–4 sessions per wk, lasting on average 40 min per session, and involving moderate to vigorous intensity physical activity.

IV. Dietary Tips

- ☆ Eat foods lower in fat, salt, and calories.
- ☆ Reduce excess body weight. Losing 3–9 per cent body weight is associated with 3 mmHg reduction in systolic blood pressure (SBP) and diastolic blood pressure (DBP).
- ☆ Eat a variety of foods.
- ☆ Increase intake of fruits and low sodium vegetables in the diet. Add a serving of vegetables at lunch and at dinner. Add a serving of fruit to meals or as a snack. Canned and dried fruits can be used but check that they don't have added sugar.
- ☆ Prefer fresh seasonal fruits and vegetables. Richly colored green, orange, and red fruits and vegetables are high in potassium and minerals that help lower blood pressure.
- ☆ Take potassium rich foods like apricots, avocados, cantaloupe, honeydew melon, prunes, grapefruit, oranges, banana, raisins, dates, tuna fish, coconut water, peas, cabbage, parsley, mushrooms, potato, tomato, spinach *etc.*
- ☆ Add dry beans to daily diet and prefer them over animal products.
- ☆ Limit intake of meat and poultry. Eat lean meat and skinless chicken. Fatty fish can be taken once or twice a week.
- ☆ Drink low-fat or skim milk and other dairy products.
- ☆ Eat foods high in dietary fiber (whole grain breads, cereals, pasta, fresh fruit, and vegetables).
- ☆ Use less oil, butter, margarine, shortening, and salad dressings.
- ☆ Limit salt intake to not more than about one teaspoon salt a day; gradually reduce to 2/3

teaspoon. These amounts include all sodium eaten, including sodium in food products as well as in what you cook with or add at the table. Although this may be an ideal goal to aim at, for many people it will require major and possibly unacceptable changes to their usual eating habits.

- Donot use table salt. Use salt-free seasoning blends or spices and herbs, vinegar, lemon or fruit juices instead of salt to flavour foods.
- Rinsing canned foods to remove some of the sodium.
- Buy foods labeled "no salt added," "sodium-free," "low sodium" or "very low sodium".
- Instead of snacking on salted chips, fried foods, fast foods or sweets, eat unsalted nuts and seeds (pumpkin, squash, and sunflower) as they are mineral-rich foods that lower blood pressure. One can take raisins, low-fat and fat-free probiotic curd, frozen yogurt and salad containing raw vegetables instead of calorie rich desserts.
- Donot take pickles, *papad*, ketchup, soy sauce, mustard, peanut butter, bakery goods and other processed foods.
- Limit caffeine consumption.
- Avoid alcohol.

Prevention of Stress and Depression

Stress can be managed through lifestyle modifications. Daily deep breathing or meditation has been shown to decrease stress, improve mental well-being, and decrease the risk for blood pressure and heart disease. Daily exercise, yoga or physical activity (whether walking, playing with kids, or gardening) and meditation also decrease stress and depression. Social or religious activities can be very helpful in promoting mental wellness and provide opportunities for meaningful social interaction. Finally, don't underestimate the importance of 8 hours of restful sleep a night which rejuvenates the mind and body.

Congestive Heart Failure

Congestive heart failure (CHF) or heart failure (HF) occurs when damage to the heart leads to reduced efficiency in pumping blood around the body with the consequent symptoms of breathlessness, excessive fatigue, fluid retention, swelling on legs, chronic coughing or wheezing, nausea or lack of appetite, high heart rate, confusion or impaired thinking. It can be caused by high blood pressure or other heart problems.

Prevalence

Heart failure is growing at an alarming rate in India and has almost attained epidemic proportions. The estimated heart failure burden in India ranges from 1.3 million to 4.6 million cases per annum. The prevalence of heart failure in India is possibly on the rise as India remains doubly burdened by the rise in the risk factors of traditional cardiovascular disease (CVD) and by the persistence of pre-transitional diseases such as RHD, endomyocardial fibrosis, tuberculous pericardial disease and anaemia. First, the population of India is ageing due to recent successes against communicable diseases such that the number of people >60 years old will increase from 62 million in 1996 to 143 million in 2021. Heart failure is predominantly a disease of the elderly, as the lifetime risk for heart failure increases with age, so the burden of this disease is likely to increase with the ageing population.

In addition to the ageing population described above, the prevalence of hypertension is projected to increase from 118 million (2000) to 214 million (2025). The annual incidence of heart failure due to obesity (body mass index [BMI] >30 kg/m^2) has been estimated to increase by 0.3 per cent in women and 0.5 per cent in men, in the Framingham Heart Study, after adjustment for age, hypertension, left ventricular hypertrophy, myocardial infarction, valve disease, diabetes and cholesterol.

Similarly, the prevalence of diabetes in India is projected to increase from 32 million (2000) to 70 million (2025).The incidence of heart failure has been demonstrated to increase from 2.3 per 1000 person-years for a HbA1c <6 per cent to 11.9 per 1000 person-years for a HbA1c >11.9 per cent. Taking the estimate of heart failure incidence based upon optimal glucose control, the annual incidence of HF due to diabetes may increase from 73 600 (2000) to 161 000 (2025).

Population, epidemiological and health transitions will continue to play an important role in the future burden of HF in India. The need of the hour is to disseminate awareness to the general public about it and take measures to prevent and control it.

Stages of Heart Failure: Goals

In 2005, the American Heart Association (AHA) and American College of Cardiology (ACC) introduced updated HF clinical practice guidelines and suggested to include four new stages of heart failure *i.e.* A through D.

Stage A represents patients who do not have structural heart disease nor do they have symptoms of heart failure but are at high risk for developing heart failure. These patients include those with hypertension, diabetes, atherosclerotic disease, obesity, metabolic syndrome, family history of cardiomyopathy, or exposure to cardiotoxic drugs (*e.g.* anthracyclines). The primary goals of treating stage A patients include treatment of hypertension and dyslipidaemia, cessation of tobacco, alcohol and illicit drug use, encouragement of exercise and management of metabolic syndrome. Anticholinesterase inhibitors (ACE-I) or angiotensin receptor blockers (ARBs) are recommended for patients with concomitant diabetes and/or vascular disease.

Stage B represents patients with evidence of structural heart disease in the absence of symptoms of heart failure (such as left ventricular hypertrophy, left ventricular dysfunction or valvular heart disease). The primary goals of treating stage B patients are similar to those for stage A patients. ACE-I or ARBs and beta-blockers are recommended for appropriate patients with left ventricular dysfunction and/or vascular disease, as well as implantable cardioverter defibrillators (ICDs), in selected patients.

Stage C represents patients with a history, symptoms and clinical signs consistent with heart failure. The primary goals of treating stage C patients include all the goals of stages A and B, as well as dietary salt restriction. Aldosterone antagonists, digoxin, hydralazine/nitrate combination therapy, and biventricular pacemaker/ICDs are recommended for selected patients.

Stage D represents patients with advanced heart failure who have marked symptoms at rest despite maximal medical therapy. These patients are often hospitalized repeatedly and cannot be discharged without specialized therapies. The primary goals of treating stage D patients include all the goals of stages A, B and C, as well as decisions regarding the appropriate level of care. Clinicians and patients in India can use this paradigm to help guide their goals and strategies, particularly in stages A and B patients where prevention of heart failure is achievable.

Prevention and Control

One of the best ways to prevent heart failure is to control conditions that cause heart failure, such as coronary artery disease, high blood pressure, diabetes or obesity; dietary and life style management as discussed earlier for these diseases should be followed for heart failure too. Basic guidelines for the prevention and control of CHF are:

1. Follow a low sodium, low cholesterol, low fat (especially saturated and trans fat), heart healthy diet. Eat more fiber from whole grains, fresh fruits, vegetables and legumes (beans and lentils).
2. Restrict the amount of fluid that you drink.
3. Weigh yourself every day.
4. Follow healthy life style.
5. Take all medications as prescribed by the doctor.

Dietary Guidelines for Heart Failure

Following a healthy diet can help to manage the symptoms of heart failure and improve quality of life of HF patients.

Sodium enhances thirst and makes the body hold onto fluids rather than urinating them out. In addition, it is important to limit the amount of fluids to drink. The amount can vary and doctor will let you know how much you should be drinking in a day. The extra fluid may make it very hard to breathe and it may be life-threatening and require hospitalization. So, following low-sodium and fluid guidelines are a vital part of the treatment for CHF.

The goal of diet therapy for congestive heart failure is to lessen the work load of heart, reduce fluid buildup and make it easier for the patient to breathe. Major highlights of dietary management of heart failure are as under:

Restricting Dietary Sodium

Sodium is a mineral that is found in all foods and is necessary in small amounts for many body functions. Sodium is consumed in the form of salt, which is made up of sodium and chloride. High sodium levels cause the body to retain fluid, which increases the heart's workload. Fluid retention can make heart failure worse and may cause symptoms such as difficulty breathing, swelling of the ankles, feet or abdomen and weight gain. Too much sodium can contribute to high blood pressure. Eating less sodium can help control blood pressure as well as prevent fluid gain.

All heart-healthy guidelines mentioned previously are important for people with CHF, but it is extra important to follow a low-sodium diet (less than 1,500 milligrams) to prevent fluid retention in the body. Limiting sodium intake will help to maximize the effects of diuretics and thus, moderate the workload on the heart by reducing the circulating volume. Low sodium diets can be very unpalatable so a compromise between avoiding an excessive salt intake while maintaining an adequate nutritional intake is required. Most people get more than enough sodium in their diet even if they do not use salt to cook or at the table for seasoning. Following tips can slash sodium content in the diet.

- ✰ A 'no added salt' diet should exclude high sources of dietary sodium by avoiding salt added at table or while cooking as most foods have more than enough sodium already in them. While cooking, if salt is to be added, use just a pinch for taste.
- ✰ Avoid herb or spice mixtures that contain salt or sodium. Avoid using the salt shaker or replace it with a sodium-free blend of herbs. Use herbs and spices such as basil, onion, garlic, oregano, parsley, citrus and other fruit juices, and vinegars to add flavor instead of salt.
- ✰ High sodium foods include cheese, lunch meat, highly processed breads and cereals, prepared items like canned and frozen foods and baked goods. One should avoid processed foods like meat and vegetable extracts, cured meat, tinned fish and meat, tinned and packet soups, cheese, baked goods like biscuits, cakes, pastries *etc.*, salted nuts and crisps, soy sauce, *papad*, pickles, salty snacks, ketchup, mayonnaise, mustard

sauce, canned food, tomato puree, monosodium glutamate *etc.*

- ✰ Avoid convenience foods, Chinese food, instant, canned, frozen and fast foods. If you have to take canned goods, rinse the foods thoroughly with plain water before cooking and eating them.
- ✰ Read food labels for sodium content before buying packaged or canned foods. Light or low fat does not always mean low sodium. Make sure that you read food labels to see how much sodium is in a food. The label on foods is the most powerful tool for following a low sodium diet.
- ✰ Eat plenty of seasonal fresh fruits and vegetables, which are naturally low in sodium.
- ✰ **Choose foods that are low in salt**, such as fresh meats, poultry, fish, dry and fresh legumes, eggs, milk and yogurt. Plain rice, whole wheat products and oatmeal are good low-sodium choices. However, the sodium content can increase if salt or other high-sodium ingredients are added during their preparation which one should avoid.
- ✰ Choose fresh foods that are low in salt, such as seasonal fresh low sodium vegetables and fruits,
- ✰ A well-balanced and colorful diet is important for overall health.

Fluid Restriction

Too much fluid in body can make it harder for already-weakened heart to pump. This can make symptoms, such as swelling and shortness of breath, worse. Most people do not need to limit their fluids until heart failure is advanced or severe. Patients with mild or moderate heart failure are unlikely to benefit from fluid restriction. Patients with severe heart failure should be advised to restrict their fluid intake to 1.5–2.0 l/day. Some foods contain a lot of fluid. Any food that will melt, that has a high water content, or that contains a lot of liquid should be measured and counted as part of fluid intake. So that means one needs to count ice cream, gelatin, ice, juicy fruits, and soup in fluid intake.

In order to catch any fluid retention or worsening of congestive heart failure early, it is very important to weigh yourself every day. Weigh yourself on the same scale every day when you wake up. Urinate first, and always wear the same amount of clothing. A gain of 500 g to 1 kg overnight, or 2.250 kg in a week, is a signal that body is retaining extra fluid. If this occurs, call your healthcare provider right away because you may need to make adjustments in the diet or medications. Noting and dealing with fluid weight gain promptly may prevent worsening heart failure and the need for hospitalization.

Obesity

Obesity is linked to the development of cardiovascular diseases, including atherosclerosis and hypertension, and to an increased risk for heart failure. Systolic and/or diastolic heart failure is more prevalent among obese and overweight individuals than among normal weight people. The risk of developing HF in the obese population was twice as that seen in the normal BMI population. The Framingham heart study suggested that there was a 5 per cent increase in men and 7 per cent increase in women for the risk of developing HF per unit of BMI. Despite this increased risk of HF in the elevated BMI population, recent studies have demonstrated that there is, in fact, a survival advantage in overweight and obese HF patients in comparison with their normal to low BMI counterparts. This observation, known as the 'obesity paradox' has been supported by many research studies. Some counter arguments have also been advanced against the phenomenon of the obesity paradox. Currently, the etiology remains unclear. In addition, there is insufficient evidence to clearly recommend weight loss in obese patients with HF (Nagarajan *et al.*, 2016. Obesity paradox in heart failure: a heavy matter. ESC Heart Fail. 2016 Dec; 3(4): 227–234.; Hamzeh *et al.*, 2017. Obesity, heart failure and obesity paradox. J. Tehran Heart Cent. 2017 Jan; 12(1): 1–5).

From prevention standpoint, intentional weight loss in obese individuals without HF can help to reduce the risk of development of HF. Patients with excess body weight (BMI > 30 kg/m^2) should be advised to reduce slowly in order to reduce progression of heart failure and improve well-being. According to few studies, there are several theoretical reasons to believe why obese patients with HF may benefit from intentional weight loss. These include the following: improved control of obesity-related co-morbidities, such as diabetes or insulin resistance, hypertension, and obstructive sleep apnea; better quality of life; better mental health; and improved candidacy for advanced HF therapies. In moderate to severe heart failure routine weight loss advice is inappropriate because unintentional weight and anorexia is common.

Based on current evidence, the American College of Cardiology/American Heart Association HF guidelines do not specifically recommend weight reduction in obese patients with HF. They do, however, provide a Class I recommendation for exercise training (or regular physical activity) as safe and effective for patients with HF who are able to participate to improve functional status (Khalid *et al.* May, 2015. Expert analysis. American College of Cardiology).

Nutritional Adequacy

In mild cases of heart failure patients, dietary recommendations regarding macro and micronutrients as suggested previously for atherosclerosis and hypertension remain the same. However, in more advanced cases, appetite can be very poor and food intake limited by symptoms. Patients should be regularly screened for under nutrition. Changes in body weight may indicate fluid retention so a holistic assessment taking into account their food intake is required. Ensuring a nutritionally adequate intake by encouraging small, frequent, nutrient-dense meals may help maintain body weight. This may conflict with the principles of the cardio-protective diet so advice must be given holistically to take into account the likely prognosis.

Alcohol

Alcohol consumption is associated with negative effect on cardiac muscle contraction, blood pressure which makes heart to work harder and is a risk factor to cause arrhythmias. Therefore, it is not a good idea to take alcohol when heart failure is there. Since one is allowed only a limited amount of fluid each day, it is best to choose healthier fluids.

Keep Food Intake (and Calories) in Check

Because CHF can make it hard to breathe, physical activity may be restricted by the doctor. As a result, it may be difficult to maintain or reach a healthy weight with weight loss. It is even more important to restrict calorie intake to prevent weight gain if exercise is limited due to heart-related breathing problems. Following is a list of food items to be taken and to be avoided in the daily diets:

Foods to be taken	*Foods to be avoided*
Meats and Meat Substitutes	**Meats and Meat Substitutes**
✰ Chicken and turkey without the skin (white meat has less cholesterol than dark meat) ✰ Fish including all fresh fish as well as canned tuna and salmon that have been packed in water. ✰ Lean lamb. Trim the fat before cooking and avoid cuts that have a lot of fat within the meat itself. ✰ Egg ✰ Dried beans and peas ✰ Tofu ✰ Low sodium peanut butter	✰ Sausage, hot dogs, lunch meats, bacon ✰ Cured meats such as ham, salami and pepperoni ✰ Frozen, canned or pre-packaged foods instant foods, canned soups, frozen pizzas *etc.* ✰ Restaurant prepared Chinese, Italian or Mexican foods ✰ Regular Peanut butter
Breads and Starches	**Breads and Starches**
✰ Whole wheat and whole grain bread or *chapati* ✰ Homemade baked goods that are made with low fat, low sodium ingredients (pancakes, muffins, breads, cookies) ✰ Low sodium crackers (the more fiber, the better) ✰ Pastas- spaghetti, macaroni *etc.* (whole wheat is best) ✰ Rice, bajra, barley, oat ✰ Other grains such as bulgur, quinoa and others	✰ Biscuits ✰ Packaged/prepared muffins, pastries, sweet rolls, coffee cakes, quick breads (banana bread, nut breads), cakes, pies, cookies ✰ Snack foods such as *samosa*, *pakora*, *tikki*, *bhujia*, all kinds of salted and fried savories like *mathi*, *matar*, *kachori etc.*, pizza, patties, buttered popcorn, salted crackers, chips, burgers *etc.* ✰ Ramen Noodles ✰ Pre-packaged mixes such as seasoned rice, upma, noodles, poha *etc.*
Vegetables and Fruits	**Vegetables and Fruits**
✰ Fresh fruits and vegetables ✰ Canned fruit that is canned in its own juice ✰ Salt-free canned vegetables ✰ Canned tomato puree and tomato paste (don't add salt) ✰ Homemade vegetable soups and curries	✰ Fried vegetables ✰ Vegetables cooked with more saturated fat ✰ Pickled vegetables (these are high in sodium) ✰ Sauerkraut ✰ Packaged potato mixes ✰ Regular canned vegetables ✰ Instant vegetable soup mixes ✰ Tomato products with salt added
Fats and Oils	**Fats and Oils**
✰ Vegetable oils liquid at room temperature such as canola, safflower, corn, sunflower, peanut, soy and olive oil ✰ Low-fat or fat-free mayonnaise or salad dressing	✰ Hydrogenated fats ✰ Butter ✰ Stick margarine ✰ Lard, salt pork, fat back, or meat drippings ✰ Creamy salad dressings that aren't light or low-fat ✰ Regular mayonnaise ✰ Gravy ✰ Cream or cheese sauces (look for these on restaurant menus) ✰ Coconut or palm oils

Moderate Amount of Unsalted Nuts	Excessive Consumption of Salted Nuts
Milk and milk products ☆ Skim milk or 1 per cent fat milk and milk products	**Milk and milk products** ☆ Whole milk and its products ☆ Cheese
Sweets and Desserts ☆ Frozen fruit bars ☆ Light ice cream, sherbet, sorbet, frozen yogurt ☆ Angel food cake ☆ Vanilla wafers and ginger snaps ☆ Homemade baked goods that have been modified with low fat ingredients ☆ Hard candies ☆ Low fat pudding	**Sweets and Desserts** ☆ Regular and pre-packaged baked goods ☆ Ice cream ☆ Store bought mixes (cakes, brownies, cookies, pie crusts)
Moderate Amount of Unsalted Nuts	***Excessive Consumption of Salted Nuts***
Beverages ☆ Water ☆ Skim milk ☆ Decaf coffee ☆ Tea (especially green) ☆ Light Soy milk ☆ Diet Soda	**Beverages** ☆ Sweet tea or regular coffee prepared with whole milk or artificial creamers ☆ Regular soda ☆ Whole milk
Condiments ☆ Salt free seasoning mixes, salt substitutes ☆ Vinegar ☆ Lemon Juice ☆ Garlic ☆ Garlic or onion powder ☆ Herbs and spices such as curry, basil, oregano, thyme, pepper, paprika, sage, onion, parsley, rosemary, mint, coriander and many others	**Condiments** ☆ BBQ sauce, steak sauce, Worcestershire sauce, cooking wine, cooking sherry ☆ Many salad dressings (look at sodium content) ☆ No more than 2 Tbsp per day of ketchup and mustard ☆ Meat tenderizer ☆ MSG ☆ Herb mixes that have sodium in them such as lemon pepper, BBQ seasoning, Jamaican Jerk seasoning and others ☆ Lite Salt ☆ Sea Salt/rock salt ☆ Soy sauce, fish sauce ☆ Pickles ☆ Olives ☆ All kinds of processed foods containing sodium as a salt or preservative

22

Diet in Gastrointestinal Disorders

The gastrointestinal tract (GIT) consists of a hollow continuous muscular tube starting from the oral cavity, where food enters the mouth, continuing through the pharynx, oesophagus, stomach and intestines to the rectum and anus, where food is expelled. There are various accessory organs that assist the tract by secreting enzymes to help break down food into its component nutrients. Thus, the salivary glands, liver, pancreas and gall bladder have important functions in the digestive system. The function of the digestive system is digestion, absorption and excretion. Digestion is the breakdown of food into small molecules, which are then absorbed into the body. In the case of gastrointestinal disease or disorders, these functions of the gastrointestinal tract are not achieved successfully. GIT diseases may affect any section of the GIT, from oesophagus to the rectum, and the accessory digestive organs. The term encompasses acute, chronic, recurrent or functional disorders and covers a wide range of diseases.

Some of the most common GI problems cover a whole range of disorders, from a simple stomachache to colorectal, gastric and pancreatic cancers. Digestive diseases include acid reflux, constipation, diarrhea, cirrhosis of the liver, celiac disease, colitis, Crohn's disease, diverticulosis and diverticulitis, gallstones, gastritis, hepatitis, inflammatory bowel disease, irritable bowel syndrome, liver failure, pancreatitis, peptic ulcers and ulcerative colitis, among other disorders. While some of these conditions have genetic roots, many are diet related. The symptoms vary widely, as do the treatments.

GI diseases may be mild, self-limiting and temporary or may persist over the long term and affect health adversely. For instance, sustained constipation can lead to diverticular disease (where the intestine becomes damaged from passing hard stools and then form pocket-like sacs) and impaction (causing the intestine and rectum to become blocked). Sustained diarrhoea can make blood pressure drop, which can lead to fainting or heart rhythm abnormalities. Nutrient and fluid loss from body due to diarrhoea or vomiting can lead to dehydration, electrolyte imbalances, vitamin deficiencies, weight loss and malnutrition. Similarly, nausea, indigestion and constipation can reduce appetite, and thus, reduce food intake leading to nutritional deficits. The nutritional effects of GI problems can be particularly serious for people who are already malnourished or ill.

Prevalence of GI Diseases

A new trend is emerging in Asia as diseases that were once considered "Western" are increasingly found in wealthy countries such as Japan and South Korea, as well as developing countries like India and China. A number of studies have highlighted the rapid growth of gastrointestinal diseases in Asian populations during the past two decades. Both incidence and prevalence rates have shown a notable rise, with scientists pointing out that GI diseases are increasingly the cause of morbidity and mortality in Asian countries. Gastrointestinal (GI) diseases in particular are becoming more common and have been linked to changing environmental factors brought on by industrialization, changes in diet, improved sanitation, and the increased use of antibiotics. The most common GI diseases like diarrhoeas, hepatitis, typhoid, TB of the intestinal tract *etc.* are still prevalent in these countries, but what is alarming is the rise of lifestyle diseases like irritable bowel syndrome, Gastroesophageal reflux disease (GERD), ulcerative colitis, inflammatory bowel disease like Crohn's and non-alcoholic fatty liver and cancers of the food pipe, colon and liver, pancreas and gall bladder.

Causes

Among the top risk factors for GI diseases, researchers writing for the *Gastroenterology and Hepatology Journal* listed smoking, urbanization, diet and genetics. People who suffer from Crohn's disease are more prone to suffer from gastrointestinal problems. Smoking rates remain persistently high throughout Asia and, as developing countries such as China and India struggle with record-setting levels of pollution, the incidence of respiratory diseases is expected to rise, too. Habits like tobacco and *pan masala* has continued to cause more GERD, dyspepsia, oral and food pipe cancers. Researchers also concluded that the increased consumption of refined sugar, fat, fast foods, fatty acids and oils, less fibre intake, irregular food habits due to erratic working hours, lack of exercise, enough stress *etc.* also led to a higher chance of developing a GI disease. Spicy food can also be more irritant for GI tract. The unclean water and sanitation can cause waterborne diseases which typically involve the GI tract and liver predominantly.

Esophagitis

Esophagitis is an inflammation of the lining of the esophagus, the tube that carries food from the throat to the stomach. It can cause painful, difficult swallowing and chest pain. Gastro-esophageal reflux disease (GERD) is the most common cause of esophagitis *i.e.* stomach acids backing up into the esophagus; other causes include infection, allergies, chemical injury or physical trauma. Besides it, there are several types of esophagitis, some are not correlated with GERD.

Causes of Esophagitis

The esophagus is a muscular tube made of both voluntary and involuntary muscles. It is responsible for peristalsis of food. It is about 8 inches long and passes through the diaphragm before entering the stomach. At the end of the esophagus is the lower esophageal sphincter, which prevents stomach acid from moving to the esophagus. Sometimes the sphincter allows acid to backflow into the esophagus if it is not strong enough. Esophagitis is caused by an infection or irritation in the esophagus. An infection can be caused by bacteria, viruses, fungi, or diseases that weaken the immune system.

I. Infections that cause esophagitis

They include:

Candida

This is a yeast infection of the esophagus caused by the same fungus that causes vaginal yeast infections. The infection develops in the esophagus when the body's immune system is weak, such as in people with diabetes or HIV. It is usually very treatable with antifungal drugs.

Herpes

Like Candida, this viral infection can develop in the esophagus when the body's immune system is weak. It is treatable with antiviral drugs.

II. Irritation Leading to Esophagitis

It may be caused by any of the following:

- **Gastroesophageal reflux disease (GERD):** It is weakness or dysfunction of the muscle that keeps the stomach closed (sphincter) can allow stomach acid to leak into the esophagus (acid reflux), causing irritation of the inner lining. Also called GERD esophagitis; in severe cases it can become erosive esophagitis (this also may be referred to as ulcerative esophagitis).
- **Vomiting:** When vomiting is frequent or chronic it can lead to acid damage to the esophagus. Excessive or forceful vomiting may cause small tears of the inner lining of the esophagus, leading to further damage.
- **Achalasia:** This is a disorder where the lower end of the esophagus does not open normally, and as a result food can get stuck in the esophagus or is regurgitated. People with achalasia have a higher than normal risk of esophageal cancer.
- **Surgery:** Surgery, including certain types of bariatric (weight loss) surgery, can lead to increased risk of esophagitis.
- **Medications** such as aspirin, steroids, antibiotics and other anti-inflammatory drugs can irritate the lining of the esophagus, and also cause increased acid production in the stomach that can lead to acid reflux.
- **Drug-induced esophagitis** occurs when large pills are taken with too little water or just before bedtime and they dissolve or get stuck in the esophagus, causing irritation. Swallowing a toxic substance.
- **Physical injury** resulting from nasogastric tubes
- **Chemical injury** by alkaline or acid solutions
- **Hernias:** This abnormality occurs when a part of the stomach moves above the diaphragm producing a small abnormal pouch, or hiatal hernia, which can lead to excess acid refluxing into the esophagus.
- **Smoking**
- **Crohn's disease**
- **Alcohol abuse**
- **Stress**

- **Vitamin and mineral supplements** (iron, vitamin C and potassium)
- **Radiation injury** can occur following radiation treatment for cancer. Radiation to the chest (thorax), for cancer treatment can cause burns leading to scarring and inflammation of the esophagus.
- **Swallowing of foreign material** or toxic substances
- **Diets high in acidic foods** or excessive caffeine

Due to inflammation, the esophagus becomes narrow, which makes swallowing of the food difficult and may result in food bolus impaction. Irritation from the backflow of acid can cause the feeling of burning in the chest and can result in the flow of acid into the throat causing acid reflux or regurgitation. If irritation, infection, and/or inflammation to the esophagus continue, it is likely that the symptoms and signs will increase.

Subtypes of Esophagitis

Esophagitis is generally categorized by the conditions that cause it. In some cases, more than one factor may be causing esophagitis.

i) Reflux Esophagitis

A valve-like structure called the lower esophageal sphincter usually keeps the acidic contents of the stomach out of the esophagus. If this valve opens when it should not or does not close properly, the contents of the stomach may back up into the esophagus (gastroesophageal reflux). Gastroesophageal reflux disease (GERD) is a condition in which this backflow of acid is a frequent or ongoing problem. A complication of GERD is chronic inflammation and tissue damage in the esophagus.

Factors that increase the risk of gastroesophageal reflux disease (GERD) and therefore, are factors in reflux esophagitis include the following:

- Eating immediately before going to bed
- Dietary factors such as excess alcohol, caffeine, chocolate, tomato based foods, citrus fruits, spicy foods, garlic, onion, and mint-flavored foods *etc.* may worsen symptoms of GERD or reflux esophagitis
- Excessively large and fatty meals
- Smoking
- Extra weight, including from pregnancy

ii) Eosinophilic Esophagitis

Eosinophils are white blood cells that play a key role in allergic reactions. Eosinophilic esophagitis occurs with a high concentration of these white blood cells in the esophagus, most likely in response to an allergy-causing agent (allergen) or acid reflux or both. Risk factors for eosinophilic esophagitis may include a history of certain allergic reactions, including allergic rhinitis, asthma and atopic dermatitis and a family history of eosinophilic esophagitis.

In many cases, this type of esophagitis may be triggered by foods such as milk, eggs, wheat, soy, peanuts, beans, rye and beef. However, conventional allergy testing does not reliably identify these culprit foods. People with eosinophilic esophagitis may have other nonfood allergies. For example, sometimes inhaled allergens, such as pollen, may be the cause.

iii) Lymphocytic Esophagitis

It is an uncommon esophageal condition in which there are an increased number of lymphocytes in the lining of the esophagus. It may be related to eosinophilic esophagitis or to GERD.

iv) Drug-Induced Esophagitis

Several oral medications may cause tissue damage if they remain in contact with the lining of the esophagus for too long. For example, if you swallow a pill with little or no water, the pill itself or residue from the pill may remain in the esophagus. Drugs that have been linked to esophagitis include:

- Pain-relieving medications, such as aspirin, ibuprofen and naproxen sodium
- Antibiotics, such as tetracycline and doxycycline
- Potassium chloride, which is used to treat potassium deficiency
- Bisphosphonates, including alendronate, a treatment for weak and brittle bones (osteoporosis)
- Quinidine, which is used to treat heart problems

v) Infectious Esophagitis

A bacterial, viral or fungal infection in tissues of the esophagus may cause esophagitis. Infectious esophagitis is relatively rare and occurs most often in people with poor immune system function, such as people with HIV/AIDS or cancer. Also, certain cancer treatments and drugs that block immune system reactions to transplanted organs (immunosuppressants) may increase the risk of infectious esophagitis.

A fungus normally present in the mouth called *Candida albicans* is a common cause of infectious esophagitis. Such infections are often associated with poor immune system function, diabetes, cancer, or the use of steroid or antibiotic medications.

Symptoms

Common signs and symptoms of esophagitis include:

- Difficult and/or painful swallowing

- Chest pain, particularly behind the breastbone, that occurs with eating
- Heartburn
- Acid regurgitation
- Mouth sores
- A feeling of something of being stuck in the throat
- Nausea
- Vomiting
- Abdominal pain
- Cough

In infants and young children, particularly those too young to explain their discomfort or pain, signs of esophagitis may include:

- Feeding difficulties
- Failure to thrive

Complications

If left untreated esophagitis caused by GERD, can lead to bleeding, ulcers, and chronic scarring. This scarring can narrow the esophagus, eventually interfering with the ability to swallow and lead to malnutrition. Hence, possible complications related to function and structure of esophagus include:

- **Stricture:** Scarring or narrowing (stricture) of the esophagus.
- **Esophageal perforation:** Tearing of the esophagus lining tissue from retching (if food gets stuck) or during endoscopy (due to inflammation) leads to holes or ulcers in the esophagus.
- **Barrett's esophagus:** A major complication, which occurs in a significant portion of people with chronic or longstanding GERD, is Barrett's esophagus. A very few of those who develop Barrett's esophagus will develop esophageal adenocarcinoma.

Treatment

Since there can be many causes underlying esophagitis, it is important to try to find the cause to help to prevent esophagitis as treatment will depend on the underlying cause of the inflammation and the severity of tissue damage. If esophagitis is diagnosed early enough, medications and dietary or lifestyle changes are often enough to allow the body to heal. If the damage is severe or leads to scar tissue causing difficulty swallowing, more invasive treatments may be necessary.

Aim of Treatment

The aims of treatment of esophagitis are:

i) To prevent esophageal reflux
ii) To prevent further inflammation of the esophagus
iii) To give relief to the patient from pain and difficulty in swallowing
iv) To provide adequate nutrition to the patient

Nutritional Requirements

Nutritional requirements remain the same as that of a normal healthy adult. However, changes would be required in the life style and dietary pattern so as to provide food items which are soothing and soft without much fibre, spices *etc.* so that they do not irritate the lining of the esophagus, do not aggravate inflammation, neutralize the acid and at the same time are nutritionally adequate.

Dietary Management

Diet is often a key to reducing the symptoms of esophagitis. In acute cases of esophagitis, person shall not be able to take solid foods. Hence, liquid diet based on soft food items (cold milk with some protein mix, banana-milk shake, milk containing plain ice cream, sherbats, barley water, glucose water, coconut water, *dal* soups, sieved vegetable soups *etc.*) shall be required and need to be given in small quantity but at frequent intervals till the patient does not feel pain or difficulty in swallowing or till the inflammation subsides. Avoid citrus juices and soups containing spices and condiments.

If the esophagitis is chronic, then the following points should be kept in mind while planning the diet:

In Case of Eosinophilic Esophagitis

- The treatment is simply to eliminate foods that cause allergies. Common dietary triggers for it are milk, soybean, eggs, wheat, peanuts, tree nuts, shell fish *etc.* should be avoided in the diet. Follow elimination diet depending upon the food allergy.

To Prevent Reflux Esophagitis

- Avoid acidic foods and drinks.
- Include high protein foods like milk, eggs *etc.*
- Avoid raw or hard foods. Avoid fibre rich foods. Take cooked foods.
- Donot take caffeine and aerated drinks.
- Avoid eating before going to bed.
- Donot lie down, strain or bend immediately after taking meal.
- Take small meals at frequent intervals.
- Avoid taking fatty meals.
- Donot take pickles, spices and condiments.
- Take smaller bites and chew food well.
- Abstain from taking alcohol.
- Donot smoke.

- Avoid wearing tight clothes.
- Practice yoga, meditation and other relaxation techniques to prevent stress.

In the following table, list of foods to be taken and avoided in the esophageal soft diet has been given for reflux esophagitis:

Food Groups	*Choose*	*Avoid*	*Tips*
Beverages	All except those on opposite list	Carbonated and iced drinks	Very hot or very cold liquids and foods may not be tolerated
Milk And Milk Products	Milk, malted milk, or milkshakes. Soft cheese such as grated Parmesan or Ricotta, cheese sauces and cottage cheese. Plain or flavored yogurt.	Yogurt with fresh fruit or seeds. To boost calories, pour cheese sauce over vegetables or noodles.	Do not add raw eggs to milkshakes or drinks for food safety reasons.
Meat and Meat Substitutes, Eggs, Beans	Ground or tender meat and poultry mixed with gravy, soft flaked fish without bones. Soft scrambled eggs and egg substitutes. Cooked dried beans and peas.	Stringy, dry or fibrous-type meats (*i.e.* steak and spare ribs). Meats containing gristle or peppercorn. Sausage and bacon. Chunky peanut butter, nuts, seeds, stringy cooked cheese such as mozzarella.	Moisten ground meats or poultry with gravy, broth or sauces.
Breads and Starches	*Roti* with vegetable curry or *dals*. Cooked cereals without nuts or dried fruits, ready to eat cereals softened in milk. Noodles, potatoes, and pasta.	Fresh or "doughy" breads may cause "sticking". Avoid all fresh bread, rolls, muffins, biscuits, and rice.	Buy or prepare products without seeds, dried fruits, or nuts
Fruits	Canned, cooked or frozen fruit (canned peaches, applesauce). Soft fresh fruit such as bananas and melon. All fruit juices.	Raw, coarse or abrasive fresh fruits. Dried fruit.	Blend fruit in milkshakes or yogurt.
Vegetables	Canned, cooked, or frozen vegetables that are soft and without skin (mashed potatoes, squash, carrots, spinach), All vegetable juices.	All raw vegetables, including salads. Cooked vegetables that are fibrous tough, "woody", or contain seeds (broccoli, tomato, okra and lotus stem).	Season vegetables with butter or margarine and ground spices. Soften vegetables in soups and gravies.
Desserts and Snacks	Puddings, soft cookies.	Popcorn, chips, desserts with nuts, seeds or coconut.	Acid or salty foods may not be tolerated.
Fats	All	None	

For the Drug-induced Esophagitis

- Drink plenty of water when taking medicines.
- Donot take medicines while lying down, before sleeping or too many at one time.
- Take an alternative drug.

For the Esophageal Ulcer

Some home remedies and lifestyle changes to help with esophagitis include losing weight, stop smoking, lowering stress, avoid sleeping/lying down after eating, raise the head while laying down, taking medicines correctly, avoiding certain medications, and avoiding foods that cause the reflux that might be causing the esophagitis.

Gastro-Esophageal Reflux Disease (GERD)

Gastroesophageal reflux disease (GERD) is a digestive disorder where stomach acid persistently and regularly flows up into the esophagus. It affects the lower esophageal sphincter (LES), the ring of muscle between the esophagus and stomach. In normal digestion, the lower esophageal sphincter (LES) opens to allow food to pass into the stomach and closes to prevent food and acidic stomach juices from flowing back into the esophagus. GERD occurs when the LES is weak or relaxes inappropriately, allowing the stomach's contents to flow up into the esophagus frequently. It is caused by a prolonged relaxation of the lower esophageal sphincter and delayed gastric emptying causing irritation of the esophagus by gastric acid, bile, and pepsin. The acid in the esophagus causes heartburn and other symptoms, as well as. In most cases, GERD can be relieved through diet and lifestyle changes; however, some people may require medication or surgery.

GER vs. GERD

Gastroesophageal reflux (GER) is a term sometimes used to describe many of the symptoms, such as heartburn, that are seen in GERD. But GER is much more common and less serious than GERD. GER happens less often and usually goes away after one takes an antacid tablet. GERD describes symptoms that are more persistent. According to doctors, GERD is acid reflux that occurs more than a couple of times per week. GERD is a chronic condition. Once it begins, it usually is life-long.

Causes

Occasional acid reflux is quite common, often occurring as a result of overeating, lying down after eating, or eating particular foods. However, recurrent acid reflux, diagnosed as GERD, typically has other causes and risk factors and can have more serious complications.

The cause of GERD is complex and may involve multiple causes. Gastroesophageal reflux disease occurs in people of all ages, and sometimes for unknown reasons while GERD occurs when the sphincter at the bottom of the esophagus becomes weak, or opens when it should not.

1. GERD occurs more commonly in people who are:
   - overweight or obese because of increased pressure on the abdomen
   - smoking, and being exposed to second-hand smoke
   - taking certain medications, including some asthma medications, calcium channel blockers, antihistamines, sedatives, and antidepressants
   - suffering from hiatal hernia
   - suffering from lower esophageal sphincter abnormalities,
   - having abnormal esophageal contractions and
   - having slow or prolonged emptying of the stomach.
2. Patients with diseases that weaken the esophageal muscles, such as scleroderma or mixed connective tissue diseases, are more prone to develop GERD.
3. Pregnant women: The elevated hormone levels of pregnancy probably cause reflux by lowering the pressure in the lower esophageal sphincter. At the same time, the growing fetus increases the pressure in the abdomen. Both of these effects would be expected to increase reflux.
4. Certain foods and beverages, including chocolate, peppermint, fried or fatty foods, spicy foods, coffee, or alcoholic beverages, may trigger reflux and heartburn.

Symptoms

Main symptom is heartburn which is a discomfort felt behind the breastbone as a burning sensation. It tends to get worse if the person lies down or bends over and also after eating food. However, not all people with GERD experience heartburn, and there are other possible symptoms:

- Regurgitation of food or sour liquid in the mouth
- Sensation of lump in the throat
- Nausea or vomiting
- Bad breath
- Respiratory problems
- Difficulty or pain when swallowing
- Chronic cough
- Acid-induced erosions of the teeth
- Hoarseness
- Laryngitis
- Ear pain
- Disrupted sleep

Complications

If GERD goes untreated, it can lead to serious complications *e.g.* strictures and inflammation in the esophagus (esophagitis). Another complication is a condition called Barrett's esophagus (BE) which can lead to esophageal adenocarcinoma over the time too. BE is more common in people who are older than 50 y, smokers and overweight.

Treatment

The treatment of GERD requires diet and lifestyle modifications, drug therapy and in severe cases, surgery. The treatment involves not only what is eaten, but also how and when food is taken. Treatment aims at:

- To decrease the amount of reflux.
- To reduce damage to the lining of the esophagus from refluxed materials by neutralizing the acid present in the stomach.
- To maintain normal weight in obese persons.

Nutritional Requirements

Nutritional requirements are the same for most of the patients as those for their healthy counterparts. But dietary modifications are needed. If the person suffering from GERD is overweight or obese, he should cut down on calorie intake to reduce weight as it will reduce the symptoms. Similarly, he will have to reduce his fat intake too as it is the concentrated source of energy. Moreover, more fat in the diet shall delay the stomach emptying. In case of complications like bleeding, foods rich in protein, iron and B-vitamin will have to be included.

Ensure adequate protein intake. Early studies indicated that protein increased the LES pressure thereby allowing the closure of the sphincter and reducing reflux. Even though more recent studies do not support this belief, however, maintaining a diet with adequate low-fat protein intake can still help in the healing of irritated mucosa or ulcers.

Dietary Modifications and Feeding Pattern

The GERD diet is aimed at reducing acid reflux, the main cause of esophagitis. Much of the nutritional research on reflux is correlational; it shows patterns of association rather than causes. The foods affect the amount of acid which the stomach produces. Eating the right kinds of food is key to controlling acid reflux or GERD, a severe, chronic form of acid reflux. Knowing what and when to eat can go a long way in preventing and controlling bothersome GERD flare-ups.

Many people with GERD find that certain foods trigger their symptoms. No single diet can prevent all symptoms of GERD, and food triggers are different for everyone. To identify the individual triggers, one should keep a food diary and track the following:

- What foods you eat
- What time of day you eat
- What symptoms you experience

Try keeping a log of what you eat and drink throughout the day. Note when symptoms seem to flare up and review the log over time to identify patterns. Keep the diary for at least a week. It shall be helpful to track your foods for a longer period if your diet varies. You can use the diary to identify specific foods and drinks that affect GERD. Studies show that eliminating trigger foods can help decrease the chances of GERD. It shall help to plan a meal for you. The purpose of this diet is to reduce the reflux of stomach fluid into the esophagus and to avoid foods that irritate the esophageal mucosa. It may be necessary to lose weight, since excess weight increases stomach pressure. The goal is to minimize and control GERD symptoms.

Dietary Management

According to research reports, no diet has been proven to prevent GERD. There is no one magic food that can treat acid reflux. Still, aside from avoiding trigger foods, a few other dietary changes can help. Certain foods may ease symptoms in some people. American Academy of Family Physicians recommends low-fat, high-protein meals. Reducing dietary fat intake can subsequently decrease the symptoms, while getting enough protein and fiber will keep one full and prevent overeating. After each meal, one may even consider chewing non-mint gum. This can help increase saliva in mouth and keep acid out of the esophagus.

Eating healthy and balanced amounts of different types of foods is good for overall health. Eat foods that help to relieve symptoms of GERD and include those in the diet. However, foods that help improve acid reflux for one person may be problematic for someone else.

In general, following foods can manage the symptoms of acid reflux:

Vegetables

Vegetables are naturally low in fat and sugar, and they help reduce stomach acid. Eat at least five servings of fresh vegetables a day. Good options include green beans, broccoli, cauliflower, leafy greens, potatoes, and cucumbers. Sauté the vegetables in very less amount of fat. Baked, boiled and mashed potato without fat can be taken but French-fried potatoes, potato chips *etc.* should not be added in the diet. Tomato and tomato based sauces, onion and garlic should not be taken. Avoid anything fibrous or filled with seeds, such as okra, artichokes, and celery. Fresh ginger has been reported to have natural anti-inflammatory properties, and it is considered as a natural treatment for heartburn and other gastrointestinal problems. But not more than 4 g of ginger should be included in the vegetable preparations as more than this can worsen the symptoms of GERD too. All the vegetables should be cooked so that they become soft and then consumed. Avoid raw vegetables.

Fruits

Non-citrus fruits, including apple, mango, melons, bananas and pears *etc.* are less likely to trigger reflux symptoms than acidic fruits. Avoid citrus fruits *e.g.* orange, lemon, lime, tangerine, pineapple, grapefruit *etc.*

Research shows that increased fiber intake, specifically in the form of fruits and vegetables, may protect against GERD. But scientists are not yet certain how fiber prevents GERD symptoms. Increasing dietary fiber is generally a good idea. In addition to helping with GERD symptoms, fiber also reduces the risk of high cholesterol, uncontrolled blood sugar and other bowel related problems.

Whole Grain Cereals

According to some research studies, whole grain cereal preparations from wheat, rice, oatmeal *etc.* are excellent sources of dietary fibre which can absorb acid in the stomach and reduce symptoms of reflux. But still studies are not clear about the beneficial effects of including dietary fibre in the diet. In case of severe esophagitis, soft diet containing refined and cooked cereals may be tolerated well by the patients. Plain bread/*roti*, plain boiled rice, breakfast cereals with skim milk, pancakes, waffles, and muffins made with low-fat ingredients are tolerated well. Once a week, one may introduce a new high fibre grain such whole wheat bread, brown rice, oatmeal or quinoa, if tolerated. But all the cereal preparations should be cooked and soft before consumption. Breads and cereals prepared with high-fat ingredients such as croissants, biscuits, doughnuts, muffins, pizza, French toast *etc.* are not tolerated well and should not be included in the diet.

Dried Legumes

Well cooked beans and peas are well tolerated. Flatulence producing pulses should be avoided. Savories containing fried beans should be avoided.

Lean Meats

Well cooked lean meats, such as chicken (without skin), turkey, fish (fresh or water-packed), and seafood, are low-fat and reduce symptoms of acid reflux. Try them grilled, broiled, baked, or poached. Avoid fried meat, poultry, fish or eggs, sausages *etc.*

Egg Whites

Egg whites are a good option. Do not take egg yolks as they are high in fat and may trigger reflux symptoms. Some people with esophagitis do well with soft scrambled eggs.

Fats

Include PUFA and MUFA rich foods such as avocados, walnuts, flaxseed, olive oil, sesame oil, and sunflower oil within the prescribed limits. Reduce intake of saturated fats and trans fats and replace them with these healthier unsaturated fats. Limit high fat foods, *e.g.* fat rich gravies, fried foods, high fat baked goods, heavy cream, butter, ice cream, high fat cheeses, margarine, cream cheeses and full fat mayonnaise for salad dressings *etc.* One significant culprit that can cause symptoms is fatty food. When dietary fat enters the GIT, it triggers relaxation of the LES. Medium and long chain fatty acids seem to create the strongest response. Fatty foods have been found to delay gastric emptying and decrease the LES pressure thereby prolonging the esophageal exposure time to stomach acid and the volume available for reflux. People who suffer from reflux may be predisposed already to delay gastric emptying. For them, more dietary fat would only make the situation worse.

Dairy Foods

Most dairy products are good for esophagitis, including milk and milk shakes. Low-fat dairy is a great source of calcium, protein and vitamin D that also will help prevent GERD symptoms when substituted for full-fat options. Skim milk instead of whole milk and low-fat buttermilk, curd and *paneer* are tolerated well. Whole milk, chocolate milk, chocolate shakes or drinks, milkshakes, whole milk buttermilk and curd, fat rich cheese *etc.* should not be given in the diet. Fat free yoghurt can be a good choice for someone with esophagitis but avoid adding fruit or seeds. It is even possible to have low-fat ice cream if cold foods do not cause irritation.

Soups

Homemade soups made with lean meat and vegetables (except tomatoes) and fat-free or low-fat milk can be tolerated but avoid prepackaged and ready to serve soups. Do not add any cream in the soups. Avoid all kinds of tomato based soups.

Desserts

Angel food cake, sponge cake, low-fat cookies, gelatin, fruit-based desserts, sherbet, fruit ice, reduced-fat ice cream, pudding or custard made with 1 per cent or 2 per cent low-fat milk are tolerated well. All other pies, cookies, and cakes, full fat ice cream, any desserts containing chocolate frosting, whole milk pudding, pastries *etc.* should be avoided.

Sweets

Sugar, honey, jam, jelly, maple syrup, hard candy, marshmallows *etc.* can be include in the diet within the prescribed limit. Avoid coconut, cream-filled candies, nuts, chocolate, spearmint, peppermint *etc.*

Miscellaneous

Salt, oregano, pepper, other spices and herbs (as tolerated), decaffeinated coffee, decaffeinated tea, non-mint tea are tolerated but avoid spices and herbs in tomato-based sauces, red chili, green chilli, black and jalapeno peppers, vinegar, carbonated beverages, caffeinated or mint-flavored coffee and/or teas, alcoholic beverages *etc.* Coffee and chocolate contain methylxanthine, which causes muscles to relax and lowers the LES pressure. In addition, coffee can stimulate gastric secretions, which can promote reflux. If the esophageal mucosa is irritated, individual tolerance of coffee (regular or decaffeinated) should be determined.

Foods and Eating Pattern which Trigger GERD

No matter what is your body weight, there are certain known trigger foods that can increase the risk for acid reflux. Food restrictions are based on personal tolerance. For some people, heartburn symptoms may be triggered by these certain foods. Generally, the following foods and eating pattern worsen the symptoms of GERD and should be avoided:

1. Avoid the following foods:
   - ✰ Caffeine (regular coffee, regular tea, chocolate)
   - ✰ Citrus fruits/juices
   - ✰ Acidic foods
   - ✰ Carbonated beverages
   - ✰ Alcohol
   - ✰ Mints (peppermint, spearmint)
   - ✰ Tomato sauce and other tomato products
   - ✰ High fat foods
   - ✰ Fried foods
   - ✰ Spicy foods
   - ✰ Hard foods such as nuts, crackers, and raw vegetables.
   - ✰ Garlic and onions
   - ✰ Processed foods
   - ✰ Fast foods
   - ✰ Very hot or very cold foods
   - ✰ Alcohol

   These foods may cause further irritation to damaged esophageal mucosa but have not been found to cause reflux themselves. By limiting or avoiding these triggers altogether, one may experience fewer symptoms.

2. Include more of plant foods in the diet but they should be cooked. Include cereal (cooked),

puddings, protein shakes, and mashed potatoes in the diet. In short, eat foods that are cooked and/or mashed so that one can swallow easily.

3. Eat soft diet that is easily digested in case of severe esophagitis. If person is not able to take food orally, intravenous nutrition is given as it will allow the esophagus to heal and to prevent dehydration and malnutrition.

Life Style Changes

1. The timing and amount of food consumed also can make a difference in how one feels. Do not take large meals at one time and opt instead for smaller, more frequent meals throughout the day. Large meals may increase stomach pressure, and therefore reflux.
2. Avoid eating irregular meals.
3. Eat slowly, until you are just satisfied rather than stuffed.
4. Maintain an upright posture while eating and for 45-60 minutes afterward.
5. Try to make mealtimes relaxing. Stress will worsen GERD.
6. Do not lie down for at least half an hour to an hour after eating. Doing so can trigger heartburn.
7. Try to stop eating 2 to 3 hours before sleeping as it may lessen reflux by allowing the acid in the stomach to decrease and the stomach to empty partially. Reflux is worse following meals. This probably is so because the stomach is distended with food at that time and transient relaxations of the lower esophageal sphincter are more frequent. Therefore, smaller and earlier evening meals may reduce the amount of reflux for two reasons. First, the smaller meal results in lesser distention of the stomach. Second, by bedtime, a smaller and earlier meal is more likely to have emptied from the stomach than is a larger one. As a result, reflux is less likely to occur when patients with GERD lie down to sleep.
8. Avoid clothing that is tight in the abdominal area.
9. Achieve and maintain a healthy weight. While heartburn can happen to anyone, GERD seems to be most prevalent in adults who are overweight or obese. Weight loss is often the first line of defense against acid reflux. Excess weight especially in the abdominal area puts more pressure on the stomach. As a result, persons with abdominal obesity are at an increased risk of stomach acids working back into the esophagus and causing heartburn. If you are obese or overweight, reduce 0.5-1 kg per week. If already the person is having optimum weight, he should maintain it with a healthy diet and regular exercise.
10. Gravity, swallowing, and saliva are important protective mechanisms for the esophagus, but they are effective only when individuals are in the upright position. At night during sleep, gravity has no effect, swallowing stops, and the secretion of saliva is reduced. Therefore, reflux that occurs at night is more likely to result in acid remaining in the esophagus longer and causing greater damage to the esophagus. Hence, elevation of the upper body at night generally is recommended for all patients with GERD. When sleeping, raise the head of the bed 6-8 inches, using wooden blocks under the bedposts. Extra pillows will not work, however, wedges that elevate the head and upper chest are effective. These maneuvers raise the esophagus above the stomach and partially restore the effects of gravity. It is important that the upper body and not just the head be elevated. Elevating only the head does not raise the esophagus and fails to restore the effects of gravity.
11. Drink plenty of water. This condition can make it hard to swallow water too. In such a case, drink from a straw as this makes it easier to swallow.
12. Sip fluids when taking solids at meals and snacks to moisten foods.
13. Stop smoking to reduce GERD symptoms. Cigarette smoking weakens the LES.
14. One novel approach to treat GERD is chewing gum. Chewing gum stimulates the production of more bicarbonate-containing saliva and increases the rate of swallowing. After the saliva is swallowed, it neutralizes acid in the esophagus. In effect, chewing gum exaggerates one of the normal processes that neutralize acid in the esophagus. It is not clear, however, how effective chewing gum is in treating heartburn. Nevertheless, chewing gum after meals is certainly worth a try.
15. Take antacids and other medications as prescribed by the physician. Avoid pain killers *e.g.* aspirin or ibuprofen.
16. Take all pills with plenty of water, and in an upright position. Don't lie down for at least 30 minutes after taking a pill.

Peptic Ulcer

An "ulcer" is an open sore. The word "peptic" means that the cause of the problem is due to acid. The term 'peptic ulcer', a disease of chronic development with activation and remission periods, is a lesion or open sore of the mucosal lining of the upper gastrointestinal tract characterized by an imbalance between aggressive and protective factors of the mucosa, having *H. pylori* as the main etiologic factor. Its pathogenesis is characterized by the disrupted balance between the factors that damage the mucosa (chloride acid, pepsin, and ulcerogenic drugs) and those that protect it (mucosal barrier, prostaglandins, and mucosal secretion). Peptic ulcers happen when the acids that help to digest the food, damage the walls of the stomach or duodenum.

Types of Peptic Ulcers

Gastritis is a superficial erosion and inflammation of the gastric mucosa. It can be either acute or chronic while peptic ulcers are deeper erosions and ulcerations that extend through the muscularis layer of the gastric or duodenal mucosa. Peptic ulcers have various names, depending on their location. An ulcer on the inside of the stomach lining is known as a **gastric ulcer** while that on the inner side of the first or upper part of the small intestine is known as a **duodenal ulcer**. A person may have one or both gastric and duodenal ulcers at the same time. Having both types is known as **gastro-duodenal ulcer**. In general, gastritis and gastric ulcers are associated with insufficient mucosal protection, whereas duodenal ulcers are associated with excess acid secretion. Rarely, a peptic ulcer may develop just above stomach in the esophagus and it is called as **esophageal ulcer**.

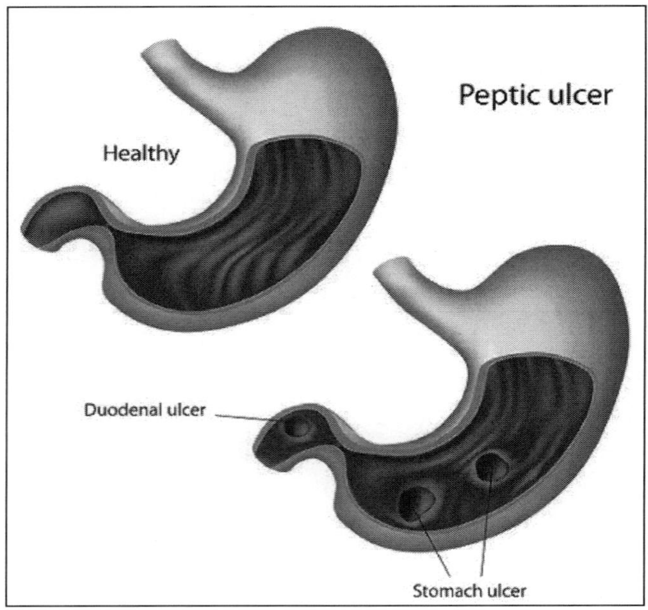

Gastric Ulcer	*Duodenal Ulcer*
Gastric ulcers occur in the mucosal membrane of the stomach.	These ulcers occur in the mucosal membrane of the duodenum *i.e.* upper part of the small intestine.
It is usually caused by an infection with a germ (bacterium) called *Helicobacter pylori* (*H. pylori*).	It is also usually caused by an infection with a germ (bacterium) called *Helicobacter pylori* (*H. pylori*).
They are associated with insufficient mucosal protection which can be due to poor nutrition, diminished mucosal blood flow and a defect in the inhibition of gastric acid and pepsin secretion.	They are due to hyper secretion of acid although mucosal tissue resistance is normal. Hyper secretion of acid is due to increased number of parietal cells and impaired rapid gastric emptying with loss of buffering action.
It is formed by the breakdown of gastric mucosa. Local epithelial tissue damage occurs because of cytokines released by *H. pylori*, and because of abnormal mucus production associated with gastritis affecting the body and antrum. Gastrin stimulates the parietal cell proliferation and increases parietal cells as a result decrease in somatostatin leads to acid hyper secretion.	It is formed as a result of increased acid secretion in the duodenum because of decreased bicarbonate secretion by *H. pylori* infection, increased parietal cell mass due to alcohol and spicy foods and Zollinger- Ellison syndrome (Increased production of gastrin), decreased inhibition of acid secretion, possibly by *H. pylori* – damaging somatostatin producing cells in the antrum *by H.pylori.* Smoking impairs the healing of gastric mucosa. The persons with Blood group "O" and genetic susceptibility factors also influences the occurrences of duodenal ulcers
Stomach pain occurs 1-2 hours after eating.	Pain occurs 3-4 hours after eating.
Gastric ulcer pain cannot be relieved by eating.	Hunger aggravates duodenal ulcers and they are usually subsided when, food is eaten.
Gastric ulcers cause hematemesis or vomiting of blood.	Duodenal ulcers cause bleeding in the stool called melena.

Etiology

In the past, it was thought that ulcers were caused by hyper acidity due to lifestyle factors such as eating habits, cigarette smoking and stress which is not tenable at present. About 70 per cent of gastric ulcers and 50 per cent of duodenal ulcers are not associated with abnormally high acid production. Now it is understood that people with ulcers have an imbalance between acid and pepsin coupled with the inability of the digestive tract to protect itself from these harsh substances. Evidence now indicates the two major risk factors for peptic ulcers are:

1. Infection with a bacterium named *Helicobacter pylori* and
2. Long term use of non-steroidal anti-inflammatory drugs (NSAIDs).

1. Bacterial Infection

Research done in 1980s showed that the most common cause of gastric and duodenal ulcers is infection with a bacterium named *Helicobacter pylori*, usually called *H. pylori* and not predominantly the life style factors as thought earlier. In fact, Dr. Barry Marshall and Dr. J. Robin Warren were awarded the 2005 Nobel Prize in Medicine for this discovery that *H. pylori* can cause ulcers as it was one of the most important medical discoveries of the late 20th century.

Infection with *H. pylori* is very common, affecting more than a billion people worldwide. This bacterium exists in the stomach and affects the mucus that protects the stomach and duodenum, allowing for stomach acid to damage the lining. Not everyone who gets an ulcer is infected with *H pylori*. Adults who have an *H. pylori* infection may get a peptic ulcer, also called an *H. pylori*-induced peptic ulcer. Among those who suffer from peptic ulcers, as many as 90 per cent of those with duodenal ulcers and 70 per cent with gastric ulcers are believed to be infected with *H. pylori*. However, most people with an *H. pylori* infection never develop a peptic ulcer. Peptic ulcers caused by *H. pylori* are uncommon in children.

The bacteria disrupt the mucosal protective barrier, making it more vulnerable to acid damage and inciting an inflammatory response. The bacteria also reduce somatostatin production, leading to increased gastrin secretion and action.

H pylori spreads through the stools of an infected person. Researchers are not certain how *H. pylori* spread. They think that the stool infected with this bacterium contaminates food or water (usually through poor personal hygiene, unclean eating utensils, contact with an infected person's saliva and other bodily fluids, including kissing *etc.*). The bacteria in the stool make their way into the digestive tracts of people who consume this food or water. This is called fecal-oral transmission and is a common way for infections to spread. Researchers have found *H. pylori* in the saliva of some infected people, which means *H. pylori* infection could spread through direct contact with saliva or other bodily fluids.

Infection with *H pylori* occurs in all ages, races, and socio-economic classes. It is more common in older adults, although it is thought that many people are infected in childhood and carry the bacteria throughout their lifetimes. It is also more common in lower socio-economic classes because these households tend to have more people living together, sharing bathrooms and kitchen facilities. People who are newly infected usually develop symptoms within a few weeks.

2. Medication (NSAIDs)

In fact, after *H. pylori*, regular use of NSAIDs (non-steroidal anti-inflammatory drugs) is the other major cause of peptic ulcers. NSAIDs can irritate and damage the stomach and intestinal lining by suppressing prostaglandin formation in the mucosa, which is normally a part of the protective mechanism of the mucosal barrier. People of any age who take NSAIDs every day or multiple times per week are more likely to develop a peptic ulcer than people who do not take them regularly. NSAIDs are a class of pain killers, such as ibuprofen, naproxen, celebrex, aspirin *etc.* and their long-term use can cause peptic ulcer disease. The chance of having a peptic ulcer caused by NSAIDs, also called an NSAID-induced peptic ulcer, is increased if someone:

- Is aged 70 or older
- is female
- is taking more than two types of NSAIDs or has taken NSAIDs regularly for a long time
- had a peptic ulcer or intestinal bleeding before
- has an upset stomach or heartburn after taking NSAIDs
- has two or more medical conditions or diseases
- is taking other medicines, such as corticosteroids and medicines to increase the bone mass
- has generalized anxiety disorder
- drinks alcohol or smokes

3. Other Risk Factors

There are other less common causative factors of peptic ulcer which can also weaken the protective mucosal barrier of the stomach, thus increasing the chances of getting an ulcer and slowing down the healing of existing ulcers. These are:

- **Increasing age:** Gastric ulcers typically occur in patients over 40 years old, and the incidence of duodenal ulcers peaks at around age 60.
- **Sex:** Between 10 and 15 per cent of the world's population suffers from peptic ulcer. Duodenal ulcers, which account for 80 per cent of peptic ulcers, are more common in men than in women, but stomach ulcers affect women more frequently.
- **Excessive alcohol drinking:** Alcohol can cause gastritis by stimulating acid secretion and damaging the mucosal barrier. No evidence for a role in ulcer formation has been found, but alcohol may delay healing of gastric ulcers.
- **Physical stress:** Physical such as severe injuries or burns and major surgery can cause ulcers. Stress causes excess production of stomach acid and *H. pylori* thrives in acidic environment. Prophylaxis for gastritis and ulcers may be administered in hospitalized patients, especially those on mechanical ventilation or those undergoing major surgery. Although some reports have argued for prophylaxis in all hospitalized patients, this claim is not backed by rigorous data.
- **Extreme psychological stress:** Emotional stress especially that associated with traumatic events does not cause ulcers but can aggravate the symptoms.
- **Caffeine:** Coffee, either in its caffeinated or decaffeinated forms, stimulates acid secretion in the stomach, and some studies have suggested a

correlation between coffee intake and symptoms in patients with duodenal ulcer.

- **Cigarette smoking:** Cigarette smoking has been found to have an adverse effect on peptic ulcers, slowing healing and promoting recurrence. Use of tobacco products lower bicarbonate secretion, increase acid secretion and reduce mucosal blood flow in the stomach and duodenum and develop inflammation which in turn damage gastric mucosa. According to Maity *et al.* (2003) nicotine present in tobacco not only induces ulceration and also potentiates ulceration caused by *H.pylori*, alcohol and NSAID. Smoking is known to delay healing of gastric ulcers, but its role in the pathogenesis of ulcers is unclear.
- **Radiation therapy:** Ulcers can occur due to the use of radiation therapy for diseases such as cancer.
- **Family history**: More than 25 per cent of ulcer patients have a family history of ulcers, compared with 5 per cent of non-ulcer patients. Genetic factors can influence, and children of parents with duodenal ulcer are three times more likely to have ulcer than the population. Duodenal ulcer is more common among persons with blood group O. The familial accumulation of peptic ulcer disease observed in several studies may be attributable to genetic effects, aggregation of environmental exposure (shared environment), or both. The intrafamilial spread of *Helicobacter pylori* infection has raised the question whether shared environment could explain the familial aggregation of peptic ulcer disease rather than genetic similarity of family members.
- **Additional medications:** Additional medications that may increase the risk for gastric and duodenal ulcers include osteoporosis treatments such as alendronate (Fosamax) and risedronate (Actonel), anticoagulants (warfarin or clopidogrel), and selective serotonin reuptake inhibitors (SSRIs).
- **Spicy foods**: Dietary factors such as consumption of spicy foods were hypothesized to cause ulcers until late in the 20th century, but have been shown to be relatively minor importance now. Now it has been observed that they do not increase the risk for ulcers or cause them. But certain foods can irritate the stomach further in certain individuals.
- **Zollinger-Ellison syndrome (ZES):** A rare condition called Zollinger-Ellison syndrome causes the development of both cancerous and non-cancerous tumors usually in the pancreas and duodenum. These tumors release hormones that cause extremely high levels of stomach acid that lead to gastric and duodenal ulcers. Anyone can have ZES, yet it is rare and only occurs in about one in every 1 million people. However, ZES is more common among men 30 to 50 years old. A child who has a parent with multiple endocrine neoplasia type 1 is also more likely to have Zollinger-Ellison syndrome.

Symptoms

Some people may have no symptoms of an ulcer. The symptoms of gastric and duodenal ulcers are similar and include:

- **Dyspepsia** is the most prominent symptom, consisting of upper abdominal pain or discomfort, usually localized in the epigastrium or right upper quadrant and sometimes radiating to the back. It is characterized by a gnawing or burning sensation or hunger-like severe and continuous pain in the mid- upper abdomen. Pain may last for minutes to hours and tends to be worse at night. It usually occurs shortly after meals with gastric ulcer and 2-3 hours afterwards with duodenal ulcer.
- Flatulence, bloating, distension
- Early satiety
- Iron deficiency anemia
- Nausea
- Recurrent vomiting
- Loss of appetite
- Weight loss

Complications

The following complications due to ulcers may rarely occur and severe symptoms include the following:

- Difficulty in breathing
- Losing consciousness
- Fatigue
- **Gasto-intestinal bleeding:** Bleeding in the stomach or duodenum in severe cases may present as hematemesis (either red blood or coffee-ground emesis), melena (black, tarry stool), guaiac-positive stools, and/or anemia in the setting of chronic bleeding.
- **Perforation of the abdominal wall:** The sudden onset of severe, diffuse abdominal pain (or significant worsening of existing pain) or peritoneal signs (abdominal rigidity, guarding, and rebound tenderness) may signify a hollow organ perforation, which is a surgical emergency.
- **Gastric outlet obstruction:** It is a rare complication that occurs in about 2 per cent of patients with peptic ulcer disease. Symptoms include early satiety, bloating, indigestion, nausea, vomiting and weight loss.

Treatment

For many years, excess acid was believed to be the major cause of ulcer disease. Accordingly, the emphasis of treatment was on neutralizing and inhibiting the secretion of stomach acid. While acid is still considered necessary for the formation of ulcers and its suppression is still the primary treatment, the two most important initiating causes of ulcers are infection of the stomach by *H. pylori* and chronic use of non-steroidal anti-inflammatory medications or NSAIDs, including aspirin. The goal of treatment is:

- ✰ To relieve the symptoms
- ✰ To heal the ulcers
- ✰ To prevent complications of ulcers
- ✰ Treatment of a peptic ulcer depends on the underlying cause. Treatment includes lifestyle changes such as quitting smoking, avoiding alcohol, aspirin, and NSAIDs; acid-blocking medications; medications that protect the lining of the stomach and duodenum; and "triple-therapy" or "dual-therapy" regimens for ulcers caused by *H pylori*. Surgery may be performed in some cases that do not respond to lifestyle changes and medical treatment. Hence, treatment is mainly based on drug therapy, rest and dietary and lifestyle modifications.

Drug Therapy

Contrary to people's initial speculation, research has proven that *H. pylori* is the bacterium responsible for the spread of gastric ulcer and not stress or foods that are spicy. The recommended treatments for *H. pylori*-induced ulcers are antibiotics and drugs that stop the secretion of stomach acid, including proton pump inhibitors and H_2 blockers. Proton pump inhibitors (PPI) help to block acid-producing cells. H_2-receptor antagonists prevent the stomach from producing excess acid. Antibiotics inhibit the growth of *H. pylori*. Antacids or alginates protect the stomach lining.

Most peptic ulcers not caused by *H. pylori* infection result from the ingestion of large quantities of NSAIDs, which often are prescribed for conditions such as rheumatoid arthritis. Withdrawal of NSAID treatment usually allows the ulcer to heal, but if this is not possible, the ulcer can be managed with H_2 blockers or with the proton pump inhibitors. Symptoms often subside quickly following treatment. However, the treatment should be continued, especially if the ulcer is due to an *H. pylori* infection. It is also important to avoid drinking alcohol, smoking tobacco, and any trigger foods during treatment.

Lifestyle and Dietary Management

Studies have reported that the damages caused by ulcer can be reverted, often after treatment of an infection caused by *H. Pylori*, by changing diet and lifestyle. The first priority should be to provide physiological rest to the affected area and to support the damaged tissue. Good dietary regime and restoration of nutritional status can help in reducing the impact of this disease. Alleviation of symptoms and reducing the acids are the most important concerns of this disease.

Diet Therapy

Diet therapy has played a key role in the prevention and treatment of peptic ulcer, with the main purpose of recovering and protecting the gastrointestinal lining from hyper gastric acid secretions, improving digestion, relieving pain in the gastric and duodenal mucosa, and contributing to a satisfactory nutritional status. In addition, nutritional therapy aims to promote healing, based on a complex sequence of events going from the initial trauma to the repair of the damaged tissue. In the early 20th century, an *ulcer diet* proposed by Sippy meant a strict bland diet with lots of milk and milk cream, devoid of spicy or greasy foods, coffee, tea, alcohol, tobacco products and combined with antacids, for treatment of GIT ulcer. That type of diet was thought to provide gastric alkalinaization and relieve pain and help an ulcer heal. While these suggestions seem reasonable, but some have not stood up well recently in controlled research investigations. For example, now we know that milk and milk products used previously for relieving ulcer symptoms have not been found to be effective. Frequent milk ingestion as previously prescribed is not encouraged. This is owing to the transient buffering effect and significant gastric acid secretion effect of milk. The fat content of milk has no influence on these effects. Although certain spices (black pepper, chili powder, red pepper) may cause dyspepsia, yet they have not been shown to contribute to either gastritis or peptic ulcer. Hence, a strict bland diet is not now recommended as the days of ulcer patients surviving on a bland diet are a thing of the past. Ulcers are not caused by emotional "stress" or by worrying.

In this era of H_2-inhibitors, the available evidence does not support the need to place peptic ulcer disease patients on restrictive diets. People diagnosed with ulcers do not need to follow a specific diet.

We now know that diet has little effect on ulcers. In some people, however, certain foods seem to aggravate stomach ulcer symptoms. The American College of Gastroenterology has also concluded that while there is no set ulcer healing diet, there are certain ulcer diet foods that can fight *Helicobacter pylori* bacteria. One should keep a food diary and note down the resulting symptoms with individual foods and avoid eating any food that aggravates the ulcer symptoms. Foods that manage the ulcer pain and allow it to heal on its own should be included in the diet. In view of this, only slight modifications in the patient's usual diet are recommended. No particular diet is helpful for all the people with peptic ulcers. Hence, diet plan has to be individualized and liberal.

Nutritional Considerations

Diet may moderate the risk for gastritis or peptic ulcer through acting on *H. pylori*, among other effects. In turn, *H. pylori* infection can affect nutritional status and may specifically interfere with various nutrients. A balanced diet is vital in the treatment of peptic ulcer. The nutritional management in peptic ulcer should be as described under:

Energy

Calories distribution for patients with peptic ulcer should be normal, with values ranging from 50-60 per cent from carbohydrates, 10-15 per cent from proteins, and 25-30 per cent from lipids, with total energy value sufficient to maintain or recover the nutritional status. In case of weight loss, energy intake should be increased to overcome weight loss and achieve ideal body weight. In acute cases, if the patient is confined to bed, he will not spend any energy for activities; in that case, normal energy intake will make up the extra requirement.

According to Vomero and Colpo (2014), daily energy intake should be sufficient to maintain or recover the nutritional status *i.e.* 20-25 Kcal per kg body weight in patients for weight loss; 25-30 Kcal per kg body weight for maintenance of body weight; and 30-35 Kcal per kg body weight for weight gain (Vomero, N. D. and Colpo, E. 2014.Arq. Bras. Cir. Dig. Oct-Dec; 27(4): 298–302.).

Protein

Ulcer is a wound which if not healed on time may get perforated and it can also bleed. Adequate amount of protein is required for healing the sore and helping in the formation of new tissues. In the acute stage (5th to 8th week), protein intake of up to 1.2g/kg body wt/ day and up to 1.5 g/kg body wt./day in the recovery stage has been recommended to be provided in the daily diets. It is important to emphasize that the reduction of the gastric acidity by antacids or antiulcers alters protein digestion and affects good digestion of foods. Therefore, diet should be adequate or even more in protein content.

Carbohydrates

Carbohydrates should be adjusted to the patient's needs, without disaccharides concentration, so as to avoid fermentation, and lipids without concentration of saturated fats. Around 50 to 60 per cent of the total daily intake should come from carbohydrates in both acute and recovery phases.

Fat

Fats help in delaying the gastric emptying. Fats also play a role in inhibiting gastric juice secretion and thus, are effective in reducing ulcers. Around 25-30 per cent calories in both acute and recovery phases should be derived from fat. Emulsified fats are better tolerated. Avoid saturated and trans fats in the diet.

Dietary Fibre

Fibers regulate the bowel function, which make them vital for the well being of healthy people and in the dietary treatment of much pathology. Some research studies have reported a large number of patients with peptic ulcer having diets poor in fibers and antioxidants. A large cohort study at the Harvard School of Public Health found that high-fiber diets were associated with reduced risk for developing duodenal ulcer. Over a 6-year period, the risk was 45 per cent lower for those with the highest fiber intake, compared with those with the lowest. Although both insoluble and soluble fibres demonstrate this association, there is a stronger association between diets high in soluble fibre and a decreased risk for developing ulcers.

Food sources of soluble fiber (found in oats, legumes, flax seeds, barley, and certain fruits and vegetables such as oranges, apples and carrots) were especially protective, resulting in a 60 per cent lower risk for this group. However, supplementation with dietary fiber in the form of wheat bran had no effect on ulcer recurrence. A diet rich in fibers for individuals with peptic ulcer is advisable (20 to 30 g/day, according to WHO), because fibers act as buffers, reducing concentrations of bile acids in the stomach and the intestinal transit time, resulting in less abdominal bloating, thus decreasing discomfort and pain in the gastrointestinal tract.

Micronutrients

In addition to protein, some specific micronutrients as described under are required to accelerate the healing process.

Zinc

This micronutrient is important for maintaining a healthy immune system, as a response to oxidative stress, and for healing wounds. The patient should be given 11 and 40 mg per day in case of acute and recovery phases, respectively. Lean meats and poultry, eggs, legumes (beans and peas), nuts, seeds, oysters, spinach and soy-products contain good amount of zinc.

Selenium

This may reduce the risk of infection complications and may also promote healing. The recommended allowances for Se in acute and recovery phases are 55 and 400 µg per day, respectively. Seafoods and organ meats are the richest sources. Brazil nuts, dairy products, cereal grains, yellow fin tuna, and halibut are also recommended for their high selenium content.

Vitamin A

A diet that is rich in vitamin A (from food and supplements) has been reported to heal certain types of ulcers when combined with a high fibre diet that included fruits and vegetables. The risk of developing peptic ulcer was 54 per cent lower among persons consuming the most vitamin A, compared with those consuming the least in the Harvard cohort study. Animal studies demonstrate that vitamin A increases the production of mucus in the gastrointestinal tract. Impaired mucosal defense can allow ulcers to develop. Therefore, vitamin A may have a protective effect against the development of ulcer disease. Good sources of vitamin A (including beta carotene) include liver, carrots, broccoli, sweet potatoes, corn, papaya, pumpkin, spinach, and collard greens *etc.*

In acute and recovery phases of peptic ulcer, 900 (acute phase) and 3000 (recovery phase) µg of vitamin A in daily diets has been recommended in some studies. Vitamin A may be used as a supplement, but the research that supports this practice is of limited effectiveness, because very high dosages do not promote cure, and excessive intake may be toxic (Arnold, M. and Barbul, A. Nutrition and wound healing. Plast. Reconstr. Surg. 2006; 117: 42–58.).

Vitamin B_{12}

Deficiency of vitamin B_{12} is common in patients with peptic ulcer due to the prolonged use of antacids, making difficult the bioavailability of this vitamin. Vitamin B_{12} can be synthesized by the intestinal microbiota in the colon, but is not absorbed. Deficiency of this vitamin causes impaired cell division and megaloblastic anemia. It is estimated that 80-90 per cent of patients lacking vitamin B_{12} develop neurological alterations, if not treated. As a result, recommendation is for 2.4 µg/day of this vitamin in both acute and recovery phases, which can be obtained from animal foods, such as milk, meat and eggs. When iron or vitamin B_{12} supplementation is paired with *H. pylori* eradication therapy, deficiencies in vitamins B_{12}, vitamin C, vitamin A, iron and folic acid improve.

Folic Acid

Absorption of folic acid can be impaired in subjects that make chronic use of aluminum-based antacids because antacids make the pH of the jejunum more alkaline. In these cases, intake of 400 µg/day of this vitamin is necessary in both acute and recovery phases, which can be supplied with ingestion of leguminous foods, such as lentils, and meats.

Iron

Use of antacids in peptic ulcer can diminish absorption of iron, causing iron-deficiency anemia. Gastrointestinal bleeding due to gastro-duodenal ulcers and infection by *H. pylori* may also be associated with the development of anemia. Gastric bleeding is a major complication of peptic ulcer which can lead to iron deficiency anaemia.

Infection by *H. pylori* can also lead to an imbalance of body iron homeostasis due to the growing demand for iron by the same. Just like other kinds of bacteria, iron is essential for the growth of *H. pylori* too. To prevent or even treat iron deficiency, an intake of 45 mg of iron daily is recommended, which can be supplied by the ingestion of animal foods especially meats, the main source of heme iron. Sources of vitamin C like citrus fruit juices should be added in the diet to enhance the absorption of non-heme iron present in plant foods.

Antioxidants

Vitamin C, a powerful antioxidant may be effective in helping to eradicate *H. pylori,* especially when taken in small doses over an extended period. Accordingly, some studies in humans used antioxidants to eradicate *H. Pylori* and observed that vitamin C has important effects in the bacteria eradication in patients with peptic ulcer. But these studies showed that smaller doses of vitamin C for a longer period of time had a better response when compared with higher doses. Thus, it is observed that patients with peptic ulcer by *H. pylori* can take up to 500 mg/day of vitamin C for a period of three months. On the basis of these studies, in acute and recovery phases of peptic ulcer, 75 and 500 mg of vitamin C per day has been recommended. Citrus fruits, aonla, guava, sprouted legumes and vegetables contain high levels of vitamin C.

Another antioxidant used to eradicate *H. pylori* is the capsaicin present in pepper and chilies. Studies on animals showed that capsaicin has effect in healing gastrointestinal lesions. Likewise, some researchers studied the effect of capsaicinoids in individuals with peptic ulcer by *H. pylori* or aspirin and showed that these substances are gastro-protective only in individuals with aspirin-induced lesions. It is worth noting that peppers may be associated with irritations in the gastric mucosa, and may not have a gastro-protective effect in some individuals with peptic ulcer.

Probiotics

One of the measures that may contribute to reduce the infection rate by *H. pylori* is dietary modulation with the addition of probiotics. Food that contains active bacterial content, such as probiotic yogurt, can help to reduce a *Helicobacter pylori* infection. There is a special interest in probiotics for treating infection by *H. pylori,* because it plays a crucial role in the pathogenesis of chronic gastritis and peptic ulcer in adults. Probiotics have therapeutic agents against *H. pylori,* which can be shown by clinical data that prove the efficacy of some probiotics in diverse gastrointestinal diseases and also due to the increasing resistance of pathogenic bacteria to antibiotics. Probiotics (*e.g., Lactobacillus casei*) interfere with *H. pylori* adhesion to epithelial cells, attenuate *H. pylori*-induced gastritis, and inhibit growth of *H. pylori* in humans, in addition to reducing the side effects of eradication treatment. In a study, combining probiotic treatment with drug therapy in *H. pylori*-infected children significantly improved the treatment effectiveness, compared with drug treatment alone.

Among the clinical applications of probiotics, reduction of the side effects associated with antibiotics is the best indication documented. According to Cats (Cats, A. 2003. Effect of frequent consumption of a *Lactobacillus casei* containing milk drink in *Helicobacter pylori*-colonized subjects. Alimentary Pharmacology and Therapeutics. 17:429–435), who conducted an intervention study, 14 patients infected with *H. pylori* receiving *L. casei* (10^8 CFU) for three weeks, it was capable of inhibiting the growth *H. pylori* in 64 per cent of the volunteers. Similarly, in a study by Wang *et al.* (Wang KY, Li SN, Liu CS, Perng DS, Su YC, Wu DC, Jan CM, Lai CH, Wang TN, Wang WM. 2004. Effects of ingesting *Lactobacillus*- and Bifidobacterium-containing yogurt in subjects with colonized *Helicobacter pylori*. 80:737–741) with 59 volunteers, they received *Bifidobacterium animalis* and *L. acidophilus* (10^{10} CFU) twice a day during six weeks, and concluded that regular intake of yogurt containing *Bifidobacterium animalis* and *L. acidophilus* can effectively suppress the infection caused by *H. pylori* in humans.

Probiotic organisms do not appear to eradicate *H. pylori*, but have the ability to reduce the bacterial load and infection in studies carried out on animals and humans. Studies on humans indicate that probiotics improve slightly the elimination rate in treatment against *H. pylori*, being useful to decrease the bacterial load and likely improve dyspeptic symptoms. Thus, an intake of 10 CFU/day of lactic acid bacteria is recommended.

Other Life Style Related Factors

The following life style factors have also been associated with reduced risk of gastritis or ulcer disease in epidemiologic studies:

Avoid Alcohol

Research studies have reported that alcohol consumption causes damages to the digestive tract with appearance of symptoms of ulcer and other alcohol-related diseases, such as esophagitis, chronic pancreatitis and gastritis, among others. The relationship between alcohol and gastritis and peptic ulcer is complex and may be related to amounts consumed. Chronic alcohol abuse favors *H. pylori* infection, and the ammonia produced by this organism contributes to gastritis. Alcohol may also slow the rate of healing in established ulcers, although this has not been demonstrated in patients on proton pump inhibitors. Excessive use of alcohol can irritate and erode the mucous lining in the stomach and intestines, causing inflammation and bleeding. Avoiding alcohol can help to reduce the risk, as it causes the body to produce more gastric acid which can lead to stomach ulcers.

However, studies have also found an inverse association between moderate alcohol consumption and *H. pylori* infection. Alcohol may have bactericidal effects on *H. pylori*, and it may prevent infection and associated gastritis through an adaptive cytoprotective response (*e.g.*, endogenous release of prostaglandins with protective effects on gastric mucosa). But, while moderate consumption is associated with the lowest odds for infection, higher intakes are associated with greater risk. In combination with smoking, alcohol increases the risk for duodenal ulcer.

Avoid Smoking

Prospective and retrospective studies show higher mortality from peptic ulcer in smokers when compared to non-smokers. Researches show that, among other tobacco constituents, nicotine accounts for most of the peptic ulcer development, because it has a harmful effect on the protective mucus of the gastric epithelium, altering bicarbonate.

Avoid smoking as it diminishes the secretion of mucus and bicarbonate, raising the duodenal and gastric flow and increasing the risk of ulcers formation. Hence, it reduces the natural defenses in the stomach and duodenum and slows down the healing process.

Avoid Coffee and Carbonated Beverages

Coffee, either in its caffeinated or decaffeinated forms, stimulates acid secretion, thus resulting in mucosal irritations. Some studies have suggested a correlation between coffee intake and symptoms in patients with duodenal ulcer.

Coffee consumption may also mediate the relationship between *H. pylori* infection and ulcer, although studies have variously found coffee consumption to be associated with both increased and decreased risk for *H. pylori* infection. Overall, there is no current evidence implicating coffee consumption in the susceptibility to, treatment of, or recovery from gastritis. The same goes for soft drinks, which, besides increasing acid production, are gaseous and cause gastric distension and is dyspepsia-related. However, it is important to take into account individual tolerances, with attention to the existence of misconceptions about foods and their actions in the body. When treating peptic ulcer, clinicians and dietitians should consider individual tolerances to various food and beverages such as coffee and carbonated drinks.

Tea

According to various research studies, regular green tea consumption is associated with a 40 per cent -50 per cent lower risk for gastritis. Cellular tests suggest that the catechins (*e.g.*, epigallocatechin-3-gallate (EGCG) in some tea varieties may suppress *H. Pylori*-induced gastritis through antioxidant and antibacterial actions. However, current evidence is not yet sufficient for recommending tea for prevention of gastritis.

Control Stress

Although stress does not cause peptic ulcers, it may worsen its signs and symptoms. Consider the sources of stress and do what you can do to address the causes. Some stress is unavoidable, but one can learn to cope with stress with exercise, socialization *etc.*

Exercise

Ulcers should not stop from engaging in regular physical exercise. Exercise has been hypothesized to influence the risk for ulcer disease or gastritis through reductions in basal or meal-stimulated acid secretion. Some evidence suggests that exercise significantly decreases the risk of duodenal ulcer and of severe gastrointestinal hemorrhage in persons with gastritis or duodenal ulcer. However, controlled clinical studies have not confirmed the ability of exercise to prevent or ameliorate gastritis. According to one research study, regular physical exercise can help to improve ulcers and overall health. Exercise also prevents ulcers from becoming complicated. In fact, some others have shown that certain kinds of exercise (*e.g.*, long-distance running) actually increase the risk for this condition. Approximately 20 per cent of long-distance runners experience GI bleeding, confirmed by endoscopy. Exercise-induced ischemia and acid secretion have been postulated as mechanisms by which runners may develop erosive gastritis, hemorrhagic gastritis, or gastric ulcer.

One should participate in physical exercises such as walking, bicycling, swimming, cycling and jogging. Regular exercise can activate the immune system and help lower inflammation throughout cells. But avoid exercises that are vigorous and strain the abdominal region. Abdominal exercises such as crunches and sit-ups increase the intra-abdominal pressure, which can worsen heartburn and other symptoms of ulcers. Vigorous abdominal exercises may also cause existing ulcers to perforate through the wall of the stomach.

Try to get Enough Sleep

Sleep can help the immune system, and therefore, counter stress. Moreover, enough sleep gives the body time to repair and regenerate damaged cells without the demands of digesting food.

Avoid Late Night Snacking

Changing the way one eats may help to reduce ulcer symptoms. For example, avoid frequent meals and or eating late night snacks just before bed time as that may trigger nocturnal acid secretion unless someone gains relief from certain foods. Do not eat within 2 h of bed time. Eating regular meals may also help to reduce the production of stomach acid.

Manage Meal Patterns

If someone is on drug therapy *e.g.* he is taking PPIs (Proton Pump Inhibitors). He must know that PPIs require a meal to activate them. He should eat a meal within 30 minutes to 1 hour after taking this medication for the acid suppression therapy to work most effectively. Waiting later than this time can decrease the positive effect of this medication. This might delay healing or even result in the failure of the ulcer to heal.

Avoid NSAIDs Pain Relievers

Donot take NSAIDs. If someone has to take pain relievers, consult the doctor whether acetaminophen may be an option.

Dietary Management

Severe Cases

In severe cases, patients have no desire to eat and 'resting' the stomach from food for 1–2 days may help alleviate pain; adequate fluid including sugar and electrolytes will minimize risk of dehydration, but clearly are not nutritionally adequate. Nutrient intake should be gradually increased over 1–3 days by providing other nourishing fluids and then diet containing non-irritating foods that the patient feels able to manage, can be started.

Mild to Moderate Symptoms

For most of the past century, health authorities thought peptic ulcers were caused by stress and dietary factors. The treatment was based on prescribing a bland food diet. Gastric acid was also blamed for ulcers and treatment was with antacids and other medications that block stomach acid. However, these were often ineffective. Now the scientists have established the link between the bacteria *Helicobacter pylori* and ulcers and this has led to the antibiotic treatment of the disease. Another cause of peptic ulcers is long-term use of nonsteroidal anti-inflammatory drugs (NSAIDs) such as aspirin and ibuprofen. A third cause is the rare Zollinger-Ellison syndrome, which results in stomach tumors. None of these now-recognized causes of peptic ulcers are due to diet. As previously thought, a bland diet and drinking milk are not effective ways to treat ulcers. Instead, they are treated by antibiotics or stopping NSAIDs. Dietary considerations should be:

1. A balanced diet that includes a variety of foods from all the food groups is vital in the treatment of peptic ulcer. Choose a healthy diet full of fruits, vegetables, legumes and whole grains which is rich in micronutrients and antioxidants especially vitamin A, C, B_{12}, folic acid, Zn, Se, iron and quality protein. Not eating protein and vitamin-rich foods in the diet may make it difficult for the body to heal ulcer.
2. Diet is considered supplemental to the primary drug therapy. Individual tolerance to foods is the basis for diet planning and dietary restrictions. No food has been shown to cause ulcers, but some may make the symptoms worse or may

make it more difficult for ulcer to heal. Listen to your body. If there are some foods or beverages that may be irritating or cause to make more stomach acid, limit or avoid them. Choose foods to help the stomach or intestinal ulcer to heal.

3. Some dietary changes may help to complement medical treatment for ulcers. For example, foods and seasonings such as caffeine, black pepper, garlic, cloves, and chili powder *etc.* if not tolerated by someone should be limited. It depends upon the individual choice. Spices, in particular black pepper, red pepper, and chili powder, may produce dyspepsia in some individuals. One study shows red chili powder to have no detrimental effect on duodenal ulcer healing. It has also been proposed that daily pepper ingestion may have a beneficial adaptive cytoprotective response. While still controversial and under evaluation, peptic ulcer patients should avoid any spice if it causes discomfort, especially during exacerbation of peptic disease. However, modifying the diet in this regard is now thought to have very little impact. One can continue to eat spicy foods if they do not cause discomfort.
4. There appears to be no evidence to restrict dietary fiber. Some fiber-containing foods may possess factors that are protective against ulcer disease. Include soluble fibre rich foods in the diet. Fruits such as plums, bananas, dried prunes, strawberries, cherries, cantaloupes and pears are all very rich in soluble fiber. Besides being good source of fibre, whole grains, legumes and nuts are rich in protein, B-vitamins, and calcium. These nutrients can reduce the symptoms of the ulcer and immensely aid in the recovery process.
5. Limit the intake of high fat milk and milk products. Do not take milk in large quantity as sometimes drinking milk will make the ulcer pain better, but then later causes excess acid secretion, which increases pain.
6. Do not take excess salt in the diet.
7. Most meat and poultry choices should be lean or very lean. Fish and chicken proteins are preferable than red meat and remove fatty skin. Include omega -3 fatty acids rich fish mainly salmon, mackerel, sardines and herring; it helps in producing prostaglandins, which in turn protects the mucosa lining of the stomach and intestines.
8. If stomach ulcer is caused by an *H. pylori* infection, foods that are rich in antioxidants may be beneficial. They could help protect and activate the immune system and help fight the infection. Foods like blueberries, cherries, strawberries, apple, bell peppers, cabbage, carrots, radish, spinach, beetroot, beans, nuts *etc.* are packed with antioxidant power and should be included in the diet. Broccoli and cauliflower contain sulforaphane, a compound that exhibits anti-*H. pylori* activity can be included in the diet.
9. Most of the fruits are alkaline and can be consumed.
10. Garlic contains specific antimicrobial elements that help in treating stomach ulcer by checking

Table 22.1: Foods to be Allowed and to be Avoided in Peptic Ulcers

Food Groups	*Allowed*	*Use with Caution*	*Prohibited*
Cereal grains	Wheat, barley, maize, oat, semolina, rice	Refined cereals	–
Dairy	Milk, low-fat cheeses, probiotic yogurt, fermented butter milk	Fatty cheeses (mascarpone, cream cheese, gorgonzola), cottage cheese	–
Oilseeds	Flaxseed, Brazilian nut, walnuts, almonds	–	–
Oils	Vegetable oils, olive oil	–	Fried foods, Food products containing saturated and trans fat
Fruits	Apple, papaya, melon, banana	Orange, pineapple, acerola, passion fruit	Lemon, unripe fruits
Vegetables	Leafy dark green vegetables, carrot, beet, green bean, spinach, kale, radish, zucchini, leek	Broccoli, cauliflower, cabbage, cucumber, onion, red pepper	Spicy peppers (black pepper, chilies), raw vegetables *etc.* if they cause discomfort to someone
Legumes	Bean soup, lentils, green gram, chickpeas, soybean *etc.* in cooked form and those which do not cause gas formation	Beans	–
Meats	Lean meat (beef, pork, chicken, fish)	Fatty meats, organ meats and sausages	–
Sweets	–	Concentrated sweets	Chocolate, cakes, pastries
Beverages	Natural juices	Citrus/acidic fruit juices	Coffee, black tea, fizzy/cola drinks
Other foods	–	Industrialized seasonings, spices and condiments (Ketchup, mayonnaise, mustard)	Mustard grain

the growth of *H. pylori*. Have 2-3 cloves of garlic every day for better results.

11. Decaffeinated green tea contains ECGC, a high level of catechin that has the potential to relieve from a stomach ulcer. Its anti-inflammatory and antioxidant properties work best on the ulcer. Have a cup of it every morning and evening.
12. Use fats and oils sparingly not generously. Make most of fats sources from fish, nuts and vegetable oils (olive and canola especially). Some research shows that olive oil has the potential of treating stomach ulcer. It contains phenols that act as an anti-bacterial agent, which prevents *H. pylori* from further spreading and affecting the stomach lining.
13. Previously recommended small frequent feedings have not been shown to be more effective than three meals per day in the treatment of chronic peptic ulcer disease. This reference cites authorities advising against extra feedings because of increased acid secretion and unnecessary complication of eating patterns. However, some patients claim to be relieved of symptoms with more frequent feedings, especially during acute phases. Meal pattern should depend upon the individual choice and comfort. Stomach distention with large quantities of food should be discouraged.
14. Avoid fried foods.
15. Avoid excessive acid production by starvation or fasting.
16. Avoid bedtime snacks to prevent increased acid secretion, unless you note that you gain relief from certain foods.
17. Do not eat within two hours of bedtime.
18. Citric acid juices may induce reflux and cause discomfort in selective patients. Those patients may avoid them.
19. It is reasonable to have peptic ulcer patients restrict decaffeinated coffee and all caffeine-containing beverages. Coffee should be avoided on the basis of its strong acid secretagogue property. Coffee can induce dyspepsia. Whether non-coffee caffeine-containing beverages (tea, soft drinks) induce peptic ulcer is unknown, but they are acid secretion stimulators. Decaffeinated coffee has an acid stimulating effect as well.
20. Use probiotic foods or supplements containing the beneficial bacteria *Lactobacillus* and *Bifidobacterium* which help to fight harmful bacteria in intestine, and improve digestive function. They may also help complement medical treatments such as antibiotic therapy.
21. Although there is now little role for dietary therapy, one should note that bland and ulcer diets probably are not detrimental to most persons if they are used for a short time and may have some psychological benefit.

Constipation

Broadly speaking constipation is a condition, a symptom or a sign of the body in which the stool becomes hard and dry; its movements may become significantly slow, and it is hard or difficult to pass the stools out. The slower the food moves through the digestive tract, the more water the colon will absorb from it. Consequently, the feces become dry and hard, thereby resulting in difficult evacuation of the intestinal contents or incomplete evacuation of stools daily. The number of visits to the toilet *i.e.* passing out stool may become very infrequent or less. Although, there is no set number of visits or frequencies which can be termed as a sign of constipation; it varies from individual to individual. However, any number less than 3 to 4 times a week is regarded as a sign of constipation.

Types of Constipation

Constipation is of three types, they are:

1. Obstructive Constipation

It is the result of actual physical obstruction of the passage of feces which may be acute or chronic, partial or complete. Causes include cystic fibrosis (in children), tumors, adhesions, scars formed in the abdomen (mostly after abdominal surgery), strictures formed within the intestine (mostly caused by inflammation, like in Crohn's disease), or damaged pelvic nerves (injuries, neurological diseases) that prevent relaxation of anal muscles. Surgical treatment is often needed.

2. Spastic Constipation

Unlike atonic constipation, which has rather a physical cause, spastic constipation tends to be caused by emotional elements. It results from colonic spasms caused by irritation of the bowel or intestines due to emotional factors like nervousness, anxiety, irritability, stress *etc.*, certain foods or medications which increase the muscle tone. It is also a symptom of constipation – dominant irritable bowel syndrome. Symptoms include constipation, abdominal cramps, bloating and occasional diarrhea. The person feels constant fatigue. Painful symptom has spastic character without any propulsive activity. In this case, a normal diet that is low in fiber is usually suitable. Milled cereals, dehusked *dhals* should be included in the diet. Excessive use of spices and herbs in diet, excessive use of alcohol, tea and coffee, mental stress *etc.* are causative factors which should be avoided.

3. Atonic Constipation

Atonic constipation, also called as colon stasis or lazy colon does not respond to the normal stimuli for evacuation due to lack or loss of muscle tone of the colon, abdominal wall or pelvic floor along with weak peristalsis. Person with this constipation simply does not feel the urge to defecate even when the colon is full of feces. As a consequence, the stomach is not properly emptied and therefore, heartburn becomes constant. It is the most common type of constipation. Atonic colon is essentially a "vicious circle" phenomenon. Constipation can lead to reduced bowel activity, which in turn can lead to a worsening of constipation, and so on. It may result in chronic constipation.

Based on the duration of constipation, they are of two types:

i) Acute Constipation (Occasional problem)

Constipation occurs suddenly and lasts for few days. Acute constipation with worrisome symptoms such as rectal bleeding, abdominal pain and cramps, nausea and vomiting, and involuntary loss of weight needs urgent diagnosis.

An acute change in bowel habits, particularly over the age of 40, is a cause for concern because it may be a symptom of a colonic neoplasm. Appropriate diagnostic studies, including proctosigmoidoscopy or flexible sigmoidoscopy plus an air-contrast barium enema or colonoscopy, should be performed when a concern regarding neoplasm exists. Other acute causes are a sudden decrease in physical activity, change in diet, particularly reducing fiber, medications and anal pain.

ii) Chronic Constipation (Persistent problem)

It is the constipation that lasts for longer durations. Chronic constipation is more common and often more difficult to manage. Once organic causes such as hypothyroidism are ruled out, chronic constipation often falls into two categories: the older patient with a well-established laxative habit and the younger patient, often female, who is constipated because of her lifestyle and often is headed toward laxative dependence. A common example of chronic constipation is the young woman who prefers an extra half hour of sleep in the morning to eating breakfast and attempting to defecate spontaneously. Indeed, the normal urge to defecate in the morning may be suppressed as a matter of convenience. Her diet is either "junk food" or regular diet which is low in fibre. She begins to lake laxatives as a matter of expediency, rapidly leading to dependency. The cycle can be broken in several places by eating a breakfast containing one of the bran cereals, taking time to defecate after breakfast, encouraging increased physical activity, and stopping the laxatives. Enemas and suppositories are lesser evils because they stimulate only the rectosigmoid or rectum, respectively, rather than the entire colon, so that these can be used to retrain the bowel.

Symptoms

There is a long list of the symptoms of constipation. That does not mean that anyone suffering from it should have all the symptoms. Again, it shall depend on case to case basis. Some people may have more signs while others may have only a few. Here are some of the most common symptoms that have been noted in case of constipation

- **Change in bowel habits:** Regular change in bowel habits *i.e.* passing fewer stools than usual is one of its most common symptoms. Irregular bowel syndrome or IBS is another among the symptoms of constipation. In case of IBS, the frequency of passing the bowels does not stay regular. At times, it may be more while at others, it may be infrequent.
- **Stomach pain, discomfort, swelling and cramps:** People often complain of intermittent abdominal pain, stomach cramps, feeling of having a swollen stomach and sickness. Constipation problem affects the digestive system and hence, colon tract may have abnormal pain signs.
- **Flatulence:** As the transit times of faeces inside the intestine is longer, it causes to produce gases which are expelled through the rectum in the form of flatulence.
- Feeling belly bloated and nauseous
- Loss of appetite
- Aggravation of hernias
- **Sense of incomplete faecal evacuation:** The stool may by very dry and hard and even lumpy, dark in colour, abnormally large or just the opposite *i.e.* too small. There is increased difficulty and straining when passing stools. This also accompanies with the symptom of feeling as if one has not completely passed the bowel movement.
- **Fatigue**, fever, sore throat, headache, cough aches, pains *etc.* have also been associated with constipation.
- **Other symptoms:** Physiological distress and/or obsession with having bowel movements, stress, anxiety, weakness, headache, coated tongue, foul breath unexplained weight loss and vomiting are some of the other most common signs of constipation.

Complications of Severe Constipation

Constipation as such is not life-threatening but of course, it is uncomfortable. However, severe constipation (If one passes one stool in a week, it is called severe constipation.) acute or chronic, if untreated, can develop into more serious conditions, including:

- **Lazy colon:** Lazy colon is a synonym for severe chronic constipation and does not mean permanent colon impairment. It does not respond to presence of the stool in the colon and gastro-colic reflex that normally stimulates the bowel movement after food enters the stomach. Often even laxatives do not help. This poor colon activity may result from long term laxatives abuse, postponing bowel movements or prolonged bed rest. One can achieve relief only by eliminating all possible causes of constipation and sticking with appropriate diet and active life style.

- **Hemorrhoids,** or swollen, inflamed blood vessels of veins may occur in the rectum.
- **Rectal pain, bleeding and anal tears and fissures:** Sometimes abdominal pain is extended to rectal pain in case of severe constipation. The stool is very hard and it is difficult to pass on through the rectum. Sometimes, the rectum may even start bleeding because of cuts or fissures.
- **Faecal impaction**, in which large, dried hard mass of stools collect in the anus and rectum; get stuck so badly in the rectum that one cannot push it out by natural effort because of size and consistency of the stool. It typically occurs in bedridden patients.

Cause of Atonic Constipation

The commonest causes of constipation are:

- It is most common among elderly (loose rectal muscles) or bedridden patients (lazy colon) or after prolonged dependence on laxatives (weak muscle tone), women after giving birth (pelvic floor weakness) and diabetics (damage to intestinal nerves).
- **Sedentary life style:** Very often the disease occurs in people, which by their nature or professional activities lead a sedentary lifestyle.
- Lack of sleep and rest.
- **Regular intake of foods devoid of fiber:** If a person prefers fast foods, takes monotonous diet composed mainly of refined foods and meat dishes, no plant foods or he is accustomed to eating on the run, in the habit to fend ready to eat snacks most of the time or fully fed only once during the day, such persons are likely to develop atonic constipation.
- Ignoring the urge to have a bowel movement
- Less consumption of water and other fluids throughout the day.
- **Dehydration:** This problem may occur when large volumes body fluid is lost; if it goes mainly through the mucosa and skin of man. This can happen when the injured person is ill, and the disease causes high temperature, as well as in the sultry season.
- Low calorie intake
- **Intake of certain group of medicines:** Decreased propulsive activity, usually from medication but occasionally from intrinsic muscle disease such as scleroderma, amyloid, and certain neurologic disorders. Medications include those with anticholinergic properties, such as the antidepressants and some antiarrhythmic drugs, iron supplements, painkillers, the opiates, certain antacids and calcium channel blockers.
- **Excessive use of laxative products for prolonged period:** If people with this problem are beginning to abuse cleansing enemas, as well as decoctions or preparations with a laxative effect, the nerve endings in the intestinal mucosa, under the constant influence of the stimulus, begin to dwindle, which invariably leads to a decrease in the tone of the intestinal walls and problems with defecation. It is called as "lazy bowel syndrome." This pathology can develop as a consequence of the inflammatory process occurring in the body of the patient.
- **Prolonged emotional stress, depression and anxiety:** Not stress itself, but not being able to cope with stress may result in constipation. Work overload, changing or losing the job, problems in relationships (common in children), anxiousness or travel may all aggravate constipation. Along with a decrease in many bodily functions, such as anorexia and insomnia, the bowel function declines. Combination of stress and certain foods may cause irritable bowel syndrome (IBS) in sensitive people.
- **Endocrine causes:** Constipation rather than lack of energy may be the presenting symptom in hypothyroidism.
- Excess body weight can also trigger atonic constipation.
- Bowel movement problems can occur during pregnancy, as well as immediately after delivery.
- Painful defecation due to hemorrhoids and fissures, hence psychological aversion and avoidance to eliminate faeces.
- Consequence of surgical intervention affecting the organs of the digestive system.
- **Colonic disorders**: Such as in diverticulosis, Crohn's disease, megacolon *etc.*, constipation may occur.
- **Hirschsprung's disease:** This congenital disease refers to a blockage in the large intestine, owing to poor muscle movement.

Common Causes of Constipation in Various Age Groups

Causes of constipation may vary a lot among various age groups:

- Among **infants**, constipation is mainly caused by congenital disorders, formula feeding or introduction of solid foods.
- In **children,** psychological factors are important; celiac disease is also not rare.
- In **adolescents**, irritable bowel syndrome and Crohn's disease are common.
- In **adults**, constipation is mainly caused by inappropriate diet, sedentary life style and stress.
- In **old people**, physical inactivity, chronic diseases, medications and dehydration are main causes.

Aim of Treatment

- Appropriate intake of diet rich in fibre
- Adequate intake of fluids
- Correction of faulty habits and regular bowel evacuation
- Physical activity

I. Dietary Management

Normal well balanced diet with modifications in dietary fibre and fluid intake is required. Requirements for rest of the nutrients remain unaltered.

Dietary Fibre

The diets with a high content of fibre (about 35 g per day), such as those rich in cereals, fruits and vegetables have a positive effect on health since their consumption has been related to a decreased incidence of several types of diseases as due to its beneficial effects like increasing the volume of fecal bulk, decreasing the time of intestinal transit, trapping substances that can be dangerous for the human organism (mutagenic and carcinogenic agents), stimulating the proliferation of the intestinal flora *etc.* Thus, the mechanism by which stool bulk and laxation is promoted varies for different fibres.

Make sure that diet is rich in both types of fiber *i.e.* soluble and insoluble. Most plant foods contain some of each kind of fiber. Insoluble fiber such as resistant starches present in mainly wheat bran, fruits and vegetables adds bulk to the stool, which hastens its passage through the intestines, reduces transit time in the colon and prevents feeling constipated. Soluble fiber allows more water to remain in the stool, making waste softer, larger, and thus, easier to pass through the intestines. Foods containing high levels of soluble fiber include dried beans, oats, oat bran, rice bran, barley, citrus fruits, prunes, figs, apples, strawberries, peas, and potatoes. The fibre present in citrus fruits and whole legumes stimulate the growth of colonic bacteria, which increase the stool weight and the amount of bacteria in the stool. Encouraging the growth of certain bacteria in colon may help promote a healthy intestine. Guar gum is readily fermented by the human fecal microbiota, improves bowel functioning and relieves constipation in patients.

Foods Rich in Fiber

- **Fruits:** Pears, berries, apples, melons, oranges, ber, guava *etc.*
- **Vegetables:** Carrots, potatoes with peel, green leafy vegetables, broccoli, cabbage, sweet corn *etc.*
- **Whole legumes:** Beans, lentils, peas *etc.*
- **Whole grains:** Whole wheat breads, brown rice, bran, oatmeal, barley *etc.*
- **Nuts and seeds:** Almonds, peanuts, sunflower seeds, flax seeds, walnuts *etc.*

Diet should contain high fibre foods like whole cereals, whole legumes, nuts, green leafy vegetables and fruits and plenty of fluids to avoid constipation. Take vegetables like radish, carrot, cucumber, tomato, onion *etc.* in the raw form as salad. Instead of fruit juices, take fruits like apple, guava, ber, pears, plums *etc.* along with skin and seeds. Donot sieve whole wheat flour to make *chapatti.* Insoluble fibre present in wheat bran is as effective as laxative. Instead of dehusked pulses, take whole beans and legumes in the diet. Incorporating some amount of caffeine in the form of bed tea or coffee can stimulate the muscles of the digestive system as it contains small amounts of soluble fibers. These soluble fibres maintain the balance of gut bacteria. Bed tea also promotes peristalsis, the coordinated contraction and relaxation of intestinal muscles that accelerates bowel movements. Try adding bulking agents to the diet to soften stools and make them easier to pass. There are several mild herbs such as psyllium, flaxseed, and fenugreek that have a laxative effect and soften stool. They can be taken as supplements. Refined cereals, legumes (without husk), pureed vegetables, fried foods and sweets should be avoided.

II. Take Lots of Fluids

Generally, constipation occurs because the colon has absorbed too much water from the food that is in the colon, making the faeces to become dry and hard or if colon's muscles are too slow leading to hardening of the stool. Moreover, fiber in the diet needs water to pass through the digestive system and form stool. Consuming enough water throughout the day is essential to keep the body hydrated. Hydration makes digestion easy and supports muscle function. Having enough water can process nutrients faster thus aiding the bowel movement on a regular basis. When the body has sufficient water level it also ensures a better functioning of the colon as it absorbs less water. Hence, water has a host of health benefits, including improving the digestive system. It softens stool thereby, making it easy to pass and preventing constipation. Make sure to drink plenty of water *i.e.* 8-10 glasses daily to prevent dehydration, which is linked to constipation.

Take foods that are high in water content, such as muskmelon, watermelon, cucumber and celery. Soups whether clear or with cream are nourishing and supply a good amount of water besides electrolytes; incorporate them in the daily diet. A glass of luke warm water containing honey and lemon juice early in the morning also stimulates the urge for defecation.

III. Correction of Faulty Habits

Ignoring the desire to have bowel movements may initiate a cycle of constipation. The fact is that the longer one holds the stool in colon, the more water is absorbed by the colon, leading to hard and compact stools. After a period of time, the person may stop feeling the desire to move the bowels. This leads to progressive constipation. For example, some people may avoid using public toilets or ignore going to the toilet because they are busy in the morning. Children should be trained from the beginning to answer nature's call regularly at the same time *i.e.* in the morning as it will allow to develop a conditioned reflex to defecate. The bowel is most active in the morning and following meals. Passing stool is the easiest when efforts combine with this physiology.

IV. Do Regular Exercise

Insufficient physical activity and excessive sedentary behaviours have been found to be positively associated with constipation with dose-response relation. Stay physically active as exercise has been shown to be helpful in improving the efficiency of the digestive tract by decreasing the amount of time it takes food to move through the large intestine. In fact, there is evidence that shows morning exercise is effective for dealing with digestive problems. Aerobic exercise, such as brisk walking, accelerates the heart and breathing rates, and helps to stimulate the natural contractions of intestinal muscles too. Try exercising most days of the week. This will help to make bodily processes more regular, including the digestive system and the passing of stools. There are several yoga poses also that can help stimulate the bowel and improve digestion.

Diarrhoea

Diarrhea is one of the most common health complaints. It can range from a mild, temporary condition, to a potentially life-threatening one. Diarrhea is a condition in which abnormally loose or watery and unformed stools are passed frequently. Diarrhoea often accompanies conditions such as ulcerative colitis, various functions of bowel and most forms of gastro-enteritis. Diarrhea follows irritation or inflammation of the mucous membrane of the bowel by physical, chemical or bacterial agents. Diarrhoea if it lasts longer causes dehydration due to fluid loss in young children. Frequent passing of formed stools is not diarrhoea, nor is the passing of loose, "pasty" stools by breastfed babies.

Types of Diarrhea

Diarrhoea can be classified in two ways *i.e.* according to the length of time that the condition lasts and the mechanism by which it occurs. On the basis of duration, it is classified as:

Acute Diarrhea

If the symptoms of diarrhoea come on suddenly but resolve in less than two weeks, it is called acute diarrhea. Acute watery diarrhea causes dehydration and contributes to malnutrition. The death of a child with acute diarrhea is usually due to dehydration. This type of diarrhea is one of the most common illnesses and, worldwide, it is among the leading causes of death in children younger than 5 years old. This type of diarrhea is typically caused by infection due to bacteria, viruses or parasites or due to side effects of various medications.

Acute Bloody Diarrhea

When loose motions are accompanied by the passage of blood with or without mucus, the condition is called dysentery. Dysentery is very dangerous because of its ability to lead to anorexia, rapid weight loss, and damage to the intestinal mucosa.

Persistent or Chronic Diarrhea

Persistent diarrhea is when the illness lasts from two to four weeks and can be triggered by parasites, bacteria or virus. Up to 20 per cent of episodes of diarrhea become persistent.

If the condition lasts longer than four weeks, it is considered chronic diarrhea. Irritable bowel syndrome happens to be the most frequent cause of chronic diarrhea in developed countries. Although chronic diarrhea is caused by digestive disorders but it can also occur due to other factors like excessive consumption of alcohol, caffeine, sugar or dairy products, surgical procedures, allergies, tumours or intake of certain medicines. Persistent and chronic diarrheas often cause nutritional problems, creating the risk of malnutrition and serious non-intestinal infection. Dehydration also occurs.

Symptoms

Major symptom of diarrhea is passage of frequent watery stools. Other possible symptoms can also include stomach pain, bloating, abdominal cramps, blood or pus in the stools, persistent vomiting, general body weakness and fatigue *etc.* Two potentially serious complications of diarrhea are dehydration with acute or chronic diarrhea and malabsorption leading to malnutrition due to chronic diarrhea.

Causes

Diarrhoea is generally not a disease by itself, but a symptom of an underlying functional or organic disease. Various factors may cause acute or persistent or chronic diarrhoea.

I. Acute Diarrhea

Common causes are:

- **Infections of GIT:** Most cases of acute diarrhoea are usually caused by bacterial (e.g, *Salmonella, Shigella, E. coli etc.*), viral (e.g, rotavirus or norovirus), or parasitic (e.g, giardia) infection of the GIT which is also known as gastroenteritis. Bacteria and parasites can be ingested through contaminated food or water. They may release toxins that cause the intestines to secrete water and salt, making the stool runny.
- **Traveller's diarrhea:** Travel to places where the climate or sanitary practices differ from one's home country may cause short-term gastroenteritis (*e.g.* from contaminated food or water) and is known as traveller's diarrhoea.
- **Other infections due to poor hygiene and environmental sanitation:** For example diarrhea associated with cholera, typhoid *etc.*
- **Ingestion of environmental toxins:** For example, ingestion of insecticides from fruits and vegetables or eating poisonous mushrooms may also cause acute diarrhoea.
- **Artificial sweeteners**: Largely used as sugar substitutes, such as sorbitol are sometimes not absorbed into the bloodstream from the intestines, lead to runny stools due to osmotic diarrhoea.
- **Food allergies, faulty feeding practices**
- **Malnutrition:** Protein energy malnutrition, deficiency of vitamin A and B- complex vitamin
- **Medications**: There are many medications that can have diarrhea as their side effect. Antibiotics often disrupt the natural balance of microorganisms in the intestines, leading to digestive problems including diarrhea. Weight loss drugs, laxatives, antidepressants, blood pressure lowering agents and cancer medications can also cause diarrhea. Diarrhea, which is a side effect of medications, resolves once the aggravating substance is discontinued.

II. Chronic Diarrhea

The most common causes of chronic diarrhoea are underlying diseases or conditions, including:

- **Celiac disease:** Inability to digest gluten present in wheat and wheat products.
- **Lactose intolerance:** It is the inability to properly digest lactose - milk sugar present in dairy products. It is the main cause of osmotic diarrhea.
- **Irritable bowel syndrome:** In the developed world, irritable bowel syndrorme is the most common cause of functional diarrhea (clear cause is unknown) which is a complex of symptoms. There is cramping abdominal pain and altered bowel habits, including diarrhea, constipation or both.
- Chronic bowel infection
- **Inflammatory bowel diseases:** It is a term used to describe either Crohn's disease or ulcerative colitis. There is often blood in stool in both conditions.
- **Bowel cancer:** Neoplastic diarrhea is associated with a number of gut cancers.
- **Microscopic colitis**: This is a persistent diarrhea that usually affects older adults, often during the night.
- **Caffeine and alcohol:** In some cases, they stimulate the passage of stool, which can move too quickly through the digestive tract to allow for normal fluid absorption, resulting in watery stools.
- **Surgical procedures:** They may cause changes within the digestive tract that can result in diarrhoea. Examples include bowel resections, dumping syndrome (a common consequence of gastric surgery) and bacterial overgrowth as a result of intestinal surgery.
- **Motility disorders**: Problems with the movement of stool through the digestive tract can also lead to diarrhea. When food moves too quickly through the digestive tract, not enough fluids are absorbed from the food, leading to watery stool. Motility disorders can be due to scleroderma, irritable bowel syndrome, liver cirrhosis, viral hepatitis, pancreatic insufficiency, tuberculosis, stress, anxiety and antacids that contain magnesium.
- **Endocrine causes**: Sometimes hormonal factors cause diarrhea, for example, in the case of overactive thyroid gland, Addison disease and carcinoid tumors.

Prevention

Diarrhoeal disease is the second leading cause of death in children under five years old. Each episode deprives the child of the nutrition necessary for growth. As a result, diarrhoea is a major cause of malnutrition, and malnourished children are more likely to fall ill from diarrhoea. Diarrhoea is both preventable and treatable as a significant proportion of diarrhoeal disease can be prevented through safe drinking-water and adequate sanitation and hygiene. Hence, key measures to prevent diarrhoea include:

- ✰ access to clean and safe drinking-water;
- ✰ use of improved sanitation systems, for example, proper wastewater, garbage and sewage disposal;
- ✰ good hygiene practices, including hand washing with soap after defecation, after cleaning a child who has defecated, after disposing of a child's stool, before preparing food, and before eating;
- ✰ exclusive breastfeeding the infant for the first six months of life;
- ✰ promoting good personal and food hygiene during preparation, processing, cooking, serving, storage *etc.*;
- ✰ health education about how infections spread; and
- ✰ rotavirus vaccination.

Treatment of Diarrhoea

It consists of:

- ✰ Rest in bed
- ✰ Correction of water and electrolyte losses
- ✰ Intake of healthy well balanced diet devoid of fibre
- ✰ Therapy with suitable drugs

Nutritional and Dietary Management

Diarrhoeal diseases may cause, precipitate, or exacerbate protein-energy and micronutrient malnutrition through five possible mechanisms:

Reduced Food Intake

Reduction of food intake during diarrhoea may be due to anorexia, maternal food-withholding behaviour, or both. Anorexia is a consequence of clinical disturbances, including dehydration, electrolyte imbalance, fever, and vomiting or abdominal discomfort.

Decreased Absorption of Nutrients

Structural damage to the intestine, as well as the physical action of increased intestinal movement and reduced fluid transit time, interact to decrease absorption of nutrients during an episode of diarrhoea.

Increased Catabolic Losses

During the inflammatory process, muscle protein is converted into glucose through gluconeogenesis by the liver. This glucose is used as a fuel by the diseased tissues, immune system, liver, kidneys, brain, and tissues generally to sustain the hypermetabolism associated with the development of fever. Diarrhoea of infectious origin induces an average daily negative nitrogen balance of 0.9 g/kg/day.

Nutrient Loss from the Intestine

In diarrhea, nutrients are lost directly from the intestinal tract. An increased stool clearance of alfa-1-antitrypsin, a serum protein marker present in the stools, has been observed in about half of all rotavirus diarrhoeas and even more frequently in shigellosis infection, in which tissue destruction is accompanied by losses of plasma, epithelia, and blood cells. Nearly two-thirds of patients with enterotoxigenic *Escherichia coli* also show an excessive loss of protein in the faeces. Losses of zinc and copper during early stages of acute diarrhoea have been documented. These losses are greater when stool losses are larger.

Metabolic Inefficiency Due to Micronutrient Deficiency

An acute depletion of plasma zinc has been shown in marasmic children recovering from malnutrition when they were fed an energy-dense diet meant to promote growth. These children were in zinc balance before their energy intake was increased. Based on these observations, it is likely that zinc deficiency may be a limiting factor for children to recover the weight lost during an acute infection.

Hence, nutritional and dietary management should be as under in acute and chronic cases of diarrhea so as to maintain good nutritional status:

I. Acute Diarrhea

Acute diarrhea, a common symptom of gastro-intestinal infection, can be life-threatening if not treated and managed properly especially among the infants and young children. Diarrhoea may drain body fluids consisting of water and salts heavily from the child's body. Children can get dehydrated even more quickly if they are vomiting too. If these fluids are not replaced quickly, the child can become dehydrated and may need to be hospitalized. Symptoms of dehydration are increased thirst, very little urine, drying up of mouth and tongue, loss of skin elasticity, fast heart beat, sunken eyes, sunken soft spot (fontanelle) on the baby's head and absence of tears. Correcting dehydration is the priority of diarrhea treatment. Hence, key measures to treat diarrhea include the following:

At present, WHO, UNICEF, and several other international agencies recommend sustained feeding of the child with diarrhea.

Rehydration

Initially during the illness, if you are breastfeeding, keep feeding the infant on demand. If the child (above 6 months age) is able to take fluid orally and does not vomit, offer the complementary foods which provide more water to him which he usually takes. He should be given plenty of fluids at frequent intervals in the form of boiled and cooled water, coconut water, barley water, butter milk, *shikanjvi* (water+ lemon+sugar+salt), glucose water, orange juice, light tea, *dal* water *etc.* to prevent dehydration. In case of formula feeding, do not dilute the formula and continue feeding it. Normally, the child recovers and does not show the symptoms of dehydration if rehydrated properly following proper food hygienic practices in the clean environment.

As soon as diarrhoea begins, treatment using home remedies to prevent dehydration must be started. If adults or children of any age have not been given extra drinks, or if in spite of this dehydration does occur, they must be treated with a special drink made with oral rehydration salts (ORS). At home, ORS can also be prepared by adding one pinch of salt and one tsp (one fist) of sugar in one glass of boiled and cooled water.

If diarrhea is prolonged and the child is dehydrated, it is better to start oral rehydration therapy (ORT) by feeding ORS recommended by WHO/UNICEF. The formula for ORS recommended by WHO and UNICEF contains:

Sodium chloride	3.5 g
Trisodium citrate dehydrate	2.9 g
Or	
Sodium bicarbonate	2.5 g
Potassium chloride	1.5 g
Glucose (anhydrous)	20g

The above ingredients are dissolved in one litre of clean water. WHO has recently recommended a change in the complete formula, replacing 2.5 gms of sodium bicarbonate with 2.9 gms of trisodium citrate dihydrate. The new formula gives the packets a longer shelf life and is at least as effective in correcting acidosis and reducing stool volume. Packets containing sodium bicarbonate are still safe and effective. Oral rehydration therapy (ORT) based on the administration of correct oral fluids while allowing food intake, provides a balanced water and electrolyte replacement at low cost. Rehydration with fluids improves appetite, allowing better feeding and continued weight gain.

Table 22.2: Feeding Schedule of ORS for the First 4 hrs of Oral Rehydration (for mild rehydration)

Babies under 6 months of age	30 to 90 mL every hour
Children 6 months to 2 years of age	90 to 125 mL every hour
Over 2 years of age	At least 125 to 250 mL every hour

If the child refuses to take the ORS by the cup or bottle, give the solution using a medicine dropper, a syringe or a small teaspoon to prevent rapid passage of stool due to hyperactive gastro-colic reflex.

If the child vomits, stop foods and other fluids, but continue to give the ORS using a spoon.

- Give 15 mL (1 tbsp.) every 10 min to 15 min until the vomiting stops.
- Increase the amounts gradually until the child is able to drink the regular amounts.

Adults suffering from acute diarrhea should also take fluids in the form of home foods and ORS frequently during the day or at least one glass of ORS for each passage of stool to prevent dehydration.

Consulting the Doctor

If the child continues vomiting along with diarrhea for longer than 4 to 6 hours, take the child to the hospital. In the hospital, loss of nutrients and fluids must be replaced parenterally along with medication if required. Saline solutions are given to correct losses of salts through intravenous drip. If parenteral feeding is to be continued beyond 48 hours, protein hydrolysates should be given.

In case of acute diarrhea due to gastro-enteritis among adults, medicines should be taken in consultation with the doctor if the episode of diarrhea does not resolve itself within a day or two.

After 4 hrs until 24 hrs (Recovery stage)

- Keep giving the child ORS until diarrhea is less frequent.
- Continue feeding him the regular diet if he is not vomiting.
- When vomiting stops, breastfeed as usual, give formula feeding or whole milk or give regular foods (mashed banana, dehusked *dal*, rice, *khichdi*, curd *etc.*) in small, frequent feedings as the case may be, depending upon the age of the child.

Providing Nutrient-Rich Foods

Diarrhoea is an important cause of undernutrition. This is because nutrient requirements are increased during diarrhoea, as during other infectious diseases, whereas nutrient intake and absorption are usually decreased. Each episode of diarrhoea can cause weight loss and growth faltering. Moreover, if diarrhoea occurs frequently, there may be too little time to "catch up" on growth between episodes, the result being a flattening of the normal growth curve among children. Children who experience frequent episodes of acute diarrhoea, or have persistent diarrhoea, are more likely to become undernourished than children who experience fewer or shorter episodes of diarrhoea. In general, the impact of diarrhoea on nutritional status is proportional to the number of days a child spends with diarrhoea each year. In turn, undernutrition contributes to the problem of diarrhoea. In children who are undernourished as a result of inadequate feeding, previous diarrhoeal episodes, or both, acute diarrhoeal episodes are more severe, longer lasting, and probably more frequent;

persistent diarrhoea is also more frequent and dysentery is more severe. The risk of dying from an episode of persistent diarrhoea or dysentery is considerably increased when a child is already undernourished. In general, these effects are proportional to the degree of undernutrition, being greatest when undernutrition is severe. Thus, diarrhoea and undernutrition combine to form a vicious circle which, if it is not broken, can eventually result in death; the final event may be a particularly severe or prolonged episode of diarrhoea or, when severe undernutrition is present, another serious infection such as pneumonia, measles *etc.* may occur. Deaths from diarrhoea are, in fact, usually associated with undernutrition. Therefore, this vicious circle present between diarrhea and malnutrition should be broken by continuing to give nutrient-rich foods.

For centuries, the dietary approach for treating diarrhoea recommended by physicians and other health workers consisted of a variable period of food reduction or, more likely, fasting, and then a gradual reintroduction of foods and nutrients during the recovery period. This practice was meant to avoid the clinical consequences of diarrhoea-induced malabsorption, based on the assumption that withholding food will reduce stool volume and duration of diarrhoea. The practice was transmitted by physicians to mothers as a need to "keep the bowel at rest" during the bout of diarrhoea, and was incorporated into traditional home-based treatments of the illness.

At present, WHO, UNICEF, and several other international agencies recommend sustained feeding of the child with diarrhea. This recommendation is based on the rationale that an important proportion of nutrients offered during the diarrhoea episode are actually absorbed; therefore, it is more important to emphasize absorption during diarrhoea, disregarding the degree of malabsorption that does occur when nutrients are offered, thus shortening the period during which nutritional requirements remain unsatisfied.

Exclusive breast milk feeding to infants up to 6 months of age must be continued during an episode, and by giving a nutritious diet along with breastfeeding after six months of life to children when they can tolerate oral foods. Non breastfed infants who receive cow's milk should continue to receive it with no dilution if they are not lactose intolerant.

During the recovery phase, choice of foods, mode of preparation, and frequency of feeding should depend on the child's age, feeding history, and physiologic status. Particular attention should be given to the nutrient density of the food used, the quality of its carbohydrate and fat, the biologic value of its protein, and, if appropriate, its osmolarity. The carbohydrates rich foods which are chosen should be easily digestible. Sucrose, the major dietary disaccharide, and processed vegetable starches usually are easily digested and absorbed by children with diarrhea. Food preparation should not be bulky but it should be nutrient dense. Ideally, foods should provide twice the protein-to-calorie ratio that is normally required. No food is likely to satisfy this ratio for all essential nutrients at all ages, so supplementation or fortification with key nutrients might be needed for nutrient replenishment. Mixing of plant and animal foods often increases the nutrient-to-calorie ratio substantially. The need for specific nutrient supplements also must be considered. Meals should be small and frequent in the early phase, because smaller meals are better tolerated. For older infants, at least six feedings a day should be attempted early in treatment; more frequent feedings are often necessary in younger infants. Flexibility in the feeding regimen promotes rapid convalescence.

In case of young children, adolescents and adults, when the patient's condition improves, initially, soft diet devoid of fibre *e.g.* bread pudding, jelly, boiled egg, dehusked *dal*, refined cereal preparation, milk, fruit juice, custard, *khichdi*, banana, sapota *etc.* should be given till the patient is able to tolerate normal diet. If diarrhea is due to lactose intolerance, do not give milk in the diet but curd or buttermilk can be given. Mixing dried-milk preparations with vegetables or cereals also reduces the lactose content of the final diet potentially without reducing the biologic value of the milk protein substantially. Avoid spices and condiments till the condition of the patient improves. If adequate food storage facilities are not available in the home, use of freshly prepared foods to minimize microbial contamination and growth should be encouraged. Reoffering of previously prepared food might be unavoidable, but reheating to a boil before feeding will reduce the hazards of consuming heavily contaminated food.

Provide Zinc Supplements

Zinc supplements or foods rich in zinc can be included as the diet as zinc reduces the duration of a diarrhoea episode by 25 per cent and are associated with a 30 per cent reduction in stool volume.

2. Chronic Diarrhoea

Persistent or chronic diarrhea should be diagnosed and any underlying causes are to be treated in addition to the symptoms of diarrhea. For nutritional management of chronic diarrhea, a soft diet, rich in protein and calories with liberal amounts of vitamins and minerals and large amounts of fluid is recommended. Foods rich in fats, fibre and milk should be avoided.

Energy

To overcome loss of body weight and weakness, energy requirement is increased. Energy intake should be 10-20 per cent more than the normal requirement.

Protein

Protein intake should be increased by 50 per

cent so as to take care of negative nitrogen balance which occurs due to increased catabolic losses during chronic diarrhea. Animal protein generally is more digestible and of higher biologic quality than plant protein. However, processing techniques can be used to improve digestibility of vegetable protein, and proteins from different sources can be mixed to improve their combined food value. Someone can have allergy to certain protein *e.g.* milk, peanut, egg *etc.*, they should avoid such foods in the diet.

Carbohydrates

Carbohydrates should provide 55-60 per cent of the calories in the diet. The intake of easily digestible carbohydrates should be increased so as to take care of more amount of energy to be provided in the diet. In case of lactose intolerance, lactose containing foods *i.e.* dairy products should not be given. Sucrose worsens diarrhea in some cases, can be avoided. The other major dietary disaccharide present in processed and cooked vegetables, pulses and cereal starches usually is easily digested and absorbed during diarrhea, can be included in the diet.

In case of infants, although the feeding of lactose to infants with gastroenteritis might be considered unwise, because of the recognized loss of intestinal lactase activity during intestinal infection, this type of acquired lactase deficiency is seldom total. Lactose at a concentration equivalent to that of half-strength cow's milk is generally well tolerated in diarrhea, especially in mixed foods, such as milk and cereal combinations.

Dietary Fibre

Amount of dietary fibre especially insoluble fibre should be minimal *i.e.* 1-2 g per day. The potentially adverse effects of dietary carbohydrate can be minimized by multiple feedings of small amounts of mixtures of carbohydrates. This strategy is the least likely to strain possibly decreased capacity for carbohydrate digestion.

Fat

Fats are not properly digested and absorbed due to increased motility of the intestines. Their intake should be limited. Medium chain triglycerides are utilized better as they can be directly absorbed through the lymphatic system. Fats like butter, coconut oil, whole milk *etc.* can be given.

Minerals

Increase the amount of foods rich in salt and potassium as they are lost during diarrhea. Acute depletion of plasma zinc occurs in chronic diarrhea. Absorption of calcium and iron is also poor.

Vitamins

Water soluble B-complex vitamins are lost along with the body fluids in the stools. Their loss should be compensated by increased intake through the diet. During diarrhoea, vitamin A absorption is reduced and greater amounts are used from body stores. Sufficient amount of vitamin A and beta carotene rich foods should be provided.

Dietary Management

In addition to the nutritional concerns described above, preparation of food should consider consistency, digestibility, and acceptability. Dilute watery paps or soups and excessively bulky foods should be avoided as primary nutrient sources if possible. When formulated appropriately, watery paps or soups can be used as adjuncts to fluid and electrolyte therapy, but never as a complete diet, because the concentrations of energy and nutrients in these preparations are usually too low.

- ☆ Foods should not be withheld during diarrhea. Continue feeding of tolerable foods during episodes of diarrhea.
- ☆ Start with fluid diet initially and then progress to soft bland diets later. *Khichdi*, puddings, curd, breads, soups, fruit cream, banana, moong *dal*, custard, smoothies *etc.* can be given.
- ☆ Increase the amount of fluids. Sip fluids slowly, but constantly, all day. Consuming large amounts of fluid at once may encourage diarrhea. Try to drink 8 to 10 glasses of fluid daily. Fluids at room temperature may be better tolerated than hot or cold fluids. Fluids in the form of water, ORS, fruit juices, coconut water, lemon water with sugar and salt, rice kanji, vegetable soups without skin and seeds, clear chicken soup, fruit nectars, light tea *etc.* can be taken. After each loose stool, replace lost fluids with at least one cup of liquid.
- ☆ Drink fluids between meals.
- ☆ Increase calories. The diet should not be bulky but energy dense. High energy foods such as fats, yogurt and cereals are quite well absorbed during diarrhoea. One can take well-cooked eggs, chicken or fish that is not fried, bananas, avocados, soups made from clear pudding, custard, curd, cottage cheese (unless lactose intolerant), pancakes, biscuits, waffles, and sherbet.
- ☆ In case of children, small, frequent feeds of energy-rich local foods familiar to the child should be given. A little vegetable oil can be added to foods such as millet or rice to increase the energy content.
- ☆ Eat small, frequent meals and snacks. Do not eat large amounts of food at once, because this can worsen diarrhea. Have small snacks often all day long.
- ☆ Eat foods with probiotics. Probiotics are found naturally in all yogurt products and have shown to be beneficial for various types of diarrhea, including chemotherapy and/or radiation induced diarrhea.

Table 22.3: Foods to be Included and Avoided in the Diet

Food Group	*Better Tolerated Foods*	*Food to Limit or Avoid*
Grains	Bread, crackers, and pasta made from white or refined flour	Whole wheat or whole grain breads, rolls, crackers, or pasta
	White rice	Brown rice
	Cream of wheat, cream of rice	Barley, oats, and other whole grains
	Cereals made from refined grains (puffed rice, corn flakes) without added fiber.	Cereals made from whole grain or bran
		Breads or cereals made with seeds or nuts
		Popcorn
		Tip: Choose grain foods with less than 2g dietary fiber per serving.
Legumes	Dehusked *dals* like moong *dal*	Whole pulses, beans, peas, lentils
Fruits and Vegetables	Fruit juice without pulp	Most raw fruits
	Ripe bananas	Dried fruits, including prunes and raisins
	Melons	Fruit juice with pulp especially citrus juices and beverages such as oranges, orange juice, grapefruits, grapefruit juice, lemons, limes, lemonade, limeade, pineapple and pineapple juice, and tomato products may irritate the stomach and worsen diarrhea.
	Fresh Apple, Pear, or Peach, if peeled	Canned fruit in heavy syrup
	Canned soft fruits	Any fruits sweetened with sorbitol
		Prune juice
	Most well-cooked vegetables without seeds or skins	Most raw vegetables
		Fried vegetables
	Strained vegetable juice	Beets, Broccoli, Brussels sprouts, Cabbage,
	Potatoes without skin	Cauliflower
		Collard, mustard, and turnip greens
		Whole corn kernels
		Potato and sweet potato skins
Meat, poultry, fish	Tender, well-cooked meat, poultry, fish, eggs *etc.* made without added fat	Fried meat, poultry, or fish
		Sausage, Bacon, Hot dogs
		Fatty, gristly meats
Nuts and oilseeds	Smooth nut butters (if tolerated)	Whole nuts, Chunky nut butters
Dairy	Buttermilk	In case of lactose intolerance, drinking milk products may aggravate diarrhea.
	Skim, low-fat and evaporated milk	Avoid whole milk ice cream, sour cream and other milk products.
	Soy/almond/rice/coconut milk	Avoid yogurts with nuts or dried fruit.
	Yogurt with live active cultures of probiotics	**Tip:** Try avoiding lactose, if someone is sensitive to dairy. Opt for lactose-free milk, yogurt, cheese, *etc.*
Beverages	Water	Beverages containing caffeine, including regular coffee, regular tea, colas, and energy drinks
	Broth	Limit or avoid beverages containing high fructose corn syrup or sorbitol
	Decaffeinated coffee	Alcoholic beverages
	Caffeine-free teas	**Tip:** Healthy people need 8 to 10 cups of fluid each day. You may need to drink more to replace fluids lost to diarrhea.
	Dilute fruit juices or nectars	
	Oral rehydration beverages	
Fats/Others	Fats include oil, butter, cream, cream cheese, margarine, mayonnaise and salad dressings.	Fried foods, sausage, pizza, and gravies. Salad dressings, mayonnaise, rich desserts, pastries, donuts, and potato chips may also aggravate diarrhea in some.
		Tip: Limit fats to less than 8 teaspoons a day.
		Avoid sugar alcohols such as xylitol and sorbitol; honey

- ☆ Avoid insoluble fibre rich foods *e.g.*, raw vegetables and the skins, seeds, and stringy fibers of unpeeled fruits.
- ☆ Increase foods high in soluble fiber found in foods like oatmeal, barley, rice, bananas, white toast, apple sauce, mashed potatoes *etc.* as these help to thicken the stool
- ☆ Include foods rich in sodium and potassium such as salted biscuits, salted buttermilk, dehusked lentils, bananas, mangoes, pineapples, coconut water, coconut milk *etc.*

- Limit foods and beverages that contain sugar, lactose, fructose, high-fructose corn syrup, and sorbitol. Avoid beverages with caffeine.
- Avoid greasy, spicy, highly seasoned and very sweet/sugary foods if they make symptoms worse. Not all diarrhea is a result of lactose-intolerance. However, sometimes cancer treatments, such as radiation can lead to changes in the digestive tract that can lead to lactose intolerance. If you notice that you are more sensitive to some dairy products, avoid dairy products. Cultured milk products such as yogurt, buttermilk and hard cheeses may still be tolerated, even though they are not lactose free.
- Avoid processed foods.
- Avoid smoking and alcoholic drinks.

Flatulence

Flatulence has been derived from the latin word 'flatus' which means 'a blowing, a breaking wind'. Flatulence is a buildup of excessive air or gas in the stomach and/or intestines associated with motility disorder of the intestinal tract that can lead to abdominal discomfort. It is commonly known as farting, passing wind, or having gas too. Gas can be expelled from the intestinal tract either as belching from the stomach or as flatus from the rectum with sound and/or odour. In the medical term, it is defined as 'flatus expelled through the anus' or the 'quality or state of being flatulent'.

It is normal for an average human to pass flatus through the rectum between 10 and 18 times a day; the volume and frequency may vary between the individuals. People often pass gas without noticing. Healthy gas is harmless and has no odour. It is also normal for intestinal gas passed through the rectum to have a characteristic feculent smell, although this too may vary in concentration. Flatus is brought to the rectum by specialized contractions of the muscles in the intestines and colon.

Causes

Causes of excess gas in the digestive tract are due to:

- **Swallowing excessive air:** It is normal to swallow some air through eating food. But excessive air is swallowed with food or drink due to faulty eating habits or through swallowing saliva, especially when excess saliva is produced due to nausea or acid reflux. Reasons that one may swallow more air than normal include chewing gum, smoking, sucking on objects such as pen tops, drinking carbonated drinks, and eating too quickly. This swallowed air consisting mainly of nitrogen and oxygen, accumulates in the gut. The swallowed air is usually released by burping. Whatever is not released by burping, enters into the small or large intestine, where it is released as flatulence.
- **Breakdown of undigested and unabsorbed foods:** The endogenous gases are produced either as a by-product of digesting certain types of food, or due to incomplete digestion. Undigested and unabsorbed food moves from the small intestine to the large intestine. Once it gets there, the bacteria go to work, making hydrogen, carbon dioxide, and methane, which then leave through anus in the form of flatus.
- **Intake of large amount of foods rich in fibre:** The fermentation of such foods occurs in the large intestine where they are acted upon by bacteria. Much of the gas generated is due to microbial breakdown of such foods so that gases, for example, hydrogen, carbon dioxide, and methane are generated; the odor is from other trace waste gases or compounds such as skatole and sulfur-containing substances.
- **Intake of certain flatulence producing foods:** Flatulence-producing foods are typically high in certain polysaccharides, especially oligosaccharides such as inulin. Those foods include beans, lentils, dairy products, onion, garlic, spring onion, turnips, radish, oats, wheat and yeast in breads. Foods belonging to genus *Brassica e.g.* cauliflower, cabbage, broccoli and other cruciferous vegetables not only increase flatulence, but are also known to increase the pungency of the flatus.
- **Food poisoning:** When certain food or water infested or infected with certain bacteria is consumed. Some infections, such as giadiasis, are also associated with flatulence.
- **Lactose intolerance:** When lactose intolerant persons consume large amount of dairy products, intestinal bacteria feeding on lactose can give rise to excessive gas production.
- **Malabsorption of certain foods:** For example in case of celiac disease and steatorrhea.
- **Other potential conditions underlying flatulence:** These may range from temporary conditions to digestive problems. Some of these conditions include constipation, gastroenteritis, irritable bowel syndrome, Crohn's disease, diabetes, eating disorders, ulcerative colitis, dumping syndrome, gastroesophageal reflux disease, autoimmune pancreatitis, peptic ulcer *etc.*

Symptoms

Flatulence is linked to a variety of symptoms, some of which may be distressing. Symptoms of excessive gas or flatulence may include belching, abdominal bloating, abdominal distension and/or abdominal pain or discomfort, excessive flatus volume, excessive flatus smell and gas incontinence. Excessive gas or flatulence usually is not a medical emergency; however medical care should be sought quickly if a person develops other symptoms along with flatulence such as severe cramps, swollen abdomen, unintentional weight loss, heartburn, diarrhea, constipation, bloody stools, fever, nausea and vomiting and/or right-sided abdominal pain *etc.*

Treatment

There are no long-term consequences for not treating flatulence. If the flatulence is due to a food intolerance or digestive issue, the problem may get worse. Other symptoms may also develop.

In some cases, prolonged excessive flatulence can lead to other issues, such as social discomfort and changes in eating habits. If it affects lifestyle a lot, it may also affect your mood. It is important to maintain a healthy diet and to see the doctor if the problem begins to negatively affect life. More serious causes of excessive flatulence (for example, irritable bowel syndrome (IBS) and small intestinal bacterial overgrowth (SIBO) may require additional medication and testing.

Dietary Considerations

Lifestyle changes can often reduce excessive flatulence. Most individuals that alter their diet can reduce or prevent excessive flatulence and thus have a good prognosis. Excessive flatulence can be reduced or prevented with several methods such as:

1. One should modify the eating habits.
2. **Look at the routine diet and keep a food diary:** As we all react to certain foods differently, sometimes a diet change is needed to prevent flatulence. Dietary changes should be made since chronic flatulence is often caused by certain foods that can be eliminated from the regular diet. Keep a food diary to identify the foods that cause excessive flatulence. This will help to identify any triggers in the diet, if present and one can avoid them. If the diet contains a large amount of carbohydrates that are difficult to digest, try to replace them. For example, some of the legume oligosaccharides of the raffinose family of sugars such as raffinose, stachyose and verbascose containing α galactosidic groupings are well known to be flatus producing in man. It has been emphasized in many research studies that removal of all raffinose oligosaccharides from beans even after cooking does not remove the problem of flatulence in man. Due to these oligosaccharides, many persons get flatulence problem especially after consuming *urd* bean, soybean *etc.* Carbohydrates that are easier to digest, such as potatoes, rice, and bananas, are good substitutes. One should avoid dairy products in case of lactose intolerance. People suffering from celiac disease are intolerant to gluten. Intolerance to the protein gluten means that some people have excessive flatulence when they consume barley, wheat, and rye and they should avoid such foods.

   Some of the common foods that cause flatulence include vegetables like cabbage, onion, cauliflower, radish, broccoli, garlic *etc.*, raisins, prunes, apples, soft drinks, fruit drinks and juices high in fructose or sorbitol, beans, whole grain foods (wheat, oats) *etc.* and avoid the food that is the causative agent of gas.

   According to the International Foundation for Functional Gastrointestinal Disorders, some foods that are less likely to cause gas include:

   - Meat, poultry, and fish
   - Eggs
   - Vegetables such as lettuce, tomatoes, cucumber, and okra
   - Fruits such as cantaloupe, grapes, berries, cherries, avocado, and olives
   - Carbohydrates such as gluten-free bread, rice bread, and rice
3. **Take less fibrous foods and fibre supplements:** Food should not contain high amount of fibrous vegetables, fruits, pulses, grains *etc.* Foods rich in non-available carbohydrates should be restricted. Unfortunately, this may also result in having fewer nutritious foods. Adding fibre supplements too rapidly to the diet can cause flatulence, especially if they contain psyllium and one should avoid them.
4. **Spices and condiments:** Avoid excessive consumption of spices and condiments in the daily diets.
5. **Avoid fried, greasy foods:** Fried or fatty foods should be avoided as they stay longer in the stomach leading to fermentation.
6. **Eat less but frequently:** Avoid overeating. Try to eat around five to six small meals a day instead of three larger ones to help the digestive process.
7. **Eat slowly and chew properly:** Donot eat fast but try to chew the food properly and slowly to avoid swallowing of air.
8. **Avoid artificial sweeteners**: Sorbitol and mannitol are found in candies, chewing-gums, and sugar-free sweet foods. Some people develop diarrhoea, gas or both when they consume these substances.
9. **Avoid air swallowing during eating and drinking:** One should avoid smoking and drinking through straw.
10. **Physical exercise:** Some people find that exercising helps to promote digestion and can prevent flatulence.

Gastritis

Gastritis is an inflammation and erosion of the stomach lining. This usually happens after the gastric mucosa has been damaged.

How Does Gastritis Occur?

The food we eat passes down the oesophagus into the stomach. The stomach creates acid which is necessary to digest the food and to kill any germs. The food mixes with the acid in the stomach and then moves into the first part of the small intestine. Since the acid is corrosive, some of the cells in the inner lining of the stomach produce mucous as a natural barrier to protect the lining of the stomach and the first part of the small intestine. Under normal conditions, there is a perfect balance between the amount of acid produced and the amount of mucus produced to protect the stomach lining. Gastritis occurs when there is an imbalance between the acid, and the mucous produced, and the acid manages to erode and damage the stomach lining.

Causes

The imbalance between the acid and mucus production can occur due to many reasons such as, bacterial infection especially caused by the *Helicobacter pylori* bacteria, regular consumption of pain killers, drinking too much alcohol regularly, overindulging in smoking, eating very spicy foods, overeating, certain food poisonings, recurring bile reflux from the small intestine, ageing, critical injury or major surgery *etc.* At times, gastritis can also be caused due to an autoimmune reaction, *i.e.* when the body's immune system instead of protecting the body, mistakenly attacks the body's own cells and tissues, in this case, the stomach lining, and cause gastritis. This is especially common among people who have an autoimmune condition such as Hashimoto's thyroid disease or diabetes type 1. In older adults, the stomach lining tends to thin with age and older people also tend to suffer from autoimmune disorder. Less common causes of chronic gastritis include Crohn's disease, IBS, sarcoidosis, food allergies and other types of fungal, bacterial, or viral infections.

Types of Gastritis

Based on the different causes there are various types of gastritis such as:

Acute Gastritis

It is a sudden inflammation or swelling in the lining of the stomach but is often temporary. It can cause indigestion, stomach irritability, severe and nagging pain but these noticeable symptoms resolve without treatment after a few days. Acute gastritis occurs due to injury, bacteria, viruses, old age, stress or ingesting irritants such as the overuse of Non-Steroidal Anti-Inflammatory (NSAID) drugs, alcohol abuse, spicy foods, corticosteroids, chemotherapy and a myocardial infarction.

Chronic Gastritis

Chronic gastritis, on the other hand, comes on more slowly and lasts longer. It occurs when stomach lining becomes inflamed repeatedly or for an extended period of time. Chronic gastritis erodes away the stomach lining. Chronic gastritis might cause more of a consistent dull ache than the more intense pain of acute gastritis. The symptoms of chronic gastritis include upper abdominal pain, bloating, vomiting, loss of appetite, and indigestion. Possible risk factors for chronic gastritis include diets high in salt or preservatives, diets high in fat and oil, especially saturated and trans fats, smoking, long term consumption of alcohol, conditions that weaken the immune system, cocaine use, long term use of NSAIDs and medications for acid reflux and indigestion.

There are different types of chronic gastritis:

Type A chronic gastritis is caused by dying stomach cells and can increase the risk of cancer, anemia, and vitamin deficiencies.

Type B chronic gastritis is caused by the *Helicobacter pylori*, and is the most common out of the three types.

Type C chronic gastritis is caused by chemical irritants such as anti-inflammatory drugs, alcohol, or bile and can lead to stomach lining erosion.

Depending upon some other causes, there can be certain other types of chronic gastritis too *e.g.* atrophic gastritis, antral gastritis (old persons), erosive gastritis, autoimmune gastritis, alcoholic gastritis and duodenitis *etc.*

Symptoms

The symptoms of gastritis may include ingestion, a burning ache or gnawing feeling in the stomach, nausea, vomiting, belching, loss of appetite, upper abdominal pain or discomfort and bloating, incessant hiccups, burping, bleeding, usually only in erosive gastritis, changes in bowel movements and the appearance of stools, such as darker stools or bloody coloured stools *etc.*

Complications

People with minor cases of gastritis that are caused by the bacterium *Helicobacter pylori* may not always notice any symptoms. If properly treated, acute cases of gastritis are rarely associated with complications. However, people may experience serious health complications if they have severe or untreated chronic gastritis. Erosive gastritis can cause peptic ulcers. Once an ulcer has formed, they can progressively degrade the surrounding tissues, widening and enlarging themselves. Severe ulcers may eventually cause internal bleeding, which can be life-threatening if left untreated. Other possible complications of chronic gastritis may include:

- ✰ Anemia caused by iron deficiency

- Anemia caused by internal bleeding
- Vitamin B_{12} deficiency
- Abnormal stomach growths, such as polyps and tumors

Treatment

For most people, gastritis is minor and symptoms of acute gastritis go away without any treatment. For chronic gastritis, medications and diet are the most common ways of treatment. Treatment for each type of chronic gastritis should focus on the cause of gastritis. Treatment may include:

- Avoid frequent use of drugs that can irritate the stomach lining such as NSAIDs or aspirin. Consume such medicines with food and water to avoid symptoms.
- Preventing acute gastritis by following food hygiene practices.
- Lifestyle changes

Nutritional Management

In case of acute gastritis, meeting the requirements of various nutrients are of secondary importance as the primary emphasis is to relieve the symptoms by withholding the foods for some time to heal the inflamed stomach mucosa. A chronic gastritis diet should also first begin with a fast for two to three days. One may only drink fruit juices in these three days as this shall allow the inflammation to subside and also help to eliminate the excess toxins from the body. After that liquid diet, bland diet without irritant fibre rich foods, high sugar, high salted and high fatty foods, spices and condiments, caffeine, carbonated and alcoholic drinks *etc.* can be given depending upon the severity of symptoms of gastritis. Once the symptoms are over, one can embark upon regular balanced diet containing various nutrients as per RDA for that age group. In the mean time, the cause of gastritis should be diagnosed and avoided.

In case of chronic gastritis, requirements for energy, protein, fat and digestible carbohydrates remain the same as that for the normal person. Indigestible carbohydrates should be avoided when the symptoms of gastritis are intense. Vitamin and mineral supplements can be added to ensure adequate intake of antioxidant vitamins- A, E and C, vitamin B_{12}, iron and folic acid.

Dietary Considerations

Since gastritis occurs in a person's stomach, diet does play a huge role in helping a person take care of this problem as well as avoid it. There is no single diet that will cure gastritis. However, embracing the dietary changes that soothe inflammation can support a treatment regimen. The foods consumed are an important part of gastritis treatment. Here are some tips on what to eat and what to avoid:

Avoiding Foods that Cause Gastritis

When suffering from gastritis, it is better to avoid certain foods that include:

- Deep fried foods, fatty foods, pickled, and salted foods. Foods that are high in fat may worsen inflammation in the lining of the stomach.
- Vinegar, acidic foods such as tomatoes and some fruits, pickles, peppermint, mustard, hot peppers, nutmeg, black pepper, red chili powder, or other spicy foods.
- Alcoholic beverages, carbonated sweet drinks and fruit juices (orange, grapefruit juices).
- Red meat and preserved meats.
- Dairy foods made from whole milk and strong or spicy cheeses, chocolates *etc.* Most people think that dairy is a good choice for soothing an upset stomach and blocking the effects of acids. However, due to its calcium and amino acid contents, may actually stimulate the release of more acid production, making the symptoms of gastritis worse.
- Strong coffee and tea. Evidence does not support the role for specific foods including spicy foods and coffee in the development of gastritis and ulcers but people should usually avoid foods if these foods trigger the problem in their cases.
- Allergenic or symptomatic foods.
- Refined and processed foods containing high fat and simple sugar contents such as sweets, pastries, rich cakes, fast foods, preserved foods, canned foods *etc.* as these can trigger food allergies and increase inflammation in the gut due to the presence of artificial colours and preservatives. Moreover, such foods with high fat act on retardation of gastric emptying and the broths with large quantities of purine raise the acid secretion. Such foods are rich in simple carbohydrates too. Therefore, concentrated carbohydrates and foods with high simple sugar and high fat are not recommended.

Each person affected by gastritis will react to foods differently, so not all foods on the above list may apply to everyone. Individual variations can be there, hence one should avoid those foods as listed above, if in their case by limiting or avoiding them, they experience improvements in the condition.

Foods that can be Included in the Diet

There are some foods that may help to manage gastritis and lessen the symptoms. These include:

- **Eat high fiber foods**: Fiber has proven to be beneficial for reducing gastritis and other digestive disorders. Opt for foods which are high in fiber such as vegetables, fruits, cereals, and whole grain. Good sources of fiber include apples, oatmeal, broccoli, carrots, and beans, nuts like almonds, seeds like chia or flax, soaked and cooked legumes/beans, and sprouted whole grains. They reduce the risk of developing ulcers and help in the healing process.

- **Take diet rich antioxidant foods**: Food items high in vitamin C, vitamin A, and flavonoids have been shown in prior research to help lower stomach inflammation and reduce the risk of digestive disorders. The best sources of antioxidants are brightly colored fresh fruits and vegetables. Foods that are especially beneficial for gastritis include onions, garlic, squash, bell peppers, leafy greens, asparagus, fennel, ginger, turmeric, cruciferous vegetables, berries, apples, and cranberries. Carrot juice (6 parts) in combination with the juice of spinach (10 parts) is considerably highly beneficial in the treatment of gastritis. Such foods hinder the growth of *H. pylori*.
- Take foods with low acidity, or those which are more alkaline, like vegetables.
- **Take probiotic foods**: Probiotic foods such as curd, buttermilk *etc.* help to reduce inflammation, regulate bowel movements, and control reactions to food allergies. Probiotic bacteria such as *Lactobacillus bulgaricus* may even be able to regulate how much stomach acid is produced, effectively reducing gastritis symptoms. Some studies show that probiotics may help with *Helicobacter pylori* (*H. pylori*) that cause an infection in the digestive system which can lead to gastritis or stomach ulcers.
- Low-fat or lean meats such as fish, chicken, and turkey breast.
- **Take healthy fats and proteins**: Healthy fats and proteins can help repair the gut wall and reduce inflammation-like symptoms. Good sources of omega-3 fatty acids such as walnuts, flax seeds, fish such as salmon or sardines should be included in the diet as they help to keep inflammation at bay and be beneficial for gastritis sufferers. Other healthy fats include olive oil, avocado *etc.* Plant based proteins like beans, tofu *etc.* should be taken. One can take soft boiled or scrambled egg.
- Drinks that are not carbonated. One can take coconut water as it provides the stomach with the necessary vitamins and minerals required for healing.
- Drinks without caffeine
- **Easily digested foods:** A diet for gastritis should include foods that can be easily digested *e.g.* steamed, well cooked and soft foods. Salt intake should be reduced as it works as an irritant.

Feeding Pattern

- Do not eat if appetite is lacking.
- For the first two days, it is better to keep fasting for healing the inflamed mucosa in chronic gastritis. Then incorporate coconut water, potato juice, rice gruel *etc.* in the diet. After the intense symptoms subside, it would be advisable to follow a regular balanced but bland diet devoid of foods that trigger the symptoms.
- Refrain from large meals, instead have small frequent (5-6) meals evenly spaced through the day.
- Too many different foods should not be mixed at the same meal.
- Meals should be taken at least two hours before going to bed at night.
- Never eat to full stomach.
- Hydrate often by frequently consuming water. Eight to ten glasses of water should be taken daily but water should not be taken with meals as it dilutes the digestive juices and delays digestion.
- And above all, haste should be avoided while eating and meals should be served in a pleasing and relaxed atmosphere.

Other Lifestyle Modifications

- Quite smoking.
- Stop or limit alcohol consumption.
- The patient should not undertake any hard physical and mental work. He should, however, undertake breathing exercises and other light exercises like walking, swimming and golf *etc.* He should avoid worries, anxiety and any other such kind of emotional stress but practice yoga and meditation.
- Medication in case of infection by *H. pylori.*
- Take probiotic, multivitamin and mineral supplements especially for vitamin B_{12}, iron and folic acid in consultation with the doctor.

Irritable Bowel Syndrome (IBS)

Irritable bowel syndrome (IBS) previously also known as spastic colon, irritable colon, mucous colitis, and spastic colitis is one of the most common ailments of the bowel (intestines). IBS is a functional disorder, meaning it is a problem caused by changes in how the gastro-intestinal tracts works but GIT is not damaged. IBS does not cause changes in the bowel tissue or increase the risk of colorectal cancer. The concept of functional disease applies to the muscular organs of the gastrointestinal tract; the esophagus, stomach, small intestine, gallbladder, and colon. In functional diseases, either the muscles of the organs or the nerves that control the organs are not working normally, and, as a result, the organs do not function normally. Hence, IBS is not a disease; it is a group of symptoms that occur together.

Symptoms

The most common signs and symptoms include cramping pain after eating certain foods, abdominal pain or discomfort, bloating, gas and changes in person's bowel habits. They may include diarrhea (IBS-D) or constipation (IBS-C), or both with one occurring after the other. The most common symptoms of IBS-D include frequent loose stools, sudden urge to have a bowel movement, nausea, gas, abdominal pain or discomfort. IBS-C stands for irritable bowel syndrome with constipation and symptoms may include infrequent stools, straining during bowel movements, feeling if one is unable to completely empty during bowel during bowel movement, abdominal pain, bloating, gas *etc.* It is not uncommon for people with IBS to have episodes of both constipation and diarrhea. Symptoms such as bloating and gas typically go away after having a bowel movement.

Although IBS is not dangerous, the symptoms can be very painful and bothersome. The symptoms vary in severity and duration from person to person. However, they last at least three months for at least three days per month.

Causes

The exact causes of IBS are not well understood. Researchers believe a combination of physical and mental health problems can lead to IBS. The possible causes of IBS may include alteration in the gastrointestinal (GI) tract motility (too fast or slow, or too strong), abnormal nervous system signals-poorly coordinated signals between the brain and the intestines, increased sensitivity to pain from a full bowel or gas, inflammation in the intestines, food intolerances or sensitivity possibly caused by poor absorption of sugars or acids in food, small intestinal bacterial overgrowth *etc.* Reproductive hormones or neurotransmitters may be off-balance in people with IBS. Many women find that signs and symptoms are worse during or around their menstrual periods.

Psychological conditions such as stress, anxiety or depression are observed in many people with IBS, though these conditions have not been found to be a direct cause of IBS. Gastroenteritis may trigger IBS symptoms. Genetics is thought to be a possible cause of IBS, but so far this hereditary link has not been proven.

Nutritional and Dietary Management

There is no known cure for this condition, but there are many treatment options to reduce or eliminate symptoms. Treatment includes dietary modifications, lifestyle changes, medication and counselling.

Follow Elimination Diet

Symptoms of IBS are not always persistent. They can resolve, only to come back. However, some people do have continuous symptoms. For some people, dietary changes can go a long way in helping ease symptoms. Because the symptoms of IBS vary among people with the condition, approaches to dietary changes need to vary. There is no specific diet for IBS, and different people react differently to different foods. But many people have worse IBS symptoms when they eat or drink certain foods or beverages, including foods high in fat, high in protein, fried foods, processed foods, wheat, rye, barley dairy products, citrus fruits, cabbage, beans, chocolate, drinks with caffeine, milk, alcohol and carbonated drinks. Keeping a diary for a few weeks may be a good way to find out if a food bothers someone. It is very important for people with IBS to identify foods that trigger their symptoms so they can avoid them.

An elimination diet should focus on avoiding certain foods for an extended period of time to see if IBS symptoms improve. The International Foundation for Functional Gastrointestinal Disorders (IFFGD) recommends cutting out these four common culprits *i.e.* coffee, chocolate, insoluble fiber, and nuts. Completely eliminate one food from diet for 12 weeks at a time. Note any differences in IBS symptoms and move on to the next food on the list of suspected foods. Some people with a sensitivity or intolerance to gluten also experience IBS. In such cases, a gluten-free diet may reduce symptoms. They should eliminate barley, rye, and wheat products from the diet. Lactose intolerant people should avoid dairy products.

Avoid High-Fat Foods

High-fat foods are generally low in fiber, which can be problematic for IBS-related constipation. According to researches, fatty foods are particularly bad for people with mixed IBS, which is characterized by a combination of constipation and diarrhea. Avoid fried and spicy foods. Embarking on a low-fat diet is good for heart and may improve uncomfortable bowel symptoms. Instead of eating fried foods and animal fats, focus on lean meats, fruits, vegetables, grains and low-fat dairy products.

Fibre Intake and IBS

In general, many people with IBS find it helpful to increase dietary fibre, drink plenty of water and eat smaller meals. Large meals can cause cramping and diarrhea, so eating smaller meals more often, or eating smaller portions, may help IBS symptoms. Eating meals that are low in fat and high in carbohydrates, such as pasta, rice, whole-grain breads and cereals, fruits, and vegetables may help.

Fibre can be helpful as it helps in the bowel movement and helps in holding water to make stools soft and easy to pass. Soluble fibre helps both diarrhea and constipation. It dissolves in water and forms a gel-like material. Many foods such as barely, oatmeal, guava, apple, berries, carrots, beans and citrus fruits contain soluble fibre. Insoluble fibre present in whole grain breads, wheat bran, nuts, tomatoes, raisins and many vegetables increase the fibre in the diet slowly and are helpful in IBS-C. Some people feel bloated and have gas if they increase their fibre intake too quickly. Gas and bloating usually improve as one gets used to eating more fibre especially soluble fibre found in fruits and vegetables instead of grains. The best way to increase fibre intake is to eat a wide variety of high fibre foods. Gradually boost the fibre intake by 2 to 3 g per day until one takes 25-38 g per day.

FODMAP Diet and IBS

One should follow FODMAP diet. This diet cuts out many common foods that may contain high FODMAP foods. The acronym stands for "fermentable oligosaccharides, disaccharides, monosaccharides, and polyols." FODMAPs are short chain carbohydrates and sugar alcohols that are difficult for the intestines to digest. They ferment in the large intestine (bowel) during digestion, drawing in water and producing carbon dioxide, hydrogen, and methane gas that causes the intestine to expand. This causes GI symptoms such as bloating and pain that are common in disorders like IBS.

FODMAPs occur in some foods naturally or as additives. They include fructose (in fruits and vegetables), fructans (like fructose, found in some vegetables and grains), lactose (dairy), galactans, (legumes), and polyols (artificial sweeteners). These foods are not necessarily unhealthy products. Some of them contain fructans, inulin, and galactooligosaccharides (GOS), which are healthy prebiotics that help stimulate the growth of beneficial gut bacteria. Many of them are otherwise good but in certain people, eating or drinking them causes gastrointestinal symptoms. Examples of foods that contain FODMAPs include:

- Wheat and rye products
- Products, including candy and gum, with sweeteners ending in "–ol," such as sorbitol, mannitol, xylitol, and maltitol
- Nuts such as cashew nuts, pistachios *etc.*
- Fruits such as apples, apricots, blackberries, cherries, mango, nectarines, pears, plums, and watermelon, or juice containing any of these fruits
- Canned fruit in natural fruit juice, or large amounts of fruit juice or dried fruit
- Vegetables such as asparagus, beans, cabbage, cauliflower, garlic and garlic salts, lentils, mushrooms and onion
- Dairy products such as milk, milk products, soft cheeses, yogurt, custard, and ice cream
- Honey and foods with high-fructose corn syrup
- It is important to note that not all carbohydrates are FODMAPs. Temporarily restricting or limiting intake of high FODMAP foods for six to eight weeks may improve symptoms of IBS. If symptoms improve by restricting certain FODMAP food, limit its intake or avoid it for few weeks and then slowly add that food back into the diet. One may be able to eat some foods with FODMAPs without having IBS symptoms. It is not meant to be a permanent solution because it is very restrictive, but it may work well enough to be a treatment for people with gastrointestinal (GI) problems.

Life Style Changes

Examples of these lifestyle changes include participating in regular physical exercise, minimizing stress, avoid smoking and big meals at a time.

Inflammatory Bowel Disease (IBD)

Inflammatory bowel diseases (IBD) are autoimmune diseases that affect both the large and small intestine, and they can manifest as either Crohn's disease or ulcerative colitis. IBD can be debilitating and sometimes leads to life-threatening complications.

Crohn's Disease

This type of IBD is characterized by inflammation of the lining of the digestive tract (from mouth to anus) which often spreads deep into affected tissues. Crohn's disease, contingent upon the location, can take shape in various forms: it can occur in the ileum-end of the small intestine (ileitis), the ileum and colon (ileo-colitis) or the colon (Crohn's colitis). Symptoms of Crohn's, if centralized in the small bowel, can include anemia, diarrhea, weight loss, pain and bleeding.

Ulcerative Colitis

Ulcerative colitis, on the other hand, affects only the colon and large intestine, and presence of the condition indicates long lasting inflammation and sores in the innermost lining of the large intestine (colon) and rectum. Signs of this may include diarrhea, bleeding, cramping and pain, all of which can become more severe over time.

Symptoms

Symptoms of inflammatory bowel disease vary, depending on the severity of inflammation and where it occurs. Symptoms may range from mild to severe. One is likely to have periods of active illness followed by periods of remission. Signs and symptoms that are common to both Crohn's disease and ulcerative colitis include severe diarrhea, fever, fatigue, abdominal pain and cramping, blood in stools, reduced appetite, unintended weight loss *etc.*

Cause

The exact cause of inflammatory bowel disease remains unknown. One possible cause is an immune system malfunction. When immune system tries to fight off an invading virus or bacterium, an abnormal immune response causes the immune system to attack the cells in the digestive tract, too. Other cause of IBD is believed to be a genetic predisposition too which may be triggered by an environmental factor such as an infection. However, most people with IBD do not have this family history.

Risk Factors

Age

IBD can happen at any age, but in most cases, it starts before the age of 35. Some people do not develop the disease until their 50s or 60s too.

Smoking

It is one of the main risk factor for developing Crohn's disease and it also aggravates the pain and other symptoms of Crohn's disease. Ulcerative colitis affects primarily non smokers and ex-smokers.

Gender

In general, IBD affects both genders equally. Ulcerative colitis is more common among men, while Crohn's disease is more common among women

Geographical Region

People who live in urban areas and industrialized countries have a higher risk of getting IBD. Those with white collar jobs are also more likely to develop the disease. This can be partially explained by lifestyle choices and diet. People who live in industrialized countries tend to eat more fat and processed food.

Family History

People who have a parent, sibling, or child with IBD are at a much higher risk for developing it themselves.

Race or Ethnicity

Although whites have the highest risk of the disease, it can occur in any race.

Nonsteroidal Anti-inflammatory Medications

These include ibuprofen, naproxen sodium, diclofenac sodium and others. These medications may increase the risk of developing IBD or worsen disease in people who have IBD.

Complications

Complications found in both conditions may include malnutrition with resulting weight loss, colon cancer, fistulas or ulcers in the bowel wall, intestinal perforation, bowel obstruction, inflammation of skin, eye and joints besides that of the colon *etc.*

Impact of IBD on Nutritional Status

i) Less Food Intake and Poor Digestion and Absorption of Nutrients

Inflammation in the small intestine of a person with Crohn's disease can interfere with the digestion and absorption of nutrients. In a person with ulcerative colitis, the small intestine works normally, but the inflamed colon does not absorb water properly, resulting in diarrhea, increased urgency to have a bowel movement and increased frequency of bowel movements.

During times of disease flares, diarrhea, urgency to have a bowel movement, abdominal pain, nausea, vomiting, blood in the stool, constipation, loss of appetite, fatigue and weight loss can negatively impact nutrition. Severe diarrhea can cause dehydration, robbing the body of fluids, nutrients and electrolytes (sodium, potassium, magnesium and phosphorus). People with Crohn's disease and ulcerative colitis often have a reduced appetite as a result of nausea, abdominal pain or altered taste sensation. This can make it difficult to consume enough calories and obtain sufficient nutrients. Additionally, the need to have numerous bowel movements in a day may cause a person with IBD to shy away from eating too much to avoid symptoms. Eating too little puts people at risk for being malnourished.

ii) Anaemia

Another possible symptom of IBD is rectal bleeding resulting from sores (ulceration) that form in the inner lining of the intestinal tract, leading to blood loss. Chronic blood loss can eventually lead to anemia, which if left unchecked, may cause fatigue. Weight loss may occur as a result of diarrhea and loss of appetite due to abdominal pain, nausea, vomiting or worsening diarrhea.

iii) Breakdown of Stored Nutrients

In addition, the inflammatory process in Crohn's disease and ulcerative colitis can result in increased consumption of stored energy and a breakdown of the body's tissues, often resulting in weight loss despite adequate caloric intake.

iv) Malabsorption of Macro- and Micronutrients

Inflammation in the small intestine of a person with Crohn's disease can interfere with the absorption of nutrients. This is called malabsorption. Amino acids (from proteins), fatty acids (from fats), sugars (from carbohydrates), vitamins and minerals are mostly absorbed from the last two sections of the small intestine (jejunum and ileum). The degree of malabsorption depends on how much of the small intestine is affected by Crohn's disease, whether the disease is active and if any portion of the intestine has been surgically removed. Generally, the malabsorption and nutrient deficiencies tend to be more significant if larger sections of the small intestine are inflamed or removed. If a significant portion of the ileum is inflamed or removed, the absorption of fat-soluble vitamins (A, D, E and K) and vitamin B_{12} will likely be affected. Bile acids or bile salts (responsible for aiding in absorbing fats and making the stool brown in color) can also be malabsorbed if the ileum is inflamed or removed. This can result in excess bile salts being transported to the colon where they cause increased fluid secretion in the colon and watery diarrhea. If a larger section of the ileum is affected, malabsorption of fatty acids may occur, resulting in abdominal cramping, diarrhea, malabsorption of fat-soluble vitamins and weight loss.

v) Growth Failure

Patients with ulcerative colitis may have less significant nutrient deficiencies; however, weight loss and anemia can be prominent due to severe diarrhea and blood loss. Some children with IBD fail to grow at a normal rate. Possible causes of poor growth include the impact of intestinal inflammation on nutrition, long-term use of corticosteroid medications (which have a negative impact on bone growth) and general poor nutrition. To minimize the negative effects of IBD on growth, good eating habits, adequate caloric intake and control of the underlying disease are essential for children.

vi Decreased Bone Density

Decreased bone mineral density (low bone mass) is common in children, adolescents and adults with IBD. Severe forms of decreased bone mineral density (osteoporosis) increases the risk for bone fracture. This may be due to insufficient calcium intake, poor absorption of calcium, vitamin D deficiency, decreased physical activity, inflammation and/or long-term use of medications such as corticosteroids.

Dietary modifications such as a low fiber or liquid diet along with medication may be necessary if the stricture is mostly inflammatory. If the narrowing is mostly scar tissue, surgery may be needed to widen the narrowed section or remove it. A low-fiber or liquid diet is often prescribed until surgery is performed.

Effect of IBD Medications on Nutritional Status

Medications tend to be more effective in people with good nutritional status. Certain medications used in IBD may also have adverse effects on nutrition. Corticosteroids may affect nutrition in many ways by increasing appetite, increasing serum glucose levels, increasing the risk for diabetes and affecting electrolyte (mineral) levels. Use of steroids for long time to reduce inflammation can slow the process of new bone formation, accelerate the breakdown of old bone and may also decrease bone mineral density. Most bone loss occurs in the first six months of corticosteroid use. Supplementation with calcium and vitamin D can help protect bone health, especially while patients are treated with steroids. Patients on sulfasalazine and methotrexate should receive supplemental folic acid. Cholestyramine, which is used sometimes to treat bile acid diarrhea, may decrease the absorption of fat-soluble vitamins.

Nutritional and Dietary Management of IBD

Inflammatory bowel disease (IBD) is associated with impairment of nutritional status both anthropometrically and biochemically, which results from both qualitative and quantitative changes in dietary intake alongside disease activity. Hence, diet and nutrition are important aspects of IBD management. Nutritional needs are specific to the individual and differ with disease state, body size and age. A nutritionist can help to estimate individual needs. During disease flares, it may be difficult to maintain adequate nutrition. However, maintaining good nutrition helps to improve overall wellness, promotes healing and immunity, increases energy levels and may alleviate some gastrointestinal symptoms.

Assessment of nutritional status is an essential part of the investigation of all patients with IBD and deficiency should be actively sought. Nutritional support does not have an established role in the treatment of active ulcerative colitis, other than in the management of malnutrition. The following are general recommendations about nutritional needs:

Calories

Having a chronic, active disease, such as Crohn's disease or ulcerative colitis, tends to increase the body's requirements for calories. Calorie needs are only slightly increased, unless weight gain is desired. Weight loss can occur due to episodes of inflammation, poor appetite and decreased intake.

Protein

Protein needs for both IBD patients between flares are the number of grams protein equal to body weight in kilograms. For example, a female of 54.5 should eat approximately 55 grams of protein each day. For weight gain and to restore losses after an acute flare, needs may be increased by 50 per cent. Needs also are increased if patient is taking corticosteroids (prednisone).

Vitamins and Minerals

People with Crohn's disease are at greater risk for deficiencies of several vitamins (as B_{12}, A, D, E, K and folate) and minerals (magnesium zinc, calcium and potassium) due to extensive inflammation or removal of large portions of the digestive tract. People with ulcerative colitis have less risk for vitamin and mineral deficiencies but are more prone to iron, fluid and electrolyte loss with bleeding, diarrhea and/or removal of the large intestine. People with ulcerative colitis and on medication generally are deficient in folate, magnesium, calcium, iron and potassium.

A standard multivitamin with minerals can be taken each day. Increased risk for deficiencies of specific nutrients should be treated with an additional amount of those nutrients.

Fluids and Electrolytes

It is important to drink adequate amounts of fluid. Fluid requirements increase during or after episodes of diarrhea and with exercise. Make sure to replenish losses of electrolytes from diarrhea. Sodium, chloride and potassium can be replenished by drinking sports drinks or ORS.

Role of Enteral Nutrition in IBD

Active Crohn's disease can be treated by exclusive enteral nutrition (elemental/polymeric/altered fat formulations all have equivalent efficacy). Enteral nutrition can maintain remission in Crohn's disease and in this context can be given alongside normal oral intake. Micronutrient deficiency (most frequently iron, vitamin B_{12}, folate and magnesium) should be replaced aggressively, parenterally if necessary. Total parenteral nutrition should not be used unless intestinal failure occurs.

Dietary Management

Dietary intervention to replace deficiency is essential and may also be used to treat active disease and to reduce symptoms. Researchers are continuing to study the interaction between diet and IBD. At present, there is no evidence to suggest that any particular food or diet causes, prevents or cures IBD. There is no one special IBD diet.

People with IBD should maintain a diverse and nutrient-rich diet to overcome malnutrition. A well-balanced diet with adequate intake of protein, carbohydrates and fat, as well as vitamins and minerals, is necessary for nutrition. This can be achieved by eating a variety of foods from all the food groups. For now, dietary recommendations are generally aimed at easing symptoms during flares and ensuring an overall adequate intake and absorption of nutrients, vitamins and minerals.

The following recommendations for dietary management are aimed at reducing uncomfortable symptoms, replacing lost fluids, preventing vitamin and mineral deficiencies, and providing adequate caloric intake.

There is no single diet or eating plan for everyone with IBD, and dietary recommendations must be individualized.

- ☆ **Planning an individual diet:** The individual diet should be based on symptoms (diarrhea, constipation, abdominal pain *etc.*), whether the person is in remission or experiencing a flare, location of disease, presence of narrowing of the small intestine (strictures), any prior surgeries, whether there are any specific nutritional deficiencies (such as iron deficiency) *etc.*

  Many people with IBD can consume a normal diet during times of disease remission but may need to alter their diet during flares. Other people with IBD, such as patients with intestinal strictures, will need to stay on a modified diet until the stricture is successfully treated medically or surgically.

- ☆ **Follow elimination diet:** During flares, certain foods or beverages may irritate the digestive tract and aggravate symptoms. Not all people with IBD are affected by the same foods, and it may be necessary to experiment to discover which foods affect symptoms the most. Keeping a food diary of possible trigger foods and food intolerances and one should avoid them. Some of the common food related triggers are:

Neither Crohn's disease nor ulcerative colitis is caused by a food allergy. Yet some people with IBD may also have food allergies. The most common foods causing an allergic reaction are milk, eggs, peanuts, tree nuts (*e.g.*, walnuts, almonds, cashews, pistachios and pecans), wheat, soy, fish and shellfish. It is important to distinguish between actual food allergy and food intolerance. A food allergy is associated with an immune system response and can cause a severe and life-threatening reaction, while a food intolerance can cause GI symptoms. Many people have food intolerances far more than have true food allergies.

Elimination diets are used to determine which foods must be avoided or minimized. This involves systematically removing foods or ingredients that may be causing symptoms. It is important to do this under the supervision of your doctor and a dietitian to be sure it is done correctly without causing poor nutrition. When eliminating foods, it is important to substitute other foods that provide the same nutrients. For example, when eliminating dairy products, be sure to obtain calcium and vitamin D from other sources.

Dietary fiber is found in plant foods, such as fruits, vegetables, nuts and grains. It is essential for health and for digestion. For many people with IBD, consuming fiber during times of disease flares or strictures can cause abdominal cramping, bloating and worsening diarrhea. But not all sources of fiber cause these problems, and some sources of fiber may help with IBD symptoms. Soluble (ability to dissolve in water) fiber helps absorb water in the gut, slowing down the transit time of food that is stored there. It can help to reduce diarrhea by forming a gel-like consistency and delaying emptying of the intestine. Insoluble fiber does not dissolve in water. It is more difficult to digest because it pulls water into the gut and makes food move more quickly through the gut. It is a harder more course fiber found in the skins of foods such as apples and seeds. Consuming insoluble fiber can aggravate IBD symptoms by causing more bloating, diarrhea, gas and pain. When there is severe inflammation or narrowing, consuming insoluble fiber can lead to worsening symptoms and a blockage in the intestinal tract. Most foods contain a combination of fibers, so cooking, peeling and removing seeds are important for patients who are in a flare and need to reduce their intake of insoluble fiber.

Some people with IBD may be sensitive to gluten and have gluten intolerance. These people may also have symptoms of abdominal bloating and diarrhea after eating gluten-containing food, and they may benefit from avoiding foods with gluten as well. A food diary can help determine the effect of gluten-containing food on symptoms. Test for celiac disease can also be got done.

Lactose intolerance is a condition in which the body does not properly digest lactose, the sugar present in milk and milk products. Some people with IBD may be lactose intolerant. Because the symptoms of lactose intolerance may mimic those of IBD, it can be difficult to recognize lactose intolerance. A simple test called a lactose breath test to diagnose this condition. Not all people with IBD are lactose intolerant. The severity of symptoms will depend on how much lactose an individual can tolerate. Some people may be able to consume small amounts of milk, while others may need to avoid it altogether. Lactase supplements can be taken along with milk to help digest it and specialty milk products that do not contain lactose are also available. Dairy products that contain yogurt and kefir may be more easily tolerated as well. Hard cheeses are generally well tolerated because of their minimal lactose content. Milk and dairy products are important sources of nutrients, particularly calcium. Therefore, people who limit or eliminate milk and dairy products from their diet must be mindful about obtaining calcium from other food sources or from supplements.

High-fat foods, such as butter, margarine and cream, may cause diarrhea and gas if fat absorption is incomplete. These symptoms tend to occur more in people who have inflammation in the small intestine or who have had large sections of the small intestine removed.

FODMAP (Fermentable Oligo-Di-Monosaccharides and Polyols) are sugars that are found in certain carbohydrates and sugar alcohols. If someone has intolerance to foods high in FODMAPs, it could result in excessive gas, bloating, diarrhea and cramping and those foods should be avoided. Sugar alcohols, such as sorbitol and mannitol often found in sugarless gums and candies cause diarrhea, bloating and gas in some people. Sorbitol is also found in ice cream and in several types of fruits, such as apples, pears, peaches and prunes, as well as the juices of these fruits. Fructose found in fruits, honey, high fructose corn syrup should be avoided in such cases. The low FODMAP diet is an intricate eating plan that should be initiated with the help of a dietitian.

Dietary Tips

There are some basic principles and guidelines to help for deciding how and what to eat, especially during flares which are:

- ✰ Increase calorie and protein intake following a flare. Abdominal pain, diarrhea and decreased appetite may have caused poor food intake. Steroids used to treat flares also can increase protein needs. One should try to maintain adequate caloric intake as meeting the body's calorie and protein demands is essential to prevent tissue wasting and weight loss. However, getting adequate calories can be challenging when appetite loss is a symptom of IBD. Caloric needs may increase during times of stress, including inflammation, fever and diarrhea. If weight begins to drop, try adding about 250 to 500 calories more per day through the foods which are tolerable.
- ✰ At times, there are very few foods that are tolerated well. During these times it is important to eat high calorie foods in tolerable amounts as frequently as possible.
- ✰ Select the right grains which are tolerable. Nutritionists often recommend eating whole grain products because the process of refining grains removes some of the iron and B vitamins. But this is not always advisable for people with IBD because the insoluble fiber in the bran and germ may be irritating, especially during a flare. During a disease flare, foods containing refined grains are generally easier to digest. Most refined grain products enriched with B vitamins and iron should be taken.
- ✰ Eat a variety of protein sources to ensure that all the required amino acids are taken. People with IBD may need to eat increased amounts of protein when experiencing inflammation or when recovering from inflammation. In general, it is best to choose lean or low-fat cuts of meat and poultry. Meat, seafood, beans, eggs, nuts and seeds are excellent sources of protein. They also provide B vitamins (niacin, thiamin, riboflavin and B_6), vitamin E, iron, zinc, magnesium and other nutrients. Before cooking meat, trim away any visible fat. Eat fish, especially oily fish that contains omega-3 fatty acids, like tuna and salmon, and try smooth nut butters.
- ✰ Eat a variety of well cooked vegetables and fruits. Tolerance for vegetables and fruits varies among people with IBD. To ease discomfort during a disease flare, select vegetables and fruits those are easier to digest, such as well cooked bottle gourd and potatoes, apple sauce and melons.

Remove the skin (the insoluble fiber part) and avoid the seeds. Avoid vegetables with a tough skin. Some vegetables, like broccoli, cauliflower and cabbage, have a tendency to produce gas. It is best to avoid eating these if gas is a problem.

- ☆ Steaming vegetables until they are very soft preserves more of the nutrients than boiling them. Vegetable stock is a good source of nutrients that can be used for making soup, or added to rice, pasta or other vegetables.
- ☆ During a disease flare, soft fleshy fruits are well tolerated. Avoid skins and seeds and cook fruits when diarrhea is severe.
- ☆ Fiber can sometimes be problematic for people with IBD, especially during disease flares. They should take low residue diet. At that time, limit foods with insoluble fiber (*i.e.*, seeds, nuts, beans, green leafy vegetables, fruit and wheat bran) but take soluble fibre rich foods. If you reduce your fiber intake during a flare, slowly increase the amount you consume when you are feeling better. Start by adding only a few grams per week.
- ☆ Limit the amount of greasy or fried foods. It is important to remember that all IBD patients have different food intolerances. One may be sensitive to spicy food while another may not be.
- ☆ Calcium consumption is especially important for people with IBD. To meet the calcium needs without a supplement, aim to eat at least three to four servings of calcium-rich foods daily. Sources of calcium include milk (regular, lactose-free, calcium fortified almond or soy), yogurt, cheese and calcium-fortified orange juice. Dark green vegetables contain less calcium but they have great nutritional value and are excellent sources of folic acid. If recommended amount of calcium is not being met through diet, one can add a calcium supplement.
- ☆ Everyone should drink plenty of fluids such as water, low-sugar sports drinks, fruit juices diluted with water *etc.* One should avoid ice-cold liquids which can cause cramps in some cases, caffeine in coffee, tea and other beverages.

  Drink beverages slowly rather than drinking fast. Also avoid using a straw. Drinking fast and using a straw may introduce air into the digestive system, which can cause discomfort.

  If someone is experiencing diarrhea, he may be at risk for dehydration. Replacing fluids and electrolytes is necessary. Rehydration drinks such as water, ORS and *shikanjvi* may be helpful for replacing lost fluids and electrolytes during times when diarrhea is severe. Excess sugar can cause more diarrhea due to the pull of water into the gut. Fruit juices used for rehydration and replenishing of vitamins and electrolytes may need to be diluted.
- ☆ During IBD flare, begin with well-tolerated liquids such as diluted juices and advance to soft solids (apple sauce, custard, mashed potato *etc.*) then solids (oatmeal, cooked eggs, white bread, plain boiled or steamed chicken or fish). Introduce one or two items every few days and avoid any foods that cause symptoms.
- ☆ Eat smaller and frequent meals. Smaller, more frequent meals are better tolerated and can maximize nutritional intake.
- ☆ There is no need to avoid foods unless they worsen the symptoms. Avoid trigger foods.
- ☆ It is best to restrict as few foods as possible to increase the chances that one can get a balanced, nutritious diet which is important for maintaining the function of digestive tract and overall health. Between flares, eat a wide variety of foods as tolerated. This includes fruits, vegetables, whole grains, lean protein, and low-fat and nonfat dairy products.
- ☆ During ulcerative flare, avoid foods that may increase stool output such as fresh fruits and vegetables, prunes and caffeinated beverages. Decrease the concentrated sweets in diet, such as juices, candy and soda, to help decrease amounts of water pulled into the intestine, which may contribute to watery stools. Decrease alcohol consumption. Try to take omega-3 fatty acids found in fatty fish as they have an anti-inflammatory effect.
- ☆ During Crohn's disease flare, follow a low residue diet to relieve abdominal pain and diarrhea. If someone is having strictures, it is especially important to avoid nuts, seeds, beans and kernels. If oily and foul-smelling stools are passed, they may be due to fat malabsorption. Follow a low fat diet in that case.
- ☆ Eat in a relaxed atmosphere.
- ☆ Consider taking nutritional supplements if appetite is poor and solid foods are not tolerated well.
- ☆ One multivitamin-mineral supplement should be taken daily.

Malabsorption Syndrome

The term malabsorption syndrome is used for conditions in which a number of nutrients such as carbohydrates, fats, proteins, minerals or protein in the diet are not digested and absorbed well across the gastrointestinal tract. Normally the human gastrointestinal tract digests and absorbs dietary nutrients with remarkable efficiency but in malabsorption syndrome, there is an impairment of any of the many steps involved in the complex process of nutrient digestion and absorption and hence, there is inadequate assimilation of dietary nutrients. Causes of malabsorption can be surgical or anatomical (Fistula, gastric surgery, diverticulosis *etc.*), enzyme deficiencies (disaccharidase deficiency, pancreatic disease, biliary obstructions), mucosal defects (celiac disease, Crohn's disease and radiation), systemic causes (Diabetes, thyroid disease, severe skin disorders, scleroderma *etc.*), drugs (antibiotics, laxatives, cholestyramine *etc.*) and infections (parasitic infections, tuberculosis and bacterial over growth *etc.*). Such impairment of GIT due to any cause can be of single or multiple nutrients depending upon the abnormality which leads to malnutrition. Malabsorption can affect macronutrients, micronutrients, or both, causing excessive fecal excretion, nutritional deficiencies and GI symptoms.

Some common well known disorders related to malabsortion are celiac disease, lactose intolerance, and steatorrhoea which have been discussed as under.

Celiac Disease

This serious autoimmune disorder can occur in genetically predisposed people where the ingestion of gluten leads to damage in the small intestine. People with conditions such as type 1 diabetes, autoimmune thyroid disease, Down's syndrome and Turner syndrome are at a higher risk than the general population of having coeliac disease (CD). It is not an allergy and should not be confused with wheat intolerance. Celiac disease sometimes called sprue or gluten-sensitive enteropathy or celiac is a serious autoimmune disorder that can occur in genetically predisposed people where the ingestion of gluten leads to damage in the small intestine. When gluten protein sensitive people eat foods such as wheat, rye, barely, triticale, spelt, and kamut, their body mounts an immune response that attacks the small intestine which is the site of absorption of nutrients. The role of oats in celiac disease has recently been questioned. Apparently, oats and oat products that are totally free of wheat, rye, and barley, are safe for celiac sufferers to consume.

When the intestinal villi are damaged by an inflammatory process after consumption of the offending grains or products, nutrients cannot be absorbed properly into the body which can result in bowel symptoms and deficiencies of vitamins, minerals and other nutrients. The loss of absorptive function along with the ongoing inflammatory process results in a severe malabsorption syndrome characterized by diarrhea, bloating, weight loss, anemia, bone pain, chronic fatigue, weakness, muscle cramps, and, in children, failure to gain weight and growth retardation, and can lead to serious other complications too.

Diagnostic Testing

The diagnosis of CD relies on both serologic (measuring total IgA level as well as either IgA tissue transglutaminase (tTGA) or IgA endomysial antibodies (EMA) and histologic studies (duodenal biopsy, which is considered to be the 'gold standard' diagnostic test), as well as a response to a gluten free diet (GFD).

The patient must continue to consume gluten in the diet at the time of investigations as withdrawal of gluten from the diet will cause false-negative serology and histology results.

- ☆ Patients with symptoms, signs, or laboratory evidence suggestive of malabsorption, such as chronic diarrhea with weight loss, steatorrhea, postprandial abdominal pain, and bloating, should be tested for celiac disease.
- ☆ Patients with a first-degree family member who has a confirmed diagnosis of celiac disease should be tested if they show possible signs or symptoms or laboratory evidence of celiac disease.
- ☆ IgA anti-TTG antibody is the preferred single test for detection of celiac disease in individuals over the age of 2 years.
- ☆ In patients in whom low IgA or selective IgA deficiency is identified, IgG-based testing (IgG DGPs and IgG TTG) should be performed.
- ☆ When screening children younger than 2 years of age for celiac disease, the IgA TTG test should be combined with DGPs (IgA and IgG).
- ☆ Celiac disease should be differentiated from non-celiac gluten sensitivity in order to identify the risk for nutritional deficiency states, complications of celiac disease, risk for celiac disease and associated disorders in family members, and to influence the degree and duration of adherence to the gluten free diet (GFD).
- ☆ A diagnosis of non-celiac gluten sensitivity should be considered only after celiac disease has been excluded with appropriate testing.
- ☆ Formal gluten challenge should be considered, where necessary, to diagnose or exclude celiac disease in patients already adhering to a GFD.
- ☆ People with celiac disease should be monitored regularly for residual or new symptoms, adherence to GFD, and assessment for complications.
- ☆ In children, special attention to assure normal growth and development is recommended.
- ☆ Monitoring of adherence to GFD should be based on a combination of history and serology [IgA TTG or IgA (or IgG) DGP antibodies].

Table 22.4: Serological Testing for Celiac Disease

Test	*Comment*
TTG	IgA Highly sensitive and specific first-line test
EMA IgA	Highly specific but operator-dependent test
Deamidated gliadin peptide IgG	Might be increased in IgA-deficient patients with CD
TTG IgG	Might be increased in IgA-deficient patients with CD
Antigliadin antibody IgA and IgG	Not as sensitive or specific as TTG
Total IgA	Screens patient for selective IgA deficiency

Pathogenesis

The cause of coeliac disease is complex, with multiple factors involved. The pathogenesis of CD is dependent on the interaction of a triad of genes, gluten, and environmental influences (Fig 22. 1). People affected by coeliac disease have a genetic predisposition, where the gene, human leukocyte antibody (HLA)-DQ2 and -DQ8 are strongly associated with the disease. Gluten is consumed as protein, but specific peptide residues are resistant to hydrolysis. There is deamidation of these peptides, where HLA DQ2 binds these peptides to T cells. This causes the subsequent release of cytokines which induces the mucosal injury (villous atrophy) of celiac disease.

Whereas predisposing genes and ingestion of gluten are both pivotal to the development of CD, most HLA-DQ2/DQ8 carriers (about 30 per cent of the population) who are exposed to gluten (>99 per cent of the population) do not have CD. In fact, only 1 per cent of the population has CD, indicating the importance of environmental factors. Various environmental factors other than gluten have been implicated in the pathogenesis of CD, from vitamin D to season of birth. In particular, early-life factors that influence the intestinal environment, such as breastfeeding, infection, and alterations in the intestinal microbiota, have gained much interest. For example, it had been suggested that infections of the gastrointestinal tract in infancy and adulthood are a risk factor for later development of CD. Infant feeding practices have also been implicated in the development of CD, with conflicting conclusions. Some studies show a protective effect of breast-feeding, especially in those infants who are breast-fed during dietary gluten introduction, whereas other studies show no such effect. The degree to which the intestinal microbiome is an environmental influence in the development of CD is also an emerging field of investigation.

The pathogenesis of CD involves a triad of predisposing genes, gluten, and environmental factors. The main genes in CD are the HLA-DQ2 and/or HLA-DQ8 haplotypes. Dietary gluten is the major exposure among patients with CD. Numerous other environmental factors influence the development of CD but are less well defined than gluten. The reaction to gluten fragments occurs at the small intestinal epithelium, with both an innate and adaptive immune response. The results are characteristic autoantibodies, histologic changes (intraepithelial lymphocytosis and villous atrophy), and clinical symptoms (eg, diarrhea or iron deficiency anemia).

Prevalence and Risk Groups

Celiac disease is an under diagnosed disease that affects at least one per cent of the population mostly

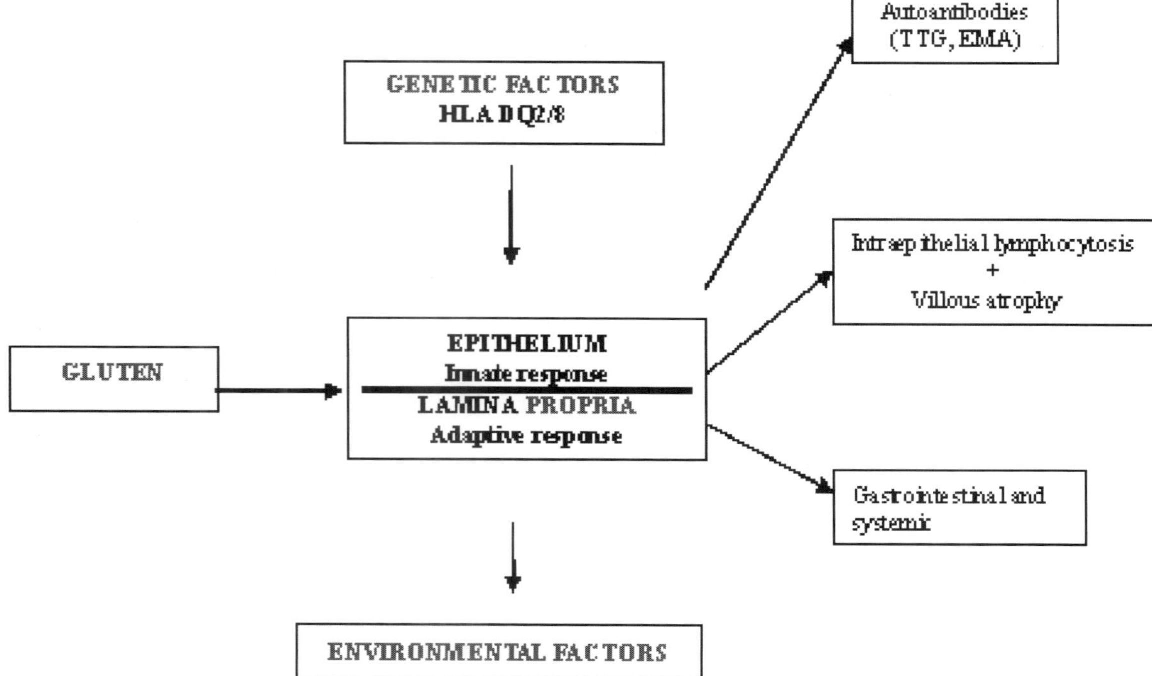

Figure 22.1: Pathogenesis of Celiac Disease.

Caucasians, Middle Eastern, South America, North Africa and West Asians. It is not common in Asians as they typically do not possess the DQ2 or DQ8 HLA profiles. One proposed reason for increasing trend is that, with a globalizing world market, developing nations that traditionally relied on gluten-free grains, such as rice and maize, are increasingly incorporating wheat-based foods into their diets. As more mass screening studies are performed in different populations, more cases of previously undiagnosed celiac disease are identified. Particularly on a global scale, there is evidence that celiac disease is a missed diagnosis in many children in whom infection and malnutrition are quite common and the presumed cause for diarrheal illness.

It is common in adults and children. Age of onset is from 6 months to 90+ years. First degree family members have a 1 in 10 risk of suffering from the disease. Sex differences also exist with respect to the rate of diagnosis of CD. One study found a female/male ratio of 2 to 3:1. The sex difference was true only in diagnoses made in adulthood, in which iron deficiency anemia was a significant presenting manifestation among women. However, there is indication that the prevalence among male and female subjects based on serologic screening is comparable at about 1 per cent.

Following are the at risk groups:

- Celiac disease in the family
- Thyroid disease
- Type 1 diabetes mellitus
- Down Syndrome
- Abnormal liver tests
- Osteoporosis
- High suspicion *e.g.* family history, anaemia
- Undefined neurological disorder/epilepsy
- Infertility/recurrent miscarriage

Celiac disease can vary in its symptoms, ranging from many to none at all. Patients may present with the following symptoms:

- Fatigue, tired all the time
- Recurrent mouth ulcers
- Iron deficiency and anaemia (Combination of iron, folic acid, or vitamin B_{12})
- Nutritional deficiencies
- Altered bowel habits (including diarrhoea and/or constipation)
- Abdominal pain or bloating
- Flatulence
- Irritability and depression
- Weight loss or gain
- Developmental delay (in children)
- Failure to thrive (in children)
- Bone or joint pain and stiffness
- Neurological symptoms such as peripheral neuropathy and ataxia
- Dental enamel defects
- Alopecia
- Itchy and blistery skin rash (dermatitis herpetiformis).

Nutritional Requirements

Nutritional requirements of such persons remain the same as for their normal healthy counterparts.

Diet plays an important role in the treatment of celiac disease and which includes exclusion of wheat, rye and barley products from the diet forever as they contain gluten protein. Gluten free diet needs to be continued for life, because the sensitivity to gluten does not disappear. Mild symptoms do not mean mild disease; all people with celiac disease require the same strict gluten free diet, irrespective of severity of symptoms.

While pure oats appear to be safely tolerated by the majority of people with celiac disease, oats should be introduced into the diet with caution and patients should be monitored closely for evidence of adverse reaction.

In North India, wheat *chapati* is a major food item consumed by most individuals on a daily basis. For this reason, compliance is a major issue. Timely dietary intervention considerably influences the achievement of appropriate growth especially among the children. The aim of effective dietary management is:

- Establishing an effective gluten free diet.
- Ensuring adequate nutrient intake.
- Ensuring good dietary compliance.

Gluten Free Diet (GFD)

There are three kinds of food categories:

- **Naturally gluten free foods:** The mainstay of treatment for celiac disease is a GFD, which entails avoidance of the gluten-containing grains *i.e.* wheat, rye, and barley. Acceptable grains include rice, oats, buckwheat, corn, millet, and quinoa. Other foods that are naturally gluten free include potato, milk, cheese, eggs, fresh fruits and vegetables, fresh meats including poultry, fresh fish and sea foods, pulses, butter, margarine, ghee and cooking oils.
- **Foods that contain obvious sources of gluten** (*e.g.* made with gluten containing cereals): Examples are bread, pasta, breakfast cereals, cakes, and biscuits.
- **Foods that may include a gluten cereal as an ingredient**, *e.g.* used as a thickener, filler, or carrier for flavours. These include some manufactured ready to serve soups, sauces, sausages, and ready meals.

For establishing an effective gluten free diet, one must have information about what foods to be taken and what foods to be avoided as given below.

Food Groups	***Avoid***	***Allowed***
Cereals	Whole wheat Whole wheat flour (*atta*) and their products like *chapatti, paratha, puri*, biscuits *etc.* Refined wheat flour (*maida*) and their products such as *nan, kulcha, bhatura, kachori, mathi, samosa*, buns, bread *pakora*, any snack containing bread crumbs *etc.* Commercial wheat starch Semolina (*suji*) Barley Rye Oats *Sevian* (vermicelli) White or brown bread Noodles, Macroni, pasta Wheat or oat porridge Wheat flakes Wafers Pastries and cakes	Rice Rice flakes (*chirwa*) Puffed rice (*murmura*) Rice noodles Cornflakes Arrowroot Maize Bajra Sorghum Ragi Tapioca Waterchestnut All savoury and sweet products incorporating above grains and millets.
Pulses	–	All pulses, lentils and soybean in any form such as *dal, roti, cheela*, or savoury and sweet products
Nuts and oilseeds	–	All salted and roasted dry nuts and peanuts.
Fats and oils	Commercial gravies, white sauce and cream sauces	Butter, fats and oils
Fruits and vegetables	Any fruit pie, fruit custard, packaged soups, pizza *etc.*	All fresh fruits and vegetables
Milk and milk products	Milk beverages that contain malt Ready to serve milk products available in the market such as milk shakes, flavoured milk, ice creams, chocolates, puddings, cheese spread *etc.*	Fresh milk Homemade milk shakes Curd, buttermilk, *khoa*, fresh *paneer etc.* prepared at home
Sweets containing milk and milk products	Generally, in milk products especially sweets, *maida* is added as a thickener or adulterant. Those sweets such as *gulabjamun, jalebi, burfi, pera*, milk cake *etc.* should be avoided.	Rice *kheer* Carrot *halwa* and *kheer* *Rasgullas* *Sandesh* Sago *kheer* *Makhana kheer* Sago *kheer* *Petha* One should try to take home made milk based sweets which are devoid of wheat flour.
Meat, fish and poultry	Fish fingers, hot dogs, sausages, *seekh kababs*, sausages, ham, commercially available ready to eat meat cutlets, burgers *etc.*	Home cooked pure meat, fish and chicken, eggs in any form
Commercial ready to eat or serve products	Malted cereal based beverages and beer, nutritional supplements (bournvita, maltova *etc.*) weaning foods (farex, cerelac *etc.*), packed and processed foods such as soups, tomato sauce and ketchup, chutney, pickles, ready to eat products, baked beans, canned and frozen foods *etc.*	Nutritional supplements (Complan, Build-up) Baby foods such as Nestum, cerelac containing rice and vegetable, Farex (rice) *etc.* Wine
Miscellaneous	Candies, chewing gums, chocolate (éclairs), custard powder, hot chocolate drinks, baking powder, mixed spices, alcoholic soft drinks containing malt extract *etc.*	Fresh fruit and vegetable juices, soft drinks, *sherbats*, squashes and homemade soups. Wine, most spirits and liqueurs and cider Chocolate drinks (pure coco) Sugar, honey and jaggery Spices and condiments Tea Coffee

A gluten free diet is not always a healthy diet as it may provide fewer amounts of fibre, iron and calcium. Some gluten free products may contain more calories and sugar. Such foods should be taken in limited amounts. For adequate nutrient intake, one should take a well balanced diet such as cereals (rice, *bajra*, corn, buckwheat, quinoa,), fruits, vegetables, potato, egg, meat, soybean, amaranth, legumes or legume flour *etc.* without additives or seasonings. Gluten free flour is not available in Indian market but it can be prepared by mixing naturally available gluten free food items. As cereals especially wheat flour *chapatti* is the main meal item in the diet of North Indians. They can prepare *chapatti, puri* or *parantha* from various other combinations such as soybean flour and rice flour (1:3), corn flour and *besan* (3:1), rice flour and *besan* (3:1), maize flour, bajra flour, *kotu ka atta* (buckwheat flour), water chest nut flour, tapioca flour, ragi and cassava flour. Various rice preparations such as boiled rice, rice *pulav*, lemon rice, tamarind rice, *dosa, idli, uttapam*, rice *papad, dahi vada, khichri etc.* can be added in the main meal. For the breakfast and evening snacks, various dishes such as milk with cornflakes, *poha, besan cheela*, potato *vada*, arrowroot and potato *tikki* or cutlet, *paneer* cutlets, fruit *chaat*, sprouted *dal chaat*, peanut *chaat*, eggs in any form, rice noodles, *paneer tikka*, fried fish or *besan pakora, bajra mathri* and *ladoos*, sago cutlets, corn cutlets, roasted dry nuts, *chikki, dal bhujia*, salted puffed rice *etc.* can be taken. For the dessert, some dishes prepared from carrot and milk, rice and milk, fruit cream, *rasgulla, besan halwa, moong dal halwa*, sweet yoghurt *etc.* can be taken. There is no restriction on oil, ghee, butter, fruits, vegetables, milk, sugar, pulses, cereals except wheat, rye, barley and their preparations. But all these foods items should be taken in balanced amounts as per recommendations of ICMR (2010) for the specific age group.

If the child is malnourished because of celiac disease, he can be given calories upto 1½ to 2 times the RDA. Calories are gradually increased from 80-100 to 150-200 Kcal per day. Calories should mainly be derived from carbohydrates and fat. They can be given energy dense foods. Similarly protein intake should also be increased till the child attains normal growth and development. Elimination of gluten from the diet and providing adequate nutrition is the only effective treatment. The diet of celiac persons is generally devoid of complex carbohydrates due to avoidance of wheat and its products. Secondary disaccharidase deficiency has also been clearly demonstrated to be present virtually in all such cases. In such case, avoid milk or low lactose diet can be given at the start of nutritional management as it will hasten the recovery. Fats may be gradually increased to provide extra calories among malnourished persons. Multivitamin and mineral deficiencies are very common in the untreated celiac disease and they should be replenished. Fat soluble vitamins should be provided in adequate amount. In severe cases, water soluble vitamins are also not absorbed properly and they should be administered vitamin C to enhance iron absorption and B-complex vitamins. Supplements of Ca and Mg should be given to prevent hyponatremia and hypokalemia as loss of these minerals is precipitated by excessive loss through stool.

Now a days, a number of gluten free food products are available in the market but one should read the labels of such foods as well as of canned, frozen and processed foods carefully. It is essential that people with CD understand food labels, so they know how to check that a food is suitable.

One should ensure good compliance of gluten free diet to prevent complications and that can be done through time to time monitoring and follow up of nutritional status of gluten sensitive persons.

Avoid Gluten Contamination

The compliance with a gluten-free diet is an extremely challenging task, given the problems related to cross-contamination and lack of clear food labeling policies. The possibility of accidental ingestion of gluten should be minimized and this is possible largely by strict supervision to avoid gluten contamination. To avoid contamination:

1. Store gluten free products and ingredients in separate, sealed containers. They are best stored on the top shelf of the pantry or on the shelf above the gluten containing food. This means that if gluten containing foods are spilled or dropped, it won't contaminate the gluten free products.
2. Prepare gluten free meals in a separate area of the kitchen or prepare the gluten free meal first and then the gluten containing one. For example first make the gluten free sandwich or *chapati* then the gluten containing one.
3. Use clean knives/spoons in spreads, dips and margarines. Never dip back into the container if there are gluten crumbs on the knife or spoon.
4. It is important when at home to use a separate *tawa, chakla-belan*, toaster and utensils *etc.* to prepare gluten-free chapatti/toast.
5. Salads or meals must not be served with or touching with gluten bread, - just a crumb is enough to cause damage.

Lifelong Compliance with Gluten Free Diet

Long-term compliance with the gluten free diet is mandatory. Failure to adhere to a strict gluten-free diet can increase risk of osteoporosis, poor growth, infertility, miscarriage, iron deficiency, twice the overall mortality rate and twice the risk if gastrointestinal tumours. The risks of these are no greater than normal when a gluten-free diet is followed.

Lactose Intolerance

Lactose intolerance is a common digestive problem where the body is unable to digest lactose, a type of sugar mainly found in milk and dairy products. The body digests lactose using an enzyme called lactase, produced in the small intestine. This enzyme is responsible for breaking down the digested lactose into two monosaccharides *i.e.* called glucose and galactose, which can be easily absorbed into the bloodstream. But people with lactose intolerance do not produce enough lactase, so unabsorbed lactose moves to the large intestine where it is fermented by bacteria present in the colon. During fermentation, these bacteria break down the unabsorbed lactose, producing fatty acid and various gases such as carbon-di-oxide, hydrogen and methane, which cause the symptoms associated with lactose intolerance.

Types of Lactose Intolerance

It can be primary, secondary, congenital or developmental.

Primary Lactose Intolerance

Primary lactase deficiency is the most common cause of lactose intolerance worldwide. This is caused by an inherited genetic defect that runs in the families. People with primary lactose intolerance start life producing plenty of lactase which is a necessity for infants, who get all nutrition through milk. Primary lactase deficiency develops when lactase production falls off sharply as the diet of child becomes less reliant on milk and dairy products and starts taking other foods. This is usually after the age of two, when breastfeeding or bottle-feeding has stopped, although the symptoms of lactose intolerance may not be noticeable until adulthood. It occurs in a large proportion of people with African, Asian or Hispanic ancestry. The condition is also common among those of Mediterranean or Southern European descent.

Secondary Lactose Intolerance

This type of lactose intolerance can occur at any age due to any problem in the small intestine due to which lactase production decreases in the small intestine. Possible causes can be illness, injury or surgery involving small intestine *e.g.* gastro-enteritis, celiac disease, Crohn's disease, ulcerative colitis, chemotherapy, long course of antibiotics, surgery, bacterial overgrowth *etc.* The decrease in the production of lactase in secondary lactase deficiency is sometimes only temporary, but it may be permanent if it is caused by a long-term condition. It is also possible to develop secondary lactase deficiency later in life, even without another condition to trigger it. This is because the body's production of lactase naturally reduces as one gets older. Treatment of the underlying disorder may restore lactase levels and improve signs and symptoms, though it can take time.

Congenital Lactose Intolerance

Congenital lactase deficiency, an inherited genetic defect is a rare condition that runs in families and is found in newborn babies. This disorder is passed from generation to generation in a pattern of inheritance called autosomal recessive, meaning that both the mother and the father must pass on the same gene variant for a child to be affected. Such babies are born with complete absence of lactase activity.

Developmental Lactose Intolerance

Premature infants (born before 37th week of pregnancy) may also have temporary lactose intolerance because of an insufficient lactase level as the small intestine may not be fully developed by the time of birth. This is known as developmental lactase deficiency and it usually improves as affected babies get older.

Lactose is a sugar found only in milk, but it is also present in dairy products and food items made with milk derivatives.

Symptoms

The signs and symptoms of lactose intolerance usually begin 30 minutes to two hours after eating or drinking foods that contain lactose. Symptoms can range from mild discomfort to a severe reaction, depending on how much lactase the person produces, and how much dairy products he consumes. Common signs and symptoms include flatulence, bloating, abdominal discomfort, cramps or pain, nausea, diarrhea, feeling sick *etc.*

Complications

Lactose intolerant persons feel discomfort when they consume milk and other dairy products. We all know that milk and other dairy products are highly nutritious as they are good sources of calcium, protein and vitamins such as vitamin A, D and $B_{12.}$ Moreover, lactose also helps the body to absorb a number of minerals such as magnesium and zinc. These vitamins and minerals are not only important for development of strong healthy bones but their deficiencies can also cause various types of malnutrition. Therefore, lactose intolerant persons are at the risk of developing osteopenia, osteoporosis and malnutrition.

Dietary and Nutritional Management

There is no known way to prevent lactose intolerance. However, one can lead a healthy and normal life by managing their daily diet routine and incorporating products that are lactose free, yet with healthy nutrients. The key is to have balance of all vital nutrients as recommended by ICMR (2010) and implementing guidelines that are meant for lactose intolerance condition.

Dietary Considerations

The amount of lactose an individual can tolerate varies from person to person. Many people with lactose intolerance can tolerate some lactose-containing foods by adjusting the type, amount and timing of these foods. Some lactose intolerant patients may need to limit or eliminate these foods completely. If one wishes to include lactose-containing foods in the diet to check whether lactose is being tolerated or not or to what extent, the following dietary suggestions given should be followed:

1. Add new foods one at a time; decrease the amount, or eliminate the food, if symptoms occur.
2. Most people with lactose intolerance do not need to avoid all dairy products, for example fresh cultured curd or yogurt contains live cultures that naturally help to digest lactose. Many people with lactose intolerance tolerate cultured yogurt well.
3. Some types of dairy are naturally low in lactose *e.g.* butter which contains only 0.1g of lactose per 20 g portion. Certain types of cheeses such as Cheddar, Swiss, Mozzarella *etc.* Curd or yoghurt containing live culture of bacteria contain small amount of lactose as bacterial lactase will hydrolyze some amount of lactose, can be taken. In case of mild lactose intolerance, such dairy products can be taken in small amount, if they are tolerated well by the person.
4. If some people wish to take milk, they can add small amount of milk in tea or coffee but not the amount one would get from the bowl of breakfast cereal with milk. Many people can tolerate up to 2 cups of milk per day when taken in smaller servings spread out over the course of the day.
5. Foods that contain lactose may be better tolerated if they are eaten with a meal.
6. In some cases of lactose intolerance, person may need to cut down or completely avoid milk (all types) and dairy products. These products are *khoa*, curd, buttermilk, hard and soft cheeses, ice cream, white sauce, butter, cream soups, some salad dressings, milk containing puddings and sweet. Certain processed foods containing milk powder, whey and cheese *etc.* should be avoided. Some of the dairy containing processed food products can be biscuits and cakes, breakfast cereals, chocolate and confectionery, breads and baked goods, sausages, desserts and custard, sauces and gravies, potato chips, nuts, ready to eat foods, packaged soups and flavoured drinks containing milk in any form *etc.* The amount of lactose in such products will depend on the amount of dairy products used.
7. While buying processed and packaged ready to eat products, one must read the food label carefully as such products should not contain milk or its product in any form. On the food labels, other names of added dairy can be described as milk, milk solids, whey, whey protein, whey protein concentrate, milk powder, skim milk powder, milk sugar, buttermilk, sour cream, caseinates, non-fat dry milk, milk by-products, cheese flavour *etc.* Such foods are to be avoided but do not get confused if a product contains lactic acid, lactalbumin or casein as these ingredients are not lactose.
8. Lactose can also be found in medications. Check the label. Ask your doctor to prescribe a lactose-free alternative if one exists.
9. There is no doubt that completely lactose free diet shall potentially be missing out on some nutrients but they can be taken from other foods instead. Many plant foods such as cereals (especially ragi), pulses (soybean, soymilk, tofu), nuts (almonds, Brazil nuts), dry figs, green leafy vegetables and oranges, *etc.* contain reasonable good amount of calcium. However, this calcium is poorly absorbed due to the presence of antinutrients such as phytates and oxalates. But after processing such as germination and fermentation of plant foods, calcium availability can be increased. Protein is not a much problem as egg, legumes, meat *etc.* are its other alternate good sources, the amount of which can be increased. For vitamin D, liver, egg yolk, fish oils, vitamin D fortified non-dairy foods can be taken. Additional calcium and vitamin D supplements can be taken for healthy and strong bones.
10. Specialty Products: If someone is not able to tolerate lactose-containing foods using the above suggestions, special products are available. Keep in mind, not everyone with lactose intolerance needs special products; many people can tolerate regular dairy products by adjusting the type and amount consumed. Now a days, some special food products are available such as:
    - ☆ 100 per cent lactose reduced milk is available in the dairy section of most grocery stores. Lactose reduced milk contains the same nutrition, including calcium and vitamin D as regular milk.
    - ☆ Available in nonfat, 1 per cent, 2 per cent, and whole milk varieties.
    - ☆ Lactose reduced milk does cost a bit more than regular milk.
    - ☆ Lactose reduced milk may taste sweeter than regular milk.
    - ☆ Soy milk, rice milk and almond milk are lactose free. If you plan to use these products

as an alternative calcium and/or vitamin D source, read labels carefully and choose a brand which specifically states it contains these nutrients and in what amounts.

11. Lactase enzyme supplements: In addition to dietary changes, lactase enzyme supplements (*e.g.*, LACTAID, Lactrase) or lactose substitutes may be helpful for some people. These are drops or tablets which one can take with the meals or drinks to improve the digestion of lactose.

Steatorrhoea

Steatorrhea is the term used for the presence of excess fat in the stool as a result of fat malabsorption. Since the small intestine is the major site of digestion of fat and the only segment of the gastrointestinal tract from which fat is absorbed; hence, the occurrence of steatorrhea invariably denotes the existence of disease affecting this organ. It means either that the small bowel itself is involved or else it is being secondarily affected by abnormalities of other organs of digestion, such as the stomach, the hepatobiliary tract, or the pancreas. Steatorrhea always indicates the presence of an organic abnormality.

Causes

In the case of steatorrhea, it is either short term or long term in duration. If it is temporary or short term, it may only be due from changes in the diet or gastrointestinal infection. However, once it persists and becomes long term then it may be a sign of serious condition.

Etiology of steatorrhoea can be broadly categorized as:

- ✰ **Pancreatic insufficiency:** Steatorrhea due to insufficient lipase and colipase to allow normal lipid hydrolysis (pancreatic exocrine insufficiency). This most commonly occurs due to chronic pancreatic inflammation and loss of acinar cells.
- ✰ **Bile salt deficiency:** It may be due to impaired production or secretion, or reduced circulating bile acids.
- ✰ **Malabsorption steatorrhea:** It is due to small intestinal disease, surgery, or medications.

Sometimes, steatorrhoea is not a major health concern as it may happen after eating foods high in fat and fiber, such as nuts with the skin, fatty fish, artificial fats, whole wheat products, excessive alcohol *etc.* Severe long term symptoms can be due to medical conditions which are known to cause steatorrhoea such as some pancreatic diseases (chronic pancreatitis), liver damage or failure, cystic fibrosis, hypothyroidism, diabetes, gastric bypass surgery, Gaucher disease, Tay-Sachs disease, celiac disease, Crohn's disease, lactose intolerance, gall bladder cancer, gallstones, gall bladder removal, lymphoma, obesity medications, parasitic infection (Giardia), tropical sprue, HIV, cholesterol medication, congestive heart failure, intestinal damage etc. In some studies, vitamin B_{12} deficiency has somehow been linked also as a cause of steatorrhea.

Aside from disease related causes, steatorrhea can also be caused by certain drugs such as Orlistat, or intake of oils such as castor oil and many more.

Symptoms

In steatorrhea, stools will be bulkier, pale, having oily appearance, foul-smelling and difficult to flush. Stools tend to float because of higher gas content. Other symptoms may also include weight loss, abdominal cramps, gas, indigestion and diarrhea.

The failure to assimilate fat normally often is associated with impaired absorption of other dietary constituents (particularly proteins, minerals, and vitamins), which may produce deficiency states of much greater clinical significance than the steatorrhea itself. Thus, the recognition of a mild impairment of absorption of fat may lead to the detection of critical deficiencies of other nutrients like poor absorption of fat soluble vitamins and other nutrients. Malnutrition, anaemia, dehydration, reduced growth in children, chronic fatigue, osteoporosis *etc.* may be caused by severe or chronic cases of steatorrhea.

Treatment

Dietary modification is a very important part in the overall treatment of steatorrhoea. Since steatorrhoea is not a disease but generally a symptom of malabsorption; the dietary treatment of steatorrhea requires knowledge of the cause of the disease associated with the steatorrhea. Once the cause is established, then an approach to the dietary management can be adopted.

What so ever may be the cause, the most basic and simple dietary modification is lessening or decreasing the amount of fat content in one's diet as it shall facilitate in maintaining the symptoms under control. In certain underlying diseases for steatorrhea such as the lactose intolerance and celiac disease, aside from a less fat or fat free diet, the person is also placed on a gluten or lactose free diet for an effective treatment. Foods that trigger the symptoms, may be avoided. For causes such as cystic fibrosis or chronic pancreatitis, medications and lifestyle changes shall be helpful.

In case of weight loss, energy intake should be increased by increasing intake of carbohydrates in the form of starches and sugars and protein (non-fat dairy products). Reduce the fibre intake. Those dietary fats should be included which can be tolerated well; foods containing short or medium chain triglycerides can be given as they are hydrolyzed by intestinal lipase, absorbed in the absence of bile acids and transported directly to the liver without getting resynthesised into triglycerides. Foods rich in vitamins and minerals or their supplements such as those incorporating fat soluble vitamins (vitamin A, D, E and K), water soluble vitamins (especially B_{12} and folic acid), iron, zinc, magnesium and calcium should be taken daily. One should remain hydrated by taking sufficient amount of water.

Sometimes, steatorrhea is caused by excessive or heavy alcohol drinking, therefore, a person's alcohol intake should be reduced, cutback or even put to an end gradually. Remember to drink moderately. But in cases where there is a great impairment or serious disease of the liver or pancreas due to chronic and heavy alcohol drinking, then the person should slowly withdraw from the alcohol until completely he/she completely stops from drinking alcohol. One should quit smoking too.

23

Diet in Diseases of Liver and Pancreas

The liver is one of the vital organs of the body, responsible for hundreds of chemical actions that the body needs to survive. The liver has two large sections, called the right and the left lobes; in fact, it is the largest internal organ in the body. It weighs between 1.44 and 1.66 kilograms and is located under the diaphragm and lies mostly in the upper right side of abdominal cavity. The gallbladder is under the liver, along with parts of the pancreas and intestines. The liver and these organs work together to digest, absorb, and process food.

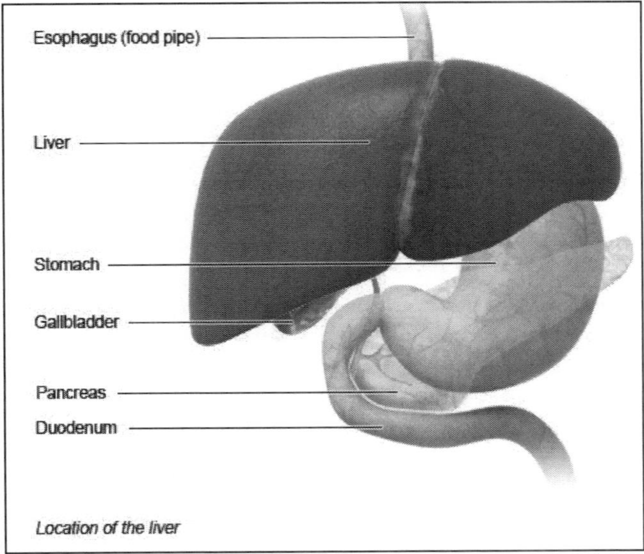

Location of the liver

The liver also actually is two different types of gland. It is a secretory gland because it has a specialized structure that is designed to allow it to make and secrete bile into the bile ducts. It also is an endocrine gland since it makes and secretes chemicals directly into the blood that have effects on other organs in the body. For these reasons the liver is both an organ and a gland.

The liver has multiple functions and its major functions include those given in the following table.

Table 23.1: Major Functions of Liver

Intermediary metabolism	**Carbohydrate metabolism:** Homeostatic regulator of blood glucose, glycogen metabolism (synthesis, storage and release), gluconeogenesis, interconversion of monosaccharides, synthesis of heparin, conversion of glucose into fatty acids *etc.*
	Protein metabolism: Deamination and transamination of amino acids, formation of urea, production of proteins and peptides (plasma albumin, alpha globulins, transferrin, ferritin, ceruloplasmin, retinol binding protein, prealbumin and other carrier proteins), synthesis of non-essential amino acids
	Lipid metabolism: β oxidation of fatty acids, synthesis of lipo-proteins, production of ketone bodies, synthesis of cholesterol,
	Bile metabolism and entero-hepatic circulation: Bile production, **absorbing and metabolizing bilirubin,** lipase activation, promote micelle formation, intestinal uptake of fat soluble vitamins, cholesterol and lipids,
Coagulation	Synthesis of most of the pro-coagulants except Factor III, VII and VIII Produces protein regulators of coagulation and fibrinolytic pathways
Heme metabolism	Homeostasis of iron, site for embryonic hematopoiesis
Xenobiotics metabolism	Drug metabolism through various reactions and their excretion from the body
Storage	Vitamin A, D, E, K dependent factors (II, VII, IX, X), vitamin B_{12}, iron, copper, glycogen

Endocrine functions	Modify amplify hormone functions *e.g.* conversion of vitamin D to 25 (OH)D in liver and conversion of 25 (OH)D to 1, 25 $(OH)_2D$ in kidney, peripheral conversion of T_4 to T_3, insulin like growth factors or sometomedins, role in cartilage function and collagen formation, removal of circulating hormones *etc.*
Immune and inflammatory response	Protection against foreign intruders, degrades toxins, process antigens and phagocytose bacteria, induce and intensify inflammation by recruiting neutrophils, release proinflammatory mediators *etc.*
Blood reservoir	Expandable organ; 10-15 per cent of total blood volume can be sequestered and quickly released after sympathetic stimulation
Excretion	Facilitates excretion of xenobiotics, lipophillic substances, bilirubin, bile salts, cholesterol and other steroids, drugs, alcohol and toxins *etc.*

The causative agents for liver diseases include malnutrition, infections (parasites and viruses, hepatitis A, B, C and D), chronic alcohol abuse, genetics (hemochromatosis, Wilson's disease, hyperoxaluria and oxalosis *etc.*), immune system abnormality, cancer and other growths, fat accumulation in the liver, blood transfusion, exposure to certain chemicals, drugs or toxins, obesity, diabetes *etc.* There are various signs and symptoms of liver diseases like weakness, fatigue, loss of appetite, weight loss, nausea, vomiting, jaundice, abdominal pain and swelling, swelling in the legs and ankles, itching, dark urine colour, pale stool colour, easy bruising of the skin *etc.* depending upon the disease.

Some of the most common types of liver diseases include the following:

Condition	***Possible Causes***
Alcohol related liver disease	Chronic alcohol abuse
Non-alcoholic fatty liver disease	Obesity, Protein deficiency in diet, High fat intake
Hepatitis	Viral infection, regularly drinking of too much alcohol
Haemochromatosis	Genetic
Primary biliary cirrhosis	Problem with the immune system

The most common liver diseases have been discussed as under.

Hepatitis

Hepatitis is an inflammation of the liver that is characterized by jaundice, liver enlargement, abdominal and gastric discomfort, abnormal liver function, and other symptoms. There are different types and causes, but the symptoms can be similar. It is commonly caused by a viral infection, but there are other possible causes too which may include other infections, autoimmune hepatitis and hepatitis that occurs as a secondary result of medications, drugs, toxins, malnutrition and alcohol.

1. Viral Hepatitis

Hepatitis caused by viruses is known as viral hepatitis which is widely prevalent in the world. There are 5 main hepatitis viruses, referred to as types A, B, C, D and E.

Hepatitis A (HAV)

It is the most common and the most infectious, spreading easily from person to person like most other viruses. This virus is present in the faeces of infected persons and is most often transmitted through consumption of contaminated food or water. It affects millions around the world and is responsible for more than 2 million deaths a year. Most people in areas of the world with poor sanitation are infected with this virus.

Hepatitis B (HBV) or Infective Hepatitis

It is acquired through exposure to infected blood, semen and other body fluids. This virus can be transmitted from infected mothers to infants at the time of birth or from family member to infant in early childhood. Transmission may also occur through transfusions of HBV-contaminated blood and blood products, contaminated injections during medical procedures, and through injection drug use.

Hepatitis C Virus (HCV)

It is mostly transmitted through exposure to infective blood. This may happen through transfusions of HCV-contaminated blood and blood products, contaminated injections during medical procedures, and through injection drug use. Sexual transmission is also possible, but is much less common. HCV infection is becoming a leading cause of cirrhosis, liver failure, and hepatocellular carcinoma, with incidence

Hepatitis D Virus (HDV)

It is unique because it can only affect those that already have hepatitis B. The dual infection of HDV and HBV can result in a more serious disease and worse outcome.

Hepatitis E Virus (HEV)

It is mostly transmitted through consumption of contaminated water or food. This virus is a common cause of hepatitis outbreaks in developing parts of the world and is increasingly recognized as an important cause of disease in developed countries.

2. Non-infectious Hepatitis

Alcohol and other Toxins

Alcoholic hepatitis is an inflammatory condition of the liver caused by heavy alcohol consumption over an extended period of time. It is also aggravated by binge drinking and ongoing alcohol use. The alcohol gets processed in the liver and produces highly toxic chemicals which injure the liver cells. Over time, alcohol intake can cause permanent damage and lead to liver failure and cirrhosis, a thickening and scarring of the liver.

Other toxic causes of hepatitis include overuse or overdose of medications and exposure to poisons.

Autoimmune System Response

In some cases, the immune system mistakes the liver as a harmful object and begins to attack it. It causes ongoing inflammation that can range from mild to severe, often hindering liver function. It is three times more common in women than in men.

Symptoms of Hepatitis

The initial phase of hepatitis is called the acute phase. Acute infection generally caused by viral infection, may occur with limited or no symptoms, or may include symptoms such as jaundice (yellowing of the skin and eyes), dark urine, extreme fatigue, nausea, vomiting, loss of appetite, abdominal pain, diarrhea, mild fever, joint or muscle pains *etc.* The bilirubin level rises in the serum. Acute phase of hepatitis is generally resolved within six months and it rarely results in acute liver failure. But if hepatitis is continued for more than six months, it is termed as chronic hepatitis. Chronic hepatitis is often asymptomatic early in its course. As the inflammation progresses, patients can develop constitutional symptoms similar to acute hepatitis, including fatigue, nausea, vomiting, poor appetite, and joint pain. Jaundice can occur as well, but much later in the disease process and is typically a sign of advanced disease. Chronic form of hepatitis may progress to scarring (fibrosis) of the liver, cirrhosis of liver, liver failure, or liver cancer.

Treatment

Treatment options are determined by the type of hepatitis and whether the infection is acute or chronic. In some cases of severe nausea, vomiting, liver damage *etc.*, hospitalization may also be required for relieving symptoms, to help in regeneration of liver tissues and prevent further liver damage.

Nutritional Management and Feeding Pattern in Acute Hepatitis

Patients with acute hepatitis are usually adequately nourished before the illness. Acute hepatitis is usually a mild disease, associated with only a few days of anorexia, nausea, and occasionally vomiting. These are usually well tolerated by the patients, who require no nutritional supplementation, and are encouraged to eat normally. Nutritional requirements remain the same as for their normal counterparts if no weight loss occurs. Old literature emphasized lipid restriction. This, however, is not true, and lipid restriction has no role in acute hepatitis unless fats aggravate nausea in an individual patient. Dietary restrictions have no place in the management of mild or moderate acute parenchymal liver disease. Usually they can take some food by mouth and enough fluids to prevent dehydration. Only small frequent meals at regular intervals should be provided and overfeeding should be avoided.

Nutritional supplementation and intravenous fluids and nutrients are reserved for the patients with excessive nausea and vomiting who cannot maintain a sufficient fluid balance. Alcohol should be avoided in acute hepatitis and for next 6 months following recovery.

Nutritional Management of Chronic Hepatitis

Malnutrition is prevalent in all forms of liver diseases; from 20 per cent in compensated liver disease to more than 80 per cent in those patients with decompensate disease. Patients with alcoholic liver disease are reported to have a greater incidence of malnutrition than those with nonalcoholic disease. Inadequate food intake is one of the primary causes of malnutrition and occurs in up to two-thirds of patients with chronic liver disease. Moreover, all the symptoms of chronic hepatitis *i.e.* anorexia, nausea, vomiting, abdominal pain, taste changes, fever, jaundice, early satiety, malabsorption, medication induced losses, altered energy and protein metabolism, restricted diet *etc.* complicate the food intake, nutrient bioavailability and make it difficult to ensure that the patient is well nourished at a time when it is essential to provide him with a highly nutritious diet to prevent liver damage. In such cases, a high carbohydrate, high protein and moderate fat is recommended.

Energy

Patients with chronic liver disease should be encouraged to maintain adequate energy consumption. A high energy diet (35-40 Kcal/kg body wt/day) is required to take care of weight loss and for ensuring maximum protein utilization. Due to bed rest, energy expenditure of the patient is low and moreover, initially due to anorexia, patient may not be able to take sufficient energy through diet and therefore, in such situations, only 1500-2000 Kcals may be sufficient. As the appetite improves, energy intake should be gradually increased *i.e.* 20-30 per cent higher than the normal recommended intake to ensure weight gain.

Excess calories should be avoided, particularly as carbohydrates, as this promotes hepatic lipogenesis, liver dysfunction, and increase CO_2 production and the work of breathing.

Carbohydrates

Carbohydrates should be sufficient to provide enough calories so as to avoid breakdown of tissue proteins for energy purpose, to build up glycogen stores, to maintain normal blood glucose levels, and should not exceed insulin reserves. They should supply 60-70 per cent of non-nitrogen calories. If the patient is able to take food orally, carbohydrates in the form of simple sugars such as glucose, sugar, honey, fruits and fruit juices and starches such as cereals and root vegetables should be provided.

Protein

A high protein intake is required to prevent negative nitrogen balance and for regeneration of damaged liver tissues. Increase in protein intake depends upon the severity of the liver disease. If liver damage is more, it shall not be able to metabolize high protein load and there is a danger of impending hepatic coma. In case of mild to moderate jaundice, diet should provide 60-80 g protein daily (1.2-1.5 g/kg body wt./day) while in severe jaundice, 40 g daily protein intake is adequate. In case of hepatic precoma and coma, protein rich foods should be limited and high carbohydrate diet should be given to provide enough calories.

Fat

Fat requirement varies depending upon the severity of hepatitis. In mild to moderate jaundice, 40-50 g while in severe jaundice, 20-30 g fat per day should be provided in the diet. Fat makes the food palatable and decreases the bulk of the diet. Moreover, it is required for the absorption of fat soluble vitamins. Emulsified fats in the form of butter, cream and egg should be given which do not require bile for emulsification. Foods containing medium chain triglycerides are also well tolerated.

Vitamins

Normally, the required amount of vitamins is supplied if the person is tolerating a well balanced diet. However, the availability of fat soluble vitamins can be low due to poor intake and absorption in chronic hepatitis. In such cases, care should be taken that diet provides vitamin A/β-carotene and vitamin K rich foods. Vitamin C (500 mg) rich foods should be provided as this vitamin is helpful in regeneration of liver. The requirement of B-complex vitamins is increased for increased energy and protein metabolism.

In case of nausea and vomiting, injections of vitamins can be given. But one must be aware that as vitamins must pass through the liver to be metabolized, if taken in excess, any vitamin has the potential to cause serious health problems. The damage is much greater, depending upon the severity of liver damage. Therefore, vitamin supplements should be taken with care.

Minerals

Patients with chronic liver disease are at increased risk for the development of osteoporosis, it is important to provide foods rich in calcium and/or to supplement their diets with calcium. If patient is not taking foods orally, care should be taken to maintain normal sodium and potassium levels. In case of oral intake, fruit juices, vegetable soups, coconut water *etc.* should be given to maintain electrolyte balance. People with chronic viral hepatitis C infections should avoid iron rich foods and supplements.

Diet and Feeding Pattern

Appetite stimulation to overcome anorexia is probably one of the most difficult challenges to overcome. One should take the favorite foods. Eat small meals at a time. Frequency of meals will depend upon the acceptability of food by the patient. In case of mild hepatitis, normal diet containing three meals and in between snacks can be given. In severe cases, small 6-8 feeds containing soft to liquid foods may be given in a day. One should try to take whole grains and healthy proteins in the form of beans, lean meat, eggs, low fat dairy products *etc.* Fresh fruits and vegetables should be a significant part of diet to help in recovering from a liver disease as besides essential nutrients, they contain antioxidants, which can protect the liver cells from damage. One should take starchy vegetables such as potatoes when on a hepatitis recovery diet. Healthy oils such as olive oil, canola oil, rice bran oil, flaxseed oil *etc.* should be taken. Avoid eating meat especially red meats due to their high sodium content. One should refrain from eating high-sodium and processed foods. Fried and spiced foods should be avoided. Strongly flavoured foods should be avoided as they cause nausea. A Hepatitis C diet plan must also be low in iron containing foods. One should drink enough water to prevent dehydration. People with chronic hepatitis should avoid alcohol since it speeds progression to cirrhosis.

Liver Cirrhosis

The term was introduced by Laennec in 1826 and refers to colour (Greek Kirrhos, tawny) of nodules of hepatocytes regenerating between stands of fibrous tissues. Cirrhosis refers to the replacement of normal liver tissue with non-living scar tissue. Cirrhosis is a chronic disease of the liver marked by degeneration of cells, inflammation, and fibrous thickening of tissue which results in irreversible damage of the hepatocytes. It is a late stage of scarring (fibrosis) of the liver caused by many forms of liver diseases and conditions and is typically a result of chronic alcoholism or hepatitis. It can take many years for liver damage to lead to cirrhosis. It is a serious and irreversible disease.

Causes

The most common causes of cirrhosis are long-term viral hepatitis C infection and chronic alcohol abuse. Obesity is also a cause of cirrhosis, although it is not as prevalent as alcoholism or hepatitis C. Obesity can be a risk factor by itself, or in combination with alcoholism and hepatitis C.

1. Chronic Alcoholism

It is the most common cause of cirrhosis. Alcoholic cirrhosis develops for 10–20 per cent of individuals who drink heavily for a decade or more. The amount of alcohol that causes cirrhosis is different for each person.

Chronic consumption of alcohol results in the secretion of pro-inflammatory cytokines [TNF-alpha, Interleukin 6 (IL6)] and Interleukin 8 (IL8), oxidative stress, lipid peroxidation and acetaldyhyde toxicity. These factors cause inflammation, apoptosis and eventually fibrosis of liver cells. Why this occurs in only a few individuals is still unclear. There are three stages to liver damage caused by alcohol:

- **Fatty liver (Steatosis):** This is when fat builds up in the cells that make up liver.
- **Steatohepatitis:** This is when the build-up of fat causes liver to become inflamed and some of the cells to die.
- **Cirrhosis:** This is when damage to liver is replaced by fibrous scar tissue.

These stages are progressive, but can also overlap. Fatty liver is reversible if one stops drinking alcohol. But if drinking is continued, the damage will get worse and one shall then be at risk of more serious liver disease, including alcohol-related hepatitis and alcohol-related cirrhosis. In both conditions, liver stops working as it should and this causes problems. Additional risk factors that play a role in someone developing alcohol-related liver disease include obesity, malnutrition, race and ethnicity, genetic factors, gender, chronic viral hepatitis especially hepatitis C *etc.*

2. Chronic Viral Hepatitis

It is another leading cause of cirrhosis. Infection with hepatitis C virus causes inflammation of the liver and a variable grade of damage to the organ. Over several decades, this inflammation and damage can lead to cirrhosis. Among patients with chronic hepatitis C, 20–30 per cent will develop cirrhosis. Chronic hepatitis B and hepatitis D also can also cause cirrhosis. Hepatitis D is dependent on the presence of hepatitis B and accelerates cirrhosis in co-infection. Globally, 57 per cent of cirrhosis is attributable to either hepatitis B (30 per cent) or hepatitis C (27 per cent).

3. Non-alcoholic Steatohepatitis (NASH)

In NASH, fat builds up in the liver that is not caused by alcohol use. This non-alcoholic fatty liver disease can lead to non-alcoholic steatohepatitis (NASH) which can cause the liver to swell and can eventually lead to scar tissue or cirrhosis. This type of hepatitis appears to be associated with obesity (40 per cent of NASH patients), diabetes, poor eating habits, protein malnutrition, high cholesterol, coronary artery disease, and treatment with steroid medications. This disorder is similar in it signs to alcoholic liver disease, but the patient does not have an alcohol history.

4. Bile Duct Diseases

There is a damage to the bile ducts, which function to drain bile. Bile duct disease limits or stops bile from flowing to the small intestine. The bile backs up in the liver causing the liver to swell and can lead to cirrhosis.

Two common bile duct diseases are primary sclerosing cholangitis (PSC) and primary biliary cirrhosis (PBC). PSC is a progressive cholestatic disorder presenting with pruritus, steartorrhea, fat-soluble vitamin deficiencies, and metabolic bone disease. There is a strong association with inflammatory bowel disease (IBD), especially ulcerative colitis. In PBC, the bile ducts are damaged by an autoimmune process, leading to secondary liver damage.

5. Food Toxins

Aflatoxin may be one of the causes of cirrhosis.

6. Autoimmune Hepatitis

Autoimmune hepatitis may have a genetic cause. This disease is caused by an attack of the liver by lymphocytes, causing inflammation and eventually scarring and cirrhosis. Findings include elevations in serum globulins, especially gamma globulins.

7. Genetic Diseases

Some genetic diseases such as Wilson disease, hemochromatosis, glycogen storage diseases, Alpha-1 antitrypsin deficiency *etc.* can lead to cirrhosis.

8. Medications

Medications including prescription and over-the-counter drugs like acetaminophen, some antibiotics, and some antidepressants, can lead to cirrhosis.

Symptoms

Cirrhosis of the liver is slow and gradual in its development. It is usually well advanced before its symptoms are noticeable enough to cause alarm. The symptoms of cirrhosis occur because the liver is unable to purify the blood, break down toxins, produce clotting proteins, and help with absorption of fats and fat-soluble vitamins. These signs and symptoms may be either a direct result of the failure of liver cells, or secondary to the resultant portal hypertension. Often there are no symptoms in its early stage (this is called compensated cirrhosis) until the disorder has progressed over time. Weakness, loss of appetite and loss of weight may be early symptoms. As the disease progresses, person may have jaundice, itchy skin, bruising and bleeding easily, develop spinder-like blood vessels on the skin *etc.* More serious symptoms include mental confusion and difficulty in thinking clearly, ascites, edema, impotence, gynecomastia *etc.* Other complications include portal hypertension, bacterial infections, splenomegaly, hepatic encephalopathy, bleeding from dilated veins in the esophagus (esophageal varices) or dilated stomach veins (gastric varices), gall stones, malnutrition, bone disease, acute –on-chronic liver failure and risk of liver cancer.

Nutritional Management

Malnutrition is common (65-100 per cent cases) and occurs in all forms of cirrhosis as shown by studies of nutritional status in cirrhosis of differing etiology and of varying degrees of liver insufficiency. They are malnourished due to presence of anorexia, vomiting and other gastrointestinal disorders. Nutritional intervention in cirrhotic patients should aim to support hepatic regeneration, prevent or correct malnutrition and prevent and/or treat the complications associated with cirrhosis. Various studies on nutritional support in liver disease concluded that aggressive nutritional support is essential to meet elevated protein requirements and reduced muscle catabolism and improve disease outcome. There is a general consensus of opinion that nutritional intervention in patients with cirrhosis improves survival, surgical outcome, liver function, and attenuates complications.

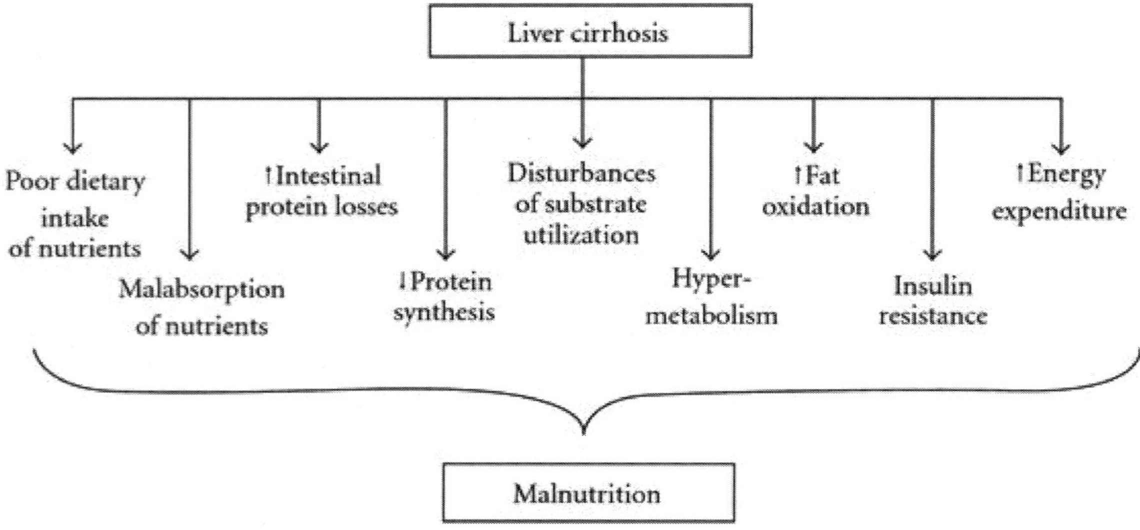

Nutritional recommendations for cirrhotic patients in general focus on suppression of hepatotoxic agents and the provision of optimal macronutrient supply in terms of energy, protein, carbohydrates and lipids together with micronutrients such as vitamins and minerals. Energy, macro- and micronutrient supplies should be based on the results of individual nutritional assessments and adjusted for weight maintenance and/or repletion. General recommendations are summarized in Table 23.2. A high energy, high protein, high carbohydrate, moderate fat, high vitamin diet has been found to be useful for regeneration of hepatic cells and in the management of ascites associated with cirrhosis. If there is ascites, sodium should be restricted. Fibre should be restricted if there is danger of oesophageal varices and portal hypertension. The diet should be attractive and palatable.

Energy and Protein

Protein-calorie malnutrition is more common in patients with cirrhosis compared to the general population, and is associated with higher in-hospital mortality rates. The severity of liver disease generally correlates with the severity of malnutrition, and protein-calorie malnutrition correlates with worsening clinical outcome. In addition, the degree of malnutrition correlates with the development of serious complications such as ascites, and hepatorenal syndrome as well as with a greater risk of post-operative complications and mortality rates in patients with cirrhosis.

Alterations of energy metabolism in chronic liver disease result in amino acid oxidation leading to protein deficiency, which occurs in all forms of cirrhosis. In addition, underlying pathophysiologic factors may cause loss of protein stores. Resting energy expenditure has been shown to be increased in cirrhotic patients and alterations in energy metabolism related to survival in these patients may even precede malnutrition in some cases. Hence, most patients with cirrhosis have high energy and protein requirements as recommended in Table 23.2.

Proteins should not be restricted in patients with liver disease unless they become protein intolerant due to encephalopathy. Protein intake should be in the range of 1-1.5 g/kg/day. Several studies have shown that a daily protein supply of 1.0-1.2g/kg/day may be sufficient to prevent negative N_2 balance in cirrhosis. With mild stress, this has to increase to 1.5 g/kg/day, and with acute exacerbations of hepatitis or decompensation to 2.0 g/kg/day.

Vegetable protein sources are good as they are higher in arginine, an amino acid that decreases blood ammonia levels through increasing urea synthesis. They are also lower in methionine and tryptophan. As per clinical studies the vegetarian protein diet increases the results of standard tests, improve nitrogen balance and electroencephalogram (EEG), and lower blood ammonia concentrations in liver cirrhotic patients. Patients who cannot achieve their protein intake goals by dietary protein may benefit from branched-chain amino acid (BCAA) supplementation. BCAA supplementation may also have a beneficial effect in patients with hepatic encephalopathy.

Table 23.2: General Recommendations for Cirrhotic Patients

Nutrient	*Recommendation*
Energy	35-50 Kcal/kg body wt Sufficient to restore/maintain nutritional status and enhance liver regeneration (adjust for obese patients)
Protein	1.0–1.8 g/kg body weight depending on the severity of malnutrition (adjust if renal disease present)
Carbohydrates	45–75 per cent of caloric intake or 4–6 meals rich in carbohydrates per day
Lipids	20-30 per cent of energy intake (adjust if steatorrhea is present)
Vitamins	B group vitamin supplements Particular attention to fat soluble vitamins Corrections of specific vitamin deficiencies
Minerals	Zinc, magnesium and selenium supplements Correct specific deficiencies

Source: Bémeur, C. and Butterworth, R.F. 2014. Nutrition in the Management of Cirrhosis and its Neurological Complications. J Clin Exp Hepatol. Jun; 4(2): 141–150.

Carbohydrates

Carbohydrates should provide 45-75 per cent of the energy intake depending upon the cause and symptoms. In infective hepatitis, high carbohydrate diet is given so as to restore the liver glycogen levels and it will help to minimize liver damage. Irritating fibres should be avoided in case of esophageal varices or gastric varices.

Fat

Lipid restriction has no scientific basis in patients with cirrhosis. Fat should be provided as polyunsaturated fatty acids, with less than 50 per cent long chain triglycerides. Fat helps make food tastier. This is important for people who suffer from a suppressed appetite due to chronic liver disease. Several investigations have concluded that excess dietary fat may encourage cirrhosis progression. High intakes of total fat, saturated fat and polyunsaturated fat have been implicated. Fat in the form of medium chain triglycerides should provide 20-30 per cent of energy intake. Medium chain triglycerides should be included in the diet of liver cirrhosis as it is better tolerated by the patients and it contains C8 to C10 which is digested and absorbed in the absence of bile. Amount of fat depends upon the symptoms. If fat malabsorption is there in cirrhosis, amount of energy from fat should be decreased.

Vitamins

In general, vitamin deficiencies in liver disease are related to disorders of hepatic function and diminished reserves and, with increasing severity of the disease, to inadequate dietary intake and/or malabsorption. Fat soluble vitamin deficiencies are common manifestations of malnutrition and cholestatic liver disease.

Vitamin A deficiency is described in cirrhosis and may be a risk factor in development of hepatitis C-related chronic liver disease. Vitamin A deficiency is also present in approximately 50 per cent of patients with alcoholic cirrhosis and patients with chronic alcoholism have been shown to have very low concentrations of hepatic vitamin A at all stages of their disease. Because high doses of vitamin A are potentially hepatotoxic, care must be taken to avoid excessive supplementation.

Chronic liver disease commonly results in vitamin D deficiency. In particular, a large proportion of patients with alcoholic liver disease have compromised vitamin D status. Vitamin D deficiency has also been linked to poor outcomes in patients with hepatitis C. An impaired immune function due to vitamin D deficiency could explain increased mortality in patients with chronic liver disease.

Vitamin E deficiency has been well documented in alcoholic liver disease. However, the beneficial effects of vitamin E supplementation in liver disease are dependent upon the nature of the disorder. Vitamin E supplementation has been reported to result in significant improvement in pathologic features of NASH including improvement in liver enzymes, as well as decreases in markers of steatosis and inflammation on liver biopsy.

Vitamin K is essential for the management of cirrhosis, because it helps in prevent bleeding of liver tissues. It also helps in conversion of glucose into glycogen, a chemical that is stored in liver. Glycogen is essential for bile excretion and healthy liver function.

Deficiency of water soluble vitamins is common in cirrhosis particularly in alcoholic cirrhosis. Thiamine deficiency is common in many forms of cirrhosis particularly alcoholic liver disease where it is caused by inadequate dietary intake, decreased hepatic storage, and impairment of intestinal thiamine absorption by ethanol. Deficiency of thiamine is well known to cause wernickes encephalopathy and korsokoffs dementia. Thiamine deficiency also occurs in HCV related cirrhosis. Riboflavin (vitamin B_2) deficiency has been described in patients with either alcoholic or non-alcoholic cirrhosis and has been explained by inadequate intake, increased utilization, deficient absorption and storage, or abnormal metabolism of the vitamin. However, a clear link between riboflavin deficiency and malnutrition in chronic liver disease has not, so far, been definitively established. Deficiencies in pyridoxine (vitamin B_6), folate (vitamin B_9) and cobalamin (vitamin B_{12}) may develop rapidly in chronic liver disease due to diminished hepatic storage. Deficiency of folate is also found in liver cirrhotic patients and an estimated 50 per cent have increased blood homocysteine concentrations which cause liver fibrosis and ultimately cirrhosis.

Minerals

Zinc plays a crucial role in the metabolism of protein, carbohydrate, lipid, nucleic acid, and ammonia. Zinc is an essential trace element required for normal cell growth, development and differentiation and zinc deficiency is common in many types of chronic liver disease. Zinc deficiency in patients with chronic alcoholism is attributed to decreased intake and absorption and diuretic-induced increased urinary excretion. Zinc supplementation reportedly reverses clinical signs of zinc deficiency in patients with liver disease and a recent randomized, double-blind, placebo-controlled clinical trial demonstrated that low dose zinc supplementation prevents deterioration of clinical status of cirrhosis. Zinc also inhibits hepatic inflammation and hepatic fibrosis. Furthermore, zinc supplementation produced metabolic effects and trended toward improvements in liver function, HE and overall nutritional status.

It has been demonstrated that alcohol impairs magnesium transport and homeostasis in brain, skeletal muscle, heart and liver. Magnesium deficiency is also associated with peripheral insulin resistance, which is common in alcoholic liver disease and, in a randomized clinical trial; magnesium treatment was reported to improve hepatic enzyme levels.

Selenium is incorporated into the active sites of multiple seleno-proteins with established antioxidant functions and several studies have shown that chronic liver disease is associated with decreases in serum, whole blood, and hepatic selenium content where selenium status correlated with severity of liver disease being most profoundly decreased in patients with decompensated cirrhosis.

Patients are greater risk for osteoporosis; they should take calcium and vitamin D rich foods or supplements.

Hepatic iron overload especially in hepatitis C produces oxidative stress and is a factor responsible for the development of HCV (hepatitis C virus) associated insulin resistance. There is a clear association between elevated iron indices and increased liver damage from HCV. Elevated iron indices have repeatedly been associated with increased serum transaminases.

Iron overload and excessive alcohol consumption might act in synergy to promote hepatic fibrogenesis. It was demonstrated that transferrin-iron saturation is associated with an increased incidence of cirrhosis, particularly in the presence of alcohol misuse. Also, untreated iron overload can lead to liver cirrhosis. Copper and copper-associated protein accumulation may be observed in chronic biliary obstructive processes and cirrhosis.

Total body manganese stores are increased in patients with liver disease, which may lead to selective manganese accumulation in several areas of the brain. Manganese deposition in basal ganglia structures of the brain has been proposed as the cause of T1-weighted magnetic resonance signal hyperintensities and cirrhosis-related Parkinsonism. Recent reports describe dysfunction of the nigrostriatal dopaminergic neuronal pathway related to manganese toxicity in patients with end-stage liver disease.

Fluid Restriction

Restriction of fluid is also important factor in nutritional therapy of cirrhotics as presence of ascites (accumulation of fluid in abdomen) which is a major complication of cirrhosis, Careful monitoring should be taken and maintenance of electrolyte and fluid balance. In liver disease, the ability of blood vessels to retain fluid is diminished because of decreased protein synthesis in liver, mainly albumin. This causes fluid leaks in the blood vessels, which in turn, causes fluid buildup in other tissues, or ascites. By limiting the amount of salt and fluid in diet, one can decrease fluid retention and swelling.

Fluid intake 30-40mL/kg/day maintain fluid balance. Dilutional hyponatremia develops due to decreased renal blood flow and greater free water accumulation. Restriction of water intake is the standard treatment of dilutional hyponatremia (serum Na < 130 mEq/l). Fluid restriction of 1 to 1.5 L fluid/day once serum sodium declines to less than 120-125 mEq/L, is considered standard practice. But generally there is no proof that water restriction in cirrhotic patients with ascites improves hyponatremia.

Sodium Restriction

It is not too much water intake that causes the buildup of ascitic fluid in the abdomen, but it is actually the body that is keeping in too much sodium (salt). Thus, it is important to cut down on salt intake, not water intake. Dietary sodium intake is usually restricted to less than 2000 mg per day. Hence, the cornerstone of ascites treatment is achievement of a negative sodium balance. This may be obtained through salt restriction in diet. A 2-g/day salt diet is palatable and an effective means in maintaining a negative sodium balance. However if applied alone, its efficacy is limited to 10 per cent of patients. Further restriction may increase efficacy but less palatable diet increases the noncompliance rates or worsens malnutrition.

Probiotic Treatment

Liver cirrhotic patients have been reported to have imbalance in bacterial gut flora which contributes significantly to ammonia production, resulting in varying degrees of encephalopathy. So these patients should take supplemental combinations of probiotics which reduce the blood ammonia concentrations.

Those patients which are treated with a combination of probiotics (*Lactobacillus plantarum*) and fiber had a lower rate of getting bacterial infections than those treated with selective intestinal decontamination, indicating a beneficial effect on the prevention of bacterial translocation. Probiotics decrease intestinal pH, inhibiting growth of pathogenic bacteria. Probiotics with fructo-oligosaccharides are equal to lactulose for hepatic encephalopathy. Generally, probiotics are safe and well tolerated.

Diet Therapy

1. Dietary restrictions should not be routinely imposed. Complications arising from cirrhosis, including ascites, oedema, encephalopathy, and steatorrhoea, may benefit from dietary modification providing it is undertaken within a context of nutritional adequacy, *i.e.* feed first, restrict second.
2. The most common and difficult to handle myth about liver disease is that there should be almost complete restriction of dietary fat and protein intake in diet, which is in contrast to the actual scientific dietary advices for such patients. Hence, we should regularly and persistently convince the patients to take high protein and fat rich diet with less salt, as decided upon degree of decompensation. Protein should be 1.0-1.8 g/kg/day in the absence of encephalopathy and restricted only in the presence of encephalopathy. Protein should be from vegetable sources.
3. Fat should not be restricted unless the patient finds it unpalatable or has clinically diagnosed steatorrhoea. For fat, medium chain triglycerides should be included as they are easily digested in the absence of bile. This fat is present in the coconut oil. Use olive oil in cooking instead of butter, shortening, margarine or vegetable oils. Unlike other oils, olive oil is an unsaturated fat, and may have a less significant impact on blood cholesterol than saturated fats. Also, saturated fats may worsen the symptoms of cirrhosis.
4. Probiotic rich foods like probiotic curd, buttermilk *etc.* should be taken.
5. **Maintain a healthy weight.** An excess amount of body fat can damage liver. Obese cirrhotic patients are often protein depleted If BMI>25, gradual weight loss of 5-10 per cent improves insulin sensitivity. Weight loss can be achieved by creating deficit of 500-1000 calories/day. They should accordingly take less energy rich foods *i.e.* foods rich in carbohydrates and fats.
6. **Eat a healthy diet.** Choose a plant-based diet that is full of fruits and vegetables. Select whole grains, beans and lean sources of protein.
7. There are several studies that support the view that a modified eating pattern with four to seven small meals rather than three big traditional meals, and including at least one late evening carbohydrate-rich snack, improves nitrogen economy in liver cirrhosis. In fact, such a modified eating pattern has been included in some international recommendations for nutritional therapy in chronic liver disease.
8. Cirrhosis is a disease of accelerated starvation, so patients should avoid long time without feeding. Patients often do better on multiple small meals with a late bed-time meal, which has been shown to reduce the need for gluconeogenesis and conserve proteins and nitrogen balance after an overnight fast, and prevent protein breakdown.
9. Increase intake of vitamin K by adding dark green vegetables such as broccoli, cabbage, avocados, spinach, kale, strawberries and eggs.
10. For vitamin A, take red, yellow and orange coloured fruits and vegetables in the diet.
11. Patients with chronic liver disease are at increased risk for the development of osteoporosis, it is important to consume foods rich in calcium and/or to supplement their diets with calcium and vitamin D.
12. Vitamins and mineral supplements should be taken.
13. Maintain a healthy lifestyle (eat a healthy diet and exercise regularly) to achieve a normal BMI (body mass index).
14. Stop drinking alcohol.
15. Stop smoking as it suppresses endogenous interferon leading to flare of HCV (hepatitis C virus). Smoking increases iron content leading to increased cirrhosis viral flare and HCC (Hepatocellular carcinoma). Smoking increases incidence of hepatic malignancies.
16. Soft foods devoid of fibre should be taken in case of oesophageal varices.
17. Patients who cannot tolerate food orally, enteral nutrition be provided in the hospital.

Foods Permitted	*Foods Avoided*
Cereals in soft form, cooked rice, *chapaties*, bread, *idli*, milk pudding, milk shakes, probiotic curd, cooked or pureed green red, yellow, orange coloured vegetables, *khichdi*, porridge, pulses, beans, meat, fruit, fruit juices, cooked beans, soft cooked eggs.	Meat, high fat fish, rich desserts, nuts, chocolate, fried foods and highly seasoned foods, pickles, rich gravies and sauces, spices and condiments, alcohol, gas forming vegetables.

Nutritional and Dietary Management of Ascites and Oedema in Cirrhosis

Fluid retention is common in cirrhosis and end-stage liver disease, and has relevance to nutrition.

Abdominal distension can impair food intake leading to an inadequate nutritional status. Energy expenditure increases due to exertion of carrying additional weight. Negative nitrogen balance can be induced by repeated large volume paracentesis, even if intravenous albumin infused. Restricting dietary sodium can help reduce or control the degree of fluid retention, and thus, improve symptoms.

Sodium and water should be restricted only in the presence of ascites and edema. A diet low in sodium can help to treat ascites and edema as it will minimize the amount of salt entering the kidney, leaving less sodium available for re-absorption, therefore, less fluid is retained. Those patients who have already poor appetite and inclusion of low salt diet will make the food unpalatable and may further reduce the food choices which can result to Protein Calorie Malnutrition in cirrhotic patients.

Diet should be fresh, perishable produce, which has to be bought, stored and prepared and many patients may not be able to do when they are already malnourished, weak and anorexic. There are also financial crises as well as issues of compliance. With these factors in mind and considering the clinical causes and significance of malnutrition, restrictions should be minimized and dietary therapy should aim to meet nutritional requirements. It would be better to use 'salt to tolerance' a reduction in salt intake that still allows adequate nutritional intake or nutritional support. More severe restriction will lead to poor compliance. Hence, a 2000 mg sodium-restricted daily diet is effective, when combined with diuretic therapy, for controlling fluid overload in 90 per cent of patients with cirrhosis and ascites.

Patients should be encouraged to eat as much as they are able and be tempted with tasty food. Small, frequent meals with snacks every 1–2 h may optimize intake. The energy density of foods may be increased by the addition of concentrated sources, *e.g.* sugar, honey, double cream, oil, and salt-free butter. Nutritional status should be monitored regularly. Body weight cannot be used; MAMC and triceps skin-fold thickness are most appropriate. Decisions about whether to restrict sodium intake must balance the potential benefits against the risk of reducing nutrient intake. Patients should be advised individually by a registered dietitian with experience in this area.

In case of low sodium diets for ascites and oedema, doctor should be consulted for prescribing diuretics to increase the urinary sodium excretion output. Limit dietary Na intake. Avoid salt at the table. Keep salt in cooking to minimum. Anything that tastes salty (such as tomato sauce, salsa, soy sauce, canned soups) probably has too much salt. Use pepper, vinegar, spices, herbs, and lemon for flavouring instead of salt. Fresh foods usually are a better bet than processed foods. Foods that are high in sodium or salt include most preserved or tinned items, such as canned soups and vegetables, bacon, ham, sausages, tinned and packet soups, stock cubes, tinned meats and fish, crisps and similar savoury snacks; cheeses; condiments; bakery goods; pickles, papads and tomato ketchup *etc.* which should be avoided. Cheese (cheddar-type) should be limited to 100 g/week. Fast foods and ready-meals should be avoided unless they are known to provide <30 mmol Na^+ per portion. Restrict bread and other bakery goods. Use salt-free butter or margarine. Breakfast cereals should be salt-free, *e.g.* Puffed Wheat, Shredded Wheat. Restrict milk to 300 ml per day. One should avoid fried foods. Avoid other sources of sodium such as antacids, some antibiotics, salt substitutes (most contain K^+ salts, but some also contain $Na^{+)}$, sodium bicarbonate *etc.*

Hepatic Encephalopathy

Hepatic encephalopathy is a common potentially reversible neuropsychiatric complication of liver dysfunction, including acute liver failure and liver cirrhosis. It is currently defined as a brain dysfunction secondary to liver insufficiency and/or portosystemic shunting that manifests as a broad spectrum of neuropsychiatric abnormalities ranging from subclinical alterations to coma. The term implies that altered brain function is due to metabolic abnormalities. The full reversibility of symptoms after improvement of liver function is considered to be direct proof of this causal relation. One of the most debilitating complications of cirrhosis, encephalopathy affects 30–45 per cent of cirrhotics. It is a significant contributor of morbidity in patients with liver disease.

Classification

Hepatic encephalopathy (HE) is generally classified according to whether its presence is overt or covert type. Covert hepatic encephalopathy includes minimal HE and grade 1 HE according to the West Haven criteria (WHC), while overt HE includes grades 2–4 (Table 23.3).

Table 23.3: Grade or Stage of HE According to the West Haven Criteria

Grade or Stage	*Clinical Features*
0	No obvious changes other than potentially mild decrease in intellectual ability and coordination
1	Minimal lack of awareness, euphoria or anxiety, shortened attention span, impairment of calculation ability, altered sleep pattern
2	Disorientation (usually to time), lethargy, inappropriate behavior, subtle personality changes,
3	Gross disorientation (time and place), marked confusion, somnolent but responsive to stimuli, incomprehensible speech
4	Coma, unresponsive to verbal or physical stimulation

In 2014, American Association for the Study of Liver Diseases/European Association for the Study of the Liver Guidelines updated the nomenclature of hepatic encephalopathy-4 axes. They have come up with joint guideline for the management of HE in chronic liver disease; the task force recommends that HE should be classified based on the underlying etiology, severity of clinical manifestations, time course and whether or not a precipitating factor is identified (Table 23.4). Each episode of HE should be characterized by one component from each of the 4 axes or columns. Also, covert HE should be further characterized as grade II, III or IV. For example: HE, Type C, Covert, Grade III, Episodic and Precipitated by dehydration.

Table 23.4: Clinical Description of Hepatic Encephalopathy (Adapted from AASLD-EASL Guideline 2014)

Type	*Grade*		*Time Course*	*Spontaneous/ Precipitated*
A	Minimal	Covert	Episodic	Spontaneous
	I			
B	II		Recurrent	
C	III			Precipitated (specify)
	IV	Overt	Persistent	

Source: Dharel, N. and Bajaj, J.S. 2015. Definition and nomenclature of hepatic encephalopathy. J. Clin. Exp. Hepatol. March: 5 (Suppl 1): S37-S41.

It should be noted that this nomenclature system primarily pertains to HE associated with chronic liver diseases. Each of these four axes bears unique importance in the clinical description, management, and prognosis of patients with HE.

Hepatic encephalopathy is further divided into 3 subtypes based on etiology, with type A associated with acute liver failure, type B with portosystemic shunts [such as transjugular intrahepatic portosystemic shunt (TIPS)] and type C with cirrhosis. Depending on its frequency, hepatic encephalopathy is classified as episodic (if one episode occurs in 6 months), recurrent (if more than one episode occurs in a 6-month period) or persistent (if the neuropsychiatric abnormalities persist and do not return to baseline). Based on presence or absence of detectable precipitating factor, HE can be Spontaneous (non- precipitated) and Precipitated (secondary to).

Etiology

While the precise molecular mechanisms that result in these morphological changes in the brain are yet to be identified, many factors have been elucidated, especially the role of ammonia, false neurotransmitters, astrocyte swelling, inflammation, and oxidative stress.

The pathogenesis is linked with ammonia and glutamine production. Ammonia, a byproduct of the metabolism of nitrogen-containing compounds is normally converted to urea in the liver and cleared out of the body through the urine. Ammonia is elevated in individuals with acute and chronic liver disease. The increase in blood ammonia in advanced liver disease is a consequence of impaired liver function and of shunting of blood around the liver. Muscle wasting, a common occurrence in these patients, also may contribute since muscle is an important site for extrahepatic ammonia removal. Ammonia is highly neurotoxic at elevated concentrations. Although ammonia is generally accepted to a play a role in hepatic encephalopathy, some individuals with elevated ammonia levels do not

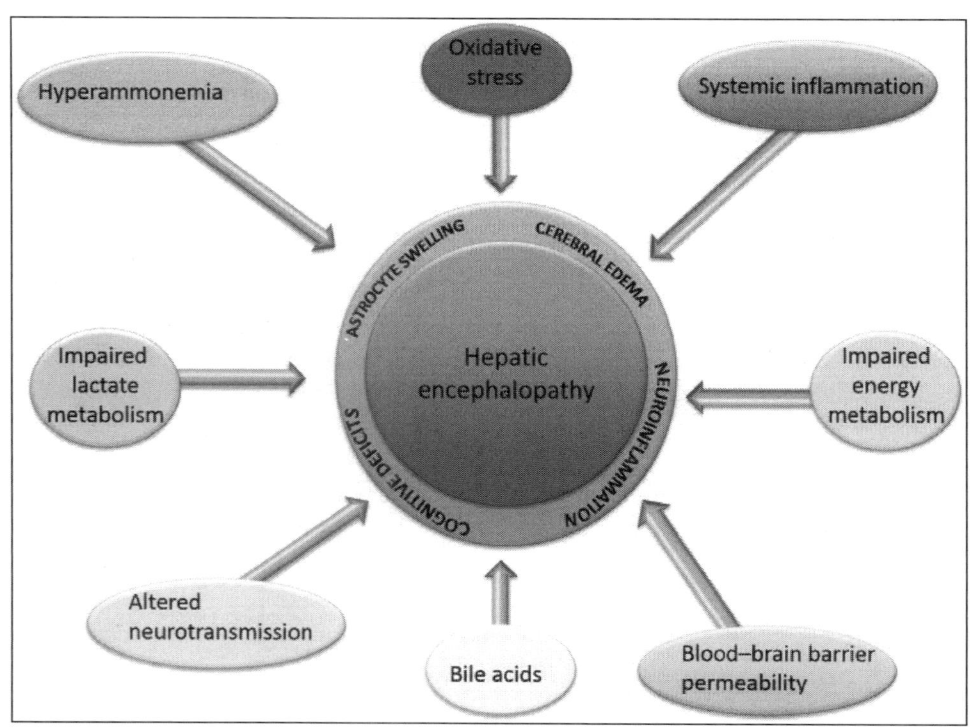

develop symptoms, suggesting that additional factors play a role in the development of the disorder.

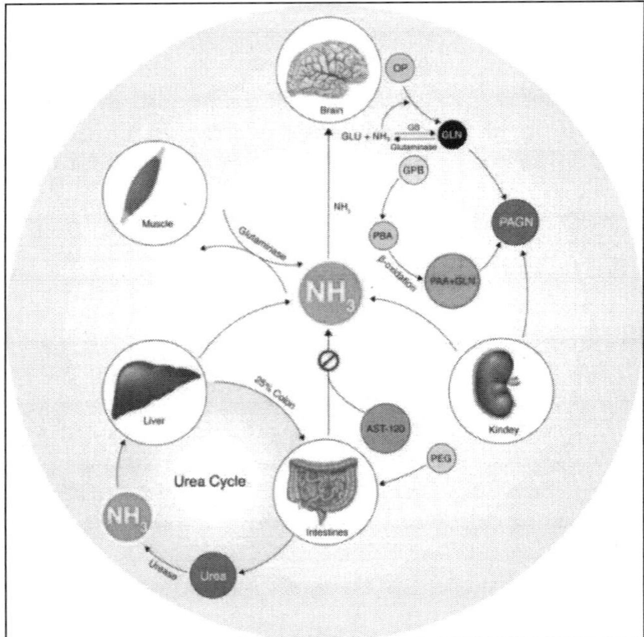

Additional factors that have been explored as potentially playing a role in the development of hepatic encephalopathy include manganese toxicity and impaired function of certain central nervous system cells called astrocytes, which play a role in regulating the blood-brain barrier and also help to detoxify certain chemicals including ammonia; dysfunction of the blood-brain barrier, which prevents dangerous substances from reaching the brain; amino acid imbalances; short chain fatty acids; infection; inflammation; and increased activity of GABA, an inhibitory neurotransmitter in the central nervous system. More research is necessary to determine the exact, underlying factors that ultimately cause the development of hepatic encephalopathy and its associated symptoms.

Episodes of hepatic encephalopathy may be triggered by certain events or occurrences such as low levels of oxygen in the body, dehydration, constipation, gastrointestinal bleeding, an alcohol binge, infection, kidney abnormalities and the use of certain drugs especially those that act on the central nervous system such as tranquilizers and other sleep medications, antidepressants, and antipsychotics. In some cases, surgery can precipitate an episode of hepatic encephalopathy. Gastrointestinal bleeding is the most common precipitating event associated with hepatic encephalopathy, most likely because individuals with cirrhosis are at a greater risk of gastrointestinal bleeding than the general population.

Symptoms

Symptoms of hepatic encephalopathy differ depending on the underlying cause of the liver damage. Symptoms and signs of moderate hepatic encephalopathy may include difficulty in thinking, personality changes, poor concentration, problems with handwriting or loss of other small hand movements, confusion, forgetfulness, poor judgment, a musty or sweet breath odor, inappropriate behavior *etc.* Symptoms of severe hepatic encephalopathy include confusion, drowsiness or lethargy, anxiety, seizures, severe personality changes, fatigue, confused speech, shaky hands, slow movements and coma *etc.*

Treatment

Hepatic encephalopathy is possibly reversible with treatment. This typically involves supportive care, addressing the triggers of the event and preventing the formation of gut derived toxins especially ammonia.

Nutritional Management

Besides medical treatment, nutrition has also a great role in the management of hepatic encephalopathy. Appropriate enteral nutrition, by mouth or nasogastric feeding tube depending upon the condition, should be administered as soon as feasible.

Patients with advanced liver disease should eat a diet that provides adequate calories, proteins, vitamins, and minerals.

Energy

The diet should include total calories providing at least 1.2 times the estimated resting energy expenditure, generally at least 35-40 kcal/kg of ideal body weight. In cirrhotic patients without ascites, actual body weight should be used to calculate basal metabolic rate using a formula such as that proposed by Harris and Benedict. In patients with ascites, the ideal weight according to body height can be used.

Carbohydrates

Even though carbohydrate continues to be the mainstay of caloric intake in cirrhotic patients, these individuals often have insulin resistance leading to decreased peripheral glucose utilization, decreased hepatic glucose production, and decreased hepatic glycogen reserves. Excess calorie intake in the form of carbohydrate should be avoided because of its tendency to promote hepatic lipogenesis, contribute to liver dysfunction, and possibly increase carbon dioxide production leading to increased work of breathing. The European Society for Parenteral and Enteral Nutrition (ESPEN) guidelines recommend that 30 per cent 35 per cent of calories be consumed as fat and the remaining 50 per cent 55 per cent as carbohydrate (Chadalavada *et al.*, 2010).

Protein

Classically, the recommendation for patients with HE with cirrhosis has been to restrict dietary protein intake, but this is a measure that nowadays is considered wrong. For several decades, it was recommended withholding all protein intake and subsequently increasing the intake to maximal tolerance; however, this recommendation has no solid support. Although avoiding intake of large amounts of protein may be positive to reduce toxins that are involved in HE, restriction may worsen liver function and increase the risk of death. Negative nitrogen balance due to protein restriction leads to protein–energy malnutrition. PEM is considered a negative prognostic factor altering immunity and is associated with life threatening complications such as refractory ascites, spontaneous bacterial peritonitis, hepatorenal syndrome, and variceal hemorrhage. According to recent research studies, cirrhotic patients exhibit increased protein requirement to achieve balanced nitrogen metabolism, and normal protein diets have been given safely to patients with hepatic encephalopathy. A positive nitrogenous balance may improve encephalopathy by promoting hepatic regeneration and increasing the capacity of muscle to detoxify ammonia. For these reasons the current recommendation is to avoid restrictions of dietary protein. The initial dose of protein in patients with acute encephalopathy is 0.5 g/kg/day – it should be carefully increased up to 1.0-1.5 g/kg/day. Most patients with mild chronic HE tolerate more than 60 to 80 g of protein per day.

Thus, restriction is rarely required but, if necessary, usually for no more than 48 hours. It should be noted that the recommended protein supplementation is based on "dry" body weight and may need alteration in edematous patients. Importantly, the risk of aspiration pneumonia in patients with advanced hepatic encephalopathy during tube feeding must be weighed against the potential complications of parenteral nutrition

Some studies suggest that the type of protein may influence the outcome. Diets containing vegetable proteins appear to be better tolerated than diets rich in animal protein. One study reported that a diet rich in vegetable protein (71 g/d) significantly improved the mental status of patients suffering from HE while increasing their nitrogen balance. Vegetable proteins may also increase intraluminal pH and decrease gastric transit time. This may be because of increased content of dietary fiber, a natural cathartic, and decreased levels of aromatic amino acids. Aromatic amino acids, as precursors for the false neurotransmitters tyramine and octopamine, are thought to inhibit dopaminergic neurotransmission and worsen HE. A daily intake of 30–40 g vegetable protein has been found to be effective in the majority of patients.

Branched Chain Amino Acids

Patients with cirrhosis have an imbalance between aromatic amino acids and branched chain amino acids. It has been suggested that this imbalance contributes to the development of HE. Branched chain amino acids are being promoted as a mechanism to correct the imbalance in the plasma amino acid profile and thought to be involved in the pathogenesis of hepatic encephalopathy. However, clinical trials using branched chain amino acids have not shown dramatically beneficial effects for episodes of HE. A new analysis of several published studies, including some more recent, supports a mild positive effect of BCAA on increasing muscle mass and minimal HE, whose clinical impact is debatable.

Table 23.5: Recommendations Regarding Energy and Protein Provision in Patients with Cirrhosis and Hepatic Encephalopathy

✰ Optimal daily energy intake should be 35–40 kcal/kg ideal body weight.
✰ Optimal daily protein intake should be 1.2–1.5 g/kg ideal body weight.
✰ Small meals evenly distributed throughout the day and a late-night snack of complex carbohydrate will minimize protein utilization.
✰ Encourage ingestion of a diet rich in vegetable and dairy protein.
✰ BCAA supplementation might allow recommended nitrogen intakes to be attained/maintained in patients who are intolerant of dietary protein.
✰ Multivitamin and mineral supplement should be given.

Source: Amodio *et al.*, 2013. The nutritional management of hepatic encephalopathy in patients with cirrhosis: International society for hepatic encephalopathy and nitrogen metabolism consensus. Hepatology. 58 (1): 325-329.

Fat

Fat should provide 10-20 per cent of the calories.

L-Ornithine-L-Aspartate (LOLA)

LOLA lowers plasma ammonia concentrations by enhancing the metabolism of ammonia to glutamine. According to one study, LOLA was significantly more effective in the improvement of HE. Two studies that compared LOLA and lactulose showed similar efficacy in the treatment of HE. Intravenous LOLA can be used as an alternative or additional agent to treat patients nonresponsive to conventional therapy.

Nonabsorbable Disaccharides as Prebiotics

Lactulose is classified, by some, as a prebiotic and its efficacy as a treatment for HE is proven. Two studies conducted in India showed marked improvement in the recurrence of HE after approximately one year of treatment with lactulose. Lactulose (β-galactosido-fructose) and other non-absorbable disaccharides are introduced with the goal of promoting the growth of *Lactobacillus bifidus*, which contains little urease activity and via this mechanism could decrease the production of ammonia in the colon. This is not its mechanism of action, which is still complex and not fully understood. The bulk of evidence links the efficacy of lactulose to an interaction with the enteric flora and to a decrease in the generation of nitrogenous compounds in the intestine

Administered orally, lactulose is not broken down by intestinal disaccharidases and reaches the cecum, where it is metabolized by enteric bacteria to lactate and acetate, causing a drop in cecal pH leading to an increase in fecal nitrogen excretion and a decrease in the amount of nitrogen that reaches portal blood.

Soluble fiber (*e.g.*, gum from pulses and pectin from fruit) is generally fermentable and, as such, has prebiotic properties. Some information available on the basis of few studies, it would appear that fermentable or prebiotic fiber may have a beneficial effect on neuropsychiatric performance in this patient population. Thus, fiber itself or fiber enriched diets, which would include diets made up of predominantly vegetable protein (*vide supra*), may benefit patients with cirrhosis and HE. However, care must be taken not to induce diarrhea, particularly in patients who are already taking lactulose.

Probiotics: The gut microbiota plays an important role in the generation of ammonia; thus, its modulation using prebiotics, probiotics, and synbiotics has been evaluated as a therapeutic option in patients with HE in several small studies.

Multiple studies have looked at the use of probiotics for the treatment of HE. A recent meta-analysis showed that compared to placebo, probiotic use was associated with improvement in covert HE, decreased hospitalization and decreased progression to overt HE in patients with minimal (*i.e.* covert) HE. However, these effects were not observed when probiotic use was compared to lactulose.

Vitamins

Supplement of vitamin B complex especially containing thiamine should be given.

Minerals

Zinc deficiency has been implicated in the development of HE because some of the urea cycle enzymes are zinc-dependent. Zinc deficiency is prevalent in patients with cirrhosis. Multiple trials have evaluated the use of zinc in the treatment of HE, yielding variable results. A meta-analysis of four RCTs showed that zinc supplementation improved the results of some psychometric tests but not of HE recurrence. In the setting of zinc deficiency or confirmed losses, zinc supplementation can be considered when treating HE.

Changes in circulating levels of calcium, magnesium, and iron, though not directly implicated in the pathophysiology of HE, can produce changes in mental function, ranging from confusion to coma or secondary dementia, which may confound its diagnosis. Hypercalcemia and hypomagnesemia should be excluded or corrected in confused or comotose patients with cirrhosis; plasma magnesium levels should be carefully monitored. Iron deficiency should be excluded or corrected in patients with persistent, otherwise unexplained, mild cognitive dysfunction.

Low circulating levels of sodium can alter brain function both directly and by interacting with the mechanisms causing HE; hyponatremia is thus a recognized risk factor for the development of HE.

Table 23.6: Nutritional Management of Patients with HE Based on Current Evidence and Consensus Opinion

	Patients With HE									
Nutritional Status	***Adequately Nourished***			***Moderately Malnourished/At Risk***				***Severely Malnourished***		
Body Weight (estimated BMI[1])	***Normal/ Over-weight (20–30)***	***Obese (30–40)***	***Obese (>40)***	***Low/ Overwt (18–30)***	***Obese (30–40)***	***Obese (>40)***	***Low/ Overwt (18–30)***	***Underwt./ Overwt. (18-30)***	***Obese (30–40)***	***Obese (>40)***
Daily energy, kcal/kg[2]	35–40	25–35	20–25	35–40	25–35[3]	20–25[3]	35–40	35-40	25–353	20–25[3]
Daily protein, g/kg[2]	1.2–1.5	1.0–1.5	1.0–1.5	1.2–1.5			1.2–1.5			
Meal patterns	Small frequent meals throughout the waking hours									
Late-evening snack	Encourage ingestion of 50 g of complex carbohydrate									
Dietary nitrogen source	Promote vegetable and dairy protein to level of tolerance				Promote high protein intake per patient preference to encourage intake					
Daily fiber[4]	Encourage ingestion of diets containing 25–45 g, especially in overweight patients									
Decompensa-ted cirrhosis	Additional medications and supplements in line with therapy goals									
HE incompletely/ poorly controlled	Consider use of probiotics and/or BCAA supplements									

1: Use estimated dry weight to calculate BMI in patients with fluid retention; 2: Use ideal body weight for calculation of requirements; 3: Achieved by reducing the carbohydrate and fat content of the diet and increasing dietary fiber; 4: Useful to aid weight loss, but care needed not to induce diarrhea in patients receiving lactulose.

Source: Monika Pazgan-Simon *et al.*, 2015. Nutrition principles and recommendations in different types of hepatic encephalopathy. Clinical and Experimental Hepatology. 4: 1–6.

Dietary sodium intake is usually reduced in patients with ascites, although the evidence for the long term utility of this manoeuver is poor and debated. However, intakes should not be reduced below 60 mmol/day because otherwise the diet becomes unpalatable, resulting in reductions in total energy and protein intakes. Hyponatremia is more likely to occur when the intake of sodium is low, while the intake of water is maintained or increased. Thus, careful monitoring of sodium and water balance is required. Hyponatremia should be corrected slowly, in patients with HE.

Manganese is an important component of a number of cerebral enzymes, in particular, glutamine synthetase. Patients with cirrhosis have elevated total body manganese levels, most likely reflecting the combined effects of hepatocellular failure, impaired biliary excretion, and the presence of portal systemic shunting of blood. However, there is no clear relationship between the presence, extent, and time course of manganese accumulation and development of HE. Nevertheless, it would seem sensible to avoid the use of nutritional supplements containing manganese in this patient group.

Dietary Management

- Diet changes have been studied to determine their effects on cognitive function in patients with MHE with cirrhosis. One study in 2016 compared patients prescribed a diet with a caloric intake of 30–35 kcal/kg and a protein intake of 1.0–1.5 g of vegetable protein/kg to a control group for six months. They found that a higher proportion of patients on the prescribed diet had reversal of HE along with a decreased incidence of overt HE in the treatment group.
- Timing of caloric ingestion is important in effecting changes in substrate utilization and hence, avoiding undue utilization of gluconeogenesis to maintain splanchnic glucose output. This is particularly important in patients with HE because utilization of amino acids for glucose production depletes tissue protein stores and produces ammonia. Patients should avoid fasting for longer than 3 6 hours during the daytime and so should be encouraged to take small, frequent meals evenly distributed throughout the day. Provision of 4-6 small meals daily with night time snack of complex carbohydrates may improve hepatic encephalopathy by avoiding protein loading and distributing protein equally.
- A late evening meal has been shown to have a positive effect on nitrogen balance in patients with cirrhosis when compared with equicaloric diet without a late evening meal.
- Ingestion of red meat protein should be discouraged.
- Patients with hepatic encephalopathy need to limit their diet to about 2 small servings a day of animal protein while most of the protein should come from vegetable sources such as dried beans and peas, soy products, other legumes and dairy. Vegetable protein is poor in the sulphated amino acids, methionine and cysteine, which are the precursors of mercaptans and indole/oxindole compounds, which, in turn, have been implicated in the genesis of HE. Conversely, they are high in ornithine and arginine, which may facilitate ammonia disposal through the urea cycle.
- Lactulose if in powder form, should be mixed with water or juice and taken once or twice a day. Lactulose syrup (2 tbsp or more once or twice daily) can also be taken.
- Avoid unnecessary dietary restrictions.
- In case of ascites, limit sodium rich foods.
- Stop alcohol intake.
- Oral intake should be encouraged in all patients. If oral intake is not possible, enteral nutrition should be started. Parenteral nutrition is less desirable than enteral nutrition.

Cholecystitis and Cholelithiasis

The gallbladder is a small organ underneath the liver on the right side of the upper abdomen. It stores a fluid called bile which is produced by the liver to help with digestion. Cholecystitis is an inflammatory condition of the gall bladder characterized by the inflammation of the gallbladder wall, which may be due to retention of bile in gallbladder or secondary to infection by microorganisms, predominantly *Escherichia coli*, *Klebsiella*, *Enterobacter*, and *Bacteroides species.*

Types of Cholecystitis

Cholecystitis can be acute or chronic. The causes of acute cholecystitis can be grouped into two categories *i.e.* calculous cholecystitis and acalculous cholecystitis.

i) Calculous Cholecystitis

It is the most common and it develops when the main opening to the gallbladder, called the cystic duct, gets blocked by a gallstone or a substance known as biliary sludge which is a mixture of bile, a liquid produced by the liver that helps digest fats, and small cholesterol and salt crystals.

Cholelithiasis refers to the presence of abnormal concretions (gallstones) in the gallbladder. Gallbladder stone disease is a chronic recurrent hepatobilliary disease, the basis for which is the impaired metabolism of cholesterol, bilirubin and bile acids, which is characterized by the formation of gall stones in the hepatic bile duct, common bile duct, or gallbladder.

The various components of gallstones are cholesterol, bile pigment, and calcium salts, in the form of phosphate, carbonate palmitate. Depending upon the composition, there are three types of gallstones:

- Pure cholesterol stones, which contain at least 90 per cent cholesterol. This type of gallstone is more common. Cholesterol gallstones are formed when the cholesterol concentration in bile exceeds the ability of bile to hold it in solution, so that crystals form and grow as stones. Cholesterol is virtually insoluble in aqueous solution, but in bile it is made soluble by association with bile salts and phospholipids in the form of mixed micelles and vesicles. Three types of abnormalities are responsible for cholesterol gallstone formation. Cholesterol super saturation, the essential requirement for cholesterol gallstone formation, might occur via excessive cholesterol biosynthesis, which is the main lithogenic mechanism in obese persons. In the non-obese, defective conversion of cholesterol to bile acids, due to a low or relatively low activity of cholesterol 7α hydroxylase, the rate limiting enzyme for bile acid biosynthesis and cholesterol elimination could result in excessive cholesterol secretion. Finally, interruption of the enterohepatic circulation of bile acids could increase bile saturation. Temporary interruption of the enterohepatic bile acid circulation during overnight fasting leads to a higher cholesterol/ phospholipid ratio in the vesicles secreted by the liver. Estrogen treatment also reduces the synthesis of bile acid in women.
- Pigment stones either brown or black, which contain at least 90 per cent bilirubin. Brown pigment stones are mainly composed of calcium bilirubinate whereas black pigment stones contain bilirubin, calcium and/or tribasic phosphate. Pigment stones occur when red blood cells are being destroyed, leading to excessive bilirubin in the bile. Black pigment stones are more common in patients with cirrhosis or chronic hemolytic conditions such as the thalassemias, hereditary spherocytosis, and sickle cell disease, in which bilirubin excretion is increased. Primary bile-duct stones, defined as stones that originate in the bile ducts, are usually brown pigment stones associated with infection. Bacteria in the biliary system release β-glucuronidases, which hydrolyze glucuronic acid from conjugated bilirubin. Primary brown pigment stones of the bile ducts often occur in Asians, associated with decreased biliary secretory Immunogloblin A (IgA.)
- Mixed composition stones, which contain varying proportions of cholesterol, bilirubin and other substances such as calcium carbonate, calcium phosphate and calcium palmitate.

ii) Acalculous Cholecystitis

It is a less common, but usually more serious, type of acute cholecystitis. It usually develops as a complication of a serious illness, infection or injury that damages the gallbladder. It can be caused by accidental damage to the gallbladder during major surgery, serious injuries or burns, sepsis, severe malnutrition or AIDS.

If cholecystitis persists over time, such as for months, with repeated attacks, or if there are recurrent problems with gallbladder function, it is known as chronic cholecystitis. Symptoms of cholecystitis often appear after eating a large, fatty meal. They include an intense pain in the middle or right side of upper abdomen, or between shoulders along with nausea and/ or vomiting, fever, indigestion especially after taking fatty meal *etc.*

Prevalence of Gallbladder Stones

Now a days, gallbladder disease is a frequent problem in India, representing a major health problem. The prevalence is seven times more common in Northern Indians than Southern Indians. According to recent study, it has been reported that gallstones are much more common in the female population (61 per cent) as compared to males (39 per cent). The age group most affected is 45–60 years (38.5 per cent) among females, and above 60 years in males (20.8 per cent). A relatively higher prevalence of 39 per cent among males when compared to reports from past studies indicates a significant shift in the pattern of prevalence of gallstone disease.

Risk Factors

Age

Increasing age especially has been associated with an increased prevalence of gallstones. Gallstones are 4-10 times more frequent in older (above 40 years of age) than younger subjects. Men older than the age of 60 and women between the ages of 20 and 60 are at increased risk of developing gallstones. Pregnant women are more likely to have gallstones with symptoms. Biliary cholesterol saturation increases with age, due to a decline in the activity of cholesterol 7α hydroxylase, the rate limiting enzyme for bile acid synthesis. Deoxycholic acid proportion in bile increases with age through enhanced 7α dehydroxylation of the primary bile acids by the intestinal bacteria.

Obesity

High-risk and most common factor for the cholesterol dominant gallstones is the obese individuals with a body mass index - 30 kg/sq.m^2, more so for women than for men. It raises the risk of cholesterol gallstones by increasing biliary secretion of cholesterol, as a result of an increase in 3-hydroxy-3-mthylglutaryl coenzyme A (HMGCoA) reductase activity. Epidemiological studies have found that the lithogenic risk of obesity is strongest in young women, and that slimness protects against cholelithiasis.

Gender, Parity and Oral Contraceptives

Increased levels of the hormone estrogen, as a result of pregnancy or hormone therapy, or the use of combined (estrogen-containing) forms of hormonal contraception, may increase cholesterol levels in bile and also decrease gallbladder movement, resulting in gallstone formation. Women during their fertile years are almost twice as likely as men to experience cholelithiasis. This preponderance persists to a lesser extent into the postmenopausal period, but the sex difference narrows with increasing age.

Genetics, Race and Ethnicity

Genetic factors play a role in gallstone disease. There is an increased risk of gallstones among first-degree relatives like a parent or sibling.

Racial differences that cannot completely be explained by environmental factors, have been shown to exist in the prevalence of gallstones. As a general rule, there appears to be higher rates of cholelithiasis in Western Caucasian, Hispanic, and Native American populations and lower rates in Eastern European, African American, and Japanese populations. Cholesterol gallstone prevalence varies widely, from extremely low (<5 per cent) in Asian and African populations, to intermediate (10-30 per cent) in European and Northern American populations, and to extremely high (30-70 per cent) in populations of Native American ancestry (Pima Indians in Arizona, Mapuche Indians in Chile).

The Pima tribe of Arizona has the highest gallstone prevalence in the world; more than 70 per cent of Pima women older than 25 years had gallstones or a history of cholecystectomy. High rates of gallstone prevalence have been also reported in other North American Indian tribes. Certain Hispanic populations in the USA are above average risk for gallbladder disease. Some studies strongly support the existence of Amerindian lithogenic genes in Mexican-Americans.

Rapid Weight Loss

Fasting in the short term increases the cholesterol saturation of gallbladder bile and in the longer term, causes gallbladder stasis which can lead to sludge, and eventually gallstone formation. Rapid weight loss is associated with occurrence of sludge and gallstones in 10-25 per cent of patients in a few weeks of initiating the slimming procedures. If a person loses weight too quickly, the liver secretes extra cholesterol; in addition there is rapid mobilization of cholesterol from adipose tissue stores. In fasting associated with severely fat restricted diets, gallbladder contraction is reduced, and the accompanying gallbladder stasis favors gallstone formation. Enhancing gallbladder emptying by inclusion of a small amount of dietary fat inhibits gallstone formation in patients undergoing rapid weight loss. A shorter overnight fasting is protective against gallstones in both sexes.

Physical Activity

Sedentary behaviour, is positively associated with the risk of cholecystectomy. Regular exercise, in addition to facilitating weight control, alone or in combination with dieting, improves several metabolic abnormalities related to both obesity and cholesterol gallstones.

Diabetes

People with diabetes generally have high levels of triglycerides which may increase the risk of gallstones. Gallbladder function is impaired in the presence of diabetic neuropathy, and regulation of hyperglycaemia with insulin seems to raise the lithogenic index.

A lack of melatonin could significantly contribute to gallbladder stones, as melatonin inhibits cholesterol secretion from the gallbladder, enhances the conversion of cholesterol to bile, and is an antioxidant, which is able to reduce oxidative stress to the gallbladder.

Diet

Increased intake of total and saturated fat, refined carbohydrates and decreased fibre content in the diet is a potent risk factor for development of gallstones. Calcium intake seems to be inversely associated with gallstone prevalence; dietary calcium decreases cholesterol saturation of gallbladder bile by preventing the reabsorption of secondary bile acids in the colon. Vitamin C influences 7α hydroxylase activity in the bile and it has been shown that ascorbic acid might reduce lithogenic risk in adults. Those who take a higher proportion of vegetable proteins have a lower incidence of gallstones.

Coffee consumption seems to be inversely correlated with gallstone prevalence, due to an increased enterohepatic circulation of bile acids; may be due to enhanced gall bladder motility, inhibition of gallbladder fluid absorption, decreased cholesterol crystallization in bile and perhaps increased intestinal motility.

Medical Conditions

Persons suffering from diseases of the gallbladder and its ducts, blood diseases with chronic hemolysis such as sickle cell anemia, liver cirrhosis, Crohn disease, surgical removal of ileum, very high triglyceride levels, hyperparathyroidism, spinal cord injury, gastric bypass, metabolic syndrome *etc.* are at risk for gallbladder stones.

Drugs

All fibric acid derivatives increase biliary cholesterol saturation while lowering serum cholesterol. Clofibrate is a potent inhibitor of hepatic acyl-CoA cholesterol acyltransferase (ACAT) which may lead to an increased availability of free or unesterified cholesterol for secretion into bile, favouring gallstone formation.

Additionally, prolonged use of proton pump inhibitors has been shown to decrease gallbladder function, potentially leading to gallstone formation.

Symptoms

Gallstones may cause no signs or symptoms. Only 25 per cent of patients have symptoms.

If a gallstone lodges in a duct and causes a blockage, the resulting signs and symptoms may include sudden and rapidly intensifying pain in the upper right portion of the abdomen or centre of the abdomen which may radiate to the right shoulder, and back, nausea, vomiting, feeling of satiety, bloating, dyspepsia *etc.* Gall stone pain may last several minutes to a few hours. Complications can be:

Acute Cholecystitis

Often patients will have history of biliary pain which lasts for > 3 hours, associated with fever and right upper quadrant tenderness.

Chronic Chlolecystitis

Patients will have episodic epigastric or right upper abdominal pain lasting for more than 30 minutes especially after eating a fatty meal.

Choledo Cholithiasis

When a stone from the gallbladder slips into the common bile duct, it can lead to biliary colic, obstructive jaundice followed by acute cholangitis or pancreatitis.

Acute Cholangitis

It is a medical emergency. Patients may present with right upper quadrant abdominal pain, fever and jaundice.

Pancreatitis

A most dreaded and life threatening complication of gallstones presents with very severe, persistent central upper abdominal pain, retching and prostration.

Dietary Treatment

Treatment of gallstones depends partly on whether they are causing symptoms or not. Recurrent episodes of upper abdominal pain related to gallstones are the most common indication for the medical treatment of gallstones which usually involves antibiotics, pain medications, and removal of gall bladder. Prior to elective surgery, obese persons should try to lose some weight.

It is important to know that there is no specific diet or food that has been proven to prevent gallbladder disease. Diet modifications may not offer any advantages for gallbladder disease before surgery as the passage of gallstones into the ducts is a random event unrelated to the type of food. However, the following set of dietary guidelines may help to keep the gallbladder healthy and fully functioning under the given situations:

Gall Bladder Stones without Symptoms

Some dietary and lifestyle changes should be made if someone has gallstones but surgery is not recommended or if one wants to avoid surgery. Rapid weight loss is not desirable in such situation as it can worsen the symptoms. As dietary fat is the most common causative agent of pain, hence one should stick to regular moderate-fat, healthy and well balanced diet in other nutrients rather than completely avoiding fat in the routine daily diets. Research shows that following a severely fat restricted diet is inappropriate for the treatment of gallstones. A diet too low in fat may actually lead to gallstone formation secondary to inadequate bile production. It is encouraged to consume a healthy diet with moderate amounts of fat; 20-30 per cent of calories from fat which equals about 30-45ml or 2-3 tbsp of fat each day. One should take healthy fat such as non-hydrogenated margarine, canola oil, olive oil, soybean oils, fish oils and nuts. One should choose low fat dairy products. Adequate amount of simple as well as complex carbohydrates rich in fibre especially soluble fibre should be taken to have adequate calories. One should take fruits, vegetables, lean protein from poultry or fish, beans, nuts, whole grains in the regular diets. Eating 140 g of nuts per week has been reported to be associated with a reduced risk of gallstone disease and gallbladder removal (cholecystectomy) in normal weight patients but the obese patients should not consume too many nuts as they are high in calories. Ensure to include foods rich in calcium and vitamin C in the daily diet. One should avoid foods rich in refined carbohydrates (white breads, biscuits, cookies, desserts, puddings cakes *etc.*) saturated fats (hydrogenated fat), animal fats (red meat, butter, ghee, lard *etc.*), fried foods and high fat foods including whole dairy products (whole milk and its products, cream, cheeses *etc.*). Limit alcohol containing beverages to 1 drink per day (150 ml wine, 360 ml beer, 45 ml hard liquor) as excessive alcohol intake can increase the risk of liver problems. One should remain physically active by doing some kind of exercise daily.

Gall Bladder Stones with Pain

In case of episode of abdomen pain, the dietary management includes fat restricted diet so as to provide rest to the gall bladder and minimize its contractions. During pain, nothing should be given orally but adequate calories should be provided through the intravenous drip. Later on, liquid diet can be given at frequent intervals for a few days. When the condition improves, soft diet with restricted amount of fat and mainly containing carbohydrates should be given. Upon recovery, eating 5-6 small normal well balanced healthy meals is recommended as this shall allow the bile in digestive tract to normalize. One big meal can cause a spasm in the gallbladder and bile ducts.

After Cholecystectomy

Before and immediately after surgery, liquids are

given through the intravenous drips, then patient is switched over to liquid diet, soft diet and finally to solid diet within a few days. Most people can return to a normal healthy, well balanced diet within a month of having gallbladder surgery.

Most people do not experience digestive problems after gallbladder removal. But if someone has, he or she may need to avoid certain foods if those cause bloating, diarrhea or gas. Depending on body's ability to adapt, one may only need to modify the diet for a short period and need not make lasting changes. Mostly, if gallbladder has been removed, there is no evidence to support the need for a fat restricted diet after surgery. The digestion of fat should not be an issue for patients following gallbladder removal and a normal diet should be tolerated soon after surgery. A normal regular well balanced healthy diet should be taken after full recovery.

Pancreatitis

Pancreas is an important organ located in the upper abdomen of the body and is assigned the dual function of producing digestive enzymes) and supply hormones like insulin and glucagon for regulating our body's energy handling mechanism and overall metabolism. The normal human pancreas secretes more than 10 different enzymes together with water, bicarbonate, and other proteins (such as secretory enzyme inhibitors). These enzymes are secreted in abundance and hydrolyze within the intestinal lumen macronutrients of which lipids, protein, and carbohydrates are of particular importance. For an undisturbed digestion, not only the quantity of the secretory response is necessary, but even the timed, controlled, and coordinated release of these enzymes into the duodenal lumen in response to a meal is required. Postprandially, regulation of human pancreatic secretion must be seen as part of an overall integrated motor and secretory response. Pancreatitis is inflammation of the pancreas. It happens when pancreatic juices become activated in the pancreas rather than the small intestine and the digestive enzymes start digesting the pancreas itself. There are two types of pancreatitis *i.e.* acute or chronic. Causes of both are similar; about 80-90 per cent are caused by alcohol abuse and gallstones (35-45 per cent) while the remaining 10-20 per cent are caused by medications, physical trauma, hereditary diseases, surgical procedures, chemical exposures, infections, genetic abnormalities in pancreas or intestine and high fat levels in the blood. Either form is serious and can lead to complications.

Acute Pancreatitis

Acute pancreatitis (AP) occurs suddenly and lasts for days. It usually goes away in a few days with treatment. But some people can go on to develop serious complications. AP can occur in different patterns, ranging from a mild inflammation to a severe necrosis of the pancreas. Gallstones and alcoholism are the long-established two most common etiological factors, and other less common risk factors such as congenital anomalies (such as pancreas divisum), infectious diseases (Coxsackie viruses, varicella virus), autoimmunity and genetic disorders, primary and secondary hypertriglyceridemia (triglycerides >1,000 mg/dl), drugs (especially azathioprine and 6 mercaptopurine), smoking, type 2 diabetes, and endoscopic retrograde cholangiopancreatography may also play a part. Common symptoms are severe pain in the abdomen, abdominal tenderness, nausea, vomiting, diarrhea, fever, rapid pulse *etc.*

Increased permeability of the gut mucosa is typical in AP. Both mild and severe pancreatitis induce bacterial translocation (BT) to the pancreatic gland in 100 per cent of the cases. In humans, one study reported a more beneficial effect of enteral nutrition (EN) vs. parenteral nutrition (PN) on gut integrity in 63 patient affected by severe AP. EN was also associated with significantly lower concentrations of plasma endotoxins at any day of observation, compared to PN. In addition, a protective effect of EN has also been reported on the integrity of enteric mucosa in a setting of AP. Enteral starvation contributes to alter gut mucosa microenvironment, its immune system, and permeability, enhancing the risk of bacterial translocation. Enteral starvation has also an impact on gut microbiota composition. Over the years it has become clear that enteral nutrition (EN) is far superior to parenteral nutrition (PN)

Treatment

Three most important issues initially are pain relief, fluid replacement and nutrition.

Nutritional Interventions in Acute Pancreatitis

The Old Paradigm

Since AP is a hypercatabolic condition, prompt and adequate provision of nutrition is essential. Traditionally, nutritional support was not part of acute pancreatitis management. According to the old idea that "to put the pancreas at rest" could be beneficial in the early phases of acute pancreatitis. Furthermore, it was believed also that enteral feeding might have some negative impact on prognosis, by stimulating exocrine pancreatic secretion and, thus, favoring the autolytic processes of the pancreas and the surrounding soft tissues. This old approach is not evidence based and may be detrimental to the patient as intestinal starvation can impair gut barrier and may favour bacterial translocation, leading to sepsis and organ failure.

New Evidenced Approach for Nutrition Support

In the past two decades, increasing interest has been developed in defining the role of nutritional support in the management of patients with acute pancreatitis. According to new approach, all patients affected by acute pancreatitis are at risk of malnutrition, and they should be screened for nutritional support according to international guidelines because of the following metabolic changes in acute pancreatitis and should be provided adequate amount of various nutrients depending upon the severity. Severity should be assessed as soon as possible for managing treatment and nutritional route of the hospitalized patient as there are different times, route and formulas for providing nutrition, depending on whether the patient has a mild or severe acute pancreatitis.

The nutritional management of patients with acute pancreatitis is now guided by five main principles:

- To start with intensive fluid replacement therapy
- To reduce pancreatic stimulation to subclinical levels to maintain gut integrity
- To attenuate the overall systemic inflammatory response
- To avoid iatrogenic complications (*e.g.* catheter-related sepsis)

Effect of Metabolic Changes on Nutritional Requirements

Knowledge of the metabolic pathophysiology of acute pancreatitis is foundation of appropriate nutritional management. During acute pancreatitis, specific and non-specific metabolic changes occur. Under the influence of inflammatory mediators and pain, the basal metabolic rate may increase leading to a higher energy consumption. These changes do not, however, occur in all the patients. If acute pancreatitis is complicated by sepsis, roughly 80 per cent of the patients are in a hypermetabolic state with an increase of the resting energy expenditure (REE). These patients have increased nutrient requirements because of increased rates of REE and protein breakdown. There is increased endogenous gluconeogenesis in patients with acute pancreatitis which can be a manifestation of the metabolic response to severe inflammation. Recommendation for total energy intake ranges from25-35 Kcal/kg/day in most patients. The upper limit of this calorie range is best in patients with sepsis coinciding with severe acute pancreatitis.

A negative nitrogen balance has been associated with an adverse clinical outcome. Net nitrogen losses are as much as 20–40 g/day in some patients with acute pancreatitis. In one study, patients with acute pancreatitis who were in a negative nitrogen balance had a 10-fold increased mortality rate than those with a positive balance. Protein calories malnutrition can also arise in patients with acute pancreatitis simply because they are subjected to prolonged periods of inadequate oral intake. Deficiencies in certain amino acids may enhance pancreatic inflammation, leading to a potential vicious circle. In general, at least 1.5g/kg/day of protein is required to improve nitrogen balance. Therefore, administration of standard exogenous amino acid mixture is the mainstay of therapy in pancreatitis patients.

Lipid metabolism seems to be altered in acute pancreatitis: there is a possibility of enhanced organ damage through high concentrations of serum triglycerides. Most prominently, hypertriglyceridemia ranging up to 80–100 mmol/l can be seen in patients with hyperlipidemia preceding pancreatitis. Increases in cholesterol and free fatty acid serum concentrations have also been reported. The mechanism of altered lipid metabolism is not entirely clear (altered lipid oxidation? lipid clearance?). It is not clear whether these represent an etiological factor or whether they are a consequence of the pancreatic disease, or possibly a combination of both. Overall, there is no proven causal relationship between infusion of exogenous fat and the development of pancreatitis. There is therefore, no firm evidence to contraindicate the use of long chain triglycerides or other fats.

Table 23.7: Metabolic Changes in Acute Pancreatitis

Increased	*Decreased*
Energy expenditure	Insulin response
Gluconeogenesis	Glutathione
Proteolysis	Vitamins A, C, and E
Urea turnover	Selenium
Lipolysis	Methionine
BCAA oxidation	Glutamine
Glutamine oxidation	

Nutritional Interventions

A. Acute Pancreatitis

Nutrition and nutritional supplements have demonstrated necessity and importance not only in restoring energy balance but also in maintaining gut barrier function and providing important immunomodulatory and antioxidant effects (Figure 23.1).

i) Guidelines for Nutrition Support in Mild and Moderate Acute Pancreatitis

Traditionally, AP patients were maintained on *nil per os* or *nothing per mouth* treatment until resolution of pain or normalization of pancreatic enzymes to allow the pancreas to rest. Clinical researches produced in recent years suggest that acute pancreatitis may benefit from early oral or enteral nutrition. Nevertheless, many clinicians still believe erroneously that fasting particularly in the early phase may reduce acute pancreatitis complications and mortality but such false belief harms the patients. Such dietary restrictions exacerbate patient's malnutrition due to imbalance between reduced food intake and higher nutritional requirements, leading to further catabolism, bacterial translocation, and ultimate mortality.

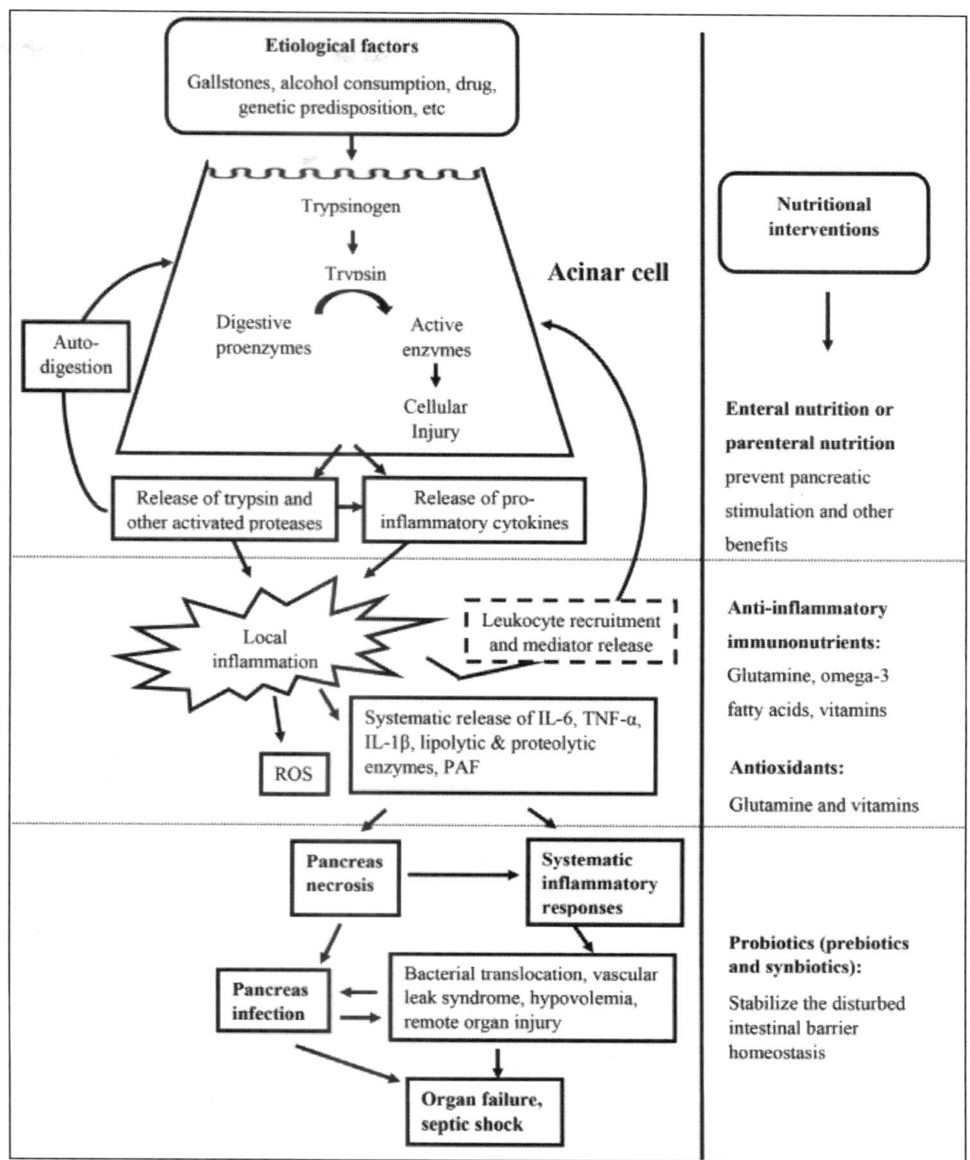

Figure 23.1: Targeted Nutritional Interventions during the Whole Episode of Acute Pancreatitis. Targeted nutritional interventions: enteral or parental nutrition and nutritional supplements including anti-inflammatory immunonutrients, antioxidants, and probiotics are presented at the administration stage.

***Source*: Pan *et al.*, 2017. Recent Advances on Nutrition in Treatment of Acute Pancreatitis. Front Immunol. 8: 762.**

Table 23.8: Nutritional Requirements in Acute Pancreatitis

Nutrients	*Amount*
Total calories	25-25 Kcal/kg/day
Carbohydrates	Less than 5 mg/kg/minute plus exogenous insulin
Fat	Less than 1.5 g/kg/day (monitor triglyceride concentration)
Protein	1.5-2g/kg/day
Vitamins, trace elements	Vitamins A, C, and E and selenium

Hence, currently, it is widely accepted that early enteral nutrition may be critical to improve acute pancreatitis-associated malnutrition and the overall outcomes, as bowel rest is associated with intestinal mucosal atrophy and increased infectious complications. Gut barrier dysfunction is found in approximately 60 per cent of patients with acute pancreatitis. Importantly, enteral nutrition exerts immunomodulatory effects to preserve gut mucosa integrity, stimulate intestinal motility, and reduce bacterial overgrowth. A randomized clinical study demonstrated that immediate oral feeding in patients with mild acute pancreatitis was feasible and safe and accelerated recovery without adverse gastrointestinal events. Another randomized controlled trial supported early-stage introduction of initial oral nutrition with either a clear liquid diet or a low-fat solid diet for patients who developed mild acute pancreatitis. In these patients, if oral intake is not tolerated, enteral feeding is recommended.

Table 23.9: Nutritional Support of Mild and Moderate Pancreatitis

1 Step (2-5 day) ↓	✰ Fasting ✰ Treat the cause of pancreatitis ✰ I.V Fluid and electrolyte replacement ✰ Analgesics
2 Step (3-7 day) No pain Enzymes regredient ↓	✰ Refeeding ✰ Diet rich in carbohydrates ✰ Moderate in protein ✰ Moderate in fat
3 Step	✰ Normal diet

Source: Meier and Sobotka. 2010. Basics in clinical nutrition: nutritional support in acute and chronic pancreatitis. Clinical nutrition (ESPEAN) 5(1); e58-e62.

According to the more rational, evidence-based new approach, in mild to moderate acute pancreatitis, patients can consume oral food when abdominal pain, nausea, and vomit are reduced, and especially when appetite returns. They are not usually malnourished and are able to eat normally within 5-7 days of disease onset, so nutritional support is not necessary. In general, patients with mild acute pancreatitis without signs of malnutrition may be treated with supportive care, including intravenous fluids and pain control, for 3 days. Fluid resuscitation is necessary in any form of acute pancreatitis. After avoidance of oral nutrition for 72 hours, a decision should be made regarding the status of the patient and what nutrition strategy is most suitable. Mostly, one should choose to initiate a clear liquid or low fat full liquid, transitioning to a low-fat, very low-fat, or no-fat regular diet in patients with mild acute pancreatitis and clinical success has been demonstrated using this strategy. Aggressive nutritional enteral or parenteral support is not required for mild to moderate forms of acute pancreatitis.

ii) Guidelines for Nutritional Interventions in Severe Acute Pancreatitis

Severe acute pancreatitis is a sepsis-like syndrome characterized by a systemic inflammation (SIRS). Patients affected by acute pancreatitis are consistently at nutritional risk. Intestinal starvation impairs gut barrier and favours bacterial translocation, leading to sepsis and organ failure. Patients with severe disease, complications or the need for surgery require early nutritional support to prevent the adverse effects of nutrient deprivation (enteral and/or parenteral nutrition is possible according to the patient condition). Severity should be assessed as soon as possible for managing treatment and nutritional route. A timely and adequate nutritional support may effectively reduce the incidence of infective and non-infective complications, mortality, length of hospital stay and hospital costs associated with severe acute pancreatitis.

The subset of patients with severe acute pancreatitis is at risk for protracted disease course, typically with an inability to achieve an oral diet by day 7. Nutritional depletion impairs host defenses, immune competence, and resistance to nosocomial infections, possibly complicating the course of severe pancreatitis. Such patients may have preexisting protein calorie malnutrition and micronutrient deficiencies. Patients who are diagnosed with severe acute pancreatitis require appropriate calories and nitrogen to prevent further nutritional and clinical compromise. In severe acute pancreatitis, latest ESPEN guidelines recommend to provide an energy supply of 25-35 kcal/kg/day, with 1.2-1.5 g/kg of protein/day (unless there are renal failure or severe hepatic failure), 3-6 g/kg of carbohydrates/day and up to 2 g/kg of lipid/day. However, plasma glucose concentration should not exceed 10 mmol/l (180 mg/dl) and plasma triglycerides 3-4 mmol/l (266 mg/dl) (Table 23.10). Limited studies have suggested that daily administration of trace elements and vitamin antioxidants should include 600 μg of selenium, 9000 IU of vitamin A, 500 mg of vitamin C and 270 IU of vitamin E. These trace elements and antioxidants can be administered within the parenteral nutrition solution, orally or through a feeding tube.

Table 23.10: Energy Requirements in Severe Acute Pancreatitis, According to ESPEN Guidelines.

Substrate	*Quantity*	*Notes*
Total calories	25-35 kcal/kg/day	–
Proteins	1.2- 1.5 g/kg/day	If not present renal failure or severe hepatic failure
Carbohydrates	3- 6 g/kg/day	Plasma glucose should be ≤ 10 mmol/l (180 mg/dl)
Triglycerides	Up to 2 g/kg/day	Plasma triglycerides should be ≤ 3 mmol/l (266 mg/dl)

Source: Rinninella *et al.*, 2017. Nutritional support in acute pancreatitis: from physiopathology to practice. An evidence-based approach. European Review for Medical and Pharmacological Sciences. 21: 421-432.

Enteral vs. Parenteral Feeding in Severe Acute Pancreatitis

All international guidelines state that nutritional support in severe acute pancreatitis should be given by enteral feeding (grade of recommendation: A). Enteral nutrition is to be preferred to parenteral nutrition even if complications such as fistulas, ascites and pseudocysts are present (grade of recommendation: C). Enteral nutrition is feasible and recommended even after surgery for pancreatitis, by intraoperative jejunostomy (grade of recommendation: C). Enteral tube feeding provides a safe nutritional support in acute pancreatitis even in cases of gastric outlet obstruction. In these cases, the tube tip should be placed distal to the obstruction (grade of recommendation: C). The only actual contraindication to enteral nutrition is prolonged paralytic ileus. However, even if this case, ESPEN (European Society for Parenteral and Enteral Nutrition) guidelines recommend to combine parenteral nutrition with a small content of an elemental or immuno-enhancing diet (10-30 ml/h) continuously perfused to the jejunum. Regarding the times of supply, continuous infusion is preferred over bolus administration (grade B recommendation).

Nasogastric vs. Nasojejunal Tube

Regarding the placement of the tube, the nasogastric tube has demonstrated to be safe and useful as well as the nasojejunal tube. No differences in terms of discharge, surgery and mortality rate have been reported between the two ways *i.e.* nasogastric and nasojejunal feeding. There were no significant differences in terms of mortality, tracheal aspiration, diarrhea, exacerbation of pain and meeting energy balance between nasogastric and nasojejunal feeding. Therefore, a post pyloric placement of the tip is no longer considered necessary (grade of recommendation: B). This evidence makes enteral nutrition more feasible in clinical practice (no more need for endoscopic or radiologic placement of the feeding tube).

Nutritional Formula

Enteral formulas are classified into elemental (monomeric), semi-elemental (oligomeric) and standard (polymeric) formulas. They differ in protein and fat contents.

- Elemental formulas contain amino acids, simple sugars, and very low fats.
- Semi-elemental formulas contain peptides of vary chain length, simple sugar, glucose polymers or starch and medium chain triglycerides (MCTs).
- Polymeric formulas contain intact proteins, complex carbohydrates and long chain triglycerides (LCTs).

Elemental and semi-elemental formulas have been preferred in many trials on acute pancreatitis, because they have a better profile of absorption than polymeric ones. However, several works have demonstrated that also standard formulations are safe and effective if administered via nasojejunal tube. In a research study, both semi-elemental and polymeric formulas were found to be well tolerated in acute pancreatitis, even if, in the semi-elemental group, the length of hospital stay was shorter. In another research study, the use of polymeric when compared with semi-elemental EN formulations, did not lead to a significantly higher risk of feeding intolerance, infectious complications or death. Moreover, semi-elemental feeds are sevenfold expensive than polymeric ones. All international guidelines recommend a small peptide and medium chain triglyceride (MCT) oil based formulation (grade B recommendation). ESPEN guidelines recommend peptide-based formulas with a grade A recommendation, even if they acknowledge that a standard formula can be tried if tolerated (grade C recommendation). The use of glutamine supplementation, immune-nutrition, prebiotics or probiotics is not supported by large-scale studies. Conversely, glutamine-supplements are effective in reducing mortality, complications, and length of stay if given in total parenteral nutrition, when such approach is inevitable.

Time of Enteral Support

Increased permeability of the gut mucosa is typical in acute pancreatitis. Both mild and severe pancreatitis induce bacterial translocation (BT) to the pancreatic gland in 100 per cent of the cases. In humans, scientists have reported a more beneficial effect of enteral nutrition (EN) vs. parenteral nutrition (PN) on gut integrity in patient affected by severe acute pancreatitis. EN was also associated with significantly lower concentrations of plasma endotoxins at any day of observation, compared to PN. A protective effect of EN on the integrity of enteric mucosa in a setting of acute pancreatitis has been reported. Enteral starvation contributes to alter gut mucosa microenvironment, its immune system, and permeability, enhancing the risk of BT.

In the light of above observations regarding gut permeability and BT as discussed above, the starting time of enteral support is crucial. On the basis of research findings, it has been reported that benefits of EN versus PN, in terms of reduction of MOF, pancreatic infectious complications and mortality rate, were statistically significant if EN was started within the first 48 hours of admission. After this time, no significant differences were observed in comparison with PN. A more recent study has demonstrated that starting EN within 24 hours after hospital admission, compared with after 24 hours is more beneficial as it was associated with lower complications.

Conclusively evidence-based data and international guidelines confirm the absolute need of an early oral or enteral feeding, depending on the grade of severity. In mild acute pancreatitis, oral nutrition should be started as soon as the patient reports to be hungry. In severe acute pancreatitis, enteral nutrition should be started within 24-48 hours from admission. This can be easily ensured in any clinical setting, either via a nasogastric or a nasojejunal tube. Enteral formulas containing small peptides and medium chain triglycerides (MCTs) should be preferred, even though polymeric formulas are equally safe. PN should be minimized for its risks of infection and other complications. Only if enteral route is not available or tolerated, PN may be considered. When side effects occur or the caloric goal cannot be achieved, in that situation also enteral nutrition should be combined with parenteral nutrition. The combined approach allows the achievement of nutritional goals most of the time. The use of intravenous lipids as part of parenteral nutrition is safe when severe hypertriglyceridaemia (<10 mmol/L) is avoided. However, the concentration of plasma triglycerides should be <3–4 mmol/L, due to metabolic problems connected with hypertriglyceridaemia. The use of lipid emulsion in TPN has been shown to be safe without aggravation of the disease.

B. Chronic Pancreatitis

Chronic pancreatitis is a syndrome defined by a combination of signs and symptoms that reflect destruction of the pancreas by a progressive, irreversible benign inflammatory process, ending in structural changes with replacement of functional exocrine and endocrine parenchyma by a fibrotic and inflammatory tissue, often evident as an inflammatory mass. Besides this, there is a calcification, stone formation, ductal dilatation, and stricture formation. In contrast to acute pancreatitis, where the gland is normal before the attack and can return to normal after resolution in chronic pancreatitis, the gland is abnormal before and after the attack in chronic pancreatitis; the gland does not heal or improve but is permanently damaged from inflammation over many years. The pathogenesis leading to chronic pancreatitis is unclear so far and is gradually evolving. Abnormal genes clearly contribute to the pathogenesis of chronic pancreatitis. Much evidence suggests that environmental factors also influence the onset and course of chronic pancreatitis. Hence, the major predisposing risk factors are toxic-metabolic, idiopathic, genetic, autoimmune, and obstructive. Smoking is an independent etiological and pathogenic factor in the development of chronic pancreatitis. Since heavy drinkers usually smoke, it is difficult to separate the effects of alcohol abuse and smoking. Cessation of smoking may help slow the progression of chronic pancreatitis. Symptoms include nausea, vomiting, weight loss, and oily stools. Treatment may also be a few days in the hospital for intravenous (IV) fluids, medicines to relieve pain, and nutritional support.

Table 23.11: Conservative Therapy in Chronic Pancreatitis

- ☆ Stop smoking and alcohol consumption
- ☆ Low fat diet and antioxidants
- ☆ Analgesics, NSAID's, codeine, antispasmodics, non-opiate, or low potency opiate analgesics
- ☆ Suppression of pancreatic secretion (octreotide)
- ☆ Pain modifying agents (antidepressants, gabapentin, and pregabalin)
- ☆ Pancreatic enzymes and adjunctive agents
- ☆ Zinc supplementation

Nutritional Support

During the course of chronic pancreatitis, enzyme secretion is decreased, resulting in maldigestion with steatorrhoea and azotorrhoea. Deficiencies of fat-soluble vitamins are the consequence of steatorrhoea. Pain and glucose intolerance are further factors, interfering with nutritional support. Insufficient pancreatic enzyme secretion and pain are the essential causes of malnutrition and weight loss.

Dietary modification, pancreatic enzyme supplements and fat-soluble vitamins are the cornerstones of nutritional management of chronic pancreatitis. It has been suggested, that with these recommendations, pain reduction can be achieved by influencing a CCK-feedback mechanism. With an adequate dietary regimen, together with enzyme preparations, in mild to moderate but not in severe disease, some pain relief has been found. Enzyme supplements are indicated when maldigestion with steatorrhoea are present. Normally 25,000–50,000 units of Lipase per main meal is necessary. The exact amount differs widely among patients.

Nutritional and Dietary Recommendations

Dietary recommendations, pancreatic enzyme supplements and fat-soluble vitamins are the cornerstones of the nutritional management of chronic pancreatitis. When weight loss is evident, a high caloric intake (30-35 Kcal/kg/day) is necessary with a diet consisting principally of carbohydrate and protein. To reach the necessary caloric goal, fat must also be given. It is crucial to supply an adequate amount of calories (2500–3000 Kcal/d). In malnourished patients, even higher amounts are necessary. In the early phase of the disease without steatorrhoea, a diet rich in carbohydrates, 70–100 g/day fat (it is tolerated if it is from vegetable source) and 1–1.5–2 g/kg/day proteins is recommended. These nutritional recommendations can be difficult to follow if overt diabetes due to endocrine insufficiency is present. Insulin treatment to control blood glucose levels is often necessary. Fat intake has to be reduced to 50–70 g/day if steatorrhoea is present. In some cases, MCT (80–120 g/day) can be tried to increase fat absorption. However, their use is often limited because they are unpalatable and because of bloating.

Fat-soluble vitamins must be given if specific deficits are documented. Antioxidants (*e.g.* Selenium, Vitamin C, E and L-Methionin) can be given. However, the beneficial effects are controversial. Six to eight meals per day together with enzyme supplements are recommended.

Enteral and Parenteral Nutrition in Chronic Pancreatitis

Only a minority of patients needs enteral or parenteral nutrition. When a patient is not able to eat (pain or pyloroduodenostenosis, due to an enlarged pancreatic head or pseudocyst formation), enteral nutrition should be instituted, using a tube introduced beyond the stenosis into the jejunum. Polymeric formulae can be used, and are usually well tolerated. Only a minority of patients need oligomeric formulae. TPN must be instituted when gastric emptying is blocked or the patient needs gastric decompression and when a double lumen tube cannot be introduced in the jejunum. In rare cases, a laparoscopically inserted needle catheter jejunostomy can be used in such patients. As patients with chronic pancreatitis often suffer not only from protein energy malnutrition but multiple electrolyte, mineral and micronutrient deficiencies, these should be corrected.

Dietary Tips

Upon the recommendation of doctor, the person should eat small meals consisting of vegetables, beans and lentils, fruits, whole grains and other plant foods frequently in the form of soft or solid diet, as the case may be depending upon toleration. These foods tend to be naturally low in fat and contain fibre and antioxidants. One should avoid red meat, full fat dairy products, beverages with added sugar, fast fried and processed foods. Saturated and trans fats should not be taken. The best cooking media for chronic pancreatitis patients is coconut oil, as it has medium chain fatty acids. Besides, eating a low-fat diet is key to avoid recurring pain. Avoid taking alcohol and smoking. If prescribed by the physician, vitamin and mineral supplements, probiotics *etc.* should be taken.

24

Diet in Diseases of Kidney

In India, Global Burden of Disease (2015) ranks chronic kidney disease as the eighth leading cause of death. The number of Indians suffering from chronic kidney ailments has doubled in the past 15 years, and at present 17 in every hundred citizens suffer from some form of kidney disease. Deaths due to renal failure constituted 2 9 per cent of all deaths in 2010–13 among 15–69 year-olds, an increase of 50 per cent from 2001–03. Diabetes was the largest contributor to renal failure deaths and next causative agent was hypertension; both account for 40-60 per cent cases of CKD.

Structure and Functions of Kidneys

The kidneys are relatively small organs, but they receive up to 25 per cent of the heart's output. The kidneys are a pair of bean-shaped organs, about 11 cm long, 6 cm wide and 3 cm thick, found in vertebrates. Each kidney weighs 125 to 170 g in males and 115 to 155 g in females. Due to the asymmetry caused by the liver, the right kidney is generally slightly smaller and lower than the left. The kidneys lie on the posterior abdominal wall, one on each side of the vertebral column, behind the peritoneum and below the diaphragm. They are embedded in, and held in position by, a mass of adipose tissue. On top of each kidney are the adrenal glands.

Externally, the kidneys are surrounded by three layers. The outermost layer is a tough connective tissue layer called the renal fascia. The second layer is called the perirenal fat capsule, which helps anchor the kidneys

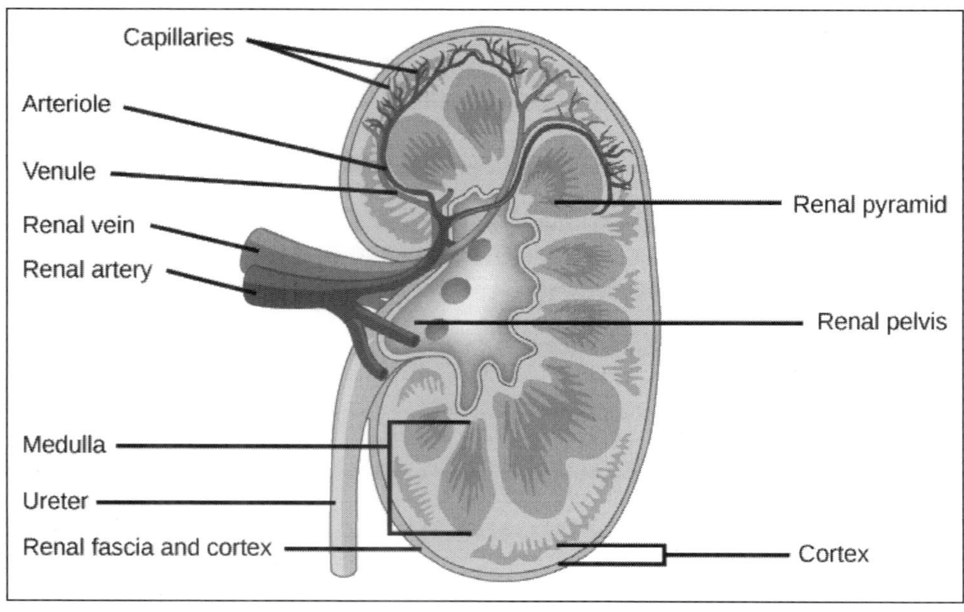

Figure 24.1: Internal Structure of Kidney.

in place. The third and innermost layer is the renal capsule. Internally, the kidney has three regions, an outer cortex, a medulla in the middle, and the renal pelvis in the region called the hilum of the kidney. The hilum is the concave part of the bean-shape where blood vessels and nerves enter and exit the kidney; it is also the point of exit for the ureters. The renal cortex is granular due to the presence of nephrons which are the functional unit of the kidney. The nephron is the structural and functional unit of the kidney. Each adult kidney contains around one million nephrons. Each nephron consists of a glomerulous which is tuft of capillaries invaginated into an epithelial sac-Bowman's capsule, from which arises a tubule. The nephron utilizes four processes to alter the blood plasma which flows to it: filtration, reabsorption, secretion, and excretion.

The medulla consists of multiple pyramidal tissue masses, called the renal pyramids. In between the pyramids are spaces called renal columns through which the blood vessels pass. The tips of the pyramids, called renal papillae, point toward the renal pelvis. There are, on average, eight renal pyramids in each kidney. The renal pyramids along with the adjoining cortical region are called the lobes of the kidney. The renal pelvis leads to the ureter on the outside of the kidney. On the inside of the kidney, the renal pelvis branches out into two or three extensions called the major calyces, which further branch into the minor calyces. The ureters are urine-bearing tubes that exit the kidney and empty into the urinary bladder.

Kidneys receive blood from the paired renal arteries; blood exits into the paired renal veins. Filtration occurs in the glomerulus; one-fifth of the blood volume that enters the kidneys is filtered. Examples of substances reabsorbed are solute-free water, sodium, bicarbonate, glucose, and amino acids. Examples of substances secreted are hydrogen, ammonium, potassium and uric acid. Kidneys filter blood and purify it. All the blood in the human body is filtered many times a day by the kidneys; these organs use up almost 25 per cent of the oxygen absorbed through the lungs to perform this function. Oxygen allows the kidney cells to efficiently manufacture chemical energy in the form of ATP through aerobic respiration. The filtrate coming out of the kidneys is called urine.

The main roles of the kidneys are as under:

Waste Excretion

A number of waste products are removed via the kidneys and expelled in the urine. Two major compounds that are removed are urea produced from the breakdown of proteins and uric acid from the breakdown of nucleic acids.

Maintenance of Homeostasis

This means that kidneys manage fluid levels, electrolyte balance, and other factors that keep the internal environment of the body within constant parameters.

Reabsorption of Nutrients

Nutrients from the blood are reabsorbed and transported to where they are needed. Other products are reabsorbed to help maintain the balance. It is called reabsorption, not absorption, because the compounds have already been absorbed once, normally in the

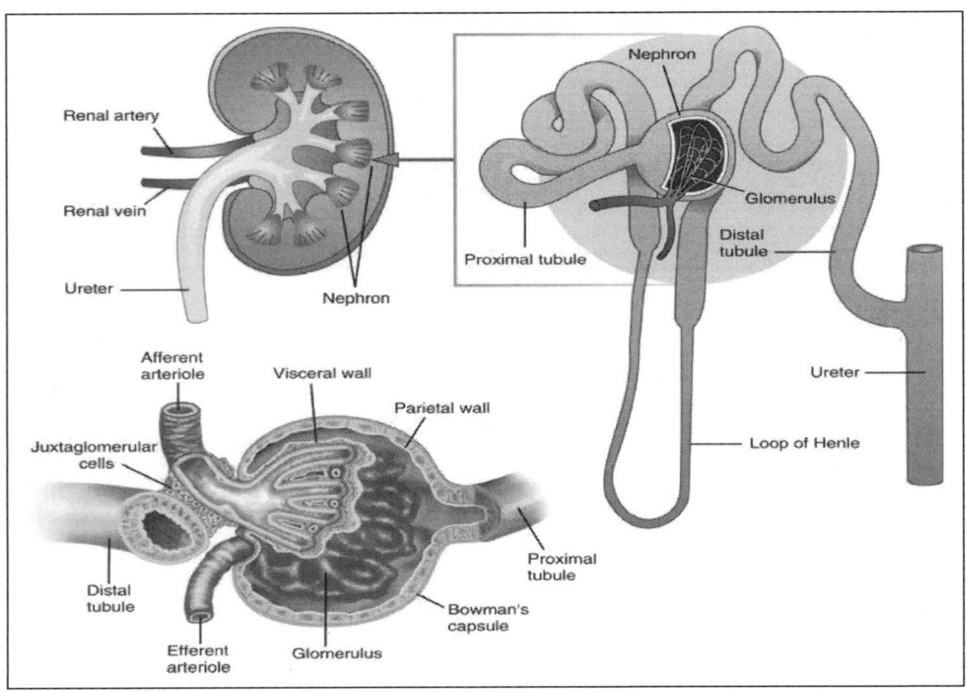

Figure 24.2: The Nephron is the Functional Unit of the Kidney. The glomerulus and convoluted tubules are located in the kidney cortex, while collecting ducts are located in the pyramids of the medulla.

intestines. Reabsorbed products include glucose, amino acids, bicarbonate, sodium, water, phosphate, chloride, magnesium and potassium ions.

Osmolality Regulation

Osmolality is a measure of the body's electrolyte-water balance. In other words, it is the ratio between fluid and minerals in the body. Dehydration is a key cause of electrolyte imbalance. If plasma osmolality rises, the hypothalamus in the brain responds by passing a message to the pituitary gland. This, in turn, releases antidiuretic hormone (ADH). In response to ADH, the kidney makes a number of changes, including increasing urine concentration, increasing water reabsorption, reopening portions of the collecting duct that are normally not permeable to water, to allow water back into the body retaining urea in the medulla of the kidney rather than excreting it, as it attracts in water. Urea can then re-enter the nephron and be excreted or recycled again depending on whether ADH is still present or not.

Regulating Blood Pressure

The kidneys regulate blood pressure when needed, but they are responsible for slower adjustments. The kidneys produce different hormones *i.e.* renin, angiotensin, aldosterone, prostaglandin *etc.* which help to regulate water and salt balance in the body, which plays vital roles in the maintenance of good blood pressure control.

The renin-angiotensin system adjusts arterial pressure over the long term by impacting the extracellular fluid compartment, or the fluid outside of cells. When blood volume or sodium levels in the body are low, or blood potassium is high, cells in the kidney release the enzyme, renin. Renin converts angiotensinogen, which is produced in the liver, to the hormone angiotensin I. An enzyme known as ACE or angiotensin-converting enzyme found in the lungs metabolizes angiotensin I into angiotensin II. Angiotensin II causes blood vessels to constrict and blood pressure to increase. Angiotensin II stimulates the release of the hormone aldosterone in the adrenal glands, which causes the renal tubules to retain sodium and water and excrete potassium. Together, angiotensin II and aldosterone work to raise blood volume, blood pressure and sodium levels in the blood to restore the balance of sodium, potassium, and fluids. If the renin-angiotensin system becomes overactive, consistently high blood pressure results.

Anything that alters blood pressure can damage the kidneys over time, including excess alcohol, smoking, and obesity.

Secretion of Active Compounds

The kidneys release a number of physiologically important products, including erythropoietin, calcitriol and rennin. Erythropoietin controls erythropoiesis, or the production of red blood cells and plays an important role in wound healing and response to neuronal injury. Renin is an enzyme which regulates angiotensin and aldosterone levels. It helps to mediate the expansion of arteries and the volume of blood plasma, lymph, and interstitial fluid; 1,25 $(OH)_2$ D_3 which is the active metabolite of vitamin D increases the level of calcium absorbed by the intestines and ups the reabsorption of phosphate in the kidney.

Definitions of Kidney Diseases

Kidney disease is defined as a heterogeneous group of disorders affecting kidney structure and function. Kidney disease differs from most other disorders of organ systems in that it is often "silent": there are few symptoms until late in the course of disease, and even then the symptoms are often nonspecific. Thus, there are few kidney-specific "clinical events," which results in a reliance on laboratory measures to define the major clinical syndromes. The duration of kidney disease must be established either by documentation or inference of the duration of abnormalities in laboratory measures.

In principle, kidney diseases can be defined and classified according to many domains: structure, function, cause, duration, and outcomes (Figure 24.3). It is recognized now that even mild abnormalities in measures of kidney structure and function are associated with increased risk for developing complications in other organ systems as well as mortality, all of which occur far more frequently than kidney failure. Duration of greater than 3 months is defined as chronic kidney disease (CKD), while duration of 3 months or fewer is termed acute kidney injury (AKI).

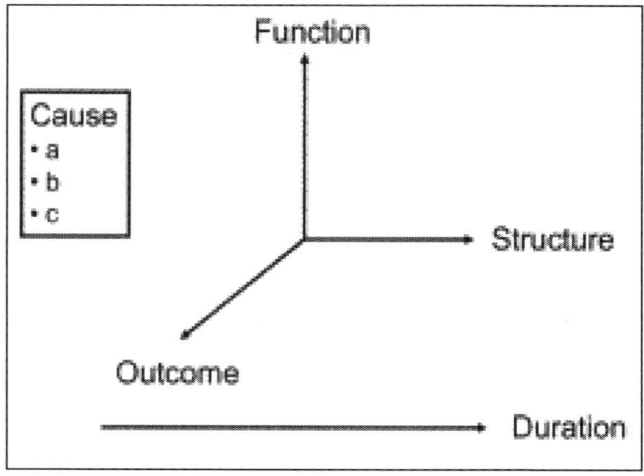

Figure 24.3: Domains for the Classification in Kidney Disease.

The current definitions for CKD and AKI have been created by independent guideline development work groups. They are based on the strong, graded, and consistent relationship of laboratory measures with important outcomes, independent of age, sex, race, location, and comorbid conditions. Criteria include measures of kidney damage (albuminuria, abnormalities in the urine sediment, imaging, or biopsy) and function (decreased glomerular filtration rate [GFR], rising serum creatinine level, or decreased urine output; Table 24.1). These criteria for AKI and CKD based on evidence and expert opinion is known as Kidney Disease: Improving Global Outcomes (KDIGO) guidelines.

Table 24.1: Criteria for the Definitions of Kidney Diseases and Disorders

	Functional Criteria	*Structural Criteria*
Acute kidney injury or impairment (AK1)	Increase in serum creatinine by 50 per cent within 7d, *or* increase in serum creatinine by 0.3 mg/dL within 2 d, *or* oliguria	No criteria
Chronic kidney disease (CKD)	Glomerula filtration rate (GFR) <60 mL/min for >3mo	Kidney damage for >3 mo
Acute kidney diseases and disorders (AKD)	AKI, *or* GFR <60 mL/min/1.73 m² for <3 mo, *or* decrease in GFR by ≥35 per cent or increase in serum creatinine by >50 per cent for <3 mo	Kidney damage for <3 mo
No known kidney disease or disorders (NKD)	GFR ≥60 mL/min/1.73 m², stable serum creatinine	No damage

Source: Levey *et al.*, 2013. Definition and Classification of Kidney Diseases. Am. J. Kidney Diseases. 61 (5): 686–688.

Risk Factors for Kidney Diseases

I. Chronic Kidney Disease (CKD)

Factors that may increase the risk of chronic kidney disease include:

- **Type 1 or Type 2 Diabetes:** Diabetes is the number one cause of kidney failure. Diabetes mellitus is the leading cause of Chronic Kidney Disease (CKD) and End Stage Renal Disease (ESRD) in both developed and developing countries. Mechanisms that lead to kidney disease in diabetes include hyperfiltration injury, advanced glycosylation end products, and reactive oxygen species. At the molecular level, numerous cytokines, growth factors and hormones such as transforming growth factor-beta and angiotensin II cause pathologic changes associated with diabetic nephropathy.

  A special diet, exercise and medicines can help to manage diabetes and prevent complications, but some people with diabetes may still develop kidney disease, even with good medical care.

- **High Blood Pressure:** Hypertension has long been a defined risk factor for both CKD and ESRD. It is number two cause of kidney failure. High blood pressure can be both a cause of kidney disease and a symptom of it because healthy kidneys help to regulate blood pressure. Essential hypertension is generally diagnosed between 25 and 45 years of age but overt kidney dysfunction does not develop unless the patient sustains at least 10 years of uncontrolled hypertension. Keeping blood pressure under control can help to prevent kidney disease, or help keep it from getting worse.

  No doubt diabetes and hypertension are the major common risk factors but there are other causes too which can lead to kidney problems. These are as under:

- **Age:** Renal function decreases with age in both men and women. Being over 60 years increases the risk of kidney diseases. Over time, the kidneys lose some function naturally. People who are over 60 are also at more likely to have diabetes and high blood pressure, the two leading causes of kidney failure.
- **Cardiovascular disease**
- **Smoking:** Smoking can increase the chronic kidney disease risk through pro-inflammatory state, oxidative stress, prothrombotic shift, endothelial dysfunction, glomerulosclerosis and tubular atrophy.
- **Obesity:** One of the strongest yet modifiable risk factors for End-stage renal disease (ESRD) in the twenty-first century is obesity. Obesity may contribute to the pathogenesis of kidney damage through inflammation, oxidative stress, endothelial dysfunction, prothrombotic state, hypervolemia, and adipokine derangements. Besides high BMI, carrying excess weight around the abdomen is linked to an increased risk of chronic kidney disease. In a study, higher waist-to-hip ratio was associated with lower glomerular filtration rate, lower effective renal plasma flow, and higher filtration fraction, even after adjustment for sex, age, mean arterial pressure, and BMI.
- **Family History of Kidney Disease:** Family members of CKD patients have a high prevalence of CKD and its risk factors.
- **Gender:** Many registries including the *Japanese Society for Dialysis Therapy* have demonstrated that End-stage renal disease (ESRD) is more frequent among men. In contrast, in another study, women were reported to have higher incidence of CKD than men.
- **Genetic:** An individual's genetic and phenotypic make-up puts him/her at risk for kidney disease. Consanguinity and genetic inbreeding increase risk of congenital anomalies of the kidney and urinary tract and obstructive or reflux nephropathy. Chronic kidney disease (CKD) has a heritable component. The involvement of renin–angiotensin system genes seems to be particularly relevant to CKD. Polycystic kidney disease is a genetic disease that causes many cysts to grow in the kidneys.

- **Ethnicity:** Certain ethnic groups are more likely to have kidney disease. Several studies performed in the United States have confirmed an increased risk for the development of ESRD in African Americans compared with Caucasians. Researchers are not exactly sure why these groups are more at risk for kidney disease. It may be because these groups tend to have higher rates of diabetes and high blood pressure, the two leading causes of kidney failure.
- **Recurrent Kidney Infection:** Recurrent kidney infection may lead to chronic kidney disease.
- **Abnormal Kidney Structure**
- **Acute Kidney Injury (AKI):** Researchers have recognized the importance of acute kidney injury (AKI) episodes in the development of CKD. Even after a single episode of experimental AKI, histologic repair can be impaired and focal tubule-interstitial fibrosis may develop.
- **Low Birth Weight:** In the 1980s, Brenner and colleagues hypothesized that intrauterine growth restriction might cause a low nephron number, which could predispose to hypertension and renal disease (also known as the Barker hypothesis). In support of this hypothesis, it has been shown that there is an increase in nephron numbers by 257,426 glomeruli per kg increase in birth weight. Low nephron number leads to intra-glomerular hypertension and hyperfiltration in the available nephrons and lower over-all glmerular filtration rate and higher urine albumin-to-creatinine ratio. In recent studies, low birth weight and intrauterine growth restriction have been significantly associated with increased risk for end stage renal disease among Norwegians.
- **Nephrotoxins:** Alcohol and recreational drugs have been linked to CKD progression as well as excessive use of analgesic drugs and exposure to heavy metals.
- **Newly Defined Risk Factors:** Obstructive sleep apnea is a disease associated with complete and partial breathing disturbances during sleep for at least five events per hour. In one study, 30.1 per cent of obstructive sleep apnea patients had CKD. Besides sharing the same risk factors, obstructive sleep apnea has an independent effect on CKD risk and progression.

  Heart rate has also been suggested as a risk factor for CKD. Subjects with higher heart rates had greater magnitude of decreasing estimated glomerular filtration rate (GFR) and higher odds ratio of developing proteinuria. Each heart rate category increment was reported to approximately 1.1 times increased risk of developing CKD and 1.2 times increased risk of developing proteinuria in middle-aged or older subjects.

  Periodontal diseases, which are initiated by gram-negative tooth-associated microbial biofilms have also been defined as a risk factor CKD. The inflammatory response in those patients has been associated with CKD.

  A slightly elevated uric acid level (7.0–8.9 mg/dl) has been associated with doubled risk for CKD and an elevated uric acid (~*9.0 mg/dl) was associated with a tripled risk.

  Hypovitaminosis A and other nutritional issues during pregnancy may cause smaller kidney volume at birth and a lower eGFR.

  Poverty, poor sanitation, pollutants, water contamination, overcrowding, and known and unknown nephrotoxins (including heavy metals and plant toxins in indigenous remedies) may lead to glomerular and interstitial kidney diseases.

II. Risk Factors for Acute Kidney Injury (AKI)

Many studies have contributed to current understanding of the risk factors for AKI and they are the following:

- **Age:** Age is one "traditional" factor affecting the incidence of AKI. Studies have reported that both elderly and very young patients are particularly inclined to suffer from this disorder. CKD is a major risk factor for AKI.
- **Morbid obesity**
- **Some Nonrenal Comobid Disease:** Diabetes mellitus, hypertension, cardiovascular disease, heart failure, chronic liver disease, and chronic obstructive pulmonary disease, have been implicated as significant predictors of AKI. Diabetics with baseline renal insufficiency represent the highest risk subgroup. Increasing evidence shows that intra-abdominal hypertension may contribute to the development of acute renal failure.
- **Many Specific Exposures:** They can modify the risk of AKI, among which the most important are sepsis, cardiac surgery, and major non-cardiac surgery. An association between sepsis, hypovolemia (shock), and acute renal failure is clear. Liver failure, rhabdomyolysis, and open-heart surgery (especially valve replacement) are clinical conditions potentially leading to acute renal failure.
- **Acute Glomerularnephritis**
- **Obstruction of Urinary Output *e.g.* tumour, renal stone disease**
- **Exposure to Nephrotoxins:** Radiocontrast, poisons, certain drugs and antimicrobial agents

are the most common causes of nephrotoxic acute renal failure.

- ☆ Certain Cancers

Classification and Staging of Kidney Diseases

Kidney diseases are mainly classified as discussed under:

Acute Kidney Injury (AKI)

Acute kidney injury (AKI) is defined as a subgroup of acute kidney diseases and disorders (AKD) in which changes in kidney function evolve within one week. They occur suddenly and there is rapid deterioration of kidney functions. It may be caused by injury or illness; often reversible.

Chronic Kidney Disease (CKD)

According to KDIGO (2012), CKD is defined as abnormalities of kidney structure or function, present for >3 months, with implications for health (not graded). There is a complex relationship between AKI and CKD; AKI can lead to CKD, and CKD increases the risk of AKI.

Markers of Kidney Damage (One or more)	*Criteria for CKD (Either of the following present for >3 months)*
	☆ Albuminuria (AER ≥30 mg/24 hours; ACR ≥30 mg/g [≥3 mg/mmol]) ☆ Urine sediment abnormalities ☆ Electrolyte and other abnormalities due to tubular disorders ☆ Abnormalities detected by histology ☆ Structural abnormalities detected by imaging ☆ History of kidney transplantation
Decreased GFR	GFR <60 ml/min/1.73 m^2 (GFR categories G3a–G5)

CKD: Chronic kidney disease; GFR: Glomerular filtration rate.

Staging of CKD

CKD is classified based on cause, GFR category, and albuminuria category (CGA) as given under in the Figure 24.4. One can assign cause of CKD based on presence or absence of systemic disease and the location within the kidney of observed or presumed pathologic-anatomic findings (not graded). Assign GFR categories as follows (Not Graded):

GFR Categories in CKD

GFR Category	*GFR (ml/min per 1.73 m^2)*	*Terms*
G1	>90	Normal or high
G2	60-89	Mildly decreased*
G3a	45-59	Mildly or moderately decreased
G3b	30-44	Moderately or severely decreased
G4	15-29	Severely decreased
G5	<15	Kidney failure

CKD: Chronic kidney disease; GFR: Glomerular filtration rate.

* Relative to young adult level

In the absence of evidence of kidney damage, neither GFR category G1 nor G2 fulfill the criteria for CKD.

Assign albuminuria[a] categories as follows (Not graded):

Albuminuria Categories in CKD

Category	*AER (mg/24h)*	*ACR (Approximate equivalent)*		*Terms*
		(mg/mmol)	*(mg/g)*	
A1	<30	<3	<30	Normal to mildly increased
A2	30-300	3-30	30-300	Moderately increased*
A3	>300	>30	>300	Severely increased**

AER: Albumin excretion rate; ACR: Albumin-to-creatinine ratio; CKD: Chronic kidney disease.

*Relative to young adult level.

** Including nephrotic syndrome (albumin excretion usually >2200 mg/24 hours [ACR >2220 mg/g; >220 mg/mmol]).

a: Note that where albuminuria measurement is not available, urine reagent strip results can be substituted.)

Markers of Kidney Function Tests

The introduction of routine reporting of estimated glomerular filtration rate and a new definition of chronic kidney disease has renewed interest in methods of measuring renal function. Coupled with this is the fact that several countries are moving towards population screening for renal impairment to try to reduce the associated increased cardiovascular risk. Accurate measurement is methodologically difficult so surrogate measures such as serum creatinine levels and prediction formulas (based on factors such as the patient's age, sex, and serum creatinine level) are more commonly used in routine practice. The routine and more specialized methods of assessing renal function are mentioned below:

1. Kidney has several interlinked functions such as excretion (nitrogenous waste, sodium, free water, potassium, phosphate and water soluble medicines *etc.*), control of blood pressure, acid-base balance, secretion of erythropoietin, hydroxylation of 25-OH vitamin D, gluconeogenesis in the fasting state and catabolism of peptide hormones including

Prognosis of CKD by GFR and albuminuria category

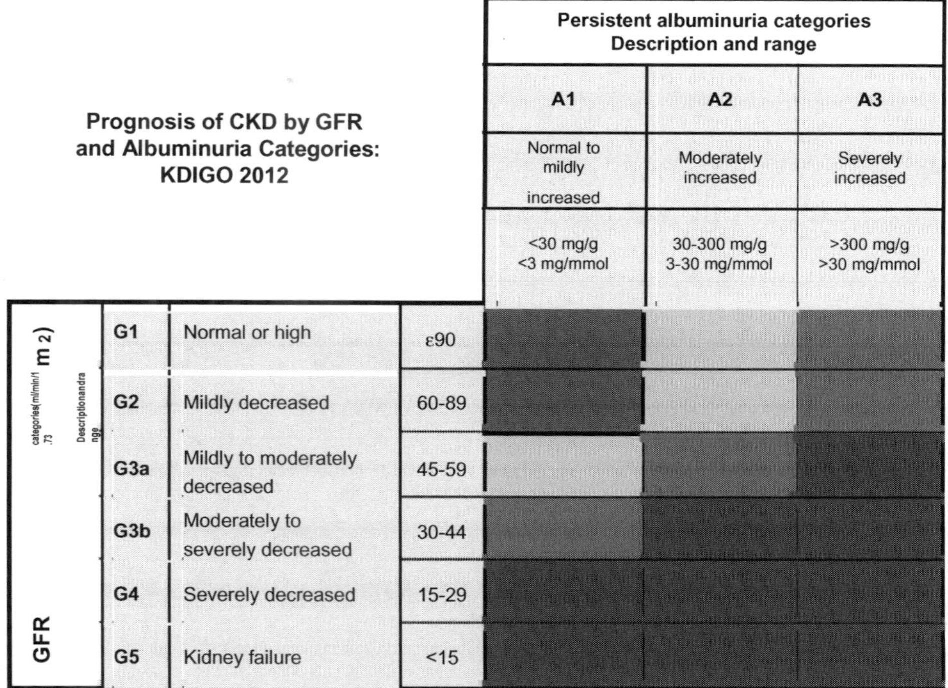

Green: low risk (if no other markers of kidney disease, no CKD); Yellow: moderately increased risk; Orange: high risk; Red, very high risk.

Figure 24.4 Prognosis of CKD.

insulin *etc.* which depend upon glomerular filtration rate (GFR), the unit measure of kidney function.

2. **Creatinine:** Creatinine is a breakdown product of creatine phosphate in muscle, and is usually produced at a fairly constant rate by the body depending on muscle mass. Creatinine is the closest to an ideal endogenous substance for measuring glomerular filtration rate. In patients with stable renal function, serum creatinine levels are usually constant (upper limit is 1.4 mg/100 ml in males and 1.3 mg/100 ml for females), with variability daily of about only 8 per cent. Creatinine is freely filtered at the glomerulus and is not reabsorbed, but up to 15 per cent is actively secreted by the tubules. In advanced renal failure, excretion of creatinine through the gastrointestinal tract increases.

   Creatinine Clearance: Measuring the creatinine clearance using serum creatinine level and a timed urine collection gives an estimate of glomerular filtration rate:

   Creatinine clearance (ClCr) = (urine creatinineX volume)/serum creatinine

   The normal creatinine clearance test value is 110-150 ml/min in male and 100-130 ml/min in female; in case of renal failure, it is reduced to 30 ml/min. As a result of tubular secretion of creatinine, creatinine clearance tends to overestimate true glomerular filtration rate. The main problem with creatinine clearance is the requirement for urine collection over 24 hours; patients find this inconvenient and therefore, collections are often inaccurate. Also a 25 per cent daily variation in the values obtained using this method has been reported. Creatinine clearance is therefore, no longer much used in clinical practice. Estimated glomerular filtration rates, calculated using the "4-v MDRD" based formula, are now routinely reported by biochemistry laboratories alongside serum creatinine results (except in non-validated patient groups). Serum creatinine levels or estimated glomerular filtration rates may be used to monitor changes in renal function in an individual patient.

3. **Urea:** Urea is major nitrogenous end product of protein and amino acid catabolism, produced by liver and distributed throughout intracellular and extracellular fluid. In kidneys, urea is filtered out of blood by glomeruli and is partially being reabsorbed with water. The most frequently determined clinical indices for estimating renal function depends upon concentration of urea in the serum. It is useful in differential diagnosis of acute renal failure and pre- renal condition where blood urea nitrogen–creatinine ratio is increased.

   Serum urea is a less reliable marker of glomerular filtration rate than creatinine because levels are more vulnerable to change for reasons unconnected to glomerular filtration rate. A high protein diet, tissue breakdown, major gastrointestinal haemorrhage, and corticosteroid therapy can lead to an increase in plasma urea whereas a low protein diet and liver disease can lead to a reduction. Also, 40-50 per cent of filtered urea may be reabsorbed by the tubules, although the proportion is reduced in advanced renal failure.

4. **Mean of Urea and Creatinine Clearance:** In advanced renal failure, the mean of urea and creatinine clearance may give a more accurate estimate of glomerular filtration rate than either clearance alone, as the effects of urea reabsorption and creatinine secretion tend to cancel each other out. It is the recommended method for estimating residual renal function in patients receiving dialysis.
5. **Markers of Tubular Function:** Tubular function tests involve evaluation of functions of the proximal tubule (*i.e.* tubular handling of sodium, glucose, phosphate, calcium, bicarbonate and amino acids) and distal tubule (urinary acidification and concentration).
6. **Measurement of Electrolytes:** The test for electrolytes includes the measurement of plasma sodium, potassium, chloride, and bicarbonate for both diagnosis and management of renal, endocrine, acid-base, water balance, and many other conditions. Potassium used as a most convincing electrolyte marker of renal failure. The combination of decreased filtration and decreased secretion of potassium in distal tubule during renal failure cause increased plasma potassium. Hyperkalemia is the most significant and life-threatening complication of renal failure
7. **Urine Volume in 24 h and Routine Urine Analysis:** Volume of urine may be increased or decreased or normal during acute or chronic renal failure. During kidney failure, some abnormal constituents like blood and protein may be present in the urine.

   Clinically the appearance of significant amount of protein in urine is one of the earliest sign of almost all renal diseases. Estimation of proteinuria helps in differentiating between tubulointerstitial and glomerular diseases and also to follow the progress of renal disease and to assess the response to therapy. Normally excretion in most healthy adults is between 20-150 mg of protein in urine over 24 hrs. Proteinuria more than 3.5 g/day is taken to be diagnostic of nephrotic syndrome. It has been recommended to use the protein/creatinine ratio as an Index of Quantitative Proteinuria in 24 hour urine collection. One of the investigations for proteinuria is semi-quantitative dipstick urinalysis as this method is relatively low cost and easily performed.
8. **Inulin Clearance:** No endogenous ideal substance exists for measuring glomerular filtration rate, so the standard method requires infusion of an exogenous agent, such as inulin. Inulin, a polymer of fructose (5200 daltons), was first used for measuring glomerular filtration rate in 1951. Rapid measurement of glomerular filtration rate by an inulin single-bolus technique would be practically useful but its use is limited because purified inulin is expensive and difficult to measure. Moreover, measuring glomerular filtration rate in this way is time consuming for both patients and clinicians. This method (nowadays often using polyfructosan) is only used in research studies when very accurate estimation of renal function is necessary.
9. **Serum Alkaline Phosphate:** Its level is increased as vitamin D synthesis is hindered.
10. **β-Trace protein (BTP):** This protein is filtered at glomerulus and then reabsorbed in proximal tubule or excreted in urine and hence, have potential to meet the criteria for use as a marker of glomerular filtration rate. β-Trace Protein is a low-molecular weight glycoprotein belonging to the lipocalin protein family with 168 amino acids and a molecular weight of 23000–29000, depending on the degree of glycosylation. It has been reported to be a better indicator of reduced glomerular filtration rate than serum creatinine. Serum β -Trace Protein has been found to be elevated in patients with renal diseases. However, when compared, Cystatin C is still a better indicator than Serum β-Trace Protein.
11. **Cystatin C:** The protease inhibitor Cystatin C is a non-glycosylated low molecular weight protein. Cystatin C has been proposed to be a marker as it is produced by all nucleated cells at a constant rate and is freely filtrated by the glomeruli and completely catabolized in the proximal tubules. It is mainly determined by glomerular filtration, which makes Cystatin C an endogenous marker of glomerular filtration rate.
12. **Radioactive Markers:** In recent decade, radioisotopes markers are being used to measure glomerular filtration rate. Some of them to mention are 125iodine (I)-iothalamate, 51CrEDTA ethylenediamine tetra acetic acid, 99mTc-DTPA (diethylene triamine penta acetic acid) and 99mTc mercapto acetyl triglycine. Renal 125iodine (I)-iothalamate clearance, is a simple and accurate test after a single subcutaneous injection, to measure glomerular filtration rate in adults.

Nutritional Assessment of Renal Disease Patients

Nutritional assessment of each patient suffering from renal disease should be undertaken to know about the nutritional status of the individual. It is generally recommended that a nutritional assessment including a nutritional care plan should be performed not later than 1-2 weeks after the start of dialysis, and should thereafter be repeated periodically.

The assessment of nutritional status is based on clinical, biophysical and biochemical parameters. Clinical assessment of subcutaneous fat, muscle mass and history of weight loss are important parts of routine nutritional assessment. Most studies have used body mass index (BMI), skin fold thickness, mid arm circumference and mid-arm muscle circumference to assess patient's nutritional status. The most commonly used laboratory parameters for routine assessment of nutritional status are plasma concentrations of albumin, transferrin, and other liver-derived proteins. In recent years, subjective global assessment of nutritional status (SGA) has been used increasingly to assess nutritional status in many studies of dialysis patients and in patients with chronic kidney disease at start of dialysis therapy. SGA correlates well with other nutritional markers in patients with chronic renal disease.

There is a general consensus that there is no single marker of nutritional status in patients with chronic kidney disease, and therefore, several nutritional markers should be evaluated together. This assessment should include:

- Biochemical and haematological data and their interpretation as shown in Table 24.2.
- Activity level
- Medical history including duration of diagnosis of CKD and all other co-morbidities
- Fluid status
- Physical examination
- **Anthropometrics:** Data regarding height, weight, BMI, alterations in lean body mass that can be masked by fluid shifts and changing fat

Table 24.2: Interpretation of Biochemical and Haematological Data

Blood Result (Normal range)	***Importance***
Potassium (K^+) 3.5–5.0 mmol/l	Hyperkalaemia can result in cardiac arrhythmia or cardiac arrest, related to diet or other reasons. ↓ values may result in general muscle weakness. Can indicate poor nutritional intake as K^+ is found in many foods.
Sodium (Na) 135–145 milliequivalents per liter (mEq/L)	Critical for blood pressure control, muscle and neurological function. Not a direct indicator of dietary intake. Most often associated with fluid status.
Urea (Ur) 2.5–7.1 mmol/l	Derived from protein degradation, influenced by diet, hydration and urine production. Increased blood urea indicates ↓ GFR, also influenced by type of protein consumed, degree of anabolism and dialysis adequacy. Data from urea kinetic modeling can be used indirectly to estimate the protein intake.
Creatinine (Cr) 0.6 to 1.2 mg/dL in adult males and 0.5 to 1.1 mg/dL in adult female	Normal waste product of muscle breakdown, not usually affected by dietary protein. Normally ↑ in CKD. Used to calculate the GFR.
Albumin (Alb) 35–55 g/l	A good predictor of mortality in CKD, however requires careful interpretation: can be influenced by fluid imbalance or an acute phase response as indicated by ↑ CRP levels (trauma, surgery, or infection). The large body pool and long half-life (14–20 d) renders albumin relatively insensitive to immediate changes in nutritional status.
Pre-albumin 150 to 360 mg/l	Shorter half-life and smaller body pool than albumin, is influenced by the acute phase response and anaemia. Levels are ↑ in CKD due to ↓ catabolism and excretion, volume changes. Not a useful nutritional marker.
Bicarbonate (HCO_3) 23–29 mmol/l	Reflects the degree of acidosis or alkalosis in the body. In CKD, kidneys are unable to maintain normal acid-base balance and HCO_3 may ↓. May also ↓ with diabetic keto-acidosis, diarrhea, hyperventilation, and lack of O_2, fever or starvation. CKD patients are often treated with sodium bicarbonate.
C-reactive protein (CRP) <3 mg/l	A non-specific positive acute phase protein that can be used to diagnose infectious and inflammatory diseases, associated with mortality in CKD. ↓ CRP is an indicator of inflammation and can result in ↑ albumin.
Protein catabolic rate (PCR) – calculated	In steady state, PCR correlates well with dietary protein intake. However, in catabolic patients, urea generated by muscle breakdown far exceeds that derived from diet. Conversely, anabolism may produce false ↓ PCR. As PCR is usually normalized for actual body wt, extremes in weight will affect nPCR.
Calcium (Ca) 8.5-*10.2 mg/dL*	Often corrected for albumin as albumin carries half the calcium in the blood. ↑ levels cause fatigue, difficulty in thinking, ↓ appetite, pain, ↑ urination, thirst, constipation, nausea and vomiting. Can be a sign of cancer. ↓ levels are most commonly seen after a surgical parathyroidectomy.
Phosphate ($PO_{4)}$ 2.5 to 4.5 mg/dL	↓ levels may indicate poor food intake or may result as part of the metabolic imbalances associated with re-feeding syndrome. ↑ levels are problematic in CKD.
Haemoglobin (Hb)	↓ levels cause anaemia, ↑ reduce quality of life due to its effect on mental function, exercise tolerance, fatigue, appetite, and sleep patterns.
Lipid profile	↓ total plasma cholesterol levels are associated with ↑ mortality, possibly as indicator of ↓ nutritional intakes.

mass over time, hand grip strength, mid-arm muscle circumference, skin fold thickness and even bioelectrical impedance analysis are all employed to more accurately monitor CKD patients, particularly those on dialysis. Waist circumference can be useful for assessing cardiovascular risk.

- Diet history
- Psycho-social factors which have influence on food intake.
- **Subjective global assessment (SGA)**: It has been found to be a useful adjunct to evaluate the nutritional status in CKD patients. The results have been found to be reproducible. Contributing factors include weight changes, dietary intake and gastrointestinal (GI) side effects, physical examination and fluid status on a 7-point scale.

Malnutrition in Renal Disease

Patients with chronic kidney disease are at substantial risk for malnutrition, characterized by protein energy wasting and micronutrient deficiency; 20–60 per cent of all patients with CKD (stages 3–5) are malnourished. The International Society of Renal Nutrition and Metabolism (ISRNM) defines protein energy wasting as a "the state of decreased body stores of protein and energy fuels (that is, body protein and fat masses)". This term was proposed by ISRNM in 2008 to specifically refer to a state of decreased body stores of protein and fat (wasting). This is to be distinguished from protein energy malnutrition, a form of protein energy wasting characterized purely by inadequate dietary intake. Moreover, unlike protein energy malnutrition, protein energy wasting cannot be corrected solely by increasing energy and protein intake.

While malnutrition in the general population might result from decreased intake but malnutrition in CKD is not entirely explained by reduced nutritional intake. The pathogenic mechanisms of malnutrition in chronic kidney disease are complex. A delicate interplay of multiple factors, including hormonal imbalances, decreased appetite and food intake, inflammation, increased catabolism, nutrient losses in dialysate and metabolic derangements predispose chronic kidney disease patients to malnutrition. Malnutrition increases the risk of morbidity, mortality and overall disease burden in these patients. A clear understanding of the pathophysiologic mechanisms of malnutrition in patients with CKD is essential for planning strategic interventions to improve growth and development and mitigate negative clinical outcomes rather than merely providing more calories and protein in the diet.

Comparison between Protein Energy Malnutrition and Protein Energy Wasting

Clinical Parameter	*Protein-Energy Malnutrition*	*Protein-Energy Wasting*
Pathogenesis	Explained by reduced nutrient and energy intake relative to metabolic demands of the body.	Not completely explained by reduced nutrient and energy intake.
Somatic mass		
Protein	Reduced	Reduced
Fat	May not be reduced	Reduced
Body mass index	May not be reduced	Reduced
Serum albumin	May be reduced	Markedly reduced
Hypermetabolism	May be present	Present
Hypercatabolism	May be present	Present
Response to therapy	Condition improved by nutrient repletion	Condition not improved solely by nutrient repletion

Source: Lorember, F.M. 2018. Malnutrition in CKD. Front Pediatr. 6: 161

Nutrition Support for CKD Patients who are Not on Dialysis

The purpose of medical nutrition therapy (MNT) for chronic kidney disease (CKD) is to maintain good nutritional status, slow progression, and to treat complications.

The key diet components to slowing progression of CKD are:

- Controlling blood pressure by reducing sodium intake
- Reducing protein intake, if excessive
- Managing diabetes

Manage with Adequate Calories, Protein and Other Nutrients

Among CKD patients, overnutrition results in sodium and volume overload, hyperkalemia, hyperphosphatemia, and accumulation of toxic metabolites of protein degradation. Undernutrition, on the other hand, exacerbates the risk for malnutrition and wasting. Hence, optimum nutrition is required for CKD patients: the optimal energy intake for non-dialyzed CKD (3-5 stages) patients, as suggested by kidney disease outcomes quality initiative (KDOQI) guidelines, is 35 kcal/kg/d for individuals under the age of 60 and 30–35 kcal/kg/d for those aged 60 years or older. However, CKD patients commonly have lower energy intake than what is recommended. High protein diet, usually defined as >1.2 g/kg/d, is known to induce significant alterations in renal function and kidney health and hence, should not be prescribed to CKD patients. High protein intake may lead to increased intraglomerular pressure and glomerular

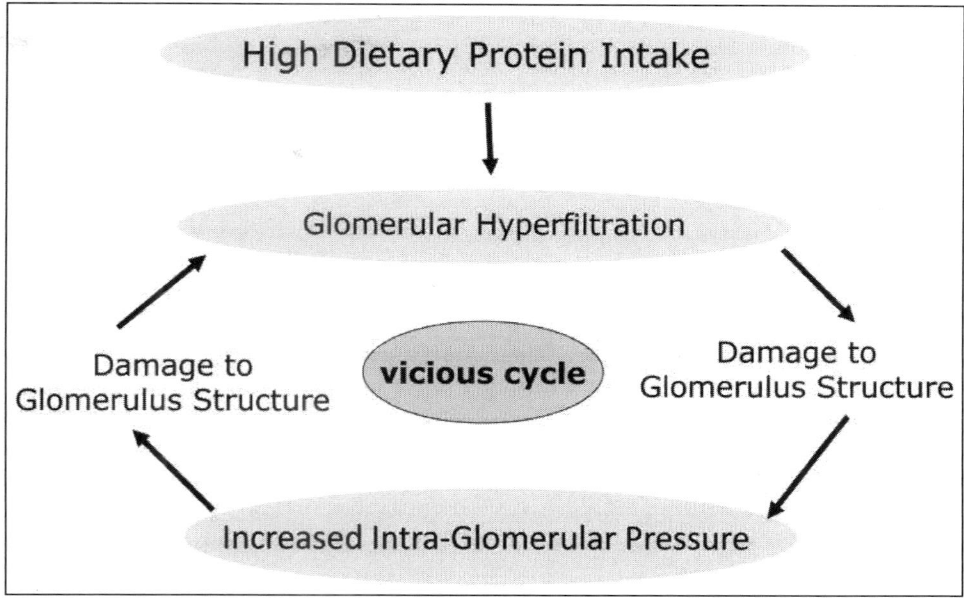

Figure 24.5: Effect of High Dietary Protein Intake on Kidney.

hyperfiltration. This can cause damage to glomerular structure leading to or aggravating chronic kidney disease (CKD).

The dietary protein requirements of non-nephrotic adult CKD patients who do not have superimposed catabolic illnesses appear to be similar to those of normal healthy adults that is about 0.8-1.0 g/kg/d, an intake that helps maintain nitrogen balance. For stable NDD-CKD patients, the recommended so-called low-protein diet (LPD) provides 0.60–0.80 g/kg/d, which represents sufficient protein intake, especially because the diet prescription includes the stipulation that ≥50 per cent of the protein should be of high biological value.

Supplemented very low protein diet (sVLPD) has also been recommended for stable CKD patients. This is a strategy used to enhance the potential beneficial effects of a protein-restricted diet; according to this strategy, reduce protein intake to 0.3 g/kg/d of mixed biological protein [very-low protein diet (VLPD)] and provide a daily supplement of 0.3 g/kg of a combination of some essential amino acids (EAA) and keto acids. The keto acids provide EAA precursors without the nitrogen load from EAAs. Hence, these sVLPDs appear to generate less toxic metabolic products than similar amounts of protein from LPDs. Therefore, the food protein component was not as limited as that in LPD.

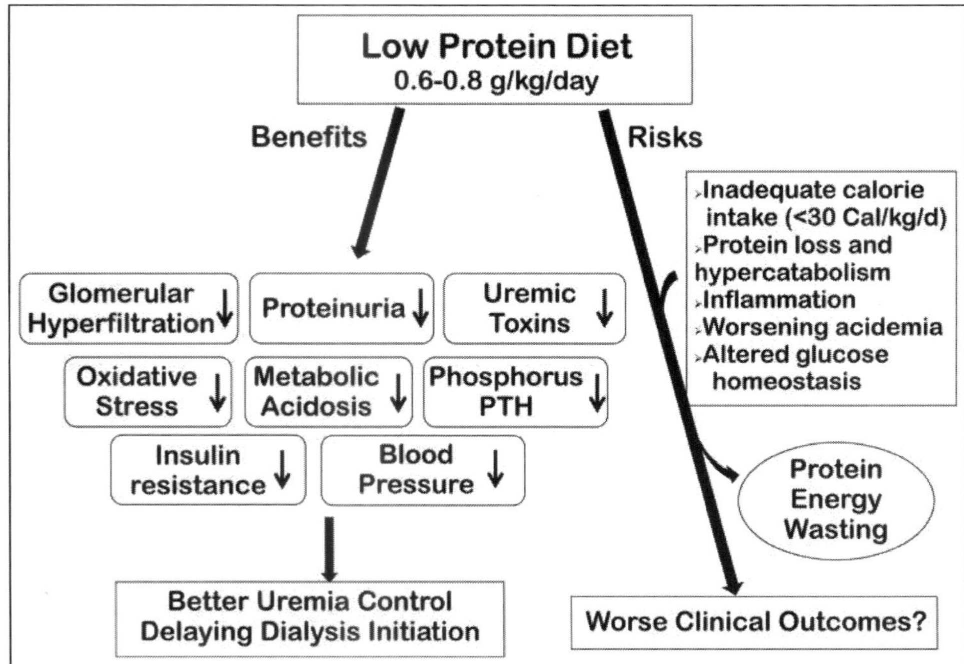

Figure 24.6: Diagram of the Role of Low Protein Diet in the Management of Chronic Kidney Disease.

Hence, a low protein diet (LPD) of 0.6–0.8 g/kg/day is often recommended for the management of CKD. One meta-analysis has suggested that a protein-restricted diet can slow down progression and loss of residual renal function in patients with advanced CKD. Kidney Disease Outcomes Quality Initiative (KDOQI) guidelines recommend a DPI of about 0.6–0.75 g/kg/d for patients with advanced CKD, defined as having a glomerular filtration rate (GFR) of less than 25 mL/min/1.73 m^2.

Low protein diet (LPD) is beneficial to improve metabolic acidosis in CKD. LPD also offers better control of CKD-mineral and bone disorder. Additionally, LPD may attenuate insulin resistance and oxidative stress, the common features that may play roles in accelerating atherosclerosis among patients with CKD. Less accumulation of protein waste products relieves clinical manifestations of uremia even in very late stage of CKD, and indeed LPD can be used effectively to delay the initiation of dialysis therapy. Some studies reported preserved body composition and skeletal muscle mass and function in CKD patients on LPD.

Renal patients who are malnourished or unable to achieve an adequate intake from food should be advised to supplement their oral intake with high energy and/or high protein foods (food fortification) or proprietary products or provided with enteral or parenteral nutrition support.

Fat

Fat should not provide more than 30 per cent of the total calories. Patients with CKD are at high risk for developing CVD; the risk increases as eGFR declines. CVD is the leading cause of mortality in CKD. If cholesterol and TG levels are high, decreasing intake of saturated and trans fats (substituting for monounsaturated and polyunsaturated fats), along with physical activity, can help control hyperlipidemia and reduce inflammation.

Phosphorus

Serum phosphorus levels may be "normal" until CKD is advanced. If serum phosphorus is elevated, dietary phosphorus restriction may be indicated. According to KDOQI recommendations, phosphorus intake should be 800–1000 mg/day (< 17 mg/kg IBW or SBW/day) while ESPEAN guidelines recommend 600-1000 mg P/day. Dietary protein restriction decreases phosphorus intake. If further restriction is needed, patients should be counseled to reduce intake of foods with added phosphorus. Limiting whole grains may help if further reduction is needed. Phosphorus binders with meals may be prescribed to limit absorption of phosphorus from food and beverages.

Sodium

Limit sodium intake to 2,300 mg a day or less. Weight reduction may be beneficial.

Potassium

Patients with CKD are at risk for hyperkalemia as a result of reduced potassium excretion, intake of high-potassium foods, metabolic acidosis, and medications that inhibit potassium excretion, such as RAAS antagonists for blood pressure control. Hyperkalemia is usually not seen until CKD is advanced, but may be seen at higher eGFRs in people with diabetes. Patients should restrict dietary potassium (take less amount of meat, poultry, fish, beans, dairy, coconut water, banana and nuts) when serum potassium level is 5.0 mEq/L or higher. Choose fruits (apple, berries, mangoes, papaya, cranberry juice, pears, peaches, watermelon *etc.*) and vegetables (bell peppers, cabbage, carrot, cauliflower, cucumber, mushroom, lettuce, okra *etc.*) low in potassium. Patients should avoid potassium-containing salt substitutes. ESPEAN guidelines recommend 1500-2000 mg K/day.

Iron

Anemia may develop early during the course of CKD due to inadequate synthesis of erythropoietin by the kidneys. Hemoglobin is used to assess anemia in CKD. Uncomplicated anemia of CKD is usually normocytic and normochromic. Both iron supplementation and injectable erythropoiesis-stimulating agents (ESAs) are used to correct anemia. The risks and benefits of these treatments in CKD are not yet defined.

Calcium and Vitamin D

Secondary hyperparathyroidism (elevated PTH) is associated with the most common cause of bone disease in CKD. Dietary phosphorus restriction and use of active vitamin D or its analogs may help control PTH levels in CKD. Calcium supplementation may help as well. Active vitamin D (calcitriol) or its analogs (doxercalciferol, paricalcitol, or alfacalcidol) may be used as eGFR declines.

B-complex Vitamins

A low protein diet (0.6 g/kg) may require supplemental riboflavin and cyanocobalamin. Supplemental pyridoxine may be beneficial due to medication-vitamin interactions. It may be prudent to provide adequate amounts of thiamin and folic acid. High doses of vitamin C are not recommended. Data do not support routine supplementation with niacin and vitamins A, E, and K. In advanced CKD, certain vitamins may accumulate due to reduced excretion (Steiber and Kopple, 2011). Haemodialysis patients should be prescribed supplements of water soluble vitamins.

Physical Exercise

Patients at all stages of CKD should be encouraged to take regular exercise. It can be a challenge to recommend safe and appropriate ways to exercise if the patient is elderly or infirm although there are multiple resources available and physiotherapy consultation can support.

Nutrition Support for Patients who are on Chronic Dialysis

Dialysis is usually initiated once GFR drops <10 ml/min (or <15 ml/min in people with diabetes) and the patient has symptoms of uraemia. Dialysis is a treatment which removes waste products normally filtered by the kidney from the blood. There are two main forms of dialysis *i.e.* haemodialysis and peritoneal dialysis. In haemodialysis, the blood is cleaned outside the body via a machine, and in the process of peritoneal dialysis the blood is filtered through the peritoneal membrane located in the abdomen. The common feature of both types of dialysis is the removal of wastes and excess fluids from the body.

When Chronic Kidney Disease (CKD) progresses on to dialysis, diet continues to play a vital role in a patient's rehabilitative care. Proper nutrition is even more important for these people as they must ensure that unnecessary waste products are not added to their system through their diet. Individuals using dialysis often consume inadequate quantities of macronutrients (energy measures as calories), fluids (water) and/or important micronutrients (vitamins, minerals and trace elements), therefore, nutrition support is very important for such patients.

Nutritional Imbalancing in Individuals Treated with Hemodialysis

Protein Energy Malnutrition

Malnutrition is common in people being treated with dialysis and close to 40 per cent suffer from varying degrees of protein-energy malnutrition especially protein energy wasting with varying causes. Evidence suggests that the buildup of toxins that can occur with renal failure can also suppress the appetite. As a result, people being treated with dialysis often reduce the amount of food they consume and suffer from protein energy wasting.

The mechanisms leading to protein energy wasting (PEW) in advanced kidney disease are still being elucidated; they cannot be attributed to any single factor in patients undergoing maintenance hemodialysis (MHD) (Figure 24.7). Still, it appears that the common pathway for all the metabolic derangements is related to exaggerated protein degradation and along with decreased protein synthesis. The observation that CKD patients spontaneously decrease their protein and energy intake as they progressively lose kidney function has led some to conclude that uremia per se causes protein catabolism stimulated by decreased nutrient intake. Protein intake may be restricted because higher intake of protein requires higher doses of dialysis. Note lower intake of protein with sufficient energy intake requires lower doses of dialysis. Hence, an additional stimulus for protein losses is the dialytic treatment per se. Both whole-body and skeletal muscle protein homeostasis are disrupted and there is a consistent finding of decreased protein synthesis at the whole-body level and an additional increase in whole-body protein breakdown. There also is evidence for a significant increase in net skeletal muscle protein breakdown. Notably, these undesirable effects persist for at least 2 hours following the completion of hemodialysis.

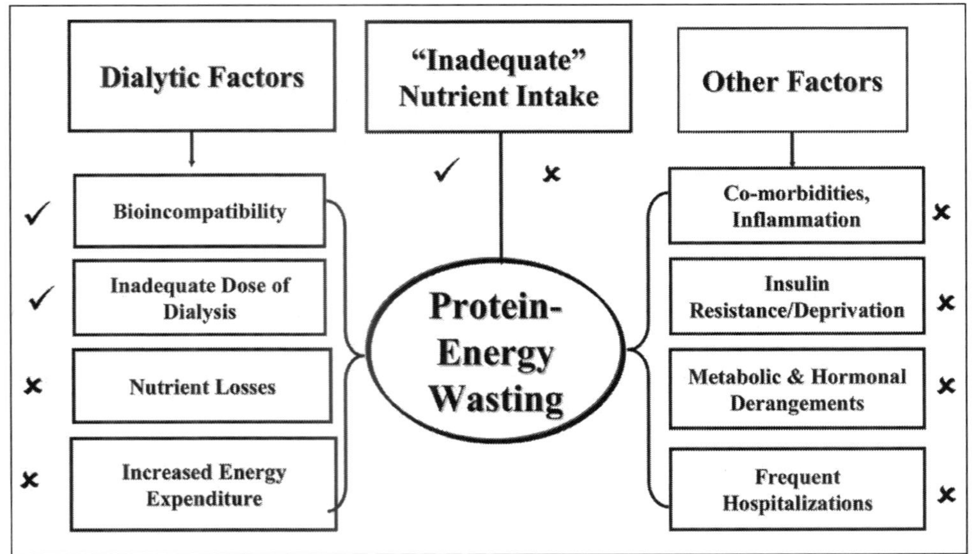

Figure 24.7: Causes of Protein-Energy Wasting in CKD Patients on Dialysis.

Note: The factors that are adequately managed are marked with a check sign and factors that need further attention are marked with an X sign. While a significant number of chronic dialysis patients report adequate intake, the optimal protein and calorie intake is still debated and marked with both a check and an X sign.

***Source*: Ikizler,T.A. 2013. Optimal nutrition in hemodialysis patients. Adv. Chronic Kidney Dis. 20 (2): 181-189.**

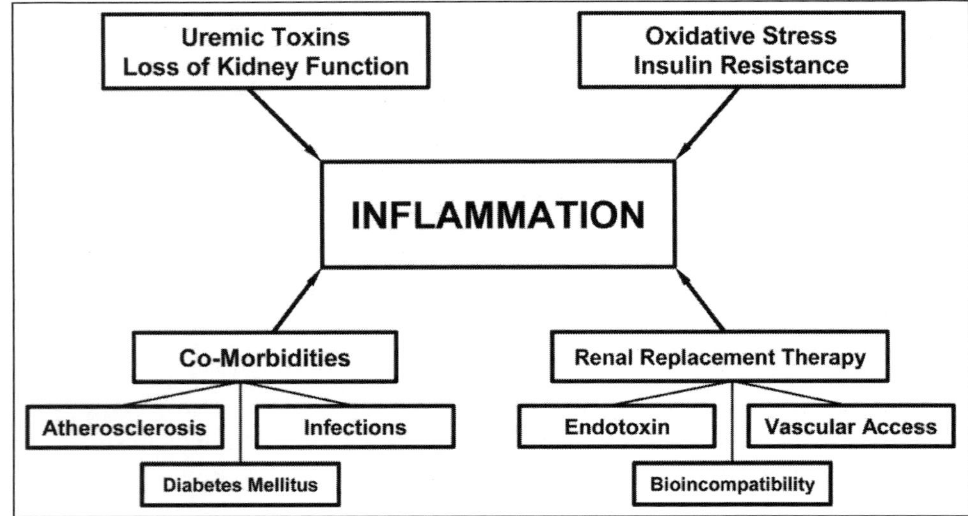

Figure 24.8: A Number of Modifiable and Non-modifiable Factors Lead to the Chronic Inflammatory State of Advanced Kidney Disease.

***Source*: Ikizler,T.A. 2013. Optimal nutrition in hemodialysis patients. Adv. Chronic Kidney Dis. 20 (2): 181-189.**

Anorexia or suppressed nutrient intake is a well-established metabolic response to inflammation in CKD patients; IL-1 and TNFα can cause anorexia through their effects on the satiety center in the central nervous system. As with inflammation, decreased insulin or decreased sensitivity to insulin can cause muscle protein losses. Several reports demonstrate that among MHD patients, muscle protein breakdown is enhanced in patients with Type II DM when compared to MHD patients without DM. In the absence of severe obesity, insulin resistance is still detectable in MHD patients and strongly associated with increased muscle protein breakdown, even after controlling for inflammation.

Metabolic acidosis, a common abnormality in patients with progressive CKD promotes PEW, by increasing protein catabolism, especially in muscle. In addition, acidosis stimulates the oxidation of essential amino acids to raise protein requirements in MHD patients.

Given the significance of the problem, as well as the complexity of the pathophysiological basis of PEW in advanced CKD, it is evident that the prevention and treatment options of PEW in MHD patients are complex. In MHD with PEW or at risk for PEW, there is no single treatment approach that will alleviate the multiple adverse consequences of PEW.

Micronutrient Deficiencies

Apart from protein-energy malnutrition, dialysis patients are also vulnerable to micronutrients deficiencies because dialysis can cause loss of water soluble vitamins such as vitamins B and C.

Micronutrient Excesses

While dialysis causes water soluble vitamins to be removed from the body more quickly and may lead to deficiencies in these vitamins. In renal failure, the body's inability to properly filter the blood of other micronutrients (which are not removed efficiently by dialysis) means that micronutrient excesses are also common in dialysis patients. The best way to correct these is by restricting intake of certain micronutrients in the diet.

Catabolic Effects of Dialysis

	Hemodialysis	*Continuous Ambulatory Peritoneal Dialysis (CAPD)*
Loss of amino acids	6-10g/dialysis	2-4 g/bag
Loss of glucose	25g/dialysis (glucose free dialysate)	uptake
Loss of protein	0	5-15g/day (higher with peritonitis)

Nutritional Assessment and Monitoring

In order to prevent and treat malnutrition, it is important firstly that people being treated with dialysis have their nutritional status assessed by a health professional (*e.g.* dietician or/doctor). Nutritional assessment allows those individuals who are not yet malnourished but at risk of becoming malnourished to be identified. There are a number of tools available to assess nutritional status in dialysis patients, including: (1) dietary intake measurements, (2) anthropometric measurements, (3) blood and urine tests, (4) nutritional scoring systems, (5) radiological or electronic-based quantitative measurements. However, because of concerns about the utility, reliability, interpretability, and practicality of these tests, malnutrition remains primarily a clinical diagnosis.

Nutritional assessment is performed by examining nutritional indicators including:

Food Intake

Dietary surveys, including short-term dietary recalls (*e.g.*, 24 hr), longer-term food records (3-7 days) and food frequency questionnaires (FFQs), can be useful in determining the risk of under- or overnutrition. They may also help track changes in dietary intake over time.

Anthropometric Measurements and Clinical Assessment

Anthropometric measurements and clinical assessment of subcutaneous fat mass and muscle mass and a history of weight loss should constitute important parts of the routine nutritional assessment of protein energy malnutrition in people on dialysis. Loss in fat deposits and muscle mass gives an indication of under nutrition. Body mass index (BMI) is also an important indicator for the assessment of malnutrition especially that of obesity. Most dialysis patients with malnutrition also have accompanying diseases, in particular cardiovascular disease and inflammation, and so their assessment is an integral part of the nutritional assessment of dialysis patients.

Subjective Global Assessment

The Subjective Global Assessment (SGA) is a subjective scoring system commonly used to assess nutritional status via a very focused nutritional history and physical examination that scores nutritional status from 1 (no nutritional loss) to 7 (severe nutritional loss). A dialysis-centric version of the SGA has also been developed.

Biochemical Assessment

This should include knowing serum creatinine and albumin levels. Creatinine levels may rise with increasing inability of the kidney to function. The normal blood creatinine level for males is usually less than 120 μmol/L and 90 μmol/L for females. Low creatinine levels may thus, reflect normal kidney function or diet patterns. Low levels of serum albumin may reflect inadequate intake of protein or calories. The Protein Equivalent of Total Nitrogen Appearance (PNA) can also help to quantify protein intake in dialysis patients.

Other Modern Tools

A number of tools exist to quantitatively measure fat and lean mass in dialysis patients, including bioelectrical impedance analysis (BIA), dual energy x-ray absorptiometry (DXA), and neutron activation analysis.

Treatment of Malnutrition

Treatment of malnutrition in dialysis patients is contingent upon whether:

1. The derangement (s) involves macronutrients (*e.g.* calories, protein, fat) and/or micronutrients (*e.g.* vitamins, minerals and trace metals), and
2. There is a deficiency or excess of the specific nutrient (s).

Table 24.3: Recommended Tests for Assessing Nutritional Status in Dialysis Patients

Tier	*Test*	*Frequency of Testing*
1	✰ Dietary intake surveys	✰ Every six months
	✰ per cent of standard weight	✰ Monthly
	✰ Body mass index	✰ Monthly
	✰ Serum creatinine	✰ Monthly
	✰ Blood urea nitrogen	✰ Monthly
2	✰ Mid-arm muscle circumference	✰ As necessary
	✰ Triceps skinfold	✰ As necessary
	✰ Subjective Global Assessment	✰ As necessary
	✰ Bioelectrical Impedance Analysis	✰ As necessary
	✰ Total Nitrogen Appearance (PNA)	✰ As necessary
3	✰ Serum albumin	✰ As necessary
	✰ Serum pre-albumin	✰ As necessary
	✰ Dual energy x-ray absorptiometry	✰ As necessary

These derangements can exist independently from one another, so it is possible for a patient to have macronutrient overnutrition (*e.g.* obesity) and a micronutrient deficiency (*e.g.* thiamine). Once the derangements are identified, a proper treatment plan can then be formulated. The next step would be to determine the route of nutrient delivery. In general, the oral or enteric routes take precedence over the parenteral route, in part to prevent mucosal atrophy and loss of the intestinal barrier. In patients who cannot tolerate nutritional feeding via the oral or enteral route, intradialytic parenteral nutrition (IDPN) is an option. This is typically given via a central venous route or dialysis shunt and usually requires the assistance of a nutritional consultant or trained dietician to calculate the appropriate prescription.

Nutritional and Dietary Management

People on dialysis may not be aware of the recommended energy and protein intake unless they receive nutritional counselling from a health professional. It is important for people who are on dialysis to access nutritional counselling so they understand the importance of different foods and the types of nutrients they add to the diet, rather than simply follow a diet which restricts intake of certain foods.

A well-balanced diet is necessary for them to stay fit as their kidneys are no longer functioning at its full capacity *i.e.* to get rid of the waste products and fluid from their blood. Monitoring diet and assessment of nutritional status plays a vital and central role in the care for dialysis patients as it is essential for dialysis patients to have the right amount of protein, calories, fluids, vitamins and minerals each day. To strive towards being healthy and happy dialysis patients, they would need to:

- ✰ Eat the right kind and right amount of food on a daily basis to maintain optimal nutritional status, prevent protein energy malnutrition and to prevent or/treat other complications and medical problems. A good meal plan choices can minimize build up of waste products and fluids between treatments and conserve muscle mass.
- ✰ Maintaining a healthy body weight.
- ✰ Take medication correctly as prescribed by the doctor
- ✰ Comply with their dialysis treatment, for instance, do not skip any dialysis session.
- ✰ Being physically active as possible.

Energy and Protein

Mostly opinion-based estimates suggest that daily protein and caloric requirements applicable to the majority of stable peritoneal and haemodialysis patients are greater than that of the general population, though certainly some dialysis patients will be able to consume lower levels without adverse consequences. Therefore, in general, management of nutritional aspects of MHD patients includes a comprehensive combination of preventive maneuvers to diminish protein and energy depletion. Getting the right amount of energy is important to overall health and well-being such as helping the body to use protein for building muscles and tissues. The energy need of each individual is different and depends on age, gender, body size, physical activity level, and medical condition. Some people need more energy to increase their body weight; others need to cut down their energy intake to reduce body weight.

During peritoneal or haemodialysis, some amount of proteins is lost and therefore, patients require a higher protein intake compared to people who are not on dialysis. Eating the right amount of protein will help dialysis patients stay fit as insufficient protein intake could result in weight loss, muscle wasting, reducing the ability to fight infections and thus, leading to malnutrition. For dialysis diet, high-quality protein, for example, meat, poultry, fish and egg whites are more desirable at every meal. Most dairy products such as yogurt, milk and cheese also contain high-quality protein. But limit the intake of dairy products *i.e.* not more than half a cup of milk or half a cup of yogurt or about 25 g cheese as most of them are rich in phosphorus. Even though nuts, dried beans, peas and lentils have protein, these foods are not generally recommended because they are high in both potassium and phosphorus. Therefore, it is very important that dialysis patients take their phosphate binders with all their meals.

The recommended dietary protein and calorie intake for (sedentary) MHD patients are 1.2–1.4 g/kg and 30–35kcal/kg, respectively. Other standard therapies for MHD patients with PEW include provision of adequate dialysis, treatment of metabolic acidosis, adjustments of dietary factors and treatment of infections.

Some patients on haemodialysis have difficulty in achieving recommended energy intakes; they should be advised to take energy rich foods and/or prescribed supplements. Patients generally have sufficient protein intake once energy requirements are met. Combined energy and protein supplements may help if intake from food is inadequate.

Overweight patients on peritoneal dialysis may require energy restriction to compensate for the additional energy absorbed from dialysate (if the dialysate contains glucose). Patient should initially be advised to reduce fat and sugar intake. High use of hypertonic dialysate to control fluid balance will increase the energy intake. Fluid management and sodium restriction should then be emphasized. Requirement for protein in patients on peritoneal dialysis are increased due to dialysate losses. High protein requirement is sometimes difficult to be met through diet alone, particularly in those with poor appetite, or during and

Table 24.4: Daily Nutritional Guidelines for Patients on Dialysis

	Haemodialysis	*Peritoneal Dialysis*	*Main food sources*
Energy	< 60 yrs old: 35 kcal/kg body weight ≟ > 60 years old: 30-35 kcal/kg body weight	< 60 yrs old: 35 kcal/kg body weight ≟ > 60 years old: 30-35 kcal/kg body weight	Cereals, bread, rice, pasta, potato, sugar, fats
Protein	1.2 g/kg ideal body weight	1.2-1.3 g/kg/ideal body weight	Meat, fish, eggs, pulses, dairy
Potassium	0.8-1 mmol/kg	Restrict only if hyperkalaemic 1mmol/kg	Fruits, vegetables, fruit juices, nuts, coffee, chocolate
Phosphate	<1000 mg	<1200 mg	Whole cereal grains, processed meat and cheese, shellfish, fish with bones, fizzy drinks with additives
Sodium	80-100 mmol (no added salt)	80-100 mmol (no added salt)	Table salt, smoked/cured foods, tinned and packet foods, ready to serve meals
Fluid	500 ml + previous day urine output	500 ml + previous day urine output + ultrafiltration	Drinks, gravies, sauces, soups, jelly, yogurt, water, ice cream, juices *etc.*

after peritonitis. Specific targets for protein rich foods and snacks, and use of prescribed supplements may be required in such cases.

Carbohydrates

Carbohydrates should provide 60-70 per cent of the energy intake; more than 70 per cent should be complex carbohydrates. Carbohydrate intake may be modified in case of patients with diabetes to achieve the goal of HgAIC <7 per cent. Fibre intake should be 20-25 g per day.

A high intake of soluble NSP is of particular importance in patients on peritoneal dialysis. The avoidance of constipation is extremely important for optimal functioning of this type of dialysis.

Fat

It should provide about 30 per cent (predominantly poly- and mono-unsaturated fatty acid sources with low saturated fats) of the total calories.

Vitamin A

Vitamin A is not removed from the body through dialysis as it is fat soluble. The kidneys play a major role in removing vitamin A and its by-products (especially retinol) from the body. Hence, in people with kidney failure, concentrations of vitamin A in the blood stream typically increase. However, the vitamin A in the blood stream is less likely to be taken up by the body's tissues and cells, where it is used for repair and maintenance.

There is a need for further research to properly understand the role of vitamin A in dialysis as there is a conflict as to whether or not people on dialysis treatment should receive vitamin A supplements. On the one hand, there are concerns that giving people on dialysis vitamin A supplements may lead to excessive concentrations in the blood stream and have toxic effects. On the other hand, there is evidence that low vitamin A levels in dialysis are associated with an increased risk of death from heart problems.

Phosphorus

Foods rich in phosphorus (*e.g.* dairy products, nuts, beans, lentils, whole grains *etc.*) increase the phosphorus level in the blood. Unfortunately, both hemodialysis and peritoneal dialysis are unlikely to remove accumulated levels of phosphorus in the blood and this can cause excessive amount of phosphorous to accumulate in the blood, which in turn cause the release of calcium from the bones. The continuous removal of calcium from the bones may eventually make them weak. It is also known that high accumulation of phosphorus in the blood results in the formation of what is called calcium-phosphorus crystals in the joints, muscle, blood and heart. The calcium-phosphorus crystals may pose problems such as bone pain, poor blood circulation and even damage to the heart. Non-dairy creamers are low in phosphorus and can be used in place of milk.

Almost all patients with high phosphate levels will also need to take phosphate binders as most foods contain some amount of phosphate. Phosphate binders prevent the body from absorbing the phosphate from the foods eaten and thus helping to prevent renal bone disease.

Potassium

Persons undertaking haemodialysis may have to limit the intake of foods rich in potassium because

Strategies for Achieving Potassium Restrictions

Limit potassium rich food sources

- **Fruits:** apricot, banana, avocado, melon, kiwi, mango, nectarine, large orange, papaya, pomegranate, all dried fruits
- **Vegetables:** spinach, mushroom, beetroot, broccoli, brussel sprouts, carrot, spinach, tomato, jacket or instant potato, chips (if not parboiled), baked beans, kidney beans, lentils, other beans and pulses
- **Drinks:** fresh fruit juices, coffee, drinking chocolate, malted drinks, blackcurrant cordials
- **Other:** salt substitutes, humus, nuts, peanut butter, chocolate, milk, yogurt, evaporated and condensed milk, yeast extract

Alter cooking methods (as K^+ is very water soluble)

- Use large volumes of water for boiling vegetables (double boiling is no longer recommended as the second boil does not remove a significant amount of additional K^+ but makes many vegetables inedible).
- Chop potatoes and other large vegetables into small to chunks to increase the surface area for K^+ removal.
- Soak beans and legumes for at least 24 h prior to cooking and change the water every 4 h.
- Rinse vegetables in warm water for a few seconds prior to boiling.
- Parboil vegetables before adding them to stews, soups, *etc.*
- Avoid pressure cooker and microwave cooking (except for re-heating previously cooked food).
- Limit portion size and quantities of fruit and vegetables.

impaired kidney function can lead to a rise in the level of potassium in the blood. Foods high in potassium include whole grain cereals, bread and biscuits, bananas, coconut water, melons, oranges, kiwi, pomegranate, avocado, dried fruits such as dates, raisins and prunes, potatoes, dark leafy vegetables, dried beans, tomatoes, milk, wine, chocolate, malted beverages, poultry, pork and fish *etc.* and they should be avoided. Canned fruits usually contain less potassium than fresh fruits. Drain all the fluid before serving. One can eat apple, berries, pear, plums, pineapple, grapes, peaches, cherries *etc.*

All vegetables have *some* potassium, but certain vegetables have more than others and should be limited or totally avoided. Choose broccoli, cabbage, carrots, cauliflower, onion, peppers, radish, cucumber, garlic, eggplant *etc.* and limit or avoid potatoes, tomato and tomato sauce, beets, pumpkin, cooked spinach *etc.* Potassium can be reduced during the preparation of vegetables by leaching or soaking.

Patients on peritoneal dialysis rarely need dietary potassium intake restriction because of continuous clearance. Hypokalaemia is often a problem requiring increased potassium diet.

Sodium

Use less salt and eat fewer salty foods as this may help to control blood pressure and reduce weight gains between dialysis sessions. High sodium diet can cause swelling of the lower joints of the feet and hands, increased weight associated with accumulated body fluids, increase in blood pressure, shortness of breath and increased heart activity. Hence, a diet high in sodium must be avoided by people on dialysis as it also increases the thirst and fluid consumption. With increased fluid consumption and impaired kidney function, a person undergoing dialysis may hold too much fluid in their body. In order to cut down on salt intake, limit its amount added to cooking. Use herbs and spices instead of common salt to flavour the food.

Fluid Intake

Dialysis patients can no longer excrete large quantities of fluids consumed; therefore, excess fluid can be retained in the body resulting in overloading of the blood circulation. Symptoms such as shortness of breath, high blood pressure and swelling of the legs can occur. Thus, both peritoneal dialysis and haemodialysis patients will need to consider urine output, remaining kidney function, and body size to restrict their fluid intake. Haemodialysis patients may keep track of weight gain in between two dialysis sessions as a gauge.

Daily allowance for fluid intake can vary from 500 ml to 1000 ml per day, depending on the patient's urine output. Limited urine output does not indicate dehydration in dialysis patients. However, it is an indication of further deterioration of kidney functions. Factors that can lead to increased intake of fluids must be watched and avoided. Daily fluid allowance includes any food or beverage that remains liquid at room temperature, for example gravy, soups, ice cream, tea, coffee, juices, porridge, ice cubes, water, fizzy drinks *etc.* Helpful tips to reduce fluid intake are:

- ✰ Choose lower-sodium foods to reduce thirst.
- ✰ Maintain good blood sugar control as high blood sugars will cause thirst.
- ✰ Choose smaller cups for drinks.
- ✰ Choose cold fluids as they are more thirst quenching.
- ✰ Sip fluids slowly so you do not drink as much.
- ✰ Drink only when you are thirsty, not out of habit or to be sociable.
- ✰ Suck on ice cubes.
- ✰ Suck on lemon or lime slices or hard, sour candy.
- ✰ Freeze allowed juices/beverages in an ice cube tray. Since frozen beverages last longer, one may drink less.
- ✰ Take sugar- free sweets or chew gums.
- ✰ Suck on allowed frozen fruit choices such as strawberries, blueberries or grapes.
- ✰ Rinse mouth with ice-cold water but do not swallow.

Micronutrient Supplementation

Kidney patients on dialysis may have to supplement their diet with vitamins and minerals in order to improve their nutritional status. If the deficiency is restricted to a single micronutrient, then the patient should be aggressively supplemented with that micronutrient. More often than not, there is concern that dialysis patients may not in general be consuming adequate micronutrients.

Potassium-restricted or protein-restricted diets may be recommended for some people on dialysis but such diets may result in thiamine and riboflavin deficiencies. Because of this and the additional concern that water soluble micronutrients, such as vitamin C and the B vitamins, may be lost in appreciable amounts with haemodialysis, many nephrologists supplement their patients with a daily dialysis "vitamin" to prevent deficiency. For vitamin B_1 a supplemental dose of 1.0-1.5 mg/day is adequate; for vitamin B_2 a supplemental of 1.0-2.0 mg/day is adequate. It is important for people on dialysis to take vitamin supplements as recommended by their doctor. However evidence suggests that many do not. For example a study conducted in New Zealand showed that many dialysis patients did not take their supplements as recommended, at least partly because their appetites were low.

Table 24.5: Recommendations for Micronutrients in Dialysis

Thiamine (B_1)	1.5 mg	Pantothenic acid	10 mg
Riboflavin (B_2)	1.7 mg	Biotin	300 µg
Pyridoxine(B_6)	10 mg	Folic acid	1000 µg
Cobalamine(B_{12})	6 µg	Ascorbic acid ©	60 mg
Nicotinamide	20 mg	Iron sulphate	150 mg

There are scattered reports of vitamin excesses (*i.e.* A, C, D) leading to clinical manifestations of hypercalcemia, hyperoxaluria, and neurologic changes for people treated with dialysis. Micronutrient toxicity should be considered when patients report the use of heavy micronutrient supplementation. The treatment is to stop the supplementation.

Nutrition in Acute Kidney Injury

AKI can be defined as an abrupt (within 48 h) reduction in kidney function with an absolute increase in serum creatinine of either ≥0.3 mg/dl (≥0.25 µmol/l) or a percentage increase of ≥50 per cent or a reduction urine output (≤0.5 ml/kg/h for >6 h). A grading system for AKI severity stratification in three stages has been also developed. The incidence of AKI is increasing, with frequencies of 3–10 per cent observed among hospitalized patients, which can rise up to 10–30 per cent in those admitted to the ICU. Up to 5 per cent of the patients with AKI in the ICU usually undergo renal replacement therapy (RRT). In the ICU, AKI usually develops in the context of multiple organ failure. Nutrition should be aimed at counteracting and attenuating the negative effects on lean body mass of both catabolism and hypermetabolism associated with critical illness. Thus, adequate nutritional support is considered a key element of overall therapeutic strategy of the syndrome.

Nutritional Assessment

Owing to the presence of many interfering factors, such as acute illness, alterations in body water distribution, external fluid balance derangements, *etc.*, the evaluation of nutritional status can be difficult in critically ill patients, especially if AKI is present, and traditional methods in this clinical setting have shown limited sensitivity and specificity. To assess the extent of prevalence of protein energy wasting which is a frequent problem in AKI, one should use four categories of diagnostic criteria: biochemical (such as albumin or prealbumin), body weight loss, decreased muscle mass and low energy and protein intakes.

Goals of Nutritional Support

The general aim of nutrition support is to maintain nutritional status whilst limiting the complications of AKI. This includes preventing or minimizing protein energy wasting, preserving lean body mass, avoiding further metabolic derangements, improving wound healing, supporting immune function and thus, reducing risk of mortality.

Goals of Nutritional Support in AKI

- ☆ To prevent protein–energy wasting (PEW)
- ☆ To preserve lean body mass and nutritional status
- ☆ To avoid further metabolic derangements
- ☆ To avoid complications
- ☆ To improve wound healing
- ☆ To support immune function
- ☆ To minimize inflammation
- ☆ To improve antioxidant activity and endothelial function
- ☆ To reduce mortality

From the nutritional point of view, dietetic classification of AKI can be divided into two groups:

i) Patients with AKI in the Non-Catabolic State

Common causes include dehydration, certain medications and urinary obstruction. Generally, these patients are stable and where required any renal replacement therapy (RRT) is usually provided by conventional haemodialysis. Oral diet alone, or with the addition of nutritionally-dense supplementary sip feeds are frequently sufficient to meet the patient's needs; if not artificial nutrition support should be implemented. Typically patients with pre-renal or post-renal injuries are present in a non-catabolic state.

ii) Patients with AKI in the Catabolic State

Causes of AKI in a catabolic state include sepsis, acidosis and trauma. Patients often have multi-organ failure and are likely to be managed on an intensive care unit. Although they could have any stages of AKI, these are likely to be predominantly patients with stage 2 and 3 AKI. PEW is a frequent finding in this group of patients and is associated with poorer patient outcomes in terms of length of hospital stay, complications and mortality rates. Protein turnover rates are increased and negative nitrogen balance results. Nitrogen requirements are increased above normal and are influenced by the mode of RRT undertaken. Although hypercatabolism cannot be reversed by simply increasing protein intake, appropriate nutrition support may reduce nitrogen losses.

Patients usually require artificial nutrition support, particularly if intubated and sedated. Wherever possible, this should be provided via the enteral route. Standard formulae can be used, but nutritionally-dense feeds with or without reduced electrolyte content are useful where the control of fluid balance and/or serum phosphate and potassium levels proves difficult.

Factors Involved in the Pathogenesis of Protein Catabolism in AKI

- ✰ Inadequate supply of nutrients
- ✰ Uraemic toxins
- ✰ Endocrine factors
- ✰ Defective response to insulin (insulin resistance)
- ✰ Increased secretion of catabolic hormones (glucagon, catecolamines, glucocorticoids, *etc.*)
- ✰ Resistance to and/or decreased/suppressed secretion of growth/anabolic factors
- ✰ Critical illness/acute phase reaction/systemic inflammatory response (cytokines)
- ✰ Metabolic acidosis
- ✰ Proteases (ubiquitine–proteasome system, *etc.*)
- ✰ Loss of nutritional substrates by renal replacement therapy

Nutritional Requirements in AKI

Energy

AKI itself has no effect on the patient's energy requirements. As generally observed in the critically ill, also in patients with AKI, energy expenditure seems to depend more on the severity of the underlying disease, pre-existing nutritional status and acute/chronic comorbidities, than on the presence of the syndrome itself; energy expenditure measured by indirect calorimetery rarely exceeds 1.3 times the basal energy expenditure (BEE) calculated by the Harris–Benedict equation, corresponding to 20–25 non-protein kcal/kg/day.

Even in individuals with AKI and multi-organ failure, measured energy requirements are only 20 per cent -30 per cent above the estimated basal metabolic rate values (KDIGO, 2012). KDIGO guidelines for AKI (2012) suggest achieving a total energy intake of 20–30 kcal/kg/day in patients with any stage of AKI. Others suggest to 25-35kcal/kg/day, in the critically ill patient with AKI (McClave *et al.*, 2016). Critical care specific equations may be used for those on ICU to avoid risks of overfeeding. In those receiving CRRT, the energy contribution from solutions that use lactate as a substrate or citrate as an anticoagulant should be taken into account.

Protein

Protein requirements are influenced by the patient's clinical condition, severity of AKI, presence or absence of a catabolic state and any renal replacement therapy (RRT) they may receive. The optimal protein intake of AKI patients is ill defined; as protein catabolic rate (PCR) in this clinical condition varies from 1.4 to 1.8 g/kg/day, an intake of at least 0.25 g of nitrogen/day is required to achieve less negative or nearly positive nitrogen balance.

Recommendation for protein requirements vary in the literature. It is essential that attention is given to individual patient condition, when considering the protein requirements as suggested in Table 24.6. For stable patients with stage 1 AKI who are not receiving RRT, protein turnover is not increased; a protein intake of 0.8–1.0 g/kg/day has been suggested (KDIGO 2012). In contrast, those with sepsis, trauma, inflammation or multiple organ failure may have increased protein requirements. Patients with stage 2 and 3 AKI receiving RRT will require a higher protein intake, with a suggested range from 1.5- 2.5g/kg/day (Gervasio *et al.*, 2011; McClave *et al.*, 2016; Patel *et al.*, 2017). It is important to remember that these recommendations are based on small sample sizes (n = 7-50) the majority of who were fed via parenteral nutrition (Brown *et al.*, 2010; Patel *et al.*, 2017).

Concerns have previously been raised regarding the safety protein intakes greater than 2.5g/kg/day, with fears the administered nitrogen may be used as an energy source or could even have a pro-inflammatory effect. Evidence from RCTs concerning advantages of very high protein intakes in critically ill patients with AKI is lacking. With 2.5 g/kg/day of protein and 35 kcal/kg/day of energy, only one-third of patients achieved a positive nitrogen balance.

Lipids

Lipids should represent ~30–35 per cent of total non-protein energy supply. In the case of parenteral nutrition, this can be obtained by giving the patient 0.8–1.2 g/kg/day of lipid from 10 to 30 per cent lipid emulsions or as a part of the commercially available three-in-one total nutrient admixtures. Lipids should be infused over 18–24 h, and serum tryglycerides should be monitored, stopping lipid administration when tryglycerides exceed 400 mg/dl. Lipid losses through the filters do not occur during haemodialysis or haemofiltration.

Electrolytes

AKI may be associated with significant electrolyte changes such as elevated potassium levels. Electrolyte intake should be individualized according to blood biochemistry, patient clinical condition as well as dietary provision. Frequent monitoring of blood biochemistry is essential.

Conservatively managed patients usually require dietary restriction of potassium, sodium, and phosphate. Low potassium and low phosphate diets can be

Table 24.6: Summary Table of Nutritional Requirements in AKI Based on Current Guidelines. These guidelines are based on low grade evidence and expert opinion, therefore their implementation in clinical practice requires an individualized and patient centred approach.

	Non-catabolic (no RRT)	*Non-catabolic (on IHD)*	*Non-catabolic (on CRRT)*	*Catabolic (no RRT)*	*Catabolic (on IHD)*	*Catabolic (on CRRT)*
Protein (g/kgBW/day)	0.8-1.0 (KDIGO, 2012)	Minimum 1.1 (Naylor *et al.*, 2013)	1.2-1.5*	At least 1.0*	1.0-1.5 (Cano *et al.*, 2006)	Up to 1.7 in hypercatabolism (KDIGO, 2012) 1.5-2.5/kg/day in critically ill patients with AKI (McClave,2016; Patel *et al.*, 2017)
Energy	Not affected by AKI. Tailored to individual requirements and critical state Some authors suggest 20-30 Kcal/kg/day even in the critical settings (KDIGO, 2012). Others instead promote 25-30 Kcal/kg/day for patients in ICU with AKI (Cano, 2009; Brown; 2010, Gervisio, 2011; McClave, 2016).					
Fluid	Fluid requirements require individual medical assessment. Standard fluid equations are unlikely to be helpful. Fluid balance and daily weights should be monitored closely.					
Electrolytes	Monitor and adjust intakes as required. Will vary depending on disease state and type of treatment.					
Micronutrients	Requirements are not well documented. Lipid soluble vitamin levels and antioxidants status are low. CRRT has negative effect on balance of some vitamins and trace elements. Whether micronutrient supplementation improves outcomes is unknown (Ch *et al.*, 2015).					

All recommendations for kg/body weight (BW) refer to actual body weight, however, special considerations to avoid overfeeding should be taken into account in obese individuals.

* There is currently no evidence to support these protein requirements and values are provided based on the expert opinion. If protein provision is increased, further dose monitoring of kidney function, patient condition and nitrogen balance will be required.

IHD: Intermittent Haemodialysis; RRT: Renal Replacement Therapy; CRRT: Continuous Renal Replacement Therapy.

implemented where serum levels are high. However, where the dietary restriction of phosphate may restrict an individual's food choice, allowing more liberal phosphate consumption to help promote food intake is a better approach. In the short-term, maintaining a good nutritional intake is of greater priority than achieving phosphate control. Avoidance of hyperkalaemia takes greater clinical priority, however not all patients with AKI and hyperkalaemia require dietary potassium restriction. Health care professionals should consider the cause of hyperkalaemia; take into account metabolic acidosis, other non-dietary related causes as well as considering dietary intake and provision via enteral and parenteral solutions if applicable. If and when kidney function recovers, serum potassium and phosphate levels normalize, any restrictions can be lifted. In the polyuric phase, oral or intravenous supplementation may be required.

Patients undergoing Continuous renal replacement therapy (CRRT) need a less restrictive diet, and many need full nutrition support. During CRRT, many AKI patients develop hypokalaemia, hypophosphataemia, and hypomagnesaemia and require supplementation. These are important metabolic derangements, especially in ventilated patients, as electrolyte depletion can lead to increasing muscle weakness, alterations in acid–base balance, and to further nephrotoxicity (especially tubular damage). Daily monitoring is necessary. At different stages of AKI, supplementation or restriction can be necessary, often determined by the catabolic state of the patient and the modality of dialysis (*e.g.* intermittent or continuous).

Micronutrients

Levels of trace elements (essential micronutrients with regulatory, immunologic and antioxidant functions) can be lower than normal in AKI. This can be due to many factors: acute phase reaction/critical illness, leading to variable protein binding, redistribution of elements between plasma and tissues, acute losses of biological fluids, dilution, varying concentrations of trace elements in dialysis/haemofiltration fluids, effects of enteral or parenteral nutrition fluid, analytical problems, *etc.*

Requirements for micronutrients are not well defined. Caution should be taken when interpreting serum micronutrient levels; many patients with AKI present in the acute phase response, which will itself alter blood levels (Gervasio *et al.*, 2011). Plasma levels of vitamins A, D, E (Druml *et al.*, 1998) and vitamin C (Story *et al.*, 1999) together with selenium and zinc (Story *et al.*, 1999; Berger *et al.*, 2004; Oh WC *et al.*, 2015) are lower in AKI than in normal subjects. However, this may be related to inflammatory processes, RRT or medication rather than true nutritional deficiencies. Micronutrient losses on CRRT may also be significant. Documented daily ultrafiltrate losses include 100 mg vitamin C (Bellomo and Boyce, 1993), 290g folate (Fortin *et al.*, 1999) and 4mg thiamine (Berger *et al.*, 2004), with some authors suggesting supplementation (Gervaso *et al.*, 2011). Megadoses of vitamin C (>250mg/day) are potentially toxic because of the danger of secondary oxalosis in kidney failure. Vitamin A toxicity has been reported in patients with AKI receiving parenteral nutrition; some authors suggest monitoring Vitamin A levels (Gervasio *et al.*, 2011). Trace elements circulate mainly bound to protein and therefore are generally unaffected by CRRT but may be affected by the acute phase response. Selenium is affected by CRRT with reported daily losses of 0.97μmol (Berger *et al.*, 2004) and some authors suggest its supplementation (Gervaso *et al.*, 2011). Since the provision of micronutrients from commonly used parenteral sources and enteral feeds may be insufficient to replace some of these losses to meet requirements, there is some suggestion that selective supplementation should be considered (Gervasio *et al.*, 2011). However, demonstrating deficiency of micronutrients in patients with AKI does not equate to demonstrating a clinical benefit from supplementation; it is not known whether micronutrient supplementation to compensate for RRT losses improves outcomes (Oh *et al.*, 2015). Current guidelines lack specific recommendations on specific micronutrient doses and duration.

Fluid

Fluid requirements require individual medical assessment. Standard fluid equations are unlikely to be helpful. Fluid balance and daily weights should be monitored closely. It is not normally restricted on CRRT. Restrictions may be necessary if the patient is receiving haemoldialysis (HD).

Monitoring

Where patients require dietary input, it is important that they receive regular dietetic review throughout the course of treatment and during recovery. Energy, protein and fluid requirements will require review as kidney function changes. Where kidney function recovers and electrolyte and fluid balance normalize, patients should be supported to relax restrictions and return to a normal balanced diet.

Glomerulonephritis

Glomerulonephritis is inflammation of the glomeruli which are a network of capillaries known as tuft, located within Bowman's capsule in the kidney at the beginning of a nephron. There are approximately 1 million glomeruli, or filters, in each kidney which remove excess fluid, electrolytes and waste from the blood stream and pass into the urine for excretion. Due to their inflammation, kidney gets injured and it cannot get rid of wastes and extra fluids in the body.

Numerous inflammatory and non-inflammatory diseases affect the glomerulus and lead to alteration in glomerular permeability, structure and function. The term glomerulonephritis (GN) implies that there is an immune pathogenesis, and non-immune-mediated conditions affecting the kidneys are also to be considered. Glomerulonephritis may be primary, restricted to the kidney, or may be part of a multisystem disease; glomerular inflammation can result in damage to any of the three major components of the glomerulus *i.e.* basement membrane, mesangium and capillary endothelium. If the severe or prolonged inflammation associated with glomeronephritis continues, the kidneys may stop working completely resulting in kidney failure.

Classification

I. Some researchers classify glomerular diseases according to whether they present as either a nephritic or a nephrotic syndrome:

A. Glomerular Diseases Presenting as Nephritic Syndromes

They are associated with a clinical presentation of:

- Hypertension
- Oedema
- Urine sample showing red blood cells, red blood cell casts, and
- Moderate degree of proteinuria.

B. Glomerular Diseases Presenting as Nephrotic Syndromes

They are characterized by:

- Heavy proteinuria (> 3.5 g/24 h)
- Hypoalbuminemia
- Hyperlipidemia and
- Oedema.

However, most of glomerular diseases can present with components of both.

II. Diseases causing glomerulonephritis can also be classified according to whether they cause only renal abnormalities (primary renal diseases) or whether the renal abnormalities result from a systemic disease (secondary renal diseases).

III. The recent classification includes following principles:

Morphological Changes	*Clinical Manifestations*	*Types of Course*	*Functional State of Kidneys*
I. Proliferative-focal or diffuse	Isolated changes in urine (haematuria, proteinuria)	I type-rare exacerbations (more rare than every 10 years)	Chronic renal failure: stage. In case of normal results special tests should be performed)
1.Extracapillary			
2. Membranous-proliferative a. with subendothelial b. dense deposits c. mixed	Nephrotic syndrome and arterial hypertension		
3. Mesangioproloferative-immune negative			
4. IgA-nephropathy (Berger's)			
5. Focal-segmental glomerulosclerosis			
II. Non-proliferative			
1. Minimal			
2. Membranous (I,II,III,IV stage)			
III. Sclerotic (fibroplastic)			
Primary (idiopathic), secondary or innate origin of glomerulonephritis also should be mentioned in diagnosis.			

IV. Acute and Chronic Glomerulonephritis

Acute glomerulonephritis is an infective renal disease characterized by sudden onset of hematuria, oliguria, edema and hypertension. It is most common in preschool and early school age children with a peak age of onset of 6-7 years; more common in boys than girls. It accounts for 90 per cent of renal diseases in childhood. Acute glomerulonephritis in all probabilities results secondary to a proceeding streptococcal (beta-hemolyticus type 12) infection of throat or skin, usually one to 3 weeks before the onset of symptoms. It may also be caused by other illnesses including lupus, Goodpasture's syndrome, Wegener's disease *etc.* Sometimes, it may get better on its own and sometimes, right treatment is required to prevent kidney damage. The early symptoms include puffiness on the face in the morning, blood in the urine, dysuira, mild edema (may appear only as rapid weight gain) malaise, mild headache, lethargy, pallor, irritability, fever, GIT disturbances (anorexia,vomiting, abdominal pain) *etc.* One may be short of breath because of extra fluid in the lungs, may have mild high blood pressure (50 per cent of the cases) during first four to five days of illness *etc.*

Chronic glomerulonephritis accounts for 13.3 per cent cases of newly diagnosed chronic kidney disease (CKD) in India. The chronic form may develop without symptoms over several years and ultimately, it often leads to complete kidney failure. Early signs and symptoms include hematuria, proteinuria, high blood pressure, oedema on the face and ankles, frequent night time urination, foamy urine *etc.* Symptoms of kidney failure include fatigue, anorexia, nausea and vomiting, insomnia, dry, itchy skin, muscle cramps at night *etc.*

Sometimes, the disease runs in the family. This kind often shows up in the young persons who may develop hearing loss and vision loss. Some forms are caused due to change in the immune system. However, in many cases, cause is not known. Sometimes, after one acute attack of the disease, it may develop into chronic form later.

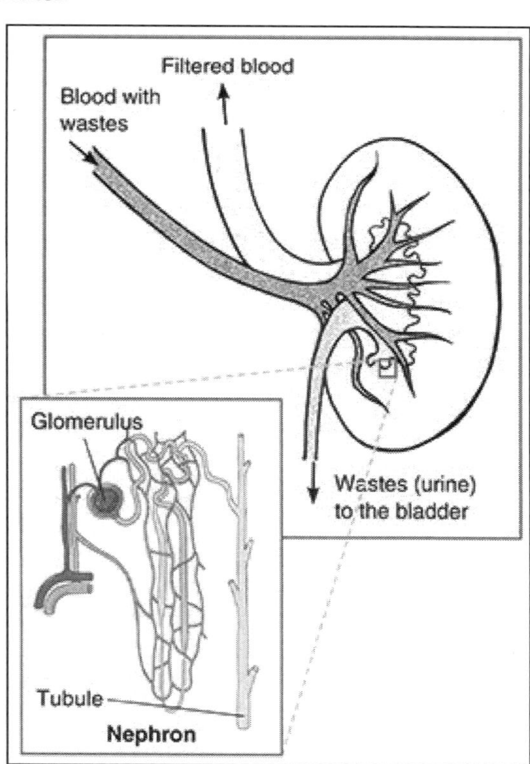

In the Nephron (left), Tiny Blood Vessels Intertwine with Fluid-collecting Tubes. Each kidney contains about 1 million nephrons.

Causes

A number of different diseases can result in glomerular disease. It may be the direct result of an infection or a drug toxic to the kidneys, or it may result from a disease that affects the entire body, like diabetes or lupus. Many different kinds of diseases can cause swelling or scarring of the nephron or glomerulus. Sometimes glomerular disease is idiopathic, meaning that it occurs without an apparent associated disease.

Conditions that can lead to inflammation of the kidneys' glomeruli include the following:

1. Infections

Acute Post-streptococcal Glomerulonephritis (PSGN)

It may develop a week or two after recovery from an episode of strep throat or, in rare cases, impetigo (a skin infection). The *Streptococcus* bacteria do not attack the kidney directly, but an infection may stimulate the immune system to overproduce antibodies, which are circulated in the blood and finally deposited in the glomeruli, causing damage. PSGN can bring on sudden symptoms of oedema, reduced urine output (oliguria), and blood in the urine *i.e.* hematuria. Tests will show large amounts of protein in the urine and elevated levels of creatinine and urea nitrogen in the blood, thus, indicating reduced kidney function. High blood pressure frequently accompanies reduced kidney function in this disease. It is most common among children between the ages 3 and 7 than adults, although can occur at any age and most often affects boys. PSGN is most common in children between the ages of 3 and 7, although it can strike at any age, and it most often affects boys. It lasts only a brief time and usually allows the kidneys to recover. In a few cases, however, kidney damage may be permanent, requiring dialysis or transplantation to replace renal function.

Bacterial Endocarditis

Bacteria occasionally can spread through bloodstream and lodge in the heart, causing an infection of one or more of the heart valves. A person having heart defect, such as a damaged or artificial heart valve, is at greater risk of this condition. Bacterial endocarditis is associated with glomerular disease, but the researchers are not sure whether the renal lesions that form after a heart infection are caused entirely by the immune response or whether some other disease mechanism contributes to kidney damage. Treating the heart infection is the most effective way of minimizing kidney damage. Endocarditis sometimes produces chronic kidney disease (CKD).

Viral Infections

Viral infections, such as the human immunodeficiency virus (HIV) that leads to AIDS, hepatitis B and hepatitis C, can trigger glomerulonephritis. Between 5 and 10 per cent of people with HIV experience kidney failure, even before developing full-blown AIDS. HIV-associated nephropathy usually begins with heavy proteinuria and progresses rapidly (within a year of detection) to total kidney failure. Researchers are looking for therapies that can slow down or reverse this rapid deterioration of renal function, but some possible solutions involving immunosuppression are risky because of the patients' already compromised immune system.

2. Autoimmune Diseases

When the body's immune system functions properly, it creates protein-like substances called antibodies and immunoglobulins to protect the body against invading organisms. In an autoimmune disease, the immune system creates auto-antibodies, which are antibodies or immunoglobulins that attack the body itself. Autoimmune diseases may be systemic and affect many parts of the body, or they may affect only specific organs or regions.

Systemic Lupus Erythematosus (SLE)

A chronic inflammatory disease, lupus can affect many parts of the body, including skin, joints, kidneys, blood cells, heart and lungs. Because women are more likely to develop SLE than men, some researchers believe that a sex-linked genetic factor may play a part in making a person susceptible, although viral infection has also been implicated as a triggering factor.

Lupus nephritis is the name given to the kidney disease caused by SLE, and it occurs when auto-antibodies form or are deposited in the glomeruli, causing inflammation. Ultimately, the inflammation may create scars that keep the kidneys from functioning properly. Conventional treatment for lupus nephritis includes immunosuppressants to suppress the immune system and steroid is given to reduce inflammation.

Goodpasture's Syndrome

It is a rare condition that affects mostly young men but also occurs in women, children and older adults. This immunological lung disorder involves an autoantibody that specifically targets the kidneys and lungs. This syndrome can mimic pneumonia; lung damage is superficial compared with progressive and permanent damage to the kidneys.

IgA Nephropathy

This primary glomerular disease results from deposits of immunoglobulin A (IgA) in the glomeruli where it causes inflammation. It is characterized by recurrent episodes of blood in the urine. IgA nephropathy can progress for years with no noticeable symptoms. The silent nature of the disease makes it difficult to determine how many people are in the early stages of IgA nephropathy, when specific medical tests are the only way to detect it. This disease is estimated to be the most common cause of primary glomerulonephritis *i.e.* glomerular disease not caused by a systemic disease like lupus or diabetes mellitus. It appears to affect men more than women. Although IgA nephropathy is found in all age groups, young people rarely display signs of kidney failure because the disease usually takes several years to progress to the stage where it causes detectable complications.

3. Genetic

The primary indicator of hereditary nephritis-Alport syndrome is a family history of chronic glomerular disease, although it may also involve hearing or vision impairment. This syndrome affects both men and women, but men are more likely to experience chronic kidney disease and sensory loss. Men with Alport syndrome usually first show evidence of renal insufficiency while in their twenties and reach total kidney failure by age 40. Women rarely have significant renal impairment, and hearing loss may be so slight that it can be detected only through testing with special equipment. Usually men can pass the disease only to their daughters. Women can transmit the disease to either their sons or their daughters.

4. Vasculitis

Polyarteritis

This form of vasculitis affects small and medium blood vessels in many parts of the body, such as heart, kidneys and intestines.

Granulomatosis with Polyangiitis

This form of vasculitis, formerly known as Wegener's granulomatosis, affects small and medium blood vessels in lungs, upper airways and kidneys.

5. Sclerotic Diseases

Glomerulosclerosis

It is scarring of the glomeruli. In several sclerotic conditions, a systemic disease like lupus or diabetes is responsible. Glomerulosclerosis is caused by the activation of glomerular cells to produce scar material. This may be stimulated by molecules called growth factors, which may be made by glomerular cells themselves or may be brought to the glomerulus by the circulating blood that enters the glomerular filter.

Diabetic Nephropathy

Kidney disease is one of several problems caused by elevated levels of blood glucose, the central feature of diabetes. In addition to scarring the kidney, elevated glucose levels appear to increase the speed of blood flow into the kidney, putting a strain on the filtering glomeruli and raising blood pressure. Diabetic nephropathy usually takes many years to develop. Good control of blood sugar levels and blood pressure might prevent or slow kidney damage.

High Blood Pressure

This can damage the kidneys and impair their ability to function normally. Glomerulonephritis can also lead to high blood pressure because it reduces kidney function and can influence how the kidneys handle sodium.

Focal Segmental Glomerulosclerosis (FSGS)

It is characterized by scattered scarring of some of the glomeruli, typically limited to one part of the glomerulus and to a minority of glomeruli in the affected region. This condition can result from another disease or occur for no known reason. Most patients with FSGS progress to kidney failure over 5 to 20 years. Some patients with an aggressive form of FSGS reach kidney failure in 2 to 3 years.

6. Other Glomerular Diseases

Membranous Nephropathy

It is also called membranous glomerulopathy. Fully 75 per cent of cases are idiopathic, which means that the cause of the disease is unknown. The remaining 25 per cent of cases are the result of other diseases like systemic lupus erythematosus, hepatitis B or hepatitis C infection, or some form of cancers.

Heavy Use of Nonsteroidal Anti-inflammatory Drugs

They may also be a risk factor. One shouldn't exceed the dosage and length of treatment.

Clinical Symptoms

- Hematuria, oliguria and proteinuria are the most classical symptoms. Other symptoms can be nausea, vomiting, shortness of breath, fever, raised blood pressure, edema *etc.* Some may develop anuria and this condition is fatal if not managed through medical treatment and dietary management.
- Clinical presentation of glomerulonephritis varies widely:
- **Asymptomatic non-nephrotic proteinuria:** Proteinuria (150 mg to 3 g per day)
- **Microscopic hematuria:** More than 2 RBC per high power field in spun urine.
- **Macroscopic hematuria:** Brown or smoky urine
- **Nephrotic syndrome:** Proteinuria greater than 3.5 g/day, hypoalbuminemia less than 3.5 g/dL, edema, hypercholesterolemia and lipiduria.
- **Nephritic syndrome:** Abrupt onset with oliguria, hematuria, proteinuria, azotemia, edema and hypertension
- **Rapidly progressive glomerulonephritis (RPGN):** Proteinuria, hematuria and renal failure developing over days to a week.
- **Chronic glomeruonephritis:** Proteinuria, hypertension, renal failure and smooth contracted kidneys on ultrasonography (USG) examination.

Nutritional and Dietary Management

The diet of the patient needs not be restricted routinely.

Regular measurement of vital signs, body weight and intake and output of fluid is essential to monitor the disease's progress and detect complications that may appear at any time during the course of the disease. Daily weight recording is important to assess the increase and decrease of edema. In case of passage of less urine and albuminuria, weight gain *etc.*, diet should be planned with restriction of protein, salt (sodium and potassium) and fluid intake, till oliguria and increased blood urea level persist.

Energy

Requirements are calculated based on age and weight (80 kcal/kg body weight for children) and an additional allowance of 10 per cent is given for infections.

Protein

Protein is to be restricted only if blood urea nitrogen is elevated and oliguria is present. If anuria develops, proteins should be stopped. In case of oliguria, dietary protein must be restricted till high blood urea persists. In that case, low protein diet is recommended to give rest to kidneys. If oliguria is not there and blood urea is normal, there is no need to restrict protein and normal recommended dietary protein intake (0.5 g/kg body weight for older children and 1 to 1.5/kg body weight per day for younger children) should be restored.

Carbohydrates

Carbohydrate containing foods should be allowed freely. Liberal intake of carbohydrates shall provide required energy as well as have protein sparing action. Both simple and complex carbohydrate food sources should be included. Starchy foods such as potato, sweet potato, colocasia, sago, arrowroot, corn flour, starchy grains (rice), jam, marmalade, honey, glucose and sugar should be included liberally in the diet.

Fat

Fat intake should not be restricted. Recommended amount for a particular age group should be given. It is better if emulsified and easily digestible fats such as butter and cream are given.

Fluid Intake

A record of daily weight is the most useful means to assess fluid balance. Sodium and water restriction is useful when the urine output is significantly reduced (< 2 to 3 dl/24hr.). In these individuals, fluid intake should be allowed in a calculated amount *i.e.* the water allowed should be equivalent to the total amount of previous day urine output in 24 hrs plus insensible losses. In general, daily fluid replacement should be 1000 or 500 ml for insensible water losses (30ml/kg body weight for infants; 20 ml/kg body weight for older children; and 10 ml/kg body weight for adults) plus urine output for 24h on the previous day. Water is calculated taking into account the water consumed with the drugs, water present in milk, tea, coffee, juices *etc.*

Calcium

The intake should be roughly 1 g/day.

Sodium

Sodium helps to regulate the amount of fluid in the body. If blood contains too much sodium, more water will be retained by the kidneys as a result of which blood volume and blood pressure rises and it puts much strain on the kidneys. Hence, sodium intake shall vary with the degree of oliguria and hypertension. Sodium is to be restricted to 500-1000 mg/day. When reducing sodium intake, avoid adding salt to foods, as well as canned foods, bakery goods, processed foods, salted snacks, dried fruits and processed meats. One should use fresh herbs and sodium-free spices in place of salt, but do not use salt substitutes, which replace sodium with potassium. Moreover, foods such as green leafy vegetables, cauliflower, radish, carrot, pulses, dairy products, cumin seeds *etc.* should be taken in restricted amount as they contain moderate amount of sodium.

Potassium

When kidneys fail, they cannot properly filter excess amounts of potassium from the blood. If potassium levels get too high, it can disrupt proper muscle and nerve function, causing irregular heart beat and even heart attack. To prevent potassium from building up too high in blood, reduce intake of potassium-rich foods, such as bananas, sweet potatoes, potato, colocasia, spinach, coriander leaves, tomatoes, mango, lemon, sweet lime, peaches, plums, peaches, aonla, coconut water, nuts, legumes and dairy products. But at the same time, all renal patients should remember that too little potassium can also be dangerous. It is restricted to 1 mmol/kg/day when urine formation is reduced.

Phosphorus

It is found in almost all foods and its intake should be restricted to 8-12 mg/kg/d. Phosphorus rich foods are milk, cheese, whole pulses, whole wheat, pearlmillet, nuts, meat, poultry, egg, mango lemon aonla, plums, peaches *etc.* and should be avoided.

Nephrotic Syndrome (Nephrosis)

What is Nephrotic Syndrome?

Nephrotic syndrome is not a specific kidney disease; instead, it is a general term for a condition where too much protein is lost in the urine. Nephrotic syndrome is caused by degenerative changes in the kidneys' capillary walls which consequently permit the passage of albumin into the glomerular filtrate and as a result, kidneys start losing large amounts of protein in urine. As kidneys get worse, extra fluids and salt build up in the body. This causes to have swelling (edema), high blood pressure and higher levels of cholesterol. People of any age can be affected by nephrotic syndrome, although children aged between 18 months and four years are at increased risk. Sometimes, nephrotic syndrome goes away after treatment. Other times, this condition may last for many years and eventually lead to kidney failure.

Etiology

Nephrotic syndrome in childhood is largely primary or idiopathic, although a small proportion of cases are secondary to infectious agents and other glomerular and systemic diseases. The etiology of nephrotic syndrome is also age dependent. Most cases appearing in the first 3 months of life are referred to as congenital nephrotic syndrome (CNS) and are due to genetic diseases. Beyond the first year of life and in the first decade, most cases are due to primary or idiopathic nephrotic syndrome, whereas the proportion of secondary nephrotic syndrome cases increases beyond the first 10 years of life. The most common histologic variant is minimal change nephritic syndrome (MCNS), which is responsible for more than 80 per cent of all cases. Other, less common histopathologic types in this age group include FSGS, MPGN, and mesangial proliferative glomerulonephritis.

Nephrotic syndrome may also be secondary to a number of systemic diseases in children. Pediatric illnesses such as systemic lupus erythematosus, especially membranous (WHO Class V) SLE; Henoch-Schönlein purpura; diabetes mellitus; and sarcoidosis may all present with nephrotic syndrome.

Infectious agents may also cause nephrotic syndrome and can be viral, bacterial, or parasitic. Although it is not yet fully understood how these agents cause nephrotic syndrome, in most cases it is probably due to an aberrant immune response to them, resulting in the formation and deposition of immune complexes in the glomerulus.

Etiologies of Nephrotic Syndrome (Beyond 3 months of age)

Idiopathic

- Minimal change nephropathy (MCN)
- Focal segmental glomerulosclerosis (FSGS)
- Membranous nephropathy (MN)
- Mesangial proliferative glomerulonephritis
- Membranoproliferative glomerulonephritis (MPGN)
- IgM nephropathy
- C1q nephropathy

Genetic

- Autosomal recessive FSGS due to mutation in gene encoding podocin (NPHS2)
- Autosomal dominant diffuse mesangial sclerosis (DMS) due to mutation in gene encoding WT1

Infections

- Hepatitis B and C
- HIV
- Malaria
- Schistosomiasis
- Filariasis

Systemic Diseases

- Systemic lupus erythematosus
- Diabetes mellitus
- Sarcoidosis

Metabolic Diseases

- Fabry's disease
- Glutaric acidemia
- Glycogen storage disease
- Mitochondrial cytopathies
- Amyloidosis

Hematologic and Oncologic Diseases

- Leukemia
- Lymphoma (Hodgkin's most likely can lead to minimal change)
- Sickle cell disease

Drugs

- Gold
- Penicillamine
- Angiotensin converting enzyme inhibitors (ACEIs)
- Pamidronate
- Interferon
- Mercury
- Heroin
- Lithium

Others

- Food allergies
- Obesity (usually with FSGS)
- Oligomeganephronia
- Pregnancy

Clinical Signs and Symptoms

Nephrotic syndrome produces a collection of symptoms and signs that occur because tiny blood vessels (the glomeruli) in the kidneys sustain damage and do not function correctly. Albumin protein is lost in the greatest quantity, but important protein-bound nutrients such as iron bound to transferrin, vitamin

D bound to vitamin D binding protein, copper bound to ceruloplasmin, and zinc bound mostly to albumin are lost as well. The syndrome is characterized by hypoalbuminemia, edema formation, hyperlipidemia, and, less well appreciated, hypovitaminosis D, resulting in hypocalcemia and rarely in iron deficiency edema. Muscle mass may also be depleted.

In general, the syndrome is characterized by:

- Proteinuria i.e high protein levels in the urine (More than 3.5 g protein per 1.73 sq m body surface area in 24 hours or >40 mg/m^2/hr or a urine protein/creatinine ratio >2.0 mg/mg)
- Hypoalbuminaemia
- Low serum protein levels (serum albumin ≤2.5 g/dl)
- Massive edema due to retention of sodium and water
- Hyperlipidemia
- Hyperlipoproteinemia
- Increased weight due to fluid retention
- Distended abdomen
- Others such as fatigue, anorexia, malaise, fever, rash, ascites *etc.*

As the condition progresses there may be:

- Susceptibility to infectious diseases due to an impaired immune system
- An increased risk of thrombosis
- Muscle wasting (due to the loss of protein)
- Malnourishment
- Anemia
- Weakening of the bones
- Abnormally low or abnormally high blood pressure
- High blood cholesterol levels
- Kidney failure

Treatment

Aim of treatment is to control oedema, reduce proteinuria and treat complications that arise including infections, hyperlipidemia and clotting problems. Diet therapy has a role to play in several of these areas. Goals of dietary treatment of nephritic syndrome are similar to as for Dietary Management to CKD stages 1-5 (non-dialyzed), plus special emphases on control of proteinuria, fluid balance and lipid abnormalities.

Dietary Treatment

The diet should be appropriate and adequate in energy for ideal body weight, moderate in protein, moderate in fat, less cholesterol and saturated fat, sodium restricted and liberal in carbohydrates (but not refined carbohydrates). Fluid intake should be restricted depending upon the severity of oedema and use of diuretics. The diet should be soft.

Energy

Aim should be to attain and maintain ideal body weight. Depending on the physical activity level, 35-45 Kcal/kg IBW/d for <60 years and 30-35 Kcal/kg IBW/d for >60 years should be provided. Energy from carbohydrates should be approximately 50-60 per cent while from fat, it should be approximately 30-35 per cent. Patients on steroid therapy may need to control body weight. Energy requirement should be modified to individual needs in case of under-nutrition or overweight/obesity.

Protein

High protein diet as recommended in the past is no longer considered beneficial as diets rich in protein increase proteinuria due to increased permeability and hyperfiltration in the glomerular basement membrane which can exacerbate proteinuria.

Protein requirement shall depend upon the degree of renal function. Usually no dietary protein modification is required; 0.8-1.0 g/kg ideal body weight/day is recommended. If dietary protein modification is required, then 0.8-1.0g/kg ideal body weight/day + g to g replacement is recommended. High biological value protein *i.e.* foods having approximately 70 per cent biological value protein such as egg and meat should be preferred should be given. For vegetarians, combination of cereal, pulse and low fat curd or buttermilk or paneer should be given in the meal.

Fat

Incidence of myocardial infarction is 5-6 greater in nephritic syndrome. The diet should not provide more than 30-35 per cent of calories. One should take monounsaturated and polyunsaturated fat rich food sources. Saturated fat should provide less than 10 per cent of the energy intake; cholesterol intake should also be less than 300 mg per day. Lipid lowering agents and bile acid sequesters are also used to control cholesterol and triglycerides. Fish oils also decrease triglycerides.

Sodium

If hypertension or oedema is present, sodium intake should be approximately 80mmol/d (no added salt in cooking or on table); salt is restricted to 2-3 g/d. In case of severe oedema, one may need to further lower sodium intake *i.e.* ~ 50mmol/d (low sodium diet). Low sodium foods should be taken liberally.

Calcium

Patients on steroid therapy require adequate calcium intake *i.e.* more than its RDI *i.e.* 800-1000mg/d.

Phosphorus

Phosphorus intake should be less than 1000 mg/d if hyperphosphataemia is there. Along with food sources of phosphorus, phosphate binders should also be given.

Potassium

In case of oliguria (depending on urinary output, oedema, dialysis and serum sodium level), potassium intake should be 35-50 mEq/d; replace diuretic losses. If diuretic are given to control oedema, then diuretic induced potassium loss should be replaced by potassium supplement.

Fluid

Fluid restriction may be necessary, depending on the response to diuretics and other treatments such as albumin infusions. Replace previous day output (urine + vomitus + diarrhea) plus 500 ml is often recommended.

Micronutrients

Hypovitaminosis D and iron deficiency can be treated by replacement. Vitamin D can result deranged calcium metabolism and hence, supplementation may be necessary. Vitamin C supplement is essential. A balanced diet rich in fruits and vegetables will help to maintain micronutrients.

Renal Transplant

Renal transplantation is considered to offer the patient with established renal failure (ERF) the best chance of rehabilitation and good quality of life. Patient's nutrition after kidney transplantation is influenced by several different factors, such as the time after transplantation, kidney function (chronic kidney disease stages 1-5T), side effects of immunosuppressive and other drugs, possible food-drug interactions and possible immunomodulatory effects of some food additives.

Importance of Good Nutrition

Before Renal Transplant

Pre-transplant nutritional status affects outcomes after transplantation. Recipients with BMI <18 (protein energy wasting) and those with BMI >35 (obesity) incur an excess of morbidity and mortality as compared to those within a normal BMI range. Wound healing and immune defence rely on enhanced synthesis of new proteins which is limited by negative energy and nitrogen balance. Presence of obesity at the time of transplantation is associated with a greater risk of delayed graft function and surgical wound complications. Pre-transplant malnutrition is associated with poor graft and patient survival. Indirect effects of abnormal nutritional status are also evident, mediated through nutrition-related disorders including diabetes, osteoporosis and cardiovascular disease. Therefore, optimal nutritional status should be maintained in all patients on the transplant waiting list.

After Renal Transplant

A patient may have spent a couple of years in the pre-dialysis stage and then several years on dialysis before receiving a transplant. Even with careful management, they will show signs of the long-term metabolic effects of CKD, including anaemia, mineral bone disease, muscle wasting and cardiovascular disease *etc.* Immunosuppressive therapy, used to prolong the life of the transplanted kidney, can exacerbate all of the above conditions as well as creating additional problems, some of which can be influenced by nutrition. Steroids cause osteoporosis, fluid retention and hypertension, dyslipidemia, new onset diabetes after renal transplant (NODAT) (~20 per cent of patients), increased appetite and weight gain. Other side effects of immunosuppressive treatment can be protein hypercatabolism, hypertension, hyperkalaemia, interference with vitamin D metabolism, increased cancer risk, increased infection risk (viral and bacterial infections that may decrease appetite and increase the nutrient requirements), gum hypertrophy *etc.* Moreover, in the early post-transplant period, adequate nutrition has to enable wound healing, prevent infection and correct clinically significant electrolyte and metabolic abnormalities caused by the restoration of kidney function.

One of the major long-term complications after kidney transplant is cardiovascular disease. Cardiovascular complications are the leading cause of death in transplant patients after the first year, with immunosuppressive drugs being unavoidable contributors. Helping patients to control cholesterol levels, weight gain, and blood pressure can be extremely beneficial in decreasing this risk and maintaining a well functioning allograft. Hence, in the long run, optimum nutrition is required to stabilize and prevent deterioration of renal function, development of obesity, dyslipidemia, anemia, diabetes/hyperglycemia, hypertension and bone disease.

Nutritional Assessment after Renal Transplant

Nutritional assessment should be performed soon after the transplantation, then monthly for the first three months and annually thereafter unless a complication occurs. Patients may have optimal post-transplant course that corrects metabolic abnormalities or may have delayed graft function, wound healing complications and gut disturbances that contributes to undernutrition. Nutritional requirements change over time after transplantation. Individual management is required.

Nutritional and Dietary Management after Renal Transplant

A. Immediately after Renal Transplant

Nutritional care should be the same as for any other post-surgical patient *i.e.* monitoring blood and urine biochemistry and urine volume, ensuring the return of normal gut function and appetite and meeting requirements with supplements if necessary. The rate at which biochemistry and urine volume return to normal can vary (from a couple of days after surgery to several weeks) and needs to be monitored closely. Treatment varies accordingly from fluid and electrolyte (Na^+ and K^+) restrictions to intravenous support if urine output is excessive and serum electrolyte levels drop below normal.

B. After Stabilization of Kidney Functions

After kidney transplant, one may be able to eat foods that were limited before the transplant. As more information becomes available regarding post-transplant nutrition, it is increasingly apparent that a well-balanced, heart healthy diet and lifestyle are valuable in the short- and especially long-term management of the renal transplant patient. Eating a healthy diet and increasing activity can lower the risk of heart disease, diabetes, high blood pressure, and osteoporosis.

The nutritional goals after renal transplantation are to:

1. Achieve or maintain optimal protein nutriture,
2. Achieve or maintain desirable body weight, and
3. Attenuate side effects of immunosuppressive agents.

Energy

Required energy intake generally after kidney transplantation is 30–35 kcal/kg IBW/d in acute and chronic phases, respectively to prevent negative nitrogen balance. For long term goal, a healthy body weight management may improve the health and life of new kidney. As the appetite may improve after transplant, it is important to maintain the optimal energy intake to avoid weight gain. One should regularly check body weight to keep it in the normal BMI range and avoid high-calorie foods such as fatty foods, sweets *etc.*

Protein

In the first two months after transplant, one needs more protein to heal after surgery and fight infection. In acute and chronic phase, protein requirements are <1.4 g/kg IBW/d and 0.75–1.0 g/kg IBW/d, respectively. One should include protein rich foods such as lean meats, poultry, fish, eggs, low fat or non-fat milk, curd, cheese, paneer, legumes, nuts *etc.*

Fat

Moderately low fat diet should be taken. Fat should contribute <30 per cent of total energy, with 8–10 per cent of total energy from n-6 polyunsaturated fat; there should be n-3 polyunsaturated fats from both plant and marine sources; approx. 20 per cent monounsaturated fat and <10 per cent saturated and trans fatty acids to prevent hyperlipidemia. Cholesterol intake should be less than 300mg/d to prevent hypercholesterolemia.

Carbohydrates

Carbohydrates should represent approx. 50 per cent of total energy. They should have low glycemic index and should contain high dietary fiber. Fiber intake should be 25-30g/day. Simple sugars should be limited.

Micronutrients

Vitamins and minerals, in particular B_6, B_{12}, magnesium and zinc intake should be according to RDA for the general population or age and gender, taking into consideration body size, nutritional status and physical activity levels. The recommended daily allowance (RDA) of calcium is 1000 - 1300 mg/d, phosphorous 1000 - 1300 mg/d and sodium 80 - 100 mmol/d (no added salt). Sodium should be limited to help control fluid retention and blood pressure; processed foods, canned foods, frozen meals, pickles, *papad*, fast foods *etc.* should be avoided.

Transplant medicines can cause a high potassium level in the blood. Potassium should be restricted if hyperkalemia persists. In that case, high potassium foods such as some fruits, (melons, bananas, avocados, oranges *etc.*), some vegetables (potatoes, tomatoes, dark leafy vegetables), chocolate, dried beans and lentils *etc.* should be avoided.

Anemia is common after renal transplantation which has been associated with development of post transplant congestive heart failure. Iron deficiency is a major cause of anemia and is independently associated with post-transplant cardiovascular events, the leading cause of death in kidney transplant recipients. Iron supplements should be given to bring the haemoglobin in the normal range.

Kidney transplant patients are at higher risk for vitamin D deficiency, for two reasons; first is the fact that they have to avoid the sun to prevent their predisposition to skin cancers because of the drugs which they take to prevent transplant rejection; other is that the drugs they take, especially glucocorticoids, increase the destruction of the active metabolite 25-hydroxyvitamin D. Hence, vitamin D deficiency generally widespread amongst renal transplant recipients (RTRs), provides one patho-mechanism that may influence the achievement of a successful degree of immunosuppression. It may also influence the development of the infectious, cardiovascular and neoplastic complications seen in RTRs. The National Kidney Foundation of USA recommends that during the first year of post renal transplant, patients should receive

50,000 IU of vitamin D once every 2 weeks to maintain a blood level of 25-hydroxyvitamin D above 30-80 ng/ml.

Fluid

If new kidney is working well, drink plenty of fluids. It should be approx. 2.0-2.5 l/d. There is no longer fluid restriction. Dehydration can be harmful to the new kidney.

Dietary Management

- **Food safety and hygiene:** Transplant drugs can lower the immune system and increase the risk for food-borne illness. One can reduce the risk by handling the food safely *i.e.* by avoiding intake of raw and undercooked meats and eggs, unpasteurized dairy products; washing fruits and vegetables well before eating; keeping cold foods cold and hot foods hot; avoiding contamination of cooking and serving utensils by washing them thoroughly *etc.* One should wash hands properly before cooking and taking meals. Strict food hygiene and safe food handling practices should be followed.
- **Weight management:** The incidence of obesity can increase dramatically post-transplantation and therefore, dietary advice should be given to prevent excessive weight gain. Advice should include healthy eating, exercise and other lifestyle improvements to keep body weight under control. One should avoid energy dense foods, fast foods, sweets, ready to serve foods *etc.* Fat rich foods should be avoided as they are energy dense. Control food portion sizes. One should always aim for healthy body mass index.
- **Follow a heart-healthy diet:** Foods high in saturated, trans fat and cholesterol should be avoided. One should choose right kind of fat for cooking. Choose cooking oils rich in MUFA such as olive oil, canola oil, blends of ricebran and soybean oil or groundnut oil and soybean or safflower oil, mustard oil *etc.* Avoid butter, shortening or hydrogenated fat for cooking. One should try to take low fat dairy products. Choose lean meats, poultry and fish. Trim all visible fat off meat before cooking. Avoid fried foods. Bake, roast, broil, boil or sauté foods instead.
- **Moderate salt restriction:** Salt should be used sparingly in cooking. Avoid table salt completely. Limit processed foods and bakery goods. Monitor blood pressure.
- **Low in sugar and refined carbohydrates:** Foods containing refined carbohydrates and simple sugars should be avoided. Donot add sugars to foods and drinks. Limit the intake of fruit juices, sodas and chocolate milk *etc.* Choose foods high in fibre such as whole grains, legumes, fruits and vegetables. Keep blood sugar under control by taking foods having low glycemic index and glycemic load.
- **Take calcium, phosphorus, iron, B12, magnesium, zinc and vitamin D rich foods:** Take foods rich in these micronutrients especially fruits and vegetables to prevent iron deficiency anaemia, osteoporosis and improve immunity. In case of need, their supplements should be taken as per medical advice. Potassium level can increase or decrease in the blood and accordingly, their food sources should be taken.
- **Avoid too much sun exposure.**
- **Avoid smoking.**
- **Do physical activity and regular exercise**

Renal Stone Disease

Kidney stone disease is one of the oldest and most common problems of the urinary system. Mankind has been afflicted by urinary stones since centuries dating back to 4000 B.C., and it is the most common disease of the urinary tract. The formation of kidney stone is also known as renal calculi or crystal. It is a serious though not life threatening disorder prevalent throughout the world. In medical terminology condition of having urinary calculi is termed as nephrolithiasis and urolithiasis where the root word "Lith" meaning "a stone". Nephrolithiasis is common worldwide, often debilitating that has different etiology and pathophysiology.

Prevalence

Recent studies have reported that the prevalence of urolithiasis has been increasing in the past decades in both developed and developing countries. This growing trend is believed to be associated with changes in lifestyle modifications such as lack of physical activity and dietary habits and global warming. Urolithiasis affects about 12 per cent of the world population at some stage in their lifetime. It affects all ages, sexes, and races but occurs more frequently in men than in women within the age of 20–49 years.

The incidences of kidney stones are rising in rural and urban societies in India too. Approximate 2 million people in India are affected with nephrolithiasis every year and some parts of country have name denoted as a stone belt that is, Gujarat, Maharashtra, Punjab, Rajasthan, Delhi, Haryana and part of states on North East side. Urinary stone are also found in south India due to high intake of tamarind in regular diet. In India, approximate 50 per cent of the population is affected with renal calculi, which may end up to renal damage or loss of kidney function. The rate of nephrolithiasis incidence, mainly staghorn calculi is very high in Manipur and some reports indicates North Western region of India have also increased prevalence.

What are Kidney Stones

Renal stone formation is a biological process that involves physicochemical changes and supersaturation of urine. Renal calculi are crystalline structures composed most commonly of calcium oxalate salts. They form when the concentrations of these ions, as well as solutes such as hydrogen ions, sodium ions and uric acid are present in the filtrate (urine) in higher than normal amounts. This condition is known as supersaturation. Hence, supersaturated solution refers to a solution that contains more of dissolved material than could be dissolved by the solvent under normal circumstances. As a result of supersaturation, solutes precipitate in urine, leads to nucleation and then crystal concretions are formed. That is, crystallization occurs when the concentration of two ions exceeds their saturation point in the solution.

Therefore, the first step in the formation of kidney stone begins by the formation of nucleus (termed as nidus) from supersaturated urine retained inside the kidneys. Supersaturated ions are more likely to come out of solution and go on crystallizing over a nucleus, which ultimately takes the shape of a stone. Kidney stones grow slowly over several days to month. Renal calculi are found in different sizes, some are as small as like grain of sand or large as pearls and big as golf boll. Kidney stones can start small but can grow larger in size, even filling the inner hollow structures of the kidney. They are smooth, spiky or asymmetric and jagged. Renal stone are mainly found in three colors as brown, yellow and reddish.

Some stones stay in the kidney, and do not cause any problems. Sometimes, the kidney stone can travel down from kidney into the ureter, the tube between the kidney and the bladder. If the stone reaches the bladder, it can be passed out of the body in urine. If the stone becomes lodged in the ureter, it blocks the urine flow from that kidney and causes pain. These stones may persist for indefinite period of time, leading to secondary complications thereby causing serious consequences to patient's life.

Risk factors for supersaturation include dehydration, high fat diet, animal protein, high salt intake and obesity. Race, gender and ethnicity play a part in who may get kidney stones. Men get kidney stones more often than women, but recent studies show the number of women getting kidney stones is rising.

Composition of Kidney Stones

Urinary calculi may have various compositions which include, in order of decreasing frequency: calcium oxalate (monohydrate or dihydrate), uric acid, struvite (magnesium ammonium phosphate), calcium phosphate, and cystine. There are other less common stones, including xanthine and drug-related stones as well. Stones are solutes that occur in amounts too high to stay dissolved (supersaturated) in urine. As a result of supersaturation, the solutes precipitate and aggregate to form concretions or stones.

The chemical compositions of urinary stones include crystals and noncrystalline phases or the organic material (the matrix). The organic matrix of urinary stones consists of macromolecules such as glycosaminoglycans (GAG's), lipids, carbohydrates, and proteins. These molecules play a significant role by promoting or inhibiting the processes of kidney stone development. The main components of the stone matrix are proteins (64 per cent), non-amino sugars (9.6 per cent), hexosamine as glucosamine (5 per cent), water (10 per cent), and inorganic ash (10.4 per cent). The matrix acts as a template participating in the assembly of kidney stones. The matrix of all stones contains phospholipids (8.6 per cent) of the total lipid, which in turn represents about 10.3 per cent of stone matrix. Cell membrane phospholipids, as part of organic matrix, promote the formation of calcium oxalate and calcium phosphate

Characteristics of Kidney Stones

- ☆ Stone formation can occur anywhere in the kidney, ureter, or bladder.
- ☆ The size of stone can vary from microscopic to the large 'staghorn' calculi and can lead to kidney failure if the kidney or urinary tract becomes obstructed.
- ☆ Stones vary in composition.
- ☆ Increased urinary concentration of calcium, sodium, oxalate, urate and citrate promote stone formation and increased concentration in urinary magnesium, pyrophosphate, citrate and nephrocalcin inhibit stone formation.
- ☆ Stone formation can also be linked to congenital kidney abnormalities, short bowel syndrome or recurrent infection with urease positive organisms.
- ☆ About 50 per cent of stone formers excrete increased urinary calcium, > 7.5 mmol/day in men or > 6.2 mmol/day in women. Hypercalciuria may be due to increased absorption of calcium from the gut, calcium resorption from the bone, or decreased ability of the kidney to reabsorb calcium.
- ☆ Stone formation is exacerbated by decreased urine output volume due to low fluid intake or increased losses *e.g.* sweat and GI tract, *e.g.* intestinal failure.

stones. Albumin is the major component of the matrix of all stone types.

Brushite stone is a hard phosphate mineral with an increasing incidence rate, and a quarter of calcium phosphate (CaP) patients form stones containing brushite. In the urinary tract, CaP may be present in the form of hydroxyapatite, carbonate apatite, or brushite (calcium monohydrogen phosphate dihydrate, CaHPO4 2H2O). Brushite is resistant to shock wave and ultrasonic lithotripsy treatment.

Types of Kidney Stones

Kidney stones come in many different types and colors. The chemical composition of kidney stones depends on the abnormalities in urine composition of various chemicals. Stones differ in size, shape, and chemical compositions (mineralogy). Based on variations in mineral composition and pathogenesis, kidney stones are commonly classified into five types as follows:

Calcium Stones (80 per cent of stones)

Calcium stones are predominant renal stones comprising about 80 per cent of all urinary calculi. There are two types of calcium stones: calcium oxalate and calcium phosphate. The proportion of calcium stones may account for pure calcium oxalate (CaOx) (50 per cent), calcium phosphate (CaP, termed as apatite) (5 per cent), and a mixture of both (45 per cent). The main constituent of calcium stones is brushite (calcium hydrogen phosphate) or hydroxyapatite. Calcium oxalate is found in the majority of kidney stones and exists in the form of CaOx monohydrate (COM, termed as mineral names: whewellite, CaC_2O_4 H_2O), and CaOx dihydrate (COD, weddellite, CaC_2O_4 $2H_2O$), or as a combination of both which accounts for greater than 60 per cent. COM is the most thermodynamically stable form of stone. COM is more frequently observed than COD in clinical stones.

It is thought that the majority of calcium oxalate stones form from an initial calcium phosphate concretion that originates near the renal calyx epithelium in the highly concentrated environment of the terminal collecting duct. The calcium phosphate concretion (called a Randall's plaque) erodes through the urothelium, is exposed to urine, and forms a nidus for calcium oxalate deposition with time. The calcium oxalate deposition grows until the stone becomes large enough to break free of its urothelial "anchor" and then may pass through the collecting system. Many factors which contribute to CaOx stone formation are hypercalciuria (resorptive, renal leak, absorptive, and metabolic diseases), hyperuricosuria, hyperoxaluria, hypocitraturia, hypomagnesuria, and hypercystinuria. Mostly, urinary pH of 5.0 to 6.5 promotes CaOx stones, whereas calcium phosphate stones occur when pH is greater than 7.5. The recurrence of calcium stone is greater than other types of kidney stones.

Calcium phosphate stones can be present in combination with calcium oxalate or struvite. Because of differences in solubility due to urine pH, calcium phosphate is not found mixed with uric acid. The two forms of calcium phosphate include apatite (sometimes reported as carbonate apatite), which is the crystal type found in bone, or calcium hydrogen phosphate (brushite); the frequency of apatite is much greater than brushite. Calcium phosphate crystals in the urine sediment are typically dark and amorphous.

Uric Acid Stones or Urate (3–10 per cent of stones)

Uric acid is a product of purine metabolism. Uric acid is 100 times more soluble at a pH > 6 compared to a pH <5.5. Uric acid crystals do not dissolve well in acidic urine and instead will form a uric acid stone. Having acidic urine may come from:

- Being overweight
- Chronic diarrhea
- Type 2 diabetes
- Gout
- A diet that is high in animal protein and low in fruits and vegetables

Diets high in purines especially those containing animal protein diet such as meat and fish, results in hyperuricosuria, low urine volume, and low urinary pH (pH < 5.05) exacerbates uric acid stone formation. People with gouty arthritis may form stones in the kidney(s). The most prevalent cause of uric acid nephrolithiasis is idiopathic, and uric acid stones are more common in men than in women.

Struvite or Magnesium Ammonium Phosphate Stones (10-15 per cent of stones)

Struvite stones are not a common type of stone. These stones have also been referred to as infection stones and triple phosphate stones. They occur among patients with chronic urinary tract infections that produce urease, the most common being *Proteus mirabilis* and less common pathogens include *Klebsiella pneumonia*, *Pseudomonas aeruginosa*, and *Enterobacter*. *Urease* is necessary to split/cleave urea to ammonia and CO_2, making urine more alkaline which elevates pH (typically > 7). Phosphate is less soluble at alkaline versus acidic pH, so phosphate precipitates on to the insoluble ammonium products, yielding to magnesium ammonium phosphate *i.e.* a large staghorn stone formation. As the bacteria that produce urease remain in urine and within the stone, they continue to produce urease, and continue to cleave urea, and so large (staghorn shaped) stones may develop quite rapidly and fill the calyceal spaces of the kidney. Hence, these stones are often large, with branches, and they often grow very fast. Women's are likely to develop this type of stone than the male. *Escherichia coli* is not capable of splitting urea and is not associated with struvite stones.

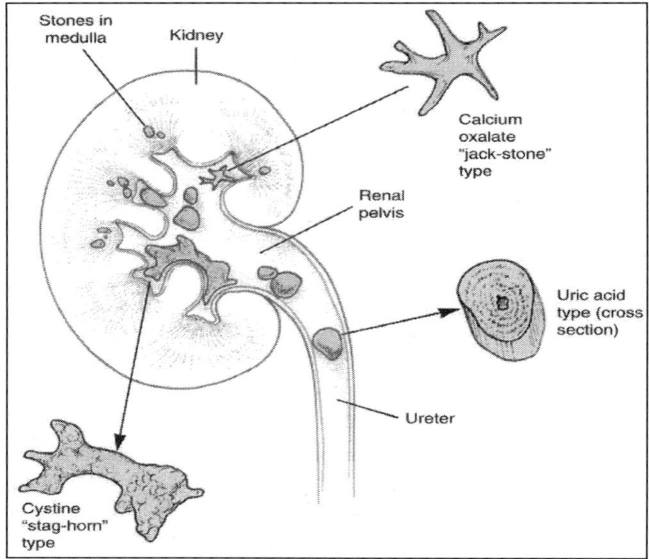

Cystine Stones (less than 2 per cent of stones)

These stones are formed in those who suffer from a genetic disorder of the transport of an amino acid cystine. It results in an excess of cystinuria (too much cystine in the urine) in urinary excretions, which is an autosomal recessive disorder caused by a defect in the rBAT gene on chromosome 2, resulting in impaired renal tubular absorption of cystine or leaking cystine into urine. Cystine does not dissolve in urine and leads to cystine stone formation. People who are homozygous for cystinuria excrete more than 600 millimole insoluble cystine per day. The development of urinary cystine is the only clinical manifestation of this cystine stone disease. Cystine stones often start to form in childhood.

Drug-induced Stones

This accounts for about 1 per cent of all stone types. Drugs such as guaifenesin, triamterene, atazanavir, and sulfa drugs induce these stones. For instance, people who take the protease inhibitor indinavir sulphate, a drug used to treat HIV infection, are at risk of developing kidney stones. Such lithogenic drugs or its metabolites may deposit to form a nidus or on renal calculi already present. On the other hand, these drugs may induce the formation of calculi through its metabolic action by interfering with calcium oxalate or purine metabolisms.

Symptoms of Kidney Stones

Stones in the kidney often do not cause any symptoms and can go undiagnosed. When a stone leaves the kidney, it travels to the bladder through the ureter. Often the stone can become lodged in the ureter. When the stone blocks the flow of urine out of the kidney, it can cause the kidney to swell (hydronephrosis), often causing a lot of pain. Common symptoms of kidney stones are:

- A sharp, cramping pain in the back and side, often moving to the lower abdomen or groin. This discomfort usually occurs only on the side of the renal calculi and does not cross over to the other side.
- Fluctuations in discomfort intensity, with periods of discomfort lasting 20-60 min.
- Discomfort waves radiating from the side and back to the lower abdomen and groin.
- Some women say the pain is worse than childbirth labor pains. The pain often starts suddenly and comes in waves. It can come and go as the body tries to get rid of the stone.
- Persistent urge to urinate.
- Urinating more often or a burning feeling during urination.
- Discomfort, pain and inflammation on urination.
- Urine that is dark or red due to blood. Sometimes urine has only small amounts of red blood cells that can't be seen with the naked eye.
- Nausea and vomiting.
- Fever and chills if an infection is present.

Risk Factors for Kidney Stone Formation

Certain environmental and lifestyle factors can increase the stone risk which have been discussed as under: A higher risk is found in professional chefs or others working in hot environments, as well as in taxi drivers who often try to minimize their fluid intake to avoid too many 'toilet stops'.

Gender

Male have 2.5 times greater risk than women to likely suffer from kidney stones.

Low Urine Volume

A major risk factor for kidney stones is constant low urine volume. Low urine volume may come from dehydration (loss of body fluids), from hard exercise, working or living in a hot place, or not drinking enough fluids. When urine volume is low, urine is concentrated and dark in color. Concentrated urine means there is less fluid to keep salts dissolved. Increasing fluid intake will dilute the salts in the urine. By doing this, one may reduce the risk of stones forming.

Obesity

Obesity is a risk factor for formation of stones. Obesity may change the acid levels in the urine, leading to stone formation.

Age

Most common in individuals aged 30-50 years.

Family History

It accounts for 45 per cent of all cases.

Lifestyle

Some of the lifestyle factors include frequent going on tropical holidays, regular strenuous exercise, and even psychological stress (although how is unclear).

Dietary Factors

Diet can also affect the chance of forming a stone. Diet-related factors that are known to increase stone risk are listed in the box given under.

Dietary Risk Factors Associated with increased stone risk

- ✰ Low fluid intake
- ✰ High intake of animal protein
- ✰ High dietary sodium
- ✰ Excessive intake of refined sugars
- ✰ Foods rich in oxalates
- ✰ High intake of grapefruit juice, apple juice and soft cola drinks

Tea or Coffee

Tea or coffee (particularly instant coffee) without milk has been shown to increase oxalate excretion, although this effect is probably offset by their diuretic action.

Dietary Factors that may Increase or Decrease Calcium Oxalate Stones

Dietary Factors	*Proposed Mechanism(s)*
Increased risk	
Oxalate	Increased urinary oxalate excretion
Sodium	Increased urinary calcium excretion
Animal protein	Increased urinary calcium and uric acid excretion, reduced urinary citrate extraction
Vitamin C	Increased oxalate generation and excretion
Carbohydrates	Increased urinary calcium excretion
Decreased risk	
Dietary calcium	Binding of dietary oxalate in gut
Potassium	Increased urinary citrate excretion, reduced urinary calcium excretion
Phytate	Inhibition of calcium oxalate absorption, inhibition of calcium oxalate crystal formation
Magnesium	Vitamin B6 deficiency may increase oxalate production and oxaluria

One of the more common causes of calcium kidney stones is high levels of calcium in the urine.

Calcium and Sodium

High urine calcium levels may be due to the way body handles calcium. It is not always due to how much calcium one eats. Lowering the amount of calcium in the diet rarely stops stones from forming. Dietary calcium has a biphasic risk curve: stone risk is greater in those on a high or low calcium diet. Studies have shown that restricting dietary calcium can be bad for bone health and may increase kidney stone risk. One should not limit dietary calcium in order to lower urine calcium. But calcium intake should not be too high. Instead of lowering dietary calcium intake, one should try to reduce urine calcium level by decreasing the sodium (salt) intake. The risk of kidney stones increases with increased daily sodium consumption. Too much salt in the diet is a risk factor for calcium stones. This is because too much salt is passing into the urine, keeping calcium from being reabsorbed from the urine and into the blood. Reducing salt in the diet lowers urine calcium, making it less likely for calcium stones to form. Because oxalate is a component of the most common type of kidney stone (calcium oxalate), eating foods rich in oxalate can raise the risk of forming these stones.

Animal Protein

A diet high in animal protein, such as beef, fish, chicken and pork, can raise the acid levels in the body and in the urine. High acid levels make it easier for calcium oxalate and uric acid stones to form. The breakdown of meat into uric acid also raises the chance that both calcium and uric acid stones will form. Diet low in fibre and fluids or other dietary patterns causing prolonged imbalances in acidity of urine can also cause stone formation.

Vitamin D

The link between vitamin D intake and renal stones is less clear; while excessive active (1,25-OH) vitamin D supplementation, by increasing intestinal calcium absorption, increases the risk of stone formation, there is no evidence that correction of native (25-OH) vitamin D deficiency has the same effect; moreover, correction, especially if there is secondary hyperparathyroidism, is likely to be a health benefit.

Vitamin C

Vitamin C excess could also increase the risk of calcium oxalate stone formation, but in practice this is rarely encountered. High dietary intake of potassium or magnesium is inversely related to stone formation because potassium promotes urinary citrate excretion, and both citrate and magnesium inhibit crystal formation. However, the impact of low urinary magnesium on stone risk is at best modest.

Other Factors

Other factors associated with increased stone risk are chronic laxative abuse, antacid excess or betel nut chewing (both forms of 'milk-alkali syndrome') *etc.*

Non-Lifestyle Factors

Some of these have been given below:

Non-Lifestyle (and non-genetic) Factors Associated Increased Stone Risk and Rates of Recurrence

Factors Associated with Increased Stone Risk	*Underlying Mechanism(s)*
Primary hyperparathyroidism	Hypercalciruria and hyper-calcaemia
Urinary tract infection	Precipitation of calcium phosphate and magnesium ammonium phosphate-struvite stones-in alkaline urine
Chronic inflammatory bowel disease	Increased oxalate absorption
Ileostomy	Bicarbonate and fluid losses causing low urine volume with an acid urine pH
Prolonged immobilization due to spinal injury or for other reasons	Hypercalciuria from bone loss and urinary stasis due to bladder catheterization in spinal injury.

Bowel Conditions

Certain bowel conditions that cause diarrhea (such as Crohn's Disease or ulcerative colitis) or surgeries (such as gastric bypass surgery) can raise the risk of forming calcium oxalate kidney stones. Diarrhea may result in loss of large amounts of fluid from the body, lowering urine volume. The body may also absorb excessive oxalate from the intestine, resulting in more oxalate in the urine. Both low urine volume and high levels of urine oxalate can help to cause calcium oxalate kidney stone formation.

Medical Conditions

Some medical conditions have an increased risk of kidney stones. People suffering from hypertension, gout, bedridden patients *etc.* have higher chances of getting them. Abnormal growth of one or more of the parathyroid glands, which control calcium metabolism, can cause high calcium levels in the blood and urine. This can lead to kidney stones. Another condition called distal renal tubular acidosis, in which there is acid build-up in the body, can raise the risk of calcium phosphate kidney stones. Some rare, inherited disorders can also make certain types of stones more likely. Examples include cystinuria, which is too much of the amino acid cystine in the urine, and primary hyperoxaluria, in which the liver makes too much oxalate.

Novel Risk and Genetic Factors

Colonization of the gut by an oxalate degrading anaerobe, *Oxalobacter formigenes*, has been linked to reduced oxalate stone risk. It has been shown that these bacteria not only degrade oxalate, but that they also interact (in an as yet unspecified way) with the chloride/oxalate anion exchanger (CFEX, SLC26A6) present in intestinal epithelial cells to increase oxalate secretion and reduce its net absorption.

Gastric bypass surgery (Roux-en-Y) is increasingly used to treat widespread obesity (which itself is a risk factor for renal stones). Recent studies have shown increased stone risk following this form of surgery, which is due to 'enteric hyperoxaluria' from a combination of disturbed enterohepatic bile circulation and loss of calcium (that normally binds dietary oxalate) through binding to fatty acids, and changes in the gut microflora affecting oxalate absorption.

A positive family history of kidney stones is strongly associated with increased stone risk. Relative risk in those with a positive family history is 2–3 times higher, and several genetic factors have been proposed that may explain some of this association. Alanine:glyoxylate aminotransferase (AGT) is a key vitamin B_6 (pyridoxine)-dependent enzyme found predominantly in liver peroxisomes, where it converts glyoxylate to glycine, reducing the formation of oxalate from glyoxylate. A polymorphism of this enzyme ($Pro^{11}Leu$) has been identified that mistargets some AGT to mitochondria rather than peroxisomes. Peroxisomes deal more effectively with plant-derived glycolate, whereas mitochondria handle meat-derived hydroxyproline, both sources of glyoxylate production. Thus, the $Pro^{11}Leu$ AGT polymorphism may be an advantage to those consuming a high animal protein diet, but a disadvantage for those on a more vegetarian diet; as yet this hypothesis remains untested. The already mentioned novel oxalate transporter CFEX is present in renal and intestinal epithelia, and CFEX null mice have a high incidence of calcium oxalate stones, primarily due to increased oxalate absorption in their small intestine. It is possible that polymorphisms of this transporter may affect oxalate absorption in humans and thereby calcium oxalate stone risk. Other gene polymorphisms that have been linked to increased renal stone risk include the renal sodium-phosphate co-transporters NaPT2a and NaPT2c, the sodium-proton exchanger regulatory factor NHERF1, as well as the renal calcium-sensing receptor, and some monogenic disorders such as primary hyperoxaluria (types I and II), Dent's disease, Bartter's syndrome (types I and II), inherited forms of distal renal tubular acidosis (dRTA), and cystinuria.

Prevention and Control of Stone Formation

Fluid intake and dietary modifications are important interventions in all kinds of stone formers and remain first-line in stone management.

The first step in prevention of kidney stones is to understand the causes and types of stone formed. Laboratory analysis of stones as it passes in urine or retrieved surgically or with a scope inserted through the urethra into the bladder or ureter gives direct indication of type of stone. Blood and urine are to be tested for unusual levels of chemicals, such as calcium, oxalate and/or sodium to determine type of kidney stone. The information related to type of stone helps to suggest diet changes that can prevent future kidney stone growth. Depending on the type of kidney stone, changes in consumption of sodium, animal protein, calcium, citrate, potassium and oxalate are to done. For example, limiting oxalate in the diet may help prevent calcium oxalate stones but may not be helpful in preventing uric acid stones. Following a special diet may be enough to prevent forming more kidney stones in early diagnosed cases and for chronic cases, medications in addition to a special diet is needed.

I. Fluid Intake

A simple and most important lifestyle change to prevent stone disease is to drink mainly more water; it is true for all kinds of stones. Regardless of the underlying etiology and drug treatment of the stone disease, patients should be instructed to increase their water intake *i.e.* 3 litres per day in order to maintain a urine output of at least 2.5 liter per day as stone formation is exacerbated by decreased urine output volume due to low fluid intake or increased losses *e.g.* sweat and GI tract, *e.g.* intestinal failure. People with cystine stones may need to drink even more (4L/d). Water is the best but other fluids such as citrus drinks also prevent kidney stones. Some studies suggest citrus drinks like lemonade and orange juice protect against kidney stones because they contain citrate, which stops crystals from growing into stones. Citrate in the urine may prevent the calcium from binding with other constituents that lead to stones. In hot weather, patients are required to drink more water to make up the fluid loss from sweating.

Controlled trials have shown that increasing urine volume to at least 2 litres a day can reduce recurrence rates by 40–50 per cent. Enough fluid intake reduces urinary saturation and dilutes promoters of CaOx crystallization. Include one beverage at each meal. Carry a water bottle with you throughout the day. One should aim to take 2-3l water/fluid (250 ml of water/fluid every 4 waking hours + 250 ml at each meal) daily. Add some lemon slice, mint leaves *etc.* to water for refreshing flavor. Avoid sugary drinks such as fruit juices, black tea and coffee as they can increase the risk for kidney stones. Tea and coffee should be taken with some mil (which binds free oxalate). Bear in mind that increasing fluid intake, and thus urine volume, may critically lower urinary citrate concentration, especially if it is already low. Citrus fruits, particularly lemons and limes, may be of some benefit, but the sugar content of many fruit juices is a disadvantage.

II. Nutritional and Dietary Management

Nutritional factors implicated in stone disease are linked with the preventive intervention. Dietary measures are the first level of intervention in primary prevention as well as in secondary prevention of recurrences. Effective kidney stone prevention depends upon addressing the cause of stone formation. Generally, to prevent the first episodes of kidney stone formation or its secondary episodes, proper management of diet and the use of medications are required. Primary prevention of kidney stone disease via dietary intervention is low-cost public health initiative with massive societal implications. Thus, nutritional management is the best preventive strategy against urolithiasis.

Most data on the relation between diet and stone disease come from observational and physiological studies rather than randomized trials on adults. The diet of people has been scrutinized with respect to intake of animal protein, sodium, calcium, oxalate and purine. It appears that even with identical dietary intakes, people with a tendency to form stone will form larger crystals than non-stone formers. However, some important following concepts should be kept in mind while formulating a therapeutic diet for renal stones.

- ☆ It is important to individualize recommendations based on stone type and urinary composition (for initial evaluation, two 24-h urine collections obtained 6 weeks after a stone episode are important).
- ☆ The impact of dietary risk factors varies with age, sex and body mass index (BMI).
- ☆ Patients must provide follow-up 24-h urine collections to evaluate the impact of dietary recommendations.

A. *Calcium Oxalate and Calcium Phosphate Stones*

Dietary recommendations should be adjusted based on individual metabolic abnormalities. In case of calcium oxalate types of kidney stones, following dietary management should be done:

Protein

On an average, 1 g protein/kg ideal body weight should be given. It may be helpful to limit animal protein, including meat, fish, seafood, poultry, and eggs. Restriction of animal proteins is encouraged since animal proteins provide an increased acid load because of its high content of sulfur-containing amino acids. Thus, high protein intake reduces urine pH and the level of citrate and enhances urinary calcium excretion via bone reabsorption. Therefore, if someone is having a very acidic urine, he or she may need to eat less meat, fish, and poultry and avoid food with vitamin D. Instead, an increase intake of fruits and vegetables rich in potassium is recommended as it increases the urinary citrate excretion. One should choose plant based protein sources but low in oxalates.

Calcium

Calcium oxalate kidney stones are the leading type of kidney stones. About 50 per cent of stone formers excrete increased urinary calcium, > 7.5 mmol/day in men or > 6.2 mmol/day in women. Hypercalciuria may be due to increased absorption of calcium from the gut, calcium resorption from the bone, or decreased ability of the kidney to reabsorb calcium.

People who form calcium stones are usually told to avoid dairy products and other foods with high calcium content. However, persons with a tendency of kidney stone formation should not be advised to restrict calcium intake unless it has been known that he/she has an excessive use of calcium. A reduced intake of calcium leads to an increased intestinal absorption of oxalate, which itself may account for an increased risk of stone formation. Calcium supplements may reduce oxalate absorption because the calcium binds dietary oxalate in the gut lumen. However, the benefit of taking calcium pills is controversial. Calcium rich foods (Green leafy vegetables, milk and milk products, small fish with bones, prawns, ragi *etc.*) should provide 700-800 mg calcium per day to an adult man. More than 50 years older people should consume 1,000 mg calcium per day, along with 800-1,000 IU of vitamin D to help the body absorb the calcium.

Hypercalciuria can be decreased by increased alkali load (*i.e.* by intake of fruits and vegetables), dietary fibre, potassium (banana, potato, oranges, prunes, dates, mushroom, peas, honeydew, cucumber, coconut water *etc.*) and phosphate (whole cereals, bran, legumes, nuts, carrot, meat, fish, egg, cheese, organ meats, oil seeds, soft drinks *etc.*) rich foods. Eating at least five servings of fruits and vegetable daily is recommended for all people who form kidney stones as fruits and vegetables provide potassium, fibre, magnesium, antioxidants, phytate and citrate, all of which may help keep stones from forming.

Oxalates

Only 10–15 per cent of urinary oxalate is derived from dietary intake while the rest is formed endogenously (from vitamin C and glycine metabolism); if 24 h urinary oxalate is > 440 mmol, check for foods rich in oxalates and megadose intakes of vitamin C; pyridoxine supplements may increase stone formation.

For absorptive hyperoxaluria, take low oxalate foods in the diet. Foods rich in oxalates as mentioned in the below box should be avoided.

Oxalate Rich Foods-Bioavailability Varies

Drinks	Beer, black tea, cocoa, juices from high oxalate fruits, instant coffee powder
Fruits	Blackberries, blueberries, gooseberries, kiwi, currants, figs, plums raspberries, strawberries, tangerines
Vegetables	Beetroot, green beans, okra, sweet potato, watercress, yam, spinach, runner beans, celery
Legumes	Baked beans, soy products *e.g.* tofu
Grains	Wheat germ, bran
Nuts and seeds	Almonds, cashewnuts, peanuts, pecans, sesame seeds, sunflower seeds
Others	Plain chocolate, soy sauce

Ratio of Dietary Oxalate and Total Calcium

The ratio of dietary oxalate/total calcium is a guide to dietary modifications; in the general population this is around 0.05. Several studies have shown that stone formers should generally be advised to maintain a normal calcium intake. A small reduction in urinary oxalate can significantly reduce the risk of forming calcium oxalate stones, therefore, avoiding oxalate-rich foods (spinach, rhubarb, beetroot, tea, soya beans and tofu, nuts, peanut butter, okra, yams, sesame seeds, and chocolate) is justified, even if the contribution of dietary oxalate to urinary oxalate excretion is limited.

Some research suggests that limiting high oxalate foods may help reduce chance of forming further oxalate stone. Eating and drinking calcium and oxalate-rich foods together during a meal is a better approach than limiting oxalate entirely because oxalate and calcium are more likely to bind in the stomach and intestines before the kidneys begin processing thus, making it less likely that kidney stones will form.

Moderate Sodium Intake

A high sodium intake boosts stone risk by reducing renal tubular calcium reabsorption and increasing urinary calcium. Instead of reducing calcium intake, focusing on limiting the sodium paired with calcium-rich and oxalate-rich foods will be beneficial. Hence, moderate sodium restriction should be done for the calcium oxalate or calcium phosphate stone prone people. Daily intake of sodium should be 90-100 mmol. One should not take more than 2300 mg of salt per day. Canned or commercially processed foods as well as restaurant-prepared and fast foods are sources of "Hidden" sodium which should be avoided or taken in moderation.

Vitamin C

It has been implicated in stone formation because of *in vivo* conversion of ascorbic acid to oxalate. Therefore, a limited amount of vitamin C rich foods should be included in the diet.

Alkalized Diet

For prevention of calcium oxalate, cystine, and uric acid stones, urine should be alkalinized by eating a diet high in fruits and vegetables, taking supplemental or prescription citrate, or drinking alkaline mineral waters.

Others

Increased intake of potassium, magnesium fibre, fruits and vegetables is associated with decreased stone risk whereas intake of refined carbohydrates may increase the risk of stone formation.

B. *Uric Acid Stones*

If 24 h urinary uric acid is > 4 mmol, check for foods containing high purine contents which should be avoided. For uric acid stone formers, gout needs to be controlled. Animal proteins, such as red meat, poultry, eggs and seafood have high concentrations of purine. High purine intake leads to a higher production of uric acid that can accumulate as crystals in the joints, or as stones in the kidneys. To prevent occurrence of common type of uric acid kidney stones, it is necessary to cut down high-purine foods and follow a healthy diet containing mostly vegetables, fruits, whole grains and low fat dairy products. Limiting sugar-sweetened foods and drinks containing high fructose corn syrup and alcohol is also helpful because these can increase uric acid levels in the blood. Eating less animal-based protein and eating more fruits and vegetables will help decrease urine acidity and this will help reduce the chance for stone formation.

Orange juice may help to prevent formation of uric acid stones as it is high in citrate without containing too much oxalate. Beer increases urate excretion due to its high content of the purine guanosine, and it should be restricted in uric acid stone formers.

Purine-Rich Foods

Meat	Liver, kidney, brain, heart, goose, partridge
Fish	Anchovies, crab, herring, mackerel, mussels, roe, sardines, scallops, shrimps, sprats, whitebait
Other	Legumes, mushroom, spinach, asparagus, cauliflower, poultry, yeast, meat extract

C. *Cystine Stones*

For cystine stone formers, sodium and protein intakes need to be restricted. Medical drug therapy is used to control infection or produce more alkaline urine.

Summary of Dietary Principles in Kidney Stone Disease

Stone Chemistry	*Nutrient Modification*
Calcium	Normal calcium, lower protein and salt
Phosphate	Low phosphorus (1000-1200 mg)
Oxalate	Low oxalate
Struvite	Low phosphorus (1000-1200 mg; associated with urinary infections)
Uric acid	Low purine
Cystine	Low methionine (low protein)

D. *Struvite Stones*

For struvite stones, acidifying the urine is the single most important step. Patients must receive careful follow-up to be sure that the infection has cleared.

For prevention of calcium phosphate and struvite stones, urine should be acidified. However, the current treatment modalities are not efficient to prevent urolithiasis, and further research is required.

25

Diet in Obesity and Underweight

India Facing Dual Burden of Underweight and Overweight/Obese Population

India is the second most populous country in the world that comprises ~17 per cent of the world's population and contributes to 16 per cent of the world's deaths. Despite rapid economic growth in India in recent decades, the nutritional status of the adult population remains poor. On the one hand, India is experiencing a steady rise in the prevalence of overweight and obesity, which is globally recognized as a strong predictor of all-cause mortality and chronic diseases such as diabetes, cardiovascular diseases and hypertension. On the other hand, India has the highest number of underweight adults in the share of global adult underweight. With 102 million men and 101 million women underweight, India leads the world in being home to over 40 per cent of the global underweight population. Almost half the nation's under-5s are underweight; the prevalence of underweight children in India is among the highest in the world.

Nutritional status of the Indian population varies significantly across the regions. Certain regions are associated with extremely high rates of childhood undernutrition (ranging from 20 per cent to 80 per cent), whereas others have a high prevalence of adult undernutrition (>50 per cent), and some have both. Data regarding the nutritional status of adults, as determined by body mass index (BMI), indicate that 50 per cent of Indian adults suffer from different types of chronic energy deficiency, in that they have a BMI<18.5 kg/m^2. While 35 per cent older adults aged 50 and above in India are underweight; underweight is concentrated in adults of low socioeconomic status and rural residents. In the same LANCET survey, it was observed that the BMI values were similar in men and women; however, there were more overweight/obese (BMI≥25 kg/m^2) women (6.6 per cent) than men (3.5 per cent). Latest data shows that the number of overweight people have almost doubled from 10.5 per cent of the country's population in 2006 to 19.6 per cent in 2016, which means that close to one in five people in India weigh far more than they should. Furthermore, underweight and overweight/obesity are both higher for women than men. This dual disease pattern in women may have an endocrine basis, but more probably has its roots in societal and cultural norms, which prevent women from leading a healthy lifestyle. India is number one on the world map for underweight population and one of the top five in obesity. Earlier, developing countries, including India, had focused scarce public health resources primarily on the high prevalence of undernutrition. However, India is currently facing the coexistence of underweight and overweight population, referred as the dual burden of malnutrition which poses a major public health challenge. Both underweight and overweight are associated with increased risks of poor health and mortality and have to be taken care of.

Overweight and Obesity

Overweight and obesity are defined as abnormal or excessive fat accumulation that presents a risk to health. Burden of overweight and obesity has increased rapidly over the past decades globally. It is endemic in many parts of the world. Most of the world's population lives in countries where overweight and obesity kills more people than underweight. According to the World Health Organization (WHO), obesity is one of the most today's most blatantly visible, yet among the most neglected, public health problems in both developed and developing countries. Today, 2.1 billion people (nearly 30 per cent of the world's population) are either obese or overweight. According to a report of the World Health Organization, 2.8 million adults die each year as a result of being overweight or obese. In addition, 44 per cent of the diabetes, 23 per cent of the ischemic heart disease, and 7–41 per cent of certain cancers are attributable to overweight and obesity.

Obesity can now be seen as the first wave of a defined cluster of non- communicable diseases called "New World Syndrome," creating an enormous socioeconomic and public health burden in developing countries too. India, once known for its malnourished population, now has more and more overweight and obese people having high risk of adverse consequences. In the twenty-first century, obesity has reached epidemic proportions in India, with morbid obesity affecting 5 per cent of the country's population. India is following a trend of other developing countries that are steadily becoming more obese; prevalence of obesity is 12.6 per cent in women and 9.3 per cent in men. A recent study demonstrated that India has raced to third place, after the United States of America and China in the highest number of obese people worldwide; the United States of America accounting for 13 per cent of obese people globally and India together with China accounting for 15 per cent of the world's obese.

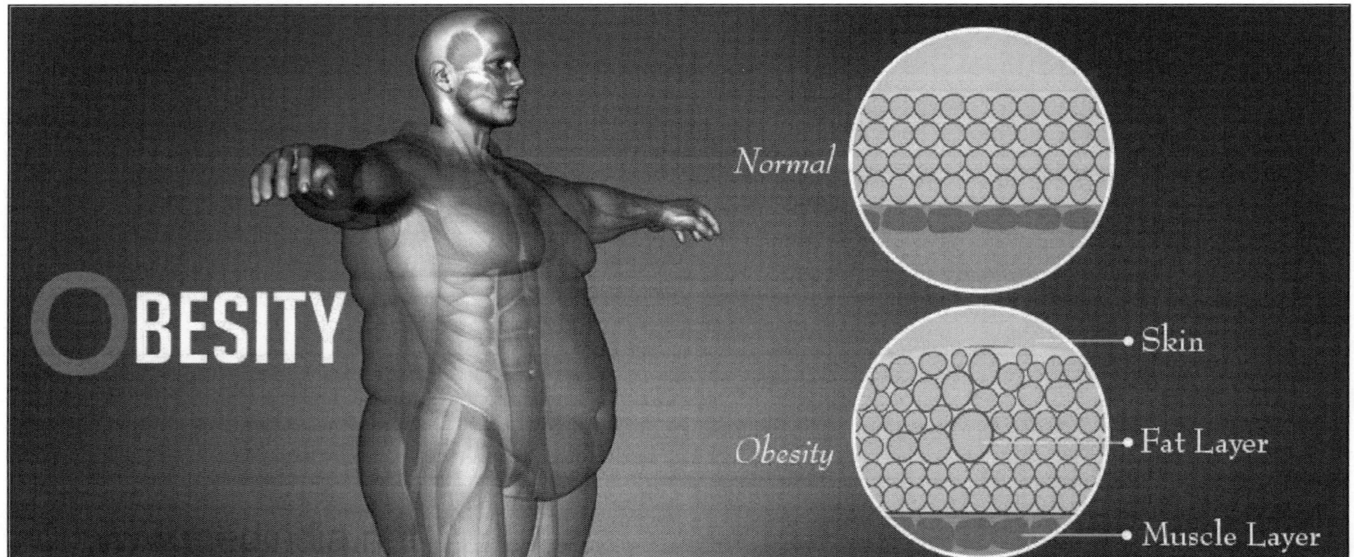

Assessment and Classification

Ideally, an obesity classification system would have the following characteristics: it would be based on a practical measurement widely available to providers regardless of their setting; it would accurately predict health risk (prognosis); and it could be used to assign treatment strategies and goals.

Body Weight

An adult having 10 per cent more body weight than the standard weight for height is termed as overweight while having 20 per cent more is obese.

Body Fat Measurements

Total body weight alone is not a clear indicator of good health because it does not distinguish between kilograms that come from body weight and those that come from lean body mass or muscle. Carrying too much body weight is a condition called obesity and puts a person at risk for many serious medical conditions including heart disease, diabetes and even certain forms of cancer. Hence, measurement of body fat mass is an important parameter to assess obesity and its implications. Body fat mass can be measured using multiple techniques. However, the measurement is typically presented as body fat percentage which is the ratio of fat weight to total body weight multiplied by 100. A classic method of measuring body fat percentage is using calipers to measure skin folds. The most accurate measures of body fat such as underwater weighing, dual-energy x-ray absorptiometry (DEXA) scanning, computed tomograpy (CT), and magnetic resonance imaging (MRI) are impractical for use in everyday clinical encounters.

However, from BMI, body fat percentage can be calculated by the given formula:

Body fat percentage= 1.39 x BMI+ 0.16 x age – 10.34 x gender - 9

where gender (sex) is 0 if female and 1 if male to account for the lower body fat percentage of men.

There is no generally accepted definition of obesity based on total body fat. Women have more body fat percentage than men. Normally, body fat percentage for men is 18 to 24 per cent and women, it is 25 to 30 per cent.

Most researchers have used >25 per cent in men, and >30 per cent in women, as cut-points to define obesity.

Asian Indians have more abdominal (visceral) fat deposition than Europids. For the same BMI, percentage body fat is 7-8 per cent points higher for Asian Indians compared to Europids. Conversely, for the same percentage body fat, BMI is 3-6 units lower for Asian Indians compared to Europids.

Where we store fat in our bodies can influence our health. On the basis of the distribution of fat, obesity is of two types *i.e.* android and gynoid obesity. In android obesity (more common among men), there is more fat around waists (being apple-shaped) which is associated with increased risk of type 2 diabetes and heart disease, even after correction for obesity. In gynoid obesity (common among women), fat is more evenly and peripherally distributed throughout the body *i.e.* around the breast, hips and thighs. In comparison, storing fat in thighs and bums (being pear-shaped) has been suggested in some research to offer some protection against diabetes and high blood pressure. Waist-to-hip ratio is a good measure of this difference in body fat distribution.

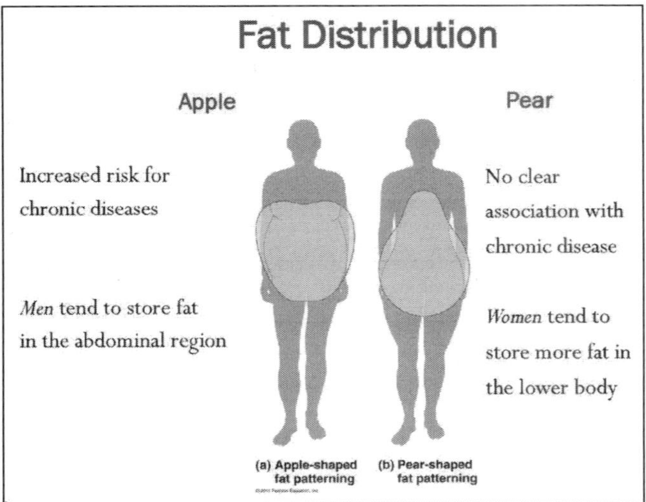

Body Mass Index (BMI)

Obesity is usually reported in terms of body mass index (BMI) and waist circumference (WC) but the cutoff points differ by ethnicity. BMI is the simplest, accepted and most widely used classification for obesity in adults. It is calculated by the following formula:

$$\text{BMI} = \frac{\text{Weight (Kg)}}{\text{Height (m)}^2}$$

For example, an adult who weighs 70 kg and whose height is 1.75 m will have a BMI of 22.9.

70 (kg)/1.75 (m²) = 22.9 BMI

The WHO classifications of overweight and obesity are intended for international use. This classification assigns increasing risk for comorbid conditions including hypertension, type 2 diabetes mellitus, and cardiovascular disease to persons with higher a BMI (Table 25.1) relative to persons of normal weight (BMI of 18.5 - 25 kg/m^2).

Table 25.1: Classification of Overweight and Obesity by BMI, Waist Circumference, and Associated Disease Risk Recommended by WHO

	BMI (kg/m²)	*Obesity Class*	*Disease Risk* (Relative to Normal Weight and Waist Circumference)*	
			Men ≤ 40 inches (≤ 102 cm) Women ≤ 35 inches (≤ 88 cm)	*> 40 in (> 102 cm) > 35 in (> 88 cm)*
Underweight	< 18.5		-	-
Normal†	18.5–24.9		-	-
Overweight	25.0–29.9		Increased	High
Obesity	30.0–34.9 35.0–39.9	I II	High Very High	Very High Very High
Extreme Obesity	≥ 40	III	Extremely High	Extremely High

* Disease risk for type 2 diabetes, hypertension, and cardiovascular disease. † Increased waist circumference can also be a marker for increased risk even in persons of normal weight.

Source: Purnell, J.Q. 2018. Definitions, classification and epidemiology of obesity. www.endotext.org

BMI Cut-Offs for Asian Populations

Adults

Earlier studies for assessment of obesity in India used the standard criteria (developed for and by Caucasians/ Europids) that have significantly underestimated the prevalence of obesity and metabolic syndrome in India (Table 25.1). There has been debate whether BMI cut-off points should be lower for Asian populations because the proportion of Asian people with a high risk of type 2 diabetes and cardiovascular disease is substantial at lower BMI ranges than the existing WHO cut-off point for non-Asian groups due to predominance of central fat distribution. Consequently, the WHO has suggested lower cutoff points for consideration of therapeutic intervention in Asians: a BMI of 18.5 to 23 kg/m^2 represents acceptable risk, 23 to 27.5 kg/m^{2}confers increased risk, and 27.5 kg/m^{2}or higher represents high risk. Recently, as per survey conducted by Ministry of Health and Family Welfare, GOI, people having BMI more than 25kg/m^2 have been considered as obese.

Children and Adolescents

The cut-off levels for categorizing overweight and obesity in children and adolescents are different. Since, growth spurt in boys and girls occurs in different age groups, age and gender specific BMI centiles of reference population is used to assess their nutritional status. Thus, children with BMI less than 5th centile are considered as undernurished and those with =5th and <85th centiles are normal, and with =85th and <95th centile are considered as overweight. More than 95th centile are considered as obese.

Limitations of BMI

Estimates of BMI though provide relevant information and are easily implemented in a variety of practice settings but do have limitations too when compared to the imaging methods. As BMI is based simply on the ratio of weight to height, it does not take account of body composition. For example, lean but well-muscled individuals may have a BMI above 25 kg/m^2 but not have an excess of body fat. However, this is an issue for a relatively small proportion of the population at the boundaries of the categories and BMI still remains practical in most situations and is widely used. If clarification is required, waist circumference can be used to identify individuals at risk (Table 25.2).

Classification of Risk of Obesity Based on Waist Circumference Measurement

In addition to an increase in total body weight, a proportionally greater amount of fat in the abdomen or trunk compared with the hips and lower extremities has been associated with increased risk for type 2 diabetes mellitus, hypertension, and heart disease in both men and women. Abdominal obesity is commonly reported as a waist-to-hip ratio. Hence, waist–hip ratio (*i.e.* the waist circumference divided by the hip circumference) has been suggested as an additional measure of body fat distribution. Waist circumference and waist–hip ratio are both related to increased risk of all cause mortality, throughout the range of adult BMIs.

A waist circumference of 102 cm (40 inches) or more in men, or 88 cm (35 inches) or more in women, is associated with health problems such as type 2 diabetes, heart disease and high blood pressure (Table 25.1). Thus, an overweight person with predominantly abdominal fat accumulation would be considered at "high" risk for these diseases even if that person is not obese by BMI criteria. These waist circumference thresholds are also used to define the "metabolic syndrome" by the most recent guidelines from the American Heart Association and the National Lipid Association. According to WHO (2008), abdominal obesity is defined as a waist–hip ratio of ≥0.90 cm (Men); ≥0.85 cm (Women) or a BMI above 30; in such individuals, the risk for metabolic complications is substantially increased.

Table 25.2: WHO Cut Off Points for Waist Circumference and Risk of Metabolic Complications*

	Waist Circumference (cm)
Men	
• Increased risk	94-102
• Substantially increased risk	>102
Women	
• Increased risk	80-88
• Substantially increased risk	>88

* World Health Organization 2008. Waist circumference and waist-hip ratio. Report of a WHO Expert Consultation, Geneva (Dec. 8-11, 2008)

There is substantial evidence of sex and age variations in waist circumference and waist–hip ratio, and some evidence for ethnic differences. Compared to Europeans, Asian populations have greater visceral

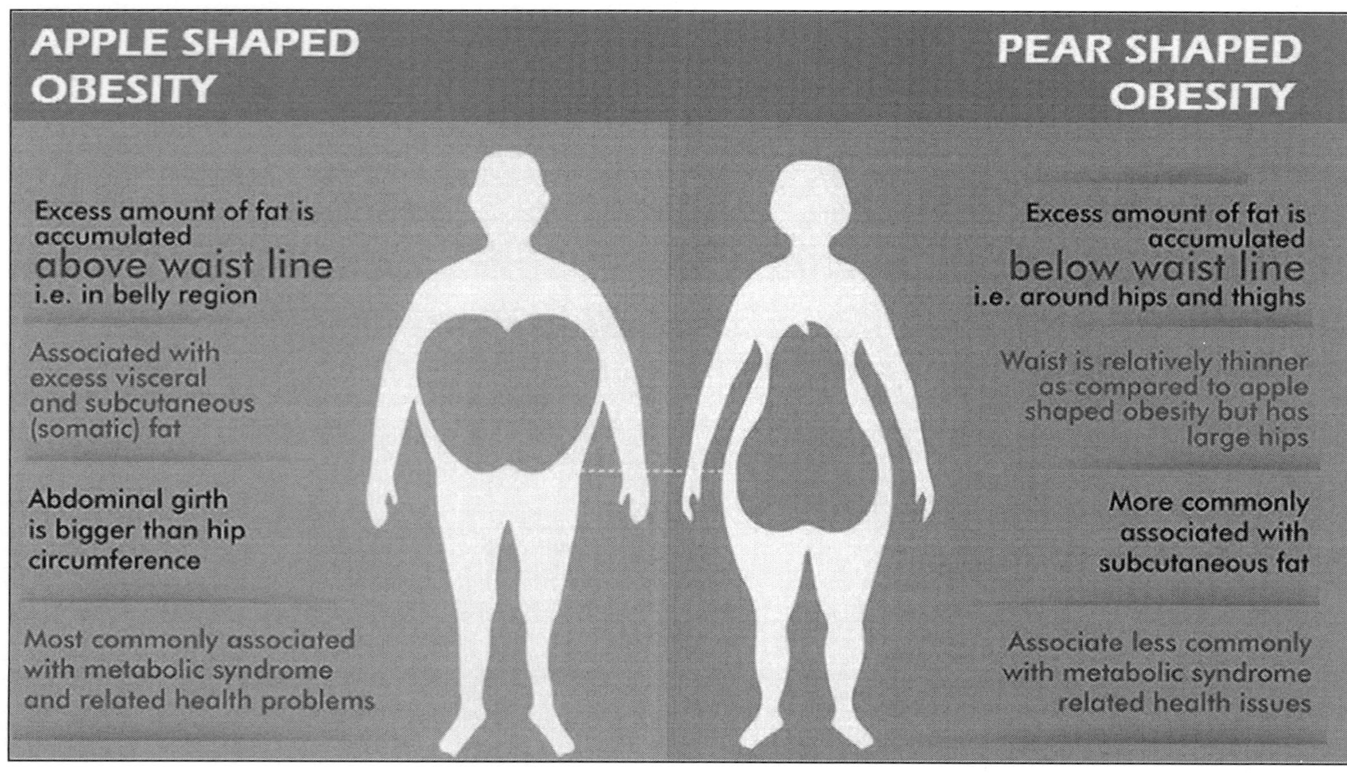

adipose tissue. On the basis of research studies, it has been reported that among Indians, due to wide prevalence of abdominal obesity even among young adults, measurement of waist to hip ratio is a better marker of obesity-related metabolic risk than BMI alone. This is based largely on the rationale that increased visceral adipose tissue is associated with a range of metabolic abnormalities, including decreased glucose tolerance, reduced insulin sensitivity and adverse lipid profiles, which are risk factors for type 2 diabetes and CVD. Waist-to-hip ratio of more than 0.9 among men and 0.8 in women and waist circumference 90cm for men and 80cm for women are associated with increased risk of several chronic diseases especially in Asian Indians.

For predicting the metabolic syndrome among population especially Indians, combining BMI with waist circumference and waist to hip ratio may be a better option so as to improve their discriminatory capability.

New Guidelines for Classification of Obesity among Indians

Alarmed by reports that India will become the global diabetes capital by 2050 if the abdominal and lower limb obesity and metabolic syndrome are not arrested, the Ministry of Health and Family Welfare, GOI has reduced the diagnostic cut-offs for body mass index to 23 kg/m^2 and the standard waist circumference to fight the battle against obesity. The standards have been set for the first time in the Ministry's consensus guidelines for prevention and management of obesity and metabolic syndrome for the country. The country's new diagnostic cut-off for the body mass index is 23 kg/m^2 as opposed to 25 kg/m^2 globally. Those with BMI of 25 kg/m^2 will be clinically termed obese (as opposed to 30 kg/m^2 at the international level) and those with BMI of 32.5 kg/m^2 will require bariatric surgery to eliminate excess flab. According to guidelines, cut-offs for waist circumstances will now be 90 cm for Indian men (as opposed to 102 cm globally) and 80 cm for Indian women (as opposed to 88 cm at the international level). The cut off levels for waist to hip ratio are 0.90 for men and 0.80 for women.

Etiology of Obesity

"The first law of thermodynamics assures that body weight cannot change if, over a specified time, energy intake and energy expenditure are equal" and hence, obesity occurs due to this imbalance of energy between calories consumed and calories expended. Both an increase in energy intake and a decrease in energy expenditure in the long run leads to excess calories being stored as fat and, ultimately, to overweight and obesity.

Research studies have indicated that energy imbalance is determined by a complex multifaceted system of determinants, where no single influence dominates. These include:

Increased Energy Intake

Food production has increased several folds over the past few years and food has become more accessible through improved agricultural practices, modern food processing techniques and efficient food transport, marketing and storage facilities in India. Food is already the largest retail consumption category in India, accounting for approximately 31 per cent of the country's consumption basket, compared to 9 per cent in the US, 17 per cent in Brazil and 25 per cent in China. Food expenditure has increased substantially. Indians have started spending a high proportion of their incomes on food and groceries, compared to consumers in other countries, and food consumption and demand are expected to grow considerably over the next few years.

In addition, changes in food basket composition are a significant component of food pattern trends happening in India. Due to globalization, industrialization, urbanization and marketing strategies of ready-to serve fast food chains, traditional food consumption patterns are slowly changing in the modern society. India's fast food industry is mounting by 40 per cent a year which statistically place India in 10th place in fast food per capita spending figures with 2.1 per cent of expenditure of annual total spending. According to the upcoming trend report of fast food market in India during 2015-2020, Indian fast food market is expected to grow at a compound annual growth rate (CAGR) of 18 per cent by 2020 due to changing consumer behaviour and demography; it is expected to be worth US$ 27.57 billion by 2020.

Easy to obtain, prepare and eat palatable energy dense fast foods also known as junk foods which are rich in fat and sugars, have provided the flexibility of having many consumption alternatives at reasonable prices. Now a days, fast food industries have acquired a place in people's life as such that ordering the fast food item or visiting the outlet of fast food items portrays a symbol of higher standard of living in the society for the all, availing their services. These growing fast food chains have been, innovative and forceful at inviting customers out of their kitchens up to fast food centres

Not only young generation but also people from all age groups now demand fast foods in their every-day eating habits, due to rapid increase in disposable income with changing attitude towards food choices. Being high in fat content, high in sugar in liquid form and low in fibre, junk foods induce gorging that leads to obesity. Eating too much junk food is one of the factors that have contributed to the children hidden obesity epidemic in India. According to the research studies, calorie content consumed by the children of out of home meals is 55 per cent higher than of in home meals. Hence, the consumption of fast foods on a regular basis leads to excess energy intake leading to an increased risk of overweight and obesity which lead to serious health issues.

Decreased Energy Expenditure

It is well known that energy expended in physical activity is an important component in controlling energy expenditure. Increases in physical activity lead to increases in energy expenditure, which can induce an energy deficit to control body weight. But in today's fast-paced environment, it is easy for people to adopt unhealthy behaviors. Behaviour, in case of obesity, relates to food choices, amount of physical activity and efforts to maintain good health. Indeed the increase in the prevalence of obesity, observed worldwide has been linked with widespread declines in the level of daily physical activity as well as increased availability of energy-dense foods and changing food habits.

A few studies conducted up to day in India about the association between metabolic and contextual causes of obesity, found some relevant links among weight increase and a general reduction of physical activities, enhanced by sedentary lifestyles. Some of the co-causing risk factors for less physical activity and decreased energy expenditure include less involvement in manual work due to availability of labour-saving devices, car ownership, more hours spent on mobile phones, watching TV and computer use, not participating in outdoor sports activities but more time on playing games on mobile phones, I-pads and computers *etc.* Rapid decline in physical activity mirrors the rise in obesity and is thus, considered the major contributing factor.

Genetic Factors

Genetic predisposition to obesity is also believed by many to play a very important role in the etiology of obesity; obesity runs in families. Some individuals are genetically more susceptible to the effects of an obesogenic environment. The recent discovery of the 'ob' gene has led to a renewed interest in understanding the patho-biological basis of genetic predisposition to obesity. The 'ob' gene codes a hormone called leptin which is an adipose derived protein hormone and has been found to have important effects on regulation of energy intake and energy expenditure. Any mutation of 'ob' gene leads to improper coding of leptin, which has been associated with obesity. Other genes are also now being identified that may contribute to obesity through dietary intake and specifically through increases in dietary fat and carbohydrate.

Although it is now widely acknowledged that obesity has a genetic component, it does not appear to be a simple genetic disorder, rather there is an underlying genetic predisposition to obesity on to which environmental factors such as a constant supply of convenient, inexpensive and palatable foods, and reduced need for physical labour in jobs, and daily life are layered. Indeed the prevalence of obesity increases so rapidly in many populations that the changes cannot be attributed to changes in genetic inheritance. Continued research is needed to fully comprehend the genetics of mechanisms responsible for energy balance, the gene-environment interactions and relate these to why certain susceptible individuals passively over consume leading to obesity.

Obesogenic Environment

Obesogenic environments are widely accepted as a driving force behind the escalating obesity epidemic today. The term 'obesogenic environment' refers to the role environmental factors may play in determining both nutrition and physical activity. The obesogenicity of an environment has been defined as 'the sum of influences that the surroundings, opportunities, or conditions of life have on promoting obesity in individuals or populations'. This includes the physical, economic, political and socio-cultural environments relating to food and physical activity. Environmental influences on diet may involve access to foods from supermarkets, or takeaways and restaurants. Environmental influences

on physical activity levels may involve access to cycle lanes, safe walking areas. Certain environments may be more 'obesogenic' than others, such that they are more likely to promote weight gain and obesity in individuals or populations, but it remains a challenge to identify the physical environmental factors with the greatest impacts on (the development of) overweight and obesity.

Psychosocial Influences

Even though caloric imbalance is likely one of the main driving forces behind the current obesity epidemic, psychosocial stress factors, which are common in modern society, have been hypothesized to contribute to the increased prevalence of obesity, through changes in lifestyle and chronic activation of the neuroendocrine system. Individuals who suffer from psychological disorders (*e.g.* depression, anxiety, and eating disorders) may have more difficulty controlling their consumption of food, exercising an adequate amount, and maintaining a healthy weight.

Foetal Programming

A growing body of empirical evidence suggests that the origins of obesity and metabolic dysfunction can be traced back to the intrauterine period of life, at which time the developing fetus is acted upon by and responds to suboptimal conditions during critical periods of cellular proliferation, differentiation, and maturation by producing structural and functional changes in cells, tissues and organ systems. These changes, in turn, may have long-term consequences to increase the individual's risk for developing a range of complex common disorders including, but not limited to, obesity and metabolic dysfunction (*i.e.*, the concept of fetal programming of health and disease risk (Glukman *et al.*, 2008).

The Barker hypothesis proposed that adverse nutrition in early life, including prenatally as measured by birth weight (<2.5 kg) i.e small size at birth, increased susceptibility to the metabolic syndrome which includes obesity, diabetes, insulin insensitivity, hypertension, and hyperlipidemia and complications that include coronary heart disease and stroke. More recent data suggest that it is not size at birth alone that best predicts disease risk. Rather, the combination of small size at birth and rapid weight gain during the first two postnatal years (known as catch-up growth) is the strongest predictor of risk for later obesity and metabolic disease. WHO estimate that 30 million babies are born with low-birth-weight annually worldwide, mostly in low-income countries, which contributes to their rising prevalence of obesity.

Metabolic Factors

There is no evidence to support the concept that a low metabolic rate is the major cause of obesity. However, people with a low metabolic rate (adjusted for body size and composition) and suffering from endocrine disorders such as Cushing's syndrome (increase in the levels of stress hormone, cortisol) and hypothyroidism, Polycystic ovarian syndrome (PCOS), Prader–Willi syndrome, and congenital leptin deficiency are prone to gain weight and become obese.

Medications

Certain medications such as antidepressants, contraceptives, antiepileptics, insulin and steroids can lead to weight gain.

Age

Obesity can occur at any age, even in young children. But with ageing, hormonal changes and a less active lifestyle can increase the risk of obesity. In addition, the amount of muscle mass decreases in the body with age. This lower muscle mass leads to decrease in metabolism. If calorie intake is not decreased with increasing age, one is likely to gain weight. Hence, older people are at a greater risk of weight gain compared to younger individuals; especially older women who are in their menopausal age.

Pregnancy

During pregnancy, a woman's weight necessarily increases. Some women find this weight difficult to lose after the baby is born. This weight gain may contribute to the development of obesity in women.

Clinical Consequences

Mortality and Co-morbidities

The effects of excess weight on morbidity and mortality have been known for more than 2000 years. Hippocrates recognized that 'sudden death is more common in those who are naturally fat than in the lean'. To achieve optimum health, the median body mass index for an adult population should be in the range of 21 to 23 kg/m2, while the goal for individuals should be to maintain body mass index in the range 18.5 to 24.9 kg/m^2. Beyond a BMI of 25, mortality rises by approximately 30 per cent for every 5kg/m^2 increase; it lowers the life expectancy of the individual. The life expectancy of men and women with a BMI of > 45 kg/m^2 aged 20–30 years is 13 and 8 years lower, respectively, than that of those with a BMI of 24 kg/m^2.

Massive global research project reveals 30 per cent of the world's population is affected by weight problems. Globally, more than 2 billion children and adults suffer from health problems related to being overweight or obese, and an increasing percentage of people die from these health conditions, according to a new study. They are dying even though they are not technically considered obese, researchers found. Of the 4.0 million deaths attributed to excess body weight in 2015, nearly 40 per cent occurred among people whose body mass index (BMI) fell below the threshold considered "obese."

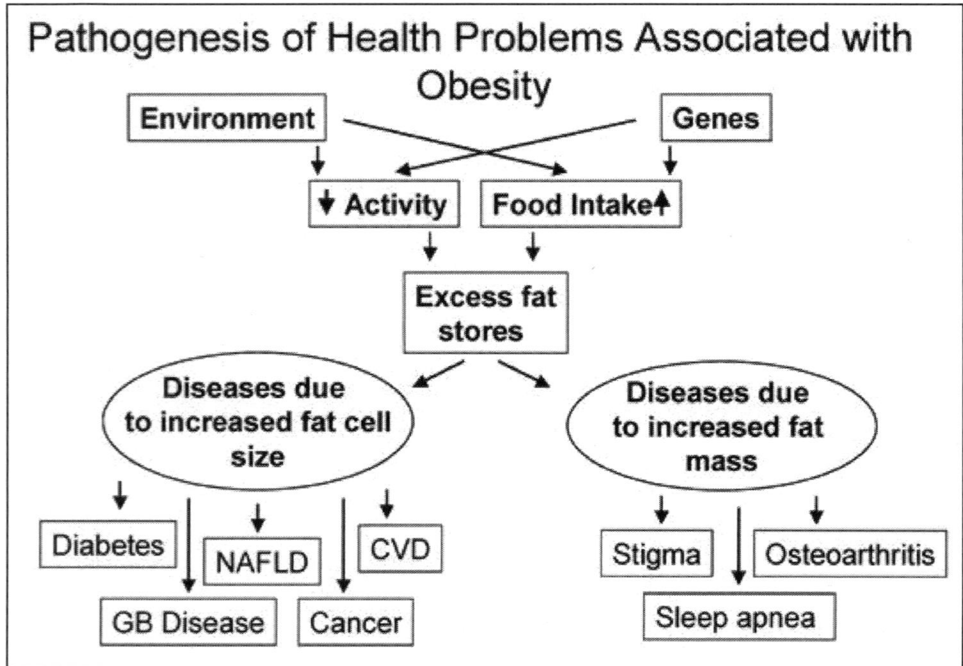

Obesity is associated with an increased risk of disease and death, particularly from CVD and cancer. Obesity increases cardiovascular mortality fourfold, and cancer-associated mortality twofold. Worldwide, at least 2.8 million people die each year as a result of being overweight or obese, and an estimated 35.8 million (2.3 per cent) of global DALYs are caused by overweight or obesity.

There is increased risk of co-morbidities for body mass index 25.0 to 29.9, and moderate to severe risk of co-morbidities for body mass index greater than 30. The pathology of obesity produces the myriad of health related problems. These health-related problems can be attributed to either the increased mass of fat or the increased release of peptides from enlarged fat cells. Some of the health problems related to obesity are:

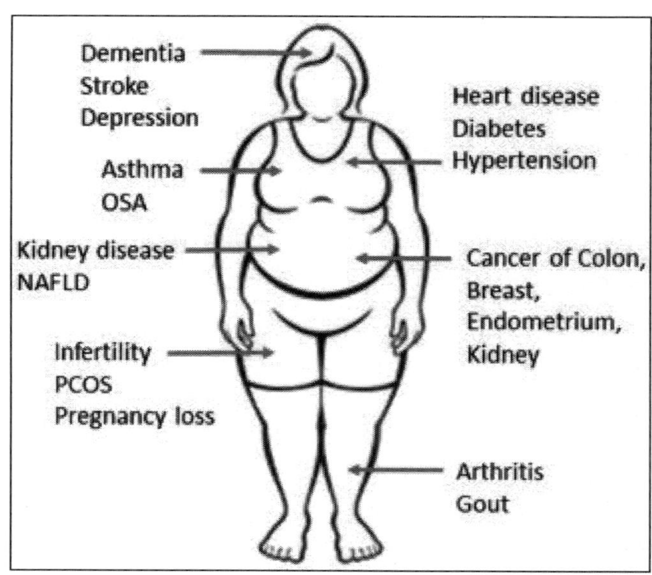

Metabolic Disorders

Overweight and obesity lead to adverse metabolic effects on blood pressure, cholesterol, triglycerides and insulin resistance. Risks of hyperlipidemia, coronary heart disease, ischemic stroke, gall stones and type 2 diabetes mellitus increase steadily with increasing body mass index (BMI). Increased steatosis is characteristic of the livers of overweight individuals and may reflect increased VLDL production associated with hyperinsulinemia. The accumulation of lipid in the liver suggests that the secretion of VLDL in response to hyperinsulinemia is inadequate to keep up with the high rate of triglyceride turnover.

Obese individuals have been shown to be twice as likely to have a stroke, either ischaemic or haemorrhagic, than people with a BMI of < 23. Raised body mass index also increases the risk of cancer of the breast, colon, prostate, endometrium, kidney and gall bladder.

Obesity can also cause infertility (men and women), and polycystic ovary syndrome. In men, obesity is associated with reduced sperm count and increased rates of erectile dysfunction. Many men have low testosterone because they have excess abdominal fat. In women, it also leads to reduced fertility, poorer outcomes after fertility treatment and more pregnancy loss. Polycystic Ovarian Syndrome (PCOS), characterized by anovulation, hyperandrogenism and a polycystic ovary, is the primary cause of female infertility and increases the rate of pregnancy complications. It has been noted that increased visceral fat assessed by waist circumference of more than 88cm is associated with hyperandrogenemia in patients with PCOS and that reduction of insulin resistance by weight loss or drugs that increase peripheral sensitivity of insulin leads to improve hormonal aberrations and ovulation.

Obesity may also affect the kidney. Glomerulopathy is significantly increased in pathological specimens compared with other forms of end-stage renal disease. There is a strong positive association between increasing obesity and GERD symptoms and esophageal erosions.

Physical Problems

Osteoarthritis, chronic back pain, respiratory problems, decreased mobility and accidents, sleep apnoea, skin problems *etc.* are some of the physical health related problems caused by obesity. Obesity is associated with significant reductions in physical activity levels and is one of the main risk factors for osteoarthritis. The greatest impact is on the knee, but there is also increased risk in hand joints, suggesting an inflammatory as well as a mechanical cause. For prevention, as little as 5.1 kg reduction over a 10-year period decreases the likelihood of developing knee osteoarthritis by 50 per cent. Radiating lower back pain is increased in obesity, as well as in those with low levels of physical activity. Gout is also more common in obese individuals.

Obesity has long been known to be a major pathogenic factor in obstructive sleep apnea in adults. Obstructive sleep apnoea (OSA) adversely affects multiple systems and is associated with hypertension, insulin resistance, liver dysfunction, systemic inflammation and dyslipidaemia. In children, it can lead to failure to thrive, behavioural problems, decreased intellectual function and a higher risk of cardiovascular morbidity.

Asthma is another condition that may occur as a complication of obesity. There is evidence that obesity increases the risk of asthma. The mechanism linking obesity and asthma includes increased airway hyper-responsiveness, decreased functional and tidal volumes, chronic systemic inflammation driven by increased inflammatory cytokines and chemokines, adipocytes derived factors leptin, adiponectin and plasminogen activator inhibitor.

Several skin changes are associated with excess weight. Stretch marks, or striae, are common and reflect the pressures on the skin from expanding lobular deposits of fat. Acanthosis nigricans with deepening pigmentation in the folds of the neck, knuckles, and extensor surfaces occurs in many overweight individuals, but is not associated with increased risk of malignancy. Hirsutism in women may reflect the altered reproductive status in these individuals.

Psychosocial Problems

The stigma of obesity is common in society and obese persons often face negative attitudes and discriminatory behavior. Obesity can lead to depression, low self-esteem, social isolation, poor employment status, impaired relationships *etc.* Obese individuals are often exposed to public disapproval and stigma due to their weight, with women experiencing more discrimination. Depression is more common in obesity, particularly in women and younger people, while weight loss is associated with improved mood. Adolescents who are obese or overweight, or perceive themselves as such, are more likely to engage in risk behaviour than those of normal weight. This can involve substance abuse, risky sexual behaviour or violence. Binge eating disorder, characterized by loss of control while eating unusually large amounts of food is quite common among obese individuals especially adolescents.

Economic Burden

Obesity imposes its cost in terms of economic burden on the individual, family and the country. Several studies have shown an increase in the healthcare costs in the overweight population. There is an increase in the outpatient visits and inpatient days along with the total cost of care of obese people. In India, it has been estimated that treatment costs of diseases directly attributable to obesity are estimated to correspond to about 2-10 per cent of the total health care expenditure. The indirect costs arising from loss of productivity due to obesity are even higher. Obese people spend 42 per cent more on healthcare costs than healthy weight people. Obesity-related conditions are estimated to have cost the Indian economy $200 billion during 2005-2015. The projections for the next 10 years can be twice or thrice the cost of the past 10 years. By 2025, obesity-related diseases will cost India a staggering $13 billion. India stands to lose $4.58 trillion before 2030 due to non-communicable diseases and obesity-related health conditions. Cardiovascular diseases, accounting for $2.17 trillion, and obesity and associated health conditions ($1.03 trillion), will lead the way in economic loss. This reiterates the importance of preventing and treating this epidemic.

Weight Management

Managing obesity can be very challenging but successful weight loss is accompanied by numerous health and social benefits. A range of approaches like dietary therapy, physical activity, behavior modification, non-clinical weight management programmes, combined therapy, pharmacotherapy, surgery (gastric band surgery, gastric bypass surgery, sleeve gastrectomy, biliopancreatic diversion (BPD) and duodenal switch (DS), electroconvulsive therapy (ECT), and gene therapy *etc.* are available in the present time for the treatment and management of obesity. These are associated with varying efficacy and the optimum choices are multicomponent and which people are willing to engage with and adhere to over time.

As obesity treatment strategies vary from person to person, hence the first essential stage of individual management approaches should be to assess:

- History of weight/BMI trajectory over time, prior attempts to lose weight, medications predisposing to weight gain and psychosocial factors:
- Family history of obesity and related conditions;
- Medical history and physical examination (underlying cause such as hypothyroidism; acanthosis nigricans, hirsutism in women; presence of any co-morbidity *e.g.*, diabetes, CVD *etc.*);
- Limited lab tests: lipid profile, glucose (or HgB A1C in adults), AST and ALT;
- Anthropometric measurements (BMI, waist circumference, waist to hip ratio *etc.*);
- Dietary history and evaluation of food intake (current dietary intake, meal pattern, food likes and dislikes, reasons for eating *etc.*);
- Physical activity status *e.g.*, current level of physical activity, reasons for being physically active or not;
- Psychosocial assessment *i.e.* to identify degree of motivation of the individual to lose weight and also barriers to weight management; and
- Preference for management therapy for weight loss.

Physical Benefits of Intentional Weight Loss

- Losing 10 kg is associated with a reduction of:
- > 20 per cent total mortality
- 10 mmHg systolic and 20 mmHg diastolic BP
- 50 per cent of fasting glucose
- 10 per cent total cholesterol and rise of 8 per cent HDL cholesterol

Management and Treatment

Goal

Weight loss goals should be realistic, achievable, individualized and aimed at the long term. General goals of weight management in obese persons are:

- Reduce body weight
- Sustain weight loss by minimizing risk of weight gain
- Prevent further weight gain.

Weight reduction should be gradual. The initial goal of weight loss therapy should be to reduce body weight by approximately 5–7 per cent from baseline. With success, further weight loss can be attempted if indicated through further assessment. A reduction of half a kg body weight per week is safe.

Goals for Weight Reduction

- **Children**: To start with, for children, decrease rate of weight gain while they continue to grow taller or 0.5 to 1.0 kg/week weight loss is reasonable.
- **Adolescents:** For adolescents at BMI ≥ 99 per cent tile, 250-500 g per week weight loss is reasonable
- **Adults:** For adults (BMI ≤ 30 kg/m^2 and no significant burden of obesity-related co morbid conditions), 10 per cent weight loss in 6 months is recommended.

Management Package

There is no single regimen for weight reduction; it has to be individualized. After assessment, management package for weight loss should be worked out with the individual's input; it should ideally include behavior modification, lifestyle changes such as dietary changes to reduce energy intake, to increase physical activity or decrease inactivity to increase energy expenditure, stress management, sleep pattern *etc.* A combination of physical activity and dietary changes has been found to be most effective for weight loss. If these measures are not successful for six months, in addition, pharmacotherapy and bariatric surgery for morbid obese adults may also be considered. Weight reduction should be gradual. Slow and steady reduction in body weight is advisable. Weight loss of 10–15 per cent using lifestyle intervention is considered a very good response and greater than 15 per cent weight loss is considered an excellent response.

A. Behaviour Change Strategies

Behavior therapies include education, which helps in deciding individuals about a balanced diet with optimal calorie as per required to achieve optimal weight. By developing a set of skills, an individual can decide how to change through the use of goal setting, self-monitoring, and problem-solving. Strategies to change behaviour relating to weight loss include cognitive behavioural therapy and a range of other psychological approaches including cognitive therapy, psychotherapy, relaxation therapy and hypnotherapy. Cognitive behavior therapy is used to reinforce the patient's lifestyle changes of diet and exercise. Eating habits like bingeing, snacking in between meals, eating large portions, eating in front of the television, eating in the car, night eating *etc.* can be addressed through behavior therapy conducted by trained personnel, psychologists or self-help groups. Slow eating techniques along with stimulus control (not distracted by television, books, or other materials) have shown the positive effect on weight loss. Self-monitoring techniques like maintaining food journals, weighing, measuring activity with pedometers, *etc.* are also helpful. Cognitive restructuring of the patient to

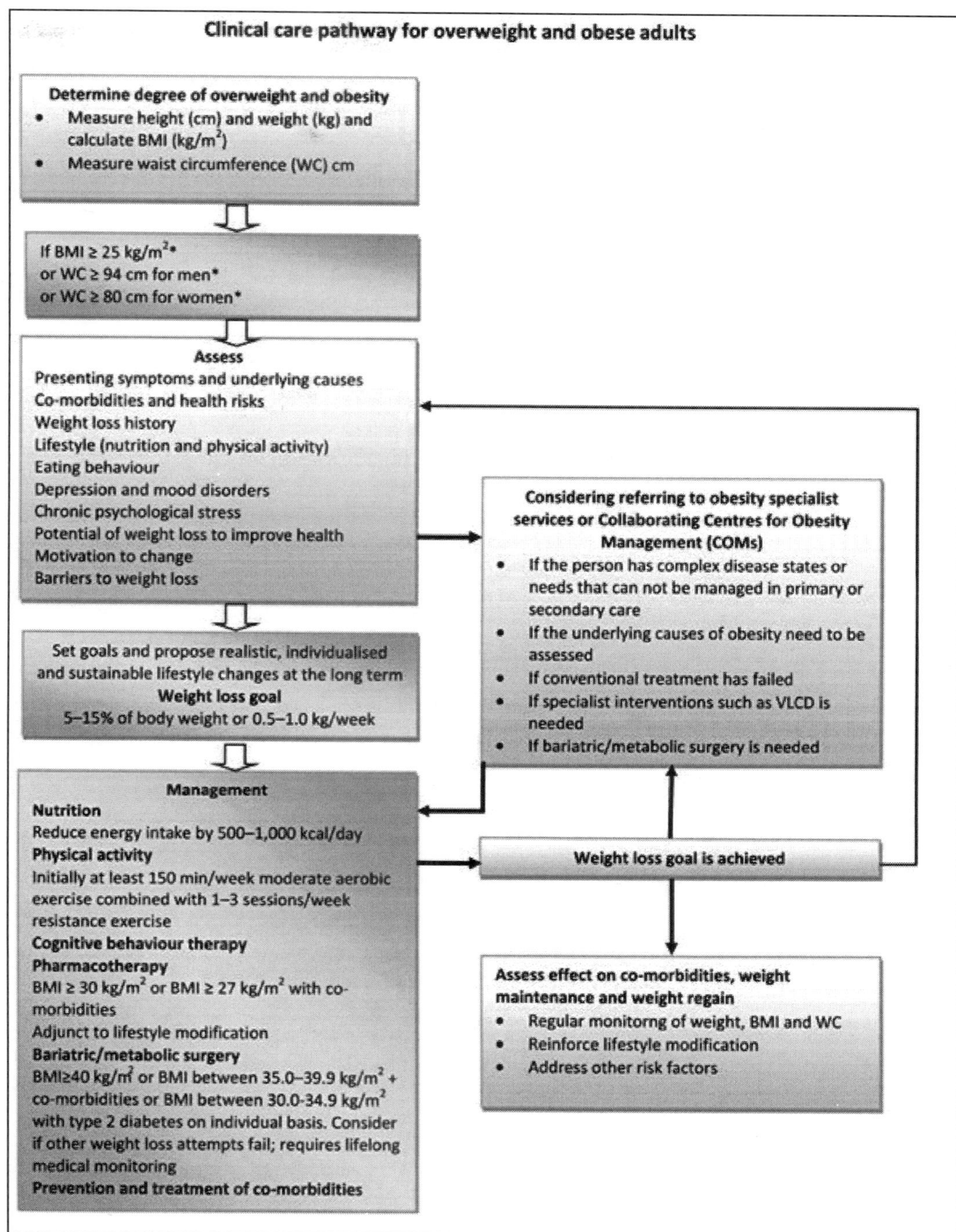

encourage positive thoughts and removing the cues or triggers that stimulate eating behaviors should be part of the therapy. Although time consuming, these activities fortify the permanent changes in lifestyle of the patients.

Systematic review studies indicate that behavior change strategies and specifically cognitive behavioural therapy are most effective when combined with physical activity and dietary modifications.

B. Dietary Therapy: Management of Obesity

The dietary therapy is based on the criteria of decreased energy intake and increased energy expenditure to produce the negative energy balance. The goal of dietary management in obesity should be to help the individual to reduce their energy intake to an acceptable level while consuming a diet that is adequate in all other nutrients, compatible with long-term good health, practical to follow and can be reconciled with their life style. Dietary advice should be individualized, tailored to food preferences and allow for flexible approaches to reducing calorie intake. One should not use unduly restrictive and nutritionally unbalanced diets because they are ineffective in the long term and can be harmful. In the longer term, people should eat a balanced diet, consistent with other healthy eating advice. Overall, energy intake should be less than energy expenditure while taking a calorie restricted balanced diet.

Various Kinds of Diets for Weight Loss

Several types of diets have been studied by the researchers for the weight loss such as vegetarian diet, low calorie diets, and very low-calorie diets, low-fat low-calorie, low-carbohydrate high-protein diet (Atkins diet), Mediterranean diet *etc.* However, adherence to a diet is more important than the type of diet. Extreme approaches should be avoided.

Various Dietary Approaches with Evidence of Weight Loss Efficacy

All of the following dietary approaches are associated with weight loss if reduction in dietary energy intake is achieved:

- A diet from the European Association for the Study of Diabetes Guidelines, which focuses on targeting food groups, rather than formal prescribed energy restriction while still achieving an energy deficit.
- Higher protein (25 per cent of total calories from protein, 30 per cent of total calories from fat, 45 per cent of total calories from carbohydrate) with provision of foods that realized energy deficit.
- Higher protein Zone type diet (5 meals per day, each with 40 per cent of total calories from carbohydrate, 30 per cent of total calories from protein, 30 per cent of total calories from fat) without formal prescribed energy restriction but realized energy deficit.
- Lacto-ovo-vegetarian-style diet with prescribed energy restriction.
- Low-calorie diet with prescribed energy restriction.
- Low-carbohydrate diet (initially <20g carbohydrate) without formal prescribed energy restriction but realized energy deficit.
- Low-fat vegan style diet (10 per cent –25 per cent of total calories from fat) without formal prescribed energy restriction but realized energy deficit.
- Low-fat diet (20 per cent of total calories from fat) without formal prescribed energy restriction but realized energy deficit
- Low-glycemic-load diet, either with or without formal prescribed energy restriction.
- Lower fat (≤30 per cent fat), high dairy (4 servings/day) diets with or without increased fiber and/or low-glycemic index or load foods (low-glycemic load), with prescribed energy restriction.
- Macronutrient-targeted diets (15 per cent or 25 per cent of total calories from protein; 20 per cent or 40 per cent of total calories from fat; 35 per cent, 45 per cent, 55 per cent, or 65 per cent of total calories from carbohydrate) with prescribed energy restriction.
- Mediterranean-style diet with prescribed energy restriction.
- Moderate protein (12 per cent of total calories from protein, 58 per cent of total calories from carbohydrate, 30 per cent of total calories from fat) with provision of foods that realized energy deficit.
- Provision of high-glycemic-load or low-glycemic-load meals with prescribed energy restriction.
- The American Heart Association style Step 1 diet (with prescribed energy restriction of 1500– 1800 kcal per day, <30 per cent of calories from fat (<10 per cent of calories from saturated fat)

Source: Jensen MD, Ryan DH, Donato KA, *et al.*, Guidelines (2013) for managing overweight and obesity in adults. Obesity 2014; 22(S2): S75.

1. Modified Healthy Eating For Weight Loss or Low Fat Diet

Initial target should be a calorie deficit of 500–1,000 kcal/day compared with the patient's habitual diet. The 500-1000 kcal deficit per day through modified healthy eating is ideal for encouraging gentle weight loss of 0.5–1.0 kg/week accompanied by a long term change in eating habits and is suitable for well-motivated individuals. A modest energy reduction of approximately 500-600 kilocalories per day through modified healthy eating can be achieved by reducing or eliminating the intake of concentrated calories particularly from fat, increasing fruit and vegetable intake to a minimum of five portions per day and maintaining intake of whole grain cereals, lean meat or fish, and low fat dairy products. Severe fasting may lead to health hazards. Hence, in any case, weight reduction diets should not be less than 1000 Kcal/day and should provide all the essential nutrients. A reduction of half a kilogram body weight per week is considered to be safe. Following dietary modifications for healthy eating should be made to promote gentle weight loss without any health hazard:

- Eat a variety of nutritious foods.
- One should avoid high calorie, low nutrient foods. Weight- reducing diets must be rich in proteins and low in carbohydrates and fats.

- ☆ One should stay within the required daily calorie need (as suggested by the dietitian) using the plate method to control calorie intake but foods should be chosen from all the food groups proportionately; take 45–65 per cent of calories from carbohydrates, 10–25 per cent from proteins and 20–25 per cent from fats, with saturated fats less than 7 per cent as part of the daily calorie intake. The percentage of monounsaturated fats should be increased up to 20 per cent of the total fat intake, while the intake of trans-fats should be restricted to the bare minimum.
- ☆ Choose smaller portion sizes of various food groups except eating more fruits and vegetables. Use smaller plates and measure foods.
- ☆ Fat is the concentrated source of calories, hence the intake of butter, ghee, full-fat dairy products, and tropical oils (palm oil and coconut oil) should be curtailed. Avoid trans-fatty rich foods (*vanaspati*, bakery products and sweets).
- ☆ Plant foods that provide complex carbohydrates and fibre may be preferred over the animal products; because plant foods reduce blood glucose, cholesterol and triglycerides.
- ☆ Eat complex carbohydrates, low glycemic foods and fibre rich diets. The diet should be rich in whole grains and millets like whole wheat, pearlmillet, oats, barely *etc.* Instead of white rice, use brown and parboiled rice. Avoid refined cereals which are devoid of fibre. Avoid white bread, *nan*, *puri* and *paranthas*. Prefer whole wheat *roti* without smearing of fat.
- ☆ Increase consumption of fruits and vegetables, legumes, whole grains and nuts.
- ☆ Take at least 5-6 servings of fruits and vegetables. Consumption of plenty of fruits and vegetables would not only result in satiety but could also help to maintain adequate micronutrient intake.
- ☆ Fruits selected should not be very sweet *e.g.* one should avoid mangoes, sapota, grapes, banana *etc.* One should take guava, *ber*, berries, watermelon, papaya, apple *etc.*
- ☆ Avoid starchy root vegetables like potato, sweet potato, colocasia *etc.* Try to incorporate more of green leafy vegetables and other seasonal vegetables which are low in calories but rich in micronutrients and antioxidants.
- ☆ Take salad before taking main meal. For salad dressing, use lemon, vinegar, herbs and spices.
- ☆ Take soups without cream and starch. Choose fat-free or low-fat milk and milk products. If lactose-intolerant, choose fat-free lactose-free milk or fat-free, plain soy milk.
- ☆ Vegetarians should include whole pulses and small amount of nuts in the diet for protein source.
- ☆ Non-vegetarians should select lean meats and skinless poultry. Two servings of fish high in omega-3 fatty acids per week can be taken.
- ☆ Instead of frying, try to bake without fat or oil, broil, steam or grill the food for cooking. Use non-stick pans for cooking.
- ☆ Sauté the vegetables in bare minimum amount of fat. It is better to take steamed vegetables without butter.
- ☆ While seasoning the cooked *dals*, avoid added fat.
- ☆ Cut down sugar, salt, fatty foods and refined foods. Avoid all kinds of Indian sweets, ice creams, desserts, fried snacks and savories, fast foods, junk foods *etc.*
- ☆ Drink lot of water instead of high caloric beverages (regular soft drinks, fruit juices, fruit punch, sweet tea and coffee, alcoholic beverages *etc.*).
- ☆ Limit eating out to rare occasions.
- ☆ Schedule regular meals and snacks throughout the day, starting with breakfast. Eat small meals regularly at frequent intervals. Avoid excessive snacking.
- ☆ Avoid eating while watching TV or reading book. In case of young children, parents should limit their screen time and exposure to food and beverage marketing.
- ☆ Limit the purchase of packaged, processed and convenience foods.
- ☆ While shopping for food, read food labels and select low fat, high fiber, low sodium and low cholesterol foods. Limit cholesterol to less than 300mg/day. Sodium intake should be less than 2,300mg/day.

2. Low Carbohydrate and High Protein Diets

The meaning of "low carbohydrate" varies from diet to diet. Some diets recommend extreme restriction of all carbohydrates, while others merely limit carbohydrates to primarily whole grains. What "low carbohydrate" means really depends on the diet and how it is followed. Two of the most popular diets, the Atkins and South Beach diets, show this contrast.

Atkins Diet

The Atkins Diet is a popular low-carbohydrate eating plan developed in the 1960s by cardiologist Robert C. Atkins. Proponents of this diet claim that one can lose weight while eating as much protein and fat as one wants, as long as foods high in carbohydrates are avoided. The main objectives of the Atkins Diet are to remove "carbohydrate cravings," "reset" the body's metabolism, and induce fat loss by eliminating carbohydrate-containing foods. The premise is that all carbohydrate-containing foods, regardless of whether they are high in fiber, contain whole grains, or have vitamins and minerals, are responsible for weight gain because of the way they affect blood sugar and the hormone insulin. Insulin regulates blood glucose levels and is released by the body after a person eats carbohydrates. The diet contends that insulin, not the types or quantity of foods, leads to an imbalanced metabolism and, ultimately, to fat storage. As a result, the diet recommends limiting breads, pasta, rice, potatoes, fruit, and starchy vegetables (corn, peas), and sugar filled sweets. The diet replaces them with high protein foods such as beef, chicken, pork, eggs, and fish and high-fat foods, such as heavy cream, butter, salad dressing, and oils because protein and fat do not stimulate the release of insulin.

Atkins diet is split in 4 phases:

Phase 1 (Induction)

In this first phase of two weeks, one has to restrict carbohydrate intake to ≤ 20 grams per day. Eat high fat, high protein and low carbohydrate-containing some vegetables and leafy greens such as asparagus, broccoli, cucumber, green beans, peppers only. This kick-starts the weight loss. One can lose 6.8 kg during this phase.

Phase 2 (Balancing)

After two weeks, one can increase the carbohydrate intake by 5 grams per week on foundation vegetables until weight loss stops. To achieve it, slowly add some nuts, berries and small amount of fruits and vegetables over time.

Phase 3 (Fine tuning)

When one is very close to goal weight, add more carbohydrates like fruits, starchy vegetables, whole grains *etc.* to diet until weight loss slows down.

Phase 4 (Maintenance)

Start to increase healthy carbohydrate intake by 10-gram increments as weight maintenance is sustained and body can tolerate without regaining weight.

One of the biggest benefits of low-carbohydrate (high-protein) diets such as Atkins diet is rapid initial weight loss. This is due to depletion of glycogen and associated water stores resulting in a loss of 1-3 kg within the first few days. Although this is not fat loss, fast weight loss of any kind, may be of muscle often appears encouraging. In fact, compared to a traditional low-fat, high-carbohydrate, low-calorie diet, these diets have been shown to result in greater weight loss in studies lasting less than six months. In a few studies carried out for one year, dieters on the Atkins diet lost more initial body weight compared to the low-fat group and at the end of six months. Although this has been attributed to the induction of ketosis (rapid fat breakdown) by Atkins and other low carbohydrate diet programs, research does not find a link between the degree of ketosis and weight loss. Weight loss on low carbohydrate diets has been associated with eating less food and kilojoules. This has been attributed to the novelty, simplicity and monotony of the diet supporting tight compliance to the program in the initial phase. There is also evidence that the higher protein content of the diet enhances feelings of fullness or satiety. This is one reason that they have been so successful. At one year, however, the Atkins diet group gained back more weight compared to the low-fat diet group with the end weight loss being similar for the two groups at the end of the studies. Many scientists argue that the short-term greater weight loss for individuals on the Atkins diet is due to fewer calories, not necessarily because of the lack of carbohydrates.

Although low-carbohydrate (high-protein) diets typically include foods high in saturated fat and cholesterol, to date, studies have indicated that they do not positively or negatively change total cholesterol, LDL cholesterol, or HDL cholesterol in short-term studies. These diets do have a positive impact on triglycerides. Levels went down in all reported studies. High triglycerides in combination with low HDL cholesterol levels are risk factors for heart disease. In this case, weight loss, not the saturated fat and cholesterol in the diet, may be the reason that cholesterol is not changed. It is unclear what would happen to cholesterol levels if the diet were followed over an extended period of time.

Foods to be Avoided when on Atkins Diet

- ☆ **Sugars:** Soft drinks, fruit juices, cakes, confectionery, ice cream
- ☆ **Grains:** Wheat, rye, barley, rice
- ☆ **Vegetable oils:** Soybean oil, corn oil, canola oil, cottonseed oil *etc.*
- ☆ **Trans fats:** Usually found in processed foods *e.g.* hydrogenated fat
- ☆ **'Diet' and 'low-fat' foods:** These are usually very high in sugar.
- ☆ **High- carbohydrate vegetables:** Carrots, turnips *etc.* (induction only)
- ☆ **High-carbohydrate fruits:** Bananas, apples, oranges, pears, grapes (induction only)
- ☆ **Starches:** Potatoes, sweet potato (induction only)
- ☆ **Legumes:** Lentils, beans, chickpeas *etc.* (induction only)

South Beach Diet

The South Beach Diet was created in the mid-1990s by Dr. Arthur Agatston, a Florida-based cardiologist. Similar to Atkins, the South Beach Diet also focuses on removing "carbohydrate cravings" and promoting weight loss by removing carbohydrates and allowing high-protein, high fat foods during the first two weeks. It is a low-carbohydrate diet that emphasizes lean meats, unsaturated fats and low glycemic index carbohydrates. Unlike the Atkins diet, the South Beach Diet does not limit carbohydrate intake or allow unlimited consumption of high-protein, high-fat foods after the first two weeks. Instead, carbohydrates are "re-introduced" with a modified diet that includes small portions of carbohydrates, such as wholegrain products (breads, cereals, and pastas), that are high in fiber and have a "low-glycemic index." The diet also encourages proper portion sizes and high-fiber vegetables and nuts. It makes no restriction on calorie intake, includes an exercise program, and is based around taking three main meals and two snacks per day. This diet has been promoted as improving risk factors associated with cardiovascular disease but the effectiveness for improving these risk factors is unclear because no evidence on its effects is available.

The diet has three phases. Phase I is most restrictive; no intake of bread, rice, potatoes, pasta or fruit, fruit juices, starchy foods, dairy products, alcohol *etc.* One can take a lot of protein, such as lean meat, poultry, seafood, eggs and cheese, some fats, including canola oil, extra-virgin olive oil and avocado and carbohydrates with the lowest glycemic index such as tomatoes, spinach, brinjal, broccoli *etc.* About 5-6 kg weight loss is expected in this phase. In phase 2, some of these foods such as healthy carbohydrates (whole grain bread, whole wheat pasta, sweet potatoes, fruit *etc.*) can be gradually added. One should expect weight loss to slow to 0.5 g to 1.0 kg a week on average. As the amount of carbohydrate increase, proportionately, the amounts of protein and fat are decreased. In phase 3, when goal weight is achieved, one can stay there for life. In case of carbohydrate craving, one should go back to phase 1 or 2 depending upon the situation.

Disadvantages and Risks of Low Carbohydrate Diets

One of the biggest concerns with restrictive, low-carbohydrate diets is the elimination of certain foods and food groups. There are definitely some foods in each food group that offer more nutrients than others. There are numerous studies showing that healthy choices and sensible portion sizes from all of the food groups of the Food Guide Pyramid, combined with physical activity, help achieve a healthy weight and promote lifelong health. Protein and fat alone cannot provide with all the nutrients one requires for good health, which is not necessarily the same thing as weight loss. Vitamin and mineral supplements taken along these diets can be just that "supplements" and should not be used as replacements. They do not offer the same benefits as vitamins and minerals obtained from "real" natural foods. Following is a list of some of the nutritional benefits that are lost as a result of restrictive carbohydrate diets:

- Whole cereal grains are excellent sources of fiber and B vitamins, like folic acid, which help promote heart health and prevent childbearing women from having babies with neural tube defects.
- Carbohydrates are the best source of energy for brain and body, especially for growing children and active adults. Protein and fat are not as efficient sources of energy; they need to be converted into energy units that one can use. Many people complain about being fatigued and getting headaches on low-carbohydrate diets because of this. Diets low in carbohydrates can cause muscle weakness, confusion, dizziness, tiredness, and dehydration–making physical activity less enjoyable for sports persons.
- These diets are not appropriate during pregnancy and lactation or in childhood or the pre-teen or teen years. During these periods of life, growth is very rapid and the body requires calories and several nutrients, such as calcium, folate, and vitamin A, to promote optimal development. Restricting calories and these nutrients could lead to poor growth during these critical years. These diets may hinder the production of breast milk during lactation
- Fruits and vegetables are rich sources of fiber, vitamins, minerals, antioxidants and phytochemicals that help protect against diseases. But low-carbohydrate, high-protein diets severely limit the amount of fruits, vegetables, and whole-grain foods which are 'allowed' to eat whereas these plant-based foods are the best choices for prevention of many diseases.
- Dairy products are excellent sources of calcium and vitamin D for strong bones and teeth, but because they also contain natural sugars, they are also avoided in many diets.
- Nuts provide healthy sources of fat, such as omega-3 fatty acids, which help protect against heart disease.
- Although many of the low-carbohydrate diets stress the importance of eating proteins and not worrying about fat, it is important to distinguish between healthier and unhealthier choices. High-fat meats, heavy cream, butter, and gravy contain saturated fat and cholesterol, which can contribute to an elevated risk of heart disease.

One should avoid saturated fat and try to incorporate healthy fats.

- Constipation is a common problem for people following low carbohydrate, high protein diets. Hence, The Atkins diet is not recommended for individuals with chronic gastrointestinal (GI) diseases such as Crohn's, irritable bowel syndrome (IBS), or diverticulitis. Persons with GI diseases need dietary fiber to maintain regular bowel function and to prevent flare ups.
- If a person already has poor kidney function as a result of untreated or poorly controlled hypertension, for example, then a high protein diet may cause further kidney damage and is not recommended.
- Lack of variety in food choices and boredom, particularly with the Atkins diet, is a common reason as to why individuals are unable to remain on the low-carbohydrate diets and keep the weight off for the long-term.
- Sudden cardiac arrest has also been reported among a few individuals on these diets.

If someone chooses to go on a low carbohydrate diet, do so for only a short period of time. Then, incorporate whole-grains, fruits, and vegetables to maintain weight, to increase the variety in the diet, and to minimize any potentially negative health effects like cancer or osteoporosis.

III. High Protein, Low Carbohydrate and Low Fat Diet (Dukan Diet)

This diet was created by Dr. P. Dukan, a French general practitioner in 1970. It is a multi-step high protein, low carbohydrate and low fat diet. The weight loss phases are primarily based on eating unlimited high protein foods and mandatory oat bran. Lean protein, oat bran, water and a daily 20 minute walk are the basic aspects of the plan. The theory is that limiting carbohydrates forces the body to burn fat. Basically, one can eat unlimited quantities of food, as long as they are on the approved food list, which includes very few carbohydrates, if any.

This diet consists of four phases *i.e.* two weight loss and two maintenance phases. How long one stays in each phase depends on how much weight one need to lose and each phase has its own dietary pattern. First phase is known as 'Attack Phase' which consists of pure protein and creates a kick-start to the diet. During this phase, one can eat unlimited amount of 68 high protein foods and 1.5 tablespoons of oat bran that produce immediate and noticeable weight loss. Second phase is known as 'Cruise Phase' to reach one's true weight. This phase adds 32 non-starchy vegetables out of 100 natural foods which can be taken to attain true weight. One will gradually but steadily lose weight by alternating pure protein days and protein + non-starchy vegetable days plus 2 tablespoons of oat bran every day. The average length of this phase is based on a schedule of 3 days for each 500 g one wants to lose. The third phase is known as 'Consolidation Phase' to prevent the rebound effect. During this phase, previously forbidden foods are added in limited amounts and allowing upto two 'celebration' meals per week. Unlimited lean protein and vegetables, some carbohydrates and fats, one day of lean protein per week and 2.5 tablespoons of oat bran per day is followed. One should follow this dietary pattern for 5 days for every 500g weight lost in phases 1 and 2. The fourth phase is called 'Stabilization Phase' which is for rest of the life. Three simple but non-negotiable rules are to be followed *i.e.* three tablespoons of oat bran daily; walk 20 minutes every day and choose to take stairs whenever possible; have a pure protein Thursday *i.e.* Attack phase menu.

Some studies have shown significant weight loss by following this diet but there is little quality research data available regarding the benefits of this diet. The limitation of this diet is that fast weight loss achieved by severe calorie restriction may tend to cause muscle loss along with the fat loss. The loss of muscle loss and severe calorie restriction may also cause the body to conserve energy, making it very easy to regain the weight after losing it.

IV. Paleo Diet

Other names for a Paleo diet include Paleolithic diet, Stone Age diet, hunter-gatherer diet and caveman diet.

A Paleo diet is a dietary plan based on foods similar to what might have been eaten during the Paleolithic era, which dates from approximately 2.5 million to 10,000 years ago. A Paleo diet typically promotes eating whole, unprocessed animal and plant foods such as lean meats, fish, fruits, vegetables, nuts and seeds foods that in the past could be obtained by hunting and gathering. A Paleo diet limits foods that became common when farming emerged about 10,000 years ago. These foods include dairy products, legumes and grains. It avoids highly processed foods, refined sugar, salt, potatoes, dairy products, legumes such as beans, lentils, peanuts and peas and grains like wheat, oats, barley *etc.* Unlike most diets, a Paleo diet does not involve counting calories. Instead, it restricts the foods which are major sources of calories in the modern diet. On an average Paleo diet provides between 25–35 per cent calories from protein and reduces carbohydrate intake because of eliminating common sources of carbohydrates such as rice, potato, bread, sugar *etc.* Reducing carbohydrate content is the best way to reduce calorie intake and hence, weight loss.

This diet can improve health by eliminating high-fat and processed foods that have little nutritional value and too many calories. This plan emphasizes loading up on fruits and vegetables that are rich in vitamins, minerals, and fiber and therefore, give a feeling of fullness faster, so one eats less, helping to curb weight gain. Moreover, processed foods are generally energy dense and they are avoided in this diet, hence leading to less calorie intake. Instead, it encourages eating lean sources of protein, fresh fruits and vegetables and healthy fats, which are lower in calories and rich in nutrients. But the primary concern is that due to absence of whole grains, legumes and dairy products, many of essential nutrients and fibre are not available. For some people, a Paleo diet may be too expensive.

A number of randomized clinical trials have compared the Paleo diet to other eating plans, such as the Mediterranean Diet or the Diabetes Diet. Overall, these trials suggest that a Paleo diet may provide some benefits when compared with diets consisting of fruits, vegetables, lean meats, whole grains, legumes and low-fat dairy products. These benefits may include more weight loss, maintenance of weight loss, improved glucose tolerance, lower triglycerides, better appetite management and better blood pressure control. However, longer trials with large groups of people randomly assigned to different diets are needed to understand the long-term, overall health benefits and possible risks of a Paleo diet.

V. Structured Weight Loss Plans

These regimes are designed by some of the healthcare professionals, dietitians, nutritionists *etc.* and provided to the participants through their centres, social media, e-mail *etc.* on daily, weekly or monthly basis. Such weight loss plans are to be followed for a limited period of time during which a detailed, prescribed menu is followed. Ideally, they should provide approximately 600 kcal per day less than the pre-dieting intake and diet given should include adequate nutrients from various food groups.

Some of the advantages for some individuals are that it is convenient and comfortable for them having a structured eating pattern especially if it is accompanied by weight loss without detrimental effect on general health. In some such programmes, regular support meetings of the participants are held at their centres where participants are weighed and motivated by their leader and fellow weight reducers. However, some of the disadvantages of structured weight loss plans are:

- Rigidity of some regimes;
- Hard to follow, particularly in the context of a busy family or professional life;
- Extra expenditure;
- Designed generally for short term weight loss; and
- Limited contribution to long term healthier eating practices if long term weight maintenance diets are not prescribed.

Examples of some of these weight loss plans are the following:

Calorie Counting

The basis of this simple concept is taking fewer calories than the energy expenditure so as to have a negative energy balance, which is required for the weight loss. Therefore, to lose weight, keep a modest and sustainable calorie deficit in the regular diet.

Step I. Calculate the Calorie Need

To start with, one should know the daily calorie intake target which can be calculated on the basis of body weight, height, activity level, gender *etc.* There are many free websites too from where one can get calorie need charts or with the help of calorie counter, one can know his or her daily calorie requirement. One can calculate calorie needs using the equations based upon total energy expenditure also.

Total Energy Expenditure (TEE) includes two major parts:

i) Resting Energy Expenditure (REE): the amount of calories needed to maintain basic body systems and body temperature at rest.
ii) Activity Energy Expenditure (AEE): the amount of calories used during activity.

Maintenance or change in body weight is summarized in the formulas below:

- **Weight maintenance:** Calorie intake = TEE
- **Weight gain:** Calorie intake > TEE
- **Weight loss:** Calorie intake < TEE.

The number of calories needed to maintain a certain body weight can be estimated by multiplying a person's REE times an appropriate Activity Factor (AF) *i.e.* 1.5 for women and 1.6 for men for doing sedentary activity.

Adult Male: REE Calories = 10 x (weight, kg) + 6.25 x (height, cm) – 5 x (age, yr) + 5.

Adult Female: REE Calories = 10 x (weight, kg) + 6.25 x (height, cm) – 5 x (age, yr) -161.

REE x AF= Estimated total energy need (kcal/day) to maintain weight.

Step II. Calculate the Daily Calorie Deficit

Ideally, one should set a goal of approximately 10 per cent weight loss from baseline over a period of six months. After achieving this goal, further weight loss can be planned.

Daily calorie deficit through diet is calculated on the basis of the fact that one pound (about 454 g) of body fat is equivalent to 3,500 kcal and to lose one pound or about 500 g body weight per week, one will have to take 500 kcal less in the diet than the calorie need per day. To reduce about one kg body weight per week, one should take 1000 kcal less in the daily diet. Modest weight loss *i.e.* 500 g-1kg per week is generally recommended and to attain that one will have to take 500-1000 kcal less than the RDA daily depending upon the set target of weight loss.

Example

Expected weight loss= 500 g per week

RDA for sedentary adult male=2320 kcal per day

1 g of body fat= 7.7 kcal

500 g body fat= 3850 kcal

Less calories per week=3850kcal

Less calories per day= 3850 ÷ 7=550 kcal

Prescribed diet for daily calories for weight loss (500 g/week) = 2320-550=1770 kcal

Hence, for an average adult man whose daily RDA for energy is 2320 kcal, if he has to lose 500 g body weight per week, he will have to follow daily 1800 kcal dietary regimen. With decrease in body size, further calorie reduction in daily diet will be required and should be calculated. Calorie intake below 1500 kcal/day is not adequate in most of the nutrients and individual will have to take vitamin and minerals supplements along with the diet. Low calorie diets provide 800-1600 kcal/day. Very low calorie diet should not be followed as it would be devoid of many essential nutrients. Monitoring of weight loss can be done weekly and monthly. In the first month, due to water loss and utilization of liver glycogen stores, one loses about 3kg.

After setting the target for weight loss, one should count and measure food calories. One can use an app or websites, devices like caltrack, food composition tables *etc.* Information on food labels should be read. In broad term, one should know that fat rich foods are concentrated sources of energy and therefore, the amount of those foods should be limited or those foods can be avoided too. Similarly, sugar rich foods provide empty calories and their intake should be restricted to lower calorie intake. Eventually, one starts remembering how many calories are there in the most common foods one eats daily. One should keep in mind that one has to eat within the calorie intake target set for the day but at the same time, daily meals should provide adequate amount of protein, vitamins, minerals and fibre.

Very Low Calorie Diets

Very low calorie diets (VLCDs) are defined as diets providing 800 kcal/day. Some diets provide only 400 Kcals/day. Zero calorie diets are also included. According to Breatharian diet, no food is consumed on the basis of belief that food is not necessary for human subsistence. There is another KE diet in which individual is fed through feeding tube and nothing is given orally.

Healthcare professional may recommend a very-low-calorie diet if the individual needs to lose weight quickly (for example, if someone is going to have joint replacement surgery), but these diets are not used for routine weight loss. One should never follow a diet that provides only 800 kcal or less a day unless this is being closely monitored by doctor.

According to National Institute for Health and Care Excellence (NICE), a diet of 800 Kcal/day should be followed for <12 weeks or intermittently, *i.e.* 2–4 times per week, interspaced with a diet providing 1000–1600 Kcal/day if the person is obese and has reached a plateau in their weight loss. Commercial VLCDs are usually liquid, milk-based drinks or soups, which aim to provide the consumer's total nutrient requirements within a day's limited energy format. Comparable diets can be made at home from milk but require adequate micronutrient supplementation.

Research studies have reported that VLCDs are effective in bringing about short-term (4–8 weeks) weight loss. These are contraindicated in individuals who have a history of, or are at risk from cardiac disease, cerebrovascular accident, renal or liver disease, hyperuricaemia, porphyria, or psychiatric disturbances and have been associated with severe adverse effects including death.

Intermittent Fasting

The concept of weight loss is the same that it requires consuming fewer calories than those which are burnt but pattern of calorie deficit is different. Intermittent fasting is an eating pattern that involves strategically skipping meals rather than focusing on calories. Calorie counting is generally not required when doing intermittent fasting, the weight loss is mostly mediated by an overall reduction in calorie intake. Missing one meal per day by fasting can save around 30 per cent of calories in one go.

There are several types of intermittent fasting one can follow:

- ☆ **The 16/8 or Leangains Method:** Fast for 16 hours per day and eat in the remaining 8-hour window. Most people will skip breakfast and eat from 1pm to 9 pm for example. Others choose to eat a meal in the morning and keep fast going until dinner time.
- ☆ **The 5:2 Diet:** Eat normally for 5 days of the week, but only eat up to 600 calories on the other two days.
- ☆ **Eat-Stop-Eat or Spontaneous Fasting:** Fast for a 12 or 24 hour period multiple times per week. There is no real schedule for this one.
- ☆ **Alternate Day Fasting:** Fast for one day and eat normally the next day. Continue to alternate days of fasting and normal eating.

Research indicates that intermittent fasting causes more fat loss than a very low calorie diet. The basis of this concept is that short-term fasting leads to several changes in the body that make fat burning easier. This includes reduced insulin, increased growth hormone, enhanced epinephrine signaling and a small boost in metabolism. According to a recent 2014 review study, intermittent fasting can lead to significant weight loss. In this review, intermittent fasting was found to reduce body weight by 3-8 per cent over a period of 3-24 weeks.

With intermittent fasting, one still needs to eat healthy and maintain a calorie deficit if one wants to lose weight. Being consistent is absolutely crucial, and strength training is important.

Pharmacotherapy: Management of Obesity

It is not the first line of treatment. Lifestyle modification remains the cornerstone of weight management but pharmacotherapy can be considered for adult persons as a part of comprehensive weight loss programme if patient's BMI is >30kg/m^2 or 28kg/m^2 with co-morbidities *e.g.* diabetes type 2, hypertension *etc.* or if patients have not met their target weight loss or have reached a plateau after losing weight on dietary, exercise and behavioral approaches.

Antiobesity drugs can be considered under two broad categories *i.e.* central-acting appetite suppressants or satiety enhancers which may also have peripheral actions; and peripherally acting agents, for example orlistat, which is an inhibitor of gastric and pancreatic lipases that blocks GI absorption of fatty acids, cholesterol and fat soluble vitamins and increasing faecal fat excretion. RCTs show that it is an effective adjunct to dietary treatment and associated with a weight loss of 2–5 kg greater than that of control groups at 2 years. Orlistat is recommended for long term use. Given the role of the brain in controlling appetite and evidence of high expression levels of 'obesity susceptibility genes' in brain, centrally acting agents appear to hold the most promise. Currently sympathomimetic drugs such as phentermine and diethylpropion are approved only in the USA for short-term (less than 3 months) treatment and thus, do not fit into a rational paradigm for treatment of a chronic disorder. However, the FDA has recently approved the serotonin agonist (lorcaserin) and a combination of low-dose phentermine/topiramate, raising the possibility of improving current paradigms for treatment of obesity. In Europe, both drugs were rejected primarily over concerns that safety had not been demonstrated, a stance that has been criticized. The development of the glucagon-like peptide 1 (GLP-1) analogue, liraglutide, at doses higher than used for managing hyperglycaemia, as an antiobesity agent is also promising.

Drugs have wide ranging side effects and contraindication for its widespread use. Despite the many deficiencies, a progressive approach to obesity pharmacotherapy perhaps offers the best opportunity to finally address the obesity crisis on a mass scale. The use of current antiobesity drugs in isolation is unlikely to be successful and alternative therapeutic paradigms should be developed.

Before Prescribing

Before prescribing antiobesity drugs, physician should discuss the following with patients:

- ✰ The prescribed drug's potential benefits and limitations.
- ✰ How the drug works and possible adverse effects.
- ✰ How they will be monitored while taking the drug.
- ✰ Agreed goal for losing weight.

Care Required whilst Prescribing

- ✰ Patients should be monitored to evaluate the effect of drug treatment and to reinforce lifestyle advice and need for adherence.
- ✰ Review weight loss from time to time.
- ✰ Consider micronutrient supplementation if required.

Stopping Prescription

- ✰ Consider withdrawing drug treatment after 3 months if the person does not lose at least 5 per cent of their body weight.
- ✰ Less strict goals should be used in diabetes type 2.

Surgical Treatment: Management of Morbid Obesity

Gastric bypass and other types of weight-loss surgery, collectively known as bariatric surgery, make surgical changes to the stomach and digestive system that limit how much food one can eat and how many nutrients one can absorb, leading to weight loss. It is the most effective treatment both in terms of amount of weight loss achieved and durability of weight loss or 'weight maintenance', as well as amelioration of obesity-related co-morbidities.

Currently, bariatric surgery is reserved for patients if:

i) BMI is ≥ 40 kg/m^2 (severe or complex obesity);

ii) BMI is ≥ 35 kg/m^2 in the presence of at least one or more obesity-related co-morbidities such as type 2 diabetes, hypertension, sleep apnea and other respiratory disorders, non-alcoholic fatty liver disease, osteoarthritis, lipid abnormalities, GI disorders, or heart disease;

iii) All appropriate non-surgical measures have been tried but failed to achieve a healthy weight loss sustained for a period of time > 6 months with prior weight loss efforts and;

iv) Patient has been receiving or will receive intensive management in a specialist obesity service, is fit for anaesthesia and surgery and commits to long-term follow-up.

It is also recommended as a first line treatment option for adults (instead of lifestyle interventions or drug treatment) if patient's BMI is ≥50 kg/m^2. It is not generally recommended for children or young people. Individual should be at least above 18 y old. Contraindications to bariatric surgery include myocardial infarction, significant chronic obstructive airways disease or respiratory dysfunction, non-compliance with medical treatment and significant psychological disorders.

Types of Bariatric Surgery

All bariatric surgical procedures have to decrease patient's GI capacity for ingesting ± absorbing food, thus decreasing available energy and promoting negative energy balance and weight loss. The procedures being performed today include vertical banded gastroplasty (VBG), laparoscopic adjustable gastric band (LAGB), biliopancreatic diversion (BPD), BPD with duodenal switch (BPD/DS), roux-en-Y gastric bypass (RYGB), long-limb RYGB, and banded RYGB.

The most common types of bariatric surgical procedures used are:

Gastric Band (Laparoscopic)

A band is placed around the stomach, so the patient does not need to eat as much to feel full. It is less invasive than gastroplasty or bypass and can be reversed. It is associated with weight loss (mean 37 kg in 58 per cent of patients at 5 years).

Gastric Bypass (Roux-en-Y)

The Roux-en-Y gastric bypass, often called gastric bypass is considered the 'gold standard' of weight loss surgery. This is the safest and most efficacious bariatric surgery and, thus, is currently the most commonly performed operation, comprising about 70-75 per cent of all bariatric procedures. In this surgery, the top part of the stomach is joined to the small intestine, so one feels fuller sooner and does not absorb as many calories from food. It is associated with significant weight loss (50–75 per cent of pre-surgery weight lost after 5 years).

Sleeve Gastrectomy

Some of the stomach is removed, so one cannot eat as much as one could before and one will feel full sooner.

All these operations can lead to significant weight loss within a few years, but each has advantages and disadvantages. The surgeon will decide what the best is for the patient in consultation with him and his family.

Information before Surgery

Before surgery, the surgeon should discuss in detail with patient and his family about the potential benefits, long-term implications and risks, including peri-operative mortality. He should provide psychological support pre- and post-surgery that would affect adherence of patient to postoperative care, *e.g.* dietary change.

Information after Surgery

The patient should be provided regular information on dietary intake appropriate for the procedure and regarding patient support groups. Monitoring of outcomes, complications, impact on quality of life, nutritional status and co-morbidities should be done from time to time.

Patient's micronutrient status should be monitored. Accordingly, patient should be provided individualized nutritional supplementation, support and guidance for weight loss and weight maintenance.

1. Pre-Operative Nutritional Assessment and Dietary Management

Proper candidate selection, adequate nutritional assessment, and behavioral dietary guidance are essential in preparing for surgery and are key in achieving optimal surgical outcomes. To date, there are no evidence-based or uniform protocols for choosing the most appropriate type of bariatric surgery. For the most part, the type of surgery chosen is determined by the patient's request together with the surgeon's experience, taking into consideration existing co-morbidities. Surgeons should take extra caution when recommending the BPD procedure to their patients due to the greater nutritional risks related to the large malabsorptive component.

Most nutritional evaluations involve a clinical interview that includes many topics related to the surgery to assess bariatric knowledge, surgery expectations, eating behaviors (*e.g.*, number and types of meals per day, liquid intake), and eating patterns (*e.g.*, grazing, binge eating). Weight-management history includes the onset of obesity, family history of obesity, previous weight-loss regimens, and physical activity habits. As part of the preoperative preparation, patients are presented with dietary recommendations for the pre- and post-surgery periods. In addition, psychosocial assessment is conducted to assess mood, social and family support, substance use, cognitive function, psycho-social status, motivation, and willingness to undertake behavioral changes. It is advised to determine the patients' awareness and understanding of the planned operation, potential operational risks, postoperative adverse effects and benefits, and lifestyle changes required to achieve the most favorable outcomes.

Weight-loss expectations should be discussed pre-surgery, and they should be pertinent to the reported average excess weight loss after each procedure.

Unrealistic expectations of weight loss are prevalent in obese patients and can negatively affect their adherence to dietary and health goals. In addition, patients should be encouraged to consider other benefits of the surgery beyond weight loss, including substantial improvement in obesity-related co-morbidities and quality of life.

2. Pre-Operative Dietary Management

Before bariatric surgery, a low-calorie diet is often recommended. However, success in preoperative weight loss should not be a condition for approving surgery. Studies have found a positive correlation between pre-operative and post-operative weight loss. A large-scale study based on data from the Scandinavian Obesity Registry showed that weight loss of 9.5 per cent before RYGB was associated with a marked reduction in important post-operative complications, such as anastomotic leakage, deep infection or abscess, and minor wound complications. These findings were even more significant in patients with BMIs >45.8. However, even more modest pre-operative weight loss has been associated with surgical advantages, such as shortening the time of the surgery. Preoperative weight loss leads to an improved glycemic state before surgery. Given the known correlation between pre-operative hyperglycemia and post-operative complications, this finding is highly relevant. Weight reduction before surgery also provides a protective effect in patients who suffer from nonalcoholic fatty liver disease, who represent 90 per cent of patients with morbid obesity.

To date, there is no consensus with regard to the recommended duration of pre-operative diet and its macronutrient composition. When several pre-operative diets were compared, a "low-carbohydrate diet" was found to be more effective than a "low-fat diet" with regard to short-term weight loss, improvement in insulin sensitivity, and reduction in lipid concentrations. The former diet showed a positive effect, especially in patients with metabolic syndrome or nonalcoholic fatty liver disease.

The "very-low-calorie diet," aims to provide patients with low energy intake (450-800 kcal/d), whilst also maintaining a high protein formula, the combination of which results in rapid weight loss, adequate satiety for a few days, and minimizes loss of fat-free mass. Commercially available 'very low calorie diet' formula *e.g.* 'Optifast', a Nestle product is available and is being used in some countries. This product replaces the typical three meals a day, and patients instead consume five shakes per day, which provide about 450 kcal, including 70 g of protein, 15 g of fat, and 100g of carbohydrates, as well as recommended daily allowance of essential vitamins, minerals and trace elements.

The "very-low-calorie diet" has been reported to a 10 per cent pre-operative weight loss, a 9 per cent reduction in BMI, and a 15–20 per cent reduction in liver volume in research studies. It has also found to be a positive factor for improving cardiovascular risk factors in diabetic patients. However, the use of a very-low-calorie diet as a sole method for multiple weeks may induce a catabolic state, which could potentially harm recovery after surgery.

According to the American Society for Metabolic and Bariatric Surgery, the maximum length of a pre-operative diet is 3 months in order to maintain high compliance rates among patients. A longer period may decrease motivation. The specific dietary components should be tailored for each patient by the bariatric medical team. Recommendations for the duration of the diet range from 2 to 6 weeks for the primary aim of reducing liver volume but also to reduce visceral adipose tissue before surgery. Despite the proven short-term benefits of pre-operative weight loss, the evidence on long-term outcomes is still inconclusive.

3. Post-Surgery Nutrition and Dietary Management

In the days immediately following bariatric surgery, the patient has limited ability to eat "normally." Most guidelines recommend a diet that is progressed from clear liquids eventually back to solid food in the days and weeks following surgery. The diet progression may vary based on the type of procedure performed. The goals of nutritional management of bariatric surgery patients is to promote intake of sufficient energy and nutrients to allow healing and preservation of lean tissue during the period of rapid weight loss after surgery and to consume foods that do not cause dumping syndrome, reflux or early satiety, while keeping caloric intake low enough to allow the patient to lose weight. Optimum diet intake will depend on the type of surgical procedure undertaken, *i.e.* capacity-limiting± malabsorptive.

During Post Operative 1-2 Days

A clear liquid diet should be initiated within the first 24 hours after the procedure so as to keep oneself well hydrated. The clear liquids, with no carbonation, no sugar, and no caffeine, should supply fluids and electrolytes with little or no energy. It is thought that a small amount of energy will encourage gut activity. This diet should be continued through days one and two after surgery. It may be helpful to avoid extremely hot or cold liquids until the stomach fully heals, usually around six weeks.

During Post Operative 3-9 Days

Full liquids should be started 3 days after surgery. These full liquids should be sugar free or low sugar. Full liquids include milk, milk products, milk alternatives and other liquids containing solutes. Also, protein supplements may be used in this phase of the diet; mixed with milk or lactose-free milk. The full liquids should have no more than 20 grams of protein per serving. Drink slowly and sip a beverage to avoid being nauseated. The patient should be encouraged to have salty fluids and solid liquids. The assumption with the diet progression is that the patient will continue to consume foods from the previous phase of the diet, in this case, clear liquids. The patient should consume a combined total of 1.4 to 1.8 litre of full and clear liquids, about half clear and half full liquid. The full liquid diet is believed to allow the healing process to continue while providing the protein and calories of a very low calorie diet. A multivitamin supplement regimen is recommended on post-operative day three. The supplement should be a chewable multivitamin, given twice daily. Additionally a chewable or liquid form of calcium citrate plus vitamin D should also be taken.

During Post Operative 10-14 Days

An increase in the total liquid intake (more than 1.4-1.8 litre per day), while replacing the full liquids with ground, diced or pureed protein-rich foods that are soft and moist, are recommended. The pureed foods should range from the consistency of a milkshake to mashed potatoes. The pureed diet allows for the gradual increase in gastric residue and tolerance of solutes and fiber. The pureed foods pass easily through the newly created pouch and into the small intestine. Protein should be the macronutrient of emphasis during this phase of the diet. Patients may only be able to tolerate a few tablespoons of food in a given eating episode, thus food should be taken 4-6 times per day. Restricting beverage intake during meals and for the half hour following an eating episode is also recommended. Beverage intake is thought to speed gastric emptying, and limiting consumption of fluid during meals may limit the sensation of fullness. Using behavioral strategies such as placing food on small plates and using small eating utensils may also help the patient to limit portion sizes.

During Post-Operative Week 3

During the third post-operative week, the patient may begin to consume protein rich foods; meat broth or gravy, *dal* soups, cottage cheese, low-fat cheese, yogurt, scrambled egg, pureed chicken, cooked and mashed fish *etc.* that are soft and moist. Protein supplements added in water or milk can be taken. Women and men require about 50-60g and 60-70 g of protein daily, respectively. Chewable vitamin and mineral supplements should be taken.

During Post-Operative Week 4

At this point, the diet should be advanced as tolerated by the patient. Protein foods should be eaten with priority, but well-cooked, soft vegetables and soft or peeled fruit (without skin and seeds) can also be added as long as the patient is consuming enough protein. Avoid breads. Consume 3 small meals each day. Chew the food thoroughly. Stop eating whenever one feels full. After 30 minutes of each meal, continue taking protein supplements to meet protein requirement. Clear liquids without sugar should be taken at least 30 minute before meal time.

During Post-Operative Week 5

The patient should continue to consume protein and small amounts of fruits or vegetables. Some patients may be able to have salads around one month into their post-operative period. Until the patient is taking at least 60 grams of protein plus fruits and vegetables per day, they should avoid rice, bread and pasta. This reinforces the priority for protein over other macronutrients. Eat three, well balanced meals each day with calorie free fluids between meals. Avoid empty calories. Avoid foods high in sugar or fat that do not contain protein, vitamins or minerals. Continue taking vitamin and mineral supplements.

Foods to be Avoided

Patients should avoid or delay consuming certain foods after surgery which include sugar/sugar containing foods and concentrated sweets, carbonated beverages, fruit juice, high saturated fat, fried foods, highly seasoned or spicy foods, bread, pasta, rice, red meat, raw vegetables, nuts and seeds, popcorn and other fibrous foods. Additionally, caffeine and alcohol should be consumed in moderation.

4. Long-Term Nutritional Requirements and Diet Recommendations

There are multiple challenges to manage the nutritional status of bariatric surgery patients. The nutritional management of bariatric surgery patients does not end with surgery. The actual surgery should be viewed as one point along the "continuum of care" for weight loss. The reduction in the size of the patient's stomach is by design, but the reduced food intake can lead to nutrient deficiencies. For procedures involving malabsorption, this concern is even greater. Patients who undergo either of these procedures must be monitored for life. The procedure results in an iatrogenic disease of the stomach and/or the intestine with digestive and nutritional consequences, namely nutrient deficiencies. Common deficiencies include iron, folate, group B vitamins, and malnutrition. In addition to the nutrient deficiencies potentially caused by bariatric surgery, many patients had nutritionally inadequate diets prior to the surgery (*e.g.*, eating calorie dense foods of low nutritional value, or "food faddism". Such deficiencies are more likely to be exacerbated by bariatric surgery, especially the malabsorptive procedures.

Nutritional Requirements

In order to account for the malabsorptive component of bariatric surgery, and prevent against the nutrient deficiencies, it is important for patients to comply with nutrition and lifestyle modifications, as recommended by the American Association of Clinical Endocrinologists. It is significant to note that these malabsorptive effects following bariatric surgery are more frequent in procedures such as roux-en-Y gastric bypass and biliary pancreatic diversion.

Energy

The calorie needs should be based on height, body weight and age.

Protein

Protein is the most significant component of the macronutrients that needs to be considered in long-term management of bariatric nutrition; 18-25 per cent of patients suffer protein deficiencies after malabsorptive bariatric procedures, compared with only 2 per cent of patients after restrictive bariatric surgery. Protein malnutrition has potential to occur in bariatric surgery due to malabsorption from bypassing a segment of small bowel, or less frequently, from dietary restriction. It is recommended that 1.1-1.5 g/kg ideal body weight/day of protein supply is needed for patients after bariatric surgery. The recommended daily intake of protein is at least 56 grams/day for men and 46 grams/day for women. When considering how this is adjusted after a bariatric procedure, protein requirement during the post-operative weight loss phase should be calculated as 1.2 grams/kg body weight for preservation of fat-free mass. The only exception to this rule is for patients who have undergone biliopancreatic diversion with duodenal switch, as this results in significant malabsorption, and therefore a higher protein intake of 1.5 to 2.0 g of protein/kg body weight per day.

Vitamins

Various vitamin deficiencies are also common after bariatric surgery. Fat soluble vitamins (A, D, E, K) are also a common deficiency following bariatric surgery, particularly in malabsorptive-type procedures such as roux-en-Y gastric bypass and biliopancreatic diversion. Following any type of bariatric procedure, it is crucial that patients continue to meet the recommended daily allowance for fat-soluble vitamins in the general population. With regards to those following significant malabsorptive surgery, if there are no other complications, these patients should be commenced on supplements for all four of the fat-soluble vitamins as there is a high incidence of their associated deficiencies occurring.

When considering water-soluble vitamins of clinical significance, note that vitamin B_{12} deficiency is uncommon in the initial 12 months post-operatively due to existing reserves within the liver. However, it is known to occur as time progresses, and therefore, should be monitored at regular intervals, and treated with supplements if necessary.

Minerals

The most critical mineral deficiency occurring post-operatively with bariatric patients is that of iron. Iron deficiency anemia is a common long-term problem after bariatric surgeries. Its pathogenesis is underpinned by the impaired ability of the intestinal tract to secrete adequate acid to reduce ferrous to ferric iron, as well as the reduced absorption of iron in the setting of a bypass, chronic use of drugs to suppress the secretion of gastric acid, due to anatomic changes after surgery, a low tolerance for red meat *etc.* The literature surrounding this is somewhat controversial despite several studies supporting the long-standing notion that bariatric surgery, particularly malabsorptive bariatric surgery, is linked with depletion of iron stores. Another school of thought, based on recent cohort studies, is that the role of surgery has been over-estimated, or that bariatric procedures may exacerbate an already pre-existing iron deficiency, but do not serve as an initiating factor. Regardless, there is an obvious need to monitor serum ferritin and iron levels in the body, and consider further iron supplementation if necessary.

Supplements

Nutritional complications include anaemia (multiple deficiencies including protein, Fe, Cu, Se, folate and vitamin B_{12}), osteoporosis (protein, vitamin D), neurological symptoms (B vitamins especially thiamine) and cardiovascular disease (folate deficiency leading to hyperhomocysteinaemia). Hence, supplements are an important component of the long term nutritional management process, because nutrient deficiencies can lead to adverse health consequences. Life-long nutritional and metabolic surveillance is required.

Table 25.3: Summary of Standard Supplement Regimen for Prevention of Nutritional Deficiencies

Supplement	*Amount Recommended**
Multivitamin-mineral	1-2 each day Two adult multivitamin-mineral supplements/d (only 1 supplement/d is required after LAGB) containing iron, folic acid, zinc, copper, selenium, and thiamin (200 per cent of the RDA)
Folic acid	400 µg/d should be included in the routine multivitamin; pregnant women or those planning to conceive should take 800–1000 µg folic acid/d included in a multivitamin supplement or separately
Vitamin B_{12}	IM: Either 1000 mcg monthly or 3000 mcg every 6 months after RYGB and BPD or Sublingual: 350 mcg per day or 1000 mcg/week
Iron (ferrous sulfate, fumarate or gluconate)	150-300 mcg per day (for menstruating women) 45–60 mg elemental Fe/d from multivitamin and supplements (after LAGB, iron supplementation is necessary if iron intake from food is insufficient)
Calcium citrate**	600-mg Ca supplement/d (1200 mg/d after BPD) or more
Copper	The routine daily multivitamin should contain 2 mg Cu
Zinc	The routine daily multivitamin should contain 15 mg/d.

Supplement	Amount Recommended*
Vitamin D***	3000 IU/d [as ergocalciferol (vitamin D_2), or cholecalciferol (vitamin D_3)]; titration should be done to reach normal concentrations of 30 ng/mL.
Vitamin C	The routine daily multivitamin should follow the DRI recommendation for vitamin C
Vitamin A****	5,000 to 10,000 units per day
Vitamin E****	400 International units per day
Vitamin K****	1 mg per day

Notes: *Patients with pre-existing nutrient deficiencies, or who develop other nutrient deficiencies after surgery may require additional supplementation. **Daily intake of calcium from food and supplements should reach 1200–1500 mg/d (1800–2400 after BPD). Single doses should not exceed 600 mg and should be separated by ≥ 2-h intervals from iron supplements or a multivitamin that contains iron. *** For treatment of deficiency, 50,000 IU vitamin D_2 or D_3 1 time/wk for 8 wk, followed by maintenance therapy of 1500–2000 IU/d to achieve normal concentrations. **** For Biliopancreatic diversion (BPD) and BPD/duodenal switch (DS) patients.

Source: Dagan, S.S., Goldenshluger, A., Globus, I. Schweiger, C., Kessler, Y., Sandbank, G.K., Ben-Portal, T. and Sinai. T. 2017. Nutritional recommendations for adult bariatric surgery patients: clinical practice. Adv. Nutr. 8 (2): 382-394.

5. Nutrition Therapy for GI Symptoms after Postoperative Surgery

Gastrointestinal symptoms are very common after bariatric surgery, although most of these can be managed effectively with proper nutritional treatment. As a part of nutritional management in these patients, is avoiding the potential complication of dumping syndrome. Dumping syndrome refers to the rapid gastric emptying that occurs after bariatric surgery. Its underlying pathogenesis is related to the delivery of energy-concentrated food to the small intestine, which exerts local osmotic effects and delayed hypoglycaemic events. The resulting clinical symptoms are uncomfortable, of which include vomiting, nausea, abdominal pain, diarrhoea, lethargy, and manifestations of post-prandial hypoglycaemia. This is more prevalent with malabsorptive surgery, as with most complications, however, it is also treatable or preventable with appropriate nutritional modifications which have been given in the Table 25.4.

Table 25.4: Prevention and Treatment Methods of Common Gastrointestinal Symptoms after Bariatric Surgery

Symptoms	Guidelines
Early dumping syndrome	✰ Avoid simple sugars and foods that have a high glycemic index. ✰ Combine complex carbohydrates, protein, and fiber in meals. ✰ Separate liquids from solid foods; a 30-min interval is recommended.
Late dumping syndrome	✰ Add 1 serving sugar/h after meals; one-half cup of fruit juice containing ~10 g sugar is a good option
Diarrhoea and Flatulence	✰ Prefer low-lactose or lactose-free milk. ✰ Reduce fat consumption; customize menus. ✰ Ensure adequate fluid intake. ✰ In acute cases, consider drug therapy and/or probiotics ✰ Eat slowly. ✰ For flatulence, avoid chewing gum and gas-producing foods.
Constipation	✰ Increase liquid consumption: hot and cold drinks in a variety of flavors; avoid carbonated and sugar-sweetened beverages. ✰ Increase consumption of foods rich in fiber, such as fruit, vegetables, and whole grains. ✰ If there is no improvement, supplements or medications should be considered
Dysphagia	✰ Ensure thorough mastication (chewing ≥15 times/bite) and eat slowly (wait a minute between swallows). ✰ Avoid hard and dry foods, such as toast or overcooked meat. ✰ If dysphagia occurs, discontinue eating to prevent regurgitation.
Vomiting	✰ Take small bites, chew thoroughly and eat slowly with a meal duration of ≥15 min. ✰ Separate liquids from solids. ✰ Eat meals at intervals of ≥2–4 h. ✰ Do not limit foods associated with vomiting; it is worthwhile to reintroduce them over time. ✰ Thiamin supplements in case of persistent vomiting. ✰ Monitor hydration and blood electrolytes.
Food intolerance	✰ Patient education and raising awareness of the phenomenon. ✰ Adhere to the nutritional recommendations after surgery. ✰ In cases of food avoidance, it is important to attempt to provide a balanced menu and, if necessary, use dietary supplements to prevent nutritional deficiencies.

Symptoms	Guidelines
Dehydration	☆ Consume ≥1.5 L liquids/d. ☆ Increase fluid intake, when needed, in cases of intense exercise, diarrhea and vomiting, pregnancy, fever, kidney disease, and fasting. ☆ Encourage liquid consumption by varying beverage temperatures and varying flavors, such as adding herbs, lemon *etc.*; avoid carbonated and sugar-sweetened beverages.

Source: Dagan, S.S., Goldenshluger, A., Globus, I. Schweiger, C., Kessler, Y., Sandbank, G.K., Ben-Portal, T. and Sinai. T. 2017. Nutritional recommendations for adult bariatric surgery patients: clinical practice. Adv. Nutr. 8 (2): 382-394.

Post-operative Meal Pattern

There are post-operative recommendations for specific nutrients and eating patterns.

Take Well Balanced Meal

As the patient's level of hunger increases and more foods are tolerated, they should resume a well-balanced, solid food diet. The patient should be sure to consume adequate protein, fruits, vegetables and whole grains. Eat foods high in protein. For optimal fibre intake, more than 5 servings of fruits and vegetables should be taken daily for promoting good colonic function. Carbohydrates (*i.e.*, bread, rice and pasta) should be avoided until the patient is consuming 60 grams of protein per day plus fruits and vegetables. Avoid foods that are high in fat and sugar as they can cause dumping syndrome.

Keep Meals Small

Eat several small meals a day. One can start with six small meals a day, then move to four meals and finally, when following a regular diet, three meals a day. Each meal should include about a half-cup to 1 cup of food.

It has been reported in the literature that higher the number of meals consumed each day, the lower the average monthly weight loss. Recommendations for meal frequency differs in the literature; some recommend eating 4-6 small meals per day vs. three small meals and one or more snacks per day. Both sets of recommendations call for the avoidance of high-energy density foods, to cut down calorie consumption as well as the possibility of dumping.

Control Portion Size

Using small plates and utensils is still recommended for the long term to help with portion control.

Chew Food Thoroughly

The new opening that leads from stomach into small intestine is now very narrow and can be blocked by larger pieces of food. Blockages prevent food from leaving the stomach and can cause vomiting, nausea and abdominal pain. Take small bites of food and chew them to a pureed consistency before swallowing.

Eat and Drink Slowly

Patients should eat slowly and chew food thoroughly to avoid dumping syndrome. To avoid dumping syndrome, take at least 30 minutes to eat the meals and 30 to 60 minutes to drink 1 cup of liquid.

Drink Liquids between Meals

Adequate hydration is essential for patients during the period of rapid weight loss. Patients should consume approximately 1.5-1.9 liters of fluid each day to maintain adequate hydration. But drinking too much liquid at or around mealtime can leave one feeling overly full and may prevent from eating enough nutrient-rich food. It is recommended that patients do not ingest liquids within 30 minutes of a meal, or during meal times. Wait 30 minutes before or after each meal to drink liquids. Avoid carbonated beverages.

Take Recommended Vitamin and Mineral Supplements

After surgery, the body would not be able to absorb enough nutrients from the food. Take a multivitamin mineral supplement every day.

IV. Long Term Nutrition Follow Up

For assessment of nutritional status, patients should undergo laboratory surveillance every 3-6 months for the first year after surgery and then annually thereafter. The tests should include a complete blood count (CBC), platelets, electrolytes, glucose, protein levels, iron studies, ferritin, serum folate, levels of vitamins A, D, E, K, vitamin B_{12} in the blood, liver function, lipid profile, 25- hydroxyvitamin D (with optional intact parathyroid hormone (PTH), thiamine, red blood cell count, and folate). In patients with RYGB, BPD or BPD/DS, bone density measurements with use of dual-energy X-ray absorptiometry (DEXA) may be indicated to monitor development or presence of osteoporosis at baseline, and follow-up at approximately 2 years. Patients found to have severe nutritional deficiencies or malnutrition should be admitted to the hospital for nutritional support.

E. Management of Obesity: Physical Activity

Physical activity and exercise are sometimes used interchangeably, however the terms have different definitions. "Physical activity" refers to any body movement that burns calories, whether it is for work or play, daily chores, or the daily commute. "Exercise," a subcategory of physical activity, refers to the planned, structured, and repetitive activities aimed at improving physical fitness and health. Researchers sometimes use the terms "leisure-time physical activity" or "recreational physical activity" as synonyms for exercise.

Worldwide, people are less active today than they were decades ago; physical activity associated with work, home, and transportation has declined due to economic growth, technological advancements, and social changes. People are following sedentary lifestyle activities *e.g.* watching TV, using more time while working on the computer and mobile phones, playing video games *etc.* and spending less time on outdoor games and activities. At the same time, energy input is more, hence globally, the more 'energy in' and less 'energy out' has tilted the energy balance equation towards weight gain and obesity.

Increasing energy expenditure through regular physical activity when combined with a reduction in energy intake is an effective way of improving health, reducing weight and preventing weight gain. Any movement *e.g.* doing household chores, walking instead of using a car or motorcycle, cycling, swimming, dancing, gardening, hiking, yoga, weight lifting, push-ups, playing outdoor sports, climbing stairs instead of using elevators *etc.* count towards energy expenditure. It increases lean body mass too. It slows down the development of abdominal obesity. Even if no weight is lost, additional benefits associated with regular exercise include increased sense of well-being and reduced risk of ill health including diabetes, cardiovascular disease, and some cancers. It not only improves physical health but also improves mental health. Physical activity reduces depression and anxiety and this mood may motivate people to stick with their exercise regimen over time. Systematic reviews have indicated that physical activity combined with dietary approaches is more effective than either intervention alone. The potential synergy between the two has been reported, and is believed that if weight loss through dietary restriction results in reduction in energy intake, while physical activity increases energy expenditure, the combination of the two could potentially lead to reduction in body mass, without subsequent reduction in resting energy expenditure thus, improving the ability to achieve long term weight loss.

Guidelines for Activity Prescription and Lifestyle Changes

Overweight or obese adults should be encouraged to perform at least 30 minutes of at least moderate-intensity physical activity on 5 or more days a week. The activity can be undertaken in one session or several lasting 10 minutes or more. For those who have already achieved this level of activity, an increase in the amount of their physical activity is a reasonable component of any strategy to lose weight. It is generally estimated that 0.450 kg of body fat loss requires about 3,500 kcal of energy consumption. As studies on the effect of prescription of muscle strengthening exercise on weight loss are limited, it seems reasonable to recommend overweight/obese people to follow the same muscle strengthening exercise recommendations for the general population for having a balanced exercise programme.

Exercise programs for patients who are overweight or obese should meet criteria for safety, effectiveness, and education/motivation. On the basis of the preparticipant evaluations, individually appropriate "prescriptions" for aerobic activity, resistance (strength) training, and daily lifestyle activities should be developed with particular attention to modifying exercise for conditions that may worsen during certain activities. A summary of main components of exercise regime for overweight and obese persons (FITT) has been given below:

Main Components of the Exercise Prescription for Overweight and Obese Persons
The FITT Principle*

Component	*Recommendation*
Frequency	✰ ≥ 5days a week of aerobic exercise. ✰ More frequent exercise is desirable, but care should be taken to first establish a regular exercise habit before recommending levels that may not be sustainable in the long run. ✰ Appropriate warm-up and cool-down periods should be emphasized.
Intensity	✰ To avoid musculoskeletal injuries and promote compliance, start at a low to moderate intensity† and gradually progress over the course of several weeks or months (4-12 weeks) to more vigorous efforts (if desired by the patient). ✰ Some individuals may prefer doing vigorous exercise as it is less time consuming, but vigorous exercise is probably not appropriate for the very obese (BMI > approximately 35 kg/m^2).
Time	✰ Emphasis should be on increasing duration rather than intensity, with the goal of optimizing energy expenditure. ✰ To prevent obesity, people may need 45 to 60 minutes of moderate intensity aerobic activity each day unless they also reduce their energy intake. ✰ Patients who are overweight or obese should be prescribed a volume of 45 to 60 minutes of moderate-intensity activity† a day (corresponding to approximately 225 to 300 minutes/week of moderate-intensity physical activity or lesser amounts of vigorous physical activity). ✰ People who have been obese and have lost weight should be advised they may need to do 60 to 90 minutes of activity a day to avoid regaining of weight. ✰ Sedentary individuals should build up to their physical activity targets over several weeks, starting with 10 to 20 minutes of physical activity every other day during the first week or two, to minimize potential muscle soreness and fatigue. ✰ Multiple short bouts produce benefits as a single long bout of the same total duration.

Component	Recommendation
Type	✰ Low impact activities (*e.g.* walking, cycling, low impact aerobics, water exercise) that are convenient, accessible, and perceived as enjoyable by the participant. ✰ Walking is an excellent form of physical activity for overweight and obese people (for obese, sedentary individuals, brisk walking often constitutes moderate-intensity physical activity.). ✰ Weight-bearing physical activity may be difficult for some individuals with BMI over approximately 35 kg/m^2, particularly for those with joint problems. For these individuals, gradually increasing non-weight bearing moderate-intensity physical activities (*e.g.* cycling, swimming, water aerobics, *etc.*) should be encouraged.

FITT: Frequency, intensity, time and type of exercise.

*Components of the exercise prescription to improve cardio-respiratory fitness in healthy, overweight, or obese adults. Specific advice on the components of frequency, intensity, time, and type of exercise provides a framework of evidence-based recommendations for a safe and effective exercise program for overweight and obese adults.

†: An example of moderate intensity can be estimated using 55 per cent to 70 per cent of the age-predicted maximal heart rate (*i.e.*, 220 - age × 0.55 – 0.70) or a rating of perceived exertion of 3 to 5 on a scale of zero (easiest) to 10 (hardest) effort

Source: Kyle, J.M. *et al.*, 2003. Counseling for physical activity in overweight and obese patients. Am Fam Physician. 67(6):1249-1256.

Management of obesity. A national clinical guideline. Edinburgh (Scotland): Scottish Intercollegiate Guidelines Network (SIGN); 2010 Feb.

Besides following regular exercise regime, obese persons should also adopt certain modifications in their daily lifestyle activities to increase energy expenditure. Physical activity should be incorporated into the daily routine through leisure activities, domestic work or travel. Obese persons should prefer climbing stairs instead of using escalators, walking greater distances, gardening, house cleaning, spending less time on computers and watching TV; all these changes in daily routine make a significant change in weight loss ultimately.

Behaviour Strategies to Promote the Adoption and Maintenance of Physical Activity

An emphasis on the health benefits of regular physical activity can be incorporated into effective counseling sessions for overweight and obese patients. Individually adapted behavior change is critical to facilitate a physically active lifestyle, but the process involves a multitude of complex variables, including personal, programmatic, social, environmental and related factors. To achieve long term changes in health-related behaviors, these and medical factors must be addressed collectively. Study have shown that physician intervention (*i.e.*, discussing the benefits of and barriers to physical activity, and patient's preferences and practices) need not take more than three to five minutes during an office visit and can play a critical role in implementation by the patient.

Following behavior management strategies should be adopted by the counselors, exercise trainers, physicians and others in the weight loss support team for promoting physical activity among the obese persons:

- ✰ Establish realistic expectations.
- ✰ Correct overly pessimistic or optimistic expectations.
- ✰ Allow patients to select their own goals and provide guidance in setting flexible, short term goals.
- ✰ Ask the patients as to what rewards will work for them.
- ✰ Use environmental cues or stimuli to remind patients of their exercise commitment.
- ✰ Establish routine times and places for exercise.
- ✰ Discuss the advantages and disadvantages of exercise.
- ✰ Discuss the benefits of social support to establish the exercise habit.
- ✰ Encourage the patient to practice self-reinforcement, including a focus on increasing self-esteem and realizing the health benefits of exercise.
- ✰ Prepare patients for situations that may create a relapse and a challenge to exercise.
- ✰ Discuss coping strategies that can help in the prevention of a complete relapse.

Underweight

Criteria of Assessing Underweight

When the body weight is less than 10-20 per cent of the average expected for height, age and sex, the person is under weight. If the individual's body weight is less by 20 per cent or more than Ideal Body Weight, that person is grossly underweight.

It is usually related with Body Mass Index (BMI). If the Body Mass Index (BMI) is less than 18.5 or a body weight below 15- 20 per cent for their age and height, they are under weight. This indicates that woman aged 25-50 years with the average height of 160 cm, the weight should be 59 kg; if it is below this weight, the person is considered as underweight.

But children and teenagers from ages 2 to 20 are still growing, their BMI scores are compared to others of the same age and gender on gender-specific BMI-for-age per centile growth charts of WHO.

Factors Contributing to Underweight

There are a variety of reasons why a person may be underweight. Sometimes, multiple underlying causes may be related. Causes of being underweight include:

Family History

Some people have a naturally low BMI due to physical characteristics that run in their family.

Poor Food Availability

It can be due to many reasons *e.g.* frequent famine conditions or due to other natural calamities/disasters, poor crop production and less food availability to the population especially to growing children, can lead to less and nutritionally poor diet which may not be adequate for normal health and well being and may cause undernutrition and underweight.

Socio-economic Status and Cultural Factors

The persons who are poor and are from low socioeconomic status are generally at risk of being underweight due to poor household food security. Contrast to this those who belong to high socio-economic status are relatively more likely to be over-nourished.

Besides socio economic status, poor cultural and some religious beliefs area are responsible for poor intake of certain nutritious foods. For example, the pure vegans, strictly avoid dairy products and eggs which are rich in protein. Unfortunately, the type of protein that cereal contains does not parallel to the proteins that animal product contain. Keeping fasts on many days a month, following certain food fads and fallacies during pregnancy, lactation *etc.* are also responsible for less intake of certain nutritious foods which have impact on the birth weight of the infant. Child nutrition is also different from family to family depending on the mother's characteristics, household ethnicity and place of residence.

Illiteracy and Ignorance

Due to illiteracy and lack of nutrition education, food taken by the families or individual is not nutritious and adequate.

Imbalanced Food Consumption

It indicates that when the quantity as well as quality of foods is less than the nutritional requirements, may be due to any reason, a person may end up with underweight.

Physical Illness or Chronic Disease

Some disease types can cause regular nausea, vomiting and diarrhoea, making it difficult to gain weight. Other conditions may decrease a person's appetite, so they do not feel like eating. Examples include cancer, diabetes, overactive thyroid disorders, celiac disease, TB, Crohn's disease or ulcerative colitis, HIV/AIDS *etc.*

Enzyme Deficiencies

Due to certain diseases, deficiency of digestive enzymes or stomach acid may also prevent or reduce digestion and absorption of food which causes weight loss.

Psychological Causes

Poor mental health can affect a person's ability to eat *e.g.* depression, anxiety, excessive stress, obsessive-compulsive disorder and eating disorders. Each of these conditions can affect a person's body image and appetite.

Depression is the major important causes of underweight other than anorexia and/or bulimia. When the patients are suffering from depression, they have low appetite, and their food intake will be very less which results in weight loss at an alarming rate.

Frequent Excessive Physical Activity

Unlimited exercise without taking the appropriated diet to the increased energy demands shall lead to loss of weight. Athletes or people who engage in high levels of physical activity, such as runners, may burn significant amounts of calories that may result in low body weight.

A High Metabolism

If a person has a high metabolism, they may not gain much weight even when eating high-energy foods.

Medication

Some medication may suppress appetite or actually cause weight loss.

Health Related Consequences of Underweight

Although overweight and obesity remain a growing concern, addressing underweight is still an unfinished agenda in India, especially in rural settings; this poses a grave challenge to India's public health-care system due to the following facts:

- The association between lower body mass index is associated with increased risk of excess mortality, poor cognition, disability, poor self-rated health and health-related quality of life leading to decreased national productivity and economic growth.
- Being underweight can lead to chronic conditions like shortness of breath, easily fatigued, lower productivity, reduced physical capacity, poor stamina, easy fractures and bruises, anaemia, and increased risk of bacterial and fungal infections.
- Being underweight can significantly affect the cognitive abilities and thus, increase the risk of dementia, which affects 50 million people worldwide and is estimated to triple by 2050. In India alone, there are over 4 million people with dementia. The deficiency of vitamin B_{12} and vitamin B_6 can be cited as one of the reasons why

underweight people could be at higher risk of dementia.

- ☆ Being underweight also leads to low haemoglobin count in the blood. Haemoglobin is essential for carrying oxygen to the brain, and being low on this parameter results in poor oxygenation of the brain, leading to regular episodes of dizziness, easy fatigability and cognitive decline. Persistent low haemoglobin in people over 40 years of age may lead to early dementia or cognitive decline.
- ☆ Since the immunity of underweight people is comparatively low, they take more time to recover from infections too. Underweight among children and adolescents is associated with higher risk of infectious diseases.
- ☆ Being underweight results in growth retardation and failure to thrive in growing children.
- ☆ An underweight adolescent's first menstruation period may be delayed by about 1-2 years or absent. Irregular or absent menstruation can cause infertility. For girls of childbearing age, underweight is associated with adverse pregnancy outcomes including maternal mortality, delivery complications, preterm birth, intrauterine growth retardation and low birth weight babies.
- ☆ Adult women who are underweight may not have regular periods, they may find early amenorrhoea. An underweight woman with short stature, iron or protein deficiency, or other nutrient deficiencies has a greater risk of adverse pregnancy outcomes such as obstructed labour or postpartum haemorrhage, and of giving birth to a baby with low birth weight and ill health.
- ☆ Underweight among men is indicative of poverty, food insecurity and inadequate health care

Management of Underweight

Goal: To restore the optimum body weight for age and height and good nutritional status.

1. Medical Treatment

First of all, the cause of underweight should be determined and then eliminated so as to improve appetite and intake of balanced diet. If low appetite, nausea, vomiting, poor digestion and absorption of food is because of some physiological condition, take medical help for the eradication of its cause.

2. Nutritional Requirements

The diet planned should be a combination of liberal amounts of carbohydrates, protein and healthy fat with sufficient amounts of vitamins and minerals taking into consideration the prevalent micronutrient deficiencies present in the underweight individual.

Energy

Recommended energy intake varies with the gender, age group, physiological condition and physical activity (ICMR, 2010). Besides it, additional 500 kcal per day should be given daily so as to increase the energy intake for weight gain (0.5 kg/week). For increasing 1.0 kg/week, one will have to take additional 1000 kcal/day. Increase in energy intake should be gradual.

Protein

Both quantity and quality of protein should be adequate. Instead of normal requirement *i.e.* 0.8-1g protein/kg body wt./day, underweight person should take 1.5 to 2.0 g protein/kg body weight per day for body building, care and repair of body tissues, to overcome infections, to gain weight *etc.* Good quality protein should be provided in liberal amounts initially.

Carbohydrates

Carbohydrates should be the basis of the diet *i.e.* these should provide 60-65 per cent of the daily calorie need. A high carbohydrate diet is needed to have protein sparing action. For taking more calories, plenty of starchy vegetables (potato, colocasia, sweet potato *etc.*), fruits, cereals (wheat, rice, *bajra etc.*), honey, sugar *etc.* should be included in the diet. High amount of dietary fibre hinders with the availability of nutrients and lowers the food intake which is not desired for underweight person. Hence, fibre intake should be just sufficient to avoid constipation.

Fats

Fats being the concentrated source of energy, reduce the bulk of the diet. Their amount should provide 25-35 per cent of the total calories. However, recommendations for the proportion of saturated (less than 7 per cent), polyunsaturated (less than 10 per cent) and monounsaturated fats (15-20 per cent) remain the same as for the normal weight person. Cholesterol intake should not be more than 300 mg/day.

Vitamins and Minerals

Liberal amount of vitamin and mineral rich foods such as fruits and vegetables, whole grains, legumes, nuts and oilseeds *etc.* should be taken. If need be, supplement of multivitamin-mineral should be taken as per prescription of the physician.

3. Dietary Management

Energy Dense Foods

To gain body weight, calorie intake is to be increased which can be done by the consumption of healthful balanced diet containing sufficient volume of nutritious energy dense foods. Some of the energy dense foods that are perfect for gaining weight include nuts and oilseeds (almonds, walnuts, peanuts, sunflower seeds, chia seeds *etc.*), fats and oils (olive oil, canola, avocados, peanut butter *etc.*) dried fruits (raisins, dates, prunes, berries, trail mixes *etc.*), high fat dairy products (whole milk, full-fat *paneer*, curd, cheese, cream *etc.*), whole grains (whole wheat, brown rice, pearl millet *etc.*) and tubers (potatoes, sweet potatoes *etc.*).

Some of the energy dense recipes include *paushtik ladoos* containing more of nuts and oilseeds (peanuts, *til*, almonds or flax seeds *etc.*), fats and oils, cereals (wheat flour), legume (*besan*); fruit juices; desserts; high energy cereal bars; porridge with nuts; malted milk; icecreams; *paranthas* instead of *chapatti*; *dal* with added ghee; peanut *chikki*; thick soups with added butter or cream; adding sesame seeds, sunflower seeds, walnuts *etc.* on salads *etc.* Instead of boiled rice, take rice *pulav*. Instead of taking milk alone, add almonds in it. Put nuts and raisins in the deserts. Take *roti* smeared with butter or any other good fat.

Take good Quality Protein Sources

Easily digestible protein foods of high biological value such as egg, milk, poultry, meat, seafoods (non-vegetarians) or a combination of cereal + legume+ dairy products, cheese, tofu, nuts *etc.* for vegetarians should be included in the diet.

Increase the Amounts of Foods

The whole day diet should contain whole grain bread, cereals, fruits and vegetables, milk, legumes, nuts and oilseeds, egg, meat, fish (if non-vegetarian) *etc.* For example, if one is taking one wheat *chapatti* at lunch time, increase it to 2 or 3 gradually. Increase the portion size of vegetables containing potato and *dal.*

Type of Fat

Healthy easily digestible fat having appropriate blend of saturated, PUFA and MUFA (Extra virgin olive oil, canola, should be provided. No trans fat should be used. Fried foods should be avoided.

Adding Snacks

In between the major meals, favourite high protein and whole grain carbohydrate snacks should be taken. Some of the energy rich snacks include egg omelets, cheese sandwich, bread with jam, whole milk yogurt with fruits, smoothies, biscuits, handful of nuts, chocolate, potato chips, pop corns *etc.* can be taken. Portion size should be small so that person can take the regular meal. Try to avoid junk and fried foods.

Avoid Empty Calories

Eating empty calorie foods that are high in sugar (carbonated beverages, synthetic soft drinks. sugary snacks and candies) and salt as besides weight gain, they may adversely affect the health. Avoid drinking too much tea or coffee as they suppress the appetite.

Eating Several Meals a Day

Sometimes, person may be underweight because he cannot take large meal at one time. Such persons should eat several nutritious small meals (at least 6 meals; 3 big and 3 small meals) throughout the day. If you have a mix of foods on plate, eat calorie-dense and protein rich foods first. Eat the vegetables last.

Use Bigger Plates and Cups

They help to eat more calories, as smaller plates cause people to automatically eat less.

Take Plenty of Water

To keep the body hydrated, take 8-10 glasses of water daily. Donot drink water just before meals to avoid feeling too full to eat.

Physical Exercise

Another way for underweight people to gain weight is by exercising. Exercise can help stimulate a person's appetite if they are not inclined to eat. Muscle hypertrophy increases body mass. Weight lifting exercises are effective in helping to improve muscle tone as well as helping with weight gain *i.e.* increasing lean body mass rather than increasing body fat. It also improves bone mineral density which underweight people have an increased risk of deficiency. Doing body resistance exercise such as situps and pushups also improve muscle strength.

26

Diet in Cancer

Non-communicable diseases are now responsible for the majority of global deaths, and cancer is the second leading cause of death globally. It is responsible for an estimated 9.6 million deaths in 2018. One in 5 men and one in 6 women worldwide develop cancer during their lifetime, and one in 8 men and one in 11 women die from the disease. Deaths from cancer worldwide are projected to continue to rise to over 13.1 million in 2030. Approximately 70 per cent of deaths from cancer occur in low and middle-income countries.

What is Cancer

Cancer in broad term, is a class of diseases characterized by out-of-control cell growth. The body is made up of billions of cells. Normally, cells multiply and die in an orderly way by the normal cell kinetics regulators, helping us to grow, replacing worn-out tissue and healing injuries. Sometimes, something goes wrong with some cells, they don't die (apoptosis) but they divide out of control and may grow into a lump (tumor) called cancer (except in the case of leukemia where cancer prohibits normal blood function by abnormal cell division in the blood stream). These changes occur due to interaction between an individual's genetic factors and exposure to certain carcinogens. Cancer can occur in any part of the body and in any organ or tissue. There are over 100 different types of cancer, and each is classified by the type of cell that is initially affected.

Even though most of the cancers are generally associated with old age, no age group is immune to this disease.

Carcinogen, Procarcinogen and Co-Carcinogen

Most of the human cancers are caused by certain carcinogens. A carcinogen is defined as any substance, radionuclide, or radiation that promotes cancer formation or carcinogenesis. This may be due to their ability to damage the genome or to the disruption of cellular metabolic processes. Many types of carcinogenic substances such as chemical, physical and biological agents are responsible for causing cancer. Radionuclides are carcinogens, whether or not they are toxic, because they emit alpha, beta, gamma, or neutron radiation that can ionize tissues. Many types of radiations such as ultraviolet rays, x-rays and gamma rays are carcinogenic. Common examples of non-radioactive carcinogens are inhaled asbestos, certain dioxins, and tobacco smoke. Chemical carcinogens can be synthetic, toxic or non-toxic. Carcinogens are not necessarily immediately toxic, thus their effect can be insidious.

Chemicals that are not carcinogenic, but become carcinogens when they are metabolized in the body are called procarcinogens. An example of a procarginogen is nitrite, which is metabolized to form carcinogenic nitrosamines.

A co-carcinogen or promoter is a chemical that does not cause cancer on its own, but promotes carcinogen activity. The presence of both chemicals together increases the likelihood of carcinogenesis. Ethanol (grain alcohol) is an example of a promoter.

Carcinogenesis

Also called tumorigenesis, carcinogenesis is the multistage molecular process by which cancer develops. Present day oncology recognizes three main phases *i.e.* initiation (alteration of DNA), promotion (cellular proliferation), and progression (caryotypic instability and frank malignancy). These stages may progress over many years.

Initiation

In the simplest terms, initiation involves one or more stable change in the genetic makeup of a cell arising spontaneously or induced by exposure to a carcinogen. At this stage, carcinogen interacts with DNA causing damage. This initial damage rarely results in cancer because the cell has in place many mechanisms to repair damaged DNA. However, if repair does not occur and the damage to DNA is in the location of a gene that regulates cell growth and proliferation, DNA repair, or a function of the immune system, then the cell is more prone to becoming cancerous. The human DNA sequences responsible for transformation are called oncogenes.

Promotion

This stage is reversible and environmentally modulated. During promotion, the mutated cell is stimulated to grow and divide faster and becomes a population of cells. Eventually a benign tumor becomes evident. In human cancers such as those of the lung, breast, prostate and GIT, promoting agents such as hormones (prolactin, estrogens and androgens.), dietary composition, UV rays (sun light), cigarette smoke, alcohol or bile acids are substances that are involved in promotion. This stage is usually reversible as evidenced by the fact that lung damage can often be reversed after smoking stops.

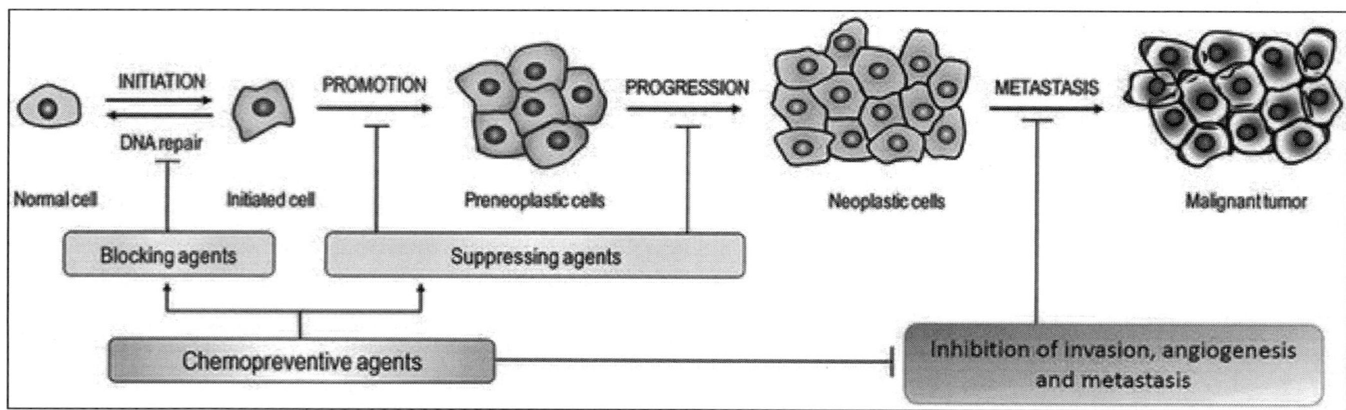

Progression

During progression, there is further growth and expansion of the tumor cells over normal cells. The genetic material of the tumor is more fragile and prone to additional mutations. These mutations occur in genes that regulate growth and cell function such as oncogenes, tumor suppressor genes, and DNA mismatch-repair genes. These changes contribute to tumor growth until conversion occurs, when the growing tumor becomes malignant and possibly metastatic. This stage is irreversible. Many of these genetic changes have been identified in the development of colon cancer and thus, it has become a model for studying multi-stage carcinogenesis.

Different types of cancer genes - oncogenes and antioncogenes (tumor suppressor genes) - are involved in cancer development. Gain of function mutations in the oncogenes, leading to abnormal cell proliferation, and loss of function mutations in the anti-oncogenes leading to suppression of cell differentiation and apoptosis, are the major events leading to cancer development. Chromosomal aneuploidy and epigenetic events are also thought to be important. Several factors like age, sex, genetic predisposition, along with extrinsic factors like diet, environmental pollutants, alcoholism and tobacco habits have a major role in determining the cancer risk. Dietary intervention as a cancer preventive measure is a primary agenda on the WHO program.

Types of Tumors

A tumour can be benign or malignant.

Benign Tumor

Benign tumors that are generally confined to one area, demonstrate limited growth and are not able to spread to other parts of the body. This is not cancer. Examples are adenomas, fibromas, lipomas, meningiomas, myomas *etc.*

Malignant Tumor and Metastasis

The word malignant is Latin for 'badly born'. This type of tumor has the ability to multiply uncontrollably, to metastasize to various parts of the body and invade surrounding tissue. Hence, these tumors can grow and interfere with any part of the body *e.g.* the digestive, nervous, and circulatory systems and they can release hormones that alter body function.

Malignant tumors are cancerous. Examples are carcinoma, sarcoma, leukemia, lymphoma *etc.*

More dangerous, or malignant, tumors form when two things occur:

1. A cancerous cell manages to move throughout the body through the bloodstream or lymphatic system, destroying healthy tissue in a process called invasion.
2. That cell manages to divide and grow, making new blood vessels to feed itself in a process called angiogenesis. When a tumor successfully spreads to other parts of the body and grows, invading and destroying other healthy tissues and forms another tumor at a new site, it is said to have metastasized and is called secondary cancer. This process itself is called metastasis, and the result is a serious condition that is very difficult to treat. Metastases are a major cause of death from cancer.

The cancer that first develops in a tissue or organ is called the primary cancer. A malignant tumor is usually named after the organ or type of cell affected. A malignant tumor that has not spread to other parts of the body is called localized cancer. A metastasis keeps the name of the original cancer. For example, lung cancer that has spread to the bones is called metastatic lung cancer, even though the person may be experiencing symptoms caused by problems in the bones.

Causes and Risk Factors for Carcinogenesis

There is no one single cause for cancer. Scientists believe that it is the interaction of many factors together that make certain people more susceptible to produce cancer *e.g.* interaction between a person's genetic factors with mainly some external agents such as physical carcinogens (UV and ionizing radiation), chemical carcinogens (components of tobacco smoke, environmental pollution, industrial effluents, exposure to industrial toxins, aflatoxin, arsenic, asbestos *etc.*), biological carcinogens (infections from certain viruses, bacteria or parasites), lifestyle factors (nutrition, dietary pattern, physical activity, alcohol consumption *etc.*) and medical treatment (radiation and medicines including chemotherapy, hormone drugs, drugs that suppress the immune system *etc.*).

Some of the intrinsic and external risk factors associated with cancer development are discussed as under:

Age and Hormonal Status

Cancer is considered to be an old age disease. Some types of cancers are almost entirely found in people above 50-55 years, *e.g.* prostate cancer. Similarly cervix cancers in women are more commonly detected at the peri- or post-menopausal ages. However, no age group is immune to this disease. Hormonal factors play an important role in the development of gender-specific cancers, *e.g.* estrogen in cancers of ovary and uterus in female. Not all fractions of estrogens seem to exert these effects. The effects of estrogen are ascribed to oestrone but not to oestriol or oestradiol. After menopause, the adrenal glands produce androstendione, a precursor of oestrone and androgens (causing the growth of facial hair in women of this age).

Family History

Some cancers are indicated to have a link with familial occurrence. For example, women whose close relatives like grandmother, mother, maternal aunt or sister has suffered from breast cancer, are found to run about 3 times higher risk of developing breast cancer than those who do not have such a family history. Similarly, cancers of the uterine cervix (females) and of prostate (males) are also thought to have a familial connection.

Genetic Predisposition

Certain genetic conditions are known to predispose the individual to cancer. Hereditary cancers are primarily caused by an inherited genetic defect. In a cancer syndrome or family cancer syndrome, inherited genetic mutations in one or more genes predispose the affected individuals to the development of cancers and may also cause the early onset of these cancers. There are more than 50 identifiable hereditary forms of cancers, less than 0.3 per cent of the populations are carriers of a cancer-related genetic mutation and these make up less than 3-10 per cent of all cancer cases. It has long been known that there are genetic factors involved in the development as cancer. As early as the first century, the Romans observed that breast cancer seemed to occur more frequently in some families than the others. Some of the examples of cancer syndromes are hereditary breast and ovarian cancer, hereditary non-polyposis colon cancer, Werner syndrome *etc.* Similarly, individuals with genetic conditions like xeroderma pigmentosum, ataxia telangiectasia, Bloom's syndrome, and Fanconi's anaemia are found to be highly susceptible to different types of cancers.

Biological Agents

Infectious diseases caused by certain bacteria, viruses and parasites cause cancer by damaging DNA or causing genomic instability. These biological agents are responsible for approximately 18 per cent of cancer cases worldwide; incidence is higher in Africa than in the developed countries. Bacteria such as *Helicobacter pylori* increase the risk of gastric carcinoma by chronic inflammation or direct action of the bacteria's virulence factors. Tuberculosis infection caused by Mycobacterium *M. tuberculosis* has been associated with the development of lung cancer.

Parasitic infections strongly associated with cancer include *Schistosoma haematobium* (squamous cell carcinoma of the bladder), *Clonorchis sinensis Opisthorchis viverrini* (cholangiocarcinoma-cancer of bile duct), *Schistosoma haematobium* (carcinoma of urinary bladder) *etc.* Inflammation triggered by the worm's eggs appears to be the cancer-causing mechanism. Certain parasitic infections can also increase the presence of carcinogenic compounds in the body, leading to the development of cancers

Viral infection is a major risk factor for cervical and liver cancer. A virus that can cause cancer is called on oncovirus. These include human papillomavirus (HPV) (cervical carcinoma), hepatitis B (HBV) and hepatitis C virus (HCV) (hepatocellular carcinoma), Human T-cell leukemia virus-1 (T-cell leukemias), HIV (Kaposi sarcoma), Epstein-Barr virus in Burkitt lymphoma and nasopharyngeal carcinoma *etc.*

Physical and Chemical Agents

Various particular physical and chemical agents, known as carcinogens due to exposure to hazardous substances, chemicals, dust or radiation (*e.g.* asbestos, uranium, coal tar pitch, wood dust, diesel exhaust, lead, arsenic, benzene, cadmium, chromium, nickel, UV radiations *etc.*) have been linked to specific types of cancer and may make some people more susceptible than others. The person who works in the chemical factories, nuclear reactors, drainage system and mining are most prone to cancer.

Radiation of certain wavelengths, called ionizing radiation, has enough energy to damage DNA and cause cancer. Ionizing radiation is an established carcinogen, having both initiating and promoting effects. Ionizing radiation includes radon, x-rays, gamma rays, and other forms of high-energy radiation. Lower-energy, non-ionizing forms of radiation, such as visible light and the energy from cell phones and electromagnetic fields, do not damage DNA and have not been found to cause cancer. Leukemia and most solid cancers have been linked with radiation. Prolonged exposure to ultraviolet radiation from the sun can lead to melanoma and other skin malignancies.

Epidemiological data indicate that women using powder (talc or asbestos) for intimate hygiene are prone to ovarian cancer, as are women who wash the trousers of men who work with asbestos. The asbestos dust spreads from the trousers during washing, penetrating the genital tract and the ovaries of the women involved.

The role of cigarette smoking in lung cancer is established. Nitrosamines and polycyclic aromatic hydrocarbons present in tobacco smoke can induce neoplastic changes in the lung cells and are associated with 80 per cent of lung cancer cases. Research studies have demonstrated the link between tobacco use and cancer in the larynx, head, neck, stomach, bladder, kidney, esophagus and pancreas too. Tobacco (cigarette smoking, tobacco intake) is responsible for about one in five of all cancer deaths worldwide. Decreasing the number of people who smoke cigarettes is the best way to prevent cancers and many other lethal diseases. Non-smoking tobacco habits, like chewing, are found to greatly increase the cancers of the upper alimentary tract and buccal mucosa. India has the highest incidence of oral cancers in the world, which is correlated with the tobacco chewing habit. Leukemia can occur from exposure to benzene at work places.

Environmental and Therapeutic Exposure Associated with Cancer Risk

Carcinogenic Agent	*Mode of Exposure*	*Type of Cancer*
Asbestos	Industrial exposure	Lungs, mesothelium
Chromium	-do-	Lungs
Iperyte	-do-	Respiratory tract
Aromatic amines	-do- (dye works, leather tannery)	Urinary bladder
Arsenic	Industrial exposure	Lungs, skin, alimentary tract, urinary bladder
Aluminum	Production, manufacture	Lungs, bladder
Coal gasication	Occupational hazard	Skin, bladder, lungs
UV radiation	Environmental exposure and occupational hazard	Melanoma
Chromium	Environmental exposure and occupational hazard	Prostate cancer
Tobacco smoke	Environmental exposure, occupational hazard, personal habit	Broncial, laryngeal and oesophageal cancers, urinary bladder cancer, head and neck region cancers
Ethanol	Environmental exposure, lifestyle factor	Oral cavity, throat, oesophagus, liver, breast cancer and colorectal cancer
Mycotoxins and aflatoxins	Infected foods	Liver
Cyclophosphamide	Therapeutic exposure	Bone marrow (leukaemia)
Dietylostilbestrol	-do-	Vagina
Diphenylhydantoine	-do-	Endometrium
Immunosuppresion	-do-	Lymphoid system and various other tissues

Certain substances such as wollastonite, glass wool, rock wool *etc.* cause cancer mainly through their physical, rather than chemical, effects on cells. For example, mesothelioma is developed from prolonged inhaling asbestos fibres.

Lifestyle Pattern

Cancer transitions are most striking in emerging economies. A recurring observation is the ongoing displacement of infection-related and poverty-related cancers (cervix, stomach and liver) by those cancers (colon, prostrate, breast, pancreatic cancers *etc.*) that already are high frequent in the most developed countries. These cancers are often ascribed to the so-called westernization of lifestyle, yet the differing cancer profiles in individual countries and between regions signify that marked geographic diversity still exists, with a persistence of local life style risk factors in populations at quite different phases of social and economic transition.

Certain life style patterns including dietary pattern, physical activity, smoking, alcohol consumption *etc.* along with obesity have been associated with risk of cancer. More than 50 per cent of all cancers are related to the diet and individual habits like alcoholism, tobacco chewing and smoking.

i) Diet

Some specific foods have been related to specific cancers *e.g.* individuals consuming red or processed meat have a higher risk of developing breast cancer, prostate cancer, and pancreatic cancer which may be partially due to the presence of carcinogens in food cooked at high temperature. Similarly, deep-fried and burnt food and preserved (high salt) food are associated with increase in gastric cancer incidence. A high salt diet is linked to gastric cancer; it is more common in Japan. Colon cancer due to increased intake of processed and red meat is very common in USA.

Other diet related risk factors include more consumption of refined carbohydrates, high fat (trans fat, animal fat, saturated fat *etc.*), less intake of dietary fibre *etc.* High fat diets particularly rich in trans fats have also been reported as culprits for causing cancers. High fat diet and obesity are associated with breast cancer. A positive correlation has been reported between age-adjusted breast cancer mortality rates and the average per capita fat consumption in a given nation on a daily basis. More consumption of unsaturated fat and lack of antioxidant rich foods in the diet can cause free radicals in the body. Free radicals react rapidly with proteins, lipids and nucleic acids, thus endangering normal cellular functions. This negative interaction with free radicals, which used to be known as oxidative stress, frequently leads to damage of the cell's genetic material. Studies have shown that the active oxygen species (H_2O_2, OH^* and O_2^* - hydrogen peroxide, hydroxy radical and hyperoxide radical) may yield as many as 104 modifications of nitrogen bases daily in a single human cell. The extent of these injuries exceeds those caused by modifications of nitrogen bases through the action of chemical carcinogens.

Processed foods that are high in sugar and low in fiber and nutrients have been linked to a higher cancer risk. In particular, researchers have found that a diet that causes blood glucose levels to spike is associated with an increased risk of several cancers, including stomach, breast and colorectal cancers. Regular consumption of food low in fibre content and rich in animal fat increased the risk of cancers of stomach and oesophagus. High intake of red meat and low fibre diet has been considered to be the cause of the high incidence of gastric cancer in the USA.

Chewing of betel nut, *pan masala etc.* have been reported to cause oral cancers. Intake of aflatoxin B1, food contaminant, more common in infected peanuts, is associated with liver cancer.

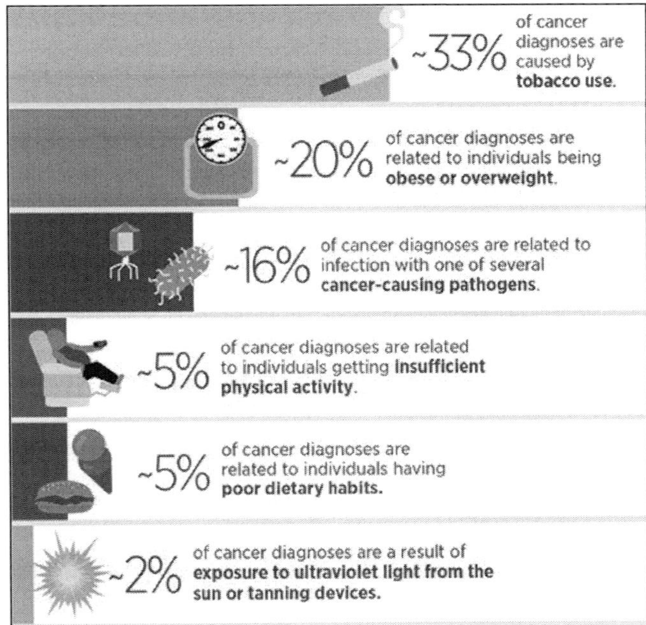

Figure 26.1: Common Risk Factors of Cancer.

ii) Alcohol

Alcohol is an example of a chemical carcinogen. The World Health Organization has classified alcohol as a Group 1 carcinogen. Numerous epidemiological studies have shown a relationship between chronic consumption of alcohol and the development of malignant tumours in humans. Alcoholism is found to increase the risk of mouth, esophagus, pharynx, larynx, stomach, liver, colon and bladder cancers. The main mechanism of cancer development involves increased exposure to acetaldehyde, a carcinogen and breakdown product of ethanol. Other mechanisms have been proposed, including alcohol-related nutritional deficiencies, changes in DNA methylation, and induction of oxidative stress in tissues.

Smoking combined with alcohol consumption poses a higher risk of cancers of the breast, oesophagus, liver, stomach and urinary bladder. The probability of developing cancer of the oral cavity or the throat is three times higher among those who abuse both alcohol and cigarettes than among abstinents. Alcoholism along with hepatitis B virus infection is a more serious risk factor in liver cancer.

Some beers contain aromatic nitrogen compounds which have carcinogenic effects. The consumption of Calvados vodka is known to promote the development of oesophageal cancer. In Japan, cohort studies have shown the highest incidence of oesophageal cancer to be among males who consumed sake daily.

iii) Obesity

Obese people often have chronic low-level inflammation, which can, over time, cause DNA damage that leads to cancer. For example, chronic local inflammation induced by gastroesophageal reflux disease or Barrett esophagus is a likely cause of esophageal adenocarcinoma. Obesity is a risk factor for gallstones, a condition characterized by chronic gallbladder inflammation, and a history of gallstones is a strong risk factor for gallbladder cancer. Chronic ulcerative colitis and hepatitis, disease of the liver causing inflammation are risk factors for different types of liver cancer. Fat tissue produces excess amounts of estrogen, high levels of which have been associated with increased risks of breast, endometrial, ovarian, and some other cancers. Obese people often have increased blood levels of insulin and insulin-like growth factor-1 (IGF-1) which may promote the development of colon, kidney, prostate, and endometrial cancers.

There is consistent evidence that large amount of body fats are associated with a number of cancers such as endometrial carcinoma, esophageal adenocarcinoma, gastric cardia cancer, cancer of liver, kidney, pancreas, gall bladder, breast, colorectal, ovarian, thyroid, multiple myeloma, meningioma *etc*.

iv) Physical Inactivity

There are strong evidence that being physically active reduces the risk of colon cancer, post menopausal breast and endometrium cancers.

v) Stress

When you experience stress, it produces certain hormones in the body, like cortisol, out of balance. When that imbalance becomes chronic, it can impair mental function, blood pressure, digestion, immune system and other bodily functions. Hence, any kind of psychological or physical stress or distress puts our immune system in jeopardy, and once the immune system is down, our body is more vulnerable to infections and chronic inflammation, and is less able to fight off diseases such as cancer. Moreover, association between stress and cancer seems to surface especially when people turn to behaviors like overindulging in alcohol, overeating unhealthy foods and smoking to cope up with the stress.

Common Cancers Prevalent in India

India ranks 3rd in cancer cases after China and the US. Every year India reports more than a million new cancer cases. Among them, half million deaths happen due to the ignorance of the disease. Cancer among women in the country is estimated to be 0.7 million reported cases. One woman dies of cervical cancer in every 8 minutes in India. For every 2 women newly diagnosed with breast cancer, one woman dies of it in the country, accordingly to National Institute of Cancer Prevention Research (NICPR).

Nearly one-third of the cancers are caused due to tobacco use. Alcohol and tobacco together pose higher risks of developing oral and other cancers. On an average, more than 1,300 Indians succumb to the dreaded cancer disease every day. With new cancer cases or its incidence in India estimated to grow by 25 per cent by 2020 (according to the cancer registry released by the Indian Council of Medical Research), cancer has become one of the major causes of death occurring in the country. According to GLOBOCAN 2018 data, in 2018

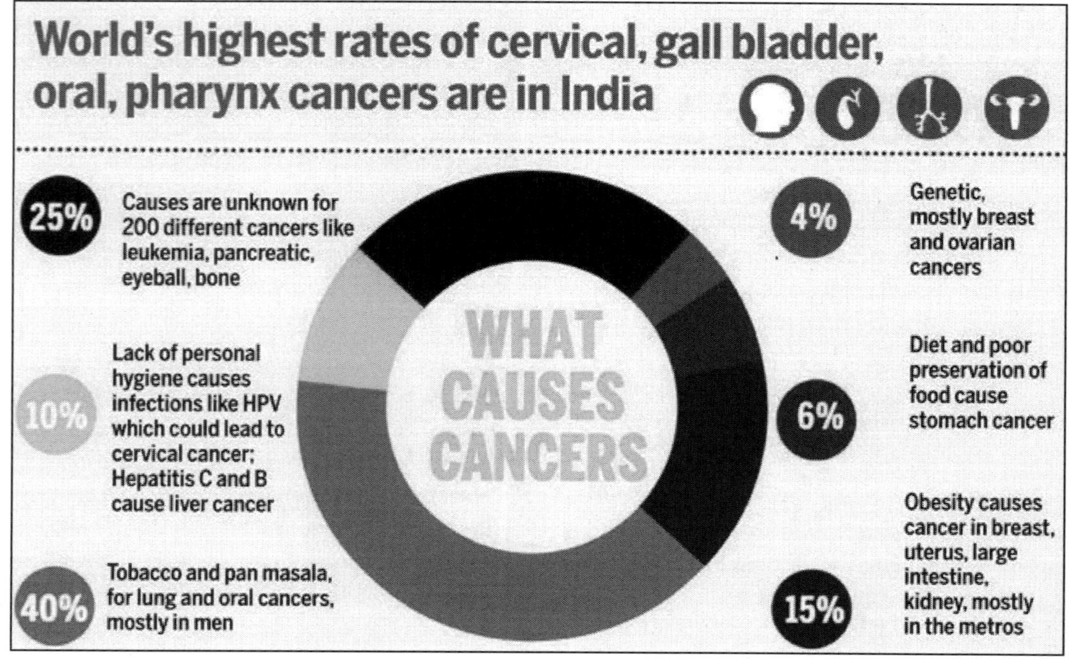

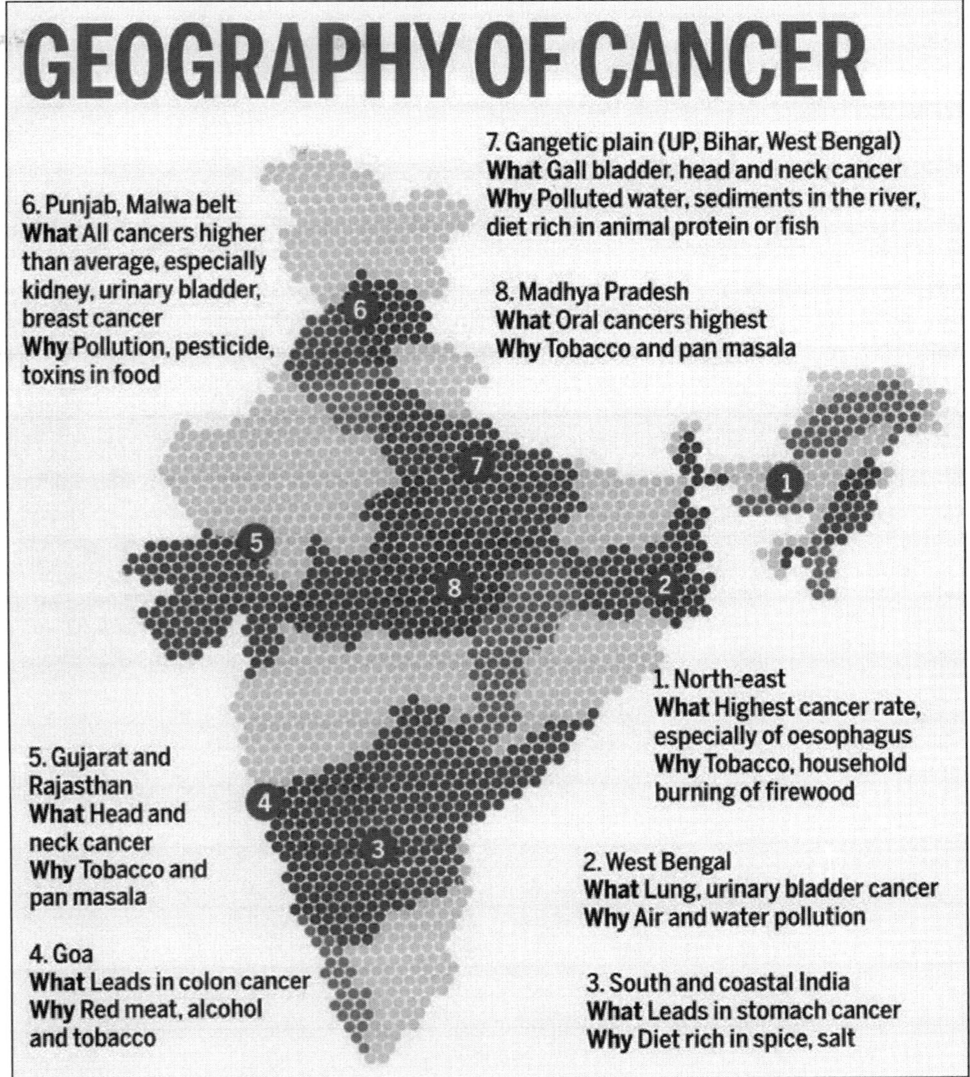

there were 11,57,294 new cancer cases in India in both men and women, 7,84,821 deaths and 22,58,208 people living with cancer (within 5 years of diagnosis).

According to WHO's Cancer Report, in India, lung, oral, lip, throat and neck cancers are the most common among men while women suffer more from cervix, breast and ovarian cancers. In the elderly, the most commonly occurring cancers are kidney, intestine and prostate cancer. Women, especially, are being increasingly diagnosed with cancer. Top 5 cancers that affect Indian population are breast, oral, cervical, gastric and lung cancers.

There are regional variations in cancer incidence and mortality is driven by multiple factors in India. Other than internal factors like genetic, hormonal and poor immune conditions, external or environmental factors like food habits, industrialization, pollution and lifestyle also cause cancer in Indian scenario. This is why a geographic pathology study can give some indication of the proportion of cancers that could be prevented by modifying specific harmful lifestyle or environmental factors.

Presently in India, it is a major cause of morbidity and mortality. Many cancers are preventable by controlling the modifiable risk factors such as tobacco use, alcohol use, unhealthy diet and physical inactivity, amongst others. In addition, a significant proportion of cancers can be cured, by surgery, radiotherapy or chemotherapy, especially if they are detected early.

Signs and Symptoms of the Most Common Cancers in India

The most common cancers in India are breast cancer, cervical cancer, lung, gastric and oral cancers. Experts suggest that if someone notices any other major changes in the way his or her body functions or feels, that person should see a doctor, especially if the changes persist for more than three weeks or gets worse. It may not be cancer but it is safer to get it checked out.

Early signs depend upon the organ affected. For example, early signs of lung cancer may include abrupt weight loss, persistent cough, abnormal voice changes, pain in the chest which becomes worth with deep breathing, laughing *etc.*, wheezing coughing up

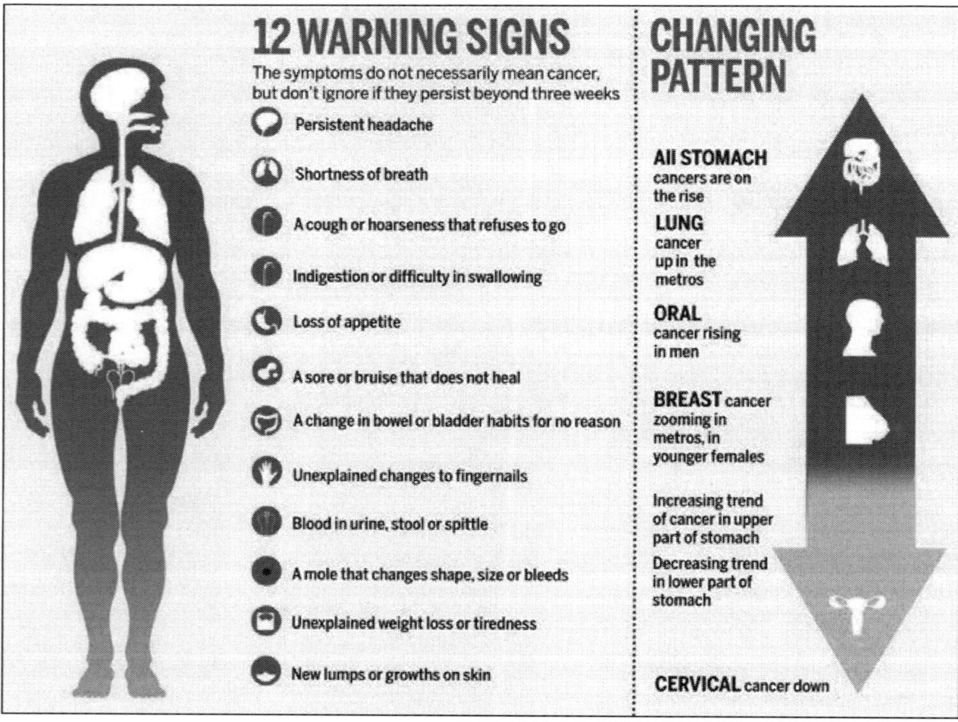

Figure 26.2: Common Symptoms of Concern.

blood, pain in bones, headache *etc.* The most common symptoms of breast cancer is formation of a lump, swelling in breast, pain in nipple or breast, skin irritation, discharge from nipples *etc.* Early signs of kidney cancer include blood in urine, loss of appetite, sudden weight loss, recurrent fever, fatigue *etc.* Unusual vaginal bleeding, pain in the pelvic area, foul smelling unusual discharge *etc.* are some of the warning signs of cervical or uterine cancer. In prostate cancer, there is an urination problem.

However, in general, some of the common possible symptoms in many types of cancers include unexplained weight loss, body ache, viral fever, fatigue, skin problems, swelling or bloating, blood in cough, pee or stool, persistent heartburn or indigestion that appear in initial stage. Below is the chart of common symptoms that may be a cause of concern.

Effect of Cancer on Nutritional Status

Changes in nutritional status may begin prior to diagnosis, when physical and psychosocial issues commonly have a negative impact on food intake. The reality is that at cancer diagnosis, half of patients present with some form of nutritional deficit. This deterioration in nutritional status has been found to poor outcomes, with as little as a 5 per cent weight loss predicting decreased response to therapy and decreased survival. Nutritional status also often declines during the natural progression of cancer and its treatment, due to treatment related side effects, and results in multiple and inter-related nutritional issues. Hence, patients with cancer face many challenges, including maintaining a good nutritional status and avoiding weight loss and malnutrition. However, research shows that the majority of patients with cancer suffer from various nutritional deficits, and up to 85 per cent of patients with certain cancer types experience some form of weight loss and malnutrition during their cancer treatment.

PEM in cancer results from multiple factors most often associated with anorexia, cachexia, and the early satiety sensation frequently experienced by individuals with cancer. These factors range from altered tastes to a physical inability to ingest or digest food, leading to reduced nutrient intake, poor absorption, altered metabolism and increased losses of nutrients.

Factors that May Lead to a Deterioration in Nutritional Status

Reduced intake due to
☆ Physical symptoms *e.g.* anorexia, nausea, vomiting, diarrhoea, constipation, taste changes, mucositis, dry mouth, fatigue and pain
☆ Psychological symptoms *e.g.* anxiety and depression
☆ Psychological issues *e.g.* reduced access to food due to poor mobility or financial concerns
☆ nil by mouth (investigations and treatment)
Reduced digestion/absorption
☆ ± Specific to site of cancer
Altered metabolism
☆ Increased protein catabolism
☆ Decreased muscle protein synthesis
☆ Increased gluconeogenesis± increased insulin resistance
☆ Increased lipolysis and increased fatty acid and glycerol turnover
☆ Increased metabolic rate 2° to tumour growth

Increased losses
☆ Vomiting
☆ Diarrhoea
☆ Fistulae
Effect of treatment
☆ Surgery
☆ Chemotherapy
☆ Radiotherapy

Some of the ill effects of cancer and its treatment on nutritional status of the individual have been highlighted in the figure given below.

Malnutrition is generally found in cancer patients which can result from the disease process, from the use of cancer therapies, or from both. The incidence of malnutrition in people with cancer ranges from 30 per cent to 87 per cent. Patients with cancer of the lung, esophagus, stomach, colon, rectum, liver, and pancreas are at greatest risk. Of people who die from cancer, up to half have been malnourished. In fact, up to 20 per cent of patients die from the effects of malnutrition rather than from the cancer itself.

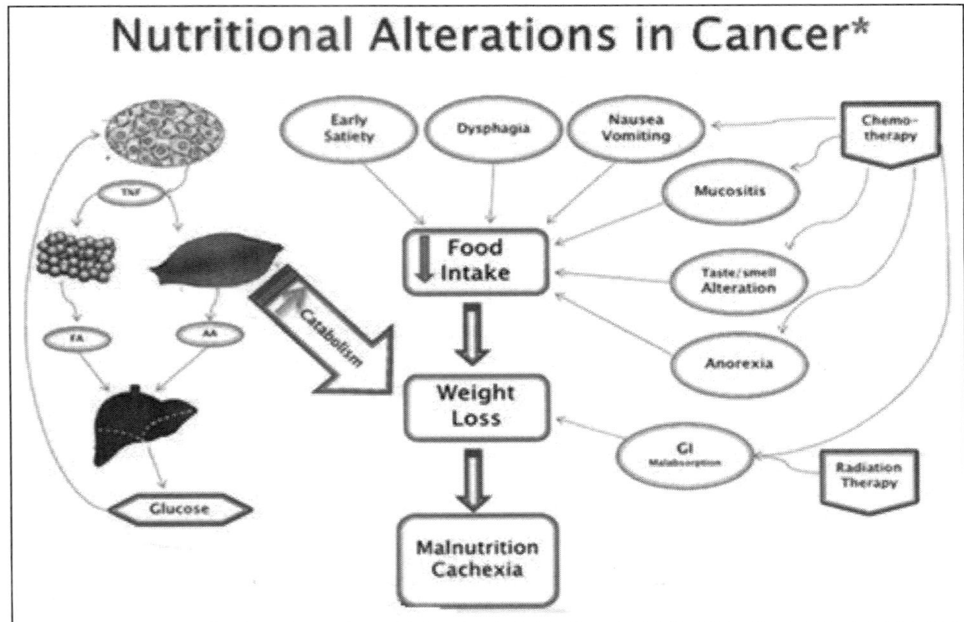

Anorexia and cachexia are common causes of malnutrition in cancer patients. Anorexia or loss of appetite or desire to eat is a common symptom in patients with cancer. Some patients already have anorexia when they are diagnosed with cancer. Most patients who have advanced cancer will have anorexia. As a result, there are many systemic effect of cancer on nutritional status. Some of the metabolic alterations take place in some of the major nutrients. Cancer related defects in carbohydrate metabolism include glucose intolerance, insulin resistance, abnormal insulin secretion, delayed glucose clearance, increased glucose production, variable increased cori cycle activity *etc.* Cancer related defects in protein metabolism are increased whole body protein turnover, persistent muscle breakdown, decreased plasma branched chain amino acids and negative nitrogen balance. Some of the cancer related defects in fat metabolism include excess body fat depletion relative to protein loss, decreased free fatty acids, and glycerol turnover, hyperlipidemia and decreased serum lipoprotein lipase activity.

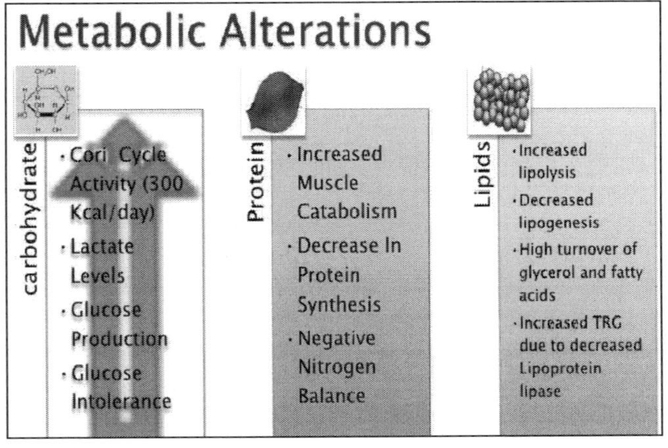

Micronutrient deficiencies also take place. Reduced levels of vitamin A in colorectal and esophageal cancer, pretreatment paediatric leukemia and lymphoma; deficiencies of vitamin E, C and β-carotene in lung, gastric, pancreas, oral and thyroid cancers; and deficiencies of vitamin D and calcium in colon cancer have been reported. Electrolyte disturbances, hyperkalemia, hyperphosphataemia and hypocalcemia occur in some of the cancers. Drug inducted metabolic alterations are also common.

For some patients, the nutritional deficits can proceed to cancer cachexia, a specific form of malnutrition characterized by loss of lean body mass, muscle wasting and impaired immune, physical, and mental function. Cytokines, *e.g.* interleukin-1, interleukin-6 and TNF- A, are released by the body in response to the tumour resulting in systemic inflammation. Metabolic changes, including an increase in protein and fat breakdown, decrease in protein synthesis and altered resting energy expenditure, result in weight loss. In addition, common symptoms experienced by those with cachexia such as anorexia, fatigue, and low mood all contribute to a reduced oral intake. Cachexia plays an important role in the development of malnutrition in people with cancer.

Furthermore, poor nutritional status, weight loss, and malnutrition can lead to poor outcomes for patients, including decreased quality of life, decreased functional status, increased complication rates, reducing the tolerance to oncology treatment and treatment disruptions. Fortunately, providing early nutrition intervention for patients can improve patients' nutritional status and help patients to maintain body weight, maintain lean body mass, better tolerate treatment, and improve quality of life. Therefore, all healthcare professionals who care for patients with cancer need to recognize the signs of malnutrition and be equipped to provide early and effective nutrition intervention to improve outcomes. Good nutrition practices can help cancer patients to maintain weight, build up body's nutritional stores, reduced post-operative infection rate, better control of cancer related symptoms, shorten the length of hospital stay, improved tolerance to treatment and improve quality of life.

Side Effects of Cancer Therapies on Nutritional Status

Side effects related to common cancer therapies, including chemotherapy, radiation, immunotherapy, and surgery, are key contributors in promoting the deterioration in nutritional status. Some of the ill effects of cancer treatment therapies have been given below in the Table:

Cancer Treatment	*General Side Effects*
Surgery	Surgery increases the requirements for nutrients and energy. However, in surgery of the mouth, stomach, intestines or throat, eating is affected. It may slow digestion and absorption.
Radiation therapy	Radiation therapy damages normal healthy cells also along with cancerous cells. The nutrition-related side-effects associated with radiotherapy depend on site of the cancer and dose prescribed. General effects can include anorexia, nausea, tiredness, and altered mood. Local effects arise from the impact of the radiation on healthy tissue near the cancer site and can cause soreness. This may have little nutritional impact if it is a discrete, non-systemic area, *e.g.* limb. However, severe problems can arise with radiotherapy to specific organs such as given under: **Problems due to radiation therapy to the head and neck (brain, neck, mouth, face, salivary glands, tonsil, larynx, nasopharynx, chest of esophagus *etc.* may cause-** ☆ Nausea, vomiting ☆ Headache, lethargy ☆ Dry mouth ☆ Sore mouth and throat ☆ Fatigue ☆ Loss of appetite ☆ Alterations in taste and smell ☆ Tooth and jaw problems ☆ Problem in swallowing food ☆ Infection, inflammation, severe pain, limited food intake *etc.* Rapid decline in nutritional status occurs. Regular nutritional assessment is required. Artificial nutrition support is often indicated in these patients. Gastrostomy tubes should ideally be placed prior to treatment commencing. **Problems due to radiation therapy to the abdominal/pelvic area (GI, colorectal, gall bladder, pancreas, cervix, prostate, rectum or pelvic area *etc.*)** ☆ Nausea ☆ Vomiting ☆ Loss of appetite ☆ Increased transit time ☆ Lactose intolerance ☆ Bloating, cramping, colitis ☆ Malabsorption of fat, vitamins and minerals ☆ Dehydration ☆ Diarrhoea

Cancer Treatment	***General Side Effects***
	✰ Hyperglycemia ✰ Fluid and electrolyte imbalance Reducing dietary fibre, fat and lactose may help some patients but should not be advised routinely to all. Individual advice from a dietitian will help identify the dietary components that exacerbate symptoms and facilitate the construction of a nutritional adequate intake compatible with avoiding these. **Lung** ✰ Shortness of breath ✰ Sore throat ✰ Swallowing difficulties ✰ Early satiety Patients who are receiving continuous hyperfractionated accelerated radiotherapy (CHART) are at greater risk, particularly as symptoms may not occur until after the treatment has been completed and the patient has been discharged home.
Chemotherapy Commonly used cytotoxic drugs include capecitabine, cisplatin, doxorubicin, 5-fluorouracil, vinblastine *etc.*	Chemotherapy works on active cells. Active cells are cells that are growing and dividing into more of the same type of cell. Cancer cells are active, but so are some healthy cells. These include cells in blood, mouth, digestive system, reproductive system and hair follicles. Side effects happen when chemotherapy damages these healthy cells. Some side effects are short-lived and minor, but others may be a sign of serious problems. Many side effects go away fairly quickly, but some might take months or even years to go away completely. Sometimes the side effects can last a lifetime, such as when chemo causes long-term damage to the heart, lungs, kidneys, or reproductive organs. The severity of side varies greatly from person to person. Every person does not get every side effect, and some people get few, if any. Some of the side effects of chemotheraptу are: ✰ Eating problems such as appetite loss, changes in the sense of taste, constipation, diarrhea, nausea, vomiting, sore throat, sore mouth, pain in swallowing *etc.* ✰ Weight gain ✰ Weight loss ✰ Fatigue ✰ Hair loss ✰ Easy bruising and bleeding ✰ Infection ✰ Anemia ✰ Urine and bladder changes and kidney problems ✰ Chemo brain can affect concentration and focus, mood changes *etc.* ✰ Fertility problems ✰ Changes in libido and sexual function ✰ Nerve and muscle problems such as numbness, tingling and pain The nutrition-related side-effects associated with chemotherapy depend on the regime and dose prescribed. Chemotherapy regimes often comprise cycles of cytotoxic drugs repeated at regular intervals over several months. Fatigue, anorexia, and an increased risk of infection are very common in people receiving chemotherapy. Eating well during the 'good times' between doses may help to maintain nutritional status, even if intake falls during and immediately after the infusion period.
Immunotherapy (Biological therapy)	✰ Immunotherapy, also called biologic therapy, is a type of cancer treatment that boosts the body's natural defenses to fight cancer. It uses substances made by the body or in a laboratory to improve or restore immune system function. Certain types of immunotherapy attack cancer or slow its spread to other parts of the body. Others make it easier for the immune system to destroy cancer cells. ✰ Problems can include: ✰ Eating problems such as changes in sense of taste, diarrhea, dry mouth, appetite loss due to flu-like symptoms such as muscle aches, fatigue and fever, sore mouth, vomiting, nausea *etc.* ✰ Skin reactions (Pain, swelling, soreness, redness, itchiness, rash *etc.*) ✰ Cough ✰ Weight gain due to oedema ✰ Headache ✰ Sinus congestion ✰ Shortness of breath ✰ Heart palpitation ✰ Risk of infections

Cancer Treatment	General Side Effects
Hormone therapy	✩ Side effects of hormonal therapy will depend mainly on the type of hormonal therapy, the dose of a drug or combination of drugs and overall health of the patient. Common side effects are: ✩ Fatigue ✩ Weight gain ✩ Nausea and vomiting ✩ Hot flashes ✩ Treatment induced menopause ✩ Fertility problems ✩ Breast tenderness ✩ Erectile dysfunction ✩ Osteoporosis ✩ Blood clots ✩ Muscle or joint pains ✩ Memory problems ✩ Hair thinning ✩ Mood swings, depression
Stem cell treatment	Side effects can happen any time during, immediately after or days to months after a stem cell transplant. Short-term (acute) side effects generally develop during the first 100 days after a stem cell transplant. Long-term (chronic) side effects generally develop 100 or more days after the transplant. Most side effects go away on their own or can be treated, but some side effects can last a long time or become permanent. Side effects of a stem cell transplant will depend mainly on: ✩ the type and dose of chemotherapy drugs used as conditioning therapy before the transplant ✩ the dose of radiation therapy if it was given before the transplant ✩ the type of transplant done ✩ how good a match the donor stem cells are ✩ Age and overall health status of the individual Chemotherapy, radiation therapy and medicines used for stem cell transplant may cause side effects that keep a patient from eating and digesting food as usual. Depending upon the therapy, they may have side effects like bleeding, anemia, low blood cell count, digestive system problems, kidney problems, lung problems, hair problems *etc.* But transplant patients have a very high risk of infection definitely. They need plenty of protein and calories to get through and recover from the treatment, prevent weight loss, fight infection from bacteria in food.

Guidelines for Nutritional Support to Cancer Patients

There are 4 steps for nutritional support to cancer patients:

1. **Step I.** Nutritional assessment and screening
2. **Step II.** It includes setting nutritional requirements and achievable goals
3. **Step III.** Intervention for symptom management and nourishment.
4. **Step IV.** Evaluation

Nutrition Screening and Assessment

Patient's nutritional status plays an integral role in determining not only risk of developing cancer related complications but also risk of therapy related toxicity and medical outcomes. Whether the goal of cancer treatment is cure or palliation, early detection of nutritional problems and prompt intervention are essential. Proactive nutritional care can prevent or reduce the complications typically associated with the treatment of cancer. Therefore, screening should be done for early assessment of nutritional risk and serial reassessments throughout cancer patient's course. The ASPEN and ESPEN guidelines for nutrition in cancer patients both recommend that nutritional screening and assessment of cancer patients should be performed frequently and nutritional intervention should be initiated early when deficits are identified.

Up to 85 per cent of all patients with cancer develop clinical malnutrition, which negatively affects patients' response to therapy, increases the incidence of treatment-related side effects and can decrease quality of life and survival rate. Therefore, screening of patients with cancer is recommended to identify those who are malnourished or at significant risk of malnutrition by clinical practice groups of oncologists. Early identification of cancer patients who are malnourished or at risk of malnutrition can promote recovery and improve prognosis. In addition, early nutritional intervention is cost effective, as it reduces complication rates and length of hospital stay. Therefore, early screening and assessment of nutritional risks and serial reassessments throughout cancer patient's course are required.

Patients identified through screening require referral to a dietician or specialist in nutrition for an in-depth nutritional assessment. Nutritional assessment is a detailed examination of nutritional status and may include exploration of nutrition impact symptoms, functional status, disease burden, metabolic impact and a physical examination; it generally involves examination of medical, dietary, psychological and social history, physical examination, anthropometry and biochemical testing. Patient-Generated Subjective Global Assessment (PG-SGA) is an outcome based assessment tool which has been developed specifically for cancer patients to assess their nutritional status. This tool is a modified version of SGA tool (developed by Detsky) adapted for oncology population. This tool is easily applicable, performs well against other tools and captures short term changes in the nutritional status due to its high sensitivity and specificity. Therefore, it is used to cross validate other screening tools. It consists of two parts: Part one is a self-completed patient history (weight loss, dietary intake, presence of nutrition impact symptoms and functional capacity). Part two assesses the metabolic demands of the disease, its impact on nutritional requirements and includes a physical examination by the healthcare team. A score is generated from the information collected. Patients are then rated as well nourished, moderately malnourished or severely malnourished. However, the PG-SGA should only be undertaken by trained individuals to maximize reliability of the assessment.

Nutritional interventions initiated after nutritional assessment should be tailored to the individual and take into consideration the patient's prognosis.

Nutritional Assessment

Criteria	*Standard*	*Interpretation*	*Remarks*
Physical examination			
Body weight	Ideal Body Weight (BMI) **Men:** 106lb for the first 5'of height plus 6lbs for each additional inch **Women:** 100lbs for the first 5' of height plus 5lbs for each additional inch	>120 per cent IBW is obese 110 per cent -120 per cent IBW is normal weight 80-90 per cent IBW is mildly underweight <80 per cent of IBW is moderately underweight	Weight loss correlates with complications and mortality. Body weight must be evaluated against normal or usual weight and not a reference standard.
Percent weight change	Amount of weight change divided by usual weight x100	Significant weight loss is greater than: 1-2 per cent in one week 5 per cent in one month 7.5 per cent in three months 10 per cent in six months	
BMI (Body Mass Index)	Weight in kg/Height $(m)^2$	<20 = Malnourished possible <18 = Malnourished likely	
Physical examination for ☆ Oedema ☆ Ascites ☆ Muscle wasting ☆ Temporal wasting (Hall mark of cachexia) ☆ Functional assessment ☆ Assessment of muscle function (grip strength)	Nil	–	Some or many of them may be present.
Biochemical and laboratory tests			
Serum albumin	3.5-5 g/dl	Mild depletion (2.8-3.4g/dl) Moderate depletion (2.1-2.7 g/dl) Severe depletion (<2.1g/dl)	Predictive of morbidity and mortality.
Total lymphocyte (mm^3)	WBC (mm^3) x per cent lymphocytes	Mild depletion (<1500 mm^3) Moderate depletion (<1200 mm^3) Severe depletion (<800 mm^3)	

Setting Nutritional Treatment Goals

The goals of nutrition therapy depend upon the individual's case and requirement. However, the general nutritional goals should be to accomplish the following:

- Stabilize or maintain the acceptable body weight for longer survival and improved quality of life compared to those who continue to lose weight. There is strong evidence to suggest that prevention of weight loss can reduce treatment complications and the length of time spent in hospital, as well as improving the patient's quality of life.
- Preserve lean body mass
- Maintain strength and energy
- Protect immune function, decreasing the risk of infection during treatment
- Prevent or reverse nutrient deficiencies
- Help patients better tolerate treatments and their side effects
- Minimize nutrition-related side effects and complications
- Aid in rapid recovery and healing
- Maximize quality of life and enhanced overall survival.
- Patients with advanced cancer can receive nutritional support even when nutrition therapy can do little for weight gain. Such support may help accomplish the following:
- Lessen side effects
- Reduce risk of infection (if given enterally)
- Reduce asthenia
- Improve well-being.

In individuals with advanced cancer, the goal of nutrition therapy should not be weight gain or reversal of malnutrition, but rather comfort and symptom relief.

Nutritional Requirements

Energy

Energy expenditure of patients with cancer varies greatly. Treatment and disease stage may alter metabolic requirements over time. However, people with cancer typically require a high calorie diet to prevent weight loss, meet increased metabolic demands and improve treatment outcomes. According to ESPEN guidelines (Arends *et al.*, 2017), they recommend, for practical purposes, that total energy expenditure of cancer patients, if not measured individually, be assumed to be rather similar to healthy subjects and ranging between 25 and 30 kcal/kg/day.

Energy requirements of cancer patients are as under:

- Normometabolic patients: 25-30 kcals per kg per day
- Moderately hypermetabolic or weight gain desired: 30-35 kcal per kg per day
- Hypermetabolic: 35-40 kcal per kg per day (More than 35 kcal per kg per may be required to maintain or promote weight gain in some situations such as bone marrow transplantation.).
- Obese patients: 21-25 kcals per kg per day (when weight maintenance is the goal, energy need may be increased when nutritional status is deteriorating.)

Protein

Protein is required for the formation and maintenance of muscles, tissues, red blood cells and hormones, fight infections, strengthen the immune system, prevent muscle wasting and maintaining optimal fluid balance. According to ESPEN guidelines (Arends *et al.*, 2017), the optimal nitrogen supply for cancer patients has not been determined and the recommendations of experts range between a minimum protein supply of 1 g/kg/day and a target supply of 1.2-2.0 g/kg/day, especially if inactivity and systemic inflammation are present. Old age, inactivity and systemic inflammation are known to induce "anabolic resistance", *i.e.* decreased responsiveness of protein synthesis to anabolic stimuli. Evidence-based recommendations for chronically ill older subjects call for a protein supply of 1.2-1.5 g/kg/d.

According to a recent literature review (Bozzetti, 2013), the dose of amino acids capable of supporting a positive protein balance in cancer patients might be close to 2 g/kg/day. In subjects with normal kidney function, intake of protein in doses up to and above 2 g/kg/d are safe; in patients with acute or chronic renal failure protein supply should not exceed 1.0 or 1.2 g/kg/d, respectively (Arends *et al.*, 2017).

Conclusively, 0.5g/kg/day is the minimum daily protein requirement. For normal maintenance level, 0.8-1.0 g/kg/day protein is needed. In case of non-stressed condition of people with cancer, protein requirement is 1-1.5 g per kg per day. However in case of hypermetabolism or protein losing conditions in cancer, it should be increased to 1.5-2.5g per kg per day.

Fat

In most patients, general recommendations (20-30 per cent of total energy) are applicable. In weight losing patients with advanced cancer, a fat intake of 35 50 per cent of total energy requirement is recommended by ESPEN guidelines.

Proportion of Carbohydrates and Fat

The optimal ratio of carbohydrates and fat in feeding cancer patients has not been determined but may be derived from pathophysiologic arguments. In patients with insulin resistance, uptake and oxidation of glucose by muscle cells is impaired; however, utilization of fat is normal or increased (Arcidiacono *et al.*, 2012); thus, suggesting a benefit for a higher fat to carbohydrate ratio. Therefore, in weight-losing cancer patients with insulin resistance, as per ESPEN guidelines (2017), an increase in the ratio of energy from fat to energy from carbohydrates is recommended. This is intended to increase the energy density of the diet and to reduce the glycemic load.

For enteral feeding, the energy density of the diet is important. This is achieved by increasing the proportion of fat. Most dietetic recommendations in anorectic cancer patients are focused on increasing the energy density of the diet and most commercially available products are touted and chosen because of their high energy density. It is well-known that low appetite, early satiety, and reduced bowel motility all conspire to limit the intake of low energy density foods.

Eicosapentaenoic Acid

Many studies have indicated that eicosapentaenoic acid (EPA), an omega-3 polyunsaturated fat supplementation reduces production of pro-inflammatory cytokines such as interleukin-6, interleukin-1and tumour necrosis factor, and in cultured cancer cell lines increases cell death rate.

In cancer patients undergoing chemotherapy at risk of weight loss, ESPEN guidelines (2017) suggest to use the supplementation with long chain n 3 fatty acids or fish oil to stabilize/improve appetite, food intake, lean body mass and body weight.

Micronutrients

Deficiencies of vitamins and minerals can occur as a result of disease and side effects of cancer treatment. In view of the restricted dietary pattern of tumor patients, the use of a multivitamin-multimineral supplement in physiological doses, *i.e.* nutrient amounts that approximately equal the recommended daily allowance, is a useful and safe measure for oral feeding. This also applies to cancer patients during chemo- and radiation therapy. However, discourage the use of high-dose micronutrients in the absence of specific deficiencies.

Similarly, vitamins and trace elements should be generally substituted in parenteral nutrition unless there are contraindications. The supplementation of vitamins and trace elements is obligatory after a parenteral nutrition of more than 1 week.

Probiotics

There is not enough data to recommend *Lactobacillus* containing probiotics to reduce radiation induced diarrhea.

Fluid

Patients with cancer are at risk for dehydration from nausea, vomiting, diarrhea, fever and medications that may be used for treatment. It can be a challenge to stay well-hydrated during cancer treatment. Fluid intake should be 30-35 ml per kg per day *i.e.* similar to other patient population without renal diseases. Fluid needs may also be greater in case of increased losses and dehydration as a result of vomiting, diarrhea *etc.*

Some caution should be used with elderly patients who may have reduced cardiac/renal function (20-25mL/kg per day suggested starting point for IV fluids).

Glutamine Supplementation

There is insufficient evidence to recommend glutamine supplementation during conventional cytotoxic or targeted therapy.

Branched Chain Amino Acids Supplementation

There is not enough clinical data to recommend the supplementation with branched chain amino acids or metabolites to improve fat free mass.

Nutrition Support, Counselling and Dietary Management in Cancer

Unfortunately, data are still lacking that define the optimal time for initiating nutritional support. However, malnutrition is associated with poorer prognosis and it is difficult to revert overt malnutrition in cancer patients with metabolic derangements. Therefore, nutritional therapy should preferably be initiated when patients are not yet severely malnourished and when the goals of care include maintaining or improving nutritional status. Nutritional support should be offered to patients who are likely to develop anorexia or gastrointestinal defects due to the side effects of treatment. Severely malnourished patients who are undergoing active treatment should be offered nutritional therapy immediately.

Nutrition therapy can help cancer patients who are malnourished or at risk of malnutrition to maintain weight and the body's nutrition stores, offering relief from nutrition impact symptoms and improving quality of life but not survival. In patients undergoing (adjuvant) radiotherapy, there is good evidence that nutritional support improves intake and weight, and some aspects of quality of life. Hence, nutritional support should be given by ensuring that the patient has access to suitable food and fluids when they are able to eat orally. Intake should be supplemented using ordinary food, proprietary products or a combination of both. If an adequate oral intake cannot be achieved and it is clinically appropriate, artificial nutrition support preferably via the gut or directly into the blood) should be undertaken.

Forms of Nutrition Support

Generally, the first form of nutritional support should be nutrition counselling to help manage symptoms and encourage the intake of energy-enriched foods and fluids that are better tolerated; a diet enriched in energy and protein is the preferred way to maintain or improve nutritional status. The additional use of ONS (Oral Nutritional Supplements) is advised when an enriched diet is not effective in reaching nutritional goals. Nutritional counselling includes nutritional history, diagnosis, and nutrition therapy. This will require ascertaining individual habits and preferences; in addition, effective counselling requires adequate communication skills to ensure high compliance with the individualized nutritional advice given. Nutrition care process undertaken by the trained nutritionist and dietitians should incorporate calculation or measurement of energy and nutrient requirements, food preparation and/or modifying of texture or nutrient content, increasing meal frequency by distribution of foods to several small meals, enriching dishes with energy- and protein-dense additives, offering oral nutritional supplements, a meal set-up plan that emphasizes supportive interventions to improve oral food intake (*e.g.* treating mucositis and other symptoms), digestion (*e.g.* pancreatic enzymes) or absorption (*e.g.* slowing of rapid gastrointestinal transit), antiemetics, and other relevant conditions.

Lifestyle Guidelines to Minimize Cancer Risk

Key recommendations for all population groups include the following:

- ☆ Choose a diet high in whole grains, legumes, fruit and vegetables, and low in fat and sugar.
- ☆ Prefer fish and white meat over red and processed meats.
- ☆ Select foods low in salt.
- ☆ Vitamin and mineral supplements should not be taken to protect against cancer.
- ☆ Drink alcohol in moderation, if at all.
- ☆ Remain physically active.
- ☆ Maintain a healthy weight and be physically active.

Mode of Nutrition

Artificial nutrition should be started if patients are unable to eat adequately (*e.g.* no food for more than one week or less than 60 per cent of requirement for more than 1-2 weeks). If a decision has been made to feed a patient, enteral nutrition is recommended if oral nutrition remains inadequate over a prolonged period despite nutritional interventions (counselling, oral nutritional supplements *etc.*). As far as possible, oral and other enteral feeding modes should be encouraged as they pose fewer problems. However, parenteral nutrition should be started if enteral nutrition is not sufficient or feasible. Increase enteral or parenteral nutrition only slowly over several days and additional precautions should be taken to prevent a refeeding syndrome.

Recommended Modes of Nutrition Support in Cancer Patients (ESPEN Guidelines, 2017)

- ☆ In surgical cancer patients at moderate or severe nutritional risk, appropriate enteral nutritional support both before and following discharge from hospital is recommended.
- ☆ In upper GI cancer patients undergoing surgical resection in the context of traditional perioperative care, oral/enteral immunonutrition has been recommended.
- ☆ Tube feeding may be done using transnasal or PEG in RT induced severe mucositis or in head neck/throracic cancers with obstructive tumor masses.
- ☆ Parenteral nutrition (PN) is not recommended in general in RT; it should only be initiated if adequate oral/enteral nutrition is not possible, *e.g.* severe RT enteritis, severe mucositis or head neck/oesophageal obstructive cancer masses.
- ☆ If artificial nutrition is required, prefer enteral tube feeding over parenteral nutrition, unless there is severe mucositis or symptomatic gastrointestinal GvHD.

Dietary Management

An individual diet plan is developed taking into consideration his or her nutritional status, type and stage of cancer, type of treatment, side effects of therapy *etc.* Eating a variety of foods and well-balanced energy-protein dense meals as per nutritional requirements can help the patient feel better and stay stronger. Eating well during treatment helps to maintain body weight, improve strength and energy, decreases the risk of infection and assists the body in healing and recovery from cancer treatments. People with cancer who are well nourished and able to maintain a healthy body weight often have a better prognosis. Given below are some common side effects of cancer treatments and suggestions to overcome them, make food more palatable and incorporate nutrition into cancer treatment.

Dietary Tips for Cancer Patients

General dietary tips
Importance of calories and protein ☆ Every person with cancer will have different nutritional needs, but people being treated for cancer sometimes need more protein and energy than usual.
Ways to increase calories and protein ☆ Eat smaller, more frequent snacks throughout the day.

- ☆ Eat favourite foods any time of the day.
- ☆ Eat every few hours – don't wait until hungry.
- ☆ Eat the largest meal when you are the most hungry.
- ☆ Try to choose high-calorie, high-protein foods at each snack.
- ☆ Exercise lightly before meals to increase the appetite.
- ☆ Drink high-calorie, high-protein shakes.
- ☆ Drink fluids between meals instead of with meals.

Eating more nutritious snacks

Snacks between meals may help people who have cancer to gain weight, increase their protein and calorie intake and boost their energy level.

Some easy and nutritious snacks are:

- ☆ Nuts, seeds and nut butters
- ☆ Cereals with milk
- ☆ Dips made with cheese (*paneer*), beans, yogurt or peanut butter
- ☆ Milkshakes or smoothies, curd
- ☆ Cereal and nutty bars
- ☆ Fruits and their juices
- ☆ Vegetables (raw, cooked, juices)
- ☆ Soups with fresh cream
- ☆ Fruit yogurt
- ☆ Sandwiches containing boiled egg slices, grilled cheese or peanut butter

Take more calories

Calories are important for maintaining weight. The patient may need to eat a diet high in calories, especially if he or she is losing weight. Calories provide the extra energy one need while recovering from surgery and fighting cancer.

Carbohydrates and fats are good sources of extra calories. Foods that are high in calories include:

- ☆ Butter, margarine and oils
- ☆ Whole milk, milkshakes and cream
- ☆ Cheese
- ☆ Yogurt with milk fat (M.F.) that is higher than 2 per cent
- ☆ Granola and granola bars
- ☆ Muffins
- ☆ Ice cream
- ☆ Mayonnaise and salad dressings

Ways to add calories to diet

- ☆ Add whole milk, cream, sour cream or yogurt to soups, sauces/ curries, vegetables *etc.*
- ☆ Add whipping cream or whipped topping to desserts, pancakes, waffles, fruit, milkshakes *etc.*
- ☆ Add butter or ghee on *chapatti*, rice, pasta, or cooked *dal* and vegetables.
- ☆ Add butter or cream to soups
- ☆ Use regular mayonnaise and salad dressings on sandwiches, salads and dips.
- ☆ Add jam, marmalade or honey to bread toasts
- ☆ Add ice cream to desserts
- ☆ Add hard cheese to scrambled eggs, sandwiches, soups, sauces *etc.*
- ☆ Add dried fruits to cereals and puddings
- ☆ Add nuts and seeds to cereal, desserts, baked goods and stir-fry vegetables.
- ☆ Eat nut trail daily

Eat more protein

People who have cancer may need to eat a diet high in protein. Protein is important for cell growth and repair and a healthy immune system. Without enough protein, the body may take longer to recover from illness or cancer treatments.

Good sources of protein include dried beans, peas and lentils, soy foods, fish, poultry, lean meat, eggs, dairy products, nuts.

Ways to add extra protein to diet

- ☆ Eat cheese on toast or *paneer* in *roti* roll
- ☆ Add grated or melted cheese/cottage cheese/soy *paneer* to soups, vegetables, scrambled sauces *etc.*
- ☆ Use milk in place of water in hot cereals, soups *etc.*
- ☆ Include cream or cheese sauces or tofu in soups
- ☆ Add yogurt, tofu or cottage cheese to fruits, dips or blended fruit smoothies.
- ☆ Add hard-boiled eggs to salads, soups, vegetable curries or sandwiches.
- ☆ Sprinkle nuts and seeds on salads, pastas, muffins, cookies, ice cream or pudding.
- ☆ Spread peanut or almond butter on toast, fruit, muffins or biscuits, or blend in milkshakes or smoothies.

Fluid

Body needs water and other fluids to stay healthy. One can easily become dehydrated due to side effects of cancer treatment such as vomiting or diarrhea if fluid intake is not sufficient. Fluid intake of normal cancer patient (no renal problem) should be 30-35 ml per kg per day.

Commercial nutrition supplements

Commercial nutrition supplements are a good source of calories and nutrients. They come in a variety of flavours and forms, including a ready-to-use beverage (similar to a milkshake), a powder, a pudding or a bar. They can be tried if someone is not eating regular meals in sufficient quantity and has less appetite.

Suggestions to Alleviate Nutrition-Related Side-Effects In Cancer

Anorexia: Try to maximize intake by taking high-energy/high-protein foods and fluids every 1-2 hours; smaller more frequent meals may help; schedule meal time; eat more when hungry; gentle exercise/ fresh air before meals may promote appetite; take soft, cold and frozen foods such as milk shakes, nutty ice creams; take high calorie, high protein snacks-boiled eggs, roasted nuts as such, nuts/seeds on salads or in custards, cheese, peanut butter sandwich, sprouts *etc.*; prefer liquids or semisolids with nutritional supplements (if there is a dislike for solids); take liquids between meals; avoid filling the stomach with a large amount of liquids before eating; drinking liquids after eating only; have food in pleasant environment; avoid smells that make someone sick.

Taste changes: Foods that look appetizing; try food chilled; eat more of the foods that still taste good; flavouring foods with new tastes or spices; counter salty taste with added sweeteners, sweet taste with added lime juice; stronger flavours like ginger, lemon or spices may help; liquids with meals to rinse away any unpleasant taste; avoid eating in rooms that are stuffy or too warm; maintain regular mouth care.

Bitter or metallic taste: Cooking in microwave oven or glass utensils may minimize metallic taste; use fresh fruits and vegetables instead of canned foods; sauces and seasonings may be added to the food may improve palatability and mask the bitter taste; maintain oral hygiene.

Dry mouth: Sip cool drinks; try using a drinking straw; suck ice chips/ popsicles; chew sugarless gum to keep the insides of the mouth moist; take soft moist foods that are cool or at room temperature; take small bites and chew food well; lots of fluids to loosen mucus; liquids with meals; sharp tastes like grapefruit or lemon may stimulate saliva; serve meals with sauce or gravy; avoid foods that stick to the roof of the mouth; limit caffeine containing drinks like tea, coffee *etc.*; keep mouth clean.
Gum and dental problem: Use soft brush for brushing teeth; avoid sugary and sticky foods; visit dentist regularly.
Nausea and vomiting: Avoid off-putting smells; plain foods in small quantities may be better tolerated; avoid lying down after eating (a gentle walk may help); sip drinks throughout day, but wait for 15 min after eating before taking more fluid; try ginger flavours, mints, toasts and plain biscuits; avoid greasy foods, too sweet foods, fried foods, spicy foods; eating slowly and chewing thoroughly; sucking on hard candy or polo if there is a bad taste in the mouth.
Difficulty in swallowing: Liquids or semisolid foods at frequent intervals; take liquids such as milk shakes, custard, pudding, cream soups, strained vegetable soup, buttermilk, ice cream *etc.*; pureed vegetables or fruits without skin and seeds; soft scrambled egg; soft well cooked, bland foods like *khichdi*, curd rice *etc.*; add butter or sauce to the food.
Mucositis (Sore mouth± Oesophagus): Frequent small meals and snacks, soft in texture; try soft, smooth foods with plenty of sauce; take soft foods like custard, milk shakes, buttermilk; avoid spicy and salty foods and sharp, citrus tastes; chilled or warm foods may be less painful than hot; coarse/crumbly food like toast, crackers, and pastry may be better avoided; avoid rough textured foods like raw fruits and raw vegetables; ensure adequate fluid intake (>1.2L/day); drink through a straw; sucking on ice cubes may alleviate pain; good oral hygiene should include brushing teeth with soft brush twice daily, flossing teeth daily and using mouth wash four times daily; avoid tobacco and alcohol.
Diarrhoea: Avoid irritating foods that exacerbate, *e.g.* pulses, onions, strong spices; consider reducing fibre-rich foods; ensure fluid intake is adequate and try and keep eating even if small quantities of smooth foods; reducing fat intake and milk products is often suggested: this may have a negative effect on nutrient intake so individual advice from a dietitian is required; avoid greasy, fried foods; visit the doctor, if required.
Constipation: This may arise due to cancer, treatment or analgesia, or simply from a poor intake and inactivity; eating more and gentle exercise may help; increase fluid intake ~2 L/day; gradually increasing intake of fruit, vegetables, and wholegrain cereals may help if tolerated; a little hot water on waking may stimulate the bowel.
Lactose intolerance: Lactose free or low lactose diet-restrict milk and milk products; include curd, buttermilk instead of milk.
Fatigue: Ignore the clock and eat when feel more awake; a variety of well-balanced snacks can be as nutritious as a main meal; ask for help with food preparation or use ready meals or home-delivered food.
Unwanted weight gain: Take fruits and vegetables; take lean meats and low fat dairy foods; cut down on added butter and ghee; avoid sweet and high fat foods; daily walks if tolerated and with doctor's approval.
Weight loss: Energy and protein dense supplements in addition to regular diet.

Precautionary Suggestions on Other Alternate Dietary Approaches Available and Used for Cancer Treatment in Literature

Depending on region and culture, different, often complex and contradictory, dietary suggestions are proclaimed to antagonize cancer growth and are proposed as anticancer diets. As many as 48 per cent of cancer patients pursue popular diets, including the alkaline, Paleolithic, ketogenic, vegan, and macrobiotic diets, with the hope that they will improve survival and prevent recurrence. No doubt, focusing on food can be a valuable and positive activity for a cancer patient and his/her carers, but alternative diet based approaches should not automatically be regarded as 'bad' or 'good'. One must know that till date, there are no diets known to reproducibly cure cancer or prevent cancer recurrence.

There is a considerable amount of information about food and cancer available to the general public in literature and some include dietary advice that is based on limited or questionable scientific evidence as given below in the Box 26.1. Whilst some is compatible with what is generally regarded as a healthy, well balanced diet, other advice is not. Some of these include regimes recommending intakes that are nutritionally inadequate and/or unpalatable or advocating complex eating patterns that require either the sourcing of unusual or hard-to-buy items or complicated preparation regimes. One should not follow all forms of diets that are not based on scientific clinical evidence, have no proven efficacy, and that potentially could be harmful. Specific harms may include secondary micronutrient deficiency, exacerbation of malnutrition, and high cost.

In many cases, the supporting arguments are neither based on scientific reasoning nor on solid evidence and the supporting information is derived from anecdote and unverifiable sources in the popular literature and internet rather than peer-reviewed literature. Some diets may be described as fad diets. It is very important to know that one should not follow dietary advice or diets, which increase the risk of inducing or aggravating malnutrition. Fad diets are generally highly restrictive in the type and quantity of specific foods *e.g.* low carbohydrate or low fat diets and, as such, generally restrict food intake. These diets increase the risk of insufficient intake of energy, fat, and protein, as well as generate risk of micronutrient deficiency. Some such diets also have low energy density and/or low protein content. In cancer patients who are already malnourished, this may be harmful and should be avoided. However, patients often are anxious to discuss dietary options and are eager to commit themselves to fighting their cancer by choosing foods that are perceived as "protective". These patients' need should be recognized and acknowledged and it should initiate an unbiased discussion and counselling on what nutrition can and cannot achieve and on the risks associated with an inadequate or restrictive diet. Such cancer patients, who are interested in following the complementary or alternative dietary regimes or so called fad diets, should be encouraged to discuss this with their doctor and dietitian so that the potential benefits and/or detrimental effects can be evaluated with respect to their individual needs. Consideration can be

BOX 26.1: Some Popular Dietary Approaches to Treat Cancer in Literature

Bristol approach to healthy eating

Bristol Cancer Help Centre is the UK's leading holistic cancer charity that has pioneered the *Bristol Approach* to cancer care. It has been developed from the old 'Bristol Diet' at Penny Brohn Cancer Care, UK.

Highlights: As such, the Bristol Diet means avoiding foods that could boost the risk of cancer. Promotes healthy eating and recommends a diet that is based primarily on plant foods including vegetables, fruits, whole grains, pulses, nuts, seeds, herbs, and spices. A smaller quantity of animal products (poultry, eggs) is recommended alongside the plant foods whilst refined grains, 'damaged fats', processed red meat, dairy products, caffeine and carbonated soft drinks, and alcohol are discouraged. Organic food, maintaining a healthy weight and enjoying meals are encouraged.

Comment: This advice includes key aspects of World Cancer Research Fund guidelines and can be nutritionally adequate for people who are eating well. During the fight against cancer, one need to fight depression and eat for strength and at that time, the Bristol Diet doesn't come across as a tempting eating plan at all. Hence, those with a poor appetite may not achieve an adequate energy or protein intake. Moreover, the scientific evidence for discouraging dairy intake is limited and not universally accepted.

Gerson therapy

Highlights: It was developed in the 1920s and 30s by a German doctor called Max Gerson. Gerson therapy is an alternative 'metabolic' therapy, using a special diet, plus supplements and a coffee enema to fight cancer. It aims to stimulate the immune system to rid the body of cancer related toxins. The diet is a very strict low salt one that includes large quantities of organic fruits and vegetables taken in the form of juice. Each day one will need to drink about 9 kgs of crushed fruits and vegetables (one glass of juice hourly, 13 times each day), have 3 or 4 coffee enemas and take potassium and other supplements, including vitamin B_{12}, pancreatic enzymes and thyroid supplements.

Comment: The diet is expensive and potentially nutritionally inadequate. Coffee enemas have been associated with infection, dehydration, heart and lung problems, constipation and inflammation of the bowel. Other side effects of this therapy like loss of appetite, diarrhea, abdominal cramps, aching, fever, sweating, dizziness and weakness *etc.* have been reported. This therapy can be very harmful for cancer patients who are already weak and ill. Recent review by the USA National Cancer Institute (2010) reported that no good quality evidence has been published to support its use.

Macrobiotics

The macrobiotic diet is a fad diet which was first developed by a Japanese philosopher called George Ohsawa and is claimed to be helpful for people with cancer and other chronic diseases. He believed in a holistic approach to health incorporating many lifestyle aspects, from diet and exercise to meditation and even the 'yin and yang' energy of particular foods. The dietary element is based on predominantly vegetarian, high carbohydrate, low fat food with regular consumption of soya and sea vegetables. Major principles of macrobiotic diets are to reduce animal products, eat locally grown foods that are in season, and consume meals in moderation.

Comment*:* The high phytoestrogen content of the diet may offer some benefits but there is no firm scientific evidence to confirm this. A low energy and protein density is a concern in patients with a poor appetite. Neither the American Cancer Society nor Cancer Research UK recommends adopting the diet.

Plant programme

Highlights: Advises the elimination of all dietary dairy produce, reducing meat intake and replacing it with soya-based foods, pulses and cereals because of a perceived causal relationship with cancer, particularly with those that are hormone-mediated, *e.g.* breast and prostate.

Comment: Published epidemiological studies do not support the proposed link. Milk and other dairy products provide useful nutrients in a palatable form and, although avoidance may not comprise intake in relatively well patients, it may be detrimental in those who are eating poorly.

Ketogenic diet

Highlights: The Ketogenic Diet (KD), a high-fat/low-carbohydrate/adequate-protein diet (75 per cent protein, 20 per cent fat and 5 per cent carbohydrates) has recently been proposed as an adjuvant therapy in cancer treatment. The rationale in providing a fat-rich, low-carbohydrate diet in cancer therapy is to reduce circulating glucose levels and induce ketosis such that cancer cells are starved of energy while normal cells adapt their metabolism to use ketone bodies and survive. Furthermore, by reducing blood glucose also levels of insulin and insulin-like growth factor, which are important drivers of cancer cell proliferation, drop. Interesting results of benefits of such diet have been obtained in *in-vitro* and in animal experiments for slowing growth of implanted tumours and prolonging survival. In case of human beings, the currently available medical literature presents strong scientific evidence for the safe application of a KD only in patients with glioblastoma. However, a clear recommendation for adjuvant use of the KD in glioblastoma patients still requires results from ongoing randomized controlled clinical trials.

Comment: Human data pertaining to KDs and cancer are mostly based on single case reports and a smattering of preliminary clinical studies with small study cohorts, heterogenous study designs, poor compliance to the diet, noncomparable regimens, or without standardized dietary guidance. A small observational series and a small randomized trial reported good tolerability of this approach in humans. Further trials are ongoing and cancer-specific recommendations await the findings of randomized controlled clinical trials in humans. Till then, beneficial effect of this diet cannot be advocated. Moreover, due to their low palatability, ketogenic diets may lead to insufficient energy intake and weight loss in cancer and this aspect needs to be looked into in future trials on human beings. Any dietary provision that may restrict energy intake in cancer patients with or at risk of malnutrition cannot be recommended.

given to combining such approaches with conventional nutritional support and healthcare professionals, nutritionists and dietitians should help provide patients with appropriate information so that they can make an informed decision. ESPEN guidelines (2017) recommend against dietary provisions like "anticancer diets" which restrict energy intake in patients with or at risk of malnutrition.

Physical Activity

Cancer patients should be advised to reduce inactivity and to avoid living a sedentary lifestyle and advice should be individualized. Physical activity in cancer patients is associated with maintenance or significant improvements in aerobic capacity, muscle strength, health-related quality of life, and self-esteem, and with reduction in fatigue and anxiety. For some patients, recommendations for physical activity should consist of motivating patients to take a daily walk in order to reduce risks of atrophy due to inactivity. Other patients would probably benefit from physical exercise programs conducted by appropriately trained experts. Additionally, physical activity will also decrease muscle catabolism and increase anabolism and also has the potential to reduce inflammation, all important pathophysiological factors in cancer cachexia.

In general, maintenance or increased level of physical activity has been recommended in cancer patients during and after treatment to support muscle mass, physical function and metabolic pattern. Individualized resistance exercise is suggested to maintain muscle strength and muscle mass during treatment. During high dose anticancer drug treatment and stem cell transplantation also, maintenance of physical activity has been recommended. Physical activity should thus probably be integrated in multimodal treatment programs.

Role of Pharmaconutrients

"Pharmaconutrients" are nutrients supplied in pharmacological doses to modulate immune and metabolic functions and exert effects on clinical outcome. However, these agents cannot substitute for conventional or specialized nutritional support. The nutritional needs of cancer patients should be adequately met, independent of pharmaconutrient or pharmacologic treatment. Specific pharmacological agents may be required or helpful in gastrointestinal disorders with relevance to food intake or absorption as well as in states typically associated with decreased appetite, like microbial infections, chronic pain, or psycho-social distress. This may include the following categories of agents which are sometimes prescribed by the medical doctors:

- Antiemetics to relieve nausea
- Antimicrobials to eliminate fungal, bacterial, or viral causes of gastrointestinal or other infections
- Analgesics to relieve chronic pain or pain associated with chewing, swallowing, or intestinal activity agents to induce saliva production in xerostomia
- Anti-secretory agents to diminish excessive saliva production or vomiting in cases of impaired intestinal transport
- Inhibitors of gastric acid secretion and other substances to treat or protect against symptomatic mucosal lesions or oesophageal reflux
- Agents to maintain or normalize intestinal motility and to treat or avoid constipation or diarrhea
- Antidepressants, agents that relieve anxiety, mood modulators

27

Food Hypersensitivity

If someone reacts to a food, he or she may have food hypersensitivity. Food hypersensitivity (FHS) reactions can be categorized as immune-mediated (food allergies) and non-immune mediated (food intolerances). Hence, food hypersensitivity is used to refer broadly to both food intolerances and food allergies. A food intolerance is not the same as a food allergy, although some of the symptoms may be similar. A true food allergy is one particular type of food intolerance that involves the body's immune system. The immune system is part of the body's defence system, as it protects against foreign organisms like bacteria and viruses. In some people, it may also react to substances in foods, or in the environment, *e.g.* pollen or animal fur. This abnormal reaction is known as an allergic reaction. People with food allergies develop antibodies against certain proteins in foods known as allergens. When they eat that food their body reacts, usually immediately or less than an hour after eating, although some babies can have very delayed reactions to milk.

Classification of Food Hypersensitivity

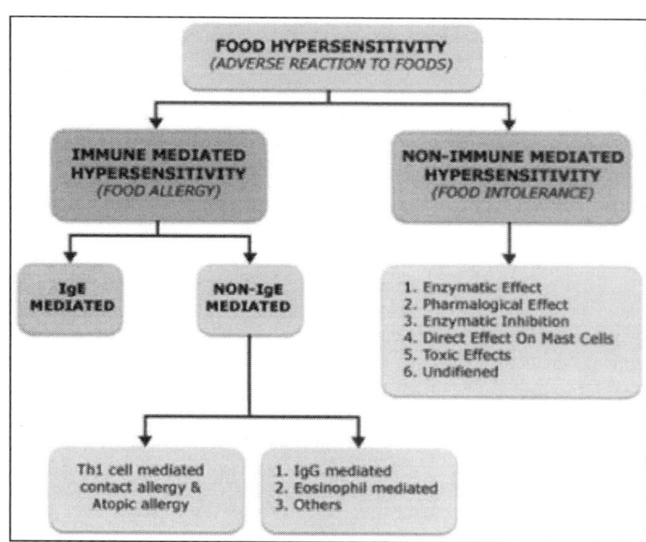

IgE-mediated Food Hypersensitivity

- A food allergy to be a hypersensitive reaction to a food that is mediated by the immune system, particularly involving IgE antibodies includes classic primary and secondary sensitization to foods *e.g.* reactions to milk, peanut, soya, eggs, tree nuts, fish, shell fish *etc.*, causing a spectrum of symptoms from mild oropharyngeal symptoms through to anaphylaxis.
- Characterized by immediate symptoms. Symptoms usually begin within a few hours of eating the food one is intolerant to.
- Prevalence of 4–8 per cent in childhood and 1.3–4 per cent in adults.

Non-IgE Mediated and Non-immune Mediated FHS

- Non-IgE food allergy includes coeliac disease, food-induced proctitis and entercolitis (FPIES) and eosinophilic oesophagitis.
- An adverse reaction that does not involve the immune system directly is considered food intolerance. It involves the digestive system generally and is different from food allergy. Non-immune mediated causes of food intolerances include enzymatic reactions such as lactose intolerance, pharmacological reactions

such as hypersensitivity to vasoactive amines, and salicylates and reactions of unknown aetiology such as those to food additives (preservatives, colouring agents, emulsifiers and taste enhanchers). Metabolic food reactions are due to inborn or acquired errors of metabolism of nutrients such as diabetes mellitus, phenylketonuria and favism. Food intolerance does not include either psychological responses or foodborne illnesses. Food intolerance may be linked to other conditions; about 70 per cent of people with irritable bowel syndrome report symptoms to specific foods.

Causes of Food Intolerance and their Mechanism of Action

There are many types of food intolerances, including:
✰ **Enzymatic**-Lactose intolerance
✰ **Intolerance to a protein, gluten** (Celiac disease)
✰ **Pharmacological reactions** Pharmacological intolerances involve reactions to certain chemical naturally occurring foods such as: ✰ Amines *e.g.* histamines containing foods such as cheese, cured and processed meat products. Low amine oxidase activity in some individuals limits detoxification of dietary histamine and increased levels of histamines can increase smooth muscle contractions. ✰ Food additives, especially the benzoate and sulphite preservatives and monosodium glutamate ✰ Caffeine in coffee, tea, cola drinks and chocolates. Caffeine stimulates central nervous system and increases gastric juice secretion and colonic motor activity possibly via gastrointestinal neuroendocrine hormones (*e.g.* cholecystokinin, exorphin, gastrin or motilin) but this is unknown. ✰ Salicylates in coffee, tea, green apples, banana, lemon, plums, grapes, carrots, peas, cucumber, grapefruit *etc.* They stimulate mast cells to overproduce leukotriene metabolites which in turn may lead to pro-inflammatory reactions and smooth muscle contraction. ✰ Glutamates (Tomato, cheese, yeast extract *etc.*). Proposed mechanism of GI symptoms provocation is unknown but exclusion of dietary additive excitotoxins (including MSG) in IBS resolved >30 per cent of symptoms in 84 per cent of patients. ✰ Toxin

- ✰ Onset of symptoms for many conditions within this FHS spectrum is usually delayed by upto 48 h and last for hours or even days, making the offending food especially difficult to pinpoint. Symptoms of food intolerance are usually less severe than those in food allergy.
- ✰ Actual prevalence for many conditions is unknown, except for well characterized disorders such as lactose intolerance and coeliac disease.
- ✰ 30 per cent or more of the population perceive themselves or their child to have FHS, whereas the actual prevalence varies greatly according to the condition.

Symptoms

It can be difficult to determine whether the person has food intolerance or food allergy because the signs and symptoms often overlap. Depending on the food involved and the sensitivity of individuals, symptoms associated with food allergy or intolerance can appear immediately (within minutes) or take several hours or days to manifest. The symptoms of food intolerance often take longer to emerge when compared to those of food allergy. When it is allergy, even small amounts of food result in symptoms, as may be the case with peanuts. Whereas, with food intolerance, tiny amounts of food usually have no effect. People can experience hypersensitive responses to more than one type of food. Some foods can trigger both an allergic reaction and intolerance *e.g.*, milk, wheat *etc.* Symptoms can range from mild irritation to severe or life threatening anaphylaxis and may affect the skin, oral cavity, gastrointestinal tract, respiratory tract or the eyes. Symptoms of food allergy can be severe and include life-threatening anaphylaxis. Anaphylaxis is a rapid and potentially life-threatening allergic reaction caused by exposure to allergens including certain foods. Symptoms experienced during anaphylactic reactions to foods such as peanut can begin in the mouth and throat within minutes of ingestion and quickly progress to affect the pharynx, skin, respiratory tract and cardiovascular system, either individually or in combination. People susceptible to anaphylactic reactions need to carry life-saving adrenaline in case of emergency. Common symptoms of food intolerance include diarrhea, bloating, rashes, nausea, fatigue, abdominal pain, headache, joint pains, constipation, flushing of the nose *etc.* Because

Table 27.1: Symptoms of Food Hypersensitivity

IgE-mediated	*Non-IgE-mediated*
✰ Pruritis, erythema, diarrhoea and abdominal pain are common to both types. ✰ Acute urticaria - localized or generalized. ✰ Acute angio-oedema - commonly the mouth, lips, face, around eyes. ✰ Oral itching, nausea, vomiting. ✰ Colicky abdominal pain. ✰ Nasal itching, sneezing, rhinorrhoea, allergic conjunctivitis. ✰ Cough, shortness of breath, wheezing and bronchospasm (or history of asthma). ✰ Other signs of anaphylaxis, feeling of impending doom, cardiovascular collapse.	✰ Pruritus, erythema, diarrhoea and abdominal pain are common to both types. ✰ Atopic eczema. ✰ Gastro-oesophageal reflux. ✰ Infantile colic. ✰ Stools: loose and/or frequent, blood and/or mucus. ✰ Constipation. ✰ Perianal redness. ✰ Pallor and tiredness. ✰ Faltering growth. ✰ Food aversion or avoidance.

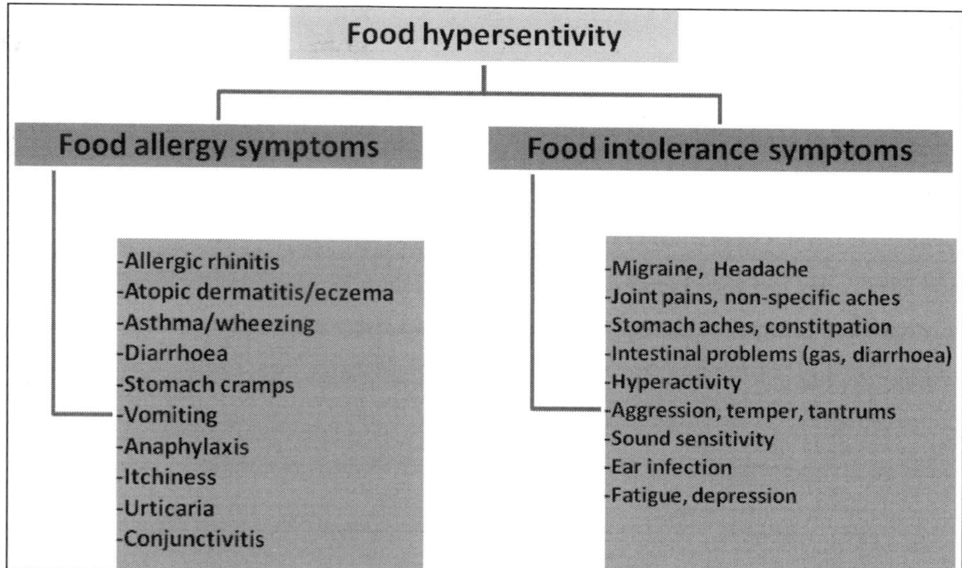

many of the symptoms associated with food allergy or intolerance can have unrelated triggers, medical advice should be sought before taking remedial action such as dietary alteration.

Methods to Diagnose Food Hypersensitivity

Medical consultation is recommended when a person believes that he is suffering from a food allergy or intolerance. Accurate medical diagnosis is particularly important for children or where the symptoms cause severe or persistent discomfort. Incorrect diagnosis can result in the actual cause of a health problem being overlooked and in unnecessary dietary restrictions. There is no perfect test to confirm or rule out food allergy. The doctors/allergist will consider a number of factors before making a diagnosis and these are as under:

Medical History and Personal Examination

The typical diagnostic routine begins with a medical history (Figure 27.1) to determine whether the symptoms are potentially related to ingestion of specific foods, whether adverse reactions are allergic in nature and, if so, the likely pathophysiologic basis. Therefore, the

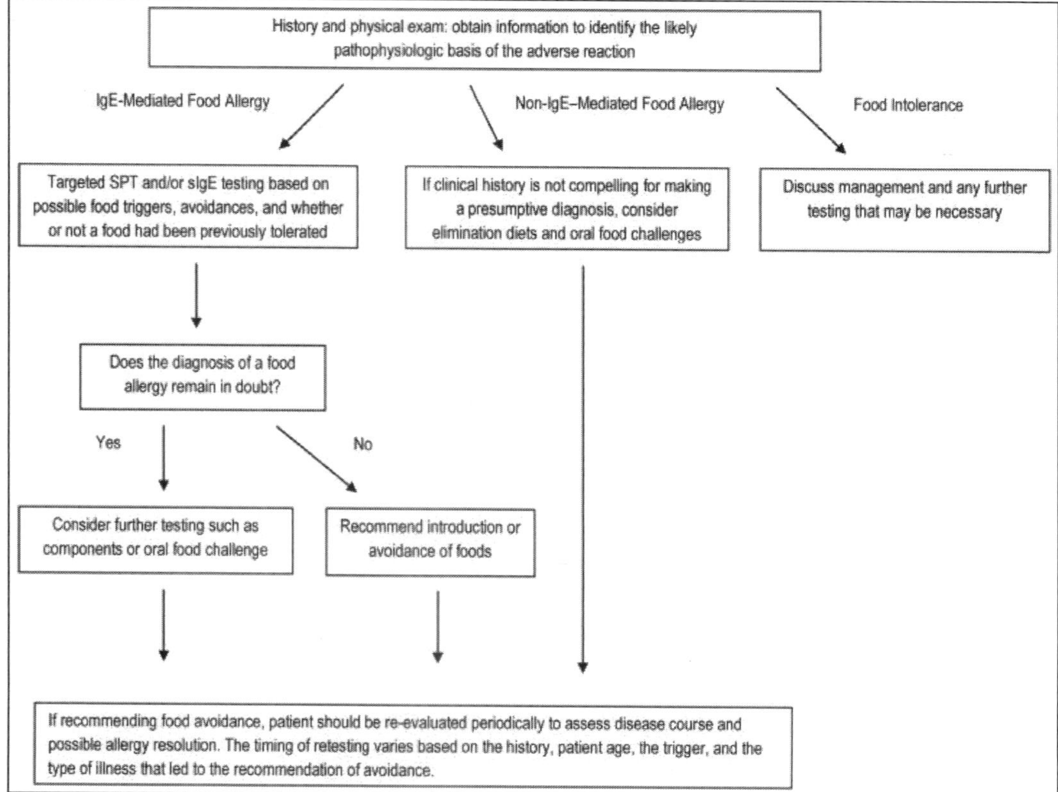

Figure 27.1: Food Allergy Diagnostic Algorithm.

first step an allergist/doctor will take to diagnose a food allergy is a thorough medical history of the patient. A detailed history about which foods, and how much, seem to cause problems, should be provided to identify the culprit food. Family history of any food allergies should also be provided.

The history is important in determining the likely pathophysiologic basis of the reaction, specifically whether the food-induced allergic reaction is IgE mediated. This is important because tests of food-specific IgE would not be diagnostic for disorders that are cell mediated/non-IgE mediated such as food protein-induced enterocolitis syndrome. Once a possible food trigger is identified, additional history can help decipher if that food is the likely culprit for the reaction. For example, a food that was not previously ingested or was ingested infrequently is more likely to have caused an acute reaction than one that had been regularly tolerated. Common ("major") allergens such as milk, egg, wheat, soy, peanut, tree nuts, fish, and shellfish are more likely to be triggers than other foods. The history is therefore, an important tool to guide allergy test selection and interpretation. After that, a careful physical examination by the medical doctor should be done as it can often identify or exclude other medical problems.

Standard Diagnostic Tests

Knowledge of the epidemiology of food allergy and details of the history may identify potential triggers to which simple tests, such as skin pricking test (SPT) and serum food specific IgE (sIgE) levels can be applied and interpreted in the context of the history and a knowledge of test limitations. When the diagnosis is uncertain, an oral food challenges (OFC) can be utilized as the diagnostic "gold standard".

Laboratory Investigations

These may include blood and faecal tests, endoscopy and/or radiological imaging to rule out any organic disease. It is not easy to determine whether somebody has a food intolerance or allergy because the signs and symptoms overlap. The doctor may recommend a skin test and/or blood test to rule out food allergy. Apart from lactose intolerance and celiac disease, there is no accurate, reliable, and validated test to identify food intolerance.

A Skin Prick Test (SPT)

This test, sometimes called scratch test also, measures the presence of IgE antibodies for the suspect food and is a good first line test for all food hypersensitivities. This test is inexpensive, no bleeding, produces immediate results (within 30 min) and provides 95 per cent negative predictive value and good sensitivity. This form of testing is often helpful to rapidly rule out an IgE-mediated food allergy. It can also help to confirm a food allergy when positive, in the setting of a recent clear history of an acute allergic reaction to the tested food (high prior probability).

Depending upon the case history, one may be tested for one food or several. During this test, a drop of solution containing the food allergen is placed on the forearm or back. (When testing fruits or vegetables, fresh food may be used instead of a solution.) Using a small plastic probe or needle, the doctor gently pricks or scratches the skin to allow a tiny amount of the solution to enter just below the skin surface. The scratching on the surface of the skin feels similar to a fingernail scratch. It is possible to test for multiple allergies at once since each one can be tested with its own scratch. Positive results are indicated by a wheal, a raised white bump surrounded by a small circle of itchy red skin. A skin wheal over a certain size is predictive at 95 or 100 per cent level for milk, egg and peanuts. In general, a large wheal is more likely to indicate a true food allergy, but size is not always an accurate predictor. If no wheal appears, it is unlikely that the person is allergic to the test food. Negative results are almost fairly accurate as SPTs seldom produce 'false negative' results. Positive tests, however, are not always accurate. About 50-60 per cent of all SPTs yield "false positive" results, meaning that the test shows positive even though the person is not really allergic to the food being tested. Hence, SPTs are considered less sensitive than some other food allergy tests. One should keep in mind that a positive reaction to this test alone is not enough to confirm food allergy.

There is debate over whether SPT is suitable if the patient is taking antihistamines, many antidepressants and some heartburn medications or has reported anaphylaxis. SPT may not be possible in extensive eczema where there is insufficient clear skin available to allow testing.

Serum IgE

Blood test measures the presence of IgE antibodies (allergy related antibody known as immunoglobulin E that triggers the food allergy symptoms) to specific foods. It involves taking a blood sample and checking for each allergen suspected. Allergens may be selected one at a time or by choosing panels such as food panels, which contain the most common adult or child food allergens. Serum food specific IgE (sIgE) -mediated hypersensitivity to milk, eggs, and peanuts tends to be more common in children, whereas peanuts, tree nuts, fish, and shellfish are more commonly associated with reactions in adults. Children are more likely to outgrow allergy to milk, soy, wheat, and eggs than allergy to peanuts, tree nuts, fish, and shellfish, only about 20 per cent of children outgrow peanut allergy.

Advantages

This test is good for validating SPT or if SPTs are not possible. It has good negative predictive value and sensitivity with cut off values that are highly positively predictive for some foods. Unlike the skin prick test, the blood test is not affected by antihistamines and can be performed for people with extensive rashes that prevent using skin tests and those who have experienced an anaphylactic reaction.

Table 27.2: Predictive sIgE Values for Reactivity to Egg, Milk, and Peanut

Food	*Mean Age 5 years, 50 per cent React*	*Mean Age 5 years, ~95 per cent React*	*Age <2 years, ~95 per cent React*
Egg (kU_A/L)	2	7	2
Milk (kU_A/L)	2	15	5
Peanut (kU_A/L)	2 (with a clinical history) 5 (without a clinical history)	14	

Source: Data from Sicherer SH, Sampson HA. Food allergy: Epidemiology, pathogenesis, diagnosis, and treatment. J Allergy Clin Immunol 2014; 133:291–307.

Limitations

No immediate results like SPT and not a good predictor for wheat (unless checking specifically for the wheat allergen omega 5 gliadin), soy, fruits and vegetables, naturally occurring substances in food followed by open or blinded challenge if symptoms improve. Another important limitation is that sIgE levels are not highly correlated with reaction severity. If a specific IgE test is negative, then chances are that the person tested is not allergic to that substance, but a positive test must be evaluated in the context of the person's clinical history. Someone can have a low level of IgE and still have a severe reaction to actual exposure to the allergen or an elevated level and never experience a reaction. Children who outgrow a food allergy may continue to have positive IgE test results for many years.

Limitations of Both SPT and sIgE

Neither SPT nor sIgE are sufficient to diagnose food allergy on their own. These tests are not predictive of reaction severity. Importantly, there are situations where clinical reactions occur despite negative tests either because the allergy is not IgE mediated, for example with food protein-induced enterocolitis syndrome, or because the test does not detect the relevant antigen. These limitations underscore the importance of the medical history-if the suspicion is high, a negative test may not be sufficient to exclude a food allergy. Cross reactivity among foods is often not clinically relevant and this must also be considered in test interpretation. For example, 50 per cent of persons with peanut allergy test positive to other legumes, but 95 per cent tolerate these legumes. Under these situations, oral food challenge 'OFC' proves to be a gold standard for diagnosis.

Oral Food Challenge

Given the limitations of current testing modalities, it is often necessary to perform an Oral Food Challenge (OFC) to make the distinction between a patient who is simply sensitized to an allergen and a patient who is clinically reactive to it. This is used to establish a diagnosis for IgE-mediated FHS if discordance between history, test results and diagnostic diet exists. It is a definitive diagnostic test for most non-IgE mediated and non-immune mediated FHS. OFCs are also helpful to determine if a child has outgrown their allergy.

Speed of onset and symptom severity normally dictate whether a challenge should be carried out in the hospital setting and whether it should be open or blinded. An OFC is generally a physician-supervised oral provocation procedure performed in a hospital or outpatient office, where a patient ingests gradually increasing amounts of a food under medical supervision until an age-appropriate serving is reached or the feeding is terminated because of symptoms. There are a number of factors that must be considered before performing an OFC, including the patient's medical history, age, past adverse food reactions, skin prick test and serum food allergen-specific IgE results, and the importance of the food to the patient, both socially and nutritionally. Indications for performing the OFC can be found in Box 27.1. The OFC is usually performed when the patient's history and test results suggest that the allergy may not exist or may have resolved. The clinician can use the medical history and test results to estimate a probability of tolerance, and the odds can then be considered in the context of the patient's needs, weighing the risks and benefits.

BOX 27.1
Indications for an Oral Food Challenge (OFC)

- ☆ Identify foods causing acute reactions for initial diagnosis of food allergy and for monitoring resolution of food allergy.
- ☆ Determine whether food allergens associated with chronic conditions such as atopic dermatitis or allergic eosinophilic esophagitis will cause immediate reactions.
- ☆ Expand the diet in persons with multiple dietary restrictions, usually because of subjective complaints such as headaches or hyperactive behavior.
- ☆ Assess the status of tolerance to cross-reactive foods.
- ☆ Assess the effect of food processing on food tolerability, *e.g.* fruits and vegetables that may be tolerated in cooked form in the pollen-food allergy syndrome.

The OFC can be performed as an open feeding or as a single blind, or double-blind, placebo-controlled food challenge (DBPCFC). In an open challenge, the patient is aware that they are ingesting the potentially problematic food, which can lead to symptoms from anxiety, biasing the test. While an open feeding is the most common approach used clinically for OFCs, this bias must be considered if the child fails the challenge. The DBPCFC is the most specific test for the diagnosis of a food allergy. In this test, the food is masked in a carrier, and the food to be tested or placebo is randomly administered at different times. While the DBPCFC is currently the gold standard test for diagnosing food allergy, due to the time and labor intensive nature of this method it is typically only performed in research studies and select cases in clinical practice. A single-blind placebo-controlled OFC may be helpful if the patient is suspected of having a psychological response. The open challenge is a reasonable first choice when the need for OFC is established, as less than 1/3 of suspected foods result in a positive challenge, and many of these positive OFCs will have objective symptoms.

How a food is processed can affect the challenge outcome. Heating can change the protein conformation of certain foods and may result in a change in allergenicity. This is seen with milk and egg, as well as with a number of fruits and vegetables implicated in pollen-food allergy syndrome. Based on studies of peanut, patients may be at risk of their allergy recurring if they fail to incorporate the food into their diet on a regular basis after passing an OFC.

Diagnostic Diets or Trial Elimination Diet

The best diagnostic tool is an exclusion diet, also known as elimination or diagnostic diet. It is useful if there is discordance between tests and clinical history and is essential if no suitable tests are available. Total exclusion, or 'few foods' diets are rarely used for IgE-mediated FHS but may be useful for other FHS. Unsupervised elimination diets may be unsafe and diagnostic diets should be managed by a dietitian to ensure dietary adequacy. No need for diagnostic avoidance diet if test result is strongly concordant with clinical history. In this method, patients are advised to keep a diary and write down which foods are eaten, what the symptoms were like, and when they appeared. The data in the diary can help a dietician or doctor to identify which foods are causing adverse reactions, and what steps to take. Exclusion diets are extremely useful in isolating the culprit foods.

In a typical exclusion diet, the suspected food group, food additive or naturally occurring substance in a food is avoided in diet for a set period, usually between 2 weeks and 2 months followed by open or blinded challenge if symptoms improve. If during this period, the adverse reactions resolve, it becomes more likely that the culprit has been found. This can be reconfirmed if it is then reintroduced and symptoms return. It is important to emphasize that elimination diets must only be undertaken for a short term, under strict medical supervision. Prolonged restricted diets can lead to problems with adequate nutrition, particularly in children. Patients should be discouraged from using the results of such tests as a basis for implementing dietary change.

Emerging Diagnostic Tests

Component Resolved Diagnostics (CRD)

Allergen CRD have garnered a lot of attention in recent years, with the hope of offering a more accurate assessment of allergic status. Instead of using crude allergen extracts consisting of a mixture of components, CRD measures IgE to individual allergen proteins. In recent years, a number of studies on a variety of food allergens have demonstrated that CRD can improve the specificity of allergy testing. The utility of CRD has been best demonstrated in studies on peanut allergy. CRD has the potential to be a very useful tool in the diagnosis of hazelnut allergy as well. While a number of studies have reported the utility of component testing for allergies to egg, milk, wheat, soy, and fruits, results from these studies have varied, likely reflecting different study populations and methods, manners of sensitization, environmental exposures, and degree of sensitization to various food components. In studies on cow's milk, hen's egg, and shrimp component testing, CRD offered increased specificity, but decreased sensitivity, when compared to traditional SPT and serum specific IgE

Key Points Regarding Diagnosing of Food Hypersensitivities

- Accurate diagnosis of food allergies is vital to identify patients who may have severe, life threatening allergic reactions, and to exclude suspected allergies that could lead to unnecessary dietary restrictions.
- Traditional tests for food allergy have a number of limitations; skin prick testing and food specific IgE levels are excellent tools for detecting sensitization to foods, but often positive tests are clinically irrelevant. While oral food challenges are the gold standard for diagnosing food allergy, they are time consuming and costly, and may result in an allergic reaction.
- Recent studies have identified a number of testing modalities that may improve the ability to identify true clinical reactivity/ severity, including component-resolved diagnostics, basophil activation studies, T-cell proliferative responses, and measurement of platelet activating factor.

Table 27.3: Emerging and Future Diagnostic Tests

Testing Modality	Potential Advantages	Pitfalls
Component Resolved Diagnostics	☆ Some components after increased specificity compared to SPT and sIgE. ☆ Requires a small amount of patient serum	☆ Decreased sensitivity for a variety of food allergens ☆ In studies for a number of foods, results vary, likely reflecting differences among study populations, manners of sensitization, and environmental exposures
Epitope Binding	☆ May be a marker of allergy persistence ☆ More specific marker of clinical reactivity	☆ Yet to be studied for a number of common food allergens
T Cell Responses	☆ Improved specificity over SPT and sIgE	☆ Difficult to measure, unlikely to become commercially available in the near future
Basophil Activation	☆ May better distinguish between milk allergic patients who can tolerate baked milk and those who cannot ☆ A potential marker of allergy severity	☆ Requires the ability to perform flow cytometry ☆ Not yet evaluated in a number of common food allergens
Platelet Activating Factor and PAF Acetylhydrolase	☆ A potential marker of allergy severity and risk for anaphylaxis	☆ Clinical utility yet to be substantiated in multiple studies ☆ Unclear if this can be used as a marker of allergy severity at baseline

testing. From a practical standpoint, this means that CRD for these foods could be used as an adjunct to currently used allergy tests to avoid performing some OFCs, but it is not ready to be used on its own in ruling out a food allergy.

A number of other emerging diagnostic tests such as epitope binding, T cell response, Basophil activation and Platelet Activating Factor and PAF Acetylhydrolase for food hypersensitivities have been highlighted along with their potential advanatages and pitfalls in Table 27.3.

Conclusively, the work up of a potential food allergy can be a complex assessment involving the clinical history, SPT, and sIgE levels, although ultimately these diagnostic tools may be inadequate to definitively diagnose a food allergy. Currently, OFCs remain the most definitive test in the diagnosis of food allergy, but they are time consuming, costly, and have the potential to elicit a severe allergic reaction. In recent years, several different testing modalities, including CRD, basophil activation studies, T-cell proliferation assays, and measurement of PAF have demonstrated potential to aid the allergist in better identifying a patient with true clinical reactivity to an allergen. While CRD is currently used clinically, the remaining tests are promising but require further evaluation before they are ready for implementation in clinical practice. The allergist should combine the test results with the information given in the medical history to provide a diagnosis finally.

Dietary and Lifestyle Management

The main management for any FHS reaction is avoidance of the trigger food(s) that cause signs and symptoms. For most foods, the avoidance advice will be similar but degree may vary depending on the severity of symptoms and type of reaction. Always read food labels to make sure that they do not contain an ingredient *e.g.* milk, eggs, peanuts, tree nuts, shellfish, soy, wheat *etc.* which causes allergy to someone. While taking food in restaurant/hotel *i.e.* outside home, one must get ensured that the ingredient causing food allergy in not present in the dishes ordered. In case of children, caregivers, teachers, baby sitters, relatives *etc.* should be instructed and informed to avoid giving allergy causing foods to the child.

However, despite best efforts, if someone comes into contact with a food that causes a reaction, take the prescribed antihistamines to reduce symptoms. Most people with severe food allergies should carry an epinephrine injection always.

Below is the list of common foods to which persons are generative hypersensitive or allergic and ways to manage such food hypersensitivities:

Cow's Milk

- ☆ Cow's milk protein allergy (CMPA), which is also commonly referred to as cow's milk allergy (CMA), is the leading cause of food allergy in up to 2-7.5 per cent of infants and children younger than three years. Cow's milk protein can be present in breast milk if the mother has cow's milk in her diet, and it is also in cow's milk protein formula. Most infants with CMA develop symptoms within the first month after introduction of CMP-based formula.
- ☆ Most children (75 per cent of cases) outgrow cow's milk allergy by the age of 3-5 years. However, in some people cow's milk allergy may not be outgrown.
- ☆ This food allergy may be immunoglobulin E (IgE) mediated with rapid onset of symptoms such as urticaria or angioedema and/or non-IgE mediated, producing more delayed symptoms such as eczema, gastro-esophageal reflux or diarrhoea.

- Gastrointestinal manifestations of CMPA are nonspecific. CMA does not include other adverse reactions to milk, such as lactose intolerance, which are non-immune mediated.
- The signs of CMPA are rather variable but most of the time are non-specific and include oral and perioral swelling; dysphagia, and food impaction (e.g, impaired esophageal motility); vomiting, regurgitation, dyspepsia, early satiety, anorexia, and food refusal (delayed gastric emptying); diarrhea (with or without malabsorption or protein loss due to enteropathy), rectal bleeding, failure to thrive, abdominal pain, severe colic, and persistent constipation often with perianal abnormalities; atopic dermatitis, swelling of lips or eyelids, urticaria unrelated to acute infection, drug intake or other causes; and respiratory symptoms such as runny nose, chronic cough and wheezing. Chronic iron-deficiency anemia may be the sole manifestation of CMPA in infants and children.
- In a small group of older children, CMPA may present with symptoms of GERD but also with dyspepsia or abdominal pain, and hence, may be easily confused with functional gastrointestinal disorders or lactose intolerance. Therefore, the challenge remains to make a correct diagnosis while minimizing the burden to patient and family.
- In comparison to cow's milk formula fed infants, exclusive breast feeding during the first 6 months of life reduces the risk for CMPA and most severe allergic manifestations during early infancy. The distinction between breast fed (Figure 27.2) and formula fed infants (Figure 27.3) reflects the importance of ensuring an adequate duration of breast feeding. Management principles also differ. The management of breast fed infants depends on reducing the maternal allergen load and strict avoidance of CMP in supplementary feeding. It is recommended that exclusive or partial breast feeding is continued, unless alarm symptoms require a different management.
- Diet should exclude all cow and other mammalian milks such as goat or sheep milk; also milk components such as milk solids, lactose, casein, and whey; foods containing milk *e.g.* flavoured crisps, baked beans and breakfast cereals.
- Infants 0–6 months should be exclusively breastfed or given an extensively hydrolyzed or amino acid formula after 6 months. Mothers should be encouraged to continue breast-feeding while avoiding all milk and milk products from their own diet. Milk formulae can be hydrolyzed in order to remove allergenic epitopes. A number of alternatives to cow's milk- based formulae exist and include amino acid formula (AAF), partially hydrolyzed formula (PHF), extensively

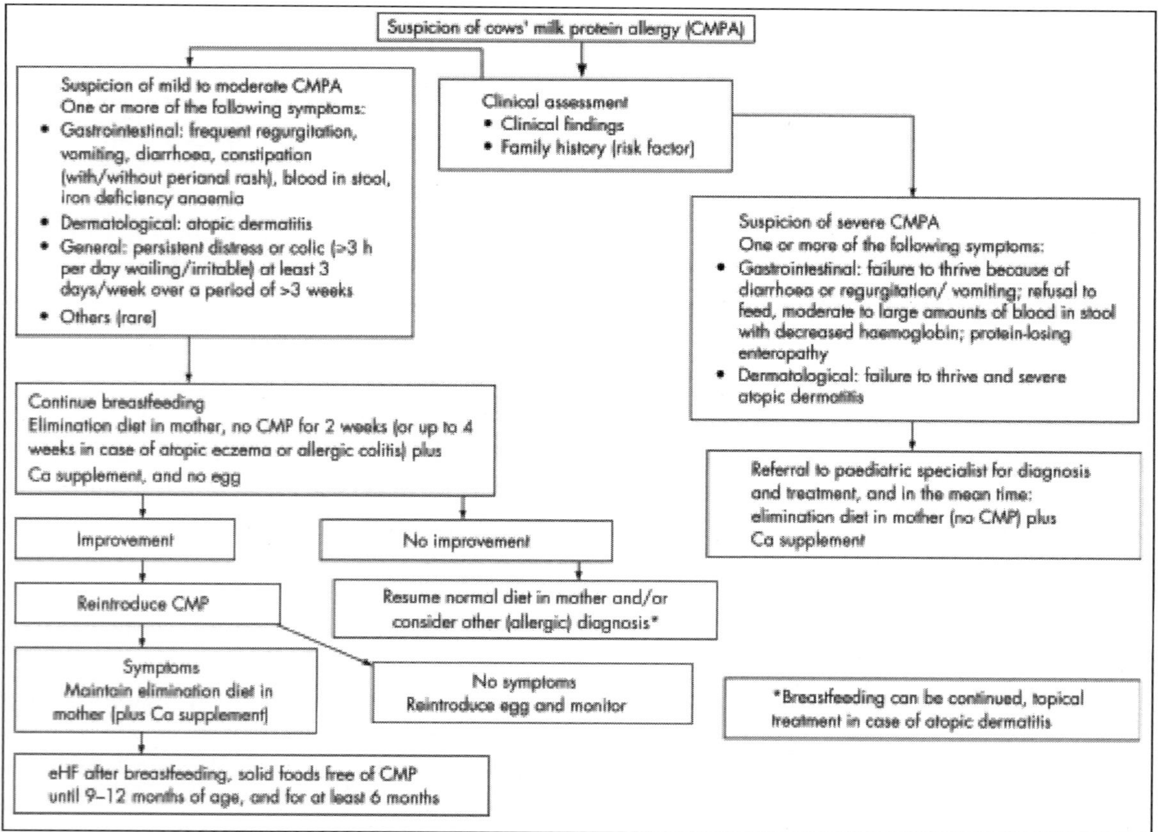

Figure 27.2: Algorithm for the Diagnosis and Management of Cow's Milk Protein Allergy (CMPA) in Exclusively Breast Fed Infants. eHF, Extensively Hydrolysed Formula.

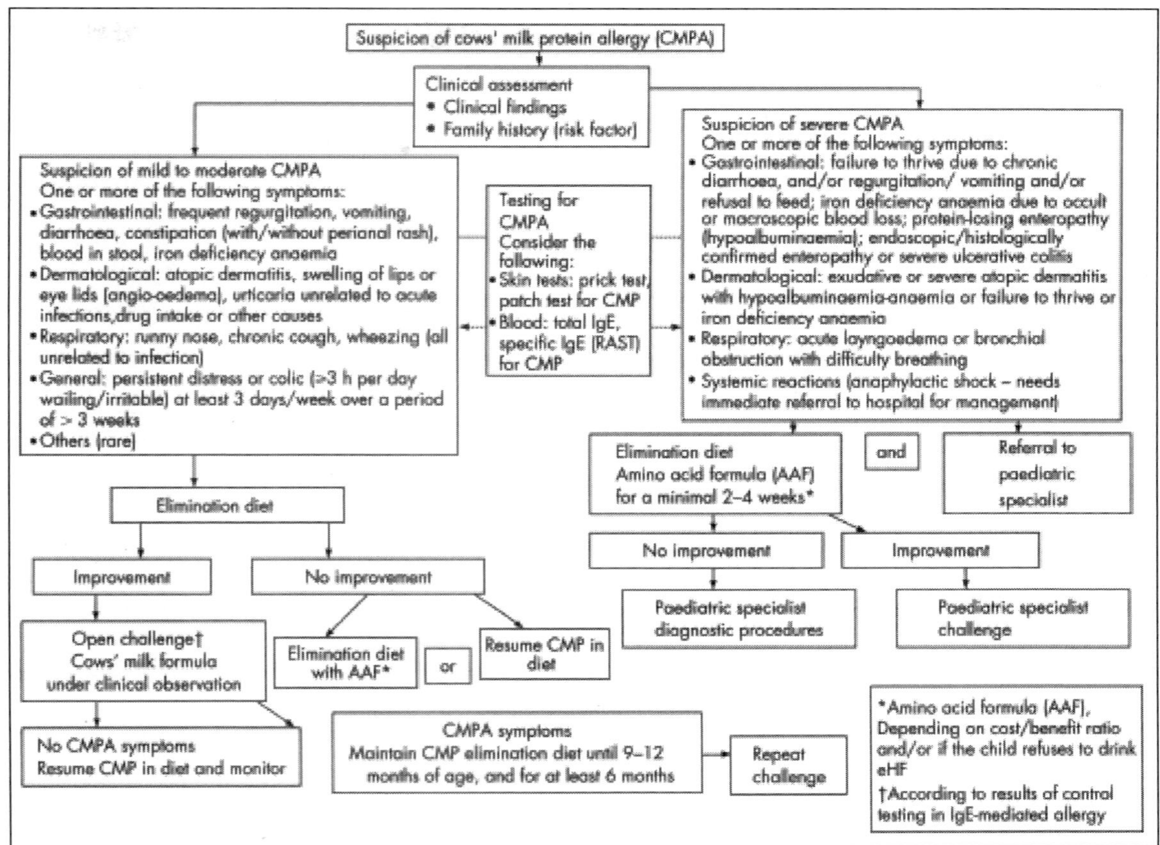

Figure 27.3: Algorithm for the Diagnosis and Management of Cow's Milk Protein Allergy (CMPA) in Formula Fed Infants.

hydrolyzed formula (eHF), casein or whey, rice partially and/or eHF, Soy formula, soy hydrolyzed formula *etc.* Milk substitutes fortified with energy, calcium, and B vitamins can be used in > 5 years; rice milk, oat milk, goat's milk, and unfortified soy milk are not suitable.

- ☆ In children older than 2 years, a nutritionally adequate elimination diet can be provided by solid foods and liquids free of CMP unless the child has multiple allergies. Goat's- and sheep's-milk protein should be strictly avoided because of the high cross-reactivity with CMP.
- ☆ Calcium and vitamin D supplements may be required by both children and adults.
- ☆ Nutritional adequacy of milk-free diets that do not include a milk replacement may be compromised.

Eggs

- ☆ After cow's milk, hen's egg allergy is the second most common food hypersensitivity in infants and young children.
- ☆ Egg allergy may be defined as an adverse reaction of immunological nature induced by egg proteins and includes IgE antibody-mediated allergy as well as other allergic syndromes such as atopic dermatitis and eosinophilic esophagitis, which are mixed IgE- and cell-mediated disorders.
- ☆ The molecules that trigger allergic reaction can be either in egg whites or egg yolks, but allergies to egg whites are more common. Most of the allergenic egg proteins found in egg white include ovomucoid (Gal d 1, 11 per cent), ovalbumin (Gal d 2, 54 per cent), ovotransferrin (Gal d 3, 12 per cent) and lysozyme (Gal d 4, 3.4 per cent). Although ovalbumin is the most abundant protein comprising hen's egg white, ovomucoid has been shown to be the dominant allergen in egg. In egg yolk, alpha-livetin (Gal d 5) is the major allergen. Several other allergens have been identified in egg yolk, including vitellenin (apovitellenin I) and apoprotein B (apovitellenin VI), although their roles in food allergy remain unclear.
- ☆ The specific molecule in eggs that triggers allergy may be present in both chicken eggs and duck eggs. Some people can therefore be allergic to both chicken and duck eggs. This is known as cross-reactivity.
- ☆ Affects 1.6–3.2 per cent of children and most commonly presents in the first year of life. Allergy to hen's egg usually presents in the second half of the first year of life, with a median

age of presentation of 10 months. This reflects the typical age of the first dietary exposure to egg. It has been shown that most reactions occur upon first known exposure to egg, particularly in sensitized children with atopic dermatitis. The development of sensitization in these patients may be due to exposure in utero or via exposure to egg proteins through maternal breast milk.

- Ingestion of raw or undercooked egg may trigger more severe clinical reactions than well-cooked egg.
- Symptoms of egg allergy range from mild reactions to severe allergic reaction (anaphylaxis) which can be life threatening.
- IgE-mediated reactions are the most common type of allergic reactions to egg on first exposure. Infants typically present with rapid-onset of urticaria or angioedema, usually within minutes to two hours after ingestion (Table 27.4). Immediate-type reactions involving the gastrointestinal (eosinophilic esophagitis) or respiratory tracts are reported as well. Other symptoms include runny nose and itchy and water eyes, cramps, nausea or vomiting. The combination of egg allergy and atopic dermatitis is a risk factor for asthma.

Table 27.4: Clinical Symptoms of Egg Hypersensitivity

IgE Mediated (Acute onset)	***Non IgE Mediated (Delayed onset)***
Skin	**Skin**
✰ Pruritus	✰ Pruritis and erythema
✰ Erythema	✰ Atopic eczema
✰ Acute urticaria (localised/ generalized)	**Gastrointestinal**
✰ Acute angioedema (face, eyes, lips)	✰ Gastro oesophageal reflux disease
Gastrointestinal	✰ Loose/frequent stools
✰ Angioedema of lips/tongue/ palate	✰ Blood and/or mucus in stools
✰ Oral pruritus	✰ Abdominal pain
✰ Colicky abdominal pain	✰ Infantile colic
✰ Vomiting and diarrhea	✰ Constipation
Respiratory	✰ Food refusal or aversion
✰ Upper respiratory (nasal itching, sneezing, rhinorrhoea, congestion)	✰ Peri-anal redness
✰ Lower respiratory (cough, wheeze, chest tightness, shortness of breath)	✰ Pallor/tiredness
	✰ Faltering growth (with one or more of above)

- The natural history of egg allergy in children is that it resolves spontaneously in the majority. Various studies given different predictions, these vary from 90 per cent at age 2 to 66 per cent at age 5 years. Parents should be reassured that resolution by age 2 is common and by school age is even more so. Most children (50 per cent), but not all, outgrow their egg allergy before adolescence and many of those with persisting allergy will be able to tolerate cooked egg as cooking reduces the allergenicity of eggs. Resolution of egg allergy occurs in stages starting with tolerance to well-cooked egg (*e.g.* cake), then lightly cooked egg (*e.g.* scrambled) followed finally by raw egg. Therefore, children who tolerate cooked egg may still react to raw or undercooked egg. As a rule, reactions do not become more severe over time and often become less severe.
- The speed with which egg allergy resolves can vary greatly between individuals, and therefore the timing and appropriateness of reintroduction should be individually assessed. Reintroduction should not be attempted within 6 months of a significant reaction to egg. According to egg ladder, start with small amount of well cooked egg like cake, biscuits *etc.*, then loosely-cooked (well cooked omelette or scrambled egg) and raw eggs in the descending manner, if no symptoms appear.
- The management of egg allergy is similar to that of other food allergies. The treatment of egg allergy consists of dietary elimination or a maternal elimination diet in breast-fed infants. It requires education on avoidance and management of allergic reactions in the event of accidental exposure.

Table 27.5: Avoid Foods Containing Eggs or other Ingredients

Avoid Foods that Contain Eggs or Any of these Ingredients given Below	***Egg Protein is Sometimes found in the Following***
Albumin	Baked goods (cakes, biscuits, pastries, bread, royal icing), egg omelette, scrambled egg, poached egg, boiled egg,
Egg (dried, powdered, solids, white, yolk)	Egg substitutes
Eggnog	Lecithin
Globulin	Macaroni
Lysozyme	Marzipan
Mayonnaise	Mashmallows
Meringue (Meringue powder)	Nougat
Ovalbumin	Dried egg pasta, egg noodles, pancake mix *etc.*
Ovovitellin	Fresh ice cream, puddings, milk custards *etc.*

☆ Hen's egg is a versatile ingredient used in foods including a wide range of manufactured food products. Read the food labels and avoid foods that contain egg or any of these ingredients given in the Table 27.5. Some of the raw or loosely cooked egg may still be present in foods *e.g.* royal icing, marzipan, meringue, mayonnaise *etc.* which should be avoided.

☆ The dietary avoidance of egg can thus, be challenging and can pose significant quality of life concerns. To ensure that elimination of egg does not result in nutritional deficiency, and in particular for those who have additional dietary limitations, *e.g.* vegetarian diet or multiple food allergies, a dietician should be involved in the care of the patient.

Fish and Shellfish

☆ Seafood can be divided into two groups *i.e.* shellfish (calamari, crayfish, cuttlefish, lobster, mussels, oysters, prawns, squid *etc.*) and fish (cod, flake, salmon, snapper, trout, tuna, whiting, barramundi *etc.*). Shrimp is behind most of the reactions, and shellfish as a group which can include oysters, mussels, scallops, squid, crab, snails and many more tends to be more problematic than finned fish.

☆ Fish and shellfish consumption has increased worldwide, and there are increasing reports of adverse reactions to fish and shellfish, with an approximate prevalence of 0.5-5 per cent.

☆ Fish allergy often develops early in life and affects children, whilst shellfish allergy tends to develop later, from adolescence onwards.

☆ Fish that commonly cause allergies are given as under. Read the food labels carefully for presence of fish.

Fish that Commonly Cause an Allergic Response

Salmon	Barramundi	Pilchards	Cod	Bream	Shark	Haddock
Mackerel	Flake	Snapper	Sardines	Flathead	Sole	Perch
Herring	Flounder	Whiting	Anchovies	Garfish	Whitebait	
Tuna	Hake	John Dory	Trout	Orange roughy	Pike	

Ingredients that Indicate the Presence of Fish

Antipasto	Dips	Soups	Oyster sauce	Fish sauce	Fish stock
Marinara sauce	Fish gelatin	Canned spreads (fish paste)	Caesar salad	Salad dressings (anchovies)	Caesar salad dressing
Pescatore sauce	Suushi	Surimi	Greek dishes	Bouillabaisse	Worchester sauce

Products fortified with omega 3 fatty acids.

Anything from a fish and chip shop (cooked in oil used to cook fish).

☆ Little is known about the natural history of these allergies, but both are thought to be persistent; they are usually lifelong.

☆ Even though fish and shellfish live in the same place, the protein allergens present in one seafood group are generally considered to be very different to those in others. That means that one can be allergic to fish only, shellfish only or members of both groups. Similarly, allergy to one crustacean usually means that all must be avoided. While those allergic to seafood from one group (*e.g.* crustaceans) can usually tolerate those from another (*e.g.* molluscs), this cannot be guaranteed without specific allergy testing.

☆ Some people who are allergic to one type of fish may be allergic to another type of fish or they may have allergies to several crustaceans, such as prawn, crab and lobster. This is known as cross-reactivity. There is high cross reactivity between different fish species including white and oily fish and fresh and sea water fish and between crustaceans (prawns, lobster and crab) and molluscs (mussels, squid, oysters, snails) but no cross-reactivity between fish and shellfish (crustaceans or molluscs). It is prudent for people to avoid all seafood if they are allergic to fish or shellfish due to cross contamination.

☆ Certain species of fish such as tuna and mackerel, and fish that is not completely fresh, can contain high levels of histamine which can precipitate a pseudo-allergy called scombroid poisoning, which may be mistaken for allergy.

☆ Shell fish that commonly cause an allergic response are given as under:

Shell Fish that Commonly Cause an Allergic Response

Prawns	Crayfish	Mussels	Yabbies	Scampi	Cockle
Shrimps	Abalone	Octopus	Sea snail	Sea slugs	Marron
Lobsters	Clams	Cutle fish	Scallops	Abalone	Molluscs
Crab	Oysters	Squid	Calamari	Barnacles	Balmain bugs

☆ Check the labels of the following foods, as they may contain shell fish.

Ingredients that Indicate the Presence of Shellfish

Bouillabaisse	Prawn crackers	Fish stock	Seafood extender	Squid ink	Scampi fries
Cuttlefish ink	Oyster sauce	Fish sauce	Marinara sauce	Pescatore sauce	Pet or fish food
Asian soups and dishes		Seafood/marinara pizza		Seafood flavouring	

☆ Some sensitive individuals may react to the fumes when seafood is being cooked, and occasionally from touch, such as when cleaning and handling fish. Occupational asthma and contact rashes can also be a problem where exposure is frequent, such as in seafood processing factories, or in some food handlers such as chefs.

☆ Allergic symptoms after contact with seafood are usually within minutes, however, delayed type reactions and particularly exercise-induced anaphylaxis is occasionally observed, particularly after ingestion of shellfish (such as oyster, abalone, squid and shrimp). Symptoms of fish or shellfish allergies vary and range from mild reactions to a severe allergic reaction (anaphylaxis) which can be life threatening. Symptoms of a reaction to seafood can include flushing and swelling of the lips and face, tingling of the throat and mouth, tightening of the throat; digestive tract symptoms such as cramps, stomach pain, nausea or vomiting and diarrhea; itching of the skin and hives; difficulty breathing; lightheadedness or faintness; a sudden drop in blood pressure and a rapid heartbeat.

☆ The clinical manifestations of shellfish allergy, in particular, may vary from local to life-threatening 'anaphylactic' reactions within an individual and between individuals. Parvalbumin and tropomyosin are the two major allergens, but several other allergens have been cloned and described. These allergens are highly heat and biochemically stable, and this may in part explain the persistence of these allergies. Diagnosis requires a thorough history, skin prick and in-vitro-specific IgE tests, and oral challenges may be needed for diagnostic confirmation. Strict avoidance of these allergens is the current standard of clinical care for allergic patients, and when indicated, an anaphylactic plan with an adrenaline auto-injector is prescribed. There are no published clinical trials evaluating specific oral immunotherapy for fish or shellfish allergy.

☆ Some allergens in fish can be denatured by intense heat, such as during the canning process, but allergens in shellfish are very robust and not destroyed by heat. Fish allergens may be present in cooking vapours.

☆ Avoiding seafood groups by such allergic persons is an important part of the management. One should check the food labels for presence of sea foods and avoid them. One should also learn other names of the foods too which contain sea foods and they should be avoided.

☆ One must be extra-careful when eating out. Waiters (and sometimes the kitchen staff) may not always know every dish ingredient on the restaurant's menu. Vapors may carry extremely small particles of shellfish protein, so even just walking into a kitchen or other area where food is being prepared can cause a dangerous reaction in sensitive individuals.

☆ For safety sake, one should always carry injection of epinephrine while going out and taking food in restaurants *etc.* so that it can be used in case of emergent problem. Repeated doses of epinephrine may be necessary.

Wheat and Other Grains

☆ Wheat contains a number of different proteins. Allergic reactions to wheat can result from reactions to one or more of these proteins. Currently, there are three major wheat related food illnesses *i.e.* celiac disease (CD), nonceliac gluten sensitivity (NCGS), and wheat allergy. Although there might be an overlap in the symptoms associated with NCGS, wheat allergy, and CD, the conditions have distinct characteristics (read Ch.22 for CD in detail).

☆ Allergic reactions can be caused by eating wheat, but also, in some cases, by inhaling wheat flour. It is one of the more common food allergies in children, and it is common for children with wheat allergy to have positive allergy tests to other cereal grains, such as rye, barley, oat and corn. It is also common for people with allergy to grass pollen to have positive allergy tests to these foods as well, because these cereal grains belong to the grass family. However, only about 20 per cent of people with an allergy to one cereal grain have allergies to other cereal grains, and only 20 per cent of people with grass pollen allergy have a true food allergy to any of the cereal grains.

- ✰ Although it is less common than celiac disease, wheat allergy occurs more frequently than previously thought. It affects a high proportion (10–20 per cent) of food allergy sufferers in some populations, especially North Europe.
- ✰ While a wheat allergy most often develops in early childhood, people can manifest with symptoms at any stage of their life, including later adulthood. Wheat if implicated in a form of food allergy that does affect adults may also include baker's asthma (occupational exposure to grain flour dust) and less frequently IgE-mediated allergy associated with exercise, is known as wheat-dependent, exercise-induced anaphylaxis (WDEIA) where the reaction only occurs when the trigger food is eaten in conjunction with exercise. This particular presentation is difficult to predict and to diagnose as the ingested wheat quantities as well as the exercise level necessary to induce the symptoms are very variable.
- ✰ Individuals allergic to the wheat related crops (barley, oats, rye) often can tolerate rice. Reactions to rice are rare in Europe and America, but may be more common in Asia.
- ✰ If one develops wheat allergy later in life, the more likely the person will be faced with a permanent condition. By contrast, most children approximately 2/3 of them with a wheat allergy will outgrow it by the age of 12 whereas celiac disease is a lifelong condition.
- ✰ Wheat is an increasingly recognized trigger for immune mediated food allergies, both immunoglobulin E (IgE) and non-IgE mediated (Figure 27.4). These reactions are typically characterized by a T helper type 2 (Th2) lymphocytic inflammation with predominant Th2 cytokines expression (*i.e.*, interleukin (IL)-4, IL-13, and IL-5). Th2 inflammation can lead B cells to produce IgE antibodies specific to certain foods (in IgE mediated food allergy), or can lead to a chronic cellular inflammation often characterized by the presence of T cell and eosinophils, which is a much less understood pathogenetic mechanism (non-IgE mediated food allergy.
- ✰ Wheat can cause a non-IgE mediated allergic disorder by inducing a Th2 lymphocytic response largely independent from IgE-specific antibodies to wheat (non-IgE mediated inflammation). The vast majority of these responses is characterized by an eosinophilic infiltration in the gastrointestinal tract, and is called eosinophilic gastrointestinal diseases (EGIDs). Based on clinical characteristics, pathogenesis, and response to therapy, EGID can be divided into four principal groups: EoE, EG, eosinophilic gastroenteritis (EGE), and eosinophilic colitis (EC). Non-IgE mediated food

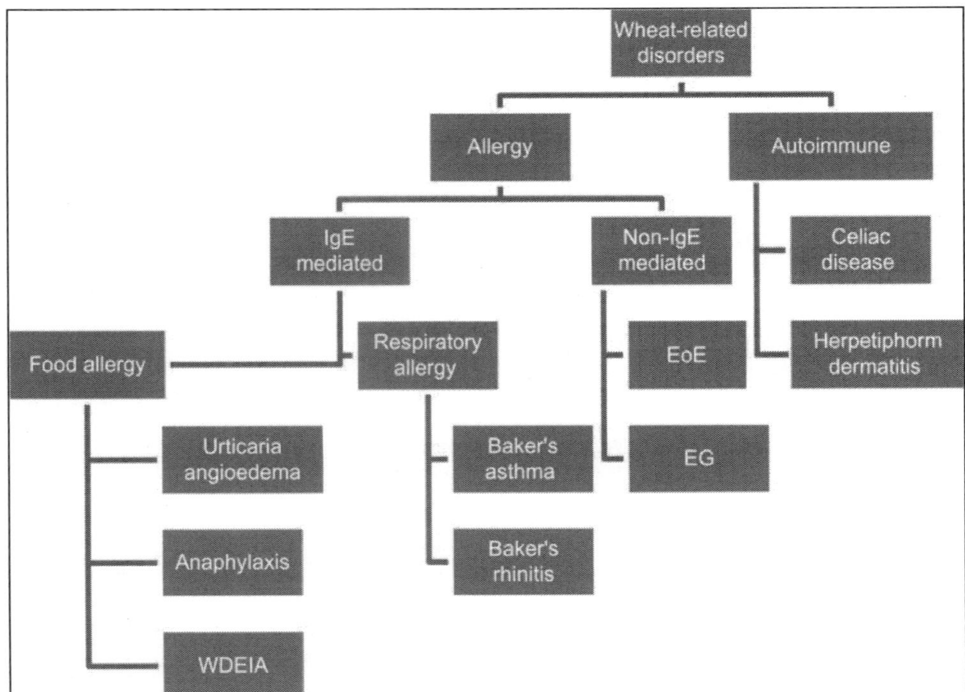

Figure 27.4: Wheat Related Disorders.

allergies to wheat are mainly EoE and *e.g.* EoE is an increasingly recognized food allergy that affects mainly Caucasian, atopic males.

- The reactions can be immediate or non-immediate according to the time interval between food ingestion and the appearance of symptoms. Immediate reactions occur within a few hours of food ingestion, and are mainly characterized by one or more of the following symptoms *i.e.* urticaria and/or angioedema, asthma, anaphylaxis, acute exacerbation of atopic dermatitis, abdominal pain, nausea, vomiting, diarrhea, allergic rhinitis, and bronchial obstruction.
- There is a high degree of cross-reactivity between wheat and grass pollen, so that 80 per cent positive SPT to wheat have no clinical significance in people with a grass pollen allergy. SPT and serum IgE are poor predictors of wheat allergy; a better choice is to test for the presence of IgE antibodies to the wheat allergen omega-5 gliadin which is highly predictive of wheat allergy in children and WDEIA triggered by wheat.
- As with all food allergies, at the moment, management of IgE mediated wheat allergy is mainly based on complete avoidance of wheat in any form both in food such as wheat bread, bread crumbs, bulgur, cereal extract, durum, all purpose flour, baked goods made from refined flour, wheat porridge, semolina, hydrolyzed wheat protein, pasta, noodles, sprouted wheat, triticale, wheat bran, wheat germ, malt, starch, wheat germ oil, wheat grass *etc.* and products made from wheat and inhaled wheat allergens.
- Wheat-allergic patients are treated the same way as patients with celiac disease, namely with the GFD. Most of these patients tolerate naturally gluten-free cereals (*i.e.* maize, rice and also oats), even though cross-allergy between wheat and these cereals has been described. At variance with celiac disease, wheat allergy is not always lifelong, especially in children.
- Read the food labels carefully to avoid taking wheat containing products.
- The promising way for treatment of a food allergy is immunotherapy. Currently, there are three techniques being studied: oral immunotherapy (OIT), sublingual immunotherapy (SLIT), and epicutaneous immunotherapy (EPIT). For baker's asthma and allergic rhinitis, like for many occupational related diseases, strict avoidance of occupational triggers, such as grain flours, remains the primary step in the management of the disease. However, as it has been done for other occupational agents such as latex, specific immunological treatments can become therapeutic options for baker's asthma. Standard subcutaneous immunotherapy has been reported to be effective in a few case series in Baker's asthma.
- In case of accidental exposure and anaphylactic reaction, epinephrine administration with a self-injector device is the lifesaving treatment.

Fruit and Vegetables

- Allergic reactions particularly in the form of the oral allergy syndrome to fruits and vegetables are frequently observed in older children, adolescents and adults and not in young children. These are likely to persist into adult life. One exception is potato allergy in young children that usually improves with time.
- These allergies can result from a primary sensitization to food allergens mediated through ingestion or from a primary sensitization to inhalant allergens such as pollens or latex.
- The reported allergies to fruits, nuts and vegetables mainly evoke Oral Allergy Syndrome (OAS) *i.e.* allergic reaction symptoms in the oral cavity subsequent to the consumption of some common fruits and vegetables in the adults who suffer from allergic rhinitis. OAS belongs to the allergy type I group, that is, allergic reactions mediated by immunoglobulin E (IgE). It has been described under various names including "pollen-food allergy syndrome," "pollen-food syndrome," and "pollen-associated food allergy syndrome". Those with oral allergy syndrome typically have allergy to birch, mugwort, ragweed, or grass pollens. PFS affects 6–47 per cent of pollen-sensitized individuals, although up to 90 per cent of those sensitized to birch pollen can have PFS.
- Unlike other food allergies, OAS is a reaction limited to the oral mucosa, lips, tongue and OAS is a form of a contact allergic reaction that occurs upon contact of the mouth and throat with fruits and vegetables. The reaction occurs because the proteins found in some fruits and vegetables are very similar to those found in pollen. These proteins can confuse the immune system and cause an allergic reaction or make existing symptoms worse, which is referred to as cross-reactivity. In case of pollen and foods, the result of cross-reactivity is called oral allergy syndrome (OA), also known as pollen fruit syndrome (PFS). Hence, OAS is not a separate food allergy, but rather represents cross-reactivity between distant remnants of tree or weed pollen still found in certain fruits and vegetables. Therefore, OAS is only seen in people with seasonal pollen allergies, and mostly people who are allergic to tree pollen. It is usually limited to ingestion of

uncooked fruits or vegetables. People affected by oral allergy syndrome can usually eat the same fruits or vegetables in cooked form because the proteins are distorted during the heating process, so that the immune system no longer recognizes the food.

- Food allergens that induce OAS rapidly dissolve in the oral cavity and are readily broken down by digestive enzymes. Preservatives in foods may also trigger the manifestation of the disease. Due to the structural similarity of individual protein molecules, a large number of allergens that exist in nature can be classified into groups as follows (Table 27.6). Among the allergens in each group, there is a possibility of cross-reactivity of IgE antibodies, that is, antibodies binding to one of two or more allergens. The reaction may start with one type of food, and subsequently allergies to other food types can develop.
- It is clear that some common fruits (apple, peach, musk melon, kiwi fruit, cherry, grape, strawberry, banana, custard apple, mango and pomegranate) and their allergens appear to be at the center of current research on food allergy.
- In Pollen Food Syndrome, fruit and vegetable allergies result from a primary sensitization to labile pollen allergens, such as Bet v 1 or profilin, and the resulting phenotype is mainly mild, consisting of local oropharyngeal reactions.
- Profilins are proteins that affect cell shape and function. Profilins have been identified in trees, grass and weed pollens as well as many fruits and vegetables. Profilins are a type of plant protein that is known to cause allergy. Fruit and vegetable allergies are commonly due to an allergy to profilins. Allergies to melon, watermelon, citrus fruits, tomato, and banana or a combination of these suggest an allergy to profilins.
- Some people may suffer from a newly recognized type of plant protein allergy called lipid-transfer protein (LTP) allergy. Lipid transfer protein (LTP) syndrome results from a primary sensitization to LTPs, which are stable plant food allergens, inducing frequent systemic reactions and even anaphylaxis. These conditions have the potential to cause more serious symptoms and there is a risk of anaphylaxis. Symptoms may include wheezing; a swelling of the throat that is serious enough to restrict breathing; and even a fall in blood pressure that can lead to collapse. Accompanying symptoms may include widespread flushing of the skin, nettle rash (otherwise known as hives or urticaria), swelling of the skin (known as angioedema) anywhere on the body, swelling of the lips, or abdominal pain, nausea and vomiting.
- Although much less prevalent, severe fruit allergies may be associated with latex. Latex allergens can also sensitize patients to cross-react to the protein found in some foods. Around 40 per cent of patients with known latex allergy have allergic symptoms after eating fruits (papaya, avocado, banana, chestnut, passion fruit, fig, melon, mango, kiwi, pineapple, peach and tomato). In some people, allergy to pollen, fruits (often peach) and latex occurs together.

Table 27.6: Types of Pollen and Food Associated with Oral Allergy Syndrome

Pollens	*Fruits*	*Vegetables*	*Nuts*	*Grains*
Birch	Kiwi, apple, pear, plum, peach, nectarine, apricot, cherry, banana, fig, avocado, strawberry, dried plum, mango	Celery, carrot, parsnip, parsley, dill, cumin, cilantro, fennel, potato, tomato, pepper (green), chicory	Hazelnut, almond, walnut, peanut	Soybeans, wheat, lentils, peas, beans
Ragweed	Banana, watermelon, melon, honey dew, cantaloupe	Squash, pepper, cucumber, artichoke, hibiscus, zucchini, chamomile tea		Sunflower seeds
Weeds	Melon, watermelon, orange, kiwi	Tomato		
Wormwood	Apple, watermelon, melon	Celery, carrot, parsley, pepper, cilantro, fennel		
Parietaria	Cherry, melon			
Grass	Fig, melon, orange, kiwi, watermelon	Tomato, potato	Peanut	Grass
Alder	Apple, cherry, peach, pear, strawberry, raspberry	Celery, parsley	Hazelnut, almond, walnut	
Japanese cedar		Tomato		
Mugwort	Mango	Celery, carrot		
Plane	Apple	Lettuce, corn	Hazelnut, peanut	Chickpea

☆ One of the most notable features of latex allergy is the patients' cross-reactivity to various fruits and vegetables, a condition often called *latex-fruit syndrome*. Commonly reported cross-reactive foods include banana, avocado, kiwi, chestnut, potato, and papaya, and numerous latex allergens cross-react with food and pollen proteins.

Table 27.7: List of Reported Cases on Various Food Items Causing OAS Symptoms

Food	*Manifestation*
Jackfruit	Itching and burning in the mouth and throat
Salami	Mild pruritus in oral mucosa
Honey	Itching in the mouth, GIT symptoms and angioedema
Tomato juice	Dyspnea, swelling of the oral and nasal mucosa, and congestion of the bulbar conjunctiva
Peanut	Lip tingling, oral itching, lip swelling, and throat itching
Cashewnut	Sialorrhea, perioral urticarial rash, tongue swelling, and immediate vomiting
Grapes	Flushing of the face and neck, followed by local itchy skin rash, itching and edema of the oral and perioral mucosa, and moderate dyspnea
Sapodilla plum	Lip edema, accompanied by itching in the lips, tongue, and throat, as well as a feeling of dryness and hardness in the throat (glottic edema)
Mango	Oropharyngeal itching, tiredness, dizziness, and swelling of the face
Cooked aubergine	Oral itching and significant perioral urticaria
Pistachio nuts	Oral and lip itching and swelling
Royal jelly	Lip, tongue edema, and palate itching

Source: Kashyap and Kashyap. 2015. Oral allergy syndrome: an update for stomatologist. J. Allergy (Cairo). Nov 8. doi: 10.1155/2015/543928.

☆ OAS symptoms may vary from person to person. Some studies have confirmed that the OAS is more common in female patients. Patients who show symptoms of OAS may have a number of other allergic reactions that start very quickly, even minutes after consuming trigger food. Reactions can range from mild to severe. Usually, it is manifested by itching and a burning sensation of the lips, mouth, ear, and throat or by the appearance of perioral erythema and generalized urticaria. Sometimes, reactions can manifest themselves in the eyes, nose, and skin. The patient may develop tingling, swelling of the lips, tongue and uvula, occasionally a sense of suffocation, and rarely anaphylaxis. Symptoms usually last for a few minutes to half an hour. In rare situations, OAS may be manifested as difficulty in breathing, appearance of a rash, or hypotension. The symptoms include itching and burning of the lips, mouth and throat. In more serious reactions, there may be swelling of the mouth, back of the throat and windpipe as well as hives. This type of allergic reaction anaphylaxis to fruit or vegetable allergens is quite uncommon when compared with nuts, egg and milk. Kiwi fruit is the fruit most likely to cause more severe reactions.

Since many of the immunogenic proteins in fruits and/or vegetables are unstable (heat-labile), patients will tolerate food cooked and canned well and fresh or raw foods badly. It has been shown that cooking food can sometimes eliminate allergens in certain species like apples, while it is impossible to destroy allergens in celery and strawberries. For some types of food (*e.g.*, nuts) that contain more than one allergen, heat treatment will destroy certain allergens, while some of them can cause a reaction even after that.

Education is the key pillar of an effective long-term elimination diet. Most patients with OAS can be treated with a combination of allergen avoidance and pharmacotherapy. The most important therapy includes antihistamines.

In patients with suspected OAS, preemptive caution is necessary because the preparation of food can be connected to reactions. Different reactions may appear at different times, such as sneezing attacks during scraping of fruits and vegetables, when particles can get into the air, or conjunctivitis if the patient touches his/her eyes after touching the fruit or vegetables. Wearing gloves and masks can help prevent contact with allergens. It is also recommended to avoid latex (rubber gloves) that can cause cross-allergic reactions to foods of plant origin. If the patient avoids areas of certain types of pollen, the syndrome usually relieves after two to three years.

In the event of a reaction, the patient is advised to stay calm, rinse their mouth with plain water, and rest. The patient can help himself/herself with hot (but not boiling) beverages that can inactivate residual allergens. This usually leads to withdrawal of the sensation of prickling, itching, and swelling, which stops within 30 minutes to an hour (before the antihistamine makes an effect).

. Subcutaneous specific immunotherapy (SIT) has been tried and has been found to significantly reduce OAS symptoms associated with ingestion of the responsible fruit and vegetables.

Peanuts and Tree Nuts

Nuts and tree nuts are among the highest producers of IgE-mediated allergic reactions following food ingestion (Table 27.8). Peanut and tree nut allergies account for just over one per cent of all allergies, with roughly 0.6 per cent having an allergy to peanuts and 0.6 per cent to tree nuts.

This type of allergy can develop at any age, but most commonly occurs in childhood. This food allergy is estimated to affect approximately 6–8 per cent of children below 4 years of age and 1 to 2 per cent of individuals over the first decade of life. Around 20 per cent of peanut allergy cases resolve, and while severity may occasionally lessen with age, around 20 per cent of cases can become worse with time. Around 10 per cent of children with an allergy to any of the tree nuts may outgrow one or all of their allergies.

Table 27.8: Nuts and Tree Nuts Implicated in IgE-mediated Reactions

Family	*Species*	*Common Name*
Leguminoseae	*Arachis hypogea*	Peanut
Betulaceae	*Corylus avellana*	Hazelnut
Anacardiaceae	*Anacardium occidentale*	Cashew
	Pistacia vera	Pistachio
Juglandaceae	*Juglans regia*	Walnut
	Carya illinoinensis	Pecan nut*
Rosaceae	*Prunus dulcis*	Almond*
Lecythidaceae	*Bertholletia excelsa*	Brazil nut
Fagaceae	*Castanea sativa*	Chestnut*
Pinaceae	*Pinus pinea*	Pine nut*
Proteaceae	*Macadamia intergrifolia*	Macadamia nut*
Palmaceae	*Cocos nucifera*	Coconut*

* Infrequent or anecdotally described as causing allergic reactions.

Table 27.9: Nut Allergens

	Allergen	*Function/Type*	*M.W. (kD)*
Cashew	Ana o 1	Vicilin (7S)	59
	Ana o 2	Legumin (11S)	
Hazelnut	Cor a 1	PR-10 (Bet v 1 homologous)	17
	Cor a 2	Profilin	14
	Cor a 8	PR-14 (LPT) 9	9
	Cor a 9	Globulin (11S)	40
	Cor a 11	Vicilin (7S)	48
Peanut	Ara h 1	Vicilin	63·5
	Ara h 2	Conglutin	17
	Ara h 3	Glicinin	60
	Ara h 4	Glicinin	37
	Ara h 5	Profilin	15
	Ara h 6	Conglutin homologous	15
	Ara h 7	Conglutin homologous	17
	Ara h 8	PR -10	17
Chestnut	Cas s 5	Chitinase Ib	
	Cas s 8	PR -14 (LPT)	9·7
Brazil nut	Ber e 1	Albumin (2S)	9
	Ber e 2	Legumin (11S)	29
Walnut	Jug r 1	Albumin (2S)	14
	Jug r 2	Vicilin (7S)	44
	Jug r 3	PR -14 (LTP)	9
	Jug r 4	Legumin (11S)	

- ✰ Peanut allergy is one of the most common food allergies. Peanut allergy often receives greater attention because it is common (3 per cent of infants), exposure is hard to avoid and in some cases even trace amounts can trigger symptoms. Peanuts are a member of the legume family like peas, lentils and chickpeas *etc.*, not related to tree nuts.
- ✰ The term "tree nuts" means any of various edible fruits or seeds of non-leguminous woody plants such as almonds, walnuts, cashews, pine nuts, pistachios, macadamia nuts, Brazil nuts *etc.* but not peanuts, soybeans, or sunflower seeds. These tree nuts contain protein that can cause an allergic reaction for around 2 per cent of children. Most people allergic to one tree nut do not have allergies to any other nuts but approximately 30 per cent will have another tree nut allergy. Common coexisting tree nut allergies are cashew-pistachio and walnut-pecan. Common nut allergens have been mentioned in Table 27.9.
- ✰ Seeds have similar allergens to tree nuts and peanuts and cross reactions may occur. Sesame seeds are the commonest seed allergens, although reactions to mustard seed and sunflower seed are becoming more common.
- ✰ The proteins in peanut are very different to those in tree nuts (such as almonds, Brazil nuts, cashews, hazelnut, macadamia nuts, pecans, pistachios or walnuts). A person can be allergic to peanuts and not be allergic to tree nuts, or they can be allergic to both. Therefore, someone allergic to peanut is not automatically going to be allergic to tree nuts.
- ✰ The allergic components in peanuts are specific identified proteins (Table 27.9). If a person is allergic to one protein present in one food (such as peanut or egg) then an allergic reaction can only occur if they eat that one particular food. Some individuals may be allergic to more than one protein in more than one food and so may be allergic to several foods.
- ✰ Foods most likely to contain nuts include almond and hazelnut paste, peanut butter, icing, glazes, marzipan, nougat; oriental and Asian cuisine such as curries, egg rolls, Szechuan and other sauces, gravy, soups; baked goods like cakes, cookies, donuts, pastries; candies, such as mandelonas, chocolates, and chocolate bars; cereals and granola, energy bars, granola bars; pesto (which commonly contains cashew nuts); chipotle sauce and other Mexican/Latin sauces; ice creams and flavoured ice water

treats, frozen desserts, frozen yogurts, sundae toppings; dried salad dressings and soup mixes; hydrolyzed plant protein/hydrolyzed vegetable protein; faux nuts made from re-formed peanut products; peanut oil; snack foods such as dried fruits, chewy fruit snacks, trail mixes, popcorn, pretzels, chips; vegetarian meat substitutes; edible fruit arrangements *etc.* Non-food sources of peanuts and tree nuts are ant bait, bird feed, mouse traps; cosmetics, hair and skin care products, massage oils, soap, sunscreen; craft materials; medications, vitamins, and health supplements; mushroom growing media; pet foods and pet toys; sunscreen and other creams; stuffing in children's toys *etc.* These foods should be avoided if someone is allergic to nuts. Read the ingredient labels every time carefully when you buy or eat a product.

- There is a misconception regarding the allergenicity of peanut oil. The vast majority of peanut oil used by foodservice and by many consumers is highly refined and processed oil, from which all of these proteins are removed. Hence, refined oils containing peanut oil are usually allowed.

Table 27.10: Common Sources of Tree Nuts

Common Sources of Tree Nut	*Common Sources of Tree Nuts (check carefully)*
Almonds	Baked goods such as cakes and slices
Almond Paste	Biscuits and cookies
Almond Meal	Breakfast cereals
Amoretto	Christmas Cakes and puddings
Artificial nuts	Chocolate
Bitter almond	Chocolate spreads
Baklava, Greek pastry commonly containing walnut	Curry pastes
Beer nuts	Dips and flavoured cheeses
Brazil nuts	Dried fruit and trail mixes
Bunya nuts	Dukkah
Candle nuts	Energy mixes or trail mix
Carponata (Italian dish that contains almonds)	Fruit crumble mix
Cashews	Flavoured coffee syrups
Gianduja	Icecream
Hazelnut	Marzipan
Indian nuts	Muesli bars and health bars
Macadamia Nuts	Natural flavourings
Macaroons	Nougat, fudge, praline, Rocky Road and Turkish delight
Macadamia (Queensland nut, candlenut)	Pasta sauces
Mixed nuts	Pesto
Non-gai nuts	Salads and salad dressings
Nut extract or flavourings	Restaurant and Takeaway meals – especially Asian, African and Indian meals
Nut meal	Stuffing
Nut oils	Vegetarian meals
Nut spreads or butters	
Pecan (Hickory Nut or Mashuga)	
Pistachios	
Shea nuts	
Walnuts	
Products at high risk of being contaminated with tree nuts	**Possible non-food sources of tree nut**
Baked goods	Animal and bird feeds
Breakfast cereal	Cosmetics and massage oils
Chocolate	
Ice creams	
Restaurant or takeaway meals	

- Allergic reactions to peanuts and tree nuts can vary from mild to severe. The clinical features of allergic reactions to nuts are acute in onset, with symptoms occurring within minutes or a few hours after ingestion. Symptoms of food allergy of peanuts and tree nuts typically include hives (urticaria), swelling around the mouth, and vomiting, usually within 30 minutes of eating a food. Other symptoms include stomach pains, or diarrhoea. In severe reactions, progressive respiratory symptoms, hypotension and dysrhythmias may take place. Cutaneous symptoms may be absent in severe cases. Symptoms of severe allergic reactions (anaphylaxis) affect breathing and/or heart and may include any of the following: difficult/noisy breathing, swelling of tongue, swelling/tightness in throat, difficulty talking/hoarse voice, wheeze or persistent cough, persistent dizziness and/or collapse, and becoming pale and floppy in young children.
- In comparison with other foods, allergic reactions to nuts seem to be particularly severe, with multisystemic or respiratory symptoms in up to 81 per cent of the cases. The role of nuts as a cause of fatal and life-threatening reactions is well documented. Peanuts most commonly are reported to provoke anaphylaxis and can be life threatening.
- Currently there is no cure for nut allergies. The only option for managing the risk is to completely avoid the specific allergen. Once the nut allergy has been diagnosed, strict avoidance of the food should be recommended. Usual dietary advice is to avoid all nuts and foods containing nuts, even

if only one nut is reported to be a trigger food, due to contamination and poor recognition of different nuts; there is usually no need to avoid other legumes in the case of peanut allergy, unless symptoms have been reported.

- Patients and families must learn how to identify nuts on food labels and how to avoid high-risk situations, such as sharing foods or ingesting products which could contain nuts as ingredients. Children with food allergy should take their own food with them to school and encouraged not to swap or share food.
- Picking nuts out of foods may leave some amount of nut protein in the product and is not recommended. One should be very careful while eating out.
- For those at risk of an anaphylactic reaction, it is essential to understand all of the facts surrounding self-management. Appropriate emergency treatment for anaphylaxis (a severe food allergy reaction) includes an injection of epinephrine, which is available in an auto-injector device. Always carry epinephrine auto-injector and it must be administered as soon as symptoms of a severe allergic reaction appear.

Food Additives

Food additives are chemicals added to foods for various reasons; for example, as antioxidants, colorings, flavorings, emulsifiers, stabilizers, taste enhancers, and preservatives. They are commonly perceived to cause food hypersensitivity (FHS), but are likely to only affect less than 0.1 per cent of the population. Contrary to common believing, the prevalence of the intolerance to food additives in the general population is rather low. People who have asthma and urticaria are more likely to report hypersensitivity to certain food additives.

The following additives are the most likely to be implicated in FHS:

Colourings

Both natural food colourings such as carmine (cochineal), annatto, turmeric and saffron and synthetic azo dyes such as tartrazine and sunset yellow. Tartrazine is a yellow dye most commonly used to color beverages, candy, ice cream, desserts, cheese, canned vegetables, salad dressings and other foods. Scientists have concluded that the color additive may cause hives in fewer than one out of 10,000 people. There is no scientific evidence that it provokes asthma attacks. Even though many studies have been done, it has been difficult to find children who consistently adversely react to tartrazine or other food dyes.

Preservatives

Benzoates added to beer, jam, fruit products, pickled foods, yoghurt, salad cream, berries, prunes and spices such as cinnamon and cloves; sulphites added to wine, cider, lager, fruit juices and squashes, meat products, dried fruits and vegetables may cause FHS in some people. Benzoate reactions are very rare. Benzoates have been linked to chronic urticaria, asthma, atopic dermatitis, rhinitis and anaphylaxis although there is limited good quality evidence to support these findings. Sulfiting agents are sometimes used to preserve the color of foods such as dried fruits and vegetables, and

Table 27.11: Foods more Likely to Contain High Levels of Natural or Added Sulphites, Benzoates and Monosodium Glutamate

Foods	*Sulphites*	*Benzoates*	*Monosodium Glutamate*
Meat, poultry and seafood	Prawns, lobster, dried salt cod, crab sticks, squid, meat burger, sausages	Dishes with a spicy sauce, ready to eat meals containing benzoates	Fish sauce
Milk and eggs		Yoghurt, cheese	Parmesan cheese
Fruits	Dried apricots, sultanas, figs, prunes, dates, dried banana, candied or glace fruit, desiccated coconut, currants	Cranberries, bilberries, prunes, papaya, dried fruit, avocado	
Vegetables, nuts, seeds and savoury snacks	Dried mushrooms and other fungi, frozen, tinned or vacuum packed potatoes, French fries, instant mash, potato cakes, potato croquettes, vegetarian burgers and sausages, tinned asparagus, broad beans, French beans, chestnuts, walnuts	Pumpkin, kidney beans, soy beans, soy flour, broccoli, spinach, baked beans in tomato/spicy sauce, dry roasted and spicy nuts, Bombay mix, crisps (except ready salted), potato or corn snacks,	Mushrooms, spinach, savoury snacks, crisps
Condiments and miscellaneous	Horseradish sauce, caramel colouring	Curry powder, allspice, mixed spice, nutmeg, clove, cinnamon, chocolate, cocoa, ketchup, soy sauce, Worcestershire sauce, salad dressing, salad cream, mayonnaise, jam, pickles	Soups, stock, gravy, rubs, coatings, ready-meals, soy sauce, black bean sauce, oyster sauce, tomato sauce, miso, marmite, instant rice and noodle dishes
Drinks	Cider, wine, beer, fruit squash and cordials, soft drinks, grape juice, fruit juice drinks, cola drinks	Tea, squash, cordial, carbonated drinks, milkshake syrup, beer, ready-to-drink alcohol and mixers, spirits with added spices	

Foods with added sulphite, benzoates and MSG will vary from country to country.

to inhibit the growth of microorganisms in fermented foods. Some beverages and foods contain naturally occurring sulfites as a result of fermentation. Sulfites can also be manufactured, and are used commercially in the pharmaceutical, cosmetic, and food industries to control microbial growth. Besides their use as preservatives, sulphites are used as bleaching agents, antioxidant, and reducing agents in the food industry. The most commonly used sulfites are sodium salts, specifically sodium sulfite, sodium bisulfite, and sodium metabisulfite (SMS). Sulfites are safe for most people. A small segment of the population, however, has been found to develop shortness of breath or fatal shock shortly after exposure to these preservatives. Sulfites can provoke severe asthma attacks in sulfite-sensitive asthmatics.

Nitrates as food additives are another cause of headache in some persons. They should avoid cured meats as they contain nitrates.

Flavour Enhancers

Originally described as 'Chinese Restaurant Syndrome' in 1968, results from studies utilizing Double-blind, placebo-controlled food challenge studies (DBPCFC) have suggested some individuals could experience symptoms from the ingestion of MSG, although only in quantities greater than the normal dietary intake. Monosodium glutamate (MSG) is used to enhance flavours in soups, sauces, dried noodle snacks, gravy, ready meals, and Chinese food *etc.* MSG is not an allergen and there is conclusive scientific evidence that MSG does not cause or exacerbate asthma. A small number of people may experience mild and transitory symptoms such as headache, a burning sensation on the back of the neck, chest tightness, nausea, diarrhea and sweating to MSG. Although the glutamate in MSG is chemically indistinguishable from glutamate present in food proteins and is metabolized in the same way, it is currently unknown whether foods high in natural bound or free form glutamate can cause the same symptoms as MSG.

- ✰ Symptoms due to, or exacerbated from, food additives usually involve non-IgE-mediated mechanisms (pseudo-allergic reactions, PAR) and are usually less severe of those induced by food allergy. Research has shown that exposure to certain food additives in processed foods can result in a variety of allergic reactions or immunological hypersensitivities such as atopic dermatitis, allergy asthma, and eosinophilic esophagitis. However, there is no evidence that any food additive causes a reaction that lasts more than one day.
- ✰ Food additives must be labeled on the processed foods.

Naturally Occurring Food Hypersensitivity Triggers

Salicylates

Salicylates are a group of chemicals derived from salicylic acid. They occur naturally in plant foods that help them to protect against insects, fungus and diseases, and have a similar chemical structure and properties to aspirin. They are found naturally in certain foods and also synthetically produced for use in products like aspirin, toothpaste and food preservatives. In natural form, foods containing the most salicylate include coffee, tea, flavoured soft drinks, fruit juices, wine, dried herbs and spices, black pepper, oil of wintergreen, spearmint, certain fruits (all fruits in varying degree except banana, limes and pears) and vegetables (all vegetables in varying degree except cabbage, celery, lettuce and potatoes). Both natural and synthetic forms can cause adverse reactions in some people.

When salicylate intolerance occurs, it can cause them to build up in the body and trigger salicylate sensitivity. Salicylate sensitivity is thought to be caused by an overproduction of leukotrienes, which are inflammatory mediators that have been linked to a variety of conditions, including asthma, rheumatoid arthritis and inflammatory bowel disease. The skin and the gastrointestinal tract may also be affected by salicylate intolerance. As the cause is not recognized and the triggering substances continue to be taken, this can lead to chronic processes, such as utricaria, colitis, or diarrhea.

Vaso-active Amines

Biogenic or vasoactive amines are produced by bacteria during fermentation, storage or decay. They include beta-phenylethylamine, tyramine, tryptamine, putrescine, cadaverine, spermine and spermidine, but histamine is the one most frequently linked to food-related symptoms. They can cause a type of FHS, which mimics IgE-mediated FHS. Histamine, the commonest vasoactive amine, is found in red wine, strong and blue cheeses, parmesan, cheddar cheese and other hard cheeses, all cured meat products, oily fish especially tuna, salmon and mackerel, spinach, aubergines, and pork products. Amines can act directly on small blood vessels to expand their capacity, perhaps accounting for their effect on flushing, migraines and nasal congestion in some patients.

When plasma histamine levels are raised above the normal range (0.3–1.0 ng/mL), this produces certain effects. For example a level of 1–2 ng/mL causes increased gastric acid secretion and heart rate, with, flushing, headache, urticaria, pruritus and tachycardia occurring at a level of 3–5 ng/mL, bronchospasm at a level of 7–12 ng/mL and cardiac arrest occurring at levels of 100 ng/mL. Thus, large amounts of ingested

Table 27.12: Foods more Likely to Contain High Levels of Vaso-active Amines and Salicylate

	Vasoacatiave Amines	*Salicylate*
Meat, poultry and seafood	All cured meat especially pork products *e.g.* ham, salami, pepperoni, game, bacon, sausages, fresh pork, fresh or canned tuna, canned sardines, anchovies, mackerel, salmon, herring, processed fish products (fish pastes, smoked, dried or pickled fish), fish sauce	
Milk and eggs	Blue cheese, parmesan, brie, camembert, emmenthal, old gouda, cheddar cheese and other hard cheeses	
Fruits	Oranges, bananas, tangerines, pineapple, grapes, strawberries	Apples, cherries, strawberries, currants, raisins, kiwi, Gala melon, peaches and nectarines, raspberries
Vegetables, nuts, seeds and savoury snacks	Tomatoes, pickled cabbage, aubergine, spinach, broad beans, peanuts, tree nuts	Asparagus, sweet corn, raw tomatoes, tomato puree
Condiments and miscellaneous	Fermented soy products including *miso* and *tempeh*	Ginger, mixed herbs, mustard, oregano, curry powder, black pepper, cardamom pods, cinnamon, cumin, fenugreek, mint, nutmeg, paprika, rosemary, thyme, turmeric, liquorice, peppermint, Worcestershire sauce, honey, tomato ketchup
Drinks	Green tea, champagne, coffee, cocoa, chocolate, wine, beer, fresh fruit juices, smoothies	Coffee, pineapple juice, cider, lemon tea, black tea, apple juice, cranberry juice, orange juice, tomato juice, fizzy drinks, Drambui liqueur, wine, rum

Source: Skypala *et al.,* 2015. Senstitivity to food additives, vaso-active amines and salicylates: a review of the evidence. Clin Transl Allergy. 5: 34.

histamine can cause significant symptoms in otherwise well individuals.

The diagnosis of sensitivity to vasoactive amines is usually made through history and dietary exclusion; however some studies have suggested that the measurement of diamine oxidase (DAO) levels may be helpful. One study found a DAO level <3 kU/mL was associated with reported symptoms to high histamine foods, whereas a level of >10 kU/mL indicated histamine intolerance was unlikely. Patients with chronic idiopathic urticaria and GI symptoms have been shown to have reduced DAO activity. Another study reported that the size of the skin prick test wheal to histamine after 50 min, the 'histamine 50-skin-prick test', was a useful diagnostic indicator; 82 per cent of subjects with histamine intolerance maintained a wheal size greater than 3 mm compared with 18 per cent of controls. A histamine-free diet is the treatment of choice for patients with histamine intolerance and chronic headache.

Other Causes of Food Reactions

Some adverse reactions to foods or food components do not involve the immune system and are known as "food intolerances". However, like true food allergies, food intolerances affect some individuals in the population but not all. Individuals suffering from food intolerances can usually tolerate small amounts of the offending food or food ingredient in their diet without ill effects.

Types of Food Intolerances

Several different types of food intolerances are also known to exist. These intolerances are food poisoning, anaphylactoid reactions, metabolic food disorders, idiosyncratic reactions and some unexplained causes of food adverse reactions.

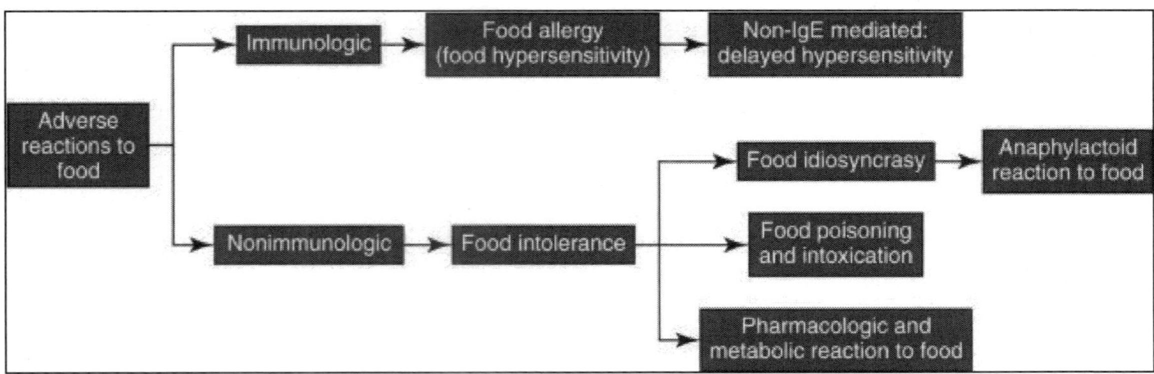

Figure 27.5: Categories of Adverse Reactions to Food.

1. Anaphylactoid Reactions (Scombroid Poisoning)

Food poisoning occurs when foods containing toxins are consumed. In some instances, food poisoning can mimic an allergic reaction. For example symptoms of flushing, sweating, urticaria, GI symptoms, palpitations and in severe cases bronchospasm may occur following the consumption of spoiled fish. This condition, known as scombroid poisoning, occurs due to the high level of histidine in certain spoiled fish species (tuna, mackrel or other fish) being converted into histamine by contaminating marine bacteria. Due to the nature of the symptoms caused, reactions involving vasoactive amines may therefore, be incorrectly diagnosed as a food allergy.

2. Metabolic Food Disorders (MFD)

Metabolic food disorders are adverse reactions to a food or food component that results from a defect in the metabolism of these foods or some substance therein or from an affect of the food or food component on the body's normal metabolic processes. Two best known examples of metabolic disorders are lactose intolerance and favism. These are genetically determined disorders which affect the host ability to metabolize food. Celiac disease could also be classified as a food allergy and/ or MFD (see chapter 22).

Lactose Intolerance

Lactose intolerance is an example of a metabolic food disorder resulting from a genetic deficiency that affects the host's ability to metabolize a food component. Lactose intolerant individuals cannot digest milk and milk products.

In lactose intolerance, there is an inherited deficiency of the intestinal enzyme, lactase (β- galactosidase), which is needed to metabolize milk sugar *i.e.* lactose to its monosaccharide components *i.e.* glucose and galactose in intestine mucosa. Since lactose cannot be absorbed in the small intestine unless it is hydrolyzed to glucose and galactose, the undigested lactose passes into the colon where it encounters large populations of bacteria. The colonic bacteria metabolize the undigested lactose into CO_2, H_2 and H_2O. Abdominal cramping, bloating, flatulence, and frothy diarrhea are the predominant symptoms of lactose intolerance and are the direct result of the action of the colonic bacteria on lactose. The symptoms vary in intensity in concert with the individual level of activity of β -galactosidase in the small intestine, GIT transit time, colonic fermentation and the amount of lactose ingested. The symptoms of lactose intolerance occur on a slightly delayed basis by comparison to cows' milk allergy.

Lactose intolerance is a fairly common metabolic food disorder on worldwide basis. Lactose intolerance is especially prevalent among some ethnic groups in the world including Greeks, Arabs, Jews, black Americans, Hispanics, Japanese, and other Asians. Only about 6-12 per cent of Caucasians are affected. Lactose intolerance can have its onset at any age, occurring as early as the age of three. However, lactose intolerance tends to worsen with advancing age and is often more common and more severe among the elderly in the affected population. The level of intestinal β -galactosidase is usually sufficient at birth to allow digestion of lactose in mother's milk, but susceptible individuals suffer as a result of decreased activity of this enzyme as life progresses. Congenital lactose intolerance is the complete absence of β–galactosidase at birth. Obviously, this condition leads to profound lactose intolerance even in infancy. However, the prevalence of this condition is quite low. Lactose intolerance may also occur on a more transitory basis on occasion, secondary to another intestinal illness or infection such as a bout of viral gastroenteritis and secondary to gastrointestinal disorders such as Crohn's disease or ulcerative colitis. Secondary lactose intolerance tends to subside rather quickly after the original illness is resolved.

Individuals with lactose intolerance are able to control their symptoms through the avoidance of dairy products (all kinds of milk-whole, skim milk, evaporated, condensed milk *etc.*, ice cream, cottage cheese, milk protein, milk solids, non-fat milk solids, whey, whey solids or protein, sweets and desserts containing milk or milk product *etc.*) containing lactose. However, many lactose-intolerant individuals can tolerate some lactose in their diets. The degree of tolerance for lactose is individualistic and variable among such individuals. There is no benefit in avoiding more lactose than is necessary to control symptoms. Systematic review has indicated that 12 g of lactose as a single dose of milk (*i.e.* ~240 ml) is tolerated by most people with lactose intolerance (15 g if consumed with other nutrients) whilst doses of 24 g are associated with substantial symptoms.

Yogurt (curd), fermented buttermilk, sour cream, and acidophilus milk that contain active cultures of bacteria with β-galactosidase activity, are better tolerated by lactose-intolerant individuals than other dairy products. Lactose-hydrolyzed milk is also available in the marketplace. And, lactose-intolerant individuals can add β -galactosidase to milk just before consumption, and this seems to be an effective practice. Certainly, the level of tolerance for dairy products is much higher with lactose intolerance than it is with IgE-mediated cows' milk allergy.

Favism

Favism is an example of a metabolic food disorder resulting from sensitivity to foodborne substances that interfere with normal metabolic processes. In favism, a genetic deficiency in glucose-6-phosphate dehydrogenase (G6PDH) in the erythrocyte results in an enhanced sensitivity to several hemolytic substances that occur naturally in fava beans. The diagnosis of G6PDH deficiency is made through an assay for enzymatic activity of G6PDH in isolated red blood cells.

Favism occurs most frequently in areas where the fava bean plant grows and the crop is harvested and sold in local markets. Symptoms occur after consumption of fava beans or the inhalation of pollen from the *Vicia faba* plant. Fava beans (broad beans) contain vicine and convicine, naturally occurring oxidants that are able to damage the erythrocyte membranes of G6PDH-deficient individuals causing hemolysis and the symptoms of hemolytic anemia. G6PDH is a critical enzyme in erythrocytes because it helps to maintain adequate levels of the reduced form of glutathione (GSH) and nicotinamide adenine dinucleotide phosphate (NADPH). GSH and NADPH help to avert oxidative damage to erythrocytes. In individuals who are G6PDH-deficient, this protective mechanism is nonfunctional and the oxidants in fava beans can exert their hemolytic effects. Sensitive individuals suffer from acute hemolytic anemia upon exposure with characteristic symptoms that include pallor, fatigue, shortness of breath, nausea, abdominal pain and/or back pain, fever and chills. In rare cases with repeated exposure, more severe symptoms can occur including hemoglobinuria, jaundice, and renal failure. The onset time after ingestion of the fava beans ranges from 5 to 24 hours. The onset time following ingestion is quite rapid, usually occurring in 5-24h. The illness is self-limited, however, with symptoms resolving promptly and spontaneously upon avoidance of further exposure.

G6PDH deficiency is the most common inherited enzymatic defect in humans and affects approximately 100 million individuals on a worldwide basis. The prevalence is highest among Oriental Jewish groups in Israel, Sardinians, Cypriot Greeks, Asians, African blacks and African Americans. This inherited trait is virtually absent among Caucasians, North American Indians, and Eskimos. It is more common in males than females and is more severe in infants and children. Despite the high prevalence of G6PDH deficiency, favism is an uncommon occurrence because fava beans are not frequently eaten except in Mediterranean and Middle Eastern locales. Obviously, the susceptible individuals can avoid favism by avoiding the ingestion of fava beans and/or the inhalation of the plant pollens.

3. Food Idiosyncrasies

These are adverse reactions to foods or food components that occur through unknown mechanisms and which can even include psychosomatic illnesses. The reaction can resemble or differ from symptoms of true food allergy. Idiosyncratic reactions to food are quantitatively abnormal responses to a food substance or additive differing in its physiologic or pharmacologic effects. This type of response resembles a hypersensitivity reaction but does not involve the immune system as seen in food allergy reactions. Psychological factors may play an important role in some cases.

Many reports, mostly anecdotal, have occurred regarding illnesses in certain individuals attributed to certain specific foods or food ingredients. Conceivably, a large number of different mechanisms could be involved in these idiosyncratic reactions. The symptoms involved in idiosyncratic reactions range from trivial to severe, life-threatening reactions. The role of specific foods or food ingredients in the causation of these idiosyncratic reactions remains to be determined in many cases. The cause-and-effect relationships can only be established through carefully controlled DBPCFCs. A positive DBPCFC would definitely confirm that the specific food or food ingredient is involved in the particular adverse reaction. The mechanism of the adverse reaction, however, cannot be determined from the positive DBPCFC alone. A negative DBPCFC indicates either that foods are not involved in causation of the reaction or at least that the specific food or food ingredient used in the challenge was wrongly incriminated. Unfortunately, DBPCFCs are rarely performed to establish convincingly that a specific food or food ingredient is associated with a particular idiosyncratic reaction.

The roles of specific foods or food ingredients are firmly established in a few of the many alleged foodborne idiosyncratic reactions. Sulfite-induced asthma is the best example of an idiosyncratic reaction that has been well documented using DBPCFC protocols to occur among small number of certain individuals, although the mechanism remains unknown.

For many other alleged idiosyncratic reactions to specific foods or food ingredients, the association with the specific food or food ingredient has not been conclusively documented through DBPCFCs. Examples would include the role of chocolate or aspartame in migraine headache; the roles of BHA, BHT, tartrazine, benzoates, or parabens in chronic urticaria; the role of tartrazine in asthma; the role of monosodium glutamate (MSG) in asthma or MSG Symptom Complex; and the role of sugar in aggressive behavior. However, very few of the clinical studies have used double blind and placebo-controlled trial designs. Furthermore, many of the studies have involved individuals with chronic, episodic symptoms such as chronic urticaria or asthma and the clinical investigators have removed critical medications from the patients before initiating the challenge trials. In such cases, the observed symptoms might be due either to the challenge substance or to the withdrawal of medications that controlled the condition. With such critical clinical design flaws, the role of these specific foods and food ingredients in these particular idiosyncratic illnesses remains unproven. Furthermore, psychological disorders may be involved in perceived reactions to specific foods or food ingredients.

28

Inherited Metabolic Disorders

Definition

Previously inherited metabolic disorders (IMD) were known as inborn errors of metabolism (term coined by a British physician, Archibald Garrod in the early 20th century (1908). Now a days, this term is often referred to as congenital metabolic diseases too. All these terms are considered synonymous.

Inherited metabolic disorders, refer to different types of medical conditions caused by genetic defects most commonly inherited from both parents that interfere with the body's metabolism. Most people with inherited metabolic disorders have a defective gene that results in an enzyme deficiency which ultimately affects the metabolism which is the complex set of chemical reactions that our body uses to maintain life, including energy production. Special enzymes break down food or certain chemicals so that the body can use them right away for fuel or store them.

Inborn errors of metabolism comprise a large class of genetic diseases involving disorders of metabolism. The majority are due to defects of single genes that code for enzymes that facilitate conversion of various substrates into other products. In most of the disorders, problems arise due to accumulation of substances which are toxic or interfere with normal function, or to the effects of reduced ability to synthesize essential compounds.

Incidence

Overall incidence of the inborn errors of metabolism is approximately 40 cases per 100 000 live births (1 in 2500). Approximately 24 children per 100 000 births (1 in 4,200) have a disease involving amino acids (*e.g.* PKU), organic acids, primary lactic acidosis, galactosemia, or a urea cycle disease. Approximately 2.3 children per 100 000 births (1 in 43,000) have some form of glycogen storage disease. Approximately 8 per 100 000 births (1 in 12,500) have a lysosomal storage disease; ~3 per 100000 births (1 in 33,000) have a respiratory chain-based mitochondrial disease; and ~3 to 4 per 100 000 of births (~1 in 30,000) have a peroxisomal disorder.

Diagnosis

It is important to make IMD diagnosis because there are many effective treatments and early diagnosis greatly enhances the chance of a better outcome. Some metabolic disorders can be diagnosed by routine screening tests done at birth. Others are identified only after a child or adult shows symptoms of a disorder. Once symptoms develop, specific blood or DNA tests are available to diagnose most genetic metabolic disorders. MR imaging is primary modality for detecting and characterizing CNS involvement of inborn errors of metabolism. MR findings are often nonspecific. Diagnosis is ultimately made from a combination of physical, biochemical, genetic analysis and imaging findings.

The severity of the disorder may vary widely depending upon the degree of enzyme deficiency. IMDs occur in many pathways of amino acid, carbohydrate, lipid, and vitamin metabolism.

Causes of Inherited Metabolic Disorders

The original cause of most genetic metabolic disorders is a gene mutation that occurred many, many generations ago. The gene mutation is passed along through the generations, ensuring its preservation. Each inherited metabolic disorder is quite rare in the general population. Considered all together, inherited metabolic disorders may affect about 1 in 1,000 to 2,500 newborns.

The code or blueprint to produce an enzyme is usually contained on a pair of genes. Most people with inherited metabolic disorders inherit two defective copies of the gene, one from each parent. Both parents are "carriers" of the bad gene, meaning they carry one defective copy and one normal copy.

In the parents, the normal gene copy compensates for the bad copy. Their enzyme levels are usually adequate, so they may have no symptoms of a genetic metabolic disorder. However, the child who inherits two defective gene copies from parents cannot produce enough effective enzyme and develops the genetic metabolic disorder.

In most inherited metabolic disorders, a single enzyme (or occasionally multiple enzymes) is either not produced by the body at all or is produced in a form that does not work in a metabolic pathway. The missing enzyme is like an absentee worker on the assembly line. Depending on that enzyme's job, the clinical consequences shall arise because:

- Substrates prior to the 'block' accumulate and can be toxic;
- Essential products beyond the 'block' are not formed;
- Other compounds may be formed via alternative pathways which may be toxic.

Symptoms

The symptoms of genetic metabolic disorders vary widely depending on the metabolism problem present. Some symptoms of inherited metabolic disorders include abdominal pain, poor appetite, lethargy, weight loss, jaundice, failure to gain weight or grow, developmental delay, seizures, coma, abnormal odour of urine, breath, sweat or saliva *etc.* The symptoms may come on suddenly or progress slowly. Symptoms may appear within a few weeks after birth in many conditions and in some disorders, it may take years for symptoms to develop.

Types of Inherited Metabolic Disorders

Major categories are:

- Disorders of carbohydrate metabolism
- Disorders of amino acid metabolism
- Lysosomal storage diseases
- Disorders of peroxisomal function
- Disorders of mitochondrial function
- Miscellaneous: Pelizaeus-Merzbacher disease, Hallervorden Spatz disease, Canavan and Alexander's diseases, Van der Knaap leukodystropy *etc.*

Table 28.1: Summary of Major Categories of Inborn Errors of Metabolism Disorders

Inborn Errors of Metabolism Categories		*Deficiency*
Disorders of carbohydrate metabolism	Von Gierke's disease	Glucose-6-Phosphatase
	Pompe's disease	Acid Maltase
	Cori's disease	Glycogen debranching enzyme
	Andersen's disease	Glycogen debranching enzyme
	McArdle's disease	Muscle glycogen phosphorylase
	Hers' disease	Liver glycogen phosphorylase
Disorders of amino acid metabolism	Alkaptonuria	Homogentisate 1,2-dioxygenase
	Phenylketonuria	Phenylalanine hydroxylase
	Maple syrup urine disease	Branched-chain alpha-keto acid dehydrogenase complex
	Glutaric acidemia type 1, type 2	Glutaryl-CoA dehydrogenase
	Urea cycle disorder	N-acetylglutamate synthase, Carbamoyl Phosphate Synthetase I, Ornithine transcarbamoylase, Argininosuccinate synthetase, argininosuccinate lyase, Arginase
Lysosomal storage diseases	Gaucher's disease	Glucocerebrosidase
	Niemann Pick disease	Sphingomyelinase
Disorder of peroxisomal function	Tay-Sachs disease	Hexosaminidase A
	Fabry's disease	α-galactosidase A
	Krabbe's disease	Galactosylceramidase (or galactocerebrosidase)
	Metachromatic leukodystrophy	Arylsulfatase A
	Mucolipidoses	
	Type I	Sialidase 1
	Type II and III	N-acetylglucosamine-1-phosphate transferase
	Type IV	Mutations in the MCOLN1 gene

Inborn Errors of Metabolism Categories		Deficiency
	Mucopolysaccharidoses	
	Hurler	Iduronidase
	Hunter	Iduronate-2-sulfatase
	Sanfilippo (A, B, C, D)	N-sulphoglucosamine sulphohydrolase ; Alpha-N-acetylglucosaminidase ; Heparan-alpha-glucosaminide N-acetyltransferase ; N-acetylglucosamine-6-sulfatase
	Morquio (A, B)	N-acetylgalactosamine-6-sulfatase ; β-galactosidase
	Maroteaux-Lamy disease	Arylsulfatase B
	Sly disease	Beta-glucuronidases
	Zellweger syndrome	Peroxisomal/peroxisome biogenesis factor
	Adrenoleukodystrophy	Transport protein
	Refsum's disease ("Phytanic acid storage disease")	Phytanoyl-CoA dioxygenase; Peroxin-7
Disorder of mitochondrial function	Leigh's disease	
	Kearns-Sayre disease	
	Mitochondrial encephalomyopathy, lactic acidosis, and stroke-like episodes (MELAS)	
	Neuropathy, ataxia, and retinitis pigmentosa (NARP)	
	Leber's hereditary optic neuropathy (LHON)	
	MERRF syndrome (Myoclonic Epilepsy with Ragged Red Fibers)	

Treatment of Inherited Metabolic Disorders

Treatment for an inherited metabolic disorder depends on the type and severity of the disorder. Because there are so many types of inherited metabolic disorders, treatment recommendations may vary considerably from dietary restrictions, certain vitamins or other therapies to medication or surgery.

Limited treatments are available for inherited metabolic disorders. The essential genetic defect causing the condition cannot be corrected with current technology. Instead, treatments try to work around the problem with metabolism. Treatments for genetic metabolic disorders follow a few general principles:

- Therapeutic diet to limit the intake of substrate that cannot be catobolized. Reduce or eliminate intake of any food or drug that cannot be metabolized properly. Special diets are recommended that eliminate certain nutrients which cannot be metabolized.
- Enzyme replacement therapy is possible in few disorders. Replace the enzyme or other chemical that is missing or inactive, to restore metabolism to as close to normal as possible.
- Provide the product that cannot be formed.
- Large doses of cofactor vitamins.
- Remove toxic products of metabolism that accumulate due to the metabolic disorder. Medicines are prescribed that conjugate with toxic metabolites so that the product is excreted in urine.
- Liver transplant for some disorders.
- IMDs are rare and complex so it is essential that patients are managed in a specialist metabolic centre by a multidisciplinary team including metabolic consultants, dietitians and nurses, with supporting specialized laboratory services.

Table 28.2: Summary of Dietary Management of some Inherited Metabolic Disorders

Disorder	Dietary Management
Amino Acid Disorders	
Classical Phenylketonuria	Low phenylalanine diet + Phenylalanine free, tyrosine enriched amino acid supplement
Maple Syrup Urine Disease	✰ Low leucine, isoleucine, valine diet + leucine, isoleucine, valine-free amino acid supplement ✰ Isoleucine and valine supplements may be needed. ✰ Amino acid supplement given during illness.
Tyrosinaemia type I and II	Low protein diet (to limit tyrosine intake) + tyrosine, phenylalanine-free amino acid supplement
Classical homocystinuria	Low methionine diet + methionine free, cystine enriched amino acid supplement

Disorder	Dietary Management
Organic acidaemias*	
Methylmalonic acidaemia *	Low protein diet (to limit methionine, threonine, valine, isoleucine intake) + methionine, threonine, valine, isoleucine-free amino acid supplement sometimes recommended
Isovaleric acidaemia *	Low protein diet (to limit isoleucine intake)
Glutaric aciduria type I *	☆ Low lysine or low protein diet (to limit lysine intake) + lysine-free, tryptophan-reduced amino acids ☆ Amino acid supplement given during illness.
Urea cycle disorders *	
Ornithine carbamoyl transferase deficiency *	Low protein diet (to limit waste nitrogen for excretion) + l-arginine supplements. Essential amino acid supplements may be needed.
Citrullinaemia *	
Argininosuccinic aciduria *	
Carbohydrate disorders	
Classical galactosaemia	Minimal galactose and lactose diet. Infant soya milk substitute. Calcium and vitamin D supplements may be needed for children
Glycogen storage disease type I * and type III *	☆ Frequent supply of exogenous glucose, provided as continuous overnight tube feeds and either 2-hourly daytime feeds or uncooked cornstarch ☆ Type III - milder disorders require less intensive dietary treatment.
Fatty acid oxidation disorders *	
Very long chain acyl-CoA dehydrogenase deficiency * Long chain 3-hydroxyacyl- CoA dehydrogenase deficiency *	Minimal long chain fat, ↑ CHO diet and medium chain triglyceride supplements. Frequent daytime feeding and continuous overnight tube feeds or uncooked cornstarch.
Medium chain acyl-CoA dehydrogenase deficiency *	Emergency regimen during illness
Lipid disorders	
Familial hypercholesterolaemia	☆ Healthy eating - restricted saturated fat, replace with poly- and monounsaturated fat
Type I hyperlipidaemia	☆ Very low long chain fat diet (to tolerance). Medium chain triglycerides can be given.
Abetali-proteinaemia	☆ Very low long chain fat diet. Vitamin A and E supplements. ☆ Medium chain triglycerides not recommended.

* Disorders requiring emergency regimen during metabolic stress such as intercurrent illnesses,*e.g.* colds, ear infections, *etc.*

Now, some of the important inherited metabolic disorders, which can be treated mainly through dietary interventions, shall be discussed in detail as under:

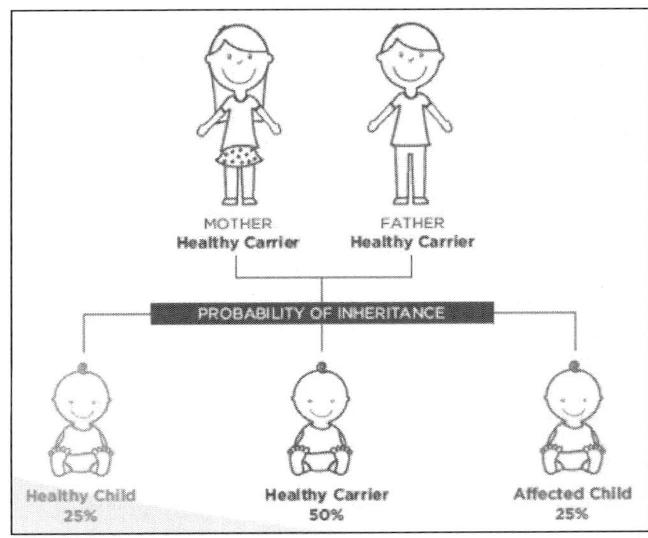

Phenylketonuria (PKU)

Phenylketonuria is the most common amino acid disorder. It is an inherited autosomal recessive disorder *i.e.* both parents are carriers of the mutation. With each pregnancy there is a 25 per cent chance of the child inheriting the condition. Both girls and boys are equally affected.

In this disorder, the essential amino acid phenylalanine cannot be metabolized due to the absence of the liver enzyme phenylalanine hydroxylase. With the development of newborn screening, it is now possible to identify children with high phenylalanine levels soon after birth.

Normally, phenylalanine is needed for protein synthesis and is converted to tyrosine by the liver enzyme phenylalanine hydroxylase (Figure 28.1). Tyrosine is necessary for the synthesis of melanin, tyroxine and a number of neurotransmitters. In the absence of phenylalanine hydroxylase, the individual with PKU is unable to convert phenylalanine to tyrosine; phenylalanine and its metabolites accumulate, and tyrosine, a precursor for neurotransmitters and melanin, becomes deficient; results in the inborn error of metabolism PKU.

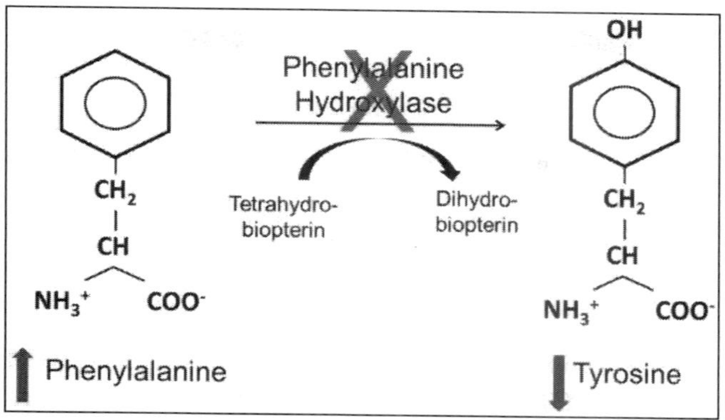

Figure 28.1: Phenylalanine Hydroxylase Enzyme Activity is Deficient in Phenylketonuria (PKU). As a result of this enzyme deficiency, phenylalanine accumulates and tyrosine becomes deficient.

Deficiency of phenylalanine hydroxylase leads to elevated levels of phenylalanine in the blood (normal range being 35-100 pmol/L). Based on their serum phenylalanine levels under normal diet, patients with hyperphenylalaninemia are classified as "classical" phenylketonuria (PKU) (>1200 pmol/L), "mild" PKU (600-1200 pmol/L), or "non-PKU hyperphenylalaninemia" (<600 pmol/L). Rare cases are attributed to defects in the biosynthesis or conservation of tetrahydrobiopterin (BH,), the cofactor for phenylalanine hydroxylase.

Screening for PKU

A newborn screening sample should be taken between 72 and 120 hours following birth and a repeat sample is done on day 10 of life to rule out PKU. Siblings of known cases generally have an elevated level of phenylalanine on the first sample if they have PKU.

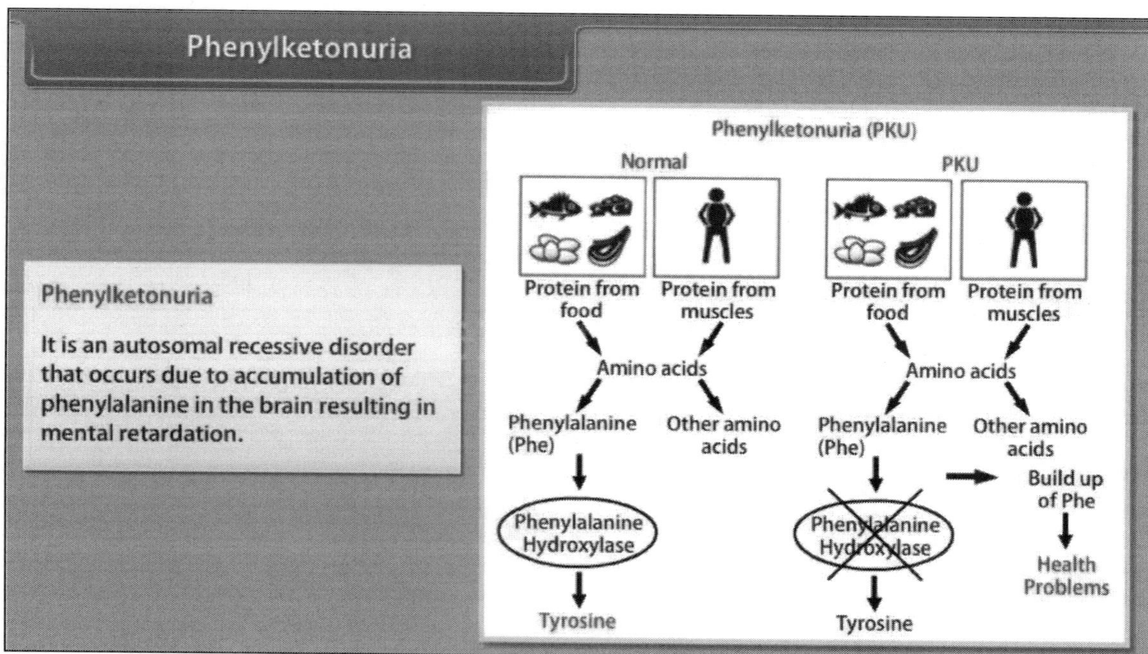

Symptoms

Symptoms are due to the accumulation of phenylalanine which inhibits brain development and these will arise within the first few months of life if the condition is undiagnosed and untreated. In the untreated patient, the accumulation of phenylalanine results in irreversible neurocognitive impairment, short attention span, poor short-term memory, mental retardation, problems with visual motor perception and motor coordination, mixed seizures, behavioural problems, growth deficiency, eczema, and skin and hair hypopigmentation. Childhood autism is very frequent. In older patients behavioral disturbances, hyperactivity, destructiveness, and schizophrenia-like symptoms are known to occur.

These complications are believed to result from a combination of factors including deficiency of tyrosine, the precursor of melanin and numerous neurotransmitters, and disruption of the transport of large neutral amino acids by the accumulation of phenylalanine.

Nutritional Management

The goal of nutritional management for those with PKU is to maintain plasma phenylalanine concentrations that support optimal growth, development, and mental functioning while providing a nutritionally complete diet. There are differences in opinion throughout the world regarding the recommended targets for plasma phenylalanine concentrations to optimize development and mental functioning for those with PKU, as summarized in table 28.3. There is consensus among European countries and the US that any newborn infant with a plasma phenylalanine concentration of >400–600 μM should be started on a low-phenylalanine diet as soon as possible, which is easily achieved in infancy and compatible with limited breastfeeding.

The low-phenylalanine PKU diet (in classical PKU, <500 mg phenylalanine per day) requires two interrelated dietary modifications to achieve metabolic control of plasma phenylalanine concentrations. First, natural foods are severely restricted to limit protein intake yet provide adequate amounts of phenylalanine, an essential AA. This generally includes elimination of all sources of animal protein, legumes and nuts, as well as limited intake of bread, pasta, rice and some vegetables. Low-protein bread and pasta products made from starch are used to provide needed energy and to increase variety. Second, consumption of an AA-based, phenylalanine-free formula (600–900 ml per day) or AA medical food is needed to provide adequate protein, vitamins, minerals and energy due to the restriction in natural foods.

The PKU diet tends to contain more carbohydrate and less total fat (20–25 per cent energy) compared to a typical omnivore diet which provides 30–35 per cent energy from fat. Moreover, the diet is devoid of meat, poultry, fish, nuts, and dairy foods which are the primary sources of long-chain polyunsaturated fatty acids (LC-PUFAs) such as arachidonic acid (20:4n6), eicosapentaenoic acid (20:5n3), and docosahexaenoic acid (DHA, 22:6n3). DHA is known to support normal brain and visual development, and some medical foods for PKU are currently supplemented with DHA. Plasma levels of the n-6 series of fatty acids, including linoleic acid and arachidonic acid, are generally normal in PKU; however, levels of the n-3 series of fatty acids, including linolenic acid and DHA, are decreased in treated children with PKU compared to control subjects. Treated adults with PKU show the same trend, although the degree of reduction in DHA is less severe compared with children. Hence, blood lipid levels of PKU patients should be monitored, intake of vegetable oils rich in linolenic acid such as canola or soybean oil should be encouraged and balanced supplements of n-3 and n-6 LC-PUFAs provided as needed for all individuals with PKU.

Table 28.3: Recommended Guidelines for Blood Phenylalanine Levels

Age	*Blood phenylalanine level, mM*			
	UK	*Germany*	*France*	*US*
Birth[1]	>400	>600	>600	>600
Infant to 10 years	120–360	40–240	120–360	120–360
10–12 years	<480	<600	<900	120–360
12–20 years	<480	<600	<900	120–600
>20 years	<700	<1,200	<900	120–900
Pregnancy				120–360

1 Phenylalanine concentration after newborn screening when countries initiate a low-phenylalanine diet.

Source: Macleod, E.L. and Ney, D.M. 2010. Nutritional management of phenylketonuria. Ann Nestle Eng. 68 (2): 58–69.

Basic Principles and Components of Dietary Management

Dietary treatment for PKU was first instituted by Bickel in the 1950's. Early diagnosis and prompt

Table 28.4: Guidelines of Daily Protein, Energy and Phenylalanine Intake for Individuals with PKU

Age	*Protein Requirement[1] g/kg*	*Minimum Phenylalanine Requirement mg/kg*	*Range of Phenylalanine Intake mg/day*	*Energy*	
				kcal/kg/day	*kcal/day*
0–6 months	3–3.5	20–70	–	95–145	
7–12 months	2.5–3(1.31)	10–35	–	80–135	
1–3 years	2–3 (1.02)	NA	200–400		900–1,800
4–6 years	2 (0.87)	13–20	210–450		1,300–2,300
7–10 years	2 (0.92)	13–20	220–500		1,650–3,300
Males					
11–14 years	2 (0.90)	NA	225–900		2,000–3,700
15–18 years	2 (0.87)	NA	295–1,100		2,100–3,900
≥ 19 years	NA (0.84)	4.6–13.6	290–1,200		2,000–3,300
Females					
11–14 years	2 (0.89)	NA	250–750		1,500–3,000
15–18 years	2 (0.84)	NA	230–700		1,200–3,000
≥ 19 years	NA (0.84)	4.6–13.6	220–700		1,400–2,500

NA = Sufficient evidence is not available for that age group.

1 Protein requirements for PKU are based on increased need with consumption of AA-based formula and values in parentheses reflect WHO safe level recommendations for the typical population.

Source: Macleod, E.L. and Ney, D.M. 2010. Nutritional management of phenylketonuria. Ann Nestle Eng. 68 (2): 58–69.

treatment with diet is needed to prevent damage to the developing brain. Dietary treatment is challenging work for the metabolic dietitian. Dietary treatment is lifelong and its aims are to:

- The basic principle of the diet is to severely restrict the substance which is not metabolized; in this case phenylalanine by strict control of natural protein intake in combination with the administration of a Phenylalanine-free protein substitute, usually based on Phenylalanine-free L-amino acids.
- The missing end product, which in PKU is tyrosine, then has to be replaced and prevent deficiencies.
- Normal growth is achieved by giving synthetically made protein devoid of phenylalanine and augmented with tyrosine. In addition, small amounts of natural protein are prescribed.
- Careful dietetic supervision is essential to ensure an adequate intake of protein, calories, minerals, vitamins and trace elements for growth and development.
- Monitoring of height and weight, regular blood tests to determine phenylalanine and tyrosine levels and nutritional status are all essential parts of the treatment.
- Catabolism associated with even mild illness/infection may cause the precursor phenylalanine to rise despite dietary compliance. In phenylketonuria, this should not have any severe pathological effect in the short-term. Patients with phenylketonuria should not be prescribed medications containing aspartame as it is a source of phenylalanine.

Phenylalanine Restricted Diet

Prior to starting the diet, it is important to rule out pterin defects (developmental delay/seizures), as these tend to be resistant to dietary interventions. This can be done by analysis of urinary pterin profile. Individuals with non-PKU hyperphenylalaninemia do not appear to be at risk for developing cognitive difficulties.

The diet is modified according to the age of the patient. It consists of restriction in phenylalanine intake. Although previously treatment was restricted to infancy and early childhood, lifelong therapy is now recommended given the emergence of numerous behavioral and psychiatric phenylalaninenotypes in patients, particularly in those who relaxed their protein restriction in late childhood or early adolescent years. Individuals may exhibit significant range in their tolerance for phenylalanine and this tolerance varies during the growth cycle. Consequently, individualized calculations for phenylalanine and tyrosine prescriptions must be made using frequent assessments of serum phenylalanine and tyrosine levels, growth parameters, and "actual" dietary intake through diet records provided by the patient.

The introduction of the phenylalanine restricted diet leads to normal, or at least, near normal development. The objective of the phenylalanine-restricted diet is to provide sufficient phenylalanine for growth, while maintaining serum phenylalanine levels within a target range of 2-6 mg/dL throughout the lifespan. The diet should provide adequate amino acids, vitamins, minerals, carbohydrates and fat. Thus, when a diet restricted in phenylalanine and supplemented with tyrosine is introduced early in infancy, permanent clinical sequelae of this disorder can be prevented.

To reduce plasma phenylalanine levels to the target range, individuals with classical forms of PKU typically require total elimination of high biological protein rich foods such as meats, eggs, milk and cheese from their diet as they contain an abundance of phenylalanine. Individuals with milder forms of hyperphenylalaninemia may be able to incorporate small amounts of high protein foods into their diet due to increased tolerance to phenylalanine. A thorough knowledge of the phenylalanine content of foods is necessary to plan an appropriate diet with variety and maintenance of essential nutrients.

Nutrient intake, specifically of vitamin B_6, B_{12}, calcium, folate and iron will be deficient if the AA formula and/or vitamin and mineral supplements are not consumed each day. Interestingly, even with adequate iron intake, studies have shown iron deficiency in those with PKU. Therefore, if AA formulas are not supplemented with micronutrients, they should be separately taken as supplements daily.

Sources of Protein in PKU Diet

Children suffering from PKU need a normal to high intake of other essential amino acids except phenylalanine, as total protein needs remain similar to that of normal children. Protein requirements reduce as the child gets older and stabilize into adulthood. However, the magnitude of natural protein restriction required to prevent elevated blood Phenylalanine concentrations could cause serious nutritional deficits in any age group without protein supplementation. In order to achieve protein sufficient diet devoid of phenylalanine, it is divided into three components:

- Synthetic protein or metabolic formulas
- Natural protein – commonly called protein exchanges
- Free foods.

Synthetic Protein

Metabolic formulas, which contain synthetic mixtures of amino acids devoid of phenylalanine, are typically required if the phenylalanine restriction results in protein intake below levels necessary for growth.

In PKU, the synthetic protein formula is devoid of phenylalanine but contains the other 20 amino acids. These special metabolic formulas are supplemented with tyrosine, the deficient product in this disorder. Formula prescriptions must be individualized and a variety of commercial products, such as coolers, gels, and ready-to-drink boxes, are available to meet patient's preferences for taste and palatability. In addition, small amounts of natural protein are prescribed.

Examples of synthetic formulas are:

- There are low phenylalanine formulae such as Lofenalac (Mead Johnson) and phenylalanine free formulae such as Phenylalaninenex 1 (Ross Lab) and Maxamaid X P.
- PKU Anamix Infant: A complete infant formula which has the same nutritional composition as all standard infant formulae except it is devoid of phenylalanine. It is made up the same way *i.e.* scoop of formula to 30 ml of water.
- For older children, there is a variety of drinks, pastes and gels.
- PKU gel and PKU Anamix Junior LQ for children: These formulae are nutritionally complete.
- PKU express and PKU Lophlex: Products for older children and adults, these formulae are nutritionally complete.
- Synthetic protein formulae are expensive and can have a bitter aftertaste. Some children and adults have difficulty complying with their prescription. The amount of synthetic protein prescribed depends on weight. Synthetic protein should be divided into three or four drinks each day.

Natural Protein (Protein exchanges)

A small amount of phenylalanine is needed for normal growth. This is provided by a system of exchanges of natural protein each containing 1g of protein (1 exchange = 1g protein). Commonly used exchange foods include measured amounts of low biological based protein foods such as cereal, pasta, beans and peas. Examples of 1g protein/1 exchange are:

- Two tablespoons of peas
- One tablespoon of baked beans
- Two tablespoons of chick peas

Most patients with phenylketonuria tolerate only a small number of exchanges (between four and eight per day). This allows for some flexibility but can be challenging to manage on a day-to-day basis. The aim of the treatment is to give sufficient phenylalanine for growth. This is achieved by keeping the blood phenylalanine levels as close to the normal physiological levels as possible. Patients are advised on the number of protein exchanges allowed based on frequent blood monitoring. It is important to note that aspartame must be avoided. This sweetener is often found in diet fizzy drinks.

During weaning, parents should be educated regarding the high, medium and low protein food. Low protein foods, including cereal, fruits, and vegetables, are given in amounts carefully calibrated using a phenylalanine equivalent system (1 equivalent = 15 mg phenylalanine).

Free Foods

Such foods are called as special low protein foods (SLPF). These contain only negligible amounts of protein. They are used to provide calories and add variety to the diet. Such foods are naturally occurring foods such as fruit and vegetables (except peas, beans, sweet corn and potatoes), sugars, jams, sweet drinks, fats and oils and specifically manufactured low protein products. Diet foods containing aspartame should be avoided as it is a phenylalanine containing peptide.

Excessive consumption of SLPF can lead to excessive energy intake with a low micronutrient contribution. So attention is required when they are consumed in excessive quantities.

Other New Strategies to Treat PKU

Two other new strategies for management of PKU include supplementation with large neutral amino acids (LNAAs) and the use of glycomacropeptide (GMP), a naturally occurring, low-phenylalanine protein source. LNAAs are proposed to interfere with the transport of phenylalanine across the intestine and blood-brain barrier since phenylalanine and other large neutral amino acids share the same transport systems. The small studies performed to date have been limited to older patients and thus, LNAAs are not considered in infants, young children, or pregnant adults with PKU since LNAAs may not be safe for use in these settings.

GMP is a naturally occurring phenylalanine-free protein source derived from whey. When isolated, GMP is contaminated with other proteins and thus, contains small amounts of phenylalanine. However, given the low phenylalanine content of GMP, GMP has been used in the manufacturing of low phenylalanine foods for patients with PKU. A small study of GMP use in patients with PKU identified no adverse events and demonstrated short-term efficacy in maintaining serum phenylalanine levels as compared to the use of traditional phenylalanine-free medical formulas. Further, long-term studies of the use of GMP in patients with PKU are necessary.

Diet Construction-Dilemmas

A regular healthy diet is composed of great food diversity and protein needs are usually supplied throughout the day, in every meal. In PKU, low-Phenylalanine containing foods (*e.g.*, low-protein bread or pasta referred to as SLPF) commonly constitute the main food sources at each meal. The protein substitute is usually given at mealtime at least three times per day in order to optimize its metabolism. There is some scientific evidence that the rate of absorption of amino acids is more rapid than intact protein sources (*e.g.*, casein). Some authors suggest that this causes higher urine nitrogen losses because rapid amino acid appearance in the blood will not allow long-term sustained anabolism. The taste and flavors of protein substitutes are also suboptimal, despite important improvements in recent years. As a consequence, adherence with protein substitute is commonly suboptimal.

It is advised to divide the total daily amount of prescribed protein substitute between several meals, in order to promote anabolism, that is, to move free amino acids toward protein synthesis rather than to the oxidative route. Anabolism promotion requires free amino acids to be given in combination with energy sources (lipids and carbohydrates). Many protein substitutes contain non-protein energy sources as well as amino acids. As recently highlighted, the composition of different protein substitutes is variable and it is important to consider the full nutritional composition to ensure that it is optimal to help promote anabolism.

Feeding Diet to Infants, Children and Adolescents Suffering from PKU

The incidence of PKU is higher among infants and children. After neonatal diagnosis, dietary treatment should be implemented immediately. In order to promote protein synthesis and provide variety within a very restrictive diet, it is best to allow the maximum amount of natural dietary protein from breast milk or infant formula in infancy, and later in life from fruits, vegetables and low protein products. However, it is unusual for the diet of an individual with classical PKU to contain more than 25 per cent of protein from natural foods (often only 5–10 g per day) given that protein in natural foods contains 2–9 per cent phenylalanine by weight.

Breast-feeding (directly from the breast) should be combined with infant Phenylalanine-free infant formula. In practice, a measured volume of an infant phenylalanine-free infant formula is given before each breast-feed. The amount of infant phenylalanine-free infant formula is titrated according to blood phenylalanine levels; if blood phenylalanine levels are high, the infant is given more infant phenylalanine-free infant formula; if blood levels are low, the volume of infant phenylalanine-free infant formula is reduced. The Phenylalanine intake is restricted/controlled to maintain the blood phenylalanine control within target ranges, but it is essential that dietary phenylalanine is given, in order to prevent phenylalanine deficiency, resulting in poor growth and other negative outcomes.

Weaning should progress as for other infants but eliminating high-protein foods and providing essential phenylalanine requirements only. In practice, dependence on fruits and vegetables is mandatory to construct the diet. However, phenylalanine intake is also derived from foods such as potatoes, sweet corn, spinach, and peas *etc.* With continued development of feeding skills, toddlers with PKU meet their phenylalanine requirements from low-protein table foods. During the toddler and early childhood years, the phenylalanine prescription changes rapidly as children grow and protein needs change. Although the PKU diet in early childhood requires a large intake of AA formula to meet protein needs (2–3 g protein/kg) and strict control of phenylalanine intake, compliance is good, presumably due to high parental control and low peer pressure.

Compliance with the PKU diet becomes more difficult as children grow older and start school, make their own food choices and develop taste preferences which are often similar to those without PKU. Hence, during this period, most difficulties are encountered for dietary management due to poor compliance, in particular during teenage years. This is like any other chronic illness such as diabetes or asthma. Independent thinking, peer pressure, time constraints, social pressures, financial limitations, and change in family relationships combine to interfere with dietary control. Added to these difficulties are the biological changes during teenage years which reduce phenylalanine tolerance. Changes in compliance are usually related to consumption of the AA-based formula. Children and adolescents may have transition from formula 3 or 4 times a day to only once or twice a day limiting the availability of AA, at a time of development when protein synthesis is essential for growth. Frequent consumption of AA formula throughout the day results in stable plasma phenylalanine concentrations and higher phenylalanine tolerance as protein retention and synthesis improve.

Consuming an AA formula at school is the most challenging aspect of compliance as AA formula looks different and has a distinctive poor odor and taste, in spite of efforts to improve formulations and packaging. Most children refuse to drink the formula at school, thus are without a good source of protein for 8 h resulting in catabolism of body protein and release of the phenylalanine into blood. Moreover, significant hunger may cause them to consume too much phenylalanine from prescribed or forbidden foods. Both of these factors lead to elevated blood phenylalanine concentrations with increased age, which negatively affects IQ. Good phenylalanine control can be achieved when children/adolescents participate actively in their own management and there is a positive parental attitude. A thorough understanding of the disease and usefulness of the diet needs to be emphasized to them.

Maternal PKU

Over the last decade, maternal PKU has emerged as a significant problem. Pregnancy provides unique challenges for the management of PKU because of the teratogenic effects of elevated maternal phenylalanine levels. Women with PKU, who when pregnant, are not on treatment or whose phenylalanine levels are poorly controlled have offspring with multisystem involvement including growth retardation, microcephaly, dysmorphic features, intellectual disability, congenital heart lesions, and developmental disabilities. This constellation of phenylalanine types has been termed maternal PKU. Thus, to prevent the development of maternal PKU, maintenance of maternal serum phenylalanine levels within the target range are necessary prior to conception and throughout pregnancy.

The treatment of maternal PKU consists of biochemical control through a phenylalanine restricted diet during pregnancy. Key to achieving metabolic control, is sufficient consumption of a phenylalanine-free AA formula. Moreover, given the increased requirements for protein, vitamins and minerals during pregnancy, consumption of AA formula is essential. For women who are unable to consume the required amount of formula and achieve plasma phenylalanine recommendations, placement of a gastrostomy tube can be an effective method of formula delivery.

Hence, close and careful monitoring of serum phenylalanine levels before, during and after pregnancy is necessary for women with PKU.

A newer strategy for PKU management involves the use of sapropterin dihydrochloride, a synthetic stereoisomer of tetrahydrobiopterin which is a cofactor for the phenylalanine hydroxylase. Sapropterin dihydrochloride is typically used in the setting of a phenylalanine-restricted diet but, in many cases, may allow for increased phenylalanine intake in the form of increased natural protein in the diet. Clinical data regarding the safety and efficacy of the use of sapropterin dihydrochloride in pregnancy is limited. However, sapropterin dihydrochloride has been used during pregnancy anecdotally, and no serious adverse events in pregnancy have been definitively linked to use during pregnancy.

Women of with PKU and their families require much psychosocial support to meet the strict requirements of a maternal PKU pregnancy, including compliance with a difficult diet. With such compliance, however, it is possible to have normal or near normal offspring.

Nutritional Outcome of Phenylalanine Restricted Diet

Normal growth and development is the positive outcome noted with early diagnosis and treatment of infants with PKU with a nutritionally adequate, low-phenylalanine diet. Evaluation of IQ scores in individuals with PKU treated shortly after birth through 12 years of age who maintained plasma phenylalanine concentrations within recommended treatment ranges show a mean IQ score of 101 ± 11. A study has reported that during the first 18 years of life for each 100 µmol increase in plasma phenylalanine concentration between 394 and 750 µ*M*, there was a 1.3- to 3.9-point decline in IQ score. Metabolic control of PKU through nutritional and dietary management is clearly effective in promoting intellectual development. Moreover, ongoing nutritional assessment over the life cycle is essential for optimal metabolic control of PKU and to ensure optimal growth.

If a patient is not consuming an adequate amount of AA formula distributed throughout the day, not only will plasma phenylalanine concentration be elevated due to protein catabolism, but growth failure may occur because formula provides the majority of energy, protein and other essential nutrients. Results of a longitudinal study of children treated early and continuously for PKU indicated that those children with plasma phenylalanine levels approximately 3-5 times normal (360-600 µmol/L or 6-10 mg/dl, levels previously considered safe) were impaired in cognitive functions dependent on the prefrontal cortex. It has been shown in numerous studies that individuals who went off the diet after 5 or 6 years of age, which was often the standard practice a few years ago, developed learning problems, lower IQ, and behavior problems. Consequently, current recommendations include lifelong adherence to the diet because formula provides the majority of energy, protein and other essential nutrients.

Nutritional Challenges in Feeding Phenylalanine Restricted Diet

The main objective in PKU treatment is to maintain blood Phenylalanine under rigorous control, especially during the first years of life. Diet is calculated to meet all energy and other nutrient requirements and should compensate for any inefficiency associated with phenylalanine-free L-amino acids. For the majority of patients, dietary adherence is not an issue in the first year of life. However, teething and childhood illnesses such as fever, gastrointestinal infections, and colic can disturb feeding and consequently adversely affect blood phenylalanine control. When protein substitute is introduced in the neonatal period, it is usually well tolerated. Nevertheless, caution is needed when formula volume or type is adjusted. Energy needs should always be respected, but care is required regarding choice of protein substitute as changes in osmolality, flavor, odor and consistency can lead to its refusal. It is also important for the clinician to differentiate between diet refusal caused by illness, flavor/odor changes of the formulation, and refusals merely as a consequence of usual child tantrums.

There are other important considerations with protein substitute administration. Nitrogen sources are known to promote satiety and giving protein substitute as close to the main meals may result in food refusal. In children particularly at risk of under nutrition, giving protein substitute just before main meals may reduce energy intakes from low-protein foods.

One of the most important messages in general nutrition for healthy individuals is that when energy needs are satisfied using a variety of foods from all foods groups, nutritional adequacy and optimal nutrient balance are easily achieved. In PKU the scenario is different. Out of the three dietary components (natural low-protein foods, protein substitutes, and SLPF), only one is natural. In fact, most natural foods are limited or controlled in variable amounts depending on individual phenylalanine tolerance. The spectrum of natural foods allowed in a PKU diet is unable to meet full micronutrient requirements.

Achieving a good balance between protein intake, energy and micronutrient supply is not always easy because protein substitutes are prescribed according to phenylalanine tolerance and weight. Therefore, micronutrient intake is influenced by the dose of protein substitute intake and the total protein equivalent from a combination of natural protein sources and protein substitute.

Nutritional imbalances (deficiencies or excesses) related to micronutrients delivered in large doses at a single time point possibly compromising their bioavailability in the gut, reduced absorption through enhanced competition for absorption, or increased fecal losses can occur in PKU individuals even if they have good dietary adherence. In addition, while SLPF are mainly responsible to provide energy intake, their micronutrient supply is often very poor in contrast to regular foods, and as a consequence, there is a reliance on protein substitutes to provide micronutrient requirements. They are usually well fortified with vitamins and minerals, and some evidence suggests that they may even contain excessive amounts of some nutrients. Vitamin B_{12} and folic acid are good examples of high amounts being added to protein substitutes and resulting in high serum concentrations. While vitamin B_{12} toxicity does not appear significant, high serum folate is a concern.

Numerous studies debating the role of various nutrients when patients are on restricted phenylalanine diet have emerged. Low levels of iron, zinc, retinol, and selenium have been reported in children maintained on a low phenylalanine diet. Reduced bone mineral density has also been described with the diet. It is unclear whether deficits in bone mineralization relate to the disease process itself or its treatment. Most individuals with PKU do not develop any of the above mentioned deficiencies and lead a healthy life.

Extra micronutrient supplementation beyond the amounts already contained in the protein substitute is not advisable as a standard recommendation in PKU, although evidence demonstrates that micronutrient imbalances continue to occur and the long-term importance of this fact remains to be clarified.

Fruits, vegetables, rice, and potatoes are good examples of regular natural foods included in the daily meals in PKU. In addition, sugars, fats, and oils can be ingested without any impact on blood phenylalanine control. Even so, classifying their intake as completely exchange-free may not be appropriate, considering the recognized health implications resulting from chronic excessive fat and sugar consumption. Very little is known about the long-term metabolic impact of diet therapy in PKU. There is an urgent need to accurately describe food patterns in PKU, although it is known that it is usually rich in carbohydrate.

Regular Measurement and Monitoring of Nutritional Status

In PKU, nutritional status has gained credence and is considered one of the more important aspects of clinical management. Considering the artificial characteristics of the dietary treatment, nutritional status monitoring is essential. Nutritional status indicators that require regular monitoring are anthropometry, body composition, biochemistry, and nutritional intake.

Growth is one of the most important outcomes to consider and its optimization constitutes a priority in PKU. World Health Organization growth standard and reference are advocated. In addition to monitoring for growth impairment, prevention of excessive weight gain with increased adiposity is an important objective in early infancy. In later childhood, information about pubertal stage and bone maturity age should be collected by the physician.

Also, clinical and a dietary history are mandatory. Social and family history is very important. Food patterns are closely linked with lifestyle, economic, cultural, and social family routines. Unfortunately, there are sometimes episodes when social actions should be taken in order to optimize treatment adherence.

Due to the potential for nutrient deficiencies with restriction of phenylalanine, comprehensive macronutrient and micronutrient analysis must compliment the physical evaluation and biochemical assessments. Normally, clinical symptoms of nutritional deficiencies are rare in patients with good dietary adherence although micronutrient imbalance does occur. Protein nutritional status should be monitored biochemically and prealbumin may constitute a good parameter to evaluate satisfaction of protein needs. Considering micronutrient status, vitamin B_{12} and folic acid blood concentrations are commonly high in patients with good adherence, while zinc and selenium concentrations are sometimes below the expected ranges. Continuous monitoring of micronutrient status is important in order to make appropriate dietary adjustments to compensate for individual imbalances.

Regular monitoring of plasma concentrations of phenylalanine and tyrosine in relation to phenylalanine intake, often provided by a 3-day record of food intake preceding blood sampling, is needed to adjust the dietary prescription for phenylalanine. When the growth rate is at its peak, as in early infancy and during the prepubertal and pubertal growth spurts, phenylalanine needs will be high relative to body mass and will then decline as the growth rate slows. The phenylalanine ingested should maintain the 2- to 4-hour postprandial plasma phenylalanine concentration within the recommended target range (The present recommended levels for metabolic control are below 400 μmol/L). During infancy, weekly adjustments in the phenylalanine prescription may be needed based on growth and assessment of plasma phenylalanine concentrations. Subsequently, phenylalanine should be assessed once or twice monthly.

Role of Nutrition Education

Although families and eventually the patient can master the basic principles of this diet, continued follow-up by health care providers is essential to allow for adjustments to the diet prescription which are necessary at various stages of development, to provide motivation to continue the diet, and to provide continued age-appropriate education. To promote long-term dietary adherence, it is necessary to help families adapt the diet to their personal life-style or cultural preferences.

For nutritional education, short- and long-terms goals should be considered. In the past, the message to adult patients was very clear that they could stop treatment in early adulthood but then treatment guidelines recommended "diet for life". As a consequence, there are many unhappy adult patients who struggle to adhere to their dietary treatment as they have become accustomed to eating higher protein foods. These and other issues illustrate the challenges in PKU adherence. The picture can differ in females because the fear of maternal PKU syndrome may contribute to higher adherence rates. The patient's outcome will also be influenced by the availability of new treatments, changes in treatment strategies, new foods, and different presentations of protein substitutes. Therefore, educational messages given to patients may significantly influence their motivation and persistence in being able to adhere to their diet.

Maple Syrup Urine Disease (MSUD)

MSUD is an autosomal recessive metabolic disorder, caused by a deficiency of the branched-chain alpha keto dehydrogenase complex (BCKDH). This enzyme deficiency leads to a buildup of the branched chain amino acids (leucine, isoleucie and valine) and their corresponding toxic by-products in the blood and urine which disrupt the metabolism of certain amino acids, causing rapid degeneration of the neurons. If left untreated, this disorder can result in poor appetite, presence of sweet-smelling urine with an odour similar to that of maple syrup, vomiting, acidosis, lethargy, lethargy, hypotonia, seizures, pancreatitis, neurological decline, coma, and death in the neonatal period. Even with treatment, severe, life-threatening metabolic decompensation can be triggered by common illness, such as upper respiratory tract infections and gastroenteritis or other forms of catabolic stress, and excess leucine consumption.

Nutrition plays an essential role in restoring and maintaining metabolic homeostasis in MSUD. The goals of medical nutrition therapy in MSUD are to rapidly reduce toxic metabolites by restricting dietary BCAA to amounts allowing individuals to achieve and maintain plasma BCAA amino acid concentrations within the targeted treatment ranges; reduce catabolism; promote anabolism; monitor nutritional status and alter intake to promote normal growth, development and health maintenance; evaluate thiamin responsiveness if the individual has residual BCKD activity; and supplement with thiamin if the individual is responsive.

Dietary Management

Of the BCAAs and ketoacids, leucine is considered the most neurotoxic, and strict, long-term control of plasma leucine levels has been associated with improved neurocognitive and neuropsychiatric outcomes in patients with MSUD. Thus, the goal of management in patients with MSUD is to maintain plasma leucine (100–200 μmol/L for infants and children <5 years and 100–300 μmol/L for those over 5 years of age) and other BCAAs within the normal range while supplying adequate nutrients necessary for growth and development. This goal is achieved with dietary restriction of BCAAs through protein restriction and the use of commercial elemental formulas which consist of synthetic amino acid mixtures devoid of the BCAAs but which contain other essential macro and micro-nutrients, dietary leucine restriction and supplemental valine and isoleucine. In addition, a trial of thiamine, the cofactor for the branched-chain alpha keto dehydrogenase complex (BCKDH), may be initiated since a subset of patients has been found to be thiamine-responsive. The levels of plasma leucine and other BCAAs are carefully monitored in all patients to monitor the efficacy of dietary therapy. Rises in plasma leucine level may occur with dietary non-compliance, over-prescription, inadequate intake of protein or energy and other catabolic stress such as illness or infection. Leucine concentrations may also rise with deficiencies of valine or isoleucine due to decreased protein synthesis or muscle catabolism. Thus, supplementation with isoleucine and/or valine is sometimes necessary in daily management of MSUD.

Comprehensive, nutrition management guidelines for MSUD recently reported by Frazier *et al.* (2014) are:

Achieving recommended BCAA blood concentrations: The goal of dietary BCAA restriction for the individual with MSUD is to achieve and maintain plasma BCAA concentrations as close to normal as possible while preventing and correcting BCAA deficiencies throughout life. Elevated leucine concentrations are most often associated with abnormal brain morphology and cognitive impairment; and impact on social and psychomotor function varies. Low plasma BCAA, particularly isoleucine and to a lesser extent valine, results in severe but reversible epithelial damage to the skin, eye and gastrointestinal tract. Poor growth is reported as a result of BCAA deficiency. Reports differ in the plasma BCAA ranges that are considered acceptable, but agree that frequent biochemical monitoring is necessary to assess the effectiveness of dietary intervention and detect deficiencies or excesses.

1. Maintain plasma leucine concentrations, with frequent monitoring, between 75 and 200 μmol/L for infants and children ≤ 5 years old and between 75 and 300 μmol/L for individuals > 5 years of age to achieve favorable cognitive outcomes.
2. Maintain plasma isoleucine and valine concentrations, with frequent monitoring, between 200 and 400 μmol/L (or slightly above the normal ranges) in all individuals to avoid metabolic instability and BCAA deficiencies.
3. Employ MSUD-specific dietary management strategies to achieve recommended BCAA concentrations (Table 28.5 and 28.6).

Thiamine Supplementation

Thiamin pyrophosphate (TPP) is a co-factor for the multi-subunit enzyme, BCKD, and thiamin is an adjunct to be considered in the treatment regime for MSUD. Patients should be continued on a BCAA-restricted diet during the assessment of the thiamin response, as well as during long-term supplementation. Recommendations are:

Table 28.5: Nutrient Recommended Intake and Sources in the Dietary Treatment of Well Individuals with MSUD

Nutrient	*Recommendation*	*Source*
Leucine (LEU)	Sufficient intake to allow adequate protein synthesis for growth, repair and health maintenance and to achieve leucine concentrations in recommended treatment range. Leucine allowance is also dependent on residual BCKD activity, age, weight, sex, life stage and health of the individual with MSUD.	Intact protein (PRO) **In infants***:* Breast milk or infant formula with known LEU content **In children and adults***:* Foods such as fruits/vegetables and some grains/cereals that are typically low in protein and for which there is a known leucine content.
Intact Protein (PRO)	RDA Plus additional 20–40 per cent if an amino acid–based medical food is used	✰ Intact PRO (as above) ✰ BCAA-free medical food[a]
Valine (VAL), Isoleucine (ILE)	VAL and ILE are essential amino acids and may need to be supplemented when BCAA are restricted to achieve appropriate LEU blood concentrations. To promote anabolism of LEU, when LEU blood concentrations are high, additional supplementation of VAL and ILE is often required.	✰ Intact PRO ✰ Supplemental VAL, ILE [b]
Energy	RDA	✰ Intact PRO ✰ BCAA-free medical food ✰ Free foods[c] ✰ Modified low-PRO food[c]
Other nutrients, minerals and vitamins[d]	RDA	✰ Intact PRO ✰ BCAA-free medical food ✰ Supplemental nutrients, vitamins and minerals[e]

* Requirements for individuals with MSUD change with catabolic illness/conditions. **a:** Medical food is a food that is formulated to be consumed or administered enterally under the supervision of a physician and that is intended for the specific dietary management of a disease or condition for which distinctive nutritional requirements, based on recognized scientific principles, are established by medical. **b:** 1 per cent solutions are convenient for adding to the medical food when supplementation is necessary. **c** Modified low-protein foods include pastas and baked goods where higher protein grains/flours are replaced by protein-free starches. **d:** Modified low-protein foods include pastas and baked goods where higher protein grains/flours are replaced by protein-free starches. **e:** Included are essential fatty acids and DHA, vitamin D, vitamin A, calcium, iron, zinc, selenium. Most BCAA-free medical foods are supplemented sufficiently with the nutrients and micronutrients that may be deficient in a diet low in BCAA. Compliance with taking the full medical food prescription is important in meeting these nutrient requirements. In addition, there are some medical foods that have been modified to improve taste, decrease calories or volume in order to increase compliance that may have insufficient supplementation of some micronutrients, vitamins and minerals.

Source: Frazier, D.M., Allgeier, C., Homer, C., Marriage, B.J., Ogata, B., Rohr, F., Splett, P.L., Stembridge, A. and Singh, R. H. 2014. Nutrition management guideline for maple syrup urine disease: an evidence-and –consensus-based approach. Molecular Genetics and Metabolism. 112 (3): 210-217.

Table 28.6: Recommended Daily Nutrient Intakes of BCAA, PRO, Energy and Fluids for Individuals with MSUD when Well

Age	*Nutrient*					
	LEU mg/kg	*ILE mg/kg*	*VAL mg/kg*	*Protein g/kg*	*Energy kcal/kg*	*Fluid mL/kg*
0 to 6 months	40–100	30–90	40–95	2.5–3.5	95–145	125–160
7 to 12 months	40–75	30–70	30–80	2.5–3.0	80–135	125–145
1–3 years	40–70	20–70	30–70	1.5–2.5	80–130	115–135
4–8 years	35–65	20–30	30–50	1.3–2.0	50–120	90–115
9–13 years	30–60	20–30	25–40	1.2–1.8	40–90	70–90
14–18 years	15–50	10–30	15–30	1.2–1.8	35–70	40–60
19 years +[a]	15–50	10–30	15–30	1.1–1.7	35–45	40–50

a: Males and non-pregnant, non-lactating females.

Source: Frazier, D.M., Allgeier, C., Homer, C., Marriage, B.J., Ogata, B., Rohr, F., Splett, P.L., Stembridge, A. and Singh, R. H. 2014. Nutrition management guideline for maple syrup urine disease: an evidence-and –consensus-based approach. Molecular Genetics and Metabolism. 112 (3): 210-217.

1. Perform a thiamine challenge for all individuals with MSUD except those known to be homozygous for the 1312T > A mutation or other mutations resulting in less than 3 per cent BCKD enzyme activity.
2. Initiate a thiamine challenge with a dosage of 50–200 mg/day.
3. Evaluate response to thiamine challenge over a one month period by assessing plasma BCAA and/or tolerance for dietary BCAA.
4. Maintain thiamine supplementation and appropriate dietary BCAA restriction in thiamine-responsive individuals.

Pregnancy and Postnatal Period in Women with MSUD

1. Monitor closely and provide individualized dietary guidance to assure that intake is adequate to meet the increased protein, BCAA and energy requirements of pregnancy.
2. Maintain plasma BCAA concentrations in the normal (MSUD) treatment range (LEU: 75–300 μmol/L; ILE and VAL: 200–400 μmol/L) throughout pregnancy.
3. Treat pregnancy-related poor appetite, nausea and vomiting aggressively to prevent or minimize endogenous protein catabolism.
4. Evaluate the need for vitamin and/or mineral supplementation based on the pregnant woman's prescribed medical food, dietary adherence and laboratory assessment.
5. Pay special attention to the prevention of catabolism during particularly vulnerable periods of labor/delivery and the first two weeks postpartum, as well as up to six weeks postpartum.
6. Provide increased nutrient intake and close clinical and biochemical monitoring for the woman with MSUD who is breastfeeding.

Liver Transplantation in individuals with MSUD

Recommendations are:

1. Liver transplantation is a viable option for individuals with MSUD and the risks and benefits should be evaluated on a case-by-case basis.
2. To be candidates for liver transplantation, individuals should be in good metabolic control through dietary management of their BCAA.
3. Prevent metabolic decompensation in the perioperative period.
4. Allow relaxation of the BCAA-restricted diet and lift precautions for severe metabolic decompensation for individuals with MSUD who have had successful liver transplantation.
5. Provide nutrition counseling to assist in dietary transition, and monitor the anthropometric and nutritional status of individuals with MSUD who have had successful liver transplantation.

Galactosemia

Galactose is a sugar found primarily in human and bovine milk and milk products as part of the disaccharide lactose. Lactose is hydrolyzed to glucose and galactose by the intestinal enzyme lactase. The galactose then is converted to glucose for use as an energy source. Free galactose also is present in some fruits and vegetables, such as tomatoes, brussels sprouts, bananas, and apples. Altered metabolism of galactose caused by deficient enzyme activity or impaired liver function results in elevated blood galactose concentration and the condition known as galactosemia.

Galactosemia is an inborn error of carbohydrate metabolism caused by an impairment of any one of the three galactose metabolizing enzymes. These enzymes are galactokinase, galactose-1-phosphate uridylyltransferase (GALT) and UDP-galactose 4-epimerase and all may lead to hypergalactosemia. The most important defect among them is severe GALT deficiency or classic galactosemia, which is often referred to as hereditary galactosemia and is inherited autosomal recessive.

GALT catalyzes the conversion of galactose-1-phosphate (GAL1P) and UDP-glucose to glucose-1-phosphate and UPD-galactose. It has been suggested that the accumulation of toxic metabolites, galactose, galactitol, and GAL1P cause permanent damage to various tissues and organs. GaL1P accumulation has been associated with a reduction in cellular phosphate levels as well as a decrease in ATP. Additionally, it may act as an inhibitor of several enzymes of glucose metabolism *in vitro* such as glucose-6-phosphatase, glucose-6- phosphate dehydrogenase, phosphoglucomutase, glycogen phosphorylase, as well as myoinositol monophosphatase enzyme, which is important in inositol signaling and recycling. It is hypothesized that the deficiency of myo-inositol due to elevated levels of Gal-1-P and galactitol in the prenatal period might be responsible for the formation of brain dysfunction in classical galactosemia.

Prevalence

The incidence of classical galactosemia in different countries has been reported to vary from 1 in 30,000 to 1 in 75,000. The prevalence of galactosemia in Indian children with suspected metabolic liver disease (MLD) has been reported to be about 20 per cent, next to Wilson's disease and glycogen storage disorders. The exact population incidence in India is not known as mutations and importance of GALT gene analysis in the diagnosis of galactosemia in Indian patients.

Symptoms and Diagnosis

Classical galactosemia symptoms begin in the neonatal period after galactose is introduced after breast or formula feeding to a newborn. They include failure to thrive, jaundice, hepatomegaly, edema, ascites, bleeding tendency and susceptibility to infection. Poor appetite, vomiting, and accompanying poor weight gain, and poor nutrition result. If the child survives infancy, cataracts and physical and mental retardation usually ensue. The majority of babies will recover. There are long-term complications of galactosaemia, despite treatment, including learning disabilities and female infertility. The majority of females with classic galactosemia also develop premature ovarian insufficiency.

(1) The occurrence of galactosemia may be suspected from the character of the symptoms; (2) previous diagnosis of galactosemia in siblings; and (3) the presence of proteinuria and galactosuria. This diagnosis can be confirmed by measuring the activity of the specific enzyme within the erythrocytes. The cord blood of newborn siblings of known galactosemic children should be assayed for the presence of galactose-1-phosphate-uridyltransferase.

Affected newborns are recognized in the first days of their life. It has been postulated that galactosaemia could be detected on newborn screening and this would prevent the immediate severe liver dysfunction and sepsis. Evaluation of GAL1P levels on a galactose-restricted diet might be helpful in providing a prognosis for galactosemia patients.

Duarte galactosemia is mild form of this carbohydrate metabolism defect. Unlike classical galactosemia, most infants with duarte galactosemia remain apparently healthy after exposure to milk, which contains high levels of galactose. However, these infants do accumulate many of the same galactose metabolites seen in classical galactosemia, although to a lesser extent. Whether older children with duarte galactosemia are at an increased risk for any of the long-term developmental complications seen in classical galactosemia has remained unclear because most are discharged from follow-up as infants or toddlers.

Principles of Dietary Management

Galactose is a monosaccharides found primarily in human and bovine milk and milk products as part of the disaccharide lactose. In the metabolic pathway, lactose is hydrolyzed to glucose and galactose by the intestinal enzyme lactase. The galactose then is converted to glucose for use as an energy source. Altered metabolism of galactose caused by deficient enzyme activity or impaired liver function results in elevated blood galactose and toxic galactose metabolites concentration.

Ingestion of galactose results in a sharp rise in the erythrocyte galactose-I-phosphate content. The aim of galactose-restricted diet is to maintain the erythrocyte galactose- I -phosphate content below 3mg/100 m1. Diet control thus, can be assessed by regular determinations of erythrocyte galactose-1-phosphate content. It is a general practice to perform this test every two to three months by obtaining a small sample of blood by venipuncture and performing the proper assay. In newly diagnosed children, this test should be performed weekly until a normal level is achieved, then monthly until the mother has demonstrated competence in dietary management.

Immediate diet intervention is necessary for infants with suspected galactosemia. Management is directed toward the rigid exclusion of galactose from the diet. Since galactose is a component of lactose, a milk sugar, this means that milk, its products, and all foods which may contain milk must be omitted completely from the diet. If this is done, most of the physical manifestations of the disease will regress with the possible exception of cataracts and brain damage. When the diagnosis is made early, so that this milk free diet regimen can be begun within the first few weeks of life, there is an excellent probability that all symptoms, including mental retardation, can be prevented. Available literature suggests that early diagnosis and treatment with lactose-free diet in initial 1-2 weeks of life reduces complications of liver failure and mortality.

Despite this metabolic handicap, the child with galactosemia needs the same kinds and amounts of nutrients that any normal child needs for growth and development. At the same time the child's diet must be free of galactose and any other substance which yields galactose during digestion. To insure an adequate intake of all nutrients, providing a well balanced diet to galactosemic child is a great challenge.

Special diets are easily available in India for galactosemic patients and are relatively much cheaper compared to diets for other metabolic liver diseases or IEMs amenable to special dietary therapy.

Medical Nutrition Therapy

Today there is inadequate evidence to support a specific age involved recommendation for the amount of galactose permitted in the diet. Worldwide there are discussions about the dietary management for galactosemia. Several European countries improved their own specific dietary management for galactosemia. The United Kingdom published its consensus on dietary guidelines for galactosemia in 1999. Following, other countries published their guidelines, *e.g.* the Netherlands in 2005 and France in 2008. Thus, each national guideline has different opinions on several points (*e.g.* acceptance of soy-based products, all vegetables allowed or not).

The Galactosemia Network (GalNet) improved recent guidelines which advise allowing any quantity and kind of fruits, vegetables, legumes, unfermented soy-based products, mature cheeses (with galactose content <25 mg/100 g), and the food additives sodium or calcium caseinate, in the diet for classical galactosemia. Though higher in galactose, all fermented soy-based products can be allowed in the small amounts that are typically used in the diet for galactosemia.

GalNet guideline advised a yearly dietary assessment of calcium and vitamin D intake with measurement of plasma total 25-OH-vitamin D levels. Both calcium and vitamin D should be added as necessary following the age specific recommendations for the general population. Therefore, the therapy of galactosemia is the lactose-free and galactose-poor diet for life. As a result of the nationwide newborn screening and the lifelong medical therapy, early treatment with galactosemia can achieve a normal life without serious complications.

Some of the major recommendations suggested by this network have been consolidated and are given below:

1. In diet therapy of patients with galactosemia; minimization of dietary galactose is accomplished by excluding milk and dairy products from the diet, however permits galactose from non-milk sources that subscribe minimal dietary galactose. This definition includes; some metabolic groups admit that small amounts of galactose are present in specific mature cheeses and caseinates.

   A person with galactosemia must avoid foods containing milk and all dairy products, such as cow's milk, butter, yogurt, cheese, ice cream.
2. Any foods or drugs which contain the following ingredients should also be avoided during infancy:
   - Casein
   - Curds
   - Whey
   - Whey solid
3. A child can start eating solid foods at around 6 months of age. A child on a galactose-restricted diet can eat most foods containing protein, such as poultry and eggs. Plain meats, poultry, fish, and eggs are excellent protein sources, but certain galactose-storing organs such as liver, pancreas, and brain must be avoided. Labels on cold-cuts and other meats commonly containing fillers should be carefully scrutinized to make sure that no milk or milk products have been added. Creamed and breaded meats, poultry, and fish should also be excluded from the diet.
4. Many breads and dry cereals have milk added. Food labels should be read carefully or enquiry about the ingredients should be ascertained from the bakery. Give only those breads which are safe to use.
5. They can also eat most types of fruits, vegetables, and grains; fruits and vegetables may be included in the galactose-restricted diet, except when these have been processed with lactose. It is recommend allowing any amount and type of fruits, vegetables, legumes, unfermented soy-based products, mature cheeses (with galactose content <25 mg/100 g), and the food additives sodium or calcium caseinate, in the diet for classical galactosemia. Although higher in galactose, all fermented soy-based products can be allowed in the small quantities that are typically used in the diet.
6. Peas, lima beans, and sugar beets contain oligosaccharides which may release galactose upon digestion, and accordingly they should be kept out of the diet.
7. Pure oils, shortenings and lard are safe for use, but margarine labels must be checked carefully to be sure the product contains no milk in any form.

8. In selecting desserts for the galactosemic patient, it must be remembered that as a general rule, commercially prepared cakes, pastries, cookies and puddings contain milk products. Parents should be instructed to be alert not only for milk, butter, cream, or cheese in foods but also for other forms of milk which contain lactose, such as nonfat dry milk solids, whey, or casein. They should also know that lactate, lactic acid and lactalbumin do not contain lactose and arc safe to use.
9. Since children with galactosemia cannot consume milk products, their calcium levels may be too low. Taking calcium supplements every day will help ensure they receive enough calcium. Vitamin D supplements may be recommended in addition to calcium. It is recommended that an annual dietary assessment of calcium and vitamin D intake with measurement of plasma total 25-OH-vitamin D levels should be done. Both calcium and vitamin D should be supplemented as necessary following the age-specific recommendations for the general population.
10. They recommend treating galactosemic patients with a life-long galactose-restricted diet that only eliminates sources of lactose and galactose from dairy products, but permits galactose from non-milk sources that contribute minimal dietary galactose. At present there is insufficient evidence to support a specific age-related recommendation for the quantity of galactose allowed in the diet.

Ensuring Nutritional Adequacy

An ongoing issue in dietary management of galactosaemia is adequate nutrient intake. Decreased height has been reported which may be related to hormonal factors but catch up has been reported beyond the age of 18 years. Full fat soy drinks are equivalent in energy and protein to full cream milk but cereal based drinks (such as rice drink) may provide inadequate protein and energy intake for young children. Cereal based drinks are not generally recommended for children under the age of 5 years. Bone density has been reported to be low in children and women with galactosaemia. Potential mechanisms are abnormal levels of sex hormones, low calcium intake and possibly an intrinsic defect in the normal galactosylation of the collagen matrix of bones. Panis *et al.* (2006) found a significant increase in bone mineral content of the lumbar spine in prepubertal children with galactosaemia who were taking calcium, vitamin K and vitamin D supplementation compared to those taking a placebo. Panis *et al.* (2007) have proposed a monitoring program for prevention of osteoporosis.

Regular Monitoring

Satisfactory erythrocyte galactose-1-phosphate levels (below 3 mg/100 ml.) do not necessarily indicate adequate intake of nutrients. A three-day record of the food ingested, recorded by the parents prior to clinic visit, is necessary also in order to permit the nutritionist or dietitian to assess the adequacy of the dietary control and relate this to subsequent erythrocyte galactose-1-phosphate content reports from the laboratory.

Dietary control is maintained by the parents. Successful management by the parents is dependent upon their understanding of the disease, the required diet, and their ability to discipline the child effectively. Successful management as the child grows up requires the child's own understanding of the diet restrictions and his self discipline. These factors relevant to management are in turn related to the help the parent receive from various professional sources.

Although responsibility for the dietary management lies with the parents, all members of the family should have a realistic understanding of the disturbance in galactosemia and of the importance of the diet. Relatives, teachers and neighbors must also be given some insight into the condition and its management. The child with galctosemia should be treated as nearly like a normal child as is possible in order to prevent behavior problems and emotional disorder.

29

Sports Nutrition

Sports nutrition has recently emerged as a recognized specialty area within the field of nutrition that partners closely with the study of the human body and exercise science. Sports nutrition can be defined as the application of nutrition knowledge to a practical daily eating plan providing the fuel for physical activity, facilitating the repair and building process following hard physical work and achieve athletic performance in competitive events, while also promoting overall health and wellness. The area of sports nutrition is often thought to be reserved only for "athletes," which insinuates the inclusion of only those individuals who are performing at the elite level. In this text, the term athlete refers to any individual who is regularly active, ranging from the fitness enthusiast to the competitive amateur or professional.

Nutrition is an important part of many sports training regimens, being popular in strength sports (such as weightlifting and body building) and endurance sports (*e.g.*, cycling, running, swimming, rowing). To fully understand and subsequently apply sports nutrition concepts, professionals instructing athletes on proper eating strategies first need to have a command of general nutrition as well as exercise science. The second step is to gain the knowledge of how nutrition and exercise science are intertwined, understanding that physical training and dietary habits are reliant on each other to produce optimal performance. The final step can be considered one of the most critical *i.e.* the practical application of sports nutrition knowledge to individual athletes participating in a sport or physical activity.

Importance of Sports Nutrition

Today the competitive sportsman has been made aware of the important role that nutrition can play in both training as well as competition, for there is clear evidence to show that improved eating habits not only benefit health but also influence an individual's endurance and capacity to perform exercise. Changing diet will not automatically make someone to run faster or jump higher. Diet is just one of the many factors involved in improving athletic performance. A proper diet *i.e.* to get a variety of foods so as to consume all the macro and micro nutrients as needed, will reduce the disturbances like way of performance, fatigue, injury and soreness. Diet significantly influences the performance of athletes. The main dietary goal for active individuals is to obtain adequate nutrition to optimize health and fitness or sports performance.

Athlete challenges their bodies on a regular basis through tough physical training and competitions. In order to keep up with demand for stamina of their activity or sport, athlete needs adequate fuel for their body on day to day basis through proper diet. Nutrition is important for an athlete because it provides energy required to perform the activity; it is important for attaining high levels of achievements in sports and athletics, physical fitness and training. Factors that may affect an athlete's nutritional needs include type of activity, gender, weight, height, body mass index, workout or activity stage (pre workout, intro workout, recovery) and time of day. Proper nutrition must be available prior, during and post competition. The food they take leaves an impact on strength, training, performance and recovery. There is strong evidence that appropriate selection of nutrients, timing of intake and proper supplement choice are associated with optimal health and exercise performances. Not only the type of food is important for sport nutrition but also the time is equally important for what they eat throughout the day. An athlete needs to pay close attention about when, what and how much does he eat or drink prior to a game or match.

Sports nutrition assumes critical importance because long before deficiency symptoms start appearing, physical performance declines. It would not be prudent to think in terms of minimum needs to keep the blood levels or enzyme levels at normal limits. Rather attempts should be made to find out the level below which physical performance starts showing changes. The level, which permits the athlete to achieve the maximum possible physical performance should be the minimum level aimed in the sports nutrition. At the time of final performance, an athlete should be supposed to be well nourished, uninjured, fit, focused and ready to compete.

Factors Affecting Sports Performance

Success in sports depends on three factors *i.e.* genetic endowments, the state of training and nutrition. Genetic make-up cannot be changed. Specialized exercise training is the major means to improve athletic performance and proper nutrition is an important component of the total training program. Athletes and fitness enthusiasts need the same essential nutrients that non-active people need with varied increases in their caloric needs as well as some increase in macro and micronutrients. Therefore, it is essential to explore and assess these increased nutritional needs of athletes before, during, and after competition for achieving optimal sports performance.

Health related fitness activities generally include cardiovascular training (aerobic activities such as jogging, swimming, cycling, and hiking), flexibility (stretching), strength (heavy resistance training), muscular endurance (extended resistance to a particular muscle), and appropriate body composition (as opposed to a general scale measurement of total weight). For example, one may set a goal to lose body fat (health-related fitness or event specific goal) and to achieve that goal, he or she would create a regular exercise program that includes all of the above components in addition to choosing more nutrient dense foods in the diet that would support increased demand from activities.

Diet and Sports Performance

The term diet refers to the collection of such type of food which helps to improve the physical condition, controls weight and helps to cure diseases by making the immune system strong. Use of diet is directly linked with physical structure and intensity of the activities which is to be performed because some activities are high powered in nature such as football, tennis, and hockey, requiring large amounts of work, rapid movements, and more energy while others are more endurance-based, such as cross-country running and triathlon training, which require continual lower force outputs for longer durations. Diet may be given to athlete according to the nature and type of activity. Athlete needs to use diet according to the physiological demands of his or her sport. For example:

1. Muscles which produce high tension need more protein for recovery.
2. Muscles which are involved in the activity of high volumes and intensity need more carbohydrates to refill glycogen stores.
3. Additionally, active bodies that produce large amounts of heat need more water to regulate the body temperature during the practice or competitive event.

Food, which selected should serve to prevent nutrient deficiencies or excesses that may decrease the risk of developing health-related problems or diseases. In moderate physical activity, a healthy diet should prevent weight gain, help to maintain appropriate body composition, and prevent any adverse health issues. The primary need for the diet of the athlete is to meet the additional nutrient requirement imposed by the training load.

Sports diet must consist of all the food nutrients in balanced amounts. Food components may be classified as macronutrients and micronutrients. It is necessary for athlete to use both macronutrients and micronutrients. Lacking of both macronutrients and micronutrients may cause poor or weak performance of the athlete during the competition. Macronutrients are essential for players in energy production, bone health, immune function and antioxidant activity. Micronutrient itself does not provide energy but helps to maintain the functional capacity of the body of athlete.

Every athlete needs to use proper diet before, during and after the activity. Lacking of proper diet not only adversely affects the performance of athlete but it significantly affects the overall functional capacity of the body of athlete. Athletes must fuel their bodies with the appropriate nutritional foods to meet their individual energy requirements in competition, training and recovery. If these nutritional needs are not met, there is an increased risk of poor performance and health issues.

Nutritional Requirements

Different kinds of physical activities demand different levels of energy and other essential nutrients. To fulfill the nutritional need of body before, during, and after the activity or training program, it is important to know the recommended intake of various nutrients.

Factors Affecting Nutritional Requirements

Factors that may affect an athlete's nutritional needs include type of activity (aerobic vs. anaerobic), gender, weight, height, body mass index, workout or activity stage (pre-workout, intro-workout, recovery), and time of day (*e.g.* some nutrients are utilized by the body more effectively during sleep than while awake). Emotional and physical stress of training and competition, hectic schedule of travelling may also affect dietary and nutrient intake. Most culprits that get in the way of performance are fatigue, injury and soreness. A proper diet based on recommended nutritional requirements will reduce these disturbances in performance.

Nutritional and Hydration Requirements and Recommendations for Sports Performance

International Life Sciences Institute-India organized a conference on Dec. 2, 2005 in Bangalore, constituted a working group to develop Guidelines on Nutrition and Hydration for Athletes who visited NIS at Patiala and had extensive discussions with sports scientists, coaches and athletes and also reviewed the institutional arrangements. Discussions were held in Delhi as well at the headquarters of Sports Authority of India (SAI). Finally, International Life Sciences Institute-India, Sports Authority of India and NIN, Hyderabad worked together and recommended nutrition and hydration guidelines for excellence in sports performance for Indian athletes.

Energy

The energy requirement of an individual has been defined by WHO/FAO/UNU (1985) as 'the level of energy intake from food that will balance energy expenditure when the individual has a body size and composition, and level of physical activity, consistent with long term good health, and that will allow for maintenance of economically necessary and socially desirable physical activity'. For the serious competitive athlete, concerns about energy go beyond health or socially desirable physical activity. It is imperative that energy intake supports the training and competitive schedules which will allow the athlete to achieve his or her personal best. Maintaining adequate energy levels, weight loss and weight gain can have profound impact on sports performance. Hence, energy requirement of a sports person depends upon the type of sports, season and period of performance besides age, sex, body composition and body weight. Now it is also evident that the energy allowances for athletes vary from phase to phase based on the intensity of training and type of event. Strenuous exercise for a long duration by a heavy person requires more energy requirement.

International nutrition standards for athletes do not exist. In our country, we follow the recommendations suggested by International Life Sciences Institute-India, SAI and NIN and these are based on one time point only, mainly keeping in view the competitive phase of training. There is an immediate need to re-evaluate/ modify the recommended energy allowances for Indian athletes from the existing norms and to incorporate phase wise allowances so as to provide optimal nutrition for maintaining desirable body weight and composition, thereby achieving optimal performance.

Carbohydrates are the major source of energy, then fats and least from proteins. The nutrition composition for athletes' diet with special reference to energy contribution from carbohydrate, protein and fat varies from event to event as reported in Tables 29.1 and 29.2. The ratio between these macronutrients has to be adjusted time to time depending on the needs of the individual athlete keeping in view event specificity so as to achieve desirable body size and composition, and optimal performance levels.

Table 29.1: Classification of Sports and Games According to Energy Expenditure

Category	*Event*
Group I	**Power Events of Higher Weight Category (80 Kg and above):** Weight Lifting, Boxing, Wrestling, Judo, Throwing Events
Group II	**Endurance Events:** Marathon, Long distance running, and walking, Road cycling, Rowing middle and Long distance swimming
Group III	**Team Events, Athletics and Power Events of Middle Weight Category (65kg):** Hockey, Foot Ball, Volley Ball, Basketball, Tennis, Sprints, Jumpers, Boxing Wrestling, Weight Lifting, Judo and Swimming group
Group IV	**Events of Light Weight Category:** Gymnastics, Table Tennis, Yachting, Boxing, Wrestling, Weight lifting and Judo
Group V	Skill Games Shooting, Archery And Equestrian

Source: National Institute of Nutrition.

Table 29.2: Average Body Weight and Energy Expenditure Levels Assumed and Allowance

Event Category	*Body Wt. (Kg)*	*Energy Allowances*		*Calories Ratio*		
		kcal/ kg/ day	*kcal/ day*	*CHO :*	*Protein :*	*Fat*
Group I	85	70	6000	55	15	30
Group II	65	80	5200	60	15	25*
Group III	65	70	4500	60	15	25
				64	15	21*
Group IV	60	60	3600	65	15	20
Group V	60	50	3000	55	15	30

* Glycogen loading

Source: National Institute of Nutrition.

The fundamental difference in dietary intake between an athlete and non-athlete is that athletes must consume adequate energy to meet the demands of intense training and competition. Determining daily energy requirements for an athlete to match the demands of training, while simultaneously maintaining/achieving body weight and body composition targets, is a primary consideration for sports nutrition professionals when providing consultation to an athlete.

Carbohydrates

Carbohydrate is the most preferred fuel for sports performance. Glucose is the simplest form of carbohydrate and starch is example of complex carbohydrates. Excess glucose is converted to glycogen in the liver and muscles. Glucose gives instant energy whereas glycogen gives sustained energy.

Carbohydrate is necessary to meet the demands of energy needed during exercise, to maintain blood glucose level and replenish muscle glycogen store. During sub-maximal exercise, carbohydrates in the body are the major source of fuel. A diet rich in carbohydrates increases both endurance and intermittent high-intensity performance because of the extra store of glycogen. Athletes need plenty of complex carbohydrates (cereals and millets like rice, wheat and wheat products, *bajra*, *jowar*, oats, barley, ragi, maize, roots and tubers, carrots *etc.*) in their diets along with proper training, as these foods help muscle and liver cells to store glycogen which is a vital energy source for most sports. When muscle cells run out of glycogen, muscle fatigue and tiredness sets in and performance suffers. Intake of complex carbohydrates improves performance and delays fatigue.

Keeping glycogen stores adequately stocked is crucial for athletes who often need to perform at high exercise intensities. Right kind of diet with right emphasis on starchy foods will result in enough stored glycogen to carry through 90 minutes of vigorous activity. It is well documented that athletes need to replenish carbohydrate stores in the body, especially during periods of intense training or competition. Consuming carbohydrates during workouts lasting over one hour can also benefit performance and delay onset of fatigue. Studies have shown that athletes who participate in intermittent sports, such as basketball and soccer, should also focus on consuming more carbohydrates during training and competition. This is not surprising since it is well-known that carbohydrates, when compared to protein and dietary fat, are the most efficiently broken down and metabolized form of energy for the body.

Well Supported Principles Regarding Carbohydrate Intake

- ☆ When it is important to train hard or with high intensity, daily carbohydrate intakes should match the fuel needs of training and glycogen restoration.
- ☆ Targets for daily carbohydrate intake are usefully based on body mass (or proxy for the volume of active muscle) and exercise load. Guidelines can be suggested (Table 29.3) but need to be fine-tuned according to the athlete's overall dietary goals and feedback from training.
- ☆ Guidelines for carbohydrate intake should not be provided in terms of percentage contributions to total dietary energy intake.
- ☆ When the period between exercise sessions is less than 8 h, athletes should consume carbohydrate as soon as practical after the first workout to maximize the effective recovery time between sessions. More importantly, in the absence of carbohydrate intake, refueling is ineffective.
- ☆ When carbohydrate intake is sub-optimal for refuelling, adding protein to a meal/snack will enhance glycogen storage.
- ☆ Early refueling may be enhanced by a higher rate of carbohydrate intake, especially when consumed in frequent small feedings.
- ☆ During longer recovery periods (24 h) when adequate energy and carbohydrate is consumed, the types, pattern, and timing of carbohydrate-rich meals and snacks can be chosen according to what is practical and enjoyable.
- ☆ Carbohydrate-rich foods with a moderate-to-high glycaemic index provide a readily available source of substrate for glycogen synthesis. This may be important when maximum glycogen storage is required in the hours after an exercise bout.
- ☆ Nutrient-rich carbohydrate foods or other foods added to recovery meals and snacks can provide a good source of protein and other nutrients.
- ☆ Adequate energy intake is needed to optimize glycogen storage; the restrained eating practices of some athletes interfere both with meeting targets for carbohydrate intake and optimizing glycogen storage from this intake.
- ☆ Although there are small differences in glycogen storage across the menstrual cycle, females can store glycogen as effectively as male athletes if they consume adequate carbohydrate and energy.
- ☆ Athletes should follow sensible practices regarding alcohol intake at all times, but particularly in the recovery period after exercise.

Recommended Intake of Carbohydrates

Generally, as the energetic demands of training or competition increase so does the dietary carbohydrate requirement. Carbohydrates should contribute 50 to 60 per cent of total energy requirement in a day. This percentage is only a guideline for estimating

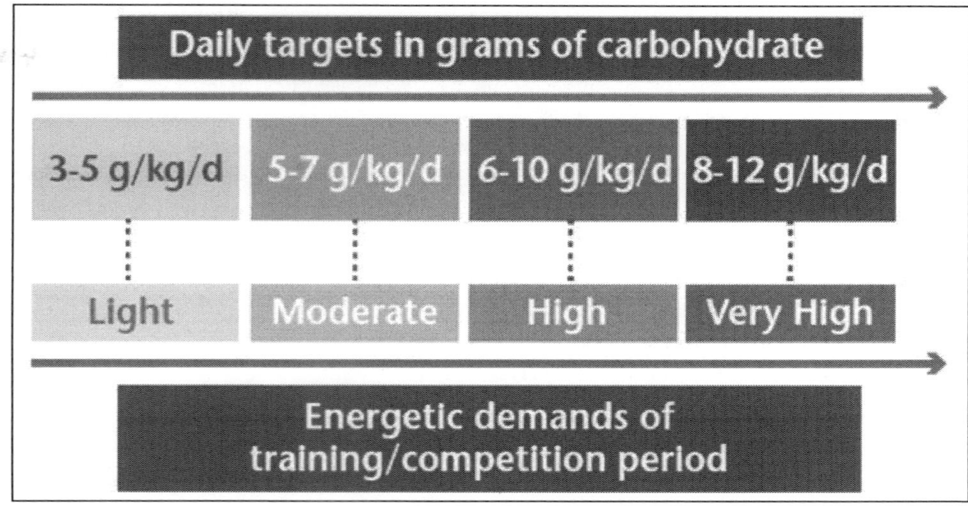

Figure 29.1: Daily Carbohydrate Intake Goals for Athletes.

Values expressed as grams of carbohydrate per kg body mass per day (g/kg/d). Light-low intensity or skill based activities; Moderate-moderate exercise programme (1 hr per day); High-endurance programme (*e.g.*, moderate to high intensity exercise of 1-3 hours per day); Very high-extreme commitment (*i.e.*, moderate to high intensity exercise of > 4-5 hours per day).

carbohydrate needs. It is now widely acknowledged that general recommendations for daily carbohydrate intake should be expressed as grams of carbohydrate per kilogram of the athlete's body mass, rather than a percentage of total dietary energy.

Depending upon the training routine, athletes should consume anywhere from 3-12 grams of carbohydrates per kilogram of bodyweight throughout the day. Depending upon the length and type of training sessions, an athlete's carbohydrate intake should adjust, with longer times and more intense trainings reflecting the higher number of grams needed.

Experts have recently refined dietary prescription regarding daily carbohydrate intakes to encourage athletes to achieve a carbohydrate intake to meet the fuel requirements of training and to optimize restoration of muscle glycogen stores between training sessions. The Figure 29.1 and Table 29.3 provide some general guidance for daily carbohydrate intake goals for athletes based on exercise context and intensity. When carbohydrate stores are inadequate, athletes cannot meet the energy needs of the activities being performed, which can result in fatigue, reduced training ability, impaired performance, and reduced immune function.

Interpretation of these guidelines into individual dietary prescription should consider the athlete's overall daily energy requirements, specific training volume and intensity, and requirements for growth and development (for children and adolescent athletes). Without an inherent knowledge of a sport, these

Table 29.3: Estimated Carbohydrate Needs of Athletes

	Situation	*Carbohydrate Targets per kg of the Athlete's Body Weight*	*Comments on Type and Timing of Carbohydrate Intake*
Daily need for fuel and recovery*			
Minimal	Light training programme (low intensity or skill-based exercise)	3-5 g/kg/day	☆ Timing of intake may be chosen to promote steady refueling, or to provide fuel intake around training sessions in the day. ☆ Otherwise, as long as total fuel needs are provided, the pattern of intake may simply be guided by convenience and individual choice. ☆ Protein and nutrient rich carbohydrate foods or meal combinations will allow the athlete to meet other acute or chronic sports nutritional goals.
Moderate	Moderate exercise programme (*i.e.* 1 hr per day)	5-7 g/kg/day	
High	Endurance programme (*i.e.*, 1-3 hr/day of moderate to high-intensity exercise)	6-10g/kg/day	
Very high	Extreme exercise (*i.e.*, >4-5 hr/day of moderate to high-intensity exercise)	8-12g/kg/day	

*Note that large athletes and athletes undertaking a weight loss programme may be better suited to reduce their fuel intake to the needs of the previous category.

These general recommendations should be fine tuned with individual consideration of total energy needs, specific training needs, and feedback from training performance.

	Situation	**Carbohydrate Targets per kg of the Athlete's Body Weight**	**Comments on Type and Timing of Carbohydrate Intake**
Special situations requiring fuel: These guidelines promote high carbohydrate availability to promote optimal performance in competition or key training sessions.			
Maximal daily refueling (General fuelling up)	-Post-event recovery or carbohydrate loading before an event -Preparation for events <90 min exercise	7-12 g/kg for each 24 h for daily fuel needs	✰ Athletes may choose carbohydrate rich sources that are low in fibre and easily consumed to ensure that fuel targets are met, and to meet goals for gut comfort or lighter 'racing weight'.
Carbohydrate loading	Preparation for events >90 min of sustained/intermittent exercise	36-48 h of 10-12g/kg per 24h	
Speedy refueling	Less than 8hr recovery between two demanding workouts	1.0-1.2 g/kg for first 4 h and then resume daily fuel needs.	✰ There may be benefits in consuming small regular snacks. ✰ Compact carbohydrate rich foods and drinks may help to ensure that fuel targets are met.
Pre-event fuelling	Before an endurance event *i.e.* > 60 min	1-4g/kg eaten 1-4 h before exercise	✰ The timing, amount and type of carbohydrate foods and drinks should be chosen to suit the practical needs of the event and individual preferences/experiences. ✰ Choice high in fat/protein/fibre may need to be avoided to reduce risk of gastrointestinal issues during the event. ✰ Low GI choices may provide a more sustained source of fuel for situations where carbohydrate cannot be consumed during exercise.
During brief exercise	<45 min	Not needed	
During sustained high-intensity exercise	45-75 min	Small amounts including mouth rinse.	✰ A range of drinks and sports products can provide easily consumed carbohydrate. ✰ Opportunities to consume foods and drinks vary according to the rules and nature of each sport. ✰ A range of everyday dietary choices and specialized sports products ranging in form from liquid to solid may be useful. ✰ The athlete should practice to find a refuelling plan that suits their individual goals including hydration needs and gut comfort.
During endurance exercise including 'stop and start' sports	1.0-2.5h	30-60 g/hr	
During ultra – endurance exercise	>2.5-3.0 h	Up to 1.5g/min (*e.g.* 60-90g/hr)	✰ As above ✰ Higher intakes of carbohydrates are associated with better performance. ✰ Products providing multiple transportable carbohydrates (glucose:fructose mixtures) will achieve high rates of oxidation of carbohydrate consumed during exercise.

Source: Burke *et al.*, 2011. Carbohydrates for training and competition. J. Sports Sci. 29:sup1, S17-S27.

guidelines can be easily misinterpreted. For example, it is common for a gymnast to train for 6-7 h per day, over two training sessions. If you consider the duration of the training sessions alone, their carbohydrate requirements would be 10-12g/kg/day. However, when the absolute amount of "activity" is calculated within these hours, the estimated amount of energy expended in that time period is quite small and suggested to be approximately 0.066 kcal/min/kg. The net result of true exercise energy expenditure in the sessions for an average 50-kg gymnast adds only 1200 kcal to their daily energy requirements. However, if the athlete consumes the amount based on training time alone (10-12 g carbohydrate/kg/day as suggested above), this would far exceed the energy expenditure of exercise, and thus, the need to "know your sport". Although there is no definitive assessment of carbohydrate usage during gymnastics training, it is likely that daily requirements are within 5-6/g/kg/day.

Carbohydrate Intake Before, During, and After Exercise

The timing of carbohydrate intake in relation to exercise is important. Carbohydrate ingestion before and during exercise has consistently been reported to increase endurance exercise capacity/performance but the mechanisms responsible and optimal dose required are still debated. Feeding carbohydrate is believed to spare muscle glycogen, spare liver glycogen, have central neural actions, and peripheral neural effects. Similarly, nutritional strategies utilizing carbohydrate during exercise may also have a positive effect on endurance performance/capacity by providing an alternative fuel source for muscle energy metabolism and therefore, limiting the amount of muscle glycogen utilized.

Pre-exercise Meal

Nutritional strategies used to increase muscle glycogen stores prior to exercise can lead to enhanced performance and an increase in exercise capacity. Hence, there are two purposes of pre-exercise or pre-training meal:

1. It keeps the athlete from feeling hungry before and during exercise, and
2. It maintains optimal levels of energy and muscle glycogen stores for the exercising muscles.

Athletes who train early in the morning, before eating or drinking, should ensure the previous night's meal contains adequate carbohydrates and then consume about 30 grams of well tolerated carbohydrates 5 minutes prior to training to help maintain blood glucose and muscle glycogen levels. While allowing for personal preferences and physiological factors, the pre-event meal should be high in carbohydrates, non-greasy, and readily digestible. Fatty foods should be limited as they delay the emptying time of the stomach and take longer to digest. The following are guidelines for the pre-event meal:

- ✰ One should avoid consuming foods closer to event and prefer to avoid high fat and high protein foods. The meal should be eaten 3-4 hours before an event.
- ✰ One should take complex carbohydrates (1-4g/kg) about 1 to 4 hrs prior to event. It shall maintain optimum glycogen stores.
- ✰ Avoid high fibre or gas forming foods.

Carbohydrates during Exercise

Carbohydrate consumption during exercise competitions (>1 hour) has become a common practice to increase performance. Indeed, consistent carbohydrate intake during sustained exercise maintains blood glucose while sparing glycogen and muscle catabolism and muscles receive adequate amounts of energy, especially during the later stages of the competition or workout. This has also been found to improve performance. However there is no one size fits all approach to carbohydrate intake during exercise.

During sustained high-intensity sports lasting ~1 h, small amounts of carbohydrate, including even mouth-rinsing, enhance performance via central nervous system effects. While 30–60 g/h is an appropriate target for sports of longer duration, events >2.5 h may benefit from higher intakes of up to 90 g/h accompanied by adequate hydration. In some individuals, consumption of carbohydrates greater than 60 g/hour may lead to gastrointestinal upset. The form of carbohydrates consumed does matter. Products containing special blends of different carbohydrates may maximize absorption of carbohydrate at such high rates.

Presently, there are numerous sports drinks, energy gels and bars that are designed to meet competitor's needs during exercise, many of which encompass multiple types of carbohydrates. Concurrent ingestion of a product containing a variety of carbohydrates (*i.e.*, sucrose, fructose, glucose, maltodexin) may lead to optimal carbohydrate fuel utilization during exercise. Use the following guidelines when consuming sports drinks with carbohydrates.

- ✰ Consume 6 to 12 ounces of a sports drink with 6-8 per cent carbohydrate concentration every 15-30 minutes during exercise. One gulp is about 2 ounces.
- ✰ Water is needed to aid in absorption of the carbohydrate. Drinks with a concentration greater than 10 per cent are often associated with abdominal cramps, nausea, and diarrhea.
- ✰ For high intensity activities, sports drinks and gels containing multiple forms of sugar can increase absorption and delivery of carbohydrates.

Sports drinks should not be confused with "energy" drinks. "Energy" drinks typically contain one or more stimulants and their carbohydrate concentration is usually greater than 10 per cent.

Adding small amounts of protein can aid in regulating energy levels by slowing down carbohydrate absorption, delivering the carbohydrates to the working muscles at a more consistent rate over time. Pay attention to salty cravings. If competing in hot/humid climates, make sure to replace electrolyte losses with salty snack foods or sport drinks with added sodium.

In real life, athletes undertake training sessions with varying carbohydrate availability.

After Exercise

Consuming a carbohydrate snack as soon as possible after training is important for the following reasons:

- ✰ To decrease muscle breakdown

- To replenish glycogen stores in the body
- To prevent gradual depletion of muscle glycogen stores over time caused by repetitive daily bouts of heavy exercise

Consuming a couple of mixed meals high in carbohydrates within six hours after training or a competition ensures that the muscles continue with glycogen restoration. Athletes who may benefit from recovery nutrition include those who are competing in tournament play or have multiple competitions over the course of one or several days, have skipped meals throughout the day, did not consume enough calories, and want to improve strength and power.

Carbohydrate intake during exercise should be scaled according to the characteristics of the event.

The recommendation is 1-1.2 grams of carbohydrates per kilogram of body weight per hour for the first four hours after exercise. Refueling may be enhanced by consuming small amounts of carbohydrate more frequently (every 15-30 minutes) for up to four hours. Add a small amount (~20 grams) of protein to the first feeding to stimulate muscle repair and rebuilding.

Post Exercise Tips

- To refill energy in the muscle with trainings less than eight hours apart, eat as soon as possible after exercise and then every 15-30 minutes for up to four hours.
- Choose higher-carbohydrate foods such as fruits, yogurts, cereal with low-fat milk, what bread with peanut butter, sports drinks, granola bars, sandwiches, fruit juice *etc*.
- Include protein to aid in muscle recovery and promote muscle growth.
- If one cannot consume solid foods as soon as possible after exercise, try 2-4 cups of a sports drink, and then consume solid food within four hours post-exercise.
- Consume 1-1.2 g of carbohydrates per kilogram body weight per hour for the first four hours after glycogen-depleting exercise. Be sure to rehydrate as well.

Eating At All-Day Events

It is important that athletes eat after competing to make sure that they will have enough energy in the muscles for the next race or competition, whether it be in the same day or the following days. The same dietary intake principles used to plan the pre-exercise meal can also apply to foods eaten at all-day events. If an athlete races at 10:00 a.m. and again after two hours, foods that are high in protein and fat will more than likely still be in the stomach potentially causing stomach or gastrointestinal (GI) distress. The following guidelines have been recommended to help athletes make wise food choices at all-day events.

One hour or less between events or heats

- Stick with carbohydrates that are in liquid form, such as sports drinks.
- If something solid needs to be eaten, try fruits like oranges, watermelon, cantaloupe, peaches, pears or bananas.

These foods consist of mostly carbohydrates and water. They are digested very fast and therefore, will not cause as much of a problem with stomach cramping or GI distress. Another key point to making food choices with limited time between events is limiting the quantity of the food eaten. The more an athlete eats, the longer it will take to digest, especially with any pre-competition nerves or stress.

Two to three hours between events or heats

- Foods containing carbohydrates and some protein can be eaten, as there is enough time to digest them before competition.
- Try eating granola bars with jerky, hot or cold cereal with nonfat milk, cake or muffins along with fruit like bananas, apples, oranges, peaches, or pears.
- Be sure to drink plenty of fluids, like water or a sports drink, for hydration, electrolyte replacement, and restoration of glycogen stores. Avoid drinks that contain caffeine, carbonation and other stimulants.

Four or more hours between events or heats

With four or more hours between events or heats, an athlete may want a meal, which should be composed of carbohydrates and some protein. Keep the meals simple. The following meal examples for this situation are appropriate:

- A chicken sandwich on two slices of whole wheat bread, yogurt/curd with fruit, and water or a sports (fluid replacement) drink; or
- Wheat bread/*chapatti* with cottage cheese/*paneer* preparation, salad with dressing, and water or a sport (fluid replacement) drink.

If there is a certain meal pattern before competition that an athlete thinks is a winning combination, then they should stick to it.

Carbohydrate Loading

Manipulating nutrition and exercise in the hours and days prior to an important exercise bout allows an athlete to commence the session with glycogen stores that are commensurate with the estimated fuel costs of the event. In the absence of severe muscle damage, glycogen stores can be normalized with 24 h of reduced training and adequate fuel intake. Events of more than 90 min duration may benefit from higher glycogen stores. Keeping it in view, to promote sufficient carbohydrate availability during competition, many recreational athletes traditionally had chosen a 'carbohydrate loading' strategy.

Carbohydrate loading is a strategy that involves changes to training and nutrition which can maximize muscle glycogen stores prior to endurance exercise lasting longer than 90 minutes. This strategy elevates muscle glycogen stores and has been found to increase endurance and exercise performance. This is also important to maintain muscle tissue stores which can be decreased with low glycogen levels.

The ability to sustain peak performance over an extended period of time (as in a marathon race) is influenced by the availability of muscle glycogen. To build up glycogen stores, two phases of preparation are recommended. First about a week before the competition, the athlete exercises vigorously to deplete glycogen stores and consumes diet high in protein and fat and restricted to about 100g carbohydrate. Second after 2 to 3 days of the glycogen depleting phase, a diet low in fat, moderate in protein and high in carbohydrate (200-500g) is consumed for 3-4 days prior to the event. Complex carbohydrates like cereals, millets that also furnish minerals and vitamins are preferred to simple sugars. Carbohydrate loading is not recommended for short term competition since it can lead to a feeling of heaviness that is a disadvantage in high intensity competition. It is also not advised for athletes in early adolescence and should be used no more than two or three times a year.

It now appears that the initial 3-4 days of depletion is not required for increased glycogen storage and carbohydrate availability (Burke *et al.,* 2011). In fact, high glycogen storage concentrations have been achieved following as little as 24 hours of rest with high carbohydrate intake. It is also important for the exercise professional to inform her/his client that carbohydrate loading does not always translate into an increase in athletic performance. Competition performance is multi-factorial involving physiological, metabolic, nutritional and psychological components, and thus, a compilation, integration and manipulation of several factors.

Carbohydrate Types for Performance

Rather than carbohydrate loading, some recreational athletes focus on choosing carbohydrate type and timing of intake. Panelists pointed out that, before and during exercise, the rapidity with which a carbohydrate source enters the bloodstream could impact exercise intensity and duration. High-carbohydrate foods and beverages that tend to be rapidly absorbed are best for providing the muscles with the energy that they need during exercise to maintain performance. Over the years, some sports nutritionists have theorized that lower glycemic index (GI) carbohydrates, those that appear in the bloodstream more slowly after ingestion and promote a blunted rise in the blood glucose response, may be preferable before exercise because they tend to "meter" the appearance of glucose in the blood.

Many recreational athletes may choose carbohydrates based on their glycemic index (GI), such as low glycemic foods/drinks prior to exercise and high glycemic foods/drinks during and following exercise. The attenuated post-prandial hyperglycaemia and hyperinsulinaemia of the low GI carbohydrate source reduces the suppression of free fatty acid oxidation that normally accompanies carbohydrate feeding, and possibly achieves better maintenance of plasma glucose concentrations and more sustained carbohydrate availability during exercise. The purported main benefit of such consideration is to achieve better maintenance of plasma glucose concentrations during and following exercise (Burke *et al.,* 2011).

However, this construct is not supported by robust data; there is inconsistent evidence that consumption of low GI foods prior to exercise results in improved athletic performance. The experts have pointed to recent research indicating no difference between pre-event consumption of a low GI meal (lentils, GI = 26) or a moderate GI meal (mashed potatoes, bread, egg whites) on the ability to maintain or improve high-intensity running performance. As per research reports, carbohydrate intake during exercise may offset the effect of the pre-exercise low glycemic carbohydrate intake. Altering the GI of meals consumed before exercise may offer benefits for some situations (when it is difficult to consume carbohydrate during exercise) or individuals (those who are sensitive to a hyperinsulinaemic response to carbohydrate feedings). However, further research is needed before systematic manipulation of the GI of carbohydrate-rich foods in the athlete's diet can be recommended. In any case, like all aspects of the pre-event meal, the type, timing, and amount of carbohydrate in the competition menu needs to be individualized to the athlete's specific event, their gut comfort, and their individual preferences.

Periodization of Carbohydrate Intake (Train Low/ Compete High – A New Paradigm for Training Adaptations?)

Apart from the recommendations for before, during and after exercise, a new concept with regard to the role of periodisation of carbohydrate intake has emerged. Based on convincing evidence of enhanced muscle markers of cellular signalling and metabolic adaptation following training in a low carbohydrate environment, a new concept of dietary periodization has been promoted whereby the athlete should "train low" to promote a greater training response, before switching to high carbohydrate availability for competition when optimal performance is required. This idea is therefore, also known as "training low and competing high".

The training-low concept generally requires an athlete to conduct heavy exercise training bouts after skipping a meal or meals to reduce carbohydrate availability and train the muscles to more readily use fat as substrate, thereby conserving limited glycogen stores and promoting greater responses in the molecular signals that lead to adaptation. When an acute bout of endurance exercise is completed in a state of low muscle glycogen, it appears that there is increased transcriptional activation of enzymes involved in carbohydrate metabolism and improvement in the adaptive responses that favour fat metabolism. Thus, the athlete should train in a depleted muscle glycogen state to elicit a greater training response, and then switch to a high carbohydrate intake when competing to ensure optimal exercise performance.

However, there are a couple of misconceptions concerning this theory. Proponents of the "train-low" concept emphasize that athletes must compete with high glycogen stores, and many recognize that "train-low" may impair the ability to use carbohydrate during competition, increase the risk of illness, reduce training intensity, and increase protein oxidation during exercise. Furthermore, all the experts have indicated that "train-low" regimens will inevitably reduce training intensity and potentially compromise performance improvements. The experts have agreed that more quality research is needed to clarify the utility of "train-low" regimens for optimal health, performance, and recovery from heavy exercise. Therefore, further research is warranted before any clear recommendations can be made in this area so as to eventually provide healthy guidelines and recommendations for serious client athletes.

Proteins

The experts have acknowledged that our understanding of dietary protein for physical activity has evolved considerably over the past decade and that science consistently demonstrates that protein is needed in greater-than-recommended daily allowance (0.8 g/kg per day) levels to enhance performance, recovery, and skeletal muscle accretion.

Dietary protein requirements depend upon the strength, speed or endurance training because protein supports muscle protein synthesis, reduces muscle protein breakdown and repairs muscle damage. There are increased requirements in the case of strength or resistance training. Endurance exercise increases leucine oxidation. Therefore, endurance athletes may have slightly higher protein requirements than their sedentary counterparts. Energy intake, exercise intensity and duration, ambient temperature, and gender and age also influence protein requirements.

Dietary protein intake should consist of high quality protein. Protein quality can be measured by the Protein Digestibility-Corrected Amino Acid Score (PDCAAS) where a score of close or equal to 1 indicates protein of high quality. Dietary protein sources with a similar score include, but are not limited to, milk (casein and whey), egg and meat products. Isolated soy protein, when all the anti-nutrient components have been removed, also has a PDCAAS score of 1. Milk protein, compared to isolated soy protein (with equivalent protein and macronutrient energy) has been shown to increase muscle protein synthesis after resistance exercise and leads to greater muscle hypertrophy. In general, the benefit of whey and milk proteins was clear in studies that compared whey, casein and soy, particularly because of the high leucine content of milk protein. The optimal timing of protein intake should also be considered when determining and prescribing protein requirements, as this can lead to faster recovery times and improved adaptation after training.

Daily Protein Requirements

The general protein requirement for a sedentary person is 0.8 g/kg BW/day. Incidentally, this requirement suffices for general fitness and can be slightly elevated to 1.0 g/kg body weight/day. The American College of Sport Nutrition (ACSM) recommends 1.2-1.7 g/kg body weight daily protein requirements for strength and endurance athletes. It is recommended that these requirements are reached through diet alone. About 60 to 80 grams of protein is sufficient for a day which can be obtained from milk, egg, meat, fish, *dhal*, cereals *etc.* but excess consumption of protein foods is not advisable as it can increase the work load on kidneys and can cause kidney damage in the later stage. It is a misconception that excess consumption of meat, milk, eggs *etc.* will give energy and stamina. Additional supplementation is not necessary, especially when the energy intake is optimal. The daily protein requirements for physical activity, as recommended by the International Society for Sport Nutrition (ISSN) have been summarized in Table 29.4.

According to International Society for Sports Nutrition (IOC), general protein guidelines for athletes are 1.3-1.8 g/kg BW and1.6-1.7 g/kg BW/day for strength-training athletes. Because of the high energy intake of these athletes, these requirements may be met easily. According to the IOC, protein intake above these guidelines does not have any additional benefit and can promote amino-acid catabolism and protein oxidation. The IOC recommends optimizing body composition in favour of losing fat and gaining muscle mass by decreasing daily carbohydrate intake (3-4 g/kg BW/day) and increasing daily protein intake (1.8-2.7 g/kg BW/day), while following a hypo-energetic diet and specified training programme (Potgieter, 2013).

Table 29.4: Daily Protein Requirements for Physical Activity

Daily or Habitual Protein Requirements		
Physical Activity Level	***g/Kg BW/day***	***Comments***
ISSN (The International Society for Sport Nutrition)		
General fitness	0.8-1.0	Focus on protein quality. Amino acid content. Whole foods. Safe, convenient supplements where needed.
Older individuals	1.0-1.2	
Moderate amount of intense training	1.0-1.5	
High volume of intense training	1.5-2.0	

Source: Kreider *et al.*, 2010. ISSN exercise and sport nutrition review: research and recommendations. Int. J. Soc. Sports Nutr.:7:7.

Potgieter, S. 2013. Sport nutrition: A review of the latest guidelines for exercise and sport nutrition from the American College of Sport Nutrition, the International Olympic Committee and the International Society for Sports Nutrition. South African Journal of Clinical Nutrition, 26(1), 6-16.

Although there appears to be consensus between the ACSM, ISSN and IOC that the daily protein requirements of athletes range between 1.2-2.0 g/kg BW/day, there are differences in the ranges provided. The ACSM provides a broad range (1.2-1.7 g/kg BW), whereas the ISSN gives recommendations according to training volume and intensity. The IOC provides specific recommendations for strength-training athletes and those who wish to prevent loss of lean body mass and to promote fat loss (Slater and Phillips, 2011; Phillips and Van Loon, 2011). It is recommended that athletes who want to increase muscle mass and reduce body fat should follow these guidelines.

Protein before Exercise

The ACSM recommends that a moderate amount of protein is added to the pre-event meal. No specific guideline on ingestion of protein before exercise is included in the consensus document. ISSN recommends that, depending on the individual's exercise duration and fitness level, protein should be included with carbohydrates in the pre-event meal before resistance exercise or when a desired change in body composition is required. This can be achieved by including 0.15-0.25 g/kg BW protein with the recommended 1-2 g/kg BW carbohydrates in the pre-event meal 3-4 hours before training or competition.

The IOC (The International Olympic Committee) states that although preliminary evidence appears to support increased muscle protein synthesis in response to resistance training when protein is given before exercise, follow-up studies have failed to confirm this finding. Therefore, the current opinion of the IOC is that protein should be ingested after exercise at a time that is associated with optimal muscle protein synthesis.

Protein during Exercise

The ACSM states that evidence pertaining to the benefit of the addition of protein to carbohydrate solutions during exercise is inconclusive. No recommendation is made in this regard in the consensus document.

According to the ISSN, the addition of protein to carbohydrates (carbohydrates to protein ratio of 3-4:1) during exercise has shown promise in recent literature. It has been demonstrated to be favourable in terms of improving endurance performance, increasing muscle glycogen stores, reducing muscle damage and promoting better training adaptations after resistance training. However, it is not known whether this addition of protein is because of the added energy that is available for substrate utilization. Although the ISSN advocates adding protein to carbohydrate intake during endurance exercise, more research in this field is needed as there is insufficient evidence to support this unequivocally.

The IOC refers to recent evidence that suggests that co-ingestion of carbohydrates and essential amino acids is beneficial before and during resistance exercise as it increases substrate availability and exercise performance, improves the anabolic hormonal environment, stimulates muscle protein synthesis and decreases muscle damage or tenderness. However, the IOC concludes that current guidelines promote the ingestion of protein at a time that is associated with maximal stimulation of muscle protein synthesis (after exercise), and do not provide any recommendations for protein intake before or during a workout.

Guidelines for protein intake before and during exercise are provided by the ISSN only, while the ACSM advocates a "moderate" intake of protein before exercise. Although there has been some evidence that supports the intake of protein before exercise, follow-up studies have failed to unequivocally support this practice. There is also inconclusive evidence to support the use of protein during endurance exercises.

Protein after Exercise

After exercise, the ACSM recommends that the primary goals of recovery should be to provide sufficient fluid, electrolytes, energy and carbohydrates to replace muscle glycogen stores and facilitate recovery. The addition of proteins can provide amino acids for the maintenance and repair of muscle protein, but no specific guideline has been provided by the ACSM to include protein as part of the recovery programme after exercise.

The ISSN recommendation for recovery is to add protein to carbohydrates at a carbohydrate to protein ratio of 3-4:1, or by supplementing with 0.2-0.5 g/kg BW protein. This results in increased glycogen resynthesis and ultimately improved performance. The consensus document also recommends that ingestion of amino acids, especially essential amino acids (EAA), stimulates muscle protein synthesis. This can be achieved by adding 6-20 g EAA to at least 30-40 g high glycaemic carbohydrates and ingesting this immediately or within three hours post exercise. This addition of protein to carbohydrates will also result in increased strength and enhanced body composition during chronic resistance training. The document also recommends adding a small amount of creatine (0.1 g/kg BW) to the carbohydrate and protein mixture post exercise to optimize the adaptations of resistance training (Kerksick *et al.*, 2008).

Current IOC guidelines also advocate the ingestion of protein after exercise, as this is when maximal stimulation of muscle protein synthesis is required. The IOC recommends that 20-25 g of high quality/or high biological value protein is included after resistance exercise (Slater and Phillips, 2011; Phillips and Van Loon, 2011). The combination of carbohydrates and protein post exercise is important to restore muscle glycogen and promote protein synthesis. Protein intake that exceeds this recommended amount does not promote muscle protein synthesis, but can lead to protein oxidation. The dietary protein form of choice is flavoured low fat milk. It shows beneficial improvements in muscle synthesis.

There is consensus from the ACSM, ISSN and IOC on the beneficial effect of the ingestion of ~20 g protein with carbohydrates within 30 minutes post exercise. This recovery strategy can be achieved through dietary sources. Additional supplementation is not warranted. Although the ISSN recommends, and the IOC recognizes, that the addition of creatine monohydrate as a supplement after exercise can increase skeletal muscle hypertrophy after resistance training, the practical application of this is questionable. Also, the risk of ingesting contaminated supplements is high. It is proposed that the ingestion of creatine after exercise is not necessary and can even be harmful to health. The same result can be achieved with the ingestion of sufficient carbohydrates and high biological value protein within 30 minutes after exercise. This can be achieved easily by the ingestion of a low-risk supplement which has the correct carbohydrate to protein ratio, such as a liquid meal replacement, especially if the athlete cannot consume a meal because of practical constraints, or because of the appetite-suppressing nature of exercise.

Fat

The fat requirements of athletes are similar, and are slightly higher than those in non-athletes. It is important to consume adequate amounts of fat to ensure optimal health, maintenance of energy balance, optimal intake of essential fatty acids and fat-soluble vitamins, as well as to replenish intramuscular triacylglycerol stores.

The amount of required fat depends largely on the training status and goals of the athletes. The ACSM recommends that daily fat intake for athletes should be 20-35 per cent of total energy intake and that fat intake should not decrease below 20 per cent of total energy intake, as the intake of fat is important for the ingestion of fat-soluble vitamins and essential fatty acids. High-fat diets for athletes are not recommended. It is not good to consume excess amount of fat in the diet as it will lead to obesity and heart diseases in the later part of life. The ISSN suggests a moderate fat intake of 30 per cent of total energy for athletes. This can increase to 50 per cent of total energy for high-volume training *i.e.* elite competitor training of 40 hours/week (like the Ironman). In order to reduce body fat or lose body weight, a fat intake of 0.5-1.0 g/kg BW/day is suggested. Optimization of the type of dietary fatty acids is important. The focus should be on increasing dietary sources of unsaturated or essential fatty acids. It is better to avoid lot of butter and ghee and take more of vegetable oils like sunflower oil, soya bean oil or rice bran oil.

The IOC recommends following a diet that does not contain less than 15-20 per cent fat of total energy.

It is suggested that athletes should be cautious of high-fat diets (> 30 per cent of total energy intake). The recommendation from the ACSM regarding fat intake should suffice for any athlete. A high-fat intake can be at the expense of carbohydrate intake and may have negative effects on training and racing performance.

Vitamins and Minerals

Vitamins and minerals are known as the micronutrients as they are found in less amounts in the body, but they play a major role in maintaining many important physiological functions. Foods rich in vitamins and minerals are also known as "protective foods" as they help to build body's resistance and immunity and help in energy metabolism, strengthening of bones and muscles, good vision, blood formation *etc.* Vitamin deficiencies of all kinds are damaging to work performance and can impair physical work capacity which is expected to have the most immediate effect. Major minerals are calcium and phosphorus which are needed for the bones and teeth and for muscle functioning. Iron deficiency is associated with decreased work capacity, poor mental performance and reduced haemoglobin levels which will lead to decreased oxygen carrying capacity and cause anemia. Low calcium levels can cause irregular muscle contractions, bone density loss *etc.*

The ACSM recommends that no additional vitamin and mineral supplementation is needed if an athlete obtains sufficient energy from a wide variety of foods. The ACSM further allows for micronutrient supplementation that is unrelated to exercise performance, such as folic acid supplementation during pregnancy. Supplementation may be individually prescribed by the attending healthcare professional for certain athletes, such as those restricting energy intake, vegetarians, people who are ill, recovering from injury or with specific medical conditions. Vegetarians may require vitamin B_{12}, iron, calcium, vitamin D, riboflavin and zinc supplementation.

According to the ISSN, specific vitamins may exhibit some health benefit, *e.g.* vitamin E, niacin, folic acid and vitamin C. However, few have been reported to provide direct ergogenic properties. Some vitamins may assist physically active individuals to endure heavy training and exercise, thereby improving exercise performance. In particular, vitamins C and E may decrease oxidative damage caused by vigorous training schedules and may also help to support a healthy immune system.

Minerals are essential nutrients too, and are important for most bodily functions. Some studies have shown mineral deficiencies in athletes. These can impact negatively on sports performance. The health and ergogenic value of some minerals has been studied. These include calcium, which reduces the risk of developing premature osteoporosis, and maintain body composition; iron, particularly in the case of athletes who are prone to iron deficiency; sodium phosphate, which increase maximal oxygen uptake, anaerobic threshold and endurance capacity; sodium chloride, to maintain fluid and electrolyte balance; and zinc, which decrease exercise-induced changes in immune functioning. However, there is little evidence to link improved sporting performance to boron, chromium, magnesium or vanadium.

The ISSN recommends that a normal nutrient-rich diet that contains a variety of food groups should provide sufficient amounts of micronutrients in most cases. Athletes who are susceptible to low energy intake, or who purposefully restrict energy intake to lose or maintain body weight, might be at risk of developing micronutrient deficiencies. Low doses of multivitamin and mineral combinations may be prescribed, or vitamin- and mineral-enriched liquid meal replacements taken in such cases. However, this should be carried out in consultation with the dietitian and in combination with altered eating habits (Kreider *et al.*, 2010).

The IOC recognized and evaluated two specific micronutrients in its latest consensus statement, namely antioxidant and vitamin D supplementation. Antioxidants have been popular in literature and research studies. There are arguments for and against the use of these by athletes. Arguments that support antioxidant supplementation propose that this group of micronutrients is able to decrease the reactive oxygen species that forms during exhaustive exercise. Free radicals can promote muscular fatigue, and in turn, decrease exercise performance (Powers *et al.*, 2011). These arguments also suggest that some athletes do not consume a healthy diet that contains antioxidant food groups, such as whole grains, fruits, vegetables, nuts and seeds, and that antioxidant supplementation is not toxic or harmful to human health. On the other hand, it is believed that although exercise can promote oxidative stress, there is no evidence to support the theory that this exercise-induced oxidative stress is detrimental to human health or performance, and that participation in regular exercise increases the body's own ability to produce endogenous antioxidants. If energy requirements are met, and the athlete is able to ingest all of the various dietary antioxidants, additional supplementation should not warranted and toxic levels may impair muscle functioning and reduce training adaptations to exercise. In light of the above controversy in the available literature, the IOC recommends that athletes should not consume antioxidant supplements and that caution should be exercised, particularly with the use of single-nutrient, high-dose antioxidant supplements.

Vitamin D supplementation is also becoming more popular, not only in the athletic arena, but also in the general population, as studies of the latter have reported vitamin D deficiencies in certain population groups (Powers *et al.*, 2011). In the case of athletes, there is particular concern about athletes who work out indoors, reside at higher latitudes and have dark pigmented skin. In sport, various factors hinder an athlete's exposure to sunlight, and therefore, the amount of vitamin D that is synthesized by athletes. Clothing, the use of sunscreen, ageing, skin pigmentation, training times during the day, seasonal factors, cloud cover and latitude may all influence vitamin D synthesis from sunlight. Vitamin D supplementation might have an ergogenic effect in athletes who do not have adequate levels of vitamin D, but there have been limited experimental studies in this regard. Some studies have found that there is a direct link between vitamin D status and jumping height, velocity, muscle tone, muscle power and hand-grip strength, while others have suggested that there is a reduction in the rate of stress fractures when vitamin D is supplemented with calcium. In general, the arguments against vitamin D supplementation are that it is not supported by a strong body of evidence, and that since vitamin D is a fat-soluble vitamin, and it can accumulate in the body and cause toxic side-effects such as nausea, vomiting, poor appetite, constipation, weakness, weight loss, mental confusion, irregularities in cardiac rhythm and calcification of soft tissues, all of which would negatively impact on exercise performance and general health. There are also large inter-individual differences in the way in which individuals respond to vitamin D supplementation (Powers *et al.*, 2011).

Current guidelines are opposed to vitamin D supplementation, unless it is medically warranted. It is also recommended that athletes have at least 5-30 minutes of direct sun exposure on the arms and legs several times per week between 10h00 and 14h00. Because vitamin D and calcium metabolism are so closely linked, it is also recommended that at the very least, DRI levels of calcium (obtained from dietary sources) must be reached in athletes. Therefore, currently, there are no clear guidelines on micronutrient supplementation in athletes and which suggest that athletes should be monitored on an individual basis (Kreider *et al.*, 2010; Rodriquez *et al.*, 2009; Powers *et al.*, 2011).

A healthy, balanced diet, which includes all the different food groups, as well as adequate exposure to sunlight, should provide sufficient micronutrients. The recommendation by the ACSM and the ISSN that groups who are at risk (including athletes on severe energy-restricting diets, and perhaps those following vegetarian diets), may benefit from taking a low-dose, multivitamin and mineral supplement, or from including micronutrient-fortified, liquid meal replacement supplements in their diet, after consultation with a medical doctor and dietitian, is supported (Kreider *et al.*, 2010).

Fluid and Electrolyte Requirements

Of all the nutrients water is probably the most essential for human life.Water constitutes 60-65 per cent of the total body weight, 70 per cent of the muscle composition, and 90 per cent of the blood plasma. During exercise, body loses water through perspiration which can lead to dehydration. Excess loss of water can lead to serious problem for sports people. It causes cramps, vomiting, delirium and lead to unconsciousness similar to sunstroke.

Endurance athletes, especially those training multiple times during a 24-hour period, are at particular risk for dehydration, primarily because of increased fluid losses from sweating as a result of prolonged and/or intensive bouts of exercise (especially when exercise is undertaken in extreme conditions such as warm weather, extreme humidity, or at altitude). Being properly hydrated improves endurance, allows the athlete to maintain their desired training intensity, and is vital to protecting against heat illness. Athletes need to know that even minimal dehydration (1 per cent -2 per cent body weight loss) can hinder their performance by reducing blood volume (and consequently the body's ability to maintain a safe core temperature) and reduce the amount of oxygenated blood pumped to working muscles. Endurance athletes need to monitor their fluid intake throughout the day to minimize acute episodes of dehydration associated with bouts of exercise. For exercise-induced losses, the endurance athlete should weigh in before and after training to determine the amount of fluid lost through sweating. For every pound "lost" after exercise, the athlete needs to drink 20 ounces (2.5 cups) of fluid over the next few hours to meet fluid needs and associated obligatory urine losses that occur during rehydration.

There is no one-size-fits-all formula for how much water to drink. How much fluid each person needs depends on the individual's age, size, level of physical activity, and environmental temperature. However, ACSM guidelines (2007 and 2009) for fluid and electrolytes have been summarized in Table 29.5.

The ISSN exercise and sport nutrition review document states that a decrease in sport performance is evident if 2 per cent or more of an athlete's body weight is lost through sweat. Athletes should not only rely on thirst as an accurate indicator of fluid needs. Body weight should be measured before and after exercise sessions to determine sweat loss. It is also recommended that, in order to maintain fluid balance and prevent hypohydration, fluid should be ingested at a rate of 0.5-2 l/hour. Also, there should be frequent (every 5-20 minutes) ingestion of small amounts of fluid (150-200 ml). Recommended fluid intake should be increased in hot and humid environments. Excessive techniques to reduce body weight, such as the use of diuretics, vomiting and saunas, are inappropriate and dangerous to human health (Kreider *et al.*, 2010).

Table 29.5: American College of Sports Medicine Guidelines on Fluid and Electrolyte Replacement for Physical Activity

Fluid and electrolyte recommendations for physical activity	
Before exercise	✰ Pre-hydration should be initiated several hours before exercise to ensure fluid absorption and normal urine output. ✰ Beverages and sodium-containing and salted snacks can increase the sensation of thirst and retain fluids.
During exercise	✰ Fluid programmes should be customized for each individual, based on body weight measurements before and after exercise. ✰ Athletes should aim to prevent > 2 per cent body weight loss during exercise. ✰ Fluids should contain carbohydrates and electrolytes to maintain fluid balance and exercise performance.
After exercise	✰ Normal meals and beverages will induce euhydration. If more rapid recovery is required, 1.5 litres of fluid per kg body weight loss during exercise should be ingested. ✰ Beverages and snacks should contain sodium to help with rapid recovery, stimulation of thirst and fluid retention.

Source: Kreider *et al.*, 2010. ISSN exercise and sport nutrition review: research and recommendations. Int. J. Soc. Sports Nutr.:7:7. Sawka *et al.*, 2007. Exercise and fluid replacement. Med Sci Sports Exerc. 39(2):377-390.

The IOC suggests that "sufficient fluid should be consumed during exercise to limit dehydration to less than approximately 2 per cent of body mass. Sodium should be included when sweat losses are high, especially if exercise takes place for more than two hours. Athletes should not drink so much that they gain weight during exercise. During recovery from exercise, rehydration should include replacement of both water and salt lost in sweat. It is also recommended that sodium should be added to fluids during exercise that lasts > 2 hours, as well as in fluids that are ingested by athletes who have lost more than 3-4 g sodium in their sweat during exercise (Shirreffs and Sawka, 2011). It is recommended that following exercise that leads to loss of body weight because of sweat loss, water and sodium are consumed in amounts that are greater than those of the losses (3- 4 g), in order to achieve optimum recovery of water and electrolyte balance. This is especially important at times when rapid recovery is required (< 24 hours), as well as in cases where a body weight loss of > 5 per cent body weight is recorded ((Shirreffs and Sawka, 2011).

The recommendations by the ACSM, ISSN and the IOC are similar, the suggestion is that athletes calculate their own individual fluid needs in order to prevent hyper- and hypohydration.

Dietary Guidelines for Sports Persons

Young athletes should follow a balanced diet and eat all types of food in moderation. This is important to improve athletic performance and maintain a healthy body.

- The daily training diet should include the following amounts of these types of foods:

  Carbohydrates should provide 55 per cent to 65 per cent of total energy (calories): Carbohydrates include foods such as breads, cereals, grains, pastas, vegetables, and fruits.

  Proteins should provide 15 per cent to 20 per cent of total energy (calories): Protein-rich foods include meats, fish, poultry, tofu, dairy foods, legumes, eggs, and nuts.

  Fats should provide 25 per cent to 30 per cent of total energy (calories): Common fats include oils, butter, ghee and margarine. Fat is also there in many protein-rich foods.
- Never skip breakfast as it the most important meal of the day.
- It is better to eat 3- 5 meals a day rather than eating three heavy meals.
- Have milk at least 2-3 glasses per day.
- Have only 1-2 eggs per day.
- Have plenty of green leafy vegetables, carrots, beetroot and other vegetables.
- Have plenty of fruits like papaya, guava, citrus fruits, watermelon *etc.* and fruit juices.
- Have sprouted pulses (*chana* or moong) and salads at least a bowl every day.
- Eat whole cereals like porridge, broken wheat *upma*, wheat and ragi porridge and oats.
- Eat dry fruits like dates, raisins and nuts like groundnuts, walnuts, almonds *etc.*
- Restrict non-vegetarian foods 2-3 times a week and reduce excess consumption of red meat.
- Drink plenty of water.
- Avoid soft drinks, alcohol and too much of coffee, tea, fatty foods and other junk foods and too many sweets.
- Never exercise immediately after a meal, or else it can lead to nausea, vomiting, distension and cramping.

Pre-game Meal Guidelines

As per George (2014), the sports persons should follow the guidelines as given below:

It is mandatory for an athlete to eat right and stay fit. Specially before participating in some game, it is extremely important for them to check upon what they consume. As what they eat affects how they feel and ultimately affects their performance. Though there is not a fixed diet chart to be recommended as a pre-game meal plan because every game and every individual has own requirements but there are obviously certain things that should be kept in mind while planning a pre-game meal.

Include Rich Carbohydrate Diet

It is very important for an athlete to get a meal that is high in carbohydrate as carbohydrates are the instant source of energy. The meal before the game should thus, contain ample amount of carbohydrate as they provide glucose to the bloodstream quickly and thus, charge the athletes with an instant source of energy. White bread, vegetables, and cereal without excess fiber can be included in the pre-game diet.

Include Fruits in the Diet

Many nutritionists are of the opinion that fruits should be included in the pregame diet as they contain natural sugar which is digested easily and proves to be a good source of instant energy. However, it should always be kept in mind that the fruits included in the pre-game diet should be light.

Include Moderate Amount of Proteins in Diet

Protein is an important part of an athlete's diet, so always remember to include rich protein contents in the diet. However, generally the food rich in protein is also high in fat content. Excess of this diet may cause sluggishness and nausea to the athlete, so remember to serve them in small servings.

Include Fluids in Diet

Include lots of fluid in the diet as they hydrate body cells and also get digested easily. They also comparatively provide energy quicker than their solid counterparts. So include glucose drinks and juices in the pre-game diet. However, consume it moderately as it can also lead to frequent urination.

Strictly Avoid Caffeine, Sweets and Fat Containing Products

Few things that one must avoid in the pre-game diet are caffeine, sweets and fatty products. Avoid caffeine because it causes excess urine production that can lead to dehydration. Sweets, on the other hand should be avoided level leaving one exhausted and lethargic. Also avoid deep fried foods, gravies, dry fruits and dairy products because they can take a toll on the energy and activeness of the athlete.

Do Not Experiment with the Meal

Make sure whatever is provided to the athlete as a pre-game meal is already consumed by him on prior occasions. In order to avoid any last moment mishap like dysentery or food allergy, remember not to experiment with food. Also, go according to the taste of the athlete, so that he eats properly.

Consume it on the Right Time

Make sure whatever is eaten as the pre-game diet, is eaten it on the right time. Dieticians say that for an afternoon or evening game, the athlete should be given a full meal about 3-4 hours before the game so that the athlete does not feel hungry during the game. However if due to any reason the meal has to happen closer to the game, try to include easily digestible small quantity of food. Thus, a well balanced diet and effective training and coaching will be the winning combination and will help the athlete to build up strength and stamina. Hence, systematized modification of diet with steady monitoring of the deficiencies if any, by a sports dietician, linked with skilled training programme and effective nutrition education can definitely lead to better performance.

Commonly Used Sports Food and Supplements

Supplements are used by athletes, bodybuilders and sports men and women to boost their strength, performance and recovery, and are available in numerous different forms ranging from multivitamins and minerals through to protein, creatine and various other 'ergogenic' aids which are intended to enhance performance. Before individuals opt to take any form of supplement they should ensure their diet is healthy, balanced and suits their sport. Those who do decide to proceed and begin taking a supplement should always consult an accredited sports dietician or a registered nutritionist who specializes in sports nutrition beforehand, as they will be able to assess your suitability for a particular supplement.

Common sports supplements include the following:

Creatine

Creatine is a high-energy compound which helps to store and provide energy. Creatine is produced within the body, occurs naturally in fish and meat and can also be taken in supplement form.

Creatine as a dietary supplement is considered to be a 'legal steroid' and is used by athletes and sports men and women to increase muscle bulk. It is intended to help the sports person train for longer and also to boost performance during frequent high intensity exercise. Concerns have been raised over the long-term health and safety implications of taking creatine for prolonged periods of time, as some research has suggested it may have undesirable effects on the digestive system and could also result in muscular and cardiovascular issues as well as potentially increasing the risk of cancer. However, more research into the possible risks of the supplement is required to substantiate these claims, and creatine currently remains a hugely popular dietary supplement.

Whey Protein

Whey protein is a natural protein present in milk, containing very little fat, carbohydrate or lactose. Whey is what is known as a naturally complete protein, meaning that it is made up of all of the essential amino acids which are needed in the average daily diet. As well as housing the perfect combination of amino acids, whey protein also contains what is known as a branch chain of amino acids (BCAAs), which are the first ones to be used during intense training. The whey protein provides the body with these amino acids and in turn they assist with repairing and rebuilding lean muscle tissue. Another benefit of whey protein is that it is extremely easy to digest, meaning it is absorbed quickly and can provide instantaneous nourishment to the muscles.

Energy Drinks

Staying well hydrated during exercise and training is extremely important, and even a small amount of dehydration could be detrimental to performance levels. Drinking water is a good way of keeping hydrated during exercise periods, and some individuals also opt for energy drinks, particularly those who undertake endurance events such as long distance running. Many energy drinks contain electrolytes such as sodium, which help to stimulate thirst and encourage drinking, as well as enhancing the body's ability to hold water. In addition, the carbohydrates contained in many energy drinks can provide individuals with extra energy which may be needed in the latter stages of training, and could also provide extra protein to help prevent muscle loss.

Sports Food and Supplement Requirements

The ACSM has concluded that "athletes should be counselled regarding the appropriate use of ergogenic aids. Such products should only be used after careful evaluation for safety, efficacy, potency, and legality". The ISSN 2010 review notes that while some supplements might have a beneficial effect on athletic performance, no amount of supplementation will compensate for inadequate dietary intake. According to this consensus document, supplements are categorized in the following manner according to safety and efficacy:

- **Apparently effective and generally safe:** These supplements include weight-gain powders, creatine, protein, EAAs, low calorie foods, ephedra (a banned substance), caffeine, water and carbohydrate-electrolyte solutions, sodium phosphate and bicarbonate and β-alanine.
- **Supplements that are possibly effective:** These include β-hydroxy-β-methylbutyrate in untrained subjects, branched chain amino acids (BCAA), calcium, conjugated linoleic acid (CLA) and green tea extract.
- **Supplements whose effectiveness is too early to tell:** The list extends to α-ketoglutarate, α-ketoisocaprate, ecdysterones, growth hormone-releasing peptides and secretogues, ornithine-α-ketoglutarate, zinc-magnesium aspartate, chitosan, phosphatidl choline, betaine, Coleus Forskolin, dehydroepiandrosterone (DHEA), psychotropic nutrients or herbs and medium-chain triglycerides.
- **Supplements which are apparently not effective or are dangerous to use:** Examples of such supplements are glutamine, smilax, isoflavones, sulphopolysaccharides, boron, chromium, CLA, gamma oryzanol, prohormones, tribulus terrestris, vanadium, calcium pyruvate, chitosan, L-carnitine, phosphates, herbal diuretics, ribose and inosine.

According to the IOC, the following supplements increase exercise performance. This is strongly supported by evidence:

- Alkalinizing agents (sodium bicarbonate and sodium citrate) increase anaerobic exercise performance.
- L-arginine boosts aerobic endurance. (There is little, but convincing evidence in this regard).
- Beta-alanine enhances anaerobic and aerobic exercise performance.
- Caffeine improves endurance and reaction time.
- Creatine increases performance in strength and power events.
- Nitrate advances aerobic endurance exercise.
- Carbohydrates, proteins, water, electrolytes and amino acids have ergogenic properties.

The IOC strongly discourages the indiscriminate use of supplements, supplementation when nutritional needs can be met via dietary intake, the use of supplements that pose a risk of a positive doping outcome and supplement used by young athletes (< 18 years). The IOC cautions against the widespread use of supplements, especially in terms of acute or long-term effects on health, positive doping outcomes and possible detrimental effects on exercise performance. Current regulations that govern supplement use are more liberal compared to those by the pharmaceutical market. Often, supplements either contain little or no active ingredient or too much of certain toxic nutrients. They may also contain harmful impurities, such as lead, broken glass and animal faeces because of poor manufacturing practices (Maughan *et al.*, 2011). The majority of products on the market fail to reach expected standards. Other involved risks include inaccurate labelling, failure to declare the ingredients on the label and cross-contamination of supplements.

Supplements and sports food are used extensively by athletes at various levels, as well as by non-athletes. Although the use of some supplements may have added benefits in terms of improving body composition, sports performance and overall health, the risk to benefit ratio needs to be carefully considered before embarking on the widespread use of supplements. Dietary supplements are poorly regulated. Although the manufacturers of these products are not allowed to state that a supplement can prevent or treat any illness or disease without sufficient scientific evidence, monitoring is not thorough. Regulation of supplements is also further complicated by the widespread sale thereof on the Internet. This promotes the use of supplements from unidentified sources.

The product safety and purity, claimed benefits and safety of the supplement for short- and long-term, needs to be considered carefully before it is taken. Poor quality control of supplements on sale in pharmacies and supermarkets can also potentially increase the likelihood of athletes obtaining negative results in doping tests (Maughan *et al.*, 2007). Poor hygiene and lack of good manufacturing practices can result in supplements containing impurities such as lead, broken glass and animal faeces, which carries obvious health risks for athletes and other users tests. Direct or deliberate, and indirect contamination of dietary supplements with undeclared and unlabelled anabolic steroids also places supplement users in a difficult position. Some supplements may not contain the exact amount of ingredients that are listed on the label as a marketing tool. Athletes may be unaware of the potential negative effects of using these supplements (Maughan *et al.*, 2007). Currently, the World Anti-Doping Agency (WADA) does not distinguish between deliberate cheating and inadvertent doping, and the responsibility and future athletic career of the individual rests solely with the athlete (Maughan *et al.*, 2011).

Carbohydrates should be the largest percentage of an athlete calorie intake, at least 50 per cent to 60 per cent in his food. This helps to meet with the demands of energy needed during exercise, maintain blood glucose and refill muscle glycogen stores. Protein is required for the hormone and enzyme production, nutrient transfer in the blood, connective tissue support, and the repair of tissue in response to periods of exercise. They should consume 10-15 per cent of total calories from protein. Fat intake is important for the energy production, protecting organs, providing insulation to the body, and facilitating fat-soluble vitamin uptake and essential fatty acid intake.

The diet which is required for an athlete before participating in any kind of training program is known as pre-training diet. Pre-event diet provides energy and strength needed to finish strong. Pre-event or training diet should be high in carbohydrate, adequate in protein, and moderate in fat and fiber. A larger food should be consumed 3-4 h prior to exercise to build and maintain energy stores, while a small snack 30-60 min before exercise will provide a last minute boost of energy to the body. Hargreaves is of the view that consumption of a high carbohydrates diet 3-7 days before the competition may be more appropriate which may contains:

- Carbohydrates 70-80 per cent
- Protein 10 per cent;
- Fats 10-20 per cent.

Diet during Training

The diet which gets utilized by an athlete during sports performance is called during competition diet. Sport activities that extend over a long period may require mid-activity refueling in order to enhance performance. Easily digestible, carbohydrate rich foods can help maintain sufficient energy levels throughout the duration of exercise. According to Bonci L, that carbohydrate is needed to provide energy during exercise. It is a fact that carbohydrates are stored mostly in the muscles and liver. And during activities the store carbohydrates are utilized for the production of energy for sports performance. During sports performance full meal cannot be consumed. However, a small, high-carbohydrate snack will need to be consumed along with adequate fluid intake from sports drinks and water. Generally, in these situations the athlete would not want to consume more than about 300 calories. The main focus is to keep the athlete hydrated and not feeling hungry, yet still leave the gastrointestinal tract empty when competition begins.

Post-event Diet or Diet after Training

The diets, which needed by athlete after performing sports activities, are known as post competition diets. Post events or training diet needed by the body to recover the body from fatigue and to adopt the load of activities. Refilling the body's energy and nutrient stores is needed immediately after high-intensity physical activity. Post-event diet should provide carbohydrate in order to restore energy losses from exercising and also protein to assist in energy restoration, in addition to muscle tissue repair and development. Fluid and electrolytes (sodium and potassium) lost in sweat should also be restored.

30

COVID-19, Nutrition, Immunity and Diet

COVID-19 (also commonly called coronavirus), a disease caused by a novel coronavirus, became a major global human threat that has turned into a pandemic. The coronavirus disease 19 (COVID-19) is a highly transmittable and pathogenic viral infection caused by severe acute respiratory syndrome coronavirus 2 (SARS-CoV-2), which emerged in Wuhan, China and spread around the world. Coronavirus is one of the major pathogens that mainly targets the human respiratory system. In late December 2019, clusters of patients were admitted to hospitals with an initial diagnosis of pneumonia of an unknown etiology. Firstly, Corona virus (2019-nCoV) was isolated from Wuhan market, China at 7 Jan. 2020.

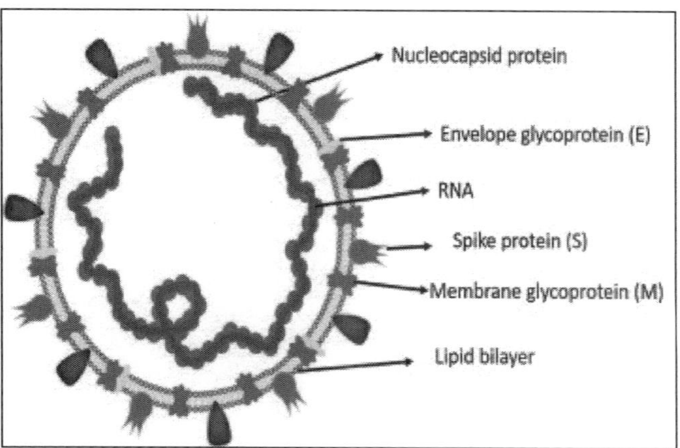

Figure 30.1: Structure of Respiratory Syndrome Causing Human Coronavirus.

A virus infects the body by entering healthy cells. There, the invader makes copies of itself and multiplies throughout the body. The new coronavirus latches its spiky surface proteins to receptors on healthy cells, especially those in lungs. Specifically, the viral proteins bust into cells through ACE2 receptors. Once inside, the coronavirus hijacks healthy cells and takes command. Eventually, it kills some of the healthy cells. There is no clinically approved antiviral drug or vaccine available to be used against COVID-19. This pandemic is posing severe threats to international health and the economy.

Symptoms

COVID-19, the illness caused by the coronavirus, starts with droplets from an infected person's cough, sneeze, or breath. They could be in the air or on a surface that one touch before touching his or her eyes, nose, or mouth. That gives the virus a passage to the mucous membranes in the throat. Within 2 to 14 days, immune system of the infected person may respond with symptoms including:

- ✰ Fever
- ✰ A cough
- ✰ Shortness of breath
- ✰ Trouble breathing
- ✰ Fatigue
- ✰ Chills, sometimes with shaking
- ✰ Body aches
- ✰ Headache
- ✰ A sore throat
- ✰ Loss of taste
- ✰ Loss of smell
- ✰ Nausea
- ✰ Diarrhea

The virus moves down the respiratory tract. That is the airway that includes mouth, nose, throat, and lungs. The lower airways have more ACE2 receptors than the rest of the respiratory tract. So COVID-19 is more likely to go deeper than viruses like the common cold. The lungs might become inflamed, making it tough for the infected person to breathe. This can lead to pneumonia, an infection of the tiny air sacs (called alveoli) inside lungs where your blood exchanges oxygen and carbon dioxide. For most people, the symptoms end with a cough and a fever. More than 8 in 10 cases are mild. But for some, the infection gets more severe. About 5 to 8 days after symptoms begin, they have shortness of breath (known as dyspnea). Acute respiratory distress syndrome (ARDS) begins a few days later. Many people who get ARDS need help breathing from a machine called a ventilator. As fluid collects in the lungs, they carry less oxygen to blood. That means blood may not supply the organs with enough oxygen to survive. This can cause your kidneys, lungs, and liver to shut down and stop working. Not everyone who has COVID-19 has these serious complications. And not everyone needs medical care. However, if symptoms include trouble breathing, get medical help right away.

Acute respiratory complications that require intensive care unit (ICU) management are a major cause of morbidity and mortality in COVID-19 patients. Patients with worst outcomes and higher mortality are reported to include immunocompromised subjects, namely older adults and polymorbid individuals and malnourished people in general.

Persons More Prone to Covid-19

About 80 per cent of people recover fully from COVID-19 without medical treatment. But COPD puts persons at higher risk to get seriously sick if they:

- ☆ Are over 60
- ☆ Undernourished people
- ☆ Have high blood pressure
- ☆ Have diabetes
- ☆ Have heart disease
- ☆ Have cancer
- ☆ Have a weakened immune system
- ☆ Smoke
- ☆ Chronic kidney disease
- ☆ Chronic obstructive pulmonary disease (COPD), like emphysema
- ☆ People with lower immune health because of a solid organ transplant
- ☆ Obesity — those with a BMI greater than 30
- ☆ Sickle cell disease
- ☆ Asthma
- ☆ Dementia
- ☆ Cerebrovascular diseases, such as stroke
- ☆ Cystic fibrosis
- ☆ Pregnancy
- ☆ Liver disease
- ☆ Scarring in the lungs (pulmonary fibrosis)
- ☆ Thalassemia (a blood disorder)

Therefore, Covid-19 does not treat us equally. Undernourished people have weaker immune systems, and may be at greater risk of severe illness due to the virus. At the same time, poor metabolic health, including obesity and diabetes, is strongly linked to worse Covid-19 outcomes, including risk of hospitalization and death. People who already suffer as a consequence of inequities including the poor, women and children, those living in fragile or conflict-affected states, minorities, refugees and the unsheltered are particularly affected by both the virus and the impact of containment measures. It is essential that they are protected, especially when responses are implemented.

The ongoing epidemic has been declared by the World Health Organization (WHO) as a global public health emergency.

Impact of Covid-19 on Food Security and Health System

Covid-19 has exposed the vulnerability and weaknesses of our already fragile food systems. Containing the virus has caused food and nutrition shortages and driven governments to reduce social services, such as mid day meal programmes, distribution of food to vulnerable groups through ICDS *etc.* that the most marginalized children and pregnant women rely upon. In the context of food and nutrition shortages, accessibility and affordability of healthy, sustainably produced food becomes even more challenging. Access to staple food distribution and local food markets is at risk. Millions of households in formerly food-secure regions of the world have fallen into severe food insecurity. Levels of hunger and malnutrition could double within the space of just a few weeks.

As measures to slow the spread of Covid-19 are enacted around the world, we must ensure that there is enough nutritious food, distributed fairly, to cover basic nutrition needs, especially for the most vulnerable. Food systems everywhere must become equitable, nutritious, efficient and inclusive. National Institute of Nutrition, ICMR has also recommended the amount of food to be distributed as free ration for the economically deprived, daily wage earners and migrant workers during the ongoing lockdown to keep them food and nutrition secure during these tough times.

Health care system should also be enhanced to address challenges faced by specific populations, especially older people and those with pre-existing conditions, such as weakened immune systems and poor metabolic health. They should specifically pay attention to women and children, especially to their nutritional well-being and healthcare. Yet even the strongest health systems are struggling with high healthcare costs and a shortage of medical personnel, equipment and facilities.

There is a need to integrate nutrition into universal health coverage as an indispensable prerequisite for improving diets, saving lives and reducing healthcare spending, while ensuring that no one is left behind. Reversing the obesity epidemic would also lessen the burden on our healthcare systems, as obesity is not only one of the costliest health conditions but also a major risk of Covid-19 hospitalizations and complications.

Prevalence

The coronavirus COVID-19 pandemic is the defining global health crisis of our time and the greatest challenge we have faced since World War II. Since its emergence in Asia late last year, the virus has spread to every continent except Antarctica. Cases are rising daily in Asia, Africa, USA, Canada and Europe. Globally, there have been 20,256,547 confirmed cases of COVID-19 in 213 countries, including 738,937 deaths till date (Aug. 9, 2020), as reported to WHO.

Prevention and Control

The COVID-19 pandemic is causing many changes in the daily lives of people around the world. The coronavirus presents many uncertainties, and none of us can completely eliminate our risk of getting infected with COVID-19. However, there are things that can be done to maintain a healthy lifestyle in these difficult times. First and foremost, everyone is encouraged to follow World Health Organization (WHO) and governmental advice to protect against COVID-19 infection and transmission. Physical distancing and good hygiene are the best protection for yourself and others against COVID-19.

In addition to the well-known personal hygiene and preventive measures against the new coronavirus (COVID-19), we can also follow some simple recommendations regarding our nutrition that strengthen our immune system and could better prepare us for an epidemic for in which the virus rapidly spreads to many people within a short period of time. One thing we can do is to eat as healthily as possible.

According to National Institute of Nutrition, ICMR, a balanced diet, comprising nutrient-rich vegetables, fruits, pulses, cereals and curd coupled with a healthy lifestyle, is the key to boost the immune system, a focal point in the fight against coronavirus.

Interaction among Corona Virus, Immunity and Nutrition

Good nutrition is an essential part of an individual's defense against Covid-19. Nutritional resilience is a key element of a society's readiness to combat the threat. Focusing on nutritional well-being provides opportunities for establishing synergies between public health and equity, in line with the 2030 Agenda for Sustainable Development.

As highlighted recently by the World Health Organization, a healthy lifestyle makes all bodily functions work better, including immunity. Although no food could prevent or treat coronavirus transmission alone; a healthy and balanced diet has been proven to strengthen the immune system along with physical activity and healthy sleeping habits. Good nutrition supports the body throughout the life course, from birth to old age. A balanced diet, including all food groups, supports an effective immune system and may provide protection against infections, cancers and other diseases. Hence, having a healthy diet especially, including lots of fruits and vegetables in the daily diets is a key component of a healthy lifestyle and plays a vital role in supporting a well-functioning and effective immune system to help protect against infection and other diseases.

How Does the Immune System Work

The immune system is involved in the protection of host against environmental agents such as pathogenic micro-organisms (bacteria, fungi, and viruses) and chemicals, thereby preserving the integrity of the body. The immune system is one of the most complex bodily systems, made up of a network of cells, molecules, tissues and organs all working together to protect the body. This complexity means that it cannot be modified acutely by a specific nutritional intervention. Rather, adhering to a healthy diet provides ongoing support to the immune system and may even delay the process of immunosenescence (the natural gradual deterioration of the immune system as we get older).

The emergence of new infectious diseases with new pathogenic properties constitutes a serious health issue worldwide. Severe acute respiratory syndrome (SARS) represents one of the most recent emerging infectious diseases, caused by a novel coronavirus member called (SARS-CoV-2), identified in Wuhan, Hubei, China in December 2019, and recognized as pandemic by the World Health Organization (WHO). The nutritional status of each COVID-19-infected patient should be assessed prior undertaking treatments. Nutritional support should be the basis of management of any infected individual. Good nutrition is key to build immunity, protect against illness and infection, and support recovery of COVID-19 patients. Not a cure for COVID-19 but healthy, balanced diets are key for boosting immunity, improve immunometabolism and preventing non-communicable diseases that are risk factors for higher COVID-19 morbidity and mortality. However, prevention measures should remain the first priority and strategy to develop throughout proper hygiene, healthy diet and staying home.

To preserve organism defense mechanisms, adequate nutritional status should be maintained with appropriate intakes of energy, vitamins, minerals and water that should be continuously provided by a healthy diet. Hence, the common denominator that drives most of the nutrition and dietary recommendations to combat viral infections, including COVID-19, lies within the link between diet and immunity. In fact, existing evidence highlights that diet has a profound effect on people's immune system and disease susceptibility. No specific one food or single supplement alone can prevent from catching COVID-19/ Corona virus. Besides following social distancing, good hygiene practices and wearing mask, there are many nutrients that are involved with the normal functioning of the immune system, which is why maintaining a healthy balanced diet is also the best way to support immune function. Hence, no one food is recommended over another but eating a variety of foods will help to maintain a healthy balanced diet.

It has been demonstrated that specific nutrients or nutrient combinations through variety of foods in the balanced diet may affect the immune system through the activation of cells, modification in the production of signaling molecules, and gene expression. Furthermore, dietary ingredients are significant determinants of gut microbial composition and consequently can shape the characteristics of immune responses in the body. Nutritional deficiencies of energy, protein, and specific micronutrients are associated with depressed immune function and increased susceptibility to infection. An adequate intake of iron, zinc, and vitamins A, E, B_6, and B_{12} is predominantly vital for the maintenance of immune function. Therefore, the key to maintaining an effective immune system is to avoid deficiencies of the nutrients that play an essential role in immune cell triggering, interaction, differentiation, or functional expression.

Creation of a pro-oxidant environment through generation of damaging reactive oxygen species is one element of innate immunity; the host needs protection against these through classic antioxidant vitamins (vitamins C and E) and the antioxidant enzymes (superoxide dismutase, catalase and glutathione peroxidase); the latter require manganese, copper, zinc, iron and selenium. Thus, the roles for nutrients in supporting the function of the immune system are many and varied and it is easy to appreciate that an adequate and balanced supply of these is essential if an appropriate immune response is to be mounted.

Important nutrients for effective immune function are:

- ✰ Copper
- ✰ Folate
- ✰ Iron
- ✰ Selenium
- ✰ Zinc
- ✰ Dietary Fibre
- ✰ Vitamins A, B_6, B_{12}, C, E and D
- ✰ Essential fatty acids

It is well known that vitamin C supports immune functions and protects against infection caused by coronavirus. The COVID-19 has been reported to infect the lower respiratory tract, and vitamin C could be one of the effective alternatives to treat COVID-19. It was reported throughout some controlled trials that vitamin C-supplemented patients presented, under certain conditions, had lower incidence of pneumonia. However, supplying patients with high-dose vitamin C has not been received any evidence-based approval.

Some nutrients, such as vitamins A and D, and their metabolites are direct regulators of gene expression in immune cells and play a key role in the maturation, differentiation and responsiveness of immune cells. Vitamin A acts in the proliferation of T lymphocytes, production of immune-reactive cytokines, and natural killer cells. Vitamins D and E regulate the immunological system. Vitamin D is known to play a role by stimulating maturation of several cells including immune ones. Vitamin D reduces the risk of acute respiratory infections. A significant number of healthy individuals have been found to be with decreased levels of vitamin D, particularly at the end of Winter season that coincides with COVID-19 discovery in Winter of 2019.

There is evidence from animal models showing the relationship between diets deficient in vitamin A, E, and D, and compromised immune response in respiratory infections and transmission caused by other coronaviruses. These deficient diets have also decreased the effectiveness of bovine coronavirus inactivated vaccines, which has made animals more susceptible to infectious diseases. Thus, these studies highlight the importance of consuming food sources of these nutrients especially during the pandemic.

Selenium, Zinc and Iron Supplementing COVID-19 affected patients with selenium could be an effective intervention for the treatment of this novel virus. COVID-19 related symptoms such as diarrhea and lower respiratory tract infection could be improved by Zinc supplementation. It has been shown that Iron deficiency constitutes a risk factor for the development of recurrent acute respiratory tract infections.

The European Food Safety Authority permits claims of 'maintenance of functions of the immune system' for vitamins A, B_6, B_{12}, C, D and folate (vitamin B_9) and for the trace elements zinc, iron, selenium and copper. Zinc and selenium act reducing the symptoms of colds. Essential fatty acids help control inflammation, infections, presenting an intrinsic relationship with the production of hormones and contribute to the production of antibodies. Dietary fibers promote a change in the microbiota with a positive effect on the immune system.

In essence, good nutrition creates an environment in which the immune system is able to respond appropriately to challenge, irrespective of the nature of the challenge. Conversely, poor nutrition creates an environment in which the immune system cannot respond well. This is amply illustrated in conditions of nutrient deficiency (either 'real life' or experimentally induced) which are accompanied by impairments of both innate and acquired immunity and increased susceptibility to, and severity of, infections. Both the immune impairments and the susceptibility to infection can be reversed by correcting the deficiency(ies) showing a causal relationship between availability of specific nutrients and immune defenses.

Effect of Coronavirus Infection on General Eating Pattern and Nutrient Intake of Masses

The COVID-19 pandemic caused by the SARS-CoV2 virus brought several individual and collective protection measures to contain the expansion of its transmission, such as social distancing and lockdown. Although extremely necessary, these measures restricted the activities in commerce, restaurants, street markets, and even the closing borders. Thus, for many individuals, usual shopping routines and eating have been entirely upended. In addition to providing a framework of economic and social instability, the lockdown could affect the food supply chain and generated a situation of food and nutritional insecurity in some regions of the world. Therefore, undernutrition and obesity might have increased, due to limited access to food, concern of running out of staple items, limited culinary ability, more sedentary lifestyle and change in food purchasing, eating behaviors, and perceptions of food safety. The higher consumption of processed foods with higher caloric content, high in saturated fats, sugars, and refined carbohydrates, greater durability, easier access and use, especially in children, could have contributed to increase the prevalence of obesity in the times of COVID-19.

During lockdown period, everybody had nothing to do but was busy in cooking and eating all the time at home especially in those families where all the family members including children were together after a long period. On social media also, people were giving demonstrations on new dishes and some were earning money through it too.

Still, staying home was a great opportunity to promote healthy eating and prevent obesity, when people are cooking more at home, buying more fresh and unpackaged foods than before the pandemic, and are more prone to change in eating habits. Some studies demonstrated that consuming homemade foods and having cooking skills are important to the ingestion of a healthy diet and the maintenance of proper nutritional status. For children, cooking may positively change intake and the preference for vegetables and fruits, by involvement and exposing them to healthier food in a fun, family, and ludic environment.

As recently discussed by the European Society for Clinical Nutrition and Metabolism (ESPEN) (Barazzoni R, Bischoff SC, Breda J, *et al.*, ESPEN expert statements and practical guidance for nutritional management of individuals with SARS-CoV-2 infection. Clin Nutr 2020; S0261- 5614(20)30140-0. DOI: 10.1016/j.clnu.2020.03.022), the obesity condition is dangerous to the severity of COVID-19 and has emerged as one of the most prominent risk factors increasing the disease mortality. In this sense, nutritional status and diets might influence the individual risk for the progression of SARS-CoV-2, but information on the impact of nutrition on COVID-19 is still arising.

Adequate dietary intake may be essential to protect against an excessive inflammatory response to SARS-Cov2 infection, preventing the evolution of the infection to severe disease or even during COVID-19, improving its outcome. Some nutrients have been reported as potentially relevant to the prevention and COVID-19 treatment. Among them are vitamins A, D, and E, minerals zinc and selenium, fiber, and essential fatty acids, as they promote an effect on the immune system. Thus, inadequate intake and status of these nutrients might lead to a decrease in the resistance to infections and an increase in disease burden.

Because of sedentary life style during lockdown period, energy intake should be less than the recommended intake for that age group. Protein intake and intake of vitamins and minerals should be as per recommendations.

- ☆ **Energy requirement:** Because of the impossibility of using indirect calorimeter, the daily requirement must be estimated through predictive equations based on body weight, such as 27–30 kcal/kg/day, adapted on the personal nutritional status, level of physical activities, clinical status, and comorbidities.
- ☆ **Protein requirement:** In case of the absence of chronic renal insufficiency, the protein intake is >1g/kg/day (up to 1.5 g/kg/day). This must be adapted to the aforementioned features as for energetic intake, in order to prevent weight loss, reduce the risk of complications, and promote a global recovery.
- ☆ **Carbohydrate and lipid requirements:** It must refer to the non-protein energy requirement, with a lipids/carbohydrates ratio ranging from 30:70 (patients without respiratory insufficiency) to 50:50 (patients with respiratory insufficiency). Low glycemic index carbohydrates have to be considered. Omega-3 and omega-6 PUFAs predominantly promote anti-inflammatory and pro-inflammatory effects. Being precursors

of resolvins /protectins and prostaglandins/ leukotrienes, respectively, omega-3 including protectin D1, serving as a novel antiviral drug, could be considered for one of the potential interventions of the novel virus COVID-19.

- **Micronutrients (Vitamins and minerals):** Deficiencies of trace elements such as iron, selenium, copper, and zinc, and vitamins A, B_6, B_{12}, folic acid, C, D and E are associated with immune dysfunction. A healthy balanced diet has the opportunity to furnish most of the essential micronutrients to exert modulatory effects on immune function, including zinc, iron, magnesium, manganese, selenium and copper that contribute to immune cells and functions sustainment and modulation. Several epidemiologic and clinical studies suggest that the risk of infection is favored by nutritional deficiencies besides poor personal hygiene, sanitation, or contaminated food and water.
- **Water requirement:** Adequate hydration must be maintained according to remote (heart or renal failure) and recent clinical history (diarrhea, vomiting, and electrolyte imbalances).

Subjects with malnutrition should ensure sufficient supplementation with vitamins and minerals. In COVID-19 non-intubated ICU patients not reaching the energy target with an oral diet, oral nutritional supplements (ONS) should be considered first and then enteral nutrition treatment. If there are limitations for the enteral route, it could be advised to prescribe peripheral parenteral nutrition in the population not reaching energy-protein target by oral or enteral nutrition.

Reducing infectious risk is achieved best by quarantine at home, which is heavily recommended during moderate disease course. However, prolonged home stay may lead to increased sedentary behaviors, such as spending excessive amounts of time sitting, reclining, or lying down for screening activities (playing games, watching television, using mobile devices); reducing regular physical activity and hence lower energy expenditure. Thus, quarantine can lead to an increased risk for and potential worsening of chronic health conditions, weight gain, loss of skeletal muscle mass and strength and possibly also loss of immune competence since several studies have reported positive impact of aerobic exercise activities on immune function. Every day > 30 min or every second day > 1h exercise

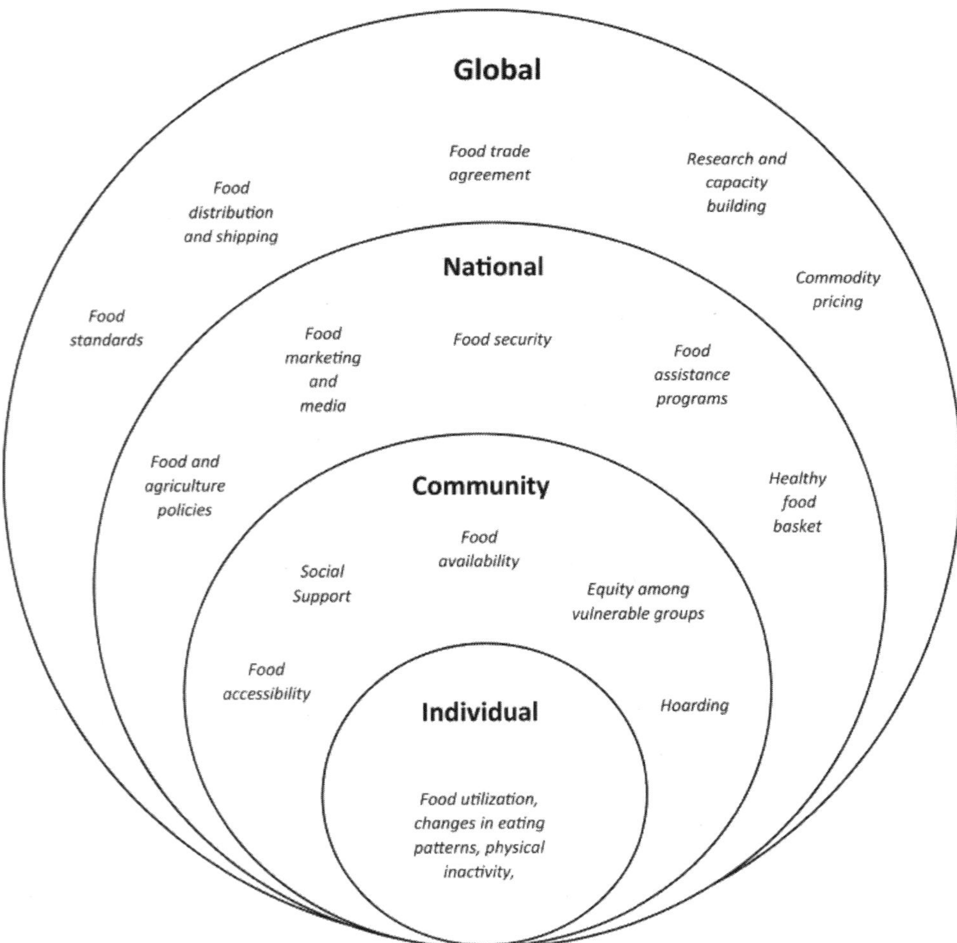

Figure 30.2: A Multi-level Framework of Action to Support Nutrition during the COVID-19 Pandemic.

***Source*: Naja, F and Hamadeh, R. 2020. Nutrition amid the COVID-19 pandemic: a multilevel framework for action. *Eur. J. Clin. Nutr.* 2020 (April, 20): 1-5.**

is recommended to maintain fitness, mental health, muscle mass and thus energy expenditure and body composition.

Nutritional Recommendations at Individual, Community, National and Global Levels amid COVID-19 Pandemic

There is currently no treatment for infection with SARS-CoV-2 or for COVID-19. Current strategies aim to limit the spread of the virus by preventing contact between people. The search for vaccines to offer immune protection against SARS-CoV-2 and for pharmacological treatments to prevent the virus from replicating is underway. In the meantime, approaches to ensure that individuals' immune systems are well supported should be taken. Nutrition should be at the forefront of these approaches.

The nutritional status of individuals has for long been considered as an indicator of resilience against destabilization. The ecology of adversity and resilience demonstrates that substantial stressors, such as inadequate nutrition, can lead to long-lasting effects that are linked to health. In fact, poor diet quality has been associated not only with physical but also mental health. Optimal nutrition and dietary intake is a resource that transcends the individual, the community to reach global influence.

A multilevel framework to support nutrition and food security during the COVID-19 pandemic, using the various levels of the ecological health model: individual, community, national and global.

i) At the Individual Level

The COVID-19 pandemic caused by the SARS-CoV2 virus brought several individual and collective protection measures to contain the expansion of its transmission, such as social distancing and lockdown all over the world. Hence, COVID-19 world pandemic imposed a new set of challenges for the individual to maintain a healthy diet. First, the state of lockdown announced in many countries around the globe led all public and private sector institutions, with the exception of health care facilities and a limited number of essential services, to close down and, if possible, carry its operations remotely (without face to face interactions). Individuals were asked to stay home and avoid contact with other people. Such measures of self-isolation and social distancing are known to be crucial in limiting the spread of the virus, flattening the curve of incidence rate, and ultimately disease containment. These measures have severe repercussions on both food access and utilization.

Food access, however, is dependent on factors that could reach beyond the individual and are more directly related to actions and policies at the community, national as well as global levels. That said, the individual remains capable of making a few choices related to food utilization. The confinement to one's home has direct effects on one's lifestyle, including dietary habits, eating, and physical activity patterns. Confinement increases sedentary behaviors that involve activities with very low energy expenditure, performed mainly in a sitting or supine position. The low physical activity levels, even for short periods, could negatively affect physical and mental health. During lockdown period, everybody had nothing to do but was busy in cooking and eating all the time at home especially in those families where all the family members including children were together after a long period and they wanted to eat something good to eat all the time to cope up with the mental stress. On social media also, people were giving demonstrations on new dishes and some were earning money through it too. This all promoted overconsumption of food in most of the well off families. The state of lockdown and confinement could also lead to irregular eating patterns and frequent snacking, both of which are associated with higher caloric intake and increased risk of obesity. The higher consumption of processed foods with higher caloric content, high in saturated fats, sugars, and refined carbohydrates, greater durability, easier access and use, especially in children, could have contributed to increase the prevalence of obesity in the times of COVID-19.

The changes in dietary patterns during the outbreak of COVID-19 could also be driven by the fear and anxiety many people around the globe are experiencing. Compelling evidence showed that dietary habits are affected by conditions of stress, distress, and emotional disturbance, whereby elevated distress levels are associated with unhealthy dietary patterns and poor quality of the diet. Furthermore, emotions like fear and sadness are associated with less desire or motivation to eat and with lessened enjoyment during eating.

More recently, in explicating a five-way model of emotions and diet, it was found that changes in food intake may be the "natural" response to stress and heightened emotional states through both psychological and physiological mechanisms.

Therefore, the responsibility of the individuals during the COVID-19 pandemic lies in making an effort to choose a healthy lifestyle, eat diets high in fruits and vegetables, exercise during free time, try to maintain a healthy weight, and get an adequate amount of sleep. In addition to taking care of one's dietary intake, the collective responsibility of individuals is to avoid the spread of misinformation related to nutrition and dietary intake, and the COVID-19. Since the outbreak, networks of social media were flooded by messages of single foods/herbs promising cure or prevention of the infection. The effects of such unfounded claims could lead to negative implications ranging from giving a false sense of protection against the infection to toxicity.

Thus, for many individuals, usual shopping routines and eating have been entirely upended. In addition to providing a framework of economic and social instability, the lockdown could affect the food supply chain and generated a situation of food and nutritional insecurity in some regions of the world. Therefore, undernutrition and obesity might have increased, due to limited access to food, concern of running out of staple items, limited culinary ability, more sedentary lifestyle and change in food purchasing, eating behaviors, and perceptions of food safety.

ii) At the Community Level

At the community level, food access and availability are particularly vulnerable to the implications of the COVID-19 outbreak, primarily because of difficulties in transportation, distribution, and delivery. This situation has led in many instances to *'hoarding'*. One way that a pandemic would indirectly impact the food supply chain is by changing consumer behavior. Pandemics create uncertainty and volatility in consumer demand, making it particularly challenging to maintain food inventories in a just in time economy. In a study of the effect of an outbreak on behavior, the most recurrent response is to stockpile supplies, food, and water. Those who can afford extra food may hoard more than they need and pose devastating consequences on at risk-populations. Hoarding could lead to extreme shortages in markets, leading to rapidly rising prices. Therefore, at a community level, it is crucial to spread awareness against *'Panic-buy'*. Furthermore, during the COVID-19 pandemic, older adults and patients with chronic diseases became particularly vulnerable and most at risk to nutrition imbalance. Firstly, available research indicated that adults 60 years and older and patients with pre-existing medical conditions, especially heart disease, lung disease, diabetes or cancer are more likely to have severe even deadly coronavirus infection than other groups. Second, the recommendations to stay home and abide by social distancing targeted these groups specifically, given their vulnerability. Third, the elderly and patients with chronic diseases may already be susceptible to malnutrition given their compromised health and limited purchasing ability. Therefore, at the level of the community, it is crucial to identify these vulnerable groups and extend assistance in food access and availability through a structured and reliable support system.

During this period, labourers were affected a lot. Because they are daily wage earners, they had no money to buy food articles during lockdown period. Though some of the social organizations tried to provide food to such labourers. But still many of them had a great sense of food insecurity and moved towards their villages in other states. This class of our society suffered the maximum during lock down period because of no work and no food.

iii) At the National Level

While governments of countries around the globe are dealing with the burden of the COVID-19 and its enormous strains on the healthcare system, they are also battling a destabilization in their economies and a rising threat of food insecurity. In light of these challenges to provide adequate and nutritious food at times of pandemics, each country is urged to define, finance, and distribute a food basket of a least-cost diet that supports the health needs of the population, ensure the use of the local agricultural produce of the country, and minimize reliance on food imports. Significant planning is needed at the national level to increase the nation's preparedness, including the formulation of policies to support the production, distribution, and access of this food basket to different communities. Among these policies are those related to mobilization of resources in order to finance food purchases and provisions, tax waiving for staple foods and commodities, and support for agricultural and food production industries. Given the effect of the COVID-19 pandemic on the demand and supply dynamics of food, price hikes became prevalent reaching at times uncontrollable levels, a situation that requires national efforts to closely monitor and inspect food prices and markets.

The COVID-19 pandemic imposed a paradigm shift on governments, whereby it became imperative to build networks with the private sector, the international agencies, and local communities. It is only through a coordinated effort of these different entities that securing essential nutritious food stocks become possible.

At the Global Level

While border protection is legitimate in safeguarding the health of citizens from external threats; however, it can severely disrupt travel, trade, and tourism, as well as infringe civil liberties. Countries that depend heavily on imported food to meet demand might face inconsistent risk from supply chain failures, especially in the face of border crossing closures. Therefore, it is essential to ensure the smooth flow of global trade and make full use of the international markets as a vital tool to secure food supply across the globe to prevent food insecurity. An important lesson of the COVID-19 relates to its global nature whereby no country is immune to its spread and infliction: a global threat requires global action. The protectionist strategies that each country is implementing should be complemented by global cooperation, solidarity, and coordination among countries to ensure that humanity emerges from this pandemic with the least possible losses.

General Dietary and Lifestyle Recommendations

Healthy Eating is Important in Quarantine Applications

Since the clearest known practice for Covid-19 is social isolation, where all kinds of contacts are minimized, many global health care providers recommend keeping at least two weeks of medicine and food for everyone

Table 30.1: Recommendations to Mitigate the Impact of COVID-19 on Nutrition and Food Security at the Individual, Community, National and Global Levels

Nutrition Recommendations during Covid-19 Pandemic	
Individual level	✰ Try to eat balanced meals, avoid irregular snacking.
	✰ Choose foods rich in Vitamins A, C, E, B_6, B_{12}, iron, zinc such as citrus fruits, dark green leafy vegetables, nuts and dairy products.
	✰ Consumption of highly processed foods should be limited and fruit juices and carbonated drinks should be avoided.
	✰ Consuming meat, poultry and eggs was not risky in present circumstances, but hand wash hygiene must be followed after handling raw meat, eggs or even vegetables. Thoroughly cooked meat and poultry products may be included in diets in moderation.
	✰ Avoid too much fat (no more than 30 gm per person per day - preferably more than two varieties of oils), salt (no more than 5 gm per person a day). Sugar is just calories with no nutrients, hence, keep it to a bare minimum,
	✰ Maintaining ideal body weight, keeping the body hydrated with adequate water intake, taking up moderate physical activity including yoga to reduce stress and build immunity.
	✰ Avoid smoking, alcohol and drugs
	✰ Most infections can be prevented by practising good personal hygiene such as washing hands before preparing or eating food, washing hands after cleaning vegetables or meat, covering mouth with a tissue or cloth while coughing or sneezing.
	✰ Stay physically active during quarantine.
	✰ Refrain from spreading misinformation related to nutrition and dietary intake and covid-19.
Community	✰ Spread awareness regarding the devastating consequences of hoarding and panic buy.
	✰ Identify and support populations at risk of malnutrition within the community especially the elderly and patients with chronic diseases.
	✰ Create a structured and reliable support system to ensure availability, access and affordability of essential food commodities to all members of the community.
National	✰ Define, finance and distribute a food basket of a least cost diet that addresses the health needs of the population, ensures the use of local agricultural produce of the country and minimizes reliance of food imports.
	✰ Mobilize resources in order to finance food purchases and provisions.
	✰ Waive taxation for staple foods and commodities.
	✰ Support agricultural and food production industries.
	✰ Closely monitor and inspect food prices and markets.
	✰ Build networks with the private sector, international agencies, and local communities.
	✰ Maintain high level of transparency, critical to build trust, support and compliance.
Global	✰ Assure continuous flow of global trade, avoiding any trade restrictions would be beneficial to keep food and feed supplies, as well as those of agricultural inputs, from worsening local conditions already strained by COVID-19 response measures.
	✰ Reduce import tariffs and other restrictions on food commodities.

as part of quarantine practices. The food needs to be nutritious, durable with a long shelf life. Balanced nutrition is highly important during these times with a diet that is rich in proteins, fibers, vitamins, minerals, and antioxidants. A Healthy Meal Plate is recommended, in which one-quarter of the plate per main meal would comprise of vegetables, the other quarter is from whole grain products and the remaining half from fruits, high-protein foods (dried legumes, meat, eggs, fish, chicken, oily seeds, *etc.*), and dairy products (milk, yogurt, cheese, *etc.*) in equal three pieces. Some amount of fats and oils should be used in cooking meals. Besides, it is recommended to ensure adequate water consumption.

Daily adequate intake of vitamin and mineral should be maintained. Finally, even if adequate intake of vitamins and minerals should be mainly provided from foods such as vegetables, fruits, whole grains, and seeds; for people who are not able to consume according to the "Healthy Food Plate"), it may be recommendable to take vitamin and mineral supplements cautiously as an alternative in this special temporary period.

One should follow the following lifestyle practices as recommended by NIN to protect from this infection:

- ✰ Maintain ideal body weight.
- ✰ Be physically active.
- ✰ Follow right eating pattern.
- ✰ Eat a variety of foods from each food group and prefer to eat whole grains, fruits, vegetables and beans.
- ✰ Eat fresh fruits and vegetables daily.
- ✰ Avoid fruit juices and carbonated drinks.
- ✰ Limit fast foods. Avoid junk foods/dairy/refined wheat flour/non-veg/deep fried/package foods as much as possible.

- ☆ Limit red and processed meat.
- ☆ Limit sugar sweetened drinks.
- ☆ Limit foods containing excessive salt.
- ☆ Do not rely on supplements.
- ☆ Limit consumption of highly processed foods.
- ☆ No risk in consuming meat, poultry and eggs in present circumstances.
- ☆ Follow hand wash hygiene after handling raw meat, eggs or even vegetables.
- ☆ Cook meat and poultry products thoroughly and include these foods in moderation in your diet.
- ☆ Avoid too much fat (no more than 30 gm per person per day), and salt (no more than 5 gm per person a day).
- ☆ Keep sugar intake to a bare minimum as it is just calories with no nutrients.
- ☆ Consume enough dietary fibre.
- ☆ Keep the body hydrated with adequate water intake.
- ☆ Taking up moderate physical activity including yoga to reduce stress and build immunity.
- ☆ Avoid smoking and alcohol consumption. Limit alcohol consumption as it has an adverse effect on vitamin and mineral absorption and is associated with sleep problems. There is no evidence that alcohol consumption prevents coronavirus infection. Additionally, the consumption of adulterated beverages containing methyl alcohol instead of ethyl alcohol can cause very serious poisoning reactions.
- ☆ Practise good personal hygiene such as washing hands before and after preparing or eating food.

Immunity Boosting Foods with Possible Anti-viral Properties

There has been much talk about one's immunity and the corona survival. Hence, the dietary management of COVID-19 should be considered in terms of improving immunity and utilizing the anti-viral properties of few nutrients. Eating a low fat, plant-based vegetarian diet may boost the immune system. Vegetarians have been shown in a few studies to have more effective white blood cells compared to non-vegetarians, because of a higher intake of vitamins and lower intake of fat (Davison *et al.*, 2016).

There are many traditional food items, which can increase the immunity with an additional benefit of some antiviral properties. Some of the foods that increase immunity and with possible anti-viral properties have been reported as under which must be included in the daily diets:

Citrus Fruits

Citrus fruit is one of the nature's best and easily available source of vitamin C, a key nutrient in supporting our immune system. Citrus fruits are known to have other benefits like antioxidant, anti-tumour, cardio protective and neuroprotective effects. They have additional fibre content also. But what makes them significant is their immune boosting potential.

Red Capsicum

Vitamin C present in red capsicum is 3 times higher than the vitamin C present in an orange. It can be included in salad every day before meals.

Other Fruits

Fruits like papaya, guava, apple, grapes, mango, tangerines, lemons, sweet lime and gooseberries, all seasonal vegetables, including green leafy, and spices, legumes, millets, flesh foods and fish are among those with rich sources of nutrients Even though almost every fruit is good for health and human immunity, it has been proved that apple, pumpkin and papaya have got antiviral effects against specific viruses (Konowalchuk and Speirs, 1978; Suchitra and Parthasarathy, 2015). Even-though extrapolation to corona virus is unscientific, the antiviral and immune boosting properties of the above said fruits is established.

Nuts and Seeds

Many nuts and seeds including almonds, pumpkin seeds, peanuts and groundnuts have high vitamin E levels. Vitamin E, a lipid-soluble antioxidant commonly present in the membrane of all cells including immune cells. This is supposed to prevent stress-induced damage to cells. Three-four spoons of pumpkin seeds daily can provide substantial quantities of Mg, zinc and healthy fats that are essential for immune functions. Eating almonds have been recently used to treat common flu symptoms. It has been suggested that almonds exhibit some antiviral actions. The peanut skin has also significant antiviral activities according to recent research (Makau *et al.*, 2018).

Curd

Curd is a source of many nutrients and it also improves gut health by regulating gut bacteria, aids immune function and reduces inflammation,

Green Tea

Green tea botanically termed as Camellia sinensis contains a group of flavonoids called catechins. These chemicals appear to inhibit viral infections by blocking the enzymes that is important in replication. Green tea has shown to be effective in inhibiting HIV, the hepatitis B and the herpes viruses (Chacko *et al.*, 2010).

Vegetables

Broccoli and other cruciferous vegetables have been proven to help boost immunity. Researchers claim that sulforaphane, a chemical found in this vegetable, switches on the antioxidant genes and enzymes in specific immune cells. This effect combats free radicals in our body and prevent the disease getting worsened. Broccoli has also been found to have anti- viral properties against influenza viruses (Antonenko *et al.*, 2013).

Garlic

Garlic has been known to have antioxidant, cardio protective and anti-tumour effects. Allicin (chemically - allyl 2-propenethiosulfinate) is the primary bioactive chemical which is present in the aqueous extract of garlic. This chemical is also found even in the raw garlic homogenate. When garlic is chopped, the enzyme alliinase is activated to produce allicin. Many studies have noted the anti-viral activity of garlic extracts against HIV, herpes, cytomegalo -virus and the flu viruses (Bayan *et al.*, 2014). The exact mechanism is unknown.

Turmeric

Turmeric is a herbaceous perennial plant (botanical name: Curcuma longa) belonging to the ginger family. The medicinal properties of turmeric, the source of bioactive compound curcumin, has been known for centuries; still the ability to know the exact mechanism of action and to determine the bioactive components are still not completely understood. The compound is known to have antioxidant, antibacterial, antiviral, cardio protective and immune stimulating properties. The bioavailability of curcumin is increased by the addition of black pepper. In a study, researchers have found that the inflammatory cytokines like the mean serum IL-1β band the vascular endothelial growth factor were found to be significantly reduced by curcumin therapy (Hewlings and Kalman, 2017). This assumes significance in the wake of corona epidemic where the cytokine surge is worsening patients rather than the virus replication.

Ginger

Ginger and its products are being used to raise the function of the immune systems. The extracts of ginger have anti-inflammatory, digestive, and anti-tumour effects. Fresh ginger but not a dried one has been shown to have anti-viral activity against Human Respiratory Syncytial Virus in a Human Respiratory Tract Cell Line study. Hence, to extrapolate for flu and a trial of such nutrient as an additive in our diet can prove useful. The ginger extract actually has been reported to stimulate the production of TNF-alpha expression by the immune system. Researchers also studied ginger along with other natural compounds in combination for inhibiting H1N1 influenza A (Chang *et al.*, 2013; Mashhadi *et al.*, 2013) and demonstrated the inhibition of viral replication.

Miscellaneous Foods

Beta-carotene is a powerful antioxidant that can reduce inflammation and boost immune function by increasing leucocytes in the body. Excellent sources of beta-carotene include sweet potatoes, carrots, and green leafy vegetables (Grune *et al.*, 2010). Coconut water is rich in vitamins like riboflavin, niacin, thiamine and folates and it possesses anti-viral and anti-bacterial properties (Chauhan *et al.*, 2014) that can help increase our body's immune system and increase our capacity to fight viral infections like flu.

Onions contain organosulfur compounds like quercetin and allicin (Sharma, 2019) which are associated with inhibition of viral infection. These bioactive compounds can hinder virus attachment to the host cell. They can alter transcription and translation of viral genome inside the host cell and hence affect the viral assembly. Inhibition of viral entry into the cell and inhibition of RNA polymerase have also been postulated as mechanism of antiviral actions of this vegetable. Tamarind leaves (Caluwé *et al.*, 2010), fruits and seeds with a multitude of uses have been also demonstrated to antiviral properties. Regular intake of probiotics allows their intimate interaction with the gut mucosa and mucosal immune system. Probiotics can modulate immune and inflammatory response in the human gut through their interaction with gut epithelial cells. The curd is a simple nutrient supplement for probiotics.

Sesame is a simple nutrient food with enough zinc as its content. Zinc has always been noted for its antibacterial and antiviral properties (Suchitra and Parthasarathy, 2020). Zinc also has a positive effect on body's defence mechanisms. Extracts of the plants and leaves of the mint family have shown anti-viral effects.

COVID-19 and Physical Activity

With the rapid coronavirus spread, the general population has been highly advised, for safety and prevention, to reduce moving and traveling and stay at home aiming to limit the COVID-19 transmission. Unfortunately, such restrictions against regular physical activities will unavoidably affect individuals' routine daily activities, psychosocial status and well-being and may increase sedentary behavior with lowering energy expenditure, favoring screening activities such as watching television, using mobile devices and playing games. Hence, regular physical activity and routinely exercising *e.g.* walking, push-up, *etc.* at least 30 min of moderate physical activity every day and/or at least 20 min of vigorous physical activity every other day in a safe home environment should be maintained to avoid stress, anxiety and depression and that constitutes a strategy for healthy living during the coronavirus crisis.

General Guidelines

- ☆ Avoid handshakes.
- ☆ Cover your mouth with a tissue or cloth while coughing or sneezing.
- ☆ Wear mask.
- ☆ Avoid crowded places and quarantine yourself.
- ☆ Avoid going out.
- ☆ Avoid public transport and unnecessary travel.
- ☆ Do not panic.
- ☆ Maintain social distancing.

31

Role of Dietitians and Nutritionists

A dietitian is a person with legally recognized qualification in nutrition/or dietetics who applies the science of nutrition to the feeding and education of groups of people and individuals in health and disease (International Congress, July 1988, Paris).

Nutritionists and dietitians both are health professionals who create diets for people. Teaching patients about healthy eating is a primary duty, as well as staying aware of developments in medicine and nutrition. However, there is a difference between dietetics and nutrition: dietetics focuses on food management, whereas nutrition focuses on the promotion of health through healthy food.

Registered Dietitian

Registered Dietitians (RDs) are the only qualified health professionals that assess, diagnose and treat dietary and nutritional problems at an individual and wider public health level.

They work with both healthy and sick people in a variety of settings. Dietitians can work in the NHS, private practice, food industry, workplace, catering, education, research, government and non-government organizations, sport and the media *etc.* Other care pathways they work in include mental health, learning disabilities, community, acute settings and public health.

They plan and supervise the preparation of therapeutic or other diets for individuals or groups in hospitals, institutions, other establishments and for workers of particular sectors, give instructions in selection and proper preparation of food according to dietetic principles, perform duties related to nutrition programmes and may be responsible for food purchasing on behalf of an organization or establishment.

Uniquely, dietitians use the most up-to-date public health and scientific research on food, health and disease which they translate into practical guidance to enable people to make appropriate lifestyle and food choices. Dietitians are the only nutrition professionals to be regulated by law, and are governed by an ethical code to ensure that they always work to the highest standard.

Nutritionists

Nutritionists generally work in different non-clinical setting roles including public health, health improvement, health policy, food industries, local and national government, in the private sector, Non Government Organizations (NGOs), international work in developing countries, media and communications, sports and in education and research. Nutritionists are qualified to provide information about food and healthy eating.

Public health nutritionists appraise various factors related to nutrition and food problems in the community, plan and execute nutrition elements of health programmes; participate in programmes of nutrition activities for particular groups or the community.

Qualifications of Registered Dietitian (RD) in India

RD stands for Registered Dietitian, who is a qualified, skilled and knowledgeable professional. A Registered Dietitian is authorized to prescribe Medical

Nutrition Therapy (MNT), which is evidence-based, and to counsel people to achieve and maintain good health.

An RD is food, nutrition and dietetics expert with:

1. 3 to 4 years' graduate degree in Nutrition and Dietetics with a UGC approved curriculum and PG Diploma or Master's in Nutrition and Dietetics as regular course.
2. Six months of internship with competency package in IDA (Indian Dietetic Association) approved hospitals under guidance of eligible Registered Dietitian.
3. Pass in RD Board Examination and acknowledged by RD Board with RD number. This is conducted by Indian Dietetic Association.
4. Life membership of IDA

After completing graduation and post-graduation/ PG diploma in Nutrition and Dietetics or PG in Foods and Nutrition, one will be termed as a Nutritionist. In order to be termed as a dietician, it is important to do internship in a hospital under RD to be eligible for the RD examination. One will have to pass the RD entrance examination conducted by IDA and then one can register with the Indian Dietetic Association as a **Registered Dietician**. With this expertise, RDs are well equipped with knowledge in the field of Clinical Dietetics and Medical Nutrition Therapy.

The RD entrance examination is divided into 2 papers. Paper-1 includes areas such as Human Physiology, Biochemistry, Physiologic and Metabolic Changes in Disease, Food Microbiology, Sanitation, and Hygiene *etc.* Paper-2 includes Human Nutrition and Meal Management, Community Nutrition, Diet Therapy (Theory), Nutrition Education and Dietetic Counselling and Food Services Management. The RD examination is held in the month of February every year. The date of examination is published on the IDA website at least 6 months in advance. The details of syllabus of RD examinations, centres to take examination and eligibility requirements *etc.* are available at the official website of IDA (http://www.idaindia.com).

Code of Ethics for Dietitians

Dietetic practitioners have voluntarily developed a code of ethics to reflect the ethical principles guiding the dietetic profession and to outline commitments and obligations of the dietetic practitioner to self, client, society and the profession.

- The dietetic practitioner provides professional services with objectivity and with respect for the unique needs and values of individuals.
- He/She avoids discrimination against other individuals on the basis of race, creed, religion, sex, age and national origin.
- He/She fulfils professional commitments in good faith.
- He/She conducts himself/herself with honesty, integrity and fairness.
- He/She remains free of conflict of interest while fulfilling the objectives and maintaining the integrity of dietetic profession.
- He/She maintains confidentiality of information.
- He/She practices dietetics based on scientific principles and current information.
- He/She organizes and exercises professional judgment within the limits of his/her qualifications and seeks counsel or makes referrals as appropriate.
- He/She assumes responsibilities and accountability for personal competence in practice.
- He/She provides sufficient information to enable clients to make their own informed decision.
- He/She does not advertise in a false or misleading manner.
- He/She promotes products in a manner that is neither false nor misleading.
- He/She permits use of his/her name for the purpose of certifying that dietetic services have been rendered only if he/she has provided or supervised the provision of those source.
- He/She presents substantiated information and interprets controversial information without personal bias, recognizing that legitimate differences of opinion exist.
- He/She makes all possible efforts to avoid bias in any kind of professional evaluation.
- The dietetic practitioner accurately presents professional qualification and credentials.
- He/She voluntarily withdraws from profession practice if he/she is engaged in substance abuse that could affect his/her practice.
- He/She complies with all applicable laws and regulations concerning the profession.
- The dietetic practitioner accepts the obligation to protect society and the profession by upholding the code of ethics for the profession of dietetics and by reporting alleged violations of code.

Role and Responsibilities of Dietitians

Dietitians have many roles to play:

- Assess patient's and client's health needs and diet
- Develop meal plans, taking both cost and client's preferences into account
- Prescribe and advise diet to the patients to treat or prevent various lifestyle related disorders and diseases.

- They supervise the quantity and quality of food consumed by patients through dietary recall method.
- They calculate the present calories intake of person and accordingly plan and recommend diet with repairing all the loop holes and thereby give a optimum nutrients filled diet.
- They also look after the RT tube feeding of the patients who are unable to chew food.
- Evaluate the effects of meal plans and change the plans as needed
- Some dietitians and nutritionists provide customized information for specific individuals. For example, a dietitian or nutritionist might teach a client with high blood pressure how to use less salt when preparing meals. Others work with groups of people who have similar needs. For example, a dietitian or nutritionist might plan a diet with limited fat and sugar to help patients lose weight. They may work with other healthcare professionals to coordinate patient care.
- Keep up with the latest nutritional science research
- They also work for food industry to keep the quality check of foods.
- They are also involved in food service management; five star hotels recruit them to take care of the health of the clients.
- They have an excellent skill of counseling their clients to change their lifestyle in order to live a long and very healthy life. They meet their patients in hospitals and solve their food related queries and myths.
- They have the ability to formulate their own personalized product and selling it among the target groups.
- They are involved in social work to help the people to recognize the importance of food and teach them recipes which they can include in their diets to increase their nutritional value.

Scope of Dietitians in India

There are different types of dietitian positions *viz.*, clinical dietitians, community dietitians, administrative dietitians, and consultant dietitians. Each of these positions offers different career options, such as food and nutrition services in hospitals, nutrition counseling in public health agencies and fitness clubs or clinical management in food service systems.

The most common settings where dietitians are employed include Health Systems (Govt. and private hospitals/nursing homes, Community Health Centres, Public Health Agencies *etc.*), Food and Agriculture (Food service/hospitality, Food producer, Food manufacturing company or association, Govt. Deptt.), Academic or research organizations, Media, Pharmaceutical companies, School Feeding Programmes, Recreation or sports facility or association, Fitness centres *etc.*

They can be employed as:

Clinical Dietitian

Clinical Dietitians are trained professionals who handle medical nutrition therapy. They work with patients in health care settings, such as hospitals, nursing care facilities and other health care facilities, private practice and other institutions *etc.* to provide nutrition therapy to patients with a variety of health conditions, and provide dietary consultations to patients and their families and help them in maintaining their nutritional intake. They confer with other health care professionals to review patients' medical charts and develop individual plans to meet nutritional requirements. They create nutritional programs based on the health needs of patients or residents and counsel patients on how to lead a healthier lifestyle. Some clinical dietitians will also create or deliver outpatient or public education programs in health and nutrition. Clinical dietitians may provide specialized services in areas of nourishment and diets, tube feedings and intravenous feedings such as total parenteral nutrition (TPN) or peripheral parenteral nutrition (PPN). They may further specialize, such as working only with patients with kidney diseases or those with diabetes. Some clinical dietitians have dual responsibilities with patient nutrition therapy and in foodservice or research.

Major responsibilities of a Clinical Dietitian are:

- Assessment of patients' nutritional status
- Determining nutritional needs of patients
- Planning normal and therapeutic diets
- Taking ward rounds to meet in-patients
- Participation in interdisciplinary team rounds/ meetings to discuss patient cases and determine nutrition risk
- Development, implementation and evaluation of nutrition intervention
- Prescribing normal, therapeutic (oral and enteral) diets
- Diet counseling during hospitalization and on discharge, based on patient's condition
- Nutrition counseling, dietary recommendation, food selection and preparation of nutrition guidelines for out-patients – both individuals and their families
- Diet counseling in Specialized Clinics like Endocrinology, Gastroenterology, Nephrology, Pediatrics, Oncology, AIDS, Obesity, Neurology, Plastic Surgery, and so on
- Diet Counseling in Preventive Health Check-up Clinics

- ☆ Liaising with the Food and Beverage team when required
- ☆ Tasting the food prepared by the kitchen and giving feedback
- ☆ Training and teaching medical students, nursing students, nutrition students and interns

Community Dietitian

Dieticians can also work with the general public in community settings, wellness programmes, health maintenance agencies, various fitness clubs, public health agencies and government projects *etc.* by assessing and promoting healthy eating to prevent diseases. They work with large groups and communities.

They often work with specific groups of people such as the elderly, adolescents, children, or other individuals with special needs or limited access to healthy food. Some community dietitians conduct home visits for patients who are too physically ill to attend consultations in health facilities in order to provide care and instruction on grocery shopping and food preparation.

Public Health Dietitian

They work in the public sector in areas including policy development, food laws and regulations, development of food guide, and regulation of food labels. They work with the government and other agencies to help implement healthy food changes.

Pediatric Dietician

They work with children of all age groups such as infants and adolescents and advise them on nutrition accordingly. They focus on early nutritional needs, and often work closely with doctors, school health services, clinics, hospitals and government agencies, in developing and implementing treatment plans for children with eating disorders, food allergies, or any condition where a child's diet factors into the equation, such as childhood obesity. They can either work in hospitals or set up an independent practice.

Neonatal Dietitian

Neonatal dietitians provide individualized medical nutrition therapy for critically ill premature newborns. They are considered a part of the Neonatal Intensive Care Unit's medical team. They perform clinical assessment of neonates, design nutrition protocols and quality improvement initiatives with the medical team, develop enteral and parenteral regimens, help establish and promote lactation/breastfeeding guidelines and often oversee the management of infection prevention in the handling, storage, and delivery of nutritional products.

Gerontological Dietitian

They are specialists in nutrition and aging. They work in nursing homes, community-based aged care agencies, government agencies in aging policy, and in higher education in the field of gerontology.

Sports and Health Dietician

There are certification/diploma programs specially designed for dietitians to become experts in sports. They work with various sports clubs, fitness centres, and athletic teams. They work with individual athletes or entire sports teams to teach them about healthy eating while being active. They provide advice on issues like maximizing energy levels, eating to optimize growth and development, eating before competition, and staying hydrated. They make diet plans for the sports personnel, both to maintain their weight and body muscle, and to keep their nutrition levels in check.

Food Service Dietitian

They are responsible for large-scale food planning and service. They coordinate, assess and plan foodservice processes in health care facilities, school food service programs, prisons, restaurants, and company cafeterias. They train and supervise other food service workers such as kitchen staff, delivery staff, and dietary assistants or aides. These dietitians may perform audits of their departments to ensure quality control and food safety standards, and launch new menus and various programs within their institution to meet health and nutritional requirements.

Research and Development

Dieticians can also work in the field of research to ensure the quality and nutritional values of food items. They work in labs and study the effects of different types of diet forms on the body. The main area of focus is the nutritive components of food. They may work on various research projects to enhance patient care, create new, healthy foods, and improve the cost-effectiveness of the foodservice industry.

They may survey foodservice systems management in order to guide quality improvement. Some research dietitians study the biochemical aspects of nutrient interaction within the body. Research dietitians can also find employment at the university level in teaching positions.

Academics

They teach in various schools, colleges and universities to pass on their knowledge of nutrition to aspiring dietitians.

Consultant Dietitian

Consultant dietitians are those who are in private practice or practice on a contractual basis with health care facilities or corporations to provide counseling and expertise on healthy eating and disease prevention to individuals or small groups. They generally provide nutrition or health related consultation and educational programs to individuals and health care facilities as well as sports teams, fitness clubs, and other health related businesses and corporations.

With experience in administrative or clinical dietetic practice, a consultant dietitian:

- Evaluates and monitors food service systems, making recommendations for a conformance level that will provide nutritionally adequate food quality.
- Develops budget proposals and recommends procedures for cost control.
- Plans, organizes and conducts orientation and in service education programmes or food service.
- Plans layout designs and determines equipment requirements for new or renovated food service facilities.
- Recommends and monitors standards for sanitation, safety and security in food service.
- Develops menu patterns.
- Assesses, develops, implements and evaluates nutritional care, plans and provides for follow-up including written reports.
- Consults with the health care team concerning the nutritional care of clients.
- Develops, maintains and uses pertinent record systems related to the needs of organization and to consultant dietitian.
- Develops, uses and evaluates educational materials related to services provided.
- Provides guidance and evaluation of the job performance of dietetic personnel.
- Interprets, evaluates and utilizes pertinent current researches relating to nutritional care.
- Maintains effective verbal and written communications and public relations, inter and intra departmentally.

Administrative Dietitian

Administrative or management dietitians oversee and direct all aspects of clinical dietetics service, food policy and/or large-scale meal service operations in hospitals, government agencies, company cafeterias, prisons, and schools.

Administrative dietitian is a member of management team and affects the nutritional care of groups through management of food service systems that provide optimal nutrition and quality of food. They recruit, train and supervise employees of dietetics departments including dietitians and other personnel. The major responsibilities of administrative dietitian are to:

- Plan, develop, control and evaluate food service systems.
- Develop short and long range department plans and programmes consistent with departmental and organizational policies.
- Manage and control fiscal resources and recommends budget programmes.
- Utilize human efforts and facilitates resources efficiently and effectively.
- Coordinate and integrate clinical and administrative aspects of dietetics to provide quality nutrition care.
- Establishe and maintain standards of food production and service, sanitation, safety and security.
- Maintain effective communication: inter and intradepartmentally.
- Complie and utilize pertinent operational data to improve efficiency and quality of food service systems.
- Plan, conduct and evaluate orientation and in-service educational programmes.
- Interpret, evaluate and utilize pertinent current research relating to nutritional care.
- Develop menu patterns and evaluates client acceptance.
- Develop specifications for the procurement of food, equipments and supplies.
- Plan or participate in the development of programme proposals for funding.
- Plan layout designs and determines equipment requirement for new or renovated food service facilities.
- Administer personnel policies as established by the dept. and organization.

Indian Dietetic Association (IDA)

In 1962, a group of like-minded nutritionists, dietitians and workers in the allied health fields resolved to form a scientific body to highlight the importance of dietetics and nutrition in the maintenance of health, and in the prevention and treatment of diseases. Thus Indian Dietetic Association (IDA) was founded, with Prof. Kalyan Bagchi as Secretary and Dr. C. Gopalan as President (Registered under Societies Registration Act, 1961).

The association was affiliated to the International Congress of Dietetics in 1975. IDA is a member of ICDA (International Confederation of Dietetic Associations) and AFDA (Asian Federation of Dietetic Associations). IDA partners with other health organizations for causes related to nutritional health of the country. IDA promotes scientific research in the field of Nutrition and Dietetics. IDA also facilitates contact and interaction within its member community and other health professionals. The

Headquarters of IDA is located in Kolkata. The major objectives of IDA are:

- ✰ To promote the cause of science by encouraging the spirit of active pursuit of knowledge and original scientific research particularly in the field of Nutrition and Dietetics.
- ✰ To facilitate social, scientific and cultural fellowship and cultivation of goodwill among its members.
- ✰ To promote close contact and interaction between persons following different branches and thus, facilitate the development of a wider outlook and the integration and application of available scientific knowledge for the welfare of society.
- ✰ To safeguard the interests of scientists generally and its members in particular and work for their welfare.

At present, there are 23 chapters of IDA in various parts of the country and this professional association is having more than 13000 members all over the country. Golden Jubilee celebrations of IDA was held in 2012 at Hyderabad. IDA has declared 10th January as Dietetics Day and a theme is decided every year. On this day, all chapters take active part in spreading the message of nutrition and dietetics to the community.

Criteria for Membership of IDA

Criteria for getting Life Membership of this association are:

A. Graduate degree (minimum three years) in Nutrition and Dietetics

OR

B. Graduate degree (minimum three years) with any of the 3 nutrition-related subjects (Human Physiology, Nutritional Biochemistry, Dietetics/ Clinical Nutrition/Diet Therapy/MNT, Food and Nutrition, Basic Nutrition, Community Nutrition, Food Science) given here in the undergraduate level from a UGC recognized university

+

PG Diploma OR PG Degree in Nutrition and Dietetics from a UGC recognised institution.

Students can become the members of IDA if they are possessing graduation (Clinical Nutrition/Nutrition and Dietetics/Home science with Nutrition Hons.)

Publication of Journal

IDA started its own scientific journal in 1963 which was renamed as JIDA (Journal of the Indian Dietetic Association) in 2003. Two issues are published every year.

Appendices

Annexure-I

Weights and Household Measures

Food Stuffs	*1 cup to gms*	*1 Tablespoon to gms*	*1 Teaspoon to gms*	*1 Cup to Tablespoons*	*1 Tablespoon to Teaspoons*
Cereals					
Whole wheat flour (*atta*)	110	–	–	11	–
Semolina	140	–	–	10	–
Refined flour (maida)	110	10	–	10	–
Rice	170	–	–	9	–
Rice flour	160	–	–	10	–
Beaten rice	50	–	–	–	–
Puffed rice	25	–	–	–	–
Corn (dry)	150	–	–	8	–
Corn flakes	40	–	–	–	–
Corn flour	90	10	2	10	4
Pulses					
Bengal gram (split)	160	–	–	7	–
Bengal gram flour	100	–	–	10	–
Dry beans	170	–	–	7	–
Gram (roasted)	100	–	–	9	–
Green gram (split)	190	–	–	9	–
Green gram (whole)	150	–	–	8	–
Kabuli gram	150	–	–	6	–
Lentils (split or whole)	160	–		8	–
Rajma	150	–	–	6	–
Red gram (split)	165	–	–	8	–
Milk/Milk products					
Butter	170	20	5	8	4
Cheese (grated)	100	–	–	–	–
Condensed milk	220	25	7	9	4
Cream	200	20	6	10	–
Curd	180	30	–	–	–
Khoa (mashed)	110	15		7	–
Milk	180 ml	20 ml	5 ml	10	–
Paneer (mashed)	125				–
Sugar/jaggery					
Sugar	190	20	5	9	4
Jaggery (grated)	200	20	–	9	4
Icing sugar	130	15	3	–	–
Fats/oils					
Dalda vanaspati (unmelted)	180	20	–	9	4
Dalda vanaspati (melted)	150	12	–	12	5
Ghee (unmelted)	190	15	–	9	4
Ghee (melted)	160	12	–	12	5
Oil	170	10	–	12	5
Miscellaneous					
Bread crumbs	135	–	–	9	4
Water	180	–	–	–	–

Annexure-II

Common Units, Measures and their Abbreviations with Conversion Factors

Mass		
1 gram (g)	=	1000 milligrams (mg)
1 milligram (mg)	=	1000 micrograms (μg)
1μg	=	1000 millimicrograms (m μg) (Nanogram)
1 m μg	=	1000 of micrograms (μg) (Picogram)
1 ounce (OZ)	=	28.4 g
1 pound (lb)	=	16 OZ = 453.6 g
Volume		
1 litre (l)	=	1000 millilitres (ml)
1 milli litre (ml)	=	1000 microlitres (μl)
1 litre	=	35.196 fluid ounce (OZ)
1 litre	=	1.76 pint (pt)
1 fluid ounce	=	28.4 ml
Conversion		
English to Metric		
Pounds x 0.454	=	Kilogram
Ounces x 28.35	=	Grams
Quarts x 0.946	=	litres
Cups x 236.6	=	millilitres
Metric to English		
Kilogram x 2.2	=	Pounds
Grams x 0.035	=	Ounces
Litres x 1.056	=	Quarts

Annexure-III

Conversions

Length
1 centimetre (cm) = 0.4 inches (in)
1 metre (m) = 100 cm = 39 in (approximately 3 feet)
1 in = 2.5 cm
1 foot = 30.5 cm
Weight
100 milligrams (mg) = 1.5 grains (gn)
1 grain = 65 mg
100 grams (g) = 3.6 ounces (oz)
1 kilogram (kg) = 2.2 pounds (lb)
Fluid measures
1 pint = 570 ml
1 teaspoon = 4 ml = 1/8 fl oz
1 tablespoon = 15 ml = ½ fl oz
Temperature
Temperature in °C = (Temperature in °F – 32) x 5/9
Temperature in °F = Temperature in °C x 9/5 + 32
Freezing point = 0°C = 32°F
Boiling point = 100°C = 212°F
Energy
1 kilocalorie (kcal) = 1 Calorie = 1000 calories (cal) = 4200 joules (J) = 4.2 kiloloules (kj)
1 kilojoule = 1000 J = 240 cal = 0.24 kcal = 0.24 Calories

Annexure-IV

Energy Rich Foods
(Values per 100 g edible portion)

Foods	*Energy (Kcal)*
Cereals	
Bajra	348
Jowar	334
Maize (dry)	334
Rice, raw, milled	356
Wheat flour, whole	320
Pulses	
Bengal gram dhal	329
Black gram, dhal	324
Field bean, dry	280
Green gram, dhal	326
Rajmah	299
Soy bean	377
Nuts and Oilseeds	
Almonds	609
Cashewnuts	583
Coconut, dry	624
Groundnut	520
Pistachio nut	539
Walnut	671
Watermelon seeds	628
Fats and Oils	
Butter	729
Ghee (cow)	900
Ghee (Buffalo)	900
Cooking oils (Groundnut, mustard, coconut, safflower *etc.*)	900
Hydrogenated fat (fortified)	900
Milk and milk products	
Paneer	258
Khoa	316
Skimmed milk powder	357
Whole milk powder (cow's milk)	496
Other foods	
Sugarcane juice	58
Honey	319
Jaggery (cane)	354

Protein Rich Foods
(Values per 100 g of edible portion)

Foods	*Protein (g)*
Cereals	
Bajra	10.96
Barley	10.94
Maize, dry	8.80
Wheat, whole	10.59
Wheat, germ	29.2
Pulses	
Bengal gram, whole	18.77
Black gram dhal	23.06
Cowpea	21.25
Fieldbean, dry	19.84
Green gram, whole	22.53
Lentil, whole	22.49
Soybean	37.80
Nuts and oilseeds	
Cashewnut	18.78
Groundnut	23.65
Pistachio nuts	23.35
Watermelon seeds	34.1
Animal sources	
Egg, hen	13.14
Goat meat	20.39
Liver, goat	20.32
Milk, buffalo	3.68
Paneer	18.86
Whole milk powder (cow's milk)	25.8
Skimmed milk powder (cow's milk)	38.0
Khoa	16.34

Calcium Rich Foods (Values per 100g edible protion)

Foods	Calcium (mg)
Cereals	
Bajra	27.35
Ragi	364
Whole wheat flour	30.94
Pulses	
Bengal gram, whole	150
Rajmah	126
Soy bean	195
Moth beans	154
Horse gram, whole	269
Black gram, dhal	55.67
Green gram, whole	92.43
Leafy vegetables	
Amaranth leaves	901
Bathua leaves	211
Bengal gram leaves	340
Carrot leaves	340
Cauliflower greens	626
Colocasia leaves	216
Fenugreek leaves	274
Radish leaves	234
Drumstick leaves	314
Mustard leaves	191
Nuts and oilseeds	
Almond	228
Coconut, dry	32
Gingelly seeds	1283
Mustard seeds	402
Linseeds	257
Niger seeds, black	572
Animal sources	
Sardine	42
Catla	43
Rohu	39
Milk (Buffalo)	121
Milk (cow)	118
Paneer	476
Curd	50
Egg, hen	50
Khoa	602
Other foods	
Jaggery (cane)	107
Jaggery (date palm)	363

Iron Rich Foods (Values per 100g of edible portion)

Foods	Iron (mg)
Cereals	
Bajra	6.42
Sorghum	3.95
Ragi	4.62
Rice, flakes	4.46
Rice puffed	4.55
Wheat flour (whole)	4.10
Pulses	
Bengal gram, whole	6.78
Bengal gram, dhal	6.08
Black gram dhal	5.97
Green gram whole	4.89
Green gram dhal	3.93
Red gram whole	5.37
Soy bean	8.22
Peas dry	7.90
Horse gram,whole	8.76
Leafy vegetables	
Amaranth leaves, green	4.64
Drumstick leaves	4.56
Fenugreek leaves	5.69
Lettuce	2.73
Cauliflower greens	2.42
Radish leaves	3.82
Parsley	5.51
Spinach	2.95
Fruits	
Dates	4.79
Tamarind pulp	9.16
Raisins dried, black	6.81
Raisins dried, golden	4.26
Animal foods	
Liver (sheep)	6.15
Chicken, liver	9.92
Egg (hen)	1.64
Fish (fresh)	1-4
Fish (dried)	20-25
Other foods	
Jaggery	2.6
Cumin seeds	20.58
Mace	22.69
Turmeric powder	46.08
Coriander seeds	17.64
Til	15.04

Vitamin A Rich Foods (Values per 100 g of edible portion)

Foods	β-Carotene (μg)
Leafy vegetables	
Amaranth leaves, green	8553
Curry leaves	7560
Drumstick leaves	17542
Spinach	2605
Colocasia leaves	5758
Lettuce	1285
Mustard leaves	2619
Parsley	2710
Radish leaves	2591
Cereals	
Bajra	28
Maize,tender, sweet	47
Pulses	
Bengal gram, whole	172
Green gram, whole	137
Bengal gram dhal	165
Peas, dry	65
Fruits	
Mango, ripe	1168
Apricot, dried	1806
Papaya, ripe	694
Tomato, ripe	1513
Dates, dry	2700
Milk and milk products	
Milk	160
Cottage cheese	366
Khoa	497
Fats and oils	
Butter	3200
Ghee (cow)	2000
Hydrogenated fat (fortified)	2500

Vitamin C Rich Foods (Values per 100 g of edible portion)

Foods	Vitamin C (mg)
Leafy vegetables	
Agathi	121
Amaranth leaves, green	83
Cabbage	40
Parsley	133
Drumstick leaves	108
Knol khol greens	71
Cauliflower leaves	52
Fruits	
Amla	600
Guava	214
Lime, sweet	47
Orange pulp	42
Lemon juice	48
Karonda	135
Papaya ripe	43
Strawberry	50
Ber	60

Source: Indian Food Composition Tables, 2017. NIN (ICMR), Hyderabad

Annexure-VI

Approximate Energy Content of some Common Cooked Indian Foods

Food Preparation/Recipe	Measure	Energy (Kcal)
Cereal		
Rice	1 cup	170
Phulka	1 No.	80
Paratha	1 No.	150
Puri	1 No.	80
Bread	2 slices	170
Poha	1 cup	270
Upma	1 cup	270
Idli	2 Nos.	150
Dosa	1 No.	125
Khichidi	1 cup	200
Wheat porridge	1 cup	220
Semolina porridge	1 cup	220
Cereal flakes with milk (corn/wheat/rice)	1 cup	220
Pulse		
Plain dhal	½ cup	100
Sambar	1 cup	110
Vegetable		
With gravy	1 cup	170
Dry	1 cup	150
Non-Vegetarian		
Boiled egg	1 No.	90
Ommelette	1 No.	160
Fried egg	1 No.	160
Mutton curry	¾ cup	260
Chicken curry	¾ cup	240
Fish fried	2 big pieces	190
Fish cutlet	2 Nos.	190
Prawn curry	¾ cup	220
Keema kofta curry	¾ cup (6 small koftas)	240
Savoury snacks		
Bajji or pakora	8 Nos.	280
Besan ka pura	1 No.	220
Chat (Dahi-pakori)	5 pieces	220
Cheese balls	2 Nos.	250
Dahi vada	2 Nos.	180
Vada	2 Nos.	140
Masala vada	2 Nos.	150
Masala dosa	1 No.	200
Pea-kachori	2 Nos.	380
Potato bonda	2 Nos.	200
Sago vada	2 Nos.	210
Samosa	1 No.	200
Sandwiches (butter - 2tbsp)	2 Nos.	200
Vegetable puff	1 No.	200
Chutneys		
Coconut/groundnuts/til	2 tbsp	120
Tomato	1 tbsp	10
Tamarind (with jaggery)	1 tbsp	60
Sweets and Desserts		
Besan barfi	2 small pieces	400
Chikki	2 pieces	290
Fruit cake	1 piece	270
Rice puttu	½ cup	280
Sandesh	2 Nos.	140
Double ka meetha	½ cup	280
Halwa (kesari)	½ cup	320
Jelly/Jam	1 tbsp	20
Custard (caramel)	½ cup	160
Srikhand	½ cup	380
Milk chocolate	25 g	140
Ice-cream	½ cup	200
Beverages		
Tea (2 tsp sugar + 1 cup 50 ml toned milk)	1 cup	75
Coffee (2 tsp sugar +100 ml)	1 cup	110
Cow's milk (2 tsp sugar)	1 cup	180
Buffalo's milk (2 tsp sugar)	1 cup	320
Lassi (2 tsp sugar)	1 cup/glass (200 ml)	110
Squash	1 cup/glass	75
Syrups (Sharabats)	1 cup/glass	200
Cold drinks	1 bottle (200 ml)	150
Fresh lime juice	1 glass	60

Source: Dietary Guidelines for Indians. A Manual, 2011. NIN, Hyderabad.

Annexure-VIII

What is Body Mass Index (BMI)?

Body Mass Index (BMI) is a ratio of body weight to height. It estimates the total body fat and assesses the risks for diseases related to increased body fat. The risk of developing many diseases increases with a higher BMI.

**BMI may not apply to athletes who tend to have large muscle mass, elderly people, pregnant women and children*

BMI Cutoff for Asians and Asian Americans	***NIH BMI Cutoff***	***Comments***
<18.5	<18.5	Your weight is **below healthy range**. This can put you at risk for developing many health problems. Talk to your healthcare provider about your ideal body weight.
18.5 - 22.9	18.5 - 24.9	Your weight is **within healthy range**. Continue exercising and eating healthfully.
23 - 26.9	25 - 29.9	Your weight is **above healthy range**. Your risk for developing type 2 diabetes and other chronic diseases are higher. Talk to your healthcare provider about your ideal body weight and how to make healthy lifestyle changes.
≥27	≥30	Your weight is **further above healthy range**. It increases the risk for developing many chronic diseases such as heart disease and type 2 diabetes, and decreases overall quality of life. Talk to your healthcare provider about your ideal body weight and how to make healthy lifestyle changes.

* Testing for diabetes should be considered for all Asian American adults who present with a BMI of ≥ 23 kg/m^2.

Hsu WC *et al.*, BMI cut points to identify at-risk Asian Americans for type 2 diabetes screening. Diabetes Care. 2015 Jan; 38(1): 150-8.

Wildman RP *et al.*, Appropriate body mass index and waist circumference cutoffs for categorization of overweight and central adiposity among Chinese adults. American Journal of Clinical Nutrition 2004; 80:1129.

Appropriate body-mass index for Asian populations and its implications for policy and intervention strategies. Lancet 2004; 363:157-63.

Target Waist Circumference for Asians and Asian Americans

Sometimes even when your BMI is within healthy range, having too much fat around the abdomen (apple-shaped body) will still put you at risk for heart disease and diabetes. Below are the target goals for waist circumference measurements.	
Men	***Women***
Equals or less than 90cm (35.5 in)	Equals or less than 80cm (31.5 in)

Annexure-IX

Glycemic Index and Glycemic Load of Common Indian Foods

Glycemic Index

The glycemic index (GI) is a measure of the blood glucose-raising potential of the carbohydrate content of a food compared to a reference food (generally pure glucose). Consumption of high-GI foods causes a sharp increase in postprandial blood glucose concentration that declines rapidly, whereas consumption of low-GI foods results in a lower blood glucose concentration that declines gradually. The glycemic index (GI) assigns a numeric score to a food based on how drastically it makes the blood sugar rise. Foods are ranked on a scale of 0 to 100, with pure glucose (sugar) given a value of 100. Carbohydrate-containing foods can be classified as high- (70), moderate- (56-69), or low-GI (£55) relative to pure glucose (GI=100).

Glycemic Load

The effect of a carbohydrate food on blood sugar levels is not only determined by its GI or carbohydrate quality but also by the amount of carbohydrate in a particular food. Glycemic Load is the term used to describe the overall effect of these two factors on blood sugar. Hence, the concept of glycemic load (GL) was developed by scientists to simultaneously describe the quality (GI) and quantity of carbohydrate in a food serving, meal, or diet. The GL of a single food is calculated by multiplying the GI by the amount of carbohydrate in grams (g) provided by a food serving and then dividing the total by 100:

Glycemic Load = GI x Carbohydrate (g) content per portion ÷ 100.

For a typical serving of a food, GL would be considered high with GL20, intermediate with GL of 11-19, and low with GL10.

It should be noted that while healthy food choices generally include low-GI foods, this is not always the case. For example, intermediate-to-high-GI foods like parsnip, watermelon, banana, and pineapple, have low-to-intermediate GLs. For example, a single apple has a GI of 38 and contains 13 grams of carbohydrates.

GL= 38 x 13/100 = 5 (GL)

A potato has a GI of 85 and contains 14 grams of carbohydrate

GL=85 x14/100 = 12 (GL)

We can therefore predict that the potato will have twice the glycemic effect of an apple.

Similarly despite similar GIs, one serving of watermelon has a GL of 8, while a medium-sized doughnut has a GL of 17. Dietary GL is the sum of the GLs for all foods consumed in the diet.

When foods have similar amounts of carbohydrate, GI will have the greatest influence on blood sugars. When the amount of carbohydrate in a portion differs the best way to predict blood sugar effects will be via the GL.

Glycemic Index of Common Foods

Food	*Glycemic Index (Glucose = 100)*
HIGH-CARBOHYDRATE FOODS	
White wheat bread	75
Whole wheat/whole meal bread	67
Specialty grain bread	53
Wheat *roti*	62
Chapatti	52
Corn tortilla	46
White rice, boiled*	73
Parboiled Rice	64
Brown rice, boiled	68
Basmati Rice	<55
Barley	28
Sweet corn	52
Spaghetti, white	49
Spaghetti, whole meal	48
Noodles	70
Rice noodles†	53
Macaroni	47
BREAKFAST CEREALS	
Cornflakes	83
Wheat flake biscuits	69
Porridge, rolled oats	55
Instant oat porridge	79
Rice porridge/congee	78
Millet porridge	67
Muesli	57
FRUIT AND FRUIT PRODUCTS	
Apple, raw	36
Orange, raw	43
Banana, raw	51
Pineapple, raw	59
Mango, raw	51
Watermelon, raw	76
Dates, raw	42
Peaches, canned	43

Food	Glycemic Index (Glucose = 100)
Strawberry jam/jelly	49
Apple juice	41
Orange juice	50
VEGETABLES	
Potato, boiled	78
Potato, instant mash	87
Potato, french fries	63
Baked potato	121
Carrots, boiled	39
Sweet potato, boiled	63
Pumpkin, boiled	64
Plantain/green banana	55
Vegetable soup	48
Cucumber	0
Green beans	0
Peas	48
Yam	51
Beetroot	64
Cabbage	10
Onion	10
DAIRY PRODUCTS AND ALTERNATIVES	
Milk, full fat	39
Milk, skim	37
Ice cream	51
Yogurt, fruit	41
Soy milk	34
LEGUMES	
Chickpeas	47
Green gram	48
Black gram	48
Kidney beans	42
Lentils	22
Soya beans	16
SNACK PRODUCTS	
Chocolate	40
Popcorn	65
Potato crisps	56
Soft drink/soda	59
Rice crackers/crisps	87
Doughnuts	76
SUGARS	
Fructose	15
Sucrose	65
Glucose	103
Honey	61

Glycemic Index and Glycemic Load of Foods

Food	GI	Serving Size (g)	GL
BAKED GOODS and CEREALS			
Corn Bread	110	60g (1 piece)	30.8
French Bread	95	64g (1 slice)	29.5
Corn Flakes	92	28g (1 cup)	21.1
Rice Krispies	82	33g (1.25 cup)	23
Corn pops	80	31g (1 cup)	22.4
Donut (lrg. glazed)	76	75g (1 donut)	24.3
Waffle (homemade)	76	75g (1 waffle)	18.7
Bagel	72	89g (1/4 in.)	33
Corn tortilla	70	24g (1 tortilla)	7.7
Whole wheat bread, average	71	30g	9
Muesli, average	66	30g	16
White Bread	70	25g (1 slice)	8.4
Kellogg's Special K	69	31g (1 cup)	14.5
Angel food cake	67	28g (1 slice)	10.7
Croissant, Butter	67	57g (1 med)	17.5
Oatmeal, Instant	65	234g (1 cup)	13.7
Coarse barley bread, 75-80 per cent kernels, average	34	30g	7
Rye bread, 100 per cent whole	65	32g (1 slice)	8.5
Blueberry Muffin	59	113g (1 med)	30
Oatmeal	58	117g (1/2 cup)	6.4
Whole wheat pita	57	64g (1 pita)	17
Oatmeal Cookie	55	18g (1 large)	6
Popcorn	55	8g (1 cup)	2.8
Sponge cake, plain	46	63g	17
Vanilla Cake and Vanilla Frosting	42	64g (1 slice)	16
Chocolate cake w/chocolate frosting	38	64g (1 slice)	12.5
GRAINS			
Quinoa	53	150g	13
Sweet corn on the cob, average	60	150g	20
White rice, average	89	150g	43
Quick cooking white basmati	67	150g	28
Brown rice, average	50	150g	16
Whole wheat kernels, average	30	50g	11
BEVERAGES			
Cola, Carbonated	63	250 ml	16
Fanta®, orange soft drink	68	250 mL	23
Orange juice, unsweetened	50	250 mL	12
Carrot juice (freshly made)	43	250g	10
Hot Chocolate Mix	51	28g (1 packet)	11.7
Pineapple Juice	46	250g (1 cup)	14.7
Apple juice, unsweetened, average	44	250 mL	30
Tomato juice, canned	38	250 mL	4

Food	*GI*	*Serving Size (g)*	*GL*
LEGUMES			
Baked Beans	48	253g (1 cup)	18.2
Chickpeas, Boiled	31	240g (1 cup)	13.3
Lentils	29	198g (1 cup)	7
Kidney Beans	27	256g (1 cup)	7
Soy Beans	20	172g (1 cup)	1.4
Peanuts	13	146g (1 cup)	1.6
VEGETABLES			
Potato	104	213g (1 med)	36.4
Carrot, raw	92	15g (1 large)	1
Corn, yellow	55	166g (1 cup)	61.5
Sweet Potato	54	133g (1 cup)	12.4
Yam	51	136g (1 cup)	16.8
Peas, Frozen	48	72g (1/2 cup)	3.4
Tomato	38	123g (1 med)	1.5
Broccoli, cooked	0	78g (1/2 cup)	0
Cabbage, cooked	0	75g (1/2 cup)	0
Cauliflower	0	100g (1 cup)	0
Green Beans	0	135g (1 cup)	0
Mushrooms	0	70g (1 cup)	0
Spinach	0	30g (1 cup)	0
FRUIT			
Watermelon	72	152g (1 cup)	7.2
Pineapple, raw	66	155g (1 cup)	11.9
Cantaloupe	65	177g (1 cup)	7.8
Raisins	64	43g (small box)	20.5
Papaya	60	140g (1 cup)	6.6
Kiwi, w/skin	58	76g (1 fruit)	5.2
Fruit Cocktail, drained	55	214g (1 cup)	19.8
Banana, ripe	62	120g (1 med)	16
Mango	51	165g (1 cup)	12.8
Orange	48	140g (1 fruit)	7.2
Grapes	43	92g (1 cup)	6.5
Strawberries	40	152g (1 cup)	3.6
Apples, w/skin	39	138g (1 med)	6.2
Dates, dried	42	60g	18
Pears	33	166g (1 med)	6.9
Apricot, dried	32	130g (1 cup)	23

Food	*GI*	*Serving Size (g)*	*GL*
Prunes	29	132g (1 cup)	34.2
Peach	28	98g (1 med)	2.2
Grapefruit	25	123g (1/2 fruit)	2.8
Plum	24	66g (1 fruit)	1.7
Sweet Cherries, raw	22	117g (1 cup)	3.7
NUTS			
Cashews, salted	27	50 g	3
Almonds	0		
Hazelnuts	0		
Macademia	0		
Pecans	0		
Walnuts	0		
Peanuts, average	7	50g	0
DAIRY			
Ice Cream (Lower Fat)	47	76g (1/2 cup)	9.4
Ice Cream, regular	38	72g (1/2 cup)	6
Pudding	44	100g (1/2 cup)	8.4
Milk, Whole	40	244g (1 cup)	4.4
Milk, skim	32	250 mL	4
Yogurt, Plain	36	245g (1 cup)	6.1
MEAT/PROTEIN			
Beef	0		
Chicken	0		
Eggs	0		
Fish	0		
Lamb	0		
Pork	0		
Veal	0		
Deer-Venison	0		
Elk	0		
Buffalo	0		
Rabbit	0		
Duck	0		
Ostrich	0		
Shellfish	0		
Lobster	0		
Turkey	0		
Ham	0		

References

1. Adhikari, S. P., Meng, S., Wu, Y. J., Mao, Y. P., Ye, R. X., Wang, Q. Z., Zhou, H. 2020. Epidemiology, causes, clinical manifestation and diagnosis, prevention and control of coronavirus disease (COVID-19) during the early outbreak period: a scoping review. Infectious Diseases of Poverty, 9(1):1–12.

2. Amodio *et al.*, 2013. The nutritional management of hepatic encephalopathy in patients with cirrhosis: International society for hepatic encephalopathy and nitrogen metabolism consensus. Hepatology. 58 (1): 325-329.

3. Anita, F.P. Clinical Dietetics and Nutrition. Oxford University Press, Delhi. Oxford, New York.

4. Arcidiacono *et al.*, 2012. Insulin resistance and cancer risk: an overview of the pathogenic mechanism. Exp Diabetes Res: 789174. http://dx.doi.org/10.1155/2012/93

5. Arends *et al.*, 2017. ESPEN guidelines on nutrition in cancer patients. Clinical Nutr. 36: 11-48.

6. ASCM's guidelines for exercise testing and prescription. 6th ed. Baltimore: Lippincott Williams and Wilkins, 2000: 243-244.

7. Bamji, M.S., Rao, P.N. and Reddy, V. Textbook of Human Nutrition. Oxford and IBH Publishing Co. Pvt. Ltd., New Delhi.

8. Barazzoni, R., Bischoff, S.C., Breda, J. *et al.*, 2020. ESPEN expert statements and practical guidance for nutritional management of individuals with SARS-CoV-2 infection. Clin. Nutr. S0261- 5614(20)30140-0. https://doi.org/10.1016/j.clnu.2020.03.022.

9. Bayan, L., Koulivand, P. H. and Gorji, A. 2014. Garlic: a review of potential therapeutic effects. Avicenna journal of phytomedicine, 4(1):1-14

10. Bellomo, R. and Boyce, N. 1993. Acute continuous hemodialfiltration: a prospective study of 110 patients and a review of the literature. Am J Kidney Dis. 21:508-518.

11. Berbel, M.A., Peron, M., Pinto, R., Ponce, D. and Balbi, A.L. 2011.Nutritional aspects in acute kidney injury. Rev. Assoc. Med. Bras. 57 (5) http://dx.doi.org/10.1590/S0104-42302011000500022

12. Berger, M.M., Shenkin, A., Revelly, J.P. *et al.*, 2004. Copper, selenium, zinc and thiamine balances during continuous venovenous hemofiltration in critically ill patients. Am J Clin Nutr. 80:410-416.

13. Bozzetti, F. 2013. Nutritional support of the oncology patient. Crit. Rev. Oncol. Hematol. 87 (2): 172-200.

14. Burke, L. M., Kiens, B., and Ivy, J. L. (2004). Carbohydrates and fat for training and recovery. Journal of Sports Sciences, 22, 15-30.

15. Caluwé, E. D., Halamová, K. and Damme, P. V. 2010. Tamarindus indica L. – A review of traditional uses, phytochemistry and pharmacology. Afrika Focus, 23(1):23–23.

16. Cano, N.J.M., Aparicio, M., Brunori, G., Carrero, J.J., Cianciaruso, B., Fiaccadori, E., Lindholm, B., Teplan, V., Fouque, D. and Guarnieri, G. 2009. ESPEN Guidelines on Parenteral Nutrition: Adult Renal Failure. Clinical Nutrition 28 (2009) 401–414.

17. Chadalavada *et al.*, 2010. Nutrition in hepatic encephalopathy. Nutr. Clin. Practice. 25 (3): 257-264.
18. Chang, J. S., Wang, K. C., Yeh, C. F., Shieh, D. E. and Chiang, L. C. 2013. Fresh ginger (Zingiber oficinale) has anti-viral activity against human respiratory syncytial virus in human respiratory tract cell lines. Journal of Ethnopharmacology, 145(1):146–151.
19. Chauhan, O. P., Archana, B. S., Singh, A., Raju, P. S. and Bawa, A. S. 2014. A refreshing beverage from mature coconut water blended with lemon juice. J. Food Sci. Technol. 51(11):3355–3361.
20. Dagan, S.S., Goldenshluger, A., Globus, I., Schweiger, C., Kessler, Y., Sandbank, G.K., Ben-Portal, T. and Sinai. T. 2017. Nutritional recommendations for adult bariatric surgery patients: clinical practice. Adv. Nutr. 8 (2): 382-394.
21. Davidson, S., Passmore, R., Brock, J.F. and Truswell, A.S. Human Nutrition and Dietetics. The English Language Society and Churchill Livingstone.
22. Dixon, M., Macdonald, A., Stafford, J., White, F. and Portnoi, P. 2015. Disorders of Carbohydrate Metabolism. Vanessa Shaw, Clinical Paediatric Dietetics in (p.527-587). West Sussex: John Wiley and Sons.
23. Druml, W. 2005. Nutritional support in acute renal failure. Handbook of Nutrition and the Kidney. 5th ed. Philadelphia: Lippincott Williams and Wilkins; pp.95-114.
24. Ferguson, M., Capra, S., Bauer, J. and Banks, M. 1999. Development of a valid and reliable malnutrition screening tool for adult acute hospital patients. Nutrition, 15: 458-64.
25. Fessler, T.A. 2014. Home Tube Feeding with Blenderized Foods. Nov.-Dec., 2014. Lifeline letter, Oley Foundation, USA.
26. Fiaccadori, E., Maggiore, U., Rotelli, C., Giacosa, R., Picetti, E., Parenti, E. *et al.*, 2005. Effects of different energy intakes on nitrogen balance in patients with acute renal failure: a pilot study. Nephrol Dial Transplant. 20:1976-80.
27. Fiaccadori, E., Parenti, E. and Maggiore, U. 2008. Nutritional support in acute kidney injury. J Nephrol. 21: 645-656.
28. Fiaccadori, E., Regolisti, G. and Cabassi, A. 2010. Specific nutritional problems in acute kidney injury, treated with non-dialysis and dialytic modalities. NDT Plus. 3(1): 1–7.
29. Fortin, M.C., Amyot, S.L. and Geadah, D. 1999. Serum concentrations and clearances of folic acid and pyridoxal-5-phosphate during venovenous continuous renal replacement therapy. Intensive Care Med 25:594-598.
30. Frazier, D.M., Allgeier, C., Homer, C., Marriage, B.J., Ogata, B., Rohr, F., Splett, P.L., Stembridge, A. and Singh, R. H. 2014. Nutrition management guideline for maple syrup urine disease: an evidence-and – consensus-based approach. Molecular Genetics and Metabolism. 112 (3): 210-217.
31. George, R. 2014. Role of nutrition and balanced diet in enhancing sports performance. 2014. ANNQUEST (December 2014): 3 (1): 9-22.
32. Gervasio, J.M., Garmon, W.P., Holowatyj, M. 2011. Nutrition support in acute kidney injury. Nutr Clin Pract. 26(4):374-81. doi: 10.1177/0884533611414029.
33. Ghosh, S. The Feeding and Care of Infants and Young Children. Voluntary Health Association of India, New Delhi.
34. Gluckman, P.D., Hanso,n M.A., Cooper, C. and Thornburg, K.L. 2008. Effect of in utero and early-life conditions on adult health and disease. N. Engl. J. Med. 359 (1): 61-73. doi: 10.1056/NEJMra0708473.
35. Hewlings, S. and Kalman, D. 2017. Curcumin: A Review of Its' Effects on Human Health. Foods, 6(10):92– 92.
36. ICMR. 2010. Nutrient Requirement and Recommended Dietary Allowances for Indian. NIN, Hyderabad.
37. Ikizler,T.A. 2013. Optimal nutrition in hemodialysis patients. Adv. Chronic Kidney Dis. 20 (2): 181-189.
38. Jensen, M.D., Ryan, D.H., Donato, K.A. *et al.*, 2014. Guidelines (2013) for managing overweight and obesity in adults. Obesity. 22(S2): S75.
39. Kashyap and Kashyap. 2015. Oral allergy syndrome: an update for stomatologist. J. Allergy (Cairo). Nov 8. doi: 10.1155/2015/543928.
40. KDIGO 2012. Clinical Practice Guideline for the Evaluation and Management of Chronic Kidney Disease. International Supplements 2, Official Journal of the International Society of Nephrology. 3 (1): 3. www.kidney-international.org
41. Kerckhove, K.V., Diels, M., Vanhaesebrouck, S. *et al.*, 2015. Consensus on the Guidelines for the Dietary Management of Classical Galactosemia. Clinical Nutrition ESPEN, 10: 1-4.
42. Kerksick, C., Harvey, T., Stout, J. *et al.*, 2008. International Society of Sports Nutrition position stand: Nutrient timing. Int J Soc Sports Nutr. 5:17.
43. Khaled, M. B. and Benajiba, N. 2020. The role of nutrition in strengthening immune system against newly emerging viral diseases: case of SARS-CoV-2. Nor. Afr. J. Food Nutr. Res. 04 (07): 240-44.
44. Kreider *et al.*, 2010. ISSN exercise and sport nutrition review: research and recommendations. Int. J. Soc. Sports Nutr. 7:7.
45. Kyle, J.M. *et al.*, 2003. Counseling for physical activity in overweight and obese patients. Am Fam Physician. 67(6):1249-1256.

46. Longvah, T., Ananthan, R., Bhaskarachary, K. and Venkalah, K. 2017. Indian Food Composition Tables. National Institute of Nutrition, ICMR, Hyderabad-500 007, India

47. Lorember, F.M. 2018. Malnutrition in CKD. Front Pediatr. 6: 161.

48. MacLeod, E.L. and Ney, D.M. 2010. Nutritional management of phenylketonuria. Ann Nestle Eng. 68 (2): 58-69.

49. Makau, J. N., Watanabe, K., Mohammed, M. M.and Nishida, N. 2018. Antiviral Activity of Peanut (Arachis hypogaea L.) Skin Extract against Human Influenza Viruses. Journal of Medicinal Food, 21(8):777–784.

50. Malone, A. M. 2005. Enteral Formula Selection: A Review of Selected Product Categories, Mt. Carmel West Hospital, Department of Pharmacy, Columbus, OH.

51. Maughan, R.J. *et al.*, 2007. The use of dietary supplements by athletes. J Sports Sci. 25 (Suppl 1):S103-S113.

52. Maughan, R.J., Greenhaff, P.L. and Hespel, P. 2011. Dietary supplements for athletes: emerging trends and recurring themes. J Sports Sci. 29(S1):S57-S66.

53. McClave, S.A., Taylor, B.E., Martindale, R.G., Warren, M.M., Johnson, D.R., Braunschweig, C., McCarthy, M.S., Davanos, E., Rice, T.W., Cresci, G.A., Gervasio, J.M., Sacks, G.S., Roberts, P.R. and Compher, C. 2016. Guidelines for the Provision and Assessment of Nutrition Support Therapy in the Adult Critically Ill Patient: Society of Critical Care Medicine (SCCM) and American Society for Parenteral and Enteral Nutrition (A.S.P.E.N.). JPEN J Parenter Enteral Nutr. 40(2):159-211.

54. Meier and Sobotka. 2010. Basics in clinical nutrition: nutritional support in acute and chronic pancreatitis. Clinical nutrition (ESPEAN) 5(1); e58-e62.

55. Nair, A.B., Madan, J. and Dhaaki, N. 2015. Formulation and Nutritional Appraisal of Low Cost Enteral Formula. International Journal of Scientific and Research Publications, 5 (9). ISSN 2250-3153.

56. Naja, F and Hamadeh, R. 2020. Nutrition amid the COVID-19 pandemic: a multilevel framework for action. Eur. J. Clin. Nutr. 2020 (April, 20): 1-5.

57. Naylor, H.L., Jackson, H., Walker, G.H., Macafee, S., Magee, S., Hooper, L., Stewart, L. and MacLaughlin, H.L. 2013. British Dietetic Association evidence-based guidelines for the protein requirements of adults undergoing maintenance haemodialysis or peritoneal dialysis J Hum Nutr Diet 26:315-328.

58. NIN. 2011. Dietary Guidelines for Indians-A Manual. National Institute of Nutrition, Hyderabad.

59. Oh *et al.*, 2015. Micronutrient loss during renal replacement therapy for acute kidney injury. Abstract – British Renal Society. Available at http://britishrenal.org/conference-2/2015-abstracts/

60. Oh, W.C., Gardner, D.S. and Devonald, M.A.J. 2015. Micronutrient and amino acid losses in acute renal replacement therapy. Curr Opin Clin Metab Care 18:593-598.

61. Pan *et al.*, 2017. Recent Advances on Nutrition in Treatment of Acute Pancreatitis. Front Immunol. 8: 762.

62. Panis, B., van Kroonenburgh, M. J. and Rubio-Gozalbo, M. E. J. 2007. Proposal for the prevention of osteoporosis in paediatric patients with classical galactosaemia. Inherit. Metab. Dis. 30(6):982.

63. Patel, J.J., McClain CJ, Sarav, M., Hamilton-Reeves, J. and Hurt, R.T. 2017. Protein requirements for critically ill patients with renal and liver failure. Nutr Clin Pract. 32 (1 suppl) 101s-111s.

64. Pazgan-Simon, M., Zuwala-Jagiello, J., Serafiñska, S. and Simon, K. 2015. Nutrition principles and recommendations in different types of hepatic encephalopathy. Clinical and Experimental Hepatology. 4: 1–6.

65. Phillips, S.M. and Van Loon, L.J. C. 2011. Dietary protein for athletes: from requirements to optimum adaptation. J Sport Sci. 29 (Suppl 1):S29-S38.

66. Potgieter, S. 2013. Sport nutrition: A review of the latest guidelines for exercise and sport nutrition from the American College of Sport Nutrition, the International Olympic Committee and the International Society for Sports Nutrition. South African Journal of Clinical Nutrition, 26(1), 6-16.

67. Powers, S., Nelson, W.B. and Larson-Meyer, E. 2011. Antioxidant and Vitamin D supplements for athletes: sense or nonsense? J Sports Sci. 29(S1):S47-S55.

68. Purnell, J.Q. 2018. Definitions, classification and epidemiology of obesity. www.endotext.org

69. Rinninella et.al. 2017. Nutritional support in acute pancreatitis: from physiopathology to practice. An evidence-based approach. European Review for Medical and Pharmacological Sciences. 21: 421-432.

70. Robinson, C.H. Normal and Therapeutic Nutrition. Oxford and IBH Publishing Company, New Delhi.

71. Rodriquez, N.R., DiMarco, N.M. and Langley, S. 2009. Position of the American Dietetic Association, Dietitians of Canada, and the American College of Sports Medicine: Nutrition and athletic performance. J Am Diet Assoc. 109 (3):509-527.

72. Sawka, M. *et al.*, 2007. Exercise and fluid replacement. Med Sci Sports Exerc. 39 (2):377-390.

73. Sen Sarma, M., Srivastava, A., Yachha, S.K., Poddar, U. and Mathias, A. 2016. Classical galactosemia among Indian children: presentation and outcome from a pediatric gastroenterology center. Indian Pediatr. 53:27-31.

74. Sharma, N. 2019. Efficacy of Garlic and Onion against virus. International Journal of Research in Pharmaceutical Sciences, 10(4):3578–3586.

75. Shirreffs, S.M. and Sawka, M.N. 2011.Fluid and electrolyte needs for training, competition, and recovery. J Sports Sci. 29(S1):S39-S46.

76. Sicherer, S.H. and Sampson, H.A. 2014. Food allergy: Epidemiology, pathogenesis, diagnosis, and treatment. J Allergy Clin Immunol. 133:291–307.

77. Sivakumar, J. 2018. Nutritional Management of Bariatric Surgical Patients in the Peri-operative Setting. J Obes Eat Disord 4:1. doi: 10.21767/2471-8203.100033

78. Skypala *et al.,* 2015. Sensitivity to food additives, vaso-active amines and salicylates: a review of the evidence. Clin Transl Allergy.5: 34.

79. Slater, G. and Phillips, S.M. 2011. Nutrition guidelines for strength sports: sprinting, weightlifting, throwing events, and bodybuilding. J Sports Sci. 29(S1):S67-S77.

80. Story, D.A., Ronco, C. and Bellomo, R. 1999. Trace element and vitamin concentrations and losses in critically ill patients treated with continuous venovenous hemofiltration. Crit Care Med. 27:220-223.

81. Suchitra, M. R. and Parthasarathy, S. 2020. Analyses of zinc content of different types of sesame seeds in the South Indian delta region. International journal of Scientific and Technology Research, (9):1867–1869.

82. Suchitra, M. R. and Parthasarathy, S. 2020. Nutrition and corona virus: Plan a diet in a pandemic Int. J. Res. Pharm. Sci. 11(SPL)(1), 110-114.

83. Suchitra, M., R. and Parthasarathy, S. 2015. Sitaphal: Reemergence. Research journal of pharmaceutical, biological and chemical sciences. 6:1560–1565.

84. Swaminathan, M. Advance Text Book on Foods and Nutrition (Volume I and II). BAPPCO, Bangalore.

85. Swaminathan, M. Principles of Nutrition and Dietetics. BAPPCO, Bangalore.

86. Tremmel, M., Gerdtham, U.G., Nilsson, P.M. and Saha, S. 2017. Economic burden of obesity: A systematic literature review. Int. J. Environ. Res. Public Health. 14 (4):435.

87. Welling, L., Bernstein, L.E., Berry, G.T., *et al.,* on behalf of the Galactosemia Network (GalNet). 2017. International Clinical Guideline for the Management of Classical Galactosemia: Diagnosis, Treatment, and Follow-up. J Inherit Metab Dis, 40 (2): 171-176.

88. World Health Organization 2008. Waist circumference and waist-hip ratio. Report of a WHO Expert Consultation, Geneva (Dec. 8-11, 2008)

89. Yumuk, V., Tsigos, C., Fried, M., Shindler, K., Busetto, L., Micic, D and Toplak, H. 2015. European guidelines for obesity management in adults. Obes Facts. 8(6): 402-424.

90. Zdenek. Zada´k, Z. and Smith, L.K. 2009. Basics in clinical nutrition: Commercially prepared formulas. Kent-Smith/e-SPEN, the European e-Journal of Clinical Nutrition and Metabolism. 4: e212-e215.

Index

D

E